SPACE SCIENCE, EXPLORATION AND POLICIES SERIES

SPACE EXPLORATION RESEARCH

SPACE SCIENCE, EXPLORATION AND POLICIES SERIES

Progress in Dark Matter Research
J. Val Blain (Editor)
2005. ISBN 1-59454-248-1

Space Science: New Research
Nick S. Maravell (Editor)
2006. ISBN 1-60021-005-8

Space Policy and Exploration
William N. Callmers (Editor)
2008. ISBN 978-1-60456-448-8

Space Commercialization and the Development of Space Law from a Chinese Legal Perspective
Yun Zhao
2009. ISBN 978-1-60692-244-6

Next Generation of Human Space Flight Systems
Alfred T. Chesley (Editor)
2009. ISBN 978-1-60692-726-7

Smaller Satellites Operations Near Geostationary Orbit
Matthew T. Erdner
2009. ISBN 978-1-60741-181-9

Environmental Satellites: Weather and Environmental Information Systems
Vincent L. Webber (Editor)
2009. ISBN 978-1-60692-984-1

Nutritional Biochemistry of Space Flight
Scott M. Smith, Sara R. Zwart, Vickie Kloeris and Martina Heer
2009. ISBN 978-1-60741-641-8

Space Exploration Research
John H. Denis and Paul D. Aldridge (Editors)
2009. ISBN: 978-1-60692-264-4

SPACE SCIENCE, EXPLORATION AND POLICIES SERIES

SPACE EXPLORATION RESEARCH

JOHN H. DENIS
AND
PAUL D. ALDRIDGE
EDITORS

Nova Science Publishers, Inc.
New York

For permission to use material from this book please contact us:
Telephone 631-231-7269; Fax 631-231-8175
Web Site: http://www.novapublishers.com

Library of Congress Cataloging-in-Publication Data

Space exploration research / edited by John H. Denis and Paul D. Aldridge.
p. cm.
Includes index.
ISBN 978-1-60692-264-4 (hardcover)
1. Astronautics--Research. 2. Outer space--Exploration. 3. Space sciences. I. Denis, John H. II. Aldridge, Paul D.
TL858.S63 2009
629.4--dc22

2009001308

Published by Nova Science Publishers, Inc. ✦ New York

CONTENTS

PREFACE

Space exploration is an immense and expanding field. The quest for knowledge about space has resulted in hundreds of very important technologies which have been incorporated into society's fabric including the biomedical field. This new book examines a multitude of issues related to space exploration including philosophy, biology, dark energy, space tourism, space station measurements, supernova, and Saturn's rings.

Chapter 1 – International law considers Space as the province of Mankind. This can only mean that humanity invests more than mere scientific and technical resources, that the stakes are more than just financial and political. Imagination also plays an important role in Mankind's march to the stars, combining the most ancient dreams with the most spectacular achievements. This is why we need to think seriously about the significance Space exploration attaches to the exploits of human astronauts, and also about the idea of founding human colonies in Space. Just how necessary is it to invest in programmes that may take on colossal proportions? And do we possess the wisdom necessary to address all the questions it raises? If humanity declares that Space is its legitimate province, this implies a commitment to understand the purpose of it all in advance, so as to be able to explore it in a responsible manner. These are the principal philosophical themes addressed in this chapter.

Chapter 2 – Space Tourism, considered not an area of priority for commercialization, has never been so heatedly discussed until the historic arrival of two "unexpected" tourists in the International Space Station (ISS). The development of space tourism no doubt calls for a legal regime, which can better regulate the market and offer clear guidance and expected outcome. It has been widely argued that the existing international space treaties are inadequate for space commercialization. The development of a clear and predictable legal regime before space tourism becomes affordable for the masses is essential. As long as the space travel technology is mature, there are always business opportunities for space tourism. This paper discusses the potential for commercial space tourism and advocates an improved but essential legal regime for space stations. In view of the many commonalities shared by aviation and space travel, the present paper takes the example of aviation and elaborates on the formulation of a legal regime for space tourism. Actually it is gradually being accepted that the most appropriate regulatory framework for space tourism is to treat it as an extension of aviation. A proper and attractive legal regime will in the end help assure the future of safe and responsible commercial space tourism.

Chapter 3 – Hypometabolism is a physiological state in which the energy requirements of the organism are drastically lowered. In hibernating animals this represents the pivotal strategy to overcome extreme environmental conditions. This paper begins with a discussion of natural hypometabolism as found in mammalian hibernation with a focus on the strategies these animals evolved to cope with life challenge in hostile environments (e.g., cold weather and/or shortage of food), covering behavioural, physiological and genetic strategies. The issue of factors required/able to induce the hypometabolic condition is then addressed, including some possible pharmacological approach. Finally, the potential impact of induced, regulated hypometabolism in the human is briefly reviewed showing why hypometabolism would be, from a resource-driven perspective, desirable during long-term space flight as well as in the broad field of biomedicine.

Chapter 4 – The well characterized nematode *C. elegans* is a superb model organism for biological experiments in space environments especially due to its similarities to humans on the most basic levels. Humans and *C. elegans* have similar numbers of genes, about 60% of those genes are homologous and both organisms have similar repair systems for radiation induced DNA damage. In addition, both have neuromuscular systems and hormonal regulatory systems. *C. elegans* can also be maintained on a chemically defined minimal medium (CeMM) so that bacteria, the nematode's normal food, need not be shipped into orbit. *C. elegans* is robust and microscopic so that nematode experiments require little oversight by astronauts and need minimal space onboard the International Space Station. Little is known about the effects of long-term exposure to different types of radiation in space and the wide range of radiation sources needed to mimic the space environment are not available on Earth. In order to address this *C. elegans* experiments have been to space on several missions starting in the early 1990's. These missions ranged from a few days to a multi-generation trip spanning several months. Five *C. elegans* mutagen testing systems have been used to analyze the effects of radiation: 1) poly-G/poly-C tract modifications; 2) assaying for dominant *unc-22* mutations; 3) fem-3 dauer analysis; 4) identifying alterations in telomere length; and 5) capturing and analyzing mutations using the eT1-system which balances approximately 1/6 of the entire genome. The eT1-system is being developed into a "biological accumulating dosimeter" to be used to capture and analyze space radiation induced mutations. Nimblegen micro-array chip analysis has proven very effective for determining the exact locations and extends of captured eT1 lethal deficiency and duplication mutations and using massively parallel sequencing instruments could give a quantitative look at the spectrum of the captured space radiation induced mutations. In addition to studying the mutational effects of radiation other experiments have focused on the effects microgravity and gene expression. These include expression of myogenic transcription factors and myosin heavy chains in *C. elegans* muscles developed during spaceflight and how checkpoint and physiological apoptosis in germ cells proceeds in space flight. These analyzes demonstrate that *C. elegans* exhibits similar molecular changes to humans in flight and should lead to the development of countermeasures necessary for continual habitation of the ISS and also for long-duration manned space-flights to the Moon and Mars

Chapter 5 – The driving force for much of the present exploration of our solar system is the search for clues to the origin of life, specifically the search for habitable environments either past or present. In short it is the search for liquid water and of most current interest is the near-surface environment of Mars. After recent intense study, from both the surface of the planet and orbit, Mars reveals itself to be relatively rich in near surface water ice mixed and overlaid by sand and dust at least down to a few meters. Despite the present cold, dry conditions Mars appears to have undergone extremely wet periods at some locations up to geologically recent times. In order to understand the transport of heat and moisture through this near surface material details of its structure, composition and transport mechanisms must be known, which are intimately linked to the nature and history of the Martian climate. Specifically dust is actively transported through the atmosphere, though the process is not understood and wind produced sand features are abundant on Mars, though it is not clear whether sand transport is still active or points towards a previous climatic environment. In this chapter the most recent laboratory simulations and modeling techniques will be presented in the context of current observations from Mars. Advances in instrumentation will also be discussed for application to future Martian surface studies.

Chapter 6 – Type~Ia supernovae, acting as standard candles, play a leading role in the exploration of the Universe evolution. Initiated by similar stellar explosions whose physics is known in detail, they provide simultaneous measurements of the (luminosity) distance versus the redshift. Observations of this type of supernovae at high redshifts, being sensitive to the Hubble expansion rate, provide the most direct evidence for the accelerating expansion of the Universe and they are consistent with cosmological models proposing a dark energy component dominating the Universe energy budget. These findings have been corroborated by several independent sources, such as measurements of the cosmic microwave background, gravitational lensing, and the large scale structure of the Cosmos.

This chapter focuses on the importance of supernova observations for exploring the nature of dark energy. It briefly outlines the procedure followed in order to extract information relevant to cosmology from measurements of supernova luminosity and spectra and addresses the statistical and systematic errors involved. A complete review is given on supernova observational evidence starting from the first observations presented in 1995 by the Supernova Cosmology Project and the High-z Supernova Search Team and reaching the recent developments by the Hubble Space Telescope, the Supernovae Legacy Survey and the ESSENCE project.

Chapter 7 – The results of precise analysis of elements and isotopes in meteorites, comets, the Earth, the Moon, Mars, Jupiter, the solar wind, solar flares, and the solar photosphere since 1960 reveal the fingerprints of a local supernova (SN)—undiluted by interstellar material. Heterogeneous SN debris formed the planets. The Sun formed on the neutron (n) rich SN core. The ground-state masses of nuclei reveal repulsive n-n interactions that can trigger axial n-emission and a series of nuclear reactions that generate solar luminosity, the solar wind, and the measured flux of solar neutrinos. The location of the Sun's high-density core shifts relative to the solar surface as gravitational forces exerted by the major planets cause the Sun to experience abrupt acceleration and deceleration, like a yoyo on a string, in its orbit about the ever-changing centre-of-mass of the solar system. Solar cycles (surface magnetic activity, solar eruptions, and sunspots) and major climate changes arise from changes in the depth of the energetic SN core remnant in the interior of the Sun.

Chapter 8 – The black hole solutions in the M-theory are provided for the intersecting M-brane solutions. To solve the bosonic part of field equations of M-theory, the Einstein equation and the Maxwell equations, the metric is assumed by the stationary flat direction with the additional spaces occupied by M-branes, which have a time-like Killing vector or null Killing vector, and the time-like or null directed M-branes for the souse of three-form fields. In null case the solutions include the BMPV solution, otherwise in time-like Killing case, the supersymmetric black ring solution is included. The BMPV solution and the other null Killing solutions are characterized by the Harmonic functions, which given by the Laplace equation, thus the solutions are easy to add the other solutions to get the new solutions. Though the black ring solution are essentially contributed by the Charn-Simons terms, which makes the Laplace equation to the Poisson equation. Therefore in general the superposition on the solutions is not allowed. These black holes are the solution of the Einstein and Maxwell equations naturally, however, taking the Killing spinor equation, which related to the gravitinofs degree of freedom, into account, the integrated constants, which related to the black hole mass or charges and angular momentums, are limited and the solutions remained supersymmetries. Finally analysis of the low-dimensional black hole solutions with torus compactification, the conserved charges in lower dimension are possible to calculation in general.

Chapter 9 – The authors use the 5D Extra Dimensional Force according to Basini-Capozziello-Ponce De Leon, Overduin-Wesson and Mashoon-Wesson-Liu to demonstrate that in flat 5D Minkowsky Spacetime, or weak Gravitational Fields, we cannot tell if we live in a 5D or a 4D Universe. But, in the extreme conditions of Strong Gravitational Fields, the authors demonstrate that the effects of the 5D Extra Dimension becomes visible and perhaps the study of the extreme conditions in Black Holes can tell if we live in a Higher Dimensional Universe. The authors also analyze the possibility of Experimental Research of Extra Dimensions On-Board International Space Station (ISS) by using a Satellite carrying a Laser device(optical Laser) on the other side of Earth Orbit targeted towards ISS. The Sun will be between the Satellite and the ISS so the Laser will pass the neighborhoods of the Sun at a distance R in order to reach ISS. The Laser beam will be Gravitationally Bent according to Classical General Relativity and the Extra Terms predicted by Kar-Sinha in the Gravitational Bending Of Light due to the presence of Extra Dimensions can perhaps be measured with precision equipment. By computing the Gravitational Bending according to Einstein the authors know the exact position where the Laser will reach the target on-board ISS. However, if the Laser arrives at ISS with a Bending different than the one predicted by Einstein and if this difference is equal to the Extra Terms predicted by Kar-Sinha then this experience would proof that we live in a Universe of more than 4 Dimensions. The authors demonstrate in this work that ISS have the needed precision to detect these Extra Terms(see eq 137 in this work). Such experience would resemble the measures of the Gravitational Bending Of Light by Sir Arthur Stanley Eddington in the Sun Eclipse of 1919 that helped to prove the correctness of General Relativity, although in ISS case would have more degrees of accuracy because we would be free from the interference of Earth's Atmosphere. The Laser Satellite could also test the Gravitational Red Shift affected by the presence of the Extra Dimensions. The authors also outline the fact that the huge number of Elementary Particles seen in 4D are as a matter of fact a small number of particles seen in 5D and an experimental proof of the Existence of Extra Dimensions can leads towards a major breakthrough in the theories of Physics unification.

Chapter 10 – The history of the discovery of Saturn's rings started right at the beginning of sky observations with a telescope after it had been invented in the Netherlands in about 1608. This discovery includes many great names of researchers in astronomy such as Galileo, Scheiner, Hevelius, Gassendi and Huygens, until Cassini finally settled the very nature of the rings.

The first person to make observations about Saturn's rings was Galileo during the time he made his sky observations that led to the acceptance of the heliocentric model. Galileo put his findings into a riddle as he was not sure enough about what he saw.

The French philosopher Gassendi described his observations in 1640 as "ansae" (Latin, English: handles), which became a scientific term for Saturn's appearance for a long time. It was not until 1656 when Huygens proved the existence of Saturn's moon Titan, using an improved telescope, that he conjectured the appearance he saw was actually a ring. This discovery was so unusual to accepted science at that time, thus again, like Galileo; he released his findings in an anagram, which he did not disclose earlier than 1659. His ideas were hardly accepted, however. While he conjectured the ring to be solid throughout, Cassini found the true nature of the ring formed by small ice particles in 1705 when he described the gap within the ring. Then finally the ring structure model was widely accepted.

The discovery of the moon Phoebe by Pickering in 1898 was both the first major discovery in astronomy made by photography as well as the first discovery of a moon from the southern hemisphere.

Chapter 11 – For the first time the role of superconductivity of the space objects within the Solar system located behind a belt of asteroids is considered. Observation of experimental data for the Saturn's rings shows that the rings particles may have superconductivity. Theoretical electromagnetic modeling demonstrates that superconductivity can be the physical reason of the origin of the rings of Saturn from the frozen particles of the protoplanetary cloud. The rings appear during some time after magnetic field of planet appears. It happened as a result of interaction of the superconducting iced particles of the protoplanetary cloud with the nonuniform magnetic field of Saturn. Finally, all the Kepler's orbits of the superconducting particles are localizing as a sombrero disk of rings in the magnetic equator plane, where the energy of particles in the magnetic field of Saturn has a minimum value. Within the sombrero disc all iced particles redistributing by the rings (strips) like it is happened for the iron particles nearby the magnet. Electromagnetism and superconductivity allow us to understand why planetary rings in the solar system appear only for the planet with the magnetic field after the belt of asteroids where the temperature is low enough and why there are no rings for the Earth, and many other phenomena.

Chapter 12 – A highly flattened, rapidly and differentially rotating disk of primarily large > cm size mutually gravitating and elastically colliding ice particles orbiting a central object is oftenly taken as an idealized model of Saturn's main A, B, and C rings. This article considers the problem of the stability of the Saturnian main ring system with special emphasis on its fine-scale of the order of 100 m density wave structure (almost regularly spaced, aligned cylindric density enhancements and rarefications). The authors attribute this periodic microstructure to the propagation of compression density waves in the ring plane. The wave propagation is a process of rotation as a solid about the center at a fixed phase velocity, despite the general differential rotation of the system; the structure consists of different material at different times. It seems likely that the key factor contributing to the generation of density waves is the classical Jeans instability of gravity perturbations (e.g., those produced by a spontaneous disturbance). This gravitational instability associated with small departures of macroscopic parameters from the dynamical equilibrium is hydrodynamical in nature and has nothing to do with any explicit resonant effects. The authors analyse Jeans' gravitational instability analytically through the use of hydrodynamic equations. It is shown that the instability in the rotating Saturnian ring layer may be stabilized by a peculiar particle motion, or "temperature" of a suitable magnitude. A stability criterion is given to suppress the instability of all perturbations including the most unstable spiral ones. The authors demonstrate that exclusively trailing spirals can be formed in Saturn's A nd B rings. The very existence and the value of the critical wavelength of the fine-scale structure is explained. Theoretical predictions are compared with numerical simulations. The stability analysis presented here would have to be regarded as an explanation of the almost regular periodic structure in the range of few tens to few hundreds meters in Saturn's A and B rings that has been recently revealed by Cassini spacecraft high-resolution measurements.

Chapter 13 – The presented review is dedicated to the consideration of the Continuous Wavelet transform from the diffusion signal and image processing point of view. Such an approach is based on the consideration of diffusion smoothing via the solution of proper partial diferential equations. Within this group of methods the real and complex wavelet transform with the wavelets of Gauss and Morlet families are considered. Especial attention is concentrated on the variety of numerical examples considering the processing of regular and irregular (random samples, chaotic ODE solutions etc.) signals. All of them are graphically illustrated.

Chapter 14 – The unique physical properties of aerogels have provided enabling technologies to a variety of both flight and proposed space science missions. The extremely low values of the density and corresponding high values of the porosity of aerogels make them suitable for stopping high velocity particles, as highly efficient thermal barriers and as a porous medium for the containment of cryogenic fluids. The use of silica aerogel as a hypervelocity particle capture and return medium for the Stardust Mission, which launched in 1999 and returned to earth in 2006, has drawn the attention of the scientific community, as well as the public, to these fascinating materials. Aerogels are currently being used as the thermal insulation material in the 2003 Mars Exploration Rovers and will be used on the 2009 Mars Science Laboratory rover, as well. The SCIM (Sample Collection for the Investigation of Mars) and the STEP (Satellite Test of the Equivalence Principle) Missions are proposed scientific missions, in which the use of aerogel is critical to their overall design and success. Composite materials comprised of silica aerogel and oxide powders are currently under development for use in a new generation of thermoelectric devices that are planned for use in future mission designs. Work is ongoing in the development and production of non-silicate and composite aerogels to extend the range of useful physical properties, and thus, the applications of aerogels in future space science missions.

Chapter 15 – The atmosphere of the Earth is very different from those of its two neighbors Venus and Mars, and furthermore, the oceans exist only on the Earth. The authors have attempted to explain these differences by undertaking a study of the origin and early evolution of the atmospheres and oceans of these planets during and right after accretion. It is conceivable that both dehydration and decarbonation of the primordial planetesimals and the surface of the growing planets had to take place when infalling materials impacted the growing planets that exceeded a certain mass. In addition, during accretion, the composition of the growing planets would become practically frozen (except for hydrogen escape) and the growing planets were covered by a “magma ocean” after the planets grew to certain masses. The proto-atmospheres, that formed after the composition became frozen, would envelope the growing planets and consist primarily of CO_2 (This CO_2 proto-atmosphere is preserved on both Venus and Mars today). At the same time, most of the H_2O would remain in the magma ocean due to the high solubility of H_2O in silicate melts at high pressures. The giant impact that formed the Moon is the most viable mechanism that drove out nearly all the H_2O from the partial melting zone of the Earth. This H_2O was then incorporated in Earth’s CO_2 proto-atmosphere to form a supercritical H_2O-CO_2 mixture. When the Earth’s surface further cooled down below 450~300 °C, the dense supercritical H_2O-CO_2 mixture would have precipitated to form the indigenous ocean. The "hot soda water" might have reacted quickly with feldspars to produce carbonate and clay minerals that precipitated on the bottom of the ocean, and effectively removed all CO_2 from the Earth's proto-atmosphere. The mantle of Venus is not yet completely solidified. Some parts of the original "magma ocean" may still be entrapped inside Venus, forming the so-called partial melting zone. Thus, unlike Earth and Mars, only a small amount of H_2O from the magma ocean was added to the Cytherean proto-atmosphere after accretion, and this H_2O has been lost due to hydrogen escape and the hot surface temperature. Mars should have been completely solidified during a certain stage of its life. Thus, H_2O should have been incorporated in the Martian atmosphere after complete solidification. In view of its small mass, there should not be a large amount of H_2O in the Martian atmosphere, and H_2O and CO_2 should act rather independently during cooling. The H_2O might have condensed to form an ocean when the Martian surface cooled to below 100 °C, but most of the H_2O from either the ocean or atmosphere should have been lost to outer-space throughout Martian life due to its small mass. The mass of the growing planets beyond which the loss of CO_2 to outer-space became negligible during accretion is calculated to be 6.4 x 10^{26} g. The CO_2 contents in the proto-atmosphere of the Earth are calculated to be 5.9 x 10^{23} g, which is equivalent to a partial pressure of 114 bar on the surface of the early Earth. The total CO_2 contents are estimated to be 5.1 x 10^{23} g for Venus, 6.3 x 10^{23} g for Earth, and 3.6 x 10^{22} g for Mars.

Chapter 16 – During their evolutions, the small bodies of our Solar System are affected by several mechanisms which can modify their properties. While dynamical mechanisms are at the origin of the orbital evolutions of asteroids, there are other mechanisms which can change their shape, spin, and their size or result in their disruption. These dynamical and disruption mechanisms have been identified and studied, both by analytical and numerical tools. It was thus found that a large fraction of the population of Near-Earth Objects (NEOs) come from several zones of the asteroid main belt where efficient dynamical mechanisms are present. Actually, most NEOs are now believed to be fragments of large parent bodies that are collisionally disrupted in the main belt. These fragments are more or less directly injected into zones where these dynamical mechanisms occur. Such mechanisms, called resonances, can increase the eccentricity of an asteroid such that it eventually crosses the Earth's orbit. The main disruption mechanisms of small bodies are collisional events, such as the ones at the origin of asteroid families and NEOs, tidal perturbations, and spin-ups. In the particular case of collisional disruption, it has been found that most large fragments generated by the disruption of at least kilometer-size bodies correspond to gravitational aggregates formed by smaller fragments which reaccumulate during their ejection from th parent body. Thus, since NEOs are likely to have undergone this process, a large fraction of them should be gravitational aggregates or rubble-piles. However, the efficiency of disruption mechanisms depends on the strength of the material constituing the small body. As there are several evidence that suggest that most asteroids larger than a few hundred meters in radius are rubble piles, i.e. cohesionless bodies, a fluid model is often used to represent them and to estimate parameters such as their tidal disruption (Roche) distance to a planet. However, even cohesionless, solid bodies do not behave like fluids. In particular, they are subjected to different failure criteria depending on the supposed strength model. This paper reviews the main studies that led us to understand the origin of NEOs and to estimate their orbital and size distribution. Then, several important aspects of material strengths that are believed to be adapted to Solar System small bodies are exposed and the most recent studies of the different disruption mechanisms of asteroids are reviewed. These studies rely on our poor understanding of the complex process of rock failure. While our knowledge of these mechanisms has improved, there is still a large debate on the appropriate strength models for Solar System small bodies and on their most likely physical properties. Current and future space missions to some of these bodies devoted to precise in-situ analysis and sample return will allow us to determine whether those models are appropriate or need to be revised.

Chapter 17 – Hilda asteroids are objects that orbit the Sun at a distance where the orbital period is exactly 2/3 of the orbital period of Jupiter. Previous analytical and numerical studies have focused their attention to the great dynamical stability within the orbital region where the Hilda asteroids are found. However the group is indeed dispersing. In this paper the authors are going to review the contribution of Hilda asteroids to the Jupiter neighborhood. Based on simulations of long term dynamical evolution of escaped Hilda asteroids, it is find that 8% of the particles leaving the resonance end up impacting Jupiter. Also they are the main source of small craters on the Jovian satellite system, overcoming the production rate by comets and Trojan asteroids. Almost all the escaped Hildas pass through the dynamical region occupied by JFCs, and the mean dynamical life time there is 1.4×10^6 years. ~ 14% of JFCs with $q > 2.5$ A.U. could be escaped Hilda asteroids. The authors analyzed also the possibility that the Shoemaker - Levy 9 was an Hilda asteroid.

Chapter 18 – Astronomy research is concerned with the precise description and understanding of Earth orientation parameters (polar motion and length of day), time series that are obtained from several terrestrial and space observational techniques. The Earth rotation time series has been extensively studied from a linear point of view, considering that the underlying dynamics were composed of well-defined oscillations above a level of noises. If this can apply to the main part of the signal, where periodic oscillations are observed, it becomes problematic when dealing with short-term fluctuations. Indeed, linear method would conclude that this part of the signal is only composed of noises. Nevertheless, as the main forcing processes of the short-term fluctuations of the Earth comes from atmospheric processes through angular momentum exchanges, and because it was demonstrated that their dynamics were mainly nonlinear, it appears necessary to investigate the research of nonlinear dynamical processes in Earth rotation data. This is a challenge that the authors have initiated and that needs to be integrated in future research.

The authors present in the chapter a review of several studies they have carried out on this topic and some perspectives of research. As an introductory part, the authors describe the main components of the Earth rotation signal and their astronomical or geophysical origin. In a second part, the authors explain the principles of nonlinear time series analysis and their theoretical links to the dynamical system theory. The authors demonstrate how these tools give the possibility to break with the traditional spectra analysis that fails to capture nonlinear processes. To illustrate those points, the authors present results from several studies (on the Earth's rotation itself and on the related atmospheric flows) where they have described chaotic components in a topological and a dynamical way. The authors explain why our results have important implications for future research in astronomy. Indeed, they show that deterministic processes can be found and have to be searched in the "noisy" part of the system. Additionally, the prediction of the Earth orientation parameters is still performed with linear methods. The authors encourage applying in the future nonlinear predicting tools for the Earth's rotation. Moreover, the oceanic processes are also acting on the Earth and an analysis of oceanic time series can precisely measure how all of those dynamics interact together.

Chapter 19 – Since the end of the final glaciation, agriculture on Earth has been developed in several areas by modifying various ecosystems. The authors have replaced natural producers with crop plants and made humans the final consumer in each ecosystem. Though terrestrial agriculture has been established after a process of trial and error, agriculture on Mars should be prepared rapidly and firmly without any failure, even under severe natural conditions completely different from those on Earth. In addition, the agriculture on Mars should supply clean water and air, as well as foods, fibers and timber necessary for sustaining human life. This engineering target can be achieved by a strategy consisting of two stages. During the first stage, air, water and regolith are brought into a pressurized dome and are regulated somewhat similar to those on Earth. In the second stage, sustainable agriculture will be managed on Mars, mainly by utilizing on-site resources and actions taken by members of the ecosystem inside the dome.

Chapter 20 – A large fraction of the venusian surface is formed by rolling plains, whose altitude is typically 1 to 3 kilometers under the planet mean radius. The plains' most interesting features are volcanic structures, winkle ridges and tectonic belts. The volcanic activity has generated huge lava fields that have inundated almost all the planet's surface concealing any previous features that possessed. This activity has resulted in the formation of the plains, volcanic buildings whose morphologies are very similar to terrestrial analogs and that usually appear in groups of several individuals, although sometimes they are isolated; may well find a shield volcanoes, slag and tephra cones and domes among others. Wrinkle ridges are interpreted as resulting from failure of the crust under compression. Tectonic belts are linear zones of concentrated deformation hundreds of kilometers long, tens of kilometers wide, which include thrusted anticlines, grabens, tight folds, and strike-slip fractures. The combination of these structures makes up belts with different kinematics: extensional belts mainly characterized by grabens, thrust and fold contractional belts, and strike-slip shear belts usually defined by the en echelon disposition of folds (transpression) or grabens (transtension). Another important characteristic of the ridged plains, indicative of a crust excess density, is that the majority of them exhibits a geoid negative anomaly and are far from isostatic equilibrium. Whereas tectonic belts have received an ample attention, there is not consensus on their origin has not been reached. Several hypotheses have been proposed, for example excess density (deduced from geoid anomaly) causing compression through the drag of vertical traction and inducing plains downwelling, push of the geoid highs over low-standing terrains, or crustal delamination. Thus, venusian plains are useful keys for the understanding of the evolution of Venus.

Chapter 21 – Europa is the smallest of the Galilean satellites of Jupiter, although it is however a remarkable body and the primary focus of astrobiological interest in the Jovian system. Geological and magnetic evidences, as well as theoretical considerations, strongly suggest that Europa has a thin ice shell, maybe a few kilometers or a few tens of kilometers thick, which floats on an internal ocean of liquid water. Some anomalies have been observed in the huge magnetic field of Jupiter whenever Europa goes through it, which seems to indicate that this ocean is made of conductive water, due to a probable high content of salts. Moreover,the low density of impact craters suggest an surface age of about 60 Myr in average, which is very young in geological terms, indicating that internal activity, driven by tidal heating, is resurfacing the moon at the present time. So, Europa shows an overprint of crisscrossing ridges and linear markings, some extending for thousand of kilometers, which are locally disrupted for chaotic terrains, building a colorful and exotic textured surface. All of these features make of Europa a theoretically good niche for life. Biological organisms might exist in the salty internal ocean, in the icy shell (active or latent) at places such as fractures, veins or liquid reservoirs, on the internal sea floor in hydrothermal vents or submarine volcanoes, or even within the rocky interior.

Chapter 22 – This article presents descriptions of instruments and measurement results of neutron dosimetry for astronaut safety inside the International Space Station (ISS). Most neutrons inside the ISS are the secondary particles, and, especially for $E > 100$ keV, are one of the major contributors to the radiation dose received by astronauts because of their high radiation quality factor. The Bonner Ball Neutron Detector (BBND) experiment was conducted onboard the US Laboratory Module of the ISS as part of the Human Research Facility project of NASA over an eight-month period in 2001, corresponding to the maximum period of solar-activity variation, in order to evaluate the neutron radiation environment in the energy range from thermal up to 15 MeV inside the ISS. The BBND experiment is the first active measurement of neutrons inside the ISS, providing new knowledge about astronaut safety concerning the radiation environment as well as important inputs for further radiation environment model calculations. However, the BBND experiment can be considered as only half of the necessary evaluation of the neutron radiation environment inside the ISS since neutrons of 10 MeV $< E <$ several hundreds MeV are considered to make a comparable contribution to the radiation dose to that in the BBND measurement energy range. A scintillation fiber neutron detector with a sensitive energy range higher than 10 MeV has been developed, which is complementary to the BBND instrument. Wide-energy range measurements of neutrons by these detectors shall permit evaluation of total contribution from neutrons to the radiation dose received by astronauts.

In: Space Exploration Research
Editors: J. H. Denis and P. D. Aldridge
ISBN: 978-1-60692-264-4

Chapter 1

SPACE: THE PROVINCE OF MANKIND A BRIEF PHILOSOPHICAL APPROACH TO SPACE EXPLORATION

Jacques Arnould
Centre National d'Etudes Spatiales (CNES), Paris, France

ABSTRACT

International law considers Space as the province of Mankind. This can only mean that humanity invests more than mere scientific and technical resources, that the stakes are more than just financial and political. Imagination also plays an important role in Mankind's march to the stars, combining the most ancient dreams with the most spectacular achievements. This is why we need to think seriously about the significance Space exploration attaches to the exploits of human astronauts, and also about the idea of founding human colonies in Space. Just how necessary is it to invest in programmes that may take on colossal proportions? And do we possess the wisdom necessary to address all the questions it raises? If humanity declares that Space is its legitimate province, this implies a commitment to understand the purpose of it all in advance, so as to be able to explore it in a responsible manner. These are the principal philosophical themes addressed in this chapter.

"All I want is for you to build a ship
and to net me some distant islands far out to sea."
(Antoine de Saint-Exupéry, *Citadelle*, Chapter CCI)

INTRODUCTION: MAN, THE WALKER

Man has always been a walker, a walker who created new worlds as he marched along. Fleeing the continent of Africa, that had suddenly grown too small for him, fleeing himself, like a stranger he knew so well that there was no room for both of them, he has always been running away from something. He flees the worlds born of his own imagination, trodden by his feet, furrowed by his ploughshares or rutted by the wheels of his chariots; he is always ready to set off and conquer new worlds. But were not those new worlds that he conquered the worlds that he himself dreamt into existence and built up from nothing? Were those new worlds anything other than the blank pages of his atlases, the frightening and fascinating uncharted *terrae incognitae* of his maps? As a sedentary creature on Earth he was a nomad only with respect to himself. As Arthur Eddington put it: “We have found a strange footprint on the shores of the unknown. We have devised many profound theories to account for its origins. At last, we have succeeded in reconstructing the creature that made the footprint. And lo! It is our own.” Man could only be certain of his own existence. He was afraid that he might be alone, for eternity. Alone with his imagination and his worlds. So he decided to conquer the Universe, without being at all sure that there really was a ‘beyond’ from which he was absent.

Gagarin was the first man to sail beyond the bar that surrounds our planetary atoll. And he reported that there was no-one out there. Where then were the angels of our ancient beliefs and the extraterrestrials of our modern-day dreams? Are we really alone? Being totally alone might be terrifying for some and reassuring for others, but it is a veritable abyss that opens up beneath the feet of our spacemen. The abyss represented by the fifteen billion years of our Universe; much too large a step for Mankind ever to catch up, becalmed in a speed of light that suddenly seems to flow not like light but like treacle. The Space we can access through our technology, the Heaven we believe in, the Cosmos we imagine, all seem petty when put into perspective by the stars and their mocking silence. Voyager 1, launched in 1977, has now reached the frontiers of the Solar System and its cosmic winds: this is the furthest that any probe from Earth has ever ventured and its signals now take half a day to reach us. Yet this is no more than a flea-hop beneath the vast marquee of the great cosmic circus.

Of course, this fifteen billion year abyss has seen an incredible story, with millions of different species, brilliant initiatives and dramatic setbacks but it is still an abyss leading back down to the enigmatic and ever-inaccessible Big Bang. Where do we come from? Where are we going? The ancient questions remain completely unanswered, despite man training his telescopes onto the blackness of the heavens and focusing his microscopes into the heart of life. And those pioneers in their spacecraft riding on the clouds are still torn between their nomadic side (the call of distant horizons) and their sedentary side (the attachment to their homeland).

When neither the eye nor even the mind’s eye can still distinguish or recognise, identify or locate the features of our world, when observation and examination must give way to suggestion and imagination, when coastlines, rivers and contours merge into fantastic shapes or into blank ignorance, then we have returned to the time of the *terrae incognitae*. These unknown regions that, ever since Man first strode the Earth, have driven him beyond the limits of the city and of his own self. These unknown regions that cartographers of past ages illustrated as being inhabited by extraordinary, frightening creatures as if to warn the readers of their charts and globes to “Stay away” from these regions, thus keeping them *incognito*. Are there no unknown parts of our Earth left today? It only needs an astronaut to photograph the Earth from the porthole of his vessel and then to forget where he was at that moment for the unknown to reappear as if by magic. And our imagination again has free rein.

This is why, despite all the knowledge accumulated by modern science, Space retains its mystery, a mystery that is only heightened by its vast dimensions and the beauty, sometimes frightening, always fascinating, of the images sent back to us. Because Space continues to ask the question of meaning: the meaning of the cosmos, of life and the life of each one of us, without ever providing a clear or definitive answer. Because today more than ever, Space calls for creativity, from the artist or the mystic, and increasingly from the engineer and the scientist. When confronted with Space in all its aspects, as a new world, a challenge, a mystery, man seems to hear the strange prayer of the Psalmist to God, which seems like a prophecy of the strivings of today's astronomers and astronauts:

> "When I consider thy heavens, the work of thy fingers,
> the moon and the stars, which thou hast ordained;
> What is man, that thou art mindful of him?
> and the son of man, that thou visitest him?
> For thou hast made him a little lower than the angels,
> and hast crowned him with glory and honour.
> Thou madest him to have dominion over the works of thy hands;
> thou hast put all things under his feet" (*Psalms* 8, 3-6)

Like our desire for Heaven, the conquest of Space is no idle dream or nostalgic reverie. Both require a commitment from the individual and society alike, from the depths of what makes us most human, most earthbound. If we wish one day to be children of the stars, to walk with angels or to talk with extraterrestrials, we must continue to examine what it means to be human. To imagine Space. And to dare to take possession of it.

1. A New Earth

On a certain night when Mankind was young, the first man to look up at the Moon thought he saw a smile, an invitation. And he thought that the stars were leaning down towards him, like the Good Fairies round the cradle of a newborn baby. Perhaps this was the moment when humanity emerged as a distinct life-form after more than three billion years, a rather exceptional species which, despite having its feet planted in the clay of this Earth, would never again cease to look up at the heavens. Konstantin Tsiolkowsky was sure of it at the end of the 19th Century in Russia, when he imagined the ancestors of our modern rockets and wrote "The Earth is the cradle of Humanity, but no-one can stay forever in his cradle."

On a certain night in December 1968, the heirs of the Russian scientist had succeeded in sending a spacecraft carrying three humans right round the Moon: the dream of several thousand years had become reality. Through their porthole, the astronauts saw a planet coming up over the grey horizon of the satellite they were orbiting: the Earth, our Earth. A cradle? More like a vessel as they describe it, a blue orb marbled with white clouds, beautiful but looking so fragile, floating in the immensity of the black, unfeeling cosmos. And then, as though to reassure themselves or to propitiate any gods whose wrath might be stirred at seeing humans pass through the gates of heaven, they read the first page of the Book of Genesis: "In the beginning God created the Heaven and the Earth." As it was Christmas Eve they dared to send back wishes of good luck to "all of you on the good Earth."

Those three men and their colleagues in Space were not the first to worry about the state of our planet. But their testimony and the images they were able to share with the rest of us who had stayed "down here" contributed then (and continue to do so today) to an increasing concern for the future of the Earth and of Mankind upon it. As Hannah Arendt wrote in The Human Condition: "Nothing, not even splitting the atom, can overshadow the launch of the first Sputnik as an event of primary importance. It confirms the prophetic inscription on the tomb of the Russian scientist Konstantin Tsiolkowsky: 'Mankind will not be bound forever by the confines of Earth'." Although the Earth once nurtured Mankind, as a mother her children, today Mankind has to protect and cherish her. Humanity has never felt so closely attached to the Earth.

On a certain night in July 1969, when Neil Armstrong made the first human footprint in the grey dust of the Moon, he wrote Mankind's destiny in the heavens once and for all, however fleeting his passage and whatever the future of the Apollo missions. He was the first, the herald who announced to humanity that they were all now citizens of the skies. Henceforth, heavenly bodies would no longer dictate human behaviour, nor reveal the future: the arrival of the white-clad traveller with the golden helmet tolled the knell of astrology and divining by himself foretelling Mankind's future among the stars. The gesture seemed so emblematic that nobody took issue with it at the time. It was not until the following year that the Nobel Prize-winning French biologist Jacques Monod put it all into perspective. "The old alliance has been broken; man is at last aware that he is alone in an immense universe that is quite indifferent to him, where he appeared by chance. No destiny awaits him, no obligations are imposed on him. He alone must choose between the Kingdom and eternal night." (Monod 1970, 194-195).

Some had perhaps already begun to doubt, even before the dark and frightening vision conjured up by Monod's words. The Apollo astronauts were forgotten, except when the troubles that beset the thirteenth in the series seemed to confirm the superstitions of another age. Many got carried away either attacking or defending the idea of a Universe with no predetermined destiny, of a Humanity with no preordained duty. Perhaps Mankind had gone mad, deluded by the promises of science that seemed to be carrying all before it, or a victim of the imperfections of his genetic heritage. The first Ethics Committees were set up with the hope of keeping the light of reason burning. We forgot the Moon.

On a certain night, Jean-Pierre Haigneré recounts that he and the other observers in the Russian Space station MIR, a few hundred kilometres above the Earth, saw a strange dark smudge spread over the planet, like oil staining a wedding-dress (Haigneré 2001, 118). It was the eclipse of 11 August 1999 crossing Europe, bringing night when there should be day and awakening ancestral fears. What if the Sun never came back? What if the heavens should fall, to punish us for our outrageous daring, for our arrogance in wishing to become the masters of the natural world, for attempting to leave the Earth? As though a childhood fear of the dark rose up in every one of us.

And on a certain night, perhaps, a spaceship will carry humans away to a new exile, in a new exodus. Banishment or privilege? The words will not matter. They will be saying a last goodbye to a planet that has become a hostile environment to them or to all humans. It will be a night of mingled doubts and hopes: will Mankind still be human, after abandoning the Earth? Tsiolkowsky, as I said, was quite sure. Those who come after him will perhaps, one day, have to make that choice.

2. NOAH'S ARK

Those in favour of Space exploration also like to quote these other words of Konstantin Tsiolkowsky's: "Mankind will not remain forever bound to the Earth. First, prudently, he will test the limits of its atmosphere, before setting off, later, to conquer the Solar System." It is a charming picture, and not far from the truth; but it is wrong and potentially dangerous. What happens to unused cradles? They end up in attics or cellars, on dumps or in second-hand shops. Will the Earth one day be stored away in some attic of the Universe, to be recycled by some monstrous black hole? This is what the astronomers predict, but not for several billion years. Even if Mankind has abandoned it by then, will he have become truly adult? There will never be a more serious question than that.

"When I was a little child," wrote Marcel Proust, "I didn't feel as sorry for any character in the Bible as I did for Noah, cooped up in that little Ark for forty days by the Flood." And indeed, what entertainment could a child expect to find in a place where you would before long have explored every nook and cranny and exhausted every pleasure? Are not legs meant for running, wings for flying, fins for swimming? Life seems too short to be shut in for forty days, even in such an extraordinary menagerie. "Later," continued Proust, "I was often ill and forced to remain in my own Ark for many long days. That was when I understood that Noah had his best view of the world from his Ark, despite being shut in and despite the night covering the face of the Earth" (*Les Plaisirs et les Jours*).

Proust is not famous for advising moderation. He does not follow Voltaire in suggesting we should each cultivate our own garden. On the contrary, he explains that Man only attains adulthood by making the edge of the world his horizon. In that case, answer Tsiolkowsky's disciples, our master was right: we must prepare to move on from Earth as we would leave a cradle that will soon be unable to hold us! Like impatient children, replies Proust calmly, they would dare to plunge headlong into the world without having first lived with themselves, gained confidence in themselves, learnt to love themselves and the world. Noah, that brilliant shipbuilder and brave navigator, did not flee the world, nor did he judge it. He drifted, buoyed up by his own lucidity and good nature, his pragmatism and optimism: he knew that one day a dove would bring him a fresh olive branch. But on one condition: he had to load the Ark with all the animals of creation. Not as owner or master, but because he was clearly responsible for them.

For a long time, Mankind dreamt and indeed believed that he stood at the centre of the Universe, although there was little evidence for it. Four hundred years later, scientific and technical progress has changed the situation: humanity now finds itself playing the role the Bible attributes to Noah, that of the captain of a ship, the curator of a menagerie, charged with exploring and conquering other worlds that are once again unknown. This is not something Mankind can afford to be too proud of. It is just that he is burdened with a conscience that will not allow him to ignore the destiny of others *a priori*. Not only are we all 'in the same boat', our human condition obliges us to take this responsibility seriously; there is no way, or at least there is no longer any way, of escaping this duty.

One day, perhaps, a few humans will board a spaceship, carrying specimens of terrestrial life with which to seed other planets. Whether the Earth they leave behind them is already dead or, on the contrary, at peace with itself and its inhabitants, we must hope that it will not have become a useless cradle to be cast aside. After being cherished by her as every child is by its mother, the future generations to spring from our loins will in turn be obliged to care for the Earth. As André Malraux might have said, our honour as men – or as Man – depends on it.

3. Riders of the Clouds

The Ancients had no one else to call riders of the clouds except the gods and, more generally, the creatures that peopled the heavens, whether they lived there or streaked through it like lightning. Living in heaven and riding the clouds meant more than just enjoying a perfect world and sipping ambrosia, the drink of the gods; it also meant acquiring and possessing immense power. In his *Icaromennipus: an Aerial Expedition*, the philosopher Lucian of Samosata ridiculed the internecine struggles between men, their contradictory prayers, the way they run busily about like a seething mass of ants. His character Menippus explains that man can rise above the clouds through the work of the spirit and broaden his knowledge in proportion to the cosmos as he becomes a citizen of the heavens. Menippus transcends his earthbound condition as a way of denouncing vanity, injustice and the absurdity of inequalities, social divisions and war. The gods and goddesses also used the clouds as a handy way of getting about and dropping in on humans, sometimes to the point that it was hard to distinguish between the passenger and the vehicle. In the Bible, the cloud that hung over the Ark of the Covenant was more than a sign or attribute of the divine presence; it was almost divine itself.

For a long time, humans on Earth dreamt of mingling with the gods in their heaven. They climbed the highest mountains, particularly those with clouds at their summits, as though this might bring them a little closer to the divinities. Olympus, Sinai, Athos: how many sacred mountains are there on our planet, promoted to divine status by the will and the faith of men? Quite apart from the artificial skyscrapers built by men themselves, from ancient pyramids to the most ambitious modern towers, via the spires of cathedrals. Whether religious or secular, they are all places where men try to acquire, steal and also disseminate the knowledge and power which they hope will make them resemble the gods. Even so, only a few can hope to meet eternity beneath a pyramid, officiate in the choir of a great cathedral or preside over human affairs from the top of a glass and steel tower. Initiations and trials, ordination and rites of passage make riding the clouds just as exclusive today as it ever was, even with one's feet firmly on the ground.

Less than two centuries after the first flight of a lighter-than-air *montgolfière,* and barely half a century after the first flight of a heavier-than-air aeroplane, a man burst through the veil of clouds that envelope the Earth: Yuri Gagarin became the first rider of the clouds with feet of clay. Did that mean that Mankind had finally usurped the place of the gods? Gagarin and his kind have a different view of the Earth and their fellow-beings to that of Menippus! They express themselves not like conquerors but like angels. One of Gagarin's successors, talking about the Mir Space station, once told me: "When I was up there I knew perfectly well that I hadn't got there alone. Thousands of people had contributed to the success of the mission. I felt that I had somehow been sent by them, an envoy of Mankind." I chose not to interrupt him, but I thought at the time: "Does he know that *envoy* is the translation of the Greek *angelos*, or angel?" I rather like the idea: astronauts as the angels of Mankind, even if we do not usually think of angels as encumbered with the heavy space-suits that make astronauts as clumsy as penguins on dry land.

But what message do these envoys carry with them in the fire and steel of their spaceships? And what message do they bring back from their time in Space? We would perhaps be wise to add: to whom do they carry this message from humanity? And where does the return message come from? Because, for the time being, there is no-one "up-there". No-one to listen to them, no-one to answer, with a message for us. But perhaps I am being unduly difficult. The man who told me he felt as though he had been sent by Mankind brought back a message in the shape of thousands of photographs, taken from Space. Not a collection of the most beautiful pictures of Earth taken from more than forty years of Space exploration, by both manned and unmanned satellites, but a graphic and dramatic portrayal of the Earth, Space and Mankind. The perils of such an adventure have been known since the dawn of humanity, as shown by the Icarus myth: it can be dangerous to use artificial wings to attain the viewpoint of the gods and examine the Heavens and the Earth. This drama had really been acted out by the cosmonaut, before he decided to lay it before the eyes of us Earthmen, for our pleasure and perhaps to make us think.

These images, this message, spark a range of feelings beyond simple aesthetic pleasure in even the most earthbound of Earthmen. Old questions arise immediately, some as old as humanity itself. Questions about how we should relate with the Earth, now that our science and technology enables us to be the masters and proprietors of nature. About weight and the graceful way living creatures interact with the world about them. About our capacity for grasping reality, even in its most complex or chaotic forms. About the shifting frontiers that sometimes separate likes and bring together unlikes. About life and death.

4. I Did It!

"I did it." When Charles Lindbergh climbed out of the *Spirit of Saint Louis*, after crossing the North Atlantic, these were the only words he could find for the admirers who crowded round him. "I did it." An artist in front of a finished work or an artiste after a performance, may have no more to say to those who would congratulate or question him: if he thought it necessary to add comments or explanations, would that not imply that he had failed in his artistic mission? Any real exploit requires no explanation; it is sufficient unto itself.

"That's one small step for a man, one giant leap for Mankind." The words uttered by Neil Armstrong, as he placed his foot on the Moon have echoed round the world; they identify what was at the time perhaps the most widely-viewed event in the history of the world. The man who conquered the Moon explained later that his words had not been prompted to him by NASA's communications department: he had merely wanted to say something not too complicated. Asked, thirty years later, what their time on the Moon had meant to him and Aldrin, he said that they had been surprised by many things (such as the trajectory of the dust raised by the LEM's thrusters or by the astronauts' own boots), but also that they had not always managed to do "a very good job" (by which he meant that they had not had time to collect enough geological samples). But whether they "did it" or they "didn't do it", Armstrong and Aldrin's exploit will surely go down in history.

Etymologically speaking, the term 'exploit' comes from the Latin *explicare* and *explicitum* thus corresponding to the idea of unfolding, developing and ultimately carrying out an action. By extension, this original sense soon changed to mean a particular feat or, on the contrary, some quite inappropriate action. We can take something from each of these three meanings, each one bringing something to the modern definition: accomplishment; exceeding expectations; adjustment. The idea of an exploit includes that of an action performed skilfully in complete liberty, independently and responsibly: I know what I want; I believe that I can; I acknowledge and evaluate the risks; and then I go to work. It represents the passage from an aspiration to an achievement; a twist to the normal run of things, a transgression of the established laws of nature, to life itself and its basic needs. Is it possible for either the human individual or the human race to grow without resorting to exploits? And, to look at it the other way around, can an achievement be qualified as an exploit if it does not provide the individual and the race with an opportunity to grow and increase its faculties?

Seen like this, it was a genuine exploit that Christopher Columbus accomplished with his three little ships. Without disregarding the catastrophic consequences of the arrival of Europeans in the New World, it cannot be denied that Columbus caused a revolution in human history by providing new prospects and pushing back the frontiers of exploration, in a way that affects us all today. After 1492 the world was never quite the same again. We can see that an exploit is only worthy of the name if it advances human history and accomplishment in some way, preferably in the direction of wellbeing and progress. I quite concede that my definition of an exploit would force the media to be a little less enthusiastic in their praise of certain accomplishments, particularly in the sports pages. What then of the mission of the astronauts on Apollo XI? There's no denying that it was an exploit in terms of the extraordinary undertaking accomplished with so much intelligence and daring; we hear that 300,000 to 400,000 people were working on it for a decade. But the exploit will only be complete when Mankind does more than just set foot on the Moon and lands there to settle.

The Space agencies sometimes agonise about which values they should be promoting. They go through the list of 'pioneer spirit', 'encouraging exploration', 'skill', 'creativity' and 'dynamism', 'professionalism' and 'technical precision', 'teamwork' and 'international cooperation', 'caution' and 'risk-management'. All values which of course have a central place in the know-how necessary for Space projects (and the ability to pass on that know-how). To that we can unhesitatingly add the idea not just of the straightforward accomplishment of a purely technical exploit but of the achievement it implies in terms of human potential and limitations.

To be an exploit an achievement must be humanist, respecting Mankind's genius and creativity, goodness and beauty, what we sometimes call *grandeur*. But is that enough to justify great enterprises, to reassure the hesitant and to silence the opposition? By no means. Although an exploit is a value in its own right it can neither be used to give a clear conscience to human activities, nor to put a brake on them. If an exploit is really a value it should, on the contrary, increase Mankind's freedom; it should support our efforts and strengthen our resolution. Even if humans now know that they are mere specks of dust in the history of the Universe and Mankind, lost in this unceasing flow of Space and time (an irresistible, irreversible, stubborn but unpredictable flow), they are discovering and coming to realise that they are capable of occupying, of living out the brief instant of their existence as though it were the beginning of a new world, of leaving a footprint as though Mankind were making many "giant leaps", each a new episode in the human saga. There will always be a new exploit tomorrow.

5. Astronauts Now, but what Next?

"…Man has never sought tranquillity alone. His nature drives him forward to fortunes which, for better or for worse, are different from those which it is in his power to pause and enjoy." This thought comes from Winston Churchill. It ranks among the store of quotations that leaders of men like to draw on when they need to raise enthusiasm, motivate their teams and overcome momentary hesitation or discouragement. It is in the same league as the speech President Kennedy gave when launching his country's participation in the Moon Race: "It is time […] for a great new American enterprise – time for this nation to take a clearly leading role in Space achievement, which in many ways may hold the key to our future on Earth. […] I believe that this nation should commit itself to achieving the goal, before this decade is out, of landing a man on the Moon and returning him safely to the Earth." And we all know the outcome of this challenge: the first steps of Neil Armstrong on the Moon in July 1969, followed by a rapid loss of interest in the Apollo programme, the serious incident with Apollo XIII and the final mission of Apollo XVII in December 1972. Then came the post-Apollo period, like an anticlimax after the golden age of Space exploration: how can we explain the meaning or the necessity of such an enterprise, especially when it results in the catastrophes of the two American Shuttles in 1986 and 2003?

Like an echo to Churchill's words, we frequently hear manned Spaceflight justified by the "human nature" argument. "It's in the DNA" say the promoters and supporters of manned flight and the notion of Man in Space. But a little thought will show that resorting to genetics to support your case can be dangerous. For one thing, it can lead to the worst kind of eugenics: if ever promoting or even defending manned flight should be outlawed, we could return to the first half of the 20th Century in the United States, when sinister agencies were charged with preventing those who were genetically inclined to poverty or murder from reproducing. And for another thing, and perhaps more seriously, we should never admit the unqualified notion that man is the consequence of his genes alone: to be human is above all to have a choice, as expressed by the French biologist and thinker René Dubos. This freedom of choice can lead to the grandeur and honour that mark out our species, but can also be its downfall.

Nobody can deny the technical, financial and human costs of manned flight. It is perfectly true that man was not designed to live in Space, that he must employ an armada of technological solutions and precautions to protect himself against the effects of weightlessness, cosmic radiation and isolation in Space. So we cannot escape the question of whether so much effort is worthwhile: what is the point of manned flight today? And the very image of astronauts and cosmonauts endlessly orbiting the Earth in an unvarying routine invites the inevitable comment: manned flights just go round and round and round…

Before we make a final judgement on whether or not to continue sending men and women into Space, I believe that justice requires us to consider one important observation that can be expressed very simply: although Mankind began visiting and working in Space as early as 1961, we have never really lived there. Yuri Gagarin opened the era of exploration and conquest and America's Skylab the era of working in Space; but I do not think we can yet claim to have entered the era of living in Space. Spacecraft, whether shuttles or stations, continue to resemble the base camps of mountaineers; their occupants wear uniforms and follow programmes and regimes worthy of explorers or soldiers, not at all like settlers.

Just because Space exploration has so far only travelled a short distance along the oft-followed path that throughout human history has led from exploration to conquest, does not prevent us from turning a critical eye on it. Once again, there is no such thing as a genetic predisposition to exploration, any more than there is for war-making; there are only beings conscious that they have to make choices, take decisions, implement those decisions and live with the consequences. The decision to send humans into Space was taken in a particular socio-political context which is now part of history, which might lead us to conclude that, with the accumulating difficulties and costs, it would be better to abandon the idea. This option has to be considered and weighed seriously, on the strict condition that we do not lose sight of the implications for the future, not just of the International Space Station, a permanent Moon base or a mission to Mars, but the future of Mankind itself.

Of course, I am not trying to say that the future of our species depends on the pursuit of Space exploration, as spelt out in the current programmes. Nor do I think that Space is the only way we can hope to survive any demographic, meteorological or ecological catastrophes that may threaten our planet. If we managed to develop the means of fleeing an Earth that had become uninhabitable, or of transforming Mars so as to make it potentially liveable, would we not also be clever enough to stop the destruction of the Earth's biosphere? The future of manned flight concerns the future of humanity not only because of its value (a value that should be constantly examined and discussed), but also because of the human qualities it reveals, identifies, puts to use, promotes, develops and so on.

It seems clear to me that this aspect is far from being properly recognised or examined. What do we know today about the human potential that might be stimulated by the adventure of manned flight? We could no doubt learn a great deal from aviation or rather from the airmen for whom, apart from the obvious commercial and military uses, flying still has an entertaining side. We could also learn from people like Antoine de Saint-Exupéry or Richard Bach who succeeded in drawing lessons about humanity from their flying experiences, without trying to use it as a justification for flight itself. Should we perhaps require that Space exploration contribute in the same way to our common culture? But if artistic genius cannot be stimulated by decree, how can we excite it, encourage it and favour it?

At the beginning of the 15th Century, when Chinese vessels had already travelled as far as the coast of Somalia, the Emperor suddenly decided to cease all exploration, to destroy his fleet and to limit sea-going to coastal navigation. It was from this point on that China turned in on itself and stayed that way until the end of the 20th Century. It would be foolhardy to make too much of this event in the context of a debate about manned flight, but we should perhaps not ignore it altogether. To suspend current programmes for manned flight would be one thing – to abandon any ambitions for them and turn in on ourselves for centuries would be another.

6. Living in Space - Tomorrow?

Even the nomad, the pilgrim or the vagabond has a roof somewhere (even if only in his imagination) to shelter him from the weather or the unwelcome attentions of others, a place to find a little comfort, to get together with others or just to be alone. At a time when the future of astronauts is under debate, perhaps we should examine the purpose of this project for living in Space, however old it may be. As soon as the first individual spacemen, followed by crews, began to visit Space, the question was raised about going to live and work there. The test pilots who made the early flights were soon joined on their Spacecraft by scientists running real and complex experiments; Space engineers were next, putting together the first elements of orbiting Space stations; and so on. "The astronaut is no longer just a pilot," wrote Gabriel Lafferranderie, "he is a scientist, an astronomer, a doctor, an engineer and a journalist and he will one day need to be a gardener, a miner and eventually a tradesman. He participates in carrying out industrial experiments. He is a jack-of-all-trades, living in a confined space, permanently observed and overheard from the ground base that sends him his instructions, wakes him up or tells him to sleep, requiring him to do physical exercises or submit to medical experiments and accomplish in Space certain movements that are totally unnatural." (Lafferranderie 1993, 255). As an astronaut-of-all-trades, can a human being be said to be living in Space? I am not so sure.

When the Americans transformed a stage of the Saturn V rocket into a living space and called it Skylab, they refused to label it a 'Space station', preferring the term 'workshop'. More realistically, Salyut, Mir and the ISS were each known as orbiting stations, but living conditions on board were still much the same. The occupants stay for longer periods, but their daily routine has hardly changed. Men and women certainly live and work in Space, but do they really reside there? I think we need to take the astronauts' own words seriously when they describe life on board these Spacecraft as 'camping'; or when they see themselves as envoys, carrying a message on behalf of Mankind. They ride the clouds, racing through the heavens and customising the little hutch which is their temporary living space with photos of their loved ones waiting back on Earth. They are orbiting gypsies, not yet Space-dwellers.

Through literature, the cinema and science fiction in general we have long been accustomed to imagining human colonies settled in Space, inside immense spacecraft or on bases, or even cities, built on (or sometimes under) the surface of other planets. Space agencies actually have designs for such bases on paper (or in the minds of their engineers). They would involve considerable scientific and technological problems and would definitely represent the kind of challenge that Eugene S. Ferguson has called *Macro-Engineering Projects* (*MEPs*), projects at the very frontier of the know-how and resources of their era (Ferguson 1978, 6-18). Some will cry "Utopia!", and they may not be entirely wrong. Yet a Utopia is precisely what we need, not just for tomorrow's adventures in Space but to help us live sustainably on today's Earth.

Let us now think about the status we would be prepared to give a Space colony, installed on the Moon or elsewhere. It would no longer be a case of individual astronauts considered as members of a crew or a team, but rather a human society. How would it be organised and governed, what would be its laws and constitution, its cultures and even its symbols? What kind of a relationship would it have with Mother-Earth: dependent, partially autonomous or totally independent? It would no longer have the sole purpose of carrying out scientific research like an Antarctic base today. Nor would it be a case of astronauts sent out into Space with the intention of returning to Earth after a reasonable period to take up again their national identities and their status as astronauts. They would be living in real colonies, as citizens of the Cosmos. Would we be prepared to create a new State, a new nationality, something different from 'internationality'? And how would the colonists' new Space culture develop, as it would certainly not spring *ex nihilo*, created out of nothing?

Finally, let us imagine that these astronaut-colonists who have now become Space-dwellers manage to acquire such a degree of autonomy, such independence from the Earth and its Earthmen that they become a different human species, by the process of divergence familiar to biologists. Have we ever wondered what kind of legal instruments we could use to govern the status of each party, the relationships between them, their respective rights and obligations? Again, the question is Utopian; but it is not so far removed from the legal challenges created in our own society by certain advances in medicine and biology. We have thankfully got beyond the stage of wondering whether or not other human populations have a soul, whether or not they can be considered as our *alter ego*. But our society still has difficulty in relating to creatures that have not yet quite become or that have ceased to be totally human.

Once again, this theoretical exercise, however Utopian or sidereal, forces us to take a close look at ourselves, our own real identity and expectations.

7. Thinking About the End

The date chosen was no coincidence. To win the X Prize (the ten-million-dollar reward for achieving the first private access to Space), SpaceShipOne needed to reach an altitude of 100 km for the second time in less than two weeks, carrying a load equivalent to two passengers. The exploit (and it was certainly an exploit in terms of the uncertain sciences of propulsion and aerodynamics) was achieved on 4 October 2004, the exact anniversary of the flight of Sputnik I, the first artificial satellite sent into Space by humanity. The success of SpaceShipOne prompts us to look back at almost half a century of Space activities which has seen so many 'firsts'. The best known of course are those that sent human beings further than ever before, as far as the surface of the Moon; from up there Mankind was able to look down for the first time on the whole of the Earth, so beautiful but so fragile set against a black velvet background speckled with stars. Other technological exploits, like the Hubble Space telescope, have already revolutionised our knowledge of the Universe and are continuing to do so. At the same time, telecommunications and observation satellites are circling the Earth in large or small networks, enhancing the quality of our lives and improving our safety and defence. Also, as some would wish to point out, SpaceShipOne's exploit brought private enterprise and tourism into the Space field, just when the President of the United States was seeking to re-focus America's Space policy on the future conquest of Mars. How far will humanity go in its race to the stars? The question inspires all those enthusiasts who revel in theories about astronomy, and love predicting technological progress. It also horrifies those who worry about humanity going too far, about the cost of all this wasted grey matter and finance, at a time when the world is undergoing a wide-spread crisis. And they seriously envisage stopping the waste and limiting our activities to a reasonable, machine-based use of our immediate Space environment.

This is a serious question and deserves one day to get the thorough democratic airing that it has so far been denied. It is easy to see why the prospect appals those who invested so much in achieving those early 'firsts' mentioned above and who dream of achieving new ones. It is certainly not my intention to try and resolve that debate here, a debate that can and no doubt should be held one day. I should simply like to take the opportunity to put the spotlight on an aspect of this situation which is often hedged around, forgotten or ignored. It is one thing to achieve a 'first', another, however difficult and dangerous, to give that achievement a real future, both sustainable and in keeping with what humanity expects, and yet another to plan for or insist on a 'last' for it.

In aviation, the hundredth anniversary of the Wright Brothers' famous flight came just months after Concorde was taken out of service, to the great disappointment of admirers of the "beautiful white bird". It inevitably reminds us of the American Space Shuttles, grounded for many months after the second shuttle disaster. What about the premature end of the unfinished Apollo programme? Or the decision by the Soviet Union to cancel the Buran shuttle programme, despite the advanced technology already developed? Or again Europe's Hermes project, that never got off the drawing-board? All potential 'firsts' that had to be clumsily abandoned. Why such a feeling of disappointment? Probably because Space exploration seemed to be picking up the baton of scientific and technological progress that had swept through the West and the 20th Century. Its roots lay deep in the past, it had broad ramifications and had accomplished much. And despite the setbacks of two World Wars during the last century, the idea of progress seemed to have found in Space an opportunity to renew its onward march. Is not Space by definition, both literally and metaphorically, infinite? After the Moon Race, Space seemed to be the arena where international tensions would be surpassed: first legislators and then politicians declared it to be part of the Common Heritage of Mankind, with a view to avoiding competition for possession or use. In short, there were many reasons to hope that Space exploration had a bright future before it and to see any interruption as a catastrophe!

Humanity is so conscious of its own faults and of the destructive power of the science that it has mastered that it needs to be convinced of the idea of progress in order to face up to present reality and imagine a future, however idealised – or however frightening. At the time, it was hoped that the First World War would be the "war to end all wars" and we know what happened next. But does this need for progress require us to deny freedom of choice, surely just as essential? Must we forget that the human adventure, wherever it may lead us, is never marked out in advance, either by its successes or its failures? The future of Space exploration, from its most banal to its most exciting achievements, is not automatically predestined to lead to victory; Mankind alone can decide to make a triumph of it, to pursue the goal or abandon it. It is a mark of human folly (but perhaps also of human grandeur) that men never cease to believe that all they have to do is choose a path for it to be crowned with success.

Conclusion: Wisdom

We have therefore seen that, as well as being the wild blue yonder, Space is also a place for action. Its importance to us will depend on what Mankind does there, what he makes of it. Without humanity and its infinitely creative imagination, Space would be just that, totally indifferent, perhaps without meaning. "Once we fully imagine Space," writes Gaston Bachelard in his *Poétique de l'espace*, "it ceases to be a neutral part of creation simply to be measured by trigonometric theoreticians. We live it, not in its straightforward reality, but with all the preconceived ideas of our imagination. It has the particular quality of almost always exercising attraction. It concentrates each creature within the boundaries of whatever protects it." (Bachelard 1994, 17). In conquering Space, Mankind is simply applying the wisdom of the ancient Greek precept: "Know thyself." And heaven knows that wisdom is a vital commodity for our decisions about Space, our Space.

Wisdom: as passed down to us by the Ancients, with advice both on how to act and how to live. The poet must work with the engineer, the artist with the inventor, the philosopher with the politician. It is no easy task: we can no longer try to read the fate of humanity in the stars, but must write it in the stars, with determination and feeling.

Wisdom remains at a premium, to introduce meaning, to give meaning to an enterprise struggling under the weight of almost half a century of human, technological and cooperative exploits. And of failures too. John F. Kennedy's speech was a long time ago. Who will give purpose and meaning to this new century? Where shall we find the inspiration to draw the new frontier, the new horizon? Some may try and avoid the responsibility: we all know only too well the painful truth of the Chinese proverb: "When the wise man points at the moon, the idiot looks at the finger." But nowadays, thanks to telescopes and probes, the wise man's finger is much further away than before and to look at it is already a way of experiencing Space.

Wisdom as well to teach humanity to renew and strengthen its ties with the Earth and the living things upon it. Voices are sometimes raised worrying about all the attention and resources that are lost by being concentrated in Space exploration. A cynic might reply (not without a modicum of good sense): "But how do you know they would be used any better on Earth?" These voices should be allowed their say, however. The "space-artist" Pierre Comte once created an installation entitled *Signature Terre* that was clearly visible to a satellite. Who else but man could dare to "sign" the Earth, thus giving it an identity and a meaning that it lacked before the appearance of the human race? The signature, inspired by Aphrodite's mirror, did not suggest that humanity possessed the earth but rather that humanity was born of the Earth. For ages past, Mankind has concluded that this relationship, both imposed and recognised, gave it both rights and obligations. What if Space provided the key to fulfilling both?

Wisdom, lastly, so that men may never forget the freedom that is their birthright. If we were to turn our backs either on Earth or on Space, it would be at the risk of transforming them into burdens encumbering our conscious or unconscious minds with remorse. If we are to accept and honour our terrestrial origins, we should constantly push back the boundaries of Space. Is there any other way of continuing the dramatic saga of Mankind?

To conclude, here are the words of a man who must have possessed wisdom to have been able to write as he did:

"We aeronauts of the spirit!

All those brave birds which fly out into the distance, into the farthest distance – it is certain that somewhere they will at last reach a place from which they will be unable to go on and will perch on a mast or a bare cliff-face – and they will even be thankful for their miserable refuge! But who would dare to infer from that, that there was not still an immense open space before them, that they had flown as far as one could fly! All our great teachers and predecessors have each one day come to a stop and when weariness brings us to a halt it is not with the noblest or most graceful of gestures. It will be the same with us! But what does that matter to you and me! Other birds will fly farther! Our insight and faith vies with them in flying up and away; it rises above our heads and above our impotence into the heights and from there surveys the distance and sees before it the flocks of birds which, far stronger than we, still strive whither we have striven, and where everything is sea, sea, sea! – And whither then would we go? Would we cross the sea? Whither does this mighty longing draw us, this longing that is worth more to us than any pleasure? Why in this particular direction, thither where all the Suns of humanity have previously gone down? Will it perhaps be said of us one day that we too, steering westward, hoped to reach a new India – but that it was our fate to be wrecked against infinity? Or what, my brothers? Or what?" (Nietzsche, *The Dawn [Thoughts on the Prejudices of Morality]*, Book V, § 575)

References

Arendt, Hannah.

Bachelard, Gaston. 1994. *La poétique de l'espace*. Paris: Presses Universitaires de France.

Ferguson, Eugene S. 1978. "Historical Perspectives on Macro-Engineering Projects", in Davidson, Frank P., Giacoletto, L. J. and Salkeld, Robert (ed.). *Macro-Engineering and the Infrastructure of Tomorrow*. Boulder (Colorado): Westview Press.

Haigneré, Jean-Pierre & Arnould, Jacques. 2001. *Chevaucheur des nuées*. Paris: Editions Solar.

Lafferranderie, Gabriel. 1993. "Espace juridique et juridiction de l'espace", in Esterle, Alain (dir.). *L'Homme dans l'espace*. Paris: Presses Universitaires de France.

Monod, Jacques. 1970. *Le hasard et la nécessité. Essai sur la philosophie naturelle de la biologie moderne*, Paris: Seuil.

In: Space Exploration Research
Editors: J. H. Denis and P. D. Aldridge

ISBN: 978-1-60692-264-4

Chapter 2

A Legal Regime for Space Tourism: The Future for Commercialization of Space Stations

Yun Zhao*

The University of Hong Kong,
Erasmus University Rotterdam, the Netherlands,
Leiden University, the Netherlands

Abstract

Space Tourism, considered not an area of priority for commercialization, has never been so heatedly discussed until the historic arrival of two "unexpected" tourists in the International Space Station (ISS). The development of space tourism no doubt calls for a legal regime, which can better regulate the market and offer clear guidance and expected outcome. It has been widely argued that the existing international space treaties are inadequate for space commercialization. The development of a clear and predictable legal regime before space tourism becomes affordable for the masses is essential. As long as the space travel technology is mature, there are always business opportunities for space tourism. This paper discusses the potential for commercial space tourism and advocates an improved but essential legal regime for space stations. In view of the many commonalities shared by aviation and space travel, the present paper takes the example of aviation and elaborates on the formulation of a legal regime for space tourism. Actually it is gradually being accepted that the most appropriate regulatory framework for space tourism is to treat it as an extension of aviation. A proper and attractive legal regime will in the end help assure the future of safe and responsible commercial space tourism.

"Just tell me the general idea you have in mind-the idea Sven and my daughter keep so mysteriously to themselves. What is this thing that's so revolutionary and so daring? Fantastic and at the same time logical? I'm quoting, of course, my daughter." He looked steadily at Lee. His eyes brightened as if an inner light had been turned on. Lee glanced

---

* Associate Professor, The University of Hong Kong; Ph.D, Erasmus University Rotterdam, the Netherlands; LL.M., Leiden University, the Netherlands; LL.M., LL.B., China University of Political Science and Law, Beijing, China.

at the architect and the girl. He found response in their faces. "I need your assistance in building a hotel in outer space," he said artlessly.[1]

1. Introduction

A story like the above must be the truly classic scene for space futurists. Outer space exhibits an unlimited source for the imaginative science fiction writers. Earlier in the mid-19th century a number of science fiction stories have been written showing the rich imagination from renowned authors.[2] Space tourism was among the most popular topics for those writers. But no one has taken this idea so seriously at that time as in the late 20th century.

The successful launch of first satellite-Sputnik I in 1957[4] and Gagarin's first manned space flight in 1961[3] marked a breakthrough in space history. The rapid development of space technology brings the dream of conquering outer space to a reality. State monopoly has been the typical characteristic of space activities since the launch of Sputnik I. The space treaties formulated by the United Nations (UN) also acknowledged this fact, which has been well justified by the large amount of investment and the long period of time needed to realize the benefits. However, private parties have increasingly shown interests in space activities, posing strong challenges to the former regime.

Space tourism, considered not an area of priority for commercialization, has never been so heatedly discussed until the historic arrival of two "unexpected" tourists-American Dennis Tito and the twenty-eight year-old South African multimillionaire Mark Shuttleworth-at the International Space Station (ISS) in April 2001[5] and April 2002[6] respectively. The ISS partners officially cleared the way for space tourism with the approval of the two visits.[7] However, these private visits were very costly except for some tycoons. As reported, Tito and Shuttleworth flew to the ISS for an amount of US $ 20 million each[8]. This amount is impossible to most people.

However, market research has clearly demonstrated that many people have strong interest in space travel if it were more affordable. These wishes can be met with the development of reusable launch vehicle (RLV) technology, which could reduce space launch costs from $ 10,000 per pound to $ 1,000 per pound.[9] Just as Bachula stated, "reliable, affordable access to space is a fundamental prerequisite if we are to realize the full potential of the outer space frontier."[10]

The development of space tourism no doubt calls for a legal regime, which can better regulate the market and offer clear guidance and expected outcomes. It has been widely argued that the existing international space treaties are inadequate for space commercialization. The development of a clear and predictable legal regime before space tourism becomes affordable for the masses is essential It is noted that the Russian Space Agency had intended to send two civilians into outer space every calendar year until the February 1, 2003 Space Shuttle Columbia disaster.[11]

Space tourism is also an interesting topic in China. It has been reported that the first Chinese space tour is expected as early as in 2006.[12] Similar reports are expected from other nations in the coming period.

With the strong demand for space tourism a strong legal regime is essential, no matter whether the RLVs can be successfully developed in the near future or not. As long as the space travel technology is mature, there are always business opportunities for space tourism.

The present paper discusses the potential for a commercial space tourism industry and advocates a legal regime. Part 2 offers a comparison of space and air travel and the rules applicable to each traveling means. This part further proposes an appropriate liability regime for space travel, trying to borrow the experience from air transportation. Part 3 examines the appropriate level of state interference through licensing measures. Part 4 specifically discusses the status of space tourists, as differed from astronauts. Part 5 concludes that current space law inadequately addresses space tourism and that the inadequacies justify an urgent need to develop an appropriate legal regime for the development of space tourism.

2. Space Travel and Air Transportation: Different Applicable Law

The Wright Brothers' successful flight at Kitty Kawk in 1903 opened a new era in transportation history. The international society was quick to respond to develop a legal regime regulating commercial air transportation. The Warsaw convention was formulated in 1929 to develop a forward-looking international aviation regime. The regime proved to be vital to the development of the air transportation industry by balancing cost prohibitive insurance premiums and liability measures protecting the struggling airline industry.

Space travel stands exactly at the same crossroad as air transportation in the early 1920s. The potential liability for accidents is a major obstacle. The legal vacuum in this respect deters the commercializing process of space travel. Insurance is not the way out since the cost prohibitive insurance premiums for space travel will be passed on to the tourists and the ticket price will go far beyond a reasonable level as to kill the whole space industry. Accordingly, the formulation of an appropriate liability regime for space travel appears all the more important.

Air transportation and space travel share a number of similarities, leading to the discussion of extending the air transportation regime to space travel. The discussion again goes back the classic question on the boundary of outer space and air space, and thus the application of air law and space law.

Outer space begins where territorial air space ends. There is currently not a clear internationally recognized boundary of outer space [13] and consequently not a proper definition despite the separate bodies of international law governing air space and outer space.[14] Striking criteria distinguishing air space and outer space for applicable law include purpose and function, technical configuration and capabilities, and the medium where the operation predominantly takes place.[15] Space travel, as denoted by its concept, has clearly classified itself as activities in outer space; furthermore, activities in sufficient distance from the Earth have no problem in justifying the application of space law for space travel.

The air transportation regime, characterized by state sovereignty over air space, substantially differs from the space travel regime where no state can claim sovereignty over outer space. This fundamental difference justifies the necessity of developing a distinct legal regime for space travel. Nevertheless, we should not neglect the fact that air transportation and space travel, though in different geographical locations, are basically transportation in essence. While the vehicles used for space tourism are rocket-powered and designed to enter outer space, they take off and land like airplane.[16] Taking space vehicles to outer space will be like taking an airplane for the travelers, although the destinations are different. Space travel, while still in its infancy is similar to the air transportation industry in its early stage. The question posed to the air transportation industry comes to space travel now: how to alleviate the liability so that the regime can effectively promote the rapid development and commercialization of the industry. In this regard, we can certainly borrow the successful legal experience of air transportation to facilitate the formulation of an appropriate regime for space travel.

2.1. Commercial Liability Regime

By referring to the liability issue in outer space, one may immediately think of the Outer Space Treaty and the 1972 Liability Convention. Article VII of the Outer Space Treaty provides that states are internationally liable for any damage caused by their objects or personnel while in space. The Liability Convention, further elaborating on Article VII of the Outer Space, envisions two situations when the launching state(s) are liable: damage caused by its space object on the surface of the earth or to aircraft in flight; damage being caused elsewhere than on the surface of the earth to a space object of one launching state or to persons or property on board such a space object by a space object of another launching state. Strict liability applies to the first situation[17] while negligence liability to the latter[18].

In view of its international nature, this Convention does not apply to two types of people, including the nationals of the launching state.[19] Furthermore, only a state may present a claim for compensation.[20] Accordingly, the Convention fails to specifically outline civilian liability in space.[21]

The ISS IGA, while incorporating the Liability Convention, has further included a provision concerning the mutual exemption of liability on board the ISS for the purpose of better cooperation among the partners. This provision applies to any claims brought by a Partner State for damage, against another Partner State, a related entity of another Partner State, and/or the employees of any of the above entities.[22] Obviously, space passengers cannot rely on this provision for any claims. The public nature of the IGA does not fit well in the present commercial regime. Liability arising out of the disputes in space tourism should be resolved according to the general international law. Thus, the space object shall be regarded as an extension of the jurisdiction of the launching state, whose law prevails.

The current liability system thereby excludes space tourism and only extends to efforts by states or international non-governmental organizations sending equipment and astronauts into space for the purpose of exploration and scientific research[23,24]

Liability to passengers is the main concern in space tourism, which is totally missing in the relevant conventions. In the air transportation legal system, domestic and international transportation are differentiated, which shall not exist in space travel. National passengers of a launching state, like that in international air transportation, should be allowed to claim compensation for damages suffered while a uniform regime should be introduced covering all the passengers, goods and related natural and legal persons directly.

In international air transportation, the Warsaw Convention[25] sets a good example in helping to establish a uniform international system, which has further enabled insurance companies to provide coverage in a new field of international transportation with confidence.[26] A negligence standard, instead of strict liability, was adopted and maximum damages to a passenger were originally set at $ 10,000.[27] However, the limitation of liability in the Convention is now considered unnecessary in view of the improving reliability of aviation; the 1999 Montreal Convention shows the sign for relaxing the above limitation.[28] Nevertheless, this Convention proves to provide the protection and freedom necessary for the air transportation industry. The industry was able to flourish and has now become the safest means of transportation.[29] The Warsaw Convention has been under revision, trying to balance the interests of the industry and other parties (including passengers and third parties). The maximum amount of damages has been changed according to social environment. However, the negligence standard remains the ground stone of the Convention.[30]

The success of the international aviation system indicates that a negligence standard should be initially introduced for space travel. The maximum damages payable to passengers should be formulated. Limiting the carrier's liability will not necessarily deter potential space tourists since they can buy additional insurance, as the case in aviation. An appropriate amount shall be determined based on several factors, including the ultimate goal of pushing the development of space travel, the financial situation of the space travel industry, and the general background of space passengers. The duration of liability should similarly be the period during which the accident takes place on board the space object or in the course of any of the operations of embarking or disembarking.[31]

Legislation providing the above propositions are indispensable for space tourism. The uncertainty concerning the liability issue can make potential investors afraid that any unknown future regulation may kill the business they are investing in. We may simply modify the Warsaw Convention for space tourism, but of course, we can formulate a new document written along similar lines.[32] By referring to the proposed document, space tourists, governments, commercial operators and insurance companies would all know in advance of possible liabilities and make sensible decisions. Thus, the international society will benefit from the transparency and legitimacy brought by such an international document.

2.2. Space Insurance

Space insurance has been available for a couple of years, especially in the field of satellite launching activities. Further development of space activities has called for more active involvement of private parties. However, a complete set of rules are still to be formulated to realize private financing for space program. In view of the high risks in space activities, the availability of insurance has been a critical element for private parties. Insurance provides relief for a whole range of liability risks currently associated with space activities, including space tourism.

Space insurance may be divided into two categories: insurance of space objects and liability insurance. Of course other categories of insurance which are related to space activities also exist, such as product liability insurance. As mentioned above, insurance of space objects have been in existence. Three types of insurance can be further differentiated: pre-launching insurance; launch failure and initial operation insurance; and insurance of the satellite itself.[33] The first satellite insurance contract providing for pre-launching insurance services was concluded in 1965 for Intelsat's "Early Bird".[34] The US Commercial Space Launch Act (CSA) requires entities that launch space vehicles to purchase $ 500 million in third-party liability insurance.[35] It is thus nothing anew. The legal basis and principle of insurance remain to be largely applicable.

However, the CSA further provides that the US government will cover excess damages up to $ 1.5 billion and that the launch entities are liable for any damage beyond the $ 2 billion.[36] Understandably, considering the high risks involved in launching activities, insurance companies are not willing to take possibly high damages. As provided in the CSA, an insurance provider may list specific exclusions to the insured's liability insurance policy.[37] To a certain extent, the US government acts as an excess insurance carrier, providing a layer on top of the required insurance. On the other hand, insurance policies for commercial launch activities have not been standardized and need to be negotiated on a case-by-case basis.[38] To a certain extent, insurance companies' confidence concerning the scale of risks involved in launching depends much on agreed standards of acceptable risk.[39]

The CSA insurance requirements do not mention space tourists.[40] Insurance for the carrier's liability for the tourists is thus something new to be added to the insurance industry with the development of space tourism. The existence of liability insurance cooperates with the commercial liability regime for space tourism. However, insurance companies might be again unwilling to undertake too much risk. As one scholar has identified, "If tourism is to become a vital part of the commercial space equation, limits on liability for the owners and operators of space facilities and vehicles will be a necessity."[41] Limits exist for liability arising out of death, personal injury or loss or damage to property; limits can also be set for each and every space flight. Several factors are relevant to the fixed limited, including the length of flight, module and model of space objects; experience of astronauts; air condition during the flight.

One the one hand, it is important to introduce insurance to the space tourism industry. At the present stage, this young industry requires support from various corners. The insurance industry is indispensable to space carriers, given the high market value of spacecraft and the great financial risks. On the other hand, it is critical to set an appropriate rate so that the insurance industry is willing to enter this potentially profitable market. Again, we can borrow successful experience from aviation. The evolution of space technology has a close relationship with the insurance industry. We are fortunate to see that the insurance industry has been mature enough to accept the risks in space industry since 1965.

The development of space tourism has introduced a new challenging market for the insurance industry. This will probably cause a temporary increase in premiums. But in the long run, with the progress of space technology and safety improvement, insurance premiums will be reduced.

2.3. Criminal Jurisdiction Regime

The criminal jurisdiction issue has so far been considered as not highly relevant to outer space activities. Currently no international treaty exists for crimes committed on private space vehicles.[42] This situation is understandable when space activities are still largely monopolized by the state. The people on board the space object are normally astronauts trained for a special mission. The commander on board the space object has the authority to enforce order and discipline during the whole flight phase.

The ISS Agreement provides the authority of a commander on board the ISS to maintain order. However, it is notable that the ISS Agreement contains a provision on criminal jurisdiction. This is necessary in view of the long-term character of the ISS and the international and multicultural character of the astronauts on board the ISS. Besides the execution of criminal jurisdiction over its nationals, article 22 further provides the jurisdiction over nationals of another Partner State whose conduct in orbit "(a) affects the life or safety of a national of another Partner State or (b) occurs in or on or causes damage to the flight element of another Partner State". It is thus obvious that the criminal jurisdiction is based on customary principles of nationality and the protective principle. That means, the criminal law of the victim's country will normally apply.

However, the development of space tourism brings new trouble. Space tourists are less prepared and controlled than astronauts, increasing the risk of criminal activities. Furthermore, the situation when a non-member country tourist becomes the target of a criminal offence, which is often the case in space tourism, will bring trouble to the above arrangement among the members. Will a non-member country's law be applied here? The ISS IGA arrangement becomes insufficient to deal with similar new problems.

In this aspect, we can refer to the similar situation in air transportation. The Tokyo Convention imposes a series of obligations upon the Contracting States that are geared towards stamping out hijacking. According to this Convention, apart from national criminal jurisdiction, each state shall take measures as may be necessary to establish its jurisdiction over the offense and any act of violence against a passenger or crew when (a) the offense is committed on board an aircraft registered in that state....[43] The provision above is claimed to establish the "semi-universal jurisdiction principle". All states have the criminal jurisdiction over any acts causing danger to the aviation industry, which has important impact on the safe operation of the industry and the confidence from the passengers.

Space travel also needs to build confidence from potential passengers, preventing the infant industry from fatal criminal activities. The "semi-universal jurisdiction principle" in the air transportation is also meaningful to space travel. Interested parties should convene to discuss the draft of a similar treaty cracking down criminal acts against space safety.

3. Registration and Licensing Regime

As mentioned above, large scale space tourism depends on the development of reusable launch vehicles (RLVs). No doubt, RLVs are space objects as identified in the Registration Convention. According to this Convention, each party is required to register and maintain a registry of its launched space objects;[44] in addition, the party must provide the UN Secretary-General information proving the establishment of a registry.[45] This requirement is reasonable for purposes like identification of space objects and determining liability. However, when space tourism develops and the launching of RLVs becomes more and more frequent, the requirement of registration appears infeasible and unnecessary. However, national registry is sufficient for the above purposes. Thus, two types of registration are suggested: that the current registration regime continues to exist; however, once space objects like RLVs are used specifically for commercial space travel, only national registry be required.

To guarantee the safety of space travel and enforcement of the international obligation of peaceful use of outer space, the state should establish a licensing regime which can provide sufficient supervisory function over space tourism. An appropriate licensing regime, as the safety valve for security in space travel is the obligation of the relevant state in guaranteeing the legitimate operation of those licensees.

The United States has established a rather complete legal framework in the licensing regime. The CSA of 1998 laid the regulatory groundwork for RLV licensing. According to the Act, prospective applicants are required to participate in pre-application consultations with the office of the Associate Administrator for Commercial Space Transportation (FAA-AST);[46] following pre-application consultations, applicants must obtain policy approval, safety approval, payload and payload reentry approval, and environmental approval.[47] The requirements have been argued to be too complicated, which will ultimately prevent private companies from getting off the ground.[48]

The 2004 Space Launch Act Amendment,[49] with the aim of regulatory reform and improved interaction with RLV developers, has further ensured its purpose to promote the development of the emerging commercial human space flight industry.[50] It represents the trend of deregulation in the field to avoid "the potential danger of industry-killing over-regulation".[51] For example, the time period needed for relevant bodies to take actions has been clearly defined, thereby preventing unduly interference from relevant bodies.[52] Considering the high risks entailed and unwavering emphasis on safety, the complex licensing process is retained. However, it has been acknowledged that a more streamlined system of requirements is needed to facilitate the licensing process.[53]

The United States example clearly shows the vital role of a licensing regime in commercial space activities and represents the first significant step towards nurturing and supporting commercial efforts in space tourism. In view of the complicated but indispensable licensing process, the FAA/AST has taken realistic measures to work directly with RLV developers, helping them better understand the process and reflecting their concerns in future space flight policy. This will help streamline the licensing requirements.

4. The Status of Space Tourists (Space Hotel Rules)

The emergence of space tourists who go to outer space for leisure pose challenges to the existing space legal regime. The astronauts as defined in the 1968 Rescue Agreement outlines procedures for astronaut rescues if the astronauts are in an accident, distress, emergency or an unintended landing are on the high seas or in any other place not under the jurisdiction of any nation.[54] According to the Agreement, nations are obliged to perform rescue duties for the personnel of a spacecraft in the event of accident, distress or emergency landing.[55] It is to be noted that "personnel of a spacecraft", instead of "astronaut", is not used in the text of the Agreement. Obviously, the term "personnel of a spacecraft" does not necessarily include astronauts.

Literally, tourists are not astronauts or the personnel of a spacecraft. If they are the same as mission specialists, like space engineers or scientists, there will not be much dispute concerning the application of Rescue Agreement.[56] For example, the first space tourist Tito spent six days on board as both a tourist and as an assistant to the crew, helping with a variety of tasks to include the transfer of supplies and scientific experiments.[57]

However, this is only an exceptional case. Space tourists generally do not play a direct role for the benefit and in the interests of all countries. Their main objective is not to contribute to the public interest, but just for their personal pleasure. In no sense do they qualify as "envoys of mankind in outer space" [58]. Accordingly, obstacles exist in applying the Rescue Agreement to space tourists.

Nevertheless, just as identified in the preface, the Rescue Agreement is prompted by sentiments of humanity. This consideration similarly applies to the rescue of tourists. Thus, ways need to be sorted out to deal with the issue of rescuing tourists in the event of accident, distress or emergency landing. Two ways can be easily identified: formulation of a new agreement with similar provisions of the Rescue Agreement, or extending the existing agreement to the application of space tourists. Considering similar measures underlying the rescue of astronauts and tourists, may logically lead to the second option.

In this respect, the 2004 Commercial Space Launch Amendments Act and the ISS IGA offer useful experience. The two documents take different approaches. The 2004 Act clearly defines two different types of people involved in space flight. It provides definitions for the terms "crew" and "space flight participants" and amends existing commercial launch legislation to include these terms alongside the inanimate payloads currently covered.[59] According to this Act, "crew" means "any employee of a licensee or transferee, or of a contractor or subcontractor of a licensee or transferee, who performs activities in the course of that employment directly relating to the launch, reentry, or other operation of or in a launch vehicle or reentry vehicle that carries human being".[60] "Space flight participant" means "an individual, who is not crew, carried within a launch vehicle or reentry vehicle".[61] So it is quite obvious space tourists are considered as space flight participants. This is a direct way to differentiate "crew" from "space tourists". However, this approach does not effectively resolve the issue of protection for space tourists as defined in the Rescue Convention.

The ISS IGA, same as the Rescue Convention, defines the term "crew" as qualified personnel.[62] But this Agreement further provides the activities of all individuals involved in outer space activities under the heading "Protected Space Operations".[63] This extensive provision validly resolves the above dilemma: the ISS IGA covers all individuals, no matter if he/she is piloting a spacecraft, conducting experiments or merely a passenger for fun. By covering anyone whoever is piloting a space object, conducting experiments or merely traveling for fun, this approach is instructive to extend application of the Rescue Agreement to space tourists.[64]

While receiving necessary protections, space tourists, as passengers of a spacecraft, should also comply with rules for good order during the journey. Basically their rights and obligations fall within the competence of the State exercising jurisdiction and control, namely, the State of registry of the RLV. The commander, providing for the safety and well-being of all persons on board, shall have sole authority throughout the flight; tourists, irrespective of their nationality, are subject to the directions of the commander.

5. Conclusion

Space tourism has received great interest from various sides. Some scholars believe that space tourism may be one of the first space industries to emerge and that it will pave the way for everything else.[65] Encouraged by the success of the first two space tourists, space tourism companies, having been set up in recent years and are actively promoting the program and soliciting support from the governments. As reported, Hong Kong Space Travel Agency has signed a cooperative agreement in early 2005 with American Space Exploration Company; more than 20 Chinese tourists will be sent to the US for training and the first Chinese tourist is scheduled to travel to space in 2006.[66] The reports released so far have sent a clear sign to the public that space tourism has come to a new era.

Drastically different from other means of transportation, such as shipping and aviation, which are governed by a comprehensive framework of national and international commercial law, space activities are supported by inter-governmental treaties negotiated during the cold-war period. While commercial space tourism is coming to reality, the legal regime is still lagging far behind. In view of the many commonalities shared by aviation and space travel, the present paper takes the example of aviation and elaborates on the formulation of a legal regime for space tourism. Actually it is gradually being accepted that the most appropriate regulatory framework for space tourism is to treat it as an extension of aviation.[67] A proper and attractive legal regime will in the end help assure the future of safe and responsible commercial space tourism.

References

[1] Kurt Siodmak, *Skyport* (Mass, 1959). See further Space Tourism in Science Fiction, available at <http://www.spacefuture.com/tourism/sciencefiction.shtml> (last visited October 11, 2004).

[2] Jules Verne, a French writer, authored several science fiction stories and the most famous amongst was entitled "De La Terre `a la Lune" in 1865.

[3] Sputnik I was launched from the Baiknur Cosmodrome in Southern Kazakstan on October 4, 1957. See Craig Covault, Policy and Technology Shape Manned Space Ops, *Aviation Week & pace Technology*, January 8, 2001, at 44.

[4] The First Cosmonaut in the World is in Space, April 1, 1998, available at <http://news.bbc.co.uk/1/hi/special_report/1998/03/98/gagarin/72182.stm> (last visited March 30, 2005).

[5] Anna Badkhen, US Tourist Arrives at Space Station: Tito is Greeted by Russians after Weekend Flight, *Boston Globe*, May 1, 2001, at C4.

[6] First African in Space, available at <http://www.africaninspace.com/home/mission/logs/1/index.shtml> (last visited March 30, 2005).

[7] Decision Paper on Russian Aviation and Space Agency (Rosaviakosmos) Request for MCB Approval of Exemption to Fly Mr. Dennis Tito Aboard the April 2001 Soyuz 2 Taxi Flight to the International Space Station, April 24, 2001, available at <http://ftp.hq.nasa.gov/pub/pao/reports/2001/tito<uscore>decision.pdf> (last visited March 19, 2003).

[8] Peter Baker, U.S., Russia Agree to Allow "Space Tourists", *Washington Post*, August 10, 2001, at A 20; see also African Space Tourist Ends $ 20 Million Odyssey, May 5, 2002, available at <http://edition.cnn.com/2002/TECH/ space/05/04/sfrica.tourist> (last visited March 30, 2005).

[9] Roscoe M. Moore, Risk Analysis and the Regulation of Reusable Launch Vehicles, 64 *Journal of Air Law & Commerce* 245, 251 (1998).

[10] Statement of Gary R. Bachula, Acting Undersecretary for Technology, US Department of Commerce, US Commercial Space Launch Industry: Hearing before the Subcommittee on Science, Technology and Space of the Senate Committee on Commerce, Science, and Transportation, 105th Congress (1998).

[11] STS-107: Columbia Disaster, February 1, 2003, available at <http://www.space.com/columbiatragedy> (last visited March 30, 2005).

[12] First Chinese to have Space Travel, available at <http://news.creaders.net/ headline/newsPool/30A220050.html> (last visited November 1, 2004).

[13] Report of the Legal Subcommittee on its 41st Session held in Vienna from April 2-12, 2002, United Nations Committee on the Peaceful Uses of Outer Space, UN Doc. A/AC.105/787, at 10 (2002).

[14] I.H.Ph. Diederiks-Verschoor, *An Introduction to Space Law* 5 (2nd ed., Kluwer, 1999).

[15] Michael Wollersheim, Considerations towards the Legal Framework of Space Tourism, 2nd International Symposium on Space Tourism, Bremen, April 21-23, 1999, available at <http://www.spacefuture.com/archive/considerations_towards_ the_legal_framework _of_space_ tourism.shtml> (last visited April 25, 2005).

[16] Recent Development: Commercialization of Space Commercial Space Launch Amendments Act of 2004, 17 *Harvard Journal of Law & Technology* 626 (Spring 2004).

[17] The Liability Convention, Article II provides, "A launching state shall be absolutely liable to pay compensation for damage caused by its space object on the surface of the earth or to aircraft in flight."

[18] The Liability Convention, Article III provides, "In the event of damage being caused elsewhere than on the surface of the earth to a space object of one launching state or to persons or property on board such a space object by a space object of another launching state, the latter shall be liable only if the damage is due to its fault or the fault of persons for whom it is responsible."

[19] The Liability Convention, Article VII provides, "The provisions of this Convention shall not apply to damage caused by a space object of a launching state to : (a) nationals of that launching state; (b) foreign nationals during such time as they are participating in the operation of that space object from the time of its launching or at any stage thereafter until its descent, or during such time as they are in the immediate vicinity of a planned launching or recovery area as the result of an invitation by that launching state."

[20] The Liability Convention, Article VIII (1) provides, "A state which suffers damage, or whose natural or juridical persons suffer damage, may present to a launching state a claim for compensation for such damage."

[21] Nandasiri Jasentuliyana, International Space Law and the United Nations 390 (1999).

[22] Article16, para. 3 (a), the 1998 IGA.

[23] Richard Berkley, Space Law Versus Space Utilization: The Inhibition of Private Industry in Outer Space, 15 *Wisconsin International Law Journal* 422 (1997); Joseph A. Bosco, International Law regarding Outer Space-An Overview, 55 *Journal of Air Law & Commerce* 614-620 (1990)

[24] Ezra J. Reinstein, Owning Outer Space, 20 *Northwestern Journal of International Law & Business* 71 (1999).

[25] Convention for the Unification of Certain Rules Relating to International Carriage by Air, opened for signature October 12, 1929, 49 Stat. 3000, 137 U.N.T.S. 11.

[26] Patrick Collins, The Regulatory Agenda for the Era of Passenger Space Transportation, Proceedings of 20th ISTS, Paper No. 96-f-13, available at <http://www.spacefuture.com/archive/the_regulatory_reform_agenda_for_the_era_of_passenger_space_transportation.shtml> (last visited April 28, 2005).

[27] James E. Dunstan, Is Launching a Rocket Still an Ultra-Hazardous Activity? Toward a Negligence Theory for Launch Activities, 9 *Space Manufacturing the High Frontier: Accession, development & Utilization* 226, 229 (1993).

[28] See for example Montreal Convention, Art. 17. The carrier is always liable to a maximum amount of 100,000 SDR; for damages exceeding this amount, the carrier is liable without limitation unless he proves that the damage was not the result of his negligence or wrongful act (nor of his servants or agents), or such damage was solely due to the negligence or wrongful act of a third party. From this provision, it is obvious that a strict liability applies to the carrier.

[29] Between 1959 and 2002, there were 1,337 accidents worldwide out of a total of 412 million departures. However, approximately 40,000 people died in automobile accidents and 1096 in railway accidents each year. See U.S. department of Transportation, A Comparison of Risk, Accidental Deaths-United States-1994-1998, at <http://hazmat.dot.gov/riskcompare.htm> (last visited November 4, 2004).

[30] The Guatemala Protocol in 1971 provides that the fault liability at present attaching to the carrier will be changed into a risk liability. However few states have ratified this Protocol to date.

[31] See Article 17 of the Warsaw Convention.

[32] Anders Lindskold, Space Tourism and Its Effects on Space Commercialization, Master of Space Studies Program 1998/99, available at <http://www.spacefuture. com/pr/archive/space_tourism_and_its_effects_on_space_commercialization.shtml>(last visited April 29, 2005).

[33] I.H.Ph. Diederiks-Verschoor, *An Introduction to Space Law* 117 (2nd Ed., Kluwer, 1999).

[34] *Id.*, at 117.

[35] 49 U.S.C. 70, 112 (2003).

[36] Id.

[37] 14 C.F.R. 440.13(a)(5).

[38] Peter D. Nesgos, The Challenges Facing the Private Practitioner: Liability and Insurance Issues in Commercial Space Transportation, 4 Journal of Law & Technology 25-26 (Winter 1989).

[39] Statement of Patricia A. Mahoney, Chair, Satellite Industry Association, Extension of Space Launch Indemnification: Hearing before the Subcommittee on Space and Aeronautics of the House Committee on Science, 106th Cong. (1999).

[40] See generally 49 U.S.C. 70, 112 (2003).

[41] Patrick Collins, The Regulatory Reform Agenda for the Era of Passenger Space Transportation, Proceedings of 20th ISTS, Paper No. 96-f-13 (1993), available at <http://www.spacefuture.com/archive/the_regulatory_reform_agenda_for_the_era_of_passenger_space_transportation.shtml> (last visited April 28, 2005).

[42] R. Thomas Rankin, Space Tourism: Fanny Packs, Ugly T-Shirts, and the Law in Outer Space, 36 *Suffolk University Law Review* 716 (2003).

[43] Tokyo convention, Article 4 (3).

[44] Registration Convention, Article 2.

[45] Registration Convention, Article 2 and 4.

[46] 14 C.F.R. 431.31.

[47] 14 C.F.R. 413.5, 415 (2001).

[48] Charity Trelease Ryabinkin, Let there be Flight: It's Time to Reform the Regulation of Commercial Space Travel, 69 *Journal of Air Law and Commerce* 129 (Winter 2004).

[49] H.R. 3752, 108th Cong. (2004).

[50] House Floor Debate on Commercial Space Launch Act of 2004, House of Representatives, November 19, 2004, available at <http://www.spaceref.com/news/viewsr.html?pid=14565> (last visited December 1, 2004).

[51] H.R. 3752, 108th Congress, 2nd Session, Report No. 108-429, available at <http://www.theorator.com/bills108/hr3752.html> (last visited December 1, 2004).

[52] For example, Art. 3(c)(7) provides that the Secretary of Transportation shall issue the experimental permit required by human space vehicle operators no later than ninety days after receipt of an application and that the Secretary of Transportation would be obliged to inform the applicant of any issues arising during the review of an application and actions to be taken to resolve them, within the first sixty days after the receipt of the application.

[53] Ryabinlin, *Supra* note, at 137.

[54] Rescue Agreement, Art. 1-4.

[55] Rescue Agreement, Art. 1. David Tan, Towards a New Regime for the Protection of Outer Space as the "Province of All Mankind", 25 *Yale Journal of International Law* 158 (2000); Ty S. Twibell, Space Law: Legal Constraints on Commercialization and Development of Outer Space, 65 *UMKC Law Review* 595 (1997).

[56] Yasuaki Hashimoto, The Space Plane and International Space Law, available at <http://www.spacefuture.com/pr/archive/the_space_plane_and_international...> (last visited October 11, 2004).

[57] Jim Banke, Space Tourist Pays His Full Fare, available at <http://www.msnbc.com/news/509288.asp> (last visited January 2, 2001).

[58] Outer Space Treaty, Art. 5.

[59] Recent Development, *supra* note, 627.

[60] H.R. 3752, 108th Congress (2004), Art. 3 (b) (2).

[61] H.R. 3752, 108th Congress (2004), Art. 3 (b) (9).

[62] ISS IGA, Art. 11 (1).

[63] ISS IGA, Art. 16 (2)(f).

[64] Lara L. Manzione, Multinational Investment in the Space Station: An Outer Space Model for International Cooperation?, *18 American University International Law Review* 521 (2002).

[65] Albert A. Harrison, Spacefaring: The Human Dimension 12 (2001).

[66] The First Chinese Tourist Expected to Explore in Space Next Year, February 28, 2005, available at <http://tour.huash.com/gb/tour/ 2005-02/28/content_ 1668183.htm> (last visited March 8, 2005).

[67] Patrick Collins & Koichi Yonemoto, Legal and Regulatory Issues for Passenger Space Travel, *Proceedings 49th Colloquium on the Law of Outer Space*, available at <http://www.spacefuture.com/pr/archive/legal_and_regulatory_issues_for_passenger_space_travel.shtml> (last visited October 11, 2004).

In: Space Exploration Research
Editors: J.H. Denis and P.D. Aldridge

ISBN: 978-1-60692-264-4

Chapter 3

HYPOMETABOLISM AS A RESOURCE FOR MANNED LONG-TERM SPACE FLIGHTS: A CHALLENGING PERSPECTIVE

Carlo Zancanaro[1], Manuela Malatesta[1], Marco Biggiogera[2] and Lorella Vecchio[2]

[1] Anatomy and Histology Section, Department of Morphological and Biomedical Sciences, University of Verona, Italy

[2] Laboratory of Cell Biology and Neurobiology, Department of Animal Biology, University of Pavia, Italy

ABSTRACT

Hypometabolism is a physiological state in which the energy requirements of the organism are drastically lowered. In hibernating animals this represents the pivotal strategy to overcome extreme environmental conditions. This paper begins with a discussion of natural hypometabolism as found in mammalian hibernation with a focus on the strategies these animals evolved to cope with life challenge in hostile environments (e.g., cold weather and/or shortage of food), covering behavioural, physiological and genetic strategies. The issue of factors required/able to induce the hypometabolic condition is then addressed, including some possible pharmacological approach. Finally, the potential impact of induced, regulated hypometabolism in the human is briefly reviewed showing why hypometabolism would be, from a resource-driven perspective, desirable during long-term space flight as well as in the broad field of biomedicine.

INTRODUCTION

Over the last several years biomimetics i.e., design inspired by nature has been attracting much interest in the space exploration field as a research discipline placed at the interface of biology and technology. The goal of biomimetics is to understand the functional principle underlying a certain physiological function and transfer it into technology. For example, biomimetics projects currently ongoing at the European Space Agency are exploring insect intelligence, dry adhesion of spiders, ground anchoring of plants. The core event of mammalian hibernation, hypometabolism, is the outcome of a complex series of physiological adaptations; therefore, it cannot be considered a biomimetics object proper. Nevertheless, it is worth considering as a source of inspiration for space application because of the enormous impact an induced, regulated, reversible hypometabolic state would have in manned space missions and in biomedicine.

NATURAL HYPOMETABOLISM: STRATEGIES OF HIBERNATION IN MAMMALS

In the wild, hypometabolism represents a common strategy adopted by many animals belonging to all Vertebrate classes - from Pisces to Amphibians, Reptiles and Birds up to the Mammals - to overcome adverse environmental conditions such as low or high temperatures, food or water scarcity. In an applied perspective for humans, mammalian hibernation represents the most suitable natural model to investigate the mechanisms of hypometabolism and to explore its potential applications in medicine and space biology.

Members of at least six mammalian orders are capable to enter hibernation. Hibernators can be found among *Monotremata* (echidna), *Marsupialia* (many dasyurids), *Crocidurinae* (tenrecs and shrews), *Insectivora* (hedgehogs), *Megachiroptera* and *Microchiroptera* (many bats), *Rodentia* (numerous sciurids, cricetids, heteromyids, murids and zapodids), and even *Primates* (dwarf and mouse lemurs), thus providing evidence that hypometabolism should be based on widespread and largely conserved mechanisms.

Mammals are endothermic organisms, being therefore able to produce endogenous heat for maintenance of a relatively high and constant body temperature. However, this independence from the environmental temperature requires high energetic costs. When the environmental conditions become particularly unfavorable, the cost for the thermoregulatory heat production can exceed the available energy, especially in small Mammals and, to overcome these energetic constraints, these organisms enter a hypometabolic state.

Natural Hypometabolism Occurs under Different Forms

Natural hypometabolism is characterised by a decrease of both body temperature (which is reduced to a new regulatory set-point) and the metabolic activity, and heart rate; as a consequence, the overall energy demand is reduced thus making survival possible, as the energy need and energy offer are both kept at the minimum. The hypometabolic state can last for variable lengths of time, from a few hours to some weeks; unlike Ectotherms, however, Mammals are able to leave the depressed metabolic state at any time, using endogenously produced heat to restore normal body temperature.

There are several forms of natural hypometabolism, characterized by different timing, duration and depth. Each state is more or less typical of a given species, in relation with its peculiar necessities [reviews in Hoffman, 1964; Nelson, 1980; Lyman et al., 1982; Wang, 1987; French, 1988; Storey and Storey, 1990].

Aestivation and hibernation are considered as seasonal torpors, being both characterized by discontinuous, successive torpor bouts of several days or weeks; these bouts are generally concentrated in one season which can even last many months. Some common groups characterized by seasonal torpor include the hedgehogs (*Erinaceous*), marmots and woodchucks (*Marmota*), ground squirrels (*Spermophilus* or *Citellus*) and bats (*Eptesicus* and *Myotis*). During the pre-hibernating phase, these animals undergo a drastic, rapid weight gain; the subsequent exposure to low ambient temperature results in the decrease of the body temperature to near 0°C during the hibernation period.

On the other hand, some species can be induced to a hypometabolic state at any time of the year by proper environmental stimuli such as cold, heat, or food shortage. A typical form of such non-seasonal hypometabolism is the daily torpor, with a duration of less than 24 hours and a body temperature ranging 10-25°C (considerably higher that that found during hibernation). Members of marsupials, insectivores, chiropters, rodents and primates exhibit daily torpor. A further type of non-seasonal torpor is typical of the Syrian (golden) hamster (*Mesocricetus auratus*), which can enter deep torpor (body temperature decreasing to about 5°C and lasting a few days), but requires a long preliminary period of cold exposure.

A particular type of natural hypometabolism is represented by the winter sleep in bears. During winter dormancy, which lasts from 3 to 7 months, bears do not eat, drink, defecate or urinate and they use fat exclusively as their energy source. The bear hibernates at a body temperature (31-35°C) very close to normal, and its metabolic depression is much less than that found even in the daily torpor (the metabolic rate decreases to 50-60% of the euthermic level and heart rate drops from 40 to 10 beats/minute). In addition, unlike other hibernating animals, when disturbed the bear is easily aroused into a mobile, reactive state. This depends on the large body size which allows great fat storage, thus making the shallow hypometabolic state of the bear optimal for this animal to meet energy requirement under fasting condition.

Despite the diversity of torpor types, the polyphyletic origin in evolution of torpor in mammals and the specific ecological features of their niches, the patterns of body temperature, metabolism, heart and respiratory rate of the natural hypometabolism are basically similar in all mammals, differing only in their quantitative aspects. Typically, a torpor bout consists of entry into, maintenance of and arousal from torpor. During entry into torpor, a progressive inhibition of heart rate, an increase in vasoconstriction to maintain blood pressure, a decrease in respiratory rate with irregular periods of apnoea and a significant decrease of oxygen consumption occur. Body temperature falls following decrease of heat production, with periodic shivering to counterbalance a too rapid cooling. During torpor, which may last from hours to weeks, all physiological functions are kept at the minimum. Heart rate may decrease to $1/30^{th}$ or less and oxygen consumption to $1/100^{th}$ or less of their respective euthermic levels. Prolonged apnoea (40-150 minutes) as well as Cheyne-Stoke breathing (apnoea followed by bursts of breathing) occurs in rodents. Body temperature can reach values close to the ambient temperature (sometimes near 0°C); however, there is a "critical" level below which a further ambient temperature decrease could result in an adaptation (metabolic rate increase or arousal) or death of the hibernator. Arousal from torpor is a chain-reaction event requiring from 20-30 minutes in small rodents and bats to a few hours in marmots: the substrate are mobilized for energy production, the cardiovascular system is stimulated for tissue perfusion and the non-shivering thermogenesis in the brown adipose tissue starts.

Natural Hypometabolism Requires Multiple Physiological and Metabolic Adaptations

The extraordinary capacity of varying body temperature and functional activities in hibernators implies peculiar regulation of metabolic and physiological pathways and structural and molecular adaptation of cell and tissue components. Numerous mammalian species have been studied under diverse experimental conditions and the scientific literature provides extensive information about the physiological and metabolic strategies adopted by hibernators to ensure survival, although the basic mechanisms of the natural hypometabolism are still unknown.

The most striking physiological feature of natural hibernators is their ability to cycle between body temperature of 37°C during euthermia to near 0°C during torpor. However, during torpor, body temperature is still controlled by a CNS regulator: thermosensitive neurons within the preoptic/anterior hypothalamic region induce cyclic bursts of heat production when the temperature drops below a certain set-point [Lyman and O'Brien, 1972; Heller and Colliver, 1974; South et al., 1975; Buck and Barnes, 2000].

The duration of the photoperiod would also play a role in the modulation of body temperature [Heldmaier et al., 1989]. The tissue mainly responsible for non-shivering thermogenesis is the brown adipose tissue, a unique lipid storing tissue, which grows substantially during the pre-hibernation period [Smith and Horwitz, 1969], probably stimulated by the short photoperiod and melatonin [Viswanathan et al., 1986]. During hibernation, brown adipose tissue is apparently quiescent, although the brown adipocyte nuclei maintain an "active" configuration [Zancanaro et al., 1993], but it switches to vigorous thermogenesis in the early phase of arousal [Haywards et al., 1965; Wang and Abbots, 1981; Horwitz et al., 1985]: the fatty acids are oxidised in specialised mitochondria where an extensive uncoupling of oxidative phosphorylation occurs with a consequent heat production [Wang and Abbots, 1981; Nedergaard and Cannon, 1984, Cannon and Nedergaard, 1985; Himms-Hagen, 1986].

Ventilation drastically decreases during hibernation, switching from continuous to intermittent with prolonged periods of apnoea [Malan, 1982]. Entrance into hibernation is also accompanied by a decrease in the respiratory exchange rate, possibly by retention of carbon dioxide in the body fluids [Snapp and Heller, 1981].

Heart rate and cardiac output also decrease dramatically, whereas stroke volume increases, however, the mechanisms regulating parasympathetic and sympathetic activities responsible for such process are still unknown. In particular, the hibernator's heart ability to cope with the lowering of the body temperature skipping fatal arrythmia remains to be explained [Johansson, 1996]. The reduction of peripheral blood flow could lead to hypoxia and, in turn, to cellular damage. However, hibernators are apparently able to down regulate their cellular metabolic activity to a new hypometabolic steady state, thereby balancing the ATP supply with the ATP demand [Boutilier, 2001]. The blood clotting time is dramatically increased during hibernation [Svihla and Bowman, 1951], thus reducing the risk of thrombus formation when blood flow is reduced. The mechanism involved could be a reduction in platelets blood count or an increase of the liver α2-macroglobulin, a universal protease inhibitor [Srere et al., 1995].

The kidney activity is markedly reduced during hibernation [Zatzman, 1984] in association with lower kidney blood flow and reduction or cessation of glomerular filtration rate [Volkert et al., 1976]; despite this, the fine architectural features of the kidney remain well preserved [Zancanaro et al, 1999], probably due to the protective role of nitric oxide [Sandovici et al., 2004].

The prolonged immobility during hibernation causes only minimal alterations in skeletal muscle: this is probably due to the maintenance of high expression levels of the major myofibrillar proteins as well as to changes in the intracellular levels of calcium; periodic arousal are also thought to play a role in avoiding muscle atrophy [Hershey et al., 2008]. Bone too seems to be only slightly affected, although some variations exist among different hibernators. In fact, in some species the bone undergoes an important resorption during the hypometabolic period and a rapid recovery at the remobilization [Donahue et al., 2003a,b], whereas in others the changes in bone mass or architecture are scarce [Lennox and Goodship, 2008].

During hibernation, animals show a remarkable suppression of responsiveness to stimuli; however, handling, re-warming, or exposure to daylight, are able to induce arousal, thus making evident that hibernation, unlike coma, does not represent a loss of function but rather results from a highly regulated process. Several brain areas seem to be involved in the regulation of metabolic depression during hibernation. The hippocampus, septum, and hypothalamus retain periodic electroencephalographic activity at temperatures below which the electroencephalograms in other structures become isoelectric [Pakotin et al., 1993]. In particular, the suprachiasmatic nucleus (containing pacemaker cells for circadian rhythms), the supraoptic and paraventricular nuclei, the lateral septal nucleus and the medial septum are thought to be involved in hibernation [Machin-Santamaria, 1978; Kilduff et al., 1982; Bitting et al., 1994; O'Hara et al., 1999]; moreover, the preoptic/anterior hypothalamic area might also play a role due to the presence of sleep promoting and temperature sensitive neurons therein [Heller, 1979].

A fundamental strategy allowing survival under natural hypometabolism is represented by the shift from carbohydrate to the lipid in the white adipose tissue, as the primary energy source [Hoffman, 1964; Lyman et al., 1982; Wang, 1987; French, 1988; Heldmaier et al., 1999]. During summer and fall, seasonal hibernators become hyperphagic, sometimes doubling the body weight they had in spring [e.g. Bintz, 1988; Vogel, 1997]. This behavioural modification probably results from the combined action of the photoperiod, ambient temperature and food availability [Bartness and Wade, 1985; Ruf et al., 1993; Körtner and Geiser, 2000], the changed activity of the enzymes responsible for lipogenesis [Mostafa et al., 1993], and the release of specific neuropeptides regulating food intake [Boswell et al., 1993]. However, some tissues, such as the brain, need carbohydrates for their metabolic functions [Musacchia, 1984]. Since during natural hypometabolism the level of the glycolytic activity remains low [Storey, 1987, 1997; Soukri et al., 1996; Nestler et al., 1997; Heldmaier et al., 1999], the gluconeogenesis (i.e., the re-synthesis, mainly in liver and kidney, of glucose from amino acids, lactate, glycerol) becomes the only means to regenerate the carbohydrate reserves. In fact, during hibernation, carbohydrate metabolism remains at levels comparable to euthermia in some tissues [e.g. Burlington and Klain, 1967; Tashima et al., 1970; Riedelsen and Steffen, 1980; Davis et al., 1990] and the adrenocortical cells which are involved in the regulation of protein and lipid catabolism for gluconeogenesis show clear signs of functional activity at light and electron microscopy [Malatesta et al., 1995; Zancanaro et al., 1997]. It is worth recalling that, upon arousal, carbohydrates become again an important energy source [Burlington and Klain, 1967; al Badry and Taha, 1983], especially for brain [Lee et al., 2002].

Mitochondria represent the sites where carbohydrates, fatty acids and amino acids are oxidized to carbon dioxide and water and the free energy released is used to convert adenosine diphosphate (ADP) and inorganic phosphate to adenosine triphosphate (ATP), the molecule responsible for most of the energy transfer involved in living processes. During hibernation mitochondria of different tissues undergo profound structural modifications [Grodums, 1977; Romita and Gatti, 1980; Brustovetsky et al., 1993b; Malatesta et al., 2000a; Kabine et al., 2003]. In parallel, several mitochondrial functions modify: the enzymes responsible for fatty acid utilization increase [Kabine et al., 2003]; many mitochondria-encoded genes are up-regulated in multiple organs to facilitate the transport rate of fatty acids [Hittel and Storey, 2001, 2002a,b] as well as the thermogenic activity, especially in brown adipose tissue [Liu et al., 1998, 2001; Boyer et al., 1998]. However, the mitochondrial proton conductance is unchanged during hibernation [Barger et al., 2003], supporting the idea that the reduced metabolism in hibernators is a partial consequence of tissue-specific depression of substrate oxidation [Martin et al., 1999]. Moreover, during hibernation calcium ions inactivate the intramitochondrial ATPase, thus preventing the exhaustion of cellular ATP in de-energized mitochondria [Bronnikov et al., 1990], and, upon arousal, remove the blockage of the respiratory chain occurring during hibernation [Gehnrich and Aprille, 1988; Brustovetsky et al., 1989, 1992a, 1993a]. A role in activation/inactivation of mitochondrial enzymes is also played by the pH [Malan et al., 1985, 1988; Malan and Mioskowski, 1988].

Natural hypometabolism implies the drastic reduction of most cellular metabolic activities [review in Kolaeva et al., 1980]. The proteo-synthetic apparatus (the RER and Golgi complex) drastically reduces in size in various cell types [Krupp et al., 1977; Reme and Young, 1977; Malatesta et al., 1998, 2001b, 2002; Popov et al., 1999; Kolomiytseva et al., 2003], according to the severe reduction of protein synthesis rate [e.g. Derij and Shtark, 1985; Bocharova et al., 1992; Frerichs et al., 1998; Koebel et al., 1991]. Such a reduction depends on a direct, active effect on protein synthesis (e.g. the inhibition of the elongation phase: Chen et al., 2001; Hittel and Storey, 2002a,b) and on indirect, passive mechanisms (e.g. temperature lowering: van Breukelen and Martin, 2001; for a comprehensive review, see also van Breukelen and Martin, 2002a); the reversible phosphorylation of several regulatory enzymes [Storey, 1987; Morano et al., 1992; MacDonald and Storey, 1999, Arendt et al., 2003] and the differential temperature-dependent enzyme control [Storey, 1997; MacDonald and Storey, 1998] are also involved in the decrease of protein synthesis. In addition, the transcriptional activity too is severely inhibited by low temperatures [van Breukelen and Martin, 2002b].

However, some proteins become more abundant during hibernation than in euthermia; for example the myoglobin of the skeletal muscle [Postnikova et al., 1999] and the intestinal stress protein GRP75 [Carey et al., 2000] are overexpressed in hibernating ground squirrels, while the phosphoprotein pp98 is specifically expressed in the brain during hibernation only [Ohtsuki et al., 1998]. Actually, a differential expression of several genes occurs [for an extensive review see van Breukelen and Martin, 2002a], so that some genes are up-regulated in some tissues whereas being down-regulated in others [e.g. Eddy and Storey, 2003]. For example, in some tissues during hibernation there is an increase in the levels of the mRNAs for the UCP 1, 2 and 3 (probably in relation with the maintenance of the thermal homeostasis: Liu et al., 1998; Boyer et al., 1998) and for the peroxisomal acyl-CoA oxidase [Kabine et al.,

2003] and the pyruvate dehydrogenase kinase isoenzyme 4 which facilitate the shift from carbohydrate oxidation to the combustion of stored fatty acids [Buck et al., 2002].

In parallel with the functional modifications, the cell nucleus of hibernating tissues undergoes structural changes in the organization of both chromatin [Kolaeva et al., 1980; Bernocchi et al., 1986] and the ribonucleoprotein (RNP)-containing domains involved in RNA transcription and maturation [Zancanaro et al., 1993; Malatesta et al., 1994a,b, 1995, 1999, 2000, 2001a, 2003; Tamburini et al., 1996].

The Mystery of Periodic Arousals

Hibernation never spans the entire dormant season, but it is interrupted by periodic arousals and brief normothermic periods [French, 1985, 1988]. The frequency and the duration of these euthermic episodes mainly depend on the animal size: large hibernators arouse more frequently than do the small ones, and the duration of their euthermic intervals is much longer [French, 1985].

The functional meaning of such periodic energy-consuming arousals is still unknown, and numerous heterogeneous hypotheses have been put forward [Lyman et al., 1982]. Some authors suggested that a trigger for the periodic arousal could be the accumulation of ketone bodies in the blood of hibernating animals, due to a shutdown in the mobilization and utilization of carbohydrates [Baumber et al., 1971]. However, other authors found that the accumulation of ketone bodies can have a positive rebound on animal survival, as high levels of ketones are associated with increased resistance to hypoxic conditions [D'Alecy et al., 1990]. It has been also suggested that periodic euthermic episodes may serve to refill the carbohydrate reserves [Galster and Morrison, 1975] or to allow "normal" sleeping [Daan et al., 1991]. Other authors hypothesized that the periodic rewarming may be related to replacement of gene products which are lost during torpor due to degradation of mRNA [Knight et al., 2000]. Recently, it has been found that periodic arousals allow for restoration of the amino acid pool reservoir through the degradation of ubiquitinated proteins that accumulate in the gut during hibernation [van Breukelen and Carey, 2002]; concomitantly, this would help in restoring the host-defence mechanisms which appear to be down-regulated during hibernation [Prendergast et al., 2002].

Hibernating to Arouse

At the tissue and cell level, there is evidence that a programmed reorganization of the subcellular components takes place, as a prerequisite for the efficient restoring of all metabolic and physiological functions upon arousal. Several examples can be made, to demonstrate this issue.

Large amounts of zymogen granules are stored in the acinar cells of the exocrine pancreas of hibernating dormice: it is likely that this accumulation of hydrolytic enzymes, which could be seen as useless and even potentially dangerous for fasting animals, has the adaptive significance to permit the immediate resumption of the digestive functions in aroused animals [Malatesta et al., 1998, 2001b].

Similarly, during hibernation there is a dramatic increase of mitochondrial matrix granules which contain inorganic (calcium, phosphorous, sodium, magnesium, chlorine) and organic (lipids, phospholipids, glycoproteins, cytochrome c oxidase) components involved in the regulation of various mitochondrial functions [Bronnikov et al., 1990; Brustovetsky et al., 1992b, 1993a; Jacob et al., 1994]; these granules suddenly decrease upon arousal in many tissues of the dormice [Malatesta et al., 2000a], suggesting that they may represent storage sites for substrates needed for the respiratory functions (and ATP synthesis) to be restored at arousal. Consistently, these matrix granules are absent in the mitochondria of the brown adipose tissue, where oxidative phosphorylation is uncoupled to allow heat production in arousing animal [Wang and Abbots, 1981; Cannon and Nedergaard, 1985; Himms-Hagen, 1986].

As a final example, the structural and molecular reorganization of cell nucleus during hibernation can be recalled. Various nuclear bodies of different type, containing RNA processing factors and even factors involved in the regulation of the circadian rhythm [Zancanaro et al., 1993; Malatesta et al., 1994a,b; 1995; 1999; 2003; Tamburini et al., 1996] accumulate during the hypometabolic period and quickly disappear at arousal [Malatesta et al., 1994a, 2001a]. These bodies would play a role as storage/assembly sites for key molecules which are necessary to cope with the dramatic increase of transcriptional and translational rate as soon as the euthermic cell function are restored. In addition, a recent study provides evidence for a differential redistribution of RNA transcription and processing factors in diverse tissues: mRNAs at early stage of maturation accumulate in the hepatocytes whereas mature mRNAs are stored in brown adipocytes, whose thermogenic role at early arousal requires the immediate full restoration of protein synthesis [Malatesta et al., 2008].

In conclusion, the plethora of heterogeneous data in the literature confirm that natural hibernation is a highly programmed hypometabolic state rather than a simple fall of metabolic and physiological functions. Unfortunately, the strategies of natural hypometabolism so far described represent only pieces of a complex mosaic whose complete design still escapes our understanding. Low body temperature certainly contributes to the reduction of all metabolic and physiological activities during hibernation, but the whole physiological regulation of the organism under hypometabolic conditions can be achieved only by means of integrated molecular mechanisms which are still poorly understood.

A fundamental point can be taken for sure: natural hypometabolism does not only imply strategies to preserve the organism integrity under conditions which would be lethal for non-hibernators, but is also essential for the complete and rapid resumption of the euthermic structural and functional features.

Factors Inducing Hypometabolism

As seen above, the necessity of a hypometabolic state occurs in several species and takes place under a plethora of aspects. Deep hibernation, torpor and aestivation are only a few. It is therefore likely that hypometabolism derives from several sources and involves multiple pathways and serial metabolic events. It is not clear, yet, what are the factors responsible for these cascade events.

Several mechanisms seem to be involved in hibernation, for instance resetting of the temperature set-point, gene expression regulation, new metabolic balance and, in particular, endogenous opioids seem to play a basic role in regulating the hibernation cycle. Neuronal complexes immunoreactive for endogeneous opiates, especially enkephalin, are involved in the neuroendocrine control of hibernation [Nurnberger, 1995]. Opioids can induce changes similar to those observed during the annual cycle of mammalian hibernators, such as an increase in feeding at low doses and anorexia at high doses [Nizielski et al., 1986], bradicardia and hypotension [Kunos et al., 1988], and lowering of set-point in thermoregulation [Burks, 1991]. Interestingly, the blockade of endogenous opioids activity by exogenous opioid antagonists shortens hibernation bouts or induces premature arousal [Wang, 1993]. Finally, proopiomelanocortin, proenkephalin and prodynorphin molecules have been found in the brain of ground squirrels [Cui et al., 1996], with specific increase in selected brain districts (hippocampus, septum, hypothalamus) during hibernation. These data pinpoint to an involvement of endogenous opioids in regulating the hibernation cycle.

A natural proto-opioid, not yet chemically characterized, has been found in blood and urine hibernators like bats, brown bears and woodchucks during lethargy but not during euthermia [Bolling et al., 1997a]. The so-called Hibernation Induction Trigger (HIT) is a powerful metabolic inhibitor since it can induce hypometabolic effects in hibernators when administered to active animals in summer, probably via the interaction with the peripheral and central opioid receptors [Benedict et al. 1999]. So far, it is the most interesting and promising candidate as regulatory factor of natural hypometabolism.

In view of inducing hypometabolism in mammals, a few selected mechanisms could be considered, although the actual lack of complete scientific data makes is still speculative.

The *decrease in body temperature* is apparently a key factor to reduce the rate of metabolic and enzymatic activities on a purely thermodynamic drive; however, during entry into torpor metabolic rates drop rapidly even before a significant decrease in body temperature [Ortmann and Heldmaier, 2000] suggesting that low temperature alone cannot explain this phenomenon. When a non hibernator is exposed to low environmental temperatures, body temperature begins to fall and hypothermia ensues; this leads to failure in the homeostatic mechanism of shivering heart fibrillation and ventilation arrest.

Hibernating animals retain the ability to sense and defend body temperature [Drew et al., 2001]; however, when they enter torpor their hypothalamic set-point for body temperature regulation is gradually lowered [Heller, 1979]. Unfortunately, the mechanism by which the set-point is determined is still a mystery [Cooper, 2002].

Ischaemic preconditioning could help the induction of a hypometabolic state. In fact, before true hibernation begins, animals go through a number of cycles where metabolic rate and body temperature drop briefly, as a sort of hypothermia preconditioning. In non hibernators one or more short periods of ischaemia results in a substantial improvement in the ability of cells and organs to tolerate a subsequent, longer period of ischaemia. For instance, repeated myocardial stunning in dogs [Di Carli et al., 2000] proved to be at the basis of prolonged and reversible reduction in systolic functions. In the same model, suspended animation by hypothermia can allow survival after a 60 min cardiac arrest [Nozari et al., 2004]. In addition, recent reports seem to indicate that induced hypothermia in swine does not affect learning and memory [Alam et al., 2002].

Low temperature could cause freezing process. However, some hibernators (non mammals and mammals) can survive body temperature as low as –3 °C by adopting the strategy of supercooling to resist freezing [Lee and Costanzo, 1998]. Frogs are able to produce and introduce glycerol in the bloodstream as cryoprotectant. On the other hand, supercooling can be dangerous because it is a metastable condition where ice nucleation starts with ease.

During torpor bouts, hibernators can down-regulate their cellular metabolic activity to $1/100^{th}$ of basal metabolic rate without damage during the prolonged cold exposure as well as during the transitions between the two conditions.

A new balance between the ATP demand and the supply is established [Boutilier, 2001] and during natural hibernation mitochondrial activity drastically decreases and lipids become the main energy source [Geiser et al., 1994]. It could then be hypothesised that a shift from carbohydrate to lipid utilization could promote a hypometabolic state. Hormones such as leptin, responsible for fatty acid mobilization [Rousseau et al., 2003], could represent a key factor. Moreover, during hibernation, several mitochondrial functions undergo modifications, in association with increased activity of the enzymes responsible for fatty acid transport and utilization [Kabine et al., 2003; Hittel and Storey, 2002a]. Changes in mitochondrial enzyme activity have been observed in non-hibernators under conditions favouring a shift from carbohydrate to fatty acid oxidation [Peters et al., 2001]; therefore, the change in metabolic fuelling could facilitate torpor entrance. However, the activation/deactivation of mitochondrial functions in hibernators is a quite complex phenomenon involving numerous and various factors. In hibernating animals all metabolic activities undergo a more or less profound depression: and the mechanisms responsible for the natural hypometabolic state identified so far are mostly based on the reversible phosphorylation of several regulatory enzymes [MacDonald and Storey, 1999; Arendt et al., 2003], as well as on the differential enzyme control at different body temperatures [van Breukelen and Martin, 2001]. Recent studies [Blackstone et al., 2005] have shown the experimentally induced suspended animation by inhibition of the cytochrome c oxydase by hydrogen sulphide. Mice exposed to up to 80 ppm H_2S had a 90% drop in the metabolic rate without showing any behavioural or functional damage at arousal.

Some seasonal hibernators can start hibernating without any external input from environmental cues; this suggests that the ability to hibernate is due to a molecular genetic mechanism rather than being an acute response to e.g., low ambient temperature. An intervention at the gene level could therefore lead to hypometabolism.

One could hypothesize, for instance, to over-express a particular factor involved in the induction of a hypometabolic state, or to increase fuel storage.

Adipose tissue mass is controlled by a hypothalamic "lipostat" that senses body lipid content and initiates compensatory changes in appetite and energy expenditure to maintain a seasonally appropriate level of adiposity, leading to fat gain during late summer and autumn, and loss during winter ("sliding set-point" hypothesis, Mrosovsky and Fisher, 1970). Molecular basis for a lipostat has come from the cloning of the *leptin (lep)* gene, (a gene defective in obese ob/ob mice, Zhang et al., 1994). In humans, mice, and rats, blood concentrations of leptin, a 16 kD protein, are proportional to total body fat, and leptin production is a peripheral signalling component of the lipostat. High levels of leptin cause decreased food intake and increased energy expenditure and low levels resulting in greater hunger and energy conservation [Stephens and Caro, 1998].

During induced torpor, blood circulation is severely slowed down, and the risk of impairing microcirculation is high. An interesting opportunity would be to express α2-

macroglobulin, a protein with an important role in preventing blood clotting [Srere et al., 1995], whose presence has been shown to enhance survival in hibernators.

Finally, the modulated expression of HIT, the elusive protein factor present in the blood and urine of natural hibernators reputed to initiate the cascade effect finally leading to hypometabolic state would be the best way to induce a hypometabolic state in non hibernators. This seems far away in the future, given the scarcity of data in our hands. However, a possible *escamotage* could be represented by an interference with primary cell functions responsible for the regulation of the whole cell metabolism.

In natural hibernators, protein synthesis and translation become depressed in the cold and reactivated during arousal. However, hibernators utilize mRNA pool storage in aid to the resumption of gene expression during the interbout arousal. Moreover, in hibernating animals, cell nuclei undergo modifications of constituents involved in RNA transcription and splicing [Zancanaro et al., 1993; Malatesta et al., 1994a,b, 1995, 1999, 2000, 2001a, 2003; Tamburini et al., 1996], which could help the transitions involved in the euthermia-hibernation-arousal cycle. HIT or other natural opioids can be considered as candidates for the role of coordinator of these activities. HIT is not yet completely characterized, and there are also contradictory results as to its efficacy. However, a HIT-like molecule acting as a trigger is likely to exist. Recently it has been shown that woodchuck plasma (containing HIT) was effective in protecting skeletal muscle from ischaemia/reperfusion in non hibernators [Hong et al., 2005].

A synthetic delta opioid [D-Ala^2, D-Leu^5] enkephalin (DADLE) has attracted some interest for its capability to mimic the effects of HIT [Bolling et al. 1997a,b]. DADLE is able to induce a hypometabolic hibernation-like state in hibernators [Oeltgen et al., 1988; Malatesta et al., 2001a], but it has given promising results also in non-hibernators.

In in vitro systems, DADLE can significantly slow down the proliferation rate of on various cell lines derived from hibernating (woodchuck) and non hibernating (rat, human) species [Kampa et al., 1997]. Moreover, it reduces both RNA transcription and export to cytoplasm and probably provokes a cascade effect on other cellular functions such as protein synthesis and cell proliferation [Vecchio et al., 2006; Baldelli et al., 2004]. No cytotoxic effects have been found and DADLE can also have antiapoptotic effects [Tsao and Su, 2001]. Finally, after removing DADLE from the culture medium, the effects rapidly disappear [Vecchio et al., 2006].

DADLE activity was reported to be due to its capability of binding delta opioid receptors [Benedict et al., 1999; Bolling et al., 1998], but it has been recently demonstrated that it can enter cells with or without opioid receptors (not only of delta type) [Baldelli et al., 2004, 2006; Vecchio et al., 2006]. DADLE, in fact, is a small, partially lipophylic molecule and like other enkephalins can interact with different subtypes of opioid receptors on the plasma membrane and in the cell nucleus, as well as with polyspecific membrane transporters [see e.g. Hu et al., 2003]; consequently, specific receptors are probably not necessarily needed for DADLE crossing the cell membrane.

Recently, Kondo and co-workers [2006] demonstrated a decrease in hibernation-specific protein (HP) complex in the blood of chipmunks during hibernation. The authors identify HP as a candidate hormone for hibernation: in chipmunks kept in constant cold and darkness, HP appears to be regulated by an individual free-running circannual rhythm that correlates with hibernation. Blocking brain HP activity using an antibody decreases the duration of hibernation. It seems therefore likely that HP, a target of endogenously generated circannual rhythm, carries hormonal signals which are essential for hibernation to the brain. It is not clear yet whether this protein is present also in other species, or signals might be generated through different, alternative pathways.

HYPOMETABOLISM: THE HUMAN IMPACT

Mammalian neonates (inclusive of the human) present tracts of the hibernating phenotype insofar they are able to sustain hypoxia and/or low temperature by physiological mechanisms; this would suggest that the ability to induce a hypometabolic state is common to all mammals [Harris et al., 2004]. At the present moment, however, we are far from the goal of detecting the genes involved and/or the physiological mechanisms that should be elicited to induce hypometabolism.

The key advantage of inducing a regulated and reversible reduction in cellular and tissue need for oxygen and nutrients is that supply matches metabolic demand, thereby preventing cell and tissue damage. It should be underlined that a full hibernation state (very low body temperature, a few heart beats per minute, striking reduction of brain electrical activity) is not needed in the human to get consistent benefits in terms of energy saving etc.: just a moderate hypometabolic state leading to a 5-6 °C decrease in body temperature would pay enough.

To artificially induce a hypometabolic state in the human has a potentially enormous impact to improve the human condition; in the medical field this would e.g., improve cadaver/organ preservation for transplant; allow for better neuro/cardioprotection following ischaemic accidents; and simplify major surgical procedures requiring extra-corporeal circulation. Inducing hypometabolism in the human would also be of great relevance to long-term manned space missions. As a matter of fact, astronauts are exposed to several different stressors; just to mention a few, altered social interactions and confinement, hypokinesia, micro-gravity. This has negative effects on the musculoskeletal apparatus [Fitts et al., 2001] as well as the psychological status [Collins, 2003]. Regulated hypometabolism would prevent most of these problems, representing an effective countermeasure.

Pharmaceutical measures are likely to play a central role in maintaining the hypometabolic state, both through initiating and regulating suppression of the metabolism. In the last several years, the synthetic opioid DADLE has attracted much interest as potential inducer of hypometabolism, because it has shown promising results in the preservation of explanted organs, increasing their survival time and improving their functional conditions and histological preservation [Oeltgen et al., 1996, Bolling et al., 1997a,b; Su, 2000]. Moreover, it has been shown that perfusion of a multiorgan block preparation (heart, liver, lung and kidney) with DADLE, results in an increase of the survival time from 8 to 46 hours [Chien et al., 1994]. This peptide can also help the functional recovery of heart tissue after a prolonged ischaemia induced in non hibernating mammals [Bolling et al. 1997 a,b] and contribute to the survival of neurons in the CNS by contrasting the effects of metamphetamine, which is responsible of the destruction of dopaminergic terminals, by acting on necrosis tumour factor p53 and of c-fos [Su, 2000], as well as the survival time after hypoxia in the rat [Mayfield and D'Alecy, 1994]. Finally, DADLE can induce a short hypothermic effect in cold-exposed rabbits [Vybiral et al., 1997] and rats [Biggiogera et al., 2006], suggesting a possible hibernation mimicry in non hibernating animals. However, the mechanisms of this molecule both at the cell level and on the whole organisms are not clear and its effects as well as the activity of analogous molecules are currently under investigation.

Provided an effective hypometabolic drug is available, several engineering and system issue must be addressed to use it in manned space missions, including design of the hibernaculum where the astronauts can enter, sustain and exit torpor safely, the hibernation control with medical sensing and administration equipment for each individual astronaut, and the life support system [Ayre et al., 2004].

Conclusion

It is evident that lowering body temperature per se does not represent the route to human hypometabolism, but it is mandatory to identify the factor(s), acting on primary cell functions, able to initiate and coordinate the cascade of metabolic events leading to a controlled and fully reversible hypometabolic state. Some candidates are at present under investigation and the ability of a primate, the Madagascan fat-tailed dwarf lemur (*Cheirogaleus medius*), to hibernate even at 30°C ambient temperature [Dausmann et al., 2004] makes human hypometabolism less speculative.

Induction of a reversible form of hypometabolism is a prized goal for biomedicine. The multidisciplinary studies tackling the problem have demonstrated its complexity and indicated that final solution is not at hand. However, the data presented above clearly show that clarifying the mechanisms of mammalian hibernation in order to mimicking controlled hypometabolism in the human is of potential great impact in several settings involving human life, especially long-term manned space missions.

References

[1] Alam, HB; Bowyer, MW; Koustova, E; Gushchin, V; Anderson, D; Stanton, K; Kreishman, P; Cryer, CM; Hancock, T; Rhee, P. Learning and memory is preserved after induced asanguineous hyperkalemic hypothermic arrest in a swine model of traumatic exsanguination. *Surgery,* 2002, 132, 278-288.

[2] al-Badry, KS; Taha, HM. Hibernation-hypothermia and metabolism in hedgehogs. Changes in some organic components. *Comp. Biochem. Physiol.* 1993, 74A, 143-148.

[3] Arendt, T; Stieler, J; Strijkstra, AM; Hut, RA; Rudiger, J; Van der Zee, EA; Harkany, T; Holzer, M; Hartig, W. Reversible paired helical filament-like phosphotylation of tau is an adaptive process associated with neuronal plasticity in hibernating animals. *J. Neurosci.* 2003, 23, 6972-6981.

[4] Ayre, M; Zancanaro, C; Malatesta, M. Morpheus - Hypometabolic Stasis for Long-term Spaceflight. *J. Brit. Interplan. Soc.* 2004, 57, 325-339.

[5] Baldelli, B; Vecchio, L; Bottone, MG; Muzzonigro, G; Biggiogera, M; Malatesta, M. The effect of the enkephalin DADLE on transcription does not depend on opioid receptors. *Histochem. Cell Biol.* 2006, 126, 189-197.

[6] Baldelli, B; Vecchio, L; Biggiogera, M; Vittoria, E; Muzzonigro, G; Gazzanelli, G; Malatesta, M. Ultrastructural and immunocytochemical analyses of opioid treatment effects on PC3 prostatic cancer cells. *Microsc. Res. Tech.* 2004, 64, 243-249.

[7] Barger, JL; Brand, MD; Barnes, BM; Boyer, BB. Tissue-specific depression of mitochondrial proton leak and substrate oxidation in hibernating arctic ground squirrels. *Am. J. Physiol. Regul. Integr. Comp. Physiol.* 2003, 284, R1306-1313.

[8] Bartness, TJ; Wade, GN. Photoperiodic control of seasonal body weight cycles in hamsters. *Neurosci. Biobehav. Rev.* 1985, 9, 599-612.

[9] Baumber, J; South, FE; Ferren, L; Zatznan, ML. A possible basis for periodic arousals during hibernation: accumulation of ketone bodies. *Life Sci.* 1971, 10, 462-467.

[10] Benedict, PE; Benedict, MB; Su, TP; Bolling, SF. Opiate drugs and delta-receptor-mediated myocardial protection. *Circulation.* 1999, 100, 357-360.

[11] Bernocchi, G; Barni, S; Scherini, E. The annual cycle of Erinaceus europaeus L. as a model for a further study of cytochemical heterogeneity in Purkinje neuron nuclei. *Neurosci.* 1986, 17, 427-443.

[12] Biggiogera, M; Fabene, P; Zancanaro, C. DADLE: a cue to human "hibernation"? *J. Brit. Interplan. Soc.* 2006, 59, 115-118.

[13] Bintz, GL. Lipid synthesis and deposition by adult Richardson's ground squirrels in the natural environment. *J. Comp. Physiol.* 1988, 158, 199-204.

[14] Bitting, L; Sutin, EL; Watson, FL; Leard, LE; O'Hara, BF; Heller, HC; Kilduff, TS. C-fos mRNA increases in the grounf squirrel suprachiasmatic nucleus during arousal from hibernation. *Neurosci. Lett.* 1994, 165, 117-121.

[15] Blackstone, E; Morrison, M; Roth, MB. H2S induces a suspended animation-like state in mice. *Science.* 2005, 308, 518.

[16] Bocharova, LS; Gordon, RY; Arkhipov, VI. Uridine uptake and RNA synthesis in the brain of torpid and awaken ground squirrels. *Comp. Biochem. Physiol.* 1992, 101B, 189-192.

[17] Bolling, K; Halldorsson, A; Allen, BS; Rahman, S; Wang, T; Kronon, M; Feinberg, H. Prevention of the hypoxic reoxygenation injury with the use of a leukocyte-depleting filter. *J. Thorac. Cardiovasc. Surg.* 1997b, 113, 1081-1089.

[18] Bolling, K; Kronon, M; Allen, BS; Wang, T; Ramon, S; Feinberg, H. Myocardial protection in normal and hypoxically stressed neonatal hearts: the superiority of blood versus crystalloid cardioplegia. *J. Thorac. Cardiovasc. Surg.* 1997a, 113, 994-1003.

[19] Bolling, SF; Benedict, MB; Tramontini, NL; Kilgore, KS; Harlow, HH; Su, TP; Oeltgen, PR. Hibernation triggers and myocardial protection. *Circulation.* 1998, 98, 220-223.

[20] Bolling, K; Halldorsson, A; Allen, BS; Rahman, S; Wang, T; Kronon, M; Feinberg, H. Prevention of the hypoxic reoxygenation injury with the use of a leukocyte-depleting filter. *J. Thorac. Cardiovasc. Surg.* 1997b, 113, 1081-1089.

[21] Bolling, K; Kronon, M; Allen, BS; Wang, T; Ramon, S; Feinberg, H. Myocardial protection in normal and hypoxically stressed neonatal hearts: the superiority of blood versus crystalloid cardioplegia. *J. Thorac. Cardiovasc. Surg.* 1997a, 113, 994-1003.

[22] Boswell, T; Richardson, RD; Schwartz, MW; D'Alessio, DA; Woods, SC; Sipols, AJ; Baskin, DG; Kenagy, GJ. NPY and galanin in a hibernator: hypothalamic gene expression and effects on feeding. *Brain Res. Bull.* 1993, 32, 379-384.

[23] Boutilier, RG. Mechanisms of cell survival in hypoxia and hypothermia. *J. Exp. Biol.* 2001, 204, 3171-3181.

[24] Boyer, BB; Barnes, BM; Lowell, BB; Grujic, D. Differential regulation of uncoupling protein gene homologues in multlipe tissue of hibernating ground squirrels. *Am. J. Physiol.* 1998, 275, R1232-1238.

[25] Bronnikov, GE; Vinogradova, SO; Chernyak, BV. Regulation of ATP hydrolysis in liver mitochondria from ground squirrel. *FEBS.* 1990, 266, 83-86.

[26] Brustovetsky, NN; Egorova, MV; Gnutov, DYu; Gogvadze, VC; Mokhova, EN; Skulachev, VP. Thermoregulatory, carboxyatractylate-sensitive uncoupling in heart and skeletal muscle mitochondria of the ground squirrel correlates with the level of free fatty acids. *FEBS.* 1992b, 305, 15-17.

[27] Brustovetsky, NN; Egorova, MV; Gnutov, DYu. The mechanism of calcium-dependent activation of respiration of liver mitochondria from hibernating ground squirrels, Citellus undulatus. *Comp. Biochem. Physiol.* 1993a, 106B, 423-426.

[28] Brustovetsky NN; Egorova MV; Grishina EV; Mayevsky EI. Analysis of the causes of the suppression of oxidative phosphorylation and energy-dependent cationic transport into liver mitochondria of hibernating gophers, Citellus undulatus. *Comp. Biochem. Physiol.* 1992a, 103B, 755-758.

[29] Brustovetsky, NN; Egorova, MV; Iliasova, EN; Bakeeva, LE. Relationship between structure and function of liver mitochondria from hibernating and active ground squirrels, Citellus undulatus. *Comp. Biochem. Physiol.* 993b, 106B, 125-130.

[30] Brustovetsky N.N., Mayevsky E.I., Grishina E.V., Gogvadze V.G., Amerikhanov Z-G. Regulation of the rate of respiration and oxidative phosphorylation in liver mitochondria from hibernating ground squirrels, Citellus undulatus. *Comp. Biochem. Physiol.* 1989, 94B, 537-541.

[31] Buck, CL; Barnes, BM. Effects of ambient temperature on metabolic rate, respiratory quotient, and torpor in an arctic hibernator. *Am. J. Physiol. Regul. Integr. Comp. Physiol.* 2000, 279, R255-262.

[32] Buck, MJ; Squire, TL; Andrews, MT. Coordinate expression of the PDK4 gene: a means of regulating fuel selection in a hibernating mammal. *Physiol. Genomics.* 2002, 8, 5-13.

[33] Burks, TF. Opioid and opioid receptors in thermoregulation. In: Schonbaum E, Lomax P, editors. *Thermoregulation: pathology, pharmacology, biosynthesis and analysis.* New York: Pergamon Press; 1991; 489-508.

[34] Burlington, RF; Klain, GJ. Gluconeogenesis during hibernation and arousal from hibernation. *Comp. Biochem. Physiol.* 1967, 22, 701-708.

[35] Cannon, B; Nedergaard, J. The biochemistry of an inefficient tissue: brown adipose tissue. *Essays Biochem.* 1985, 20, 110-164.

[36] Carey, HV; Frank, CL; Seifert, JP. Hibernation induces oxidative stress and activation of NK-kappa B in ground squirrel intestine. *J. Comp. Physiol.* 2000, 170B, 551-559.

[37] Chen, Y; Matsushita, M; Nairn, AC ; Damuni, Z; Cai, D; Frerichs, KU; Hallenbeck, JM. Mechanisms for increased levels of phosphorylation of elongation factor-2 during hibernation in ground squirrels. *Biochemistry.* 2001, 40, 11565-11570.

[38] Chien, SF; Oeltgen, PR; Diana, JN; Salley, RK; Su, TP. Extension of tissuesurvival time in multiorgan block preparation using a delta opioid DADLE. *J. Thorac. Cardiovasc. Surg.* 1994, 107, 964-967.

[39] Collins, DL. Psychological issues relevant to astronaut selection for long-duration space flight: a review of the literature. *Hum. Perf. Extrem. Environ.* 2003, 7, 43-67.

[40] Cooper, KE. Molecular biology of thermoregulation – Some historical perspectives on thermoregulation. *J. Appl. Physiol.* 2002, 92, 1717-1724.

[41] Cui, Y; Lee, TF; Wang, LCH. State-dependent changes of brain endogenous opioids in mammalian hibernation. *Brain Res. Bull.* 1996, 40, 129-133.

[42] D'Alecy, LG; Lundy, EF; Kluger, MJ; Harker, CT; LeMay, DR; Shlafer, M. Beta-hydroxybutyrate and response to hypoxia in the ground squirrel, Spermophilus tridecimlineatus. *Comp. Biochem. Physiol.* 1990, 96B, 189-193.

[43] Daan, S; Bernes, BM; Strijkstra, AM. Warming up for sleep? Ground squirrels sleep during arousal from hibernation. *Neurosci. Lett.* 1991, 128, 265-268.

[44] Dausmann, KH; Glos, J; Ganzhorn, JU; Heldmaier, G. Hibernation in a tropical primate. *Nature.* 2004, 429, 825-826.

[45] Davis, WL; Goodman, DB; Crawford, LA; Cooper, OJ; Matthews, JL. Hibernation activates glyoxylate cycle and gluconeogenesis in black bear brown adipose tissue. *Biochim. Biophys. Acta.* 1990, 1051, 276-278.

[46] Derij, LV; Shtark, MB. Hibernators' brain: protein synthesis in the neocortex and the hippocampus. *Comp. Biochem. Physiol.* 1985, 80B, 927-934.

[47] Di Carli, MF; Prcevski, P; Singh, TP; Janisse, J; Ager, J; Muzik, O; Vander Heide, R. Myocardial blood flow, function, and metabolism in repetitive stunning. *J. Nucl. Med.* 2000, 41, 1227-1234.

[48] Donahue, SW; Vaughan, MR; Demers, LM; Donahue, HJ. Bone formation is not impaired by hibernation (disuse) in black bears Ursus americanus. *J. Exp. Biol.* 2003a, 206, 4233-4239.

[49] Donahue, SW; Vaughan, MR; Demers, LM; Donahue, HJ. Serum markers of bone metabolism show bone loss in hibernating bears. *Clin. Orthop. Relat. Res.* 2003b, 408, 295-301.

[50] Drew, KL; Rice, ME; Kuhn, TB; Smith, MA. Neuroprotective adaptations in hibernation: therapeutic implications for ischaemia-reperfusion, traumatic brain injury and neurodegenerative diseases. *Free Radic. Biol. Med.* 2001, 31, 563-573.

[51] Eddy, SF; Storey, KB. Differential expression of Akt, PPARgamma, and PGC-1 during hibernation in bats. *Biochem. Cell Biol.* 2003, 81, 269-274.

[52] Fitts, RH; Riley, DR; Widrick, JJ. Functional and structural adaptations of skeletal muscle to microgravity. *J. Exp. Biol.* 2001, 204, 3201-3208.

[53] French, AR. Allometries of the duration of torpid and euthermic intervals during mammalian hibernation: a test of the theory of metabolic control of the timing of changes in body temperature. *J. Comp. Physiol.* 1985, 156B, 13-19.

[54] French, AR. The patterns of mammalian hibernation. *Am. Sci.* 1988, 76, 569-575.

[55] Frerichs, KU; Smith, CB; Brenner, M; DeGracia, DJ; Krause, GS; Marrone, L; Dever, TE; Hallenbeck, JM. Suppression of protein synthesis in brain during hibernation involves inhibition of protein initiation and elongation. *Proc. Natl. Acad. Sci. U.S.A.* 1998, 95, 14511-14516.

[56] Galster, W; Morrison, PR. Gluconeogenesis in arctic squirrels between periods of hibernation. *Am. J. Physiol.* 1975, 228, 325-330.

[57] Gehnrich, SC; Aprille, JR. Hepatic gluconeogenesis and mitochondrial function during hibernation. *Comp. Biochem. Physiol.* 1988, 91B, 11-16.

[58] Geiser, F; McAllan, BM; Kenagy, GJ. The degree of dietary fatty acid unsaturation affects torpor patterns and lipid composition of a hibernator. *J. Comp. Physiol.* 1994,164, 299-305.

[59] Grodums, EJ. Ultrastructural changes in the mitochondria of brown adipose cells during the hibernation cycle of Citellus lateralis. *Cell Tissue Res.* 1977, 185, 231-237.

[60] Harris, M; Olson, L; Milsom, W. The origin of mammalian heterothermy: a case for perpetual youth. In: Barnes B, Carey H, editors. *Life in the cold: evolution, mechanisms, adaptation*. Alaska: Institute of Arctic Biology; 2004; 41-50.

[61] Haywards, JS; Lyman, CP; Taylor, C. The possible role of brown fat as source of heat during arousal from hibernation. *Ann. N.Y. Acad. Sci.* 1965, 131, 441-446.

[62] Heldmaier, G; Klingenspor, M; Werneyer, M; Lampi, BJ; Brooks, SP; Storey, KB. Metabolic adjustements during daily torpor in the Djungarian hamster. *Am. J. Physiol.* 1999, 276, E896-906.

[63] Heldmaier, G; Steinlechner, S; Ruf, T; Wiesinger, H; Klingenspor, M. Photoperiod and thermoregulation in vertebrates: body temperature rhythms and thermogenic acclimation. *J. Biol. Rhytms.* 1989, 4, 251-265.

[64] Heller, HC; Colliver, GW. CNS regulation of body temperature during hibernation. *Am. J. Physiol.* 1974, 227, 583-589.

[65] Heller, HC. Hibernation: neural aspects. *Annu. Rev. Physiol.* 1979, 41, 305-321.

[66] Hershey, JD; Robbins, CT; Nelson, OL; Lin, DC. Minimal seasonal alterations in the skeletal muscle of captive brown bears. *Physiol. Biochem. Zool.* 2008, 81, 138-147.

[67] Himms-Hagen, J. Brown adipose tissue and cold acclimatation. In: Trayhurn P, Nicholls DG, editors. *Brown adipose tissue*. London: Edward Arnold; 1986; 214-268.

[68] Hittel, DS; Storey, KB. Differential expression of adipose-and heart-type fatty acid proteins in hibernating ground squirrels. *Biochim. Biophys. Acta.* 2001, 1522, 238-243.

[69] Hittel, DS; Storey, KB. Differential expression of mitochondria-encoded genes in a hibernating mammal. *J. Exp. Biol.* 2002a, 205, 1625-1631.

[70] Hittel, DS; Storey, KB. The translation state of differentially expressed mRNAs in the hibernating 13-lined ground squirrel (Spermophilus tridecemlineatus). *Arch. Biochem. Biophys.* 2002b, 401, 244-254.

[71] Hoffman, RA. Terrestrial animals in the cold: hibernators. In: Dill DC, Adolph EF, Wilber CG, editors. *Handbook of Phisiology, Section 4: Adaptation to the Environment.* Washington DC: American Physiological Society; 1964; 379-403.

[72] Hong, J; Sigg, DC; Coles, JA Jr; Oeltgen, PR; Harlow, HJ; Soule, CL; Iaizzo, PA. Hibernation induction trigger reduces hypoxic damage of swine skeletal muscle. *Muscle Nerve.* 2005, 32, 200-207.

[73] Horwitz, BA; Hamilton, JS; Kott, KS. GDP binding to hamster brown fat mitochondria is reduced during hibernation. *Am. J. Physiol.* 1985, 249, R689-693.

[74] Hu, H; Miyauchi, S; Bridges, CC; Smith, SB; Ganapathy, V. Identification of a novel Na+- and Cl--coupled transport system for endogenous opioid peptides in retinal pigment epithelium and induction of the transport system by HIV-1 Tat. *Biochem. J.* 2003, 375, 17-22.

[75] Jacob, WA; Bakker, A; Hertsens, RC; Biermans, W. Mitochondrial matrix granules: their behavior during changing metabolic situations and their relationship to contact sites between inner and outer mitochondrial membranes. *Microsc. Res. Tech.* 1994, 27, 307-318.

[76] Johansson, BW. The hibernator heart – Nature's model of resistance to ventricular fibrillation. *Cardiov. Res.* 1996, 31, 826-832.

[77] Kabine, M; Clemencet, MC; Bride, J; El Kebbaj, M-S; Latruffe, N; Cherkaoui-Malki, M. Changes of peroxisomal fatty acid metabolism during cold acclimatization in hibernating jerboa (Jaculus orientalis). *Biochimie.* 2003, 85, 707-714.

[78] Kampa, M; Bakogeorgou, E; Hatzoglou, A; Damianaki, A; Martin, P-M; Castanas, E. Opioid alkaloids and casomorphin peptides decrease the proliferation of prostatic cancer cell lines (LNCaP, PC3 and DU145) through a partial interaction with opioid receptors. *Eur. J. Pharmacol.* 1997, 335, 255-265.

[79] Kilduff, TS; Sharp, FR; Heller, HC. [14C]2 deoxyglucose uptake in ground squirrel brain during hibernation. *J. Neurosci.* 1982, 2, 143-157.

[80] Knight, JE; Narus, EN; Martin, SL; Jacobson, A; Barnes, BM; Boyer, BB. mRNA stability and polysome loss in hibernating Arctic ground squirrels (Spermophilus parryii). *Mol. Cell Biol.* 2000, 20, 6374-6379.

[81] Koebel, DA; Miers, PG; Nelson, RA; Steffen, JM. Biochemical changes in skeletal muscles of denning bears (Ursus americanus). *Comp. Biochem. Physiol. B*, 1991, 100, 377-380.

[82] Kolaeva, SG; Kramarova, LI; Ilyasova, EN; Ilyasova, FE. The kinetics and metabolism of the cells of hibernating animals during hibernation. *Int. Rev. Cytol.* 1980, 66, 148-169.

[83] Kolomiytseva, IK; Perepelkina, NI; Patrushev, IV; Popov, VI.: Role of lipids in the assembly of endoplasmic reticulum and dictysomes in neuronal cells from the cerebral cortex of Yakutian ground squirrel (Citellus undulatus) during hibernation. *Biochemistry.* 2003, 68, 783-794.

[84] Kondo, N; Sekijima, T; Kondo, J; Takamatsu, N; Tohya, K; Ohtsu, T. Circannual control of hibernation by HP complex in the brain. *Cell.* 2006, 125, 161-172.

[85] Körtner, G; Geiser, F. The temporal organization of daily torpor and hibernation: ciecadian and circannual rhythms. *Chronobiol. Int.* 2000, 17, 103-109.

[86] Krupp, PP; Young, RA; Frink, R. The thyroid gland of the woodchuck, Marmota monax: a morphological study of seasonal variations in the follicular cells. *Anat. Rec.* 1977, 187, 495-514.

[87] Kunos, G; Mosqueda-Garcia, R; Mastrianni, JA. Endophinergic neurons in the brainstem and the control of blood pressure and heart rate In: Illes P, Farsang C, editors. *Regulatory roles of opioid peptides* Weinheim: VCH; 1988; 460-470.

[88] Lee, M; Choi, I; Park, K. Activation of stress signaling molecules in bat brain during arousal from hibernation. *Neurochem.* 2002, 82, 867-873.

[89] Lee, RE Jr; Costanzo, JP. Biological ice nucleation and ice distribution in cold-hardy ectothermic animals. *Annu. Rev. Physiol.* 1998, 60, 55-72.

[90] Lennox, AR; Goodship, AE. Polar bears (Ursus maritimus), the most evolutionary advanced hibernators, avoid significant bone loss during hibernation. *Comp. Biochem. Physiol. A Mol. Integr. Physiol.* 2008, 149, 203-208.

[91] Liu, X; Li, Q; Lin, Q; Sun, R. Uncoupling protein 1 mRNA, mitochondrial GTP-binding, and T4 5'- deiodinase of brown adipose tissue in euthermic Daurian ground squirrel during cold exposure. *Comp. Biochem. Physiol. A Mol. Integr. Physiol.* 2001, 128, 827-835.

[92] Liu, XT; Lin, QS; Li, QF; Huang, CX; Sun, RY. Uncoupling protein mRNA, mitochondrial GTP-binding, and T4 5'- deiodinase activity of brown adipose tissue in Daurian ground squirrel during hibernation and arousal. *Comp. Biochem. Physiol. A Mol. Integr. Physiol.* 1998, 120, 745-752.

[93] Lyman, CP; O'Brien, RC. Sensitivity to low temperature in hibernating rodents. *Am. J. Physiol.* 1972, 222, 864-869.

[94] Lyman, CP; Willis, JS; Malan, A; Wang, LCH. Hibernation and topor in mammals and birds. 1st edition. New York: Accademic Press; 1982.

[95] MacDonald, JA; Storey, KB. cAMP-dependent protein kinase from brown adipose tissue: temperature effects on kinetic properties and enzyme role in hibernating ground squirrels. *J. Comp. Physiol.* 1998, 168, 513-525.

[96] MacDonald, JA; Storey, KB. Regulation of ground squirrel Na+K+-ATPase activity by reversible phosphorylation during hibernation. *Biochem. Biophys. Res. Commun.* 1999, 254, 424-429.

[97] Machin-Santamaria, C. Ultrastructure of the hypothalamic neurosecretory nuclei of the dormouse (Elyomis quercinus L.) in the awakening and hibernating states. *J. Anat.* 1978, 127, 239-249.

[98] Malan, A; Mioskowski, E; Calgari, C. Time-course of blood acid-base state during arousal from hibernation in the European hamster. *J. Comp. Physiol.* 1988,158B, 495-500.

[99] Malan, A; Mioskowski, E. pH-temperature interactions on protein function and hibernation: GDP binding to brown adipose tissue mitochondria. *J. Comp. Physiol.* 1988, 158B, 487-493.

[100] Malan, A; Rodeau, JL; Daull, F. Intracellular pH in hibernation and respiratory acidosis in the European hamster. *J. Comp. Physiol.* 1985, 156B, 251-258.

[101] Malan, A. Respiration and acid-base state in deep hibernation. In: Lyman CP, Willis JS, Malan A, Wang LCH, editors. *Hibernation and torpor in mammals and birds*. New York: Academic Press; 1982; 237-283.

[102] Malatesta, M; Biggiogera, M; Baldelli, B; Barabino, SM; Martin, TE; Zancanaro, C. Hibernation as a far-reaching program for the modulation of RNA transcription. *Microsc. Res. Tech.* 2008, 71, 564-572.

[103] Malatesta, M; Baldelli, B; Rossi, L; Serafini, S; Gazzanelli, G. Fine distribution of clock protein in hepatocytes of hibernating dormice. *Eur. J. Histochem.* 2003, 47, 233-240.

[104] Malatesta, M; Battistelli, S; Rocchi, MBL; Zancanaro, C; Fakan, S; Gazzanelli, G. Fine structural modifications of liver, pancreas and brown adipose tissue mitochondria from hibernating, arousing and euthermic dormice. *Cell Biol. Int.* 2000a, 25, 131-138.

[105] Malatesta, M; Gazzanelli, G; Battistelli, S; Martin, TE; Amalric, F; Fakan, S. Nucleoli undergo structural and molecular modifications during hibernation. *Chromosoma.* 2000, 109, 506-513.

[106] Malatesta, M; Luchetti, F; Marcheggiani, F; Fakan, S; Gazzanelli, G. Disassembly of nuclear bodies during arousal from hibernation: an in vitro study. *Chromosoma.* 2001a, 110, 471-477.

[107] Malatesta, M; Zancanaro, C; Tamburini, M; Martin, TE; Fu, X-D; Vogel, P; Fakan, S. Novel nuclear ribonucleoprotein structural components in the dormouse adrenal cortex during hibernation. *Chromosoma.* 1995, 104, 121-128.

[108] Malatesta, M; Zancanaro, C; Balzelli, B; Gazzanelli, G. Quantitative ultrastructural changes of hepatocyte constituents in euthermic, hibernating and arousing dormice (Muscardinus avellanarius). *Tissue Cell.* 2002, 34, 397-405.
[109] Malatesta, M; Zancanaro, C; Marcheggiani, F; Cardinali, A; Rocchi, MBL; Capizzi, D; Vogel, P; Fakan, S; Gazzanelli, G. Ultrastructural, morphometrical and immunocytochemical analyses of the exocrine pancreas in a hibernating dormouse. *Cell Tissue Res.* 1998, 292, 531-541.
[110] Malatesta, M; Baldelli, B; Rossi, L; Serafini; S; Gazzanelli, G. Fine distribution of clock protein in hepatocytes of hibernating dormice. *Eur. J. Histochem.* 2003, 47, 233-240.
[111] Malatesta, M; Cardinali, A; Battistelli, S; Zancanaro C; Martin, TE; Fakan, S; Gazzanelli, G. Nuclear bodies are usual constituents in tissues of hibernating dormice. *Anat. Rec.* 1999, 254, 389-395.
[112] Malatesta, M; Gazzanelli, G; Marcheggiani, F; Zancanaro, C; Rocchi, MBL. Ultrastructural characterisation of periinsular pancreatic acinar cells in the hibernating dormouse Muscardinus avellanarius. *It. J. Zool.* 2001b, 68, 101-106.
[113] Malatesta, M; Zancanaro, C; Tamburini, M; Martin, TE; Fu, X-D; Vogel, P; Fakan, S. Novel nuclear ribonucleoprotein structural components in the dormouse adrenal cortex during hibernation. *Cromosoma.* 1995, 104, 121-128.
[114] Malatesta, M; Zancanaro, C; Martin, TE; Chan, EKL; Amalric, F; Lührmann, R; Vogel, P; Fakan, S. Is the coiled body involved in nucleolar functions? *Exp. Cell Res.* 1994a, 211, 415-419.
[115] Malatesta, M; Zancanaro, C; Martin, TE; Chan, EKL; Amalric, F; Lührmann, R; Vogel, P; Fakan, S. Cytochemical and immunocytochemical characterization of nuclear bodies during hibernation. *Eur. J. Cell Biol.* 1994b, 65, 82-93.
[116] Martin, SL; Maniero, GD; Carey, C; Hand, SC. Reversible depression of oxygen consumption in isolated liver mitochondria during hibernation. *Physiol. Biochem. Zool.* 1999, 72, 255-264.
[117] Mayfield, KP; D'Alecy, LG. Delta 1 opioid receptor dependence of acute hypoxic adaptation. *J. Pharmacol. Exp. Ther.* 1994, 268, 74-77.
[118] Morano, I; Adler, K; Agostini, B; Hasselbach, W. Expression of myosin heavy and light chains and phosphorylation of the phosphorylatable myosin light chain in the heart ventricle of the European hamster during hibernation and in summer. *Muscle Res. Cell. Motil.* 1992, 13, 64-70.
[119] Mostafa, N; Everett, DC; Chou, SC; Kong, PA; Florant, GL; Coleman, RA. Seasonal changes in critical enzymes of lipogenesis and triacylglycerol syntesis in the marmot (Marmota flaviventris). *Comp. Physiol.* 1993, 163, 463-469.
[120] Mrosovsky, N; Fisher, KC. Sliding set points for body weight in ground squirrels during the hibernation season. *Can. J. Zool.* 1970, 48, 241-247.
[121] Musacchia, XJ. Comparative physiological and biochemical aspects of hypothermia as a model for hibernation. *Cryobiology.* 1984, 21, 583-592.
[122] Nedergaard, J; Cannon, B. Preferential utilization of brown adipose tissue lipids during arousal from hibernation in hamsters. *Am. J. Physiol.* 1984, 247, R506-512.
[123] Nelson, RA. Protein and fat metabolism in hibernating bears. *Fed. Proc.* 1980, 39, 2955-2958.

[124] Nestler, GR; Peterson, SJ; Smith, BD; Heathcock, RB; Johanson, CR; Sarthou, JC; King, JC. Glycolytic enzyme binding during and trans to daily torpor in deer mice (Peromyscus maniculatus). *Physiol. Zool.* 1997, 70, 61-67.

[125] Nizielski, SE; Levine, AS; Morley, GE; Hall, KA; Gosnell, BA. Seasonal variation in opioid modulation of feeding in the 13-lined ground squirrel. *Physiol. Behav.* 1986, 37, 5-9.

[126] Nozari, A; Safar, P; Wu, X; Stezoski, WS; Henchir, J; Kochanek, P; Klain, M; Radovsky, A; Tisherman, SA. Suspended animation can allow survival without brain damage after traumatic exsanguination cardiac arrest of 60 minutes in dogs. *J. Trauma.* 2004, 57, 1266-1275.

[127] Nurnberger, F. The neuroendocrine system in hibernating mammals: present knowledge and open questions. *Cell Tissue Res.* 1995, 281, 391-412.

[128] O'Hara, BF; Watson, FL; Srere, HK; Kumar, H; Wiler, SW; Welch, SK; Bitting, L; Heller, HC; Kilduff, TS. Gene expression in the brain across the hibernation cycle. *Neurosci.* 1999, 19, 3781-3790.

[129] Oeltgen, PR; Horton, ND; Bolling, SF; Su, TP. Extended lung preservation with the use of hibernation trigger factors. *Ann. Thorac. Surg.* 1996, 61, 1488-1493.

[130] Oeltgen, PR; Nichols, PA; Nilekani, WA; Spurrier, WA; Su, TP. Further studies on opioids and hibernation: delta opioid receptor ligand selectively induced hibernation in summer active ground squirrels. *Life Sci.* 1988, 43, 1565-1574.

[131] Otsuki, T; Jaffe, H; Brenner, M; Azzam, R; Frerichs, KU; Hallenbeck, JM. Stimulation of tyrosine phosphporylation of a brain protein by hibernation. *Cereb. Blood Flow Metab.* 1998,18, 1040-1045.

[132] Ortmann, S; Heldmaier, G. Regulation of body temperature and energy requirements of hibernating Alpine marmots (Marmota marmota). *Am. J. Physiol. Regul. Integr. Comp. Physiol.* 2000, 278, R689-R704.

[133] Pakotin, PL; Pakotina, ID; Belusov, AB. The study of brain slices from hibernating mammals in vitro and some approaches to the analysis of hibernation problems in vivo. *Progr. Neurobiol.* 1993, 40, 123-161.

[134] Peters, SJ; Harris, RA; Wu, P; Pehleman, TL; Heigenhauser, GJ; Spriet, LL. Human skeletal muscle PDH kinase activity and isoform expression during a 3-day high-fat/low-carbohydrate diet. *Am. J. Physiol. Endocrinol. Metab.* 2001, 281, E1151-1158.

[135] Popov, VI; Ignat'ev, DA; Lindemann, B. Ultrastructure of taste receptor cells in active and hibernating ground squirrels. *J. Electron. Microsc.* 1999, 48, 957-969.

[136] Postnikova, GB; Tselikova, SV; Kolaeva, SG; Solomonov, NG. Myoglobin content in skeletal muscles of hibernating ground squirrels rises in autumn and winter. *Comp. Biochem. Physiol. A Mol. Integr. Physiol.* 1999, 124, 35-37.

[137] Prendergast, BJ; Freeman, DA; Zucker, I; Nelson, RJ. Periodic arousal from hibernation is necessary for initation of immune responses in ground squirrels. *Am. J. Physiol. Regul. Integr. Comp. Physiol.* 2002, 282, R1054-1062.

[138] Reme, CE; Young, RW. The effects of hibernation on cone visual cells in the ground squirrel. *Invest. Ophthalmol.* 1977, 16, 815-840.

[139] Riedelsen, ML; Steffen, JM. Protein metabolism and urea recycling in rodent hibernators. *Fed. Proc.* 1980, 39, 2959-2963.

[140] Romita, G; Gatti R. Histochemical and ultrastructural aspects of liver in Chiroptera (Vesperugo savi and Rinolophus f.e.) during different year periods. *Acta Bio-Med. Ateneo Parmense.* 1980, 51, 203-237.

[141] Rousseau, K; Archa, Z; Loudon, AS. Leptin and seasonal mammals. *Neuroendocrinol.* 2003, 15, 409-414.

[142] Ruf, T; Stieglitz, A; Steinlechner, S; Blank, JL; Heldmaier, G. Cold exposure and food restriction facilitate physiological responses to short photoperiod in Djungarian hamsters (Phodopus sungorus). *J. Exp. Zool.* 1993, 267, 104-112.

[143] Sandovici, M; Henning, RH; Hut, RA; Strijkstra, AM; Epema, AH; Goor, H; Deelman, LE. Differential regulation of glomerular and interstitial endothelial nitric oxide synthase expression in the kidney of hibernating ground squirrel. *Nitric Oxide.* 2004, 11, 194-200.

[144] Smith, RE; Horwitz, BA. Brown fat and thermogenesis. *Physiol. Rev.* 1969, 49, 330-425.

[145] Snapp, BD; Heller, HC. Suppression of metabolism during hibernation in ground squirrel (Citellus lateralis). *Physiol. Zool.* 1981, 54, 297-307.

[146] Soukri, A; Valverde, F; Hafid, N; Elkebbaj, MS; Serrano, A. Occurrence of a differential expression of the glyceraldehydes-3-phosphatedehydrogenase gene in muscle and liver from euthermic and induced hibernating jerboa (Jaculus orientalis). *Gene.* 1996, 181, 139-145,

[147] South, FE; Hartner, WC; Luecke, RH. Responses to preoptic temperature manipulation in the awake and hibernating marmot. *Am. J. Physiol.* 1975, 229, 150-160.

[148] Srere, HK; Belke, D; Wang, LC; Martin, SL. α2-Macroglobulin gene expression during hibernation in ground squirrel (Citellus tridecemlineatus). *Am. J. Physiol.* 1995, 268, R1507-R1512.

[149] Stephens, TW; Caro, JF. To be lean or not to be lean: Is leptin the answer? *Exp. Clin. Endocrinol. Diabetes.* 1998, 106, 1-15.

[150] Storey, KB; Storey, JM. Metabolic rate depression and biochemical adaptation in anaerobiosis, hibernation and estivation. *Q. Rev. Biol.* 1990, 65, 145-174.

[151] Storey, KB. Metabolic regulation in mammalian hibernation:enzyme and protein adaptations. *Comp. Biochem. Physiol. A Physiol.* 1997, 118, 1115-1124.

[152] Storey, KB. Regulation of liver metabolism by enzyme phosphorylation during mammalian hibernation. *Biol. Chem.* 1987, 262, 1670-1673.

[153] Su, TP. Delta opioid peptide[D-Ala(2),D- Leu(5)]enkephalinpromots cell survival. *J. Biomed. Sci.* 2000, 7, 195-199.

[154] Svihla, A; Bowman, HR. Prolongation of blood clotting time in the dormant hamster. *Science.* 1951, 115, 298-299.

[155] Tamburini, M; Malatesta, M; Zancanaro, C; Martin, TE; Fu, XD; Vogel, P; Fakan, S. Dense granular bodies: a novel nucleoplasmic structure in hibernating dormice. *Histochem. Cell Biol.* 1996, 106, 581-586.

[156] Tashima, LS; Aldelstein, SJ; Lyman CP. Radioglucose utilization by active, hibernating, and arousing ground squirrels. *Am. J. Physiol.* 1970, 218, 303-309.

[157] Tsao, LI; Su, TP. Hibernation-induction peptide and cell death: [D-Ala2,D-Leu5]enkephalin blocks Bax-related apoptotic processes. *Eur. J. Pharmacol.* 2001, 428, 149-151.

[158] van Breukelen, F; Martin, SL. Translational initiation is uncoupled from elongation at 18 degrees C during mammalian hibernation. *Am. J. Physiol. Regul. Integr. Comp. Physiol.* 2001, 281, R1374-1379.

[159] van Breukelen, F; Carey, V. Ubiquitin conjugate dynamics in the gut and liver of hibernating ground squirrels. *J. Comp. Physiol.* 2002, 172B, 269-273.

[160] van Breukelen, F; Martin SL. Invited review: molecular adaptations in mammalian hibernators: unique adaptations or generalized responses? *J. Appl. Physiol.* 2002a, 92, 2640-2647.

[161] van Breukelen, F; Martin, SL. Reversible depression of transcription during hibernation. *J. Comp. Physiol.* 2002b, 172B, R1374-1379.

[162] Vecchio, L; Soldani, C; Bottone, MG; Malatesta, M; Martin, TE; Rothblum, LI; Pellicciari, C; Biggiogera, M. DADLE induces a reversible hibernation-like state in HeLa cells. *Histochem. Cell Biol.* 2006, 125, 193-201.

[163] Viswanathan, M; Hissa, R; George, JC. Effects of short photoperiod and melatonin treatment on thermogenesis in the Syrian hamster. *J. Pineal Res.* 1986, 3, 311-321.

[164] Vogel, P. Hibernation of recently captured Muscardinus, Elyomis and Myoxus: comparative study. *Nat. Croat.* 1997, 6, 217-231.

[165] Volker, WA; Tempel, GE; Musacchia, XJ. Renal function in hypothermic hamsters. In: Jansky L, Musacchia XJ, editors. *Regulation of depressed metabolism and thermogenesis*. Springfield: Thomas; 1976; 258-273.

[166] Vybiral, S; Jansky L. Hibernation triggers and cryogens: do they play a role in hibernation? *Comp. Biochem. Physiol. A Physiol.* 1997, 118, 1125-1133.

[167] Wang, LCH; Abbots, B. Maximum thermogenesis in hibernators: magnitudes and seasonal variations. In: Musacchia XJ, Jansky L, editors. *Survival in the cold.* Amsterdam: Elsevier/North Holland; 1981; 77-97.

[168] Wang, LCH. Mammalian hibernation. In: Grout BWW, Morris GJ, editors. *The effects of low temperature on biological systems*. London: Edward Arnold; 1987; 349-386.

[169] Wang, LCH. Is endogenous oipoid involved in hibernation? In: Cynthia C et al., editors. *Life in the cold: Ecological, physiological, and molecular mechanisms.* Colorado: Westview Press, 1993; 297-304.

[170] Zancanaro, C; Malatesta, M; Mannello, F; Vogel, P; Fakan, S. The kidney during hibernation and arousal from hibernation. A natural model of organ preservation during cold ischaemia and reperfusion. *Nephrol. Dial. Transpl.* 1999, 14, 1982-1990.

[171] Zancanaro, C; Malatesta, M; Vogel, P; Fakan, S. Ultrastructure of the adrenal cortex of hibernating, arousing and euthermic dormouse, Muscardinus avellanarius. *Anat. Rec.* 1997, 249, 359-364.

[172] Zancanaro, C; Malatesta, M; Vogel, P; Osculati, F; Fakan, S. Ultrastructural and morphometrical analyses of the brown adipocyte nucleus in a hibernating dormouse. *Biol. Cell.* 1993, 79, 55-61.

[173] Zatzman, ML. Renal and cardiovascular effect of hibernation. *Cryobiology.* 1984, 21, 593-614.

[174] Zhang, YY; Proenca, R; Maffei, M; Barone, M; Leopold, L; Friedman, JM. Positional cloning of the mouse obese gene and its human homologue. *Nature.* 1994, 372, 425-432.

In: Space Exploration Research
Editors: J. H. Denis and P. D. Aldridge

ISBN: 978-1-60692-264-4

Chapter 4

C. ELEGANS: A MODEL SYSTEM FOR STUDYING BIOLOGICAL EFFECTS OF THE SPACE ENVIRONMENT

Robert Johnsen[1], Martin Jones[1], Nathaniel Szewczyk[2] and David Baillie[1]

[1] Department of Molecular Biology and Biochemistry,
Simon Fraser University, Burnaby, British Columbia, Canada, V5A 1S6
[2] School of Graduate Entry Medicine and Health,
University of Nottingham, Derby, United Kingdom, DE22 3DT

ABSTRACT

The well characterized nematode *C. elegans* is a superb model organism for biological experiments in space environments especially due to its similarities to humans on the most basic levels. Humans and *C. elegans* have similar numbers of genes, about 60% of those genes are homologous and both organisms have similar repair systems for radiation induced DNA damage. In addition, both have neuromuscular systems and hormonal regulatory systems. *C. elegans* can also be maintained on a chemically defined minimal medium (CeMM) so that bacteria, the nematode's normal food, need not be shipped into orbit. *C. elegans* is robust and microscopic so that nematode experiments require little oversight by astronauts and need minimal space onboard the International Space Station. Little is known about the effects of long-term exposure to different types of radiation in space and the wide range of radiation sources needed to mimic the space environment are not available on Earth. In order to address this *C. elegans* experiments have been to space on several missions starting in the early 1990's. These missions ranged from a few days to a multi-generation trip spanning several months. Five *C. elegans* mutagen testing systems have been used to analyze the effects of radiation: 1) poly-G/poly-C tract modifications; 2) assaying for dominant *unc-22* mutations; 3) fem-3 dauer analysis; 4) identifying alterations in telomere length; and 5) capturing and analyzing mutations using the eT1-system which balances approximately 1/6 of the entire genome. The eT1-system is being developed into a "biological accumulating dosimeter" to be used to capture and analyze space radiation induced mutations. Nimblegen micro-array chip analysis has proven very effective for determining the exact locations and extends of captured eT1 lethal deficiency and duplication mutations and using massively parallel sequencing instruments could give a quantitative look at the spectrum of the captured space radiation induced mutations. In addition to studying the mutational effects of radiation other experiments have focused on the effects microgravity and gene expression. These include expression of myogenic transcription factors and myosin heavy chains in *C. elegans* muscles developed during spaceflight and how checkpoint and physiological apoptosis in germ cells proceeds in space flight. These analyzes demonstrate that *C. elegans* exhibits similar molecular changes to humans in flight and should lead to the development of countermeasures necessary for continual habitation of the ISS and also for long-duration manned space-flights to the Moon and Mars

INTRODUCTION

We are entering an exciting time for space exploration; we are on the verge of sending manned missions to explore the solar system. Both China and the United States have announced plans for manned lunar landings. Long Lehao, deputy chief architect of the Chinese lunar probe project set 2024 as the year of China's first moonwalk and Sun Laiyan, administrator of the China National Space Administration, stated that China would start deep space exploration focusing on Mars. The first Chinese unmanned Mars exploration program should take place between 2014-2033, followed by a manned phase in 2040-2060. The United States plans to spend $100 billion to put humans back on the Moon by 2018, and President George W. Bush called the return to the moon as the "first major step in a broader space exploration vision aimed at extending the human presence throughout the solar system."

Unfortunately, travel in deep space will be hazardous and we do not even know how hazardous. We must develop much better understandings on the effects of micro-gravity, radiation, and in addition, the psychological effects of being in closed cramped space for long time periods in order to insure the survival, safety, and health of the astronauts. We need to know more about the extent of the degeneration of the neuromuscular system and effects on other physiological systems, the effects on gene expression, and we need to know the effects of long-term exposure to the wide spectra and doses of space radiation that the spacefarers will encounter. NASA requested The Aeronautics and Space Engineering Board of the NRC to establish a committee [1] to evaluate our current understanding, and identify gaps in our knowledge, of various aspects of the radiation environments astronauts will encounter on trips to the Moon and also on future Mars missions. The committee's mandate is not only to assess the effects and health risks of radiation but also to assess our current state of knowledge and review approaches necessary to meet the radiation shielding requirements for lunar and Mars missions in order to meet NASA's radiation exposure guidelines. The committee will also develop a strategic plan and recommend technologies NASA should invest in to develop the necessary radiation mitigation capabilities to ensure the health of the astronauts.

We must do space-based experiments in order to analyze physiological effects of micro-gravity and the effects of exposure to radiation because we cannot simulate prolonged micro-gravity nor are the wide range of radiation sources needed to reproduce space environments available on Earth. Attempts have been made to simulate the radiation exposure astronauts will be subject to during a round trip to Mars. For example, New York's Brookhaven National Laboratory used their particle accelerator and found an expected exposure of approximately 130,000 Millirem. This is over two orders of magnitude greater than the 350 Millirems average annual dose for Americans. The human LD50 (dose expected to kill 50%

of a normal human population) is about 450,000 Millirems [2]. In addition to the expected exposure the spacefarers may also be subject to solar flare radiation which could increase the dose several fold.

A round trip to Mars will likely expose the spacefarers to over one-quarter the LD50 dose – and possibly much more - which means, without adequate protection, that a large fraction of the crew could die from radiation poisoning while the rest would probably become quite sick. This is much higher than allowable by NASA's radiation exposure guidelines which must be met for deep space missions. The Brookhaven results are from a simulation, we need space based studies to get better data so that we can design and develop appropriate countermeasures to protect the space travelers. Manned spacecraft have been equipped with excellent radiation detection devices, unfortunately that is not enough. We not only need to know the types and quantities of radiation but we need to know what effects those radiations have on biologically systems especially humans. What we need is a good model system to analyze the biological effects of radiation in space and the effects of micro-gravity. For this we are using a species of tiny nematodes called *Caenorhabditis elegans* which has proven to be an excellent model system for numerous studies of human biological systems and human diseases.

C. Elegans – A Model System

The nematode worm *C. elegans* has sufficient similarities to us to make it an excellent model system for human biology. The nematode is the simplest multi-cellular organism with a completely known genomic DNA sequence. It has over 20,000 genes which is the same number, within a couple thousand, as humans and approximately 60% of *C. elegans*' genes have human homologues. Not only does *C. elegans* have a similar number of genes to us, but many genes are effectively the same doing equivalent functions in the nematode and in humans including similar DNA repair systems. *C. elegans* also has many similar tissue types to humans these include a neuromuscular system, gut, reproductive system and integument. These similarities allow *C. elegans* to be a valid model system for use in predicting possible biological damage that can occur in humans in the space environment.

C. elegans has many more advantages which make it an excellent model system for studying the effects of micro-gravity and radiation in space. One is that *C. elegans* is robust and can survive harsh conditions. Under conditions of starvation and overcrowding it goes into a dispersive life stage called dauer in which it can survive for several months without food. When conditions improve worms exit dauer and resume their normal life cycle. In addition, live worms were recovered from the disaster of the Columbia in which they survived an impact 2,295 times Earth's gravity [3]. Another advantage is its small size - an adult *C. elegans* is about 1mm long – about as long as a gram of table salt is wide. Large numbers of *C. elegans* can be grown in small spaces therefore experiments can be very compact. This is important because it is expensive to launch mass into orbit and spacecraft, even the International Space Station (ISS), are space limited. BioServe Space Technologies has developed a small portable automated growth chamber that can be used to maintain and feed the worms thus minimizing the time and effort of direct involvement of an astronaut in worm experiments. In addition, *C. elegans* has a generation time of just a few days, it has a two-three week lifespan in which a single hermaphrodite produces about 300 offspring, it has a limited number of cells and is easy to maintain. Worms can also be frozen alive in liquid nitrogen which allows for long-term strain storage and strain analysis at opportune times. These advantages allow experimental procedures to be short, flexible, statistically significant, and cost effective. An exceptional characteristic of this organism is that any perturbation in its development is easy to note and characterize. The reason for this is that wild-type adult hermaphrodites contain a small number of cells (exactly 959 somatic cells) with the position of every cell constant and known [4; 5], and the organism is transparent making it relatively straightforward to track cells and follow cell lineages.

C. elegans is being studied in numerous laboratories around the world and there are many sophisticated techniques in use to analyze the worms. These include sophisticated genetic tools such as the eT1-system [6; 7; 8] that has been to space several times. There is also a complete genome cDNA based microarray technology [9] which can also utilize the newly developed Nimblegen DNA-array chip technology. One of the chip designs includes bits of DNA from *C. elegans*' 22,000 genes, so by comparing the space radiation exposed worms to those chips deletion and duplication mutations (the most common types induced by radiation) can be identified rapidly [10; 11]. There is also an available library of promoter::GFP (green fluorescent protein) fusions for over 2,000 *C. elegans* genes [12; 13]. The promoter is the region of DNA that controls expression of its gene while GFP is a protein that fluoresces green when exposed to blue light (figure 1a). So an activated promoter results in glowing green worm cells. Which cells fluoresce depends on the innate expression pattern of the particular promoter. Some genes are turned on by DNA damage such as caused by radiation. In future experiments, which include fluorescent detection systems, these promoter::GFP fusions could prove useful for analyzing biological damage in space without returning the experiments to Earth. Another new technique, which uses massively parallel sequencing instruments, allows rapid and inexpensive large scale DNA sequencing. *C. elegans* was resequenced using a Solexa (Illumigen) Sequence Analyzer [14]. This technique could be used to give a quantitative look at the spectrum of mutations that the experimental worms have picked up in space versus a ground control (pers comm. Steve Jones) fairly inexpensively.

C. elegans has been to space several times, including flights in 1993 and 1996 which carried experiments for Greg Nelson of the Jet Propulsion Laboratory (JPL); the devastating crash in Texas in 2003 of Columbia, which carried experiments for Nate Szewczyk of the University of Pittsburgh and Catharine Conley of NASA's Ames laboratory; the 2004 Delta mission, a four country collaboration (Canada, France, Japan and USA), which carried the First International C. elegans (ICE-FIRST) experiment; and a six-month mission which flew to the ISS on STS-116 in December 2006. The two JPL experiments, ICE-FIRST, and the six-month mission made use of the mutagen testing *C. elegans* eT1-system developed in David Baillie's laboratory at Simon Fraser University [6; 7; 8]. The eT1-system, allows the capture of mutations in two large regions that together cover over one sixth of the entire *C. elegans* genome. *eT1* is a reciprocal translocation between the right half of Chromosome III and the left half of Chromosome V. There is a complete suppression of recombination in the translocated regions thus any mutations that occur in those regions will not cross away [15]. Once mutations have been 'captured' in space and are back on the ground they can be analyzed.

Changes in Gene Expression

During the ICE-FIRST experiment, gene expression changes were assayed using a near full genome cDNA based microarray developed at Stanford [9]. Three populations of mixed developmental stage worms were assessed and compared to three sister populations on Earth that were subject to similar culturing conditions on a twelve hour delay [16; 17]. The gene expression changes confirmed pre-flight data demonstrating animals grown in the spaceflight hardware were stressed. Increased expression of hypoxia and stress induced genes were noted in two of the three samples cultured in the flight hardware on the ground. These gene expression changes were not noted in the third sample, presumably due to this sample having increased access to oxygen as a result of being immediately next to the oxygen exchanging membrane of the flight hardware. These results reinforce the idea that growth of animals in spaceflight hardware (see figure 1a and 1b) is significantly different than growth of animals cultured in a laboratory on Earth and is a potentially confounding variable when assessing the biological effects of spaceflight. When the authors subtracted the genes showing changes on Earth from the set showing changes in response to spaceflight alone, they found reproducibly increased expression of only nineteen genes and reproducibly decreased expression of only ten genes. The genes with increased expression include eight known metabolic genes (encoding a P450, a delta 6 fatty acid desaturase, an omega 3 fatty acid desaturase, an Acyl-CoA synthetase, two Acyl-CoA oxidases, and two UDP-glucuronosyltransferases), four genes encoding secreted proteins (*phat-3*, C49G7.3, F15A4.6, and C36C5.12), three stress response genes (*hsp-70*, *hsp-16.49*, *hsp-16.41*), two uncharacterized genes (C52D10.1 and R06F6.7), one apparent transcription factor (F19B2.5), and one organic toxin induced gene (F17H10.1). These gene changes suggest the animals were experiencing molecular stress and altering lipid metabolism in response to spaceflight. Both sets of gene changes could be the result of altered Insulin signalling as the identified genes are largely regulated by Insulin signalling. Remarkably, the set of genes identified as down regulated in response to spaceflight is also enriched for Insulin regulated genes. These genes are mostly uncharacterized; ZK6.7 is predicted to encode a lipase, and decreased expression of *dct-16* is linked to decreased tumour formation [18]. If Insulin signalling is indeed altered, this may have implications for the damaging effects of radiation as decreased Insulin signalling appears to result in cells having increased sensitivity to the negative effects of radiation [19].

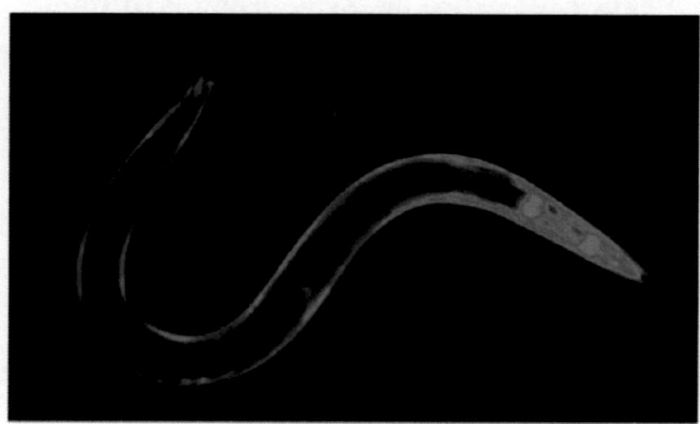

Figure 1a. Example of GFP expression in C.elegans. The promoter of the gene BO511.10 was PCR stitched to a GFP cassette so that tissues in which the gene expresses also express GFP. Photo by Allan Mah.

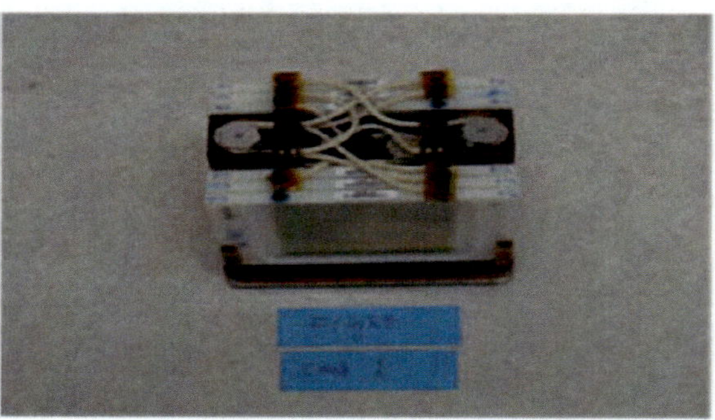

Figure 1B. C.elegans space habitat (Chab1) containing six opticalls. C.eleganse T1 lived in here for several months aboard the 1SS during the six month mission. Photo courtesy of Louis Stodieck (BioServe Space Technologies).

Figure 1C. The C.elegans habitats used on the space experiments, which flew to the ISS on the STS-116 for the six month mission, and their corresponding ground controls. The opticals can maintain thousands of worms. Photo courtesy of Louis Stodieck (BioServe Space Technologies).

The approach, to analysis of gene expression changes in response to spaceflight, taken by the authors focused upon genes displaying reproducible changes across multiple replicates and multiple populations. As a result the identified genes show reproducible changes in a fourth population of animals [20] when assayed using a different microarray platform, Affymetrix [21]. However, it is clear that the stringent analysis applied to gene expression analysis lead the authors not to report, as significant, a number of other genes that did show significant responses to spaceflight.

Changes in Gene Expression in Both Worms and Flies

On the Soyuz taxi flight before ICE-FIRST, a full genome microarray analysis was conducted on the fruit fly *Drosophila melanogaster* as the GENE experiment of the Spanish Cervantes mission. Accordingly, a comparison of gene changes between the two species was undertaken [22]. The most immediate similarities noted were that both worms and flies experienced altered expression of genes as a result of the abnormal culturing hardware required for spaceflight and altered expression of metabolic genes in response to spaceflight. Intriguingly, there were very few specific genes that showed identical changes in both species. There were six genes that showed decreased expression in both species (*C. elegans* names will be used here): T04F8.2 is uncharacterized and appears to be a membrane protein; *spon-1* is a fat-spondin like protein involved in cell adhesion; C01B7.4 is a membrane-associated guanylate kinase; F35H12.4 is a phosphatidylinositol 4-kinase; *nac-1* is a sodium coupled citrate transporter; *ldh-1* is a lactate dehydrogenase.

The authors concluded that they could not find any evidence for master genes regulating the response to spaceflight. While this is true, the results would appear to reinforce past suggestions that altered metabolism, as the result of altered fluid dynamics, underlie the response to spaceflight across species [23; 24]. Perhaps there are no master genes that regulate the response to spaceflight, but rather individual specific adaptations that must take place to allow metabolism to function properly during spaceflight.

Changes in Muscle Gene Expression

Aside from radiation exposure, muscle and skeletal atrophy is the most significant bio-medical challenge of spaceflight [23-25]. Previously, it had been shown that gene transcription is decreased in muscle cells cultured in space [26]. Thus, a decision was made to examine *C. elegans* for alterations of muscle gene expression. This effort was enabled by the fact that *C. elegans* is a major model for systems biology [27]. As a result, there have been a number of attempts to uncover tissue specific or developmental stage enriched or specific genes, these efforts included identifying muscle genes [15, 16]. Following spaceflight, significant changes in approximately 100 muscle genes were found and many, if not all, corresponded to changes in expression of gene protein products [17; 20]. Notably, a post-flight movement defect and decreased expression of the muscle specific transcription factor MyoD and three pharyngeal muscle specific transcription factors were noted. These results suggest that some factor associated with spaceflight results in decreased expression of the transcription factors that control the expression of muscle genes and consequently, the decreased expression of muscle genes. These changes may result in increased difficulty in healing muscle damaged during spaceflight as it was also shown that these changes exacerbate muscular defects in mutationally damaged muscle [30].

Intriguingly, many of the gene changes noted in space flown worm muscle have previously been observed in space flown rodent and human muscle and in various human muscle diseases [31]. In the case of spaceflight it is tempting to suggest that altered insulin signalling underlies the changes in muscle as untreated diabetes leads to muscle wasting and altered insulin signalling appears to occur in both spaceflown worms and human beings [17; 32].

Apoptosis

One of the consequences of radiation exposure is damage to DNA which can be repaired or else result in cell death [19]. Animals have evolved a process of controlled cell death, possibly because uncontrolled cell death releases the contents of dying cells with toxic consequences for surrounding cells. This process was first discovered in *C. elegans*. Because of the unique tools available it was decided to examine if this process was defective in response to spaceflight. These experiments were deemed important as if this process, apoptosis, is impaired in response to spaceflight then animals will lack a normal adaptive process by which to cope with radiation induced cellular damage. Cell death as a result of normal developmental programming, for example cells expected to die in order for proper development such as death of the interdigital webbing cells allowing for proper finger and toe development, occurred normally. Cell death as a result of improper development, for example cells dying as a result of various defects in DNA replication, also occurred normally. Thus, no changes were noted in the apoptotic processes. Furthermore, no changes in the expression of genes associated with these cell death processes were detected [17; 33]. Together these results demonstrate that animals are indeed able to remove cells damaged by radiation during spaceflight in a controlled manner that prevents further toxic effects. However, it remains unclear if the limits of this system to clear radiation damaged cells are the same or different during spaceflight.

SPACE-BASED *C. ELEGANS* MUTAGENESIS TESTING SYSTEMS

Astronauts will be irradiated by large doses of sizeable ranges of two main classes of space radiation (galactic cosmic radiation and solar particles including those from solar storms) on voyages to the Moon and Mars and we do not know, in detail, the adverse biological effects of the irradiation. A report [1], commissioned by NASA found that the:

> "lack of knowledge about the biological effects of and responses to space radiation is the single most important factor limiting prediction of radiation risk associated with human space exploration".

The biological effects, of long-term exposure to the various classes and doses of radiation in space, which include health risks and the effects on the astronauts' abilities to perform their operational tasks, are major concerns of extended deep space travel. In addition, in times of solar radiation storms, astronauts may be confined to highly shielded areas and thus be unable to do planned tasks. We must develop a much deeper understanding of the biological effects of the totality of radiation astronauts will encounter including the secondary radiation produced when the primary radiation interacts with material near the astronauts such as the hulls of spacecraft or the ground and material of their outposts on the Moon and Mars. This understanding is needed so that we can develop effective countermeasures to protect astronauts and to know when they must find shelter and when it is safe for them to leave those highly shielded areas.

To develop the necessary understanding we must use good model systems in space-based analyses because we cannot replicate the large range and types of radiation on Earth that the astronauts will encounter on their voyages. The microscopic nematode *C. elegans* is an excellent model organism for biological experiments due to its basic similarities to humans, including a comparable number of genes to (with approximately 60% orthologous) and conserved DNA repair systems which recognize and repair radiation induced DNA damage. In addition, *C. elegans*, which normally feeds on bacteria, can live on an anexic media

(CeMM) [34] so that bacteria need not be introduced to space environments when sending *C. elegans* experiments to Earth orbit or beyond.

C. elegans based experiments have been to space on several missions ranging from a few days to a multi-generation trip spanning several months. Johnson and Nelson [35] first proposed using *C. elegans* as a model system for space biology studies. Since then *C. elegans* has flown on several missions to low Earth orbit (LEO) and has proven to be an excellent model system for studying biological effects of travel in space. The eT1-system captures acquired mutations over one-sixth of the *C. elegans* genome (the eT1-balanced regions of Chromosomes III and V) and has proven to be an excellent mutagen testing system [6; 12; 36; 37]. It has been developed into an accumulating biological dosimeter for assaying genetic damage in space [38; 39].

In the 1990s the eT1-system; *unc-22* (a highly mutable gene with approximately 20 Kbp of coding sequence that, when mutated, gives raise to dominate "twitcher" alleles [40]); and *fem-3* dauers were used to do short term (few days) investigations of space radiation induced mutations in *C. elegans* in low Earth orbit [41; 42; 43]. From the dauer experiments, on Space Shuttle flight STS-76, the researchers found a relatively low 3.3-fold increased mutation rate in the spaceflight samples than the ground control spontaneous rate. The recovered mutations were similar (generally homozygous inviable mutations suggesting large deficiencies) to those induced by accelerated iron ions at the Brookhaven National Laboratory, and they concluded that the increased mutation rate was likely the direct effect of high energy space radiation and not microgravity. In 2004 the collaborative ICE-First mission lasted a little longer - 11 days in space. This short experiment also yielded few mutations, not much above the number of spontaneous mutations we would expect on Earth. These results are probably due to do fact that these missions were in low Earth orbit, which is protected by the Earth's magnetosphere. In the ICE-First mission, Zhao et *al.* [39] used four *C. elegans* mutagen testing systems to analyze the effects of radiation: 1) poly-G/poly-C tract modifications; 2) assaying for dominant *unc-22* mutations; 3) identifying alterations in telomere length; and 4) capturing and analyzing mutations using the eT1-system. In all four tests there were little or no significant differences between the ground control and the low Earth orbit results. For the poly-G/poly-C tract analysis no deletions were detected in either the 37 G-tracts analyzed from spaceflight worms or the 24 G-tracts sites sampled in the control worms. In the *unc-22* analysis progeny from 1000 worms ($3X10^5$ F_1s) from the spaceflight and 500 ($1.5X10^5$ F_1s) from the ground control were scored for dominate twitchers. No twitchers were found. These results suggest that much larger sample sizes are required to distinguish different mutation rates for this single gene and for poly-G/poly-C tract analysis. The results from identifying alterations in telomere length suggest that telomeres are elongated during the spaceflight. While the eT1-system showed no statistically significant differences among the mutation rates. The eT1-system has proved to be well suited to be a biological accumulating dosimeter for capturing and analyzing space radiation induced essential mutations.

C. elegans worms flew to space on STS-116 on December 9, 2006 and returned to Earth, from the ISS, June 22, 2007 on STS-117 for a total time of 195 day. This multi-generational flight was the longest, by far, utilizing *C. elegans*. Experiments included both wild type (CC1) worms and the eT1-system. This was a collaborative effort among the Malaysian Space Agency, Nathaniel Szewczyk (University of Pittsburgh), Louis Stodieck (BioServe Space Technologies), and David Baillie's laboratory (Simon Fraser University). The Malaysian Space Agency objective was to look at mutations and also gene expression profiling using a microarray approach (Affymetrix), BioServe developed the Commercial Generic Bioprocessing Apparatus (CGBA), a temperature controlled, suitcase-sized, automated culturing system that was used to maintain the *C. elegans* experiments. BioServe also downlinked video, still images and data from the experiments. In addition, Louis Stodieck of Bioserve and Nate Szewczyk coordinated, with Orions Quest, a K-12 educational project involving an estimated 1,000 students in schools in Canada, Malaysia and the United States. They utilized video and still images of *C. elegans*, especially *eT1* because it has three different phenotypes whereas wild type has only one phenotype, to monitor population dynamics, morphology, and movement. The conclusion is that there are no major gravity-dependent processes associated with spaceflight that preclude essentially normal *C. elegans* growth and development for at least ten generations.

The multi-generational flight was planned for three months not six months and so the food supply was exhausted long before the worms returned to Earth. Fortunately, *C. elegans* is robust and newly hatched eggs can enter dauer stage in situations of starvation and overcrowding. Even so most of the approximately 20,000 - 30,000 returned *eT1* worms proved to be sterile. Still, 948 were fertile showing that *C. elegans* is an excellent testing system in the harsh environment in space. Three of the 948 carried lethal eT1-balanced mutations (unpublished results) and these are being analyzed for their class, location and size. So far, one is a 40,000 base pair duplication of a region on Chromosome III that inserted into an essential gene on Chromosome V. One is on an eT1-translocated chromosome while the other is in the eT1 balanced region of either Chromosome III or Chromosome V. A ground control is being done, in the Baillie laboratory, to establish and spontaneous mutation rate. So far two mutants have been isolated from 7,300 screened worms. The three mutants isolated from 948 space flight worms gives a higher rate than the ground control but not excessively higher. We postulate that the mutation rate would be dramatically higher outside the protective envelop of the Earth's magnetosphere.

Array Comparative Genomic Hybridization (ACGH) as an Aid to High Resolution Genomic Analysis of Mutated Genomes

High Resolution Analysis – Genomic Analysis of Space Worms

The development of systems to capture genetic damage in *C. elegans* (such as the *eT1* system described above) is an excellent strategy to study damage acquired by animals exposed to a variety of mutagenic sources [44; 36; 45]. Though these systems are reliable methods for detecting the relative amounts of damage between test and reference samples they are limited in that they require that genetic damage occur in genes which give an easily detectable phenotype. Consequently, in-depth analysis of the type and range of damage that has occurred at the molecular level requires a significant input of time and resources. Initially, analysis of captured mutations relied on indirect genetic methods, such as 2 or 3-factor mapping and deficiency mapping to locate damage to a sub-chromosomal region of the genome, followed by the use of other techniques such as cosmid and fosmid transgenic rescue to locate a defined region which can then be sequenced cost effectively. Not only are these methods time-consuming, they also rely upon genetic damage occurring within genes whose disruption results in detectable phenotypes and therefore are not suitable for estimating the true amount and variety of genetic damage occurring throughout the captured region of the genome.

Recent years have seen a significant advance in molecular techniques suitable for the identification and evaluation of small differences at the level of DNA. These high-resolution technologies, such as DNA micro-array [46] and next-generation sequencing methodologies [47] represent an alternative to traditional mutation identification strategies. Although the reduced cost of sequencing and the development of alternative sequencing methodologies have brought the age of genome re-sequencing to the doorstep of geneticists and molecular biologists this type of analysis is not yet practical for use in the analysis of large numbers of test samples in an individual laboratory setting. The use of array-based nucleotide hybridisation platforms therefore, represents the most promising current technology for isolating genomic changes between two closely related genomes to a high resolution in an unbiased way.

Techniques using comparative analysis of closely related genomic samples have been well documented for their use for characterizing copy number variations (CNVs). Such applications include: investigation of genomic instability in cancers [48]; study of genomic evolution [49]; and genotyping certain genetic disorders [50]. In these comparative genomic hybridization (CGH) analyzes the DNA of test and reference genomes are directly compared through comparison of the ratio of fluorescence intensities from samples differentially labeled with fluorescent probes competitively binding a target substrate (figure 2). Initially, this substrate took the form of metaphase chromosomes, allowing a whole genome to be screened in a single experiment, with the caveat that relatively low resolution in the detection of altered copy number in the kilobase range was attainable [51]. With advances in microarray technologies array-CGH (aCGH) experiments have been developed which perform at higher resolution in a digital manner. Each spot on the array corresponding to a discrete genetic segment, at one time provided by the use of BACs, cosmids and fosmids and PCR products. The introduction of technologies to allow high-density microarray development has dramatically increased the content possible. These developments have led to BAC and cosmid or PCR based arrays being superseded by DNA oligonucleotide probes also known as DNA oligoarrays [46]. DNA oligoarrays offer the advantage of having discrete custom synthesized probes spaced across the target genome or genomic region to a very high resolution (figure 2). Short oligonucleotide probes, can be either synthesized and arrayed onto the slide by ink jet disposition [52] or synthesized on the array in situ by the use of photolithography [53]. In situ microarray synthesis offers several advantages over spotted oligoarrays as the array design can be customized, tailoring the probe density and coverage, allowing both specific experimental questions to be addressed and efficient use of array capacity.

With the use of high density arrays, probe resolution at the individual nucleotide is possible, though this limits the genomic region which can be covered on an individual array, requiring the use of multiple arrays to achieve complete coverage of larger genomes [54]. With the current technology, a standard oligoarray can contain up to 10^4 probes [55], though arrays which can contain 10^5 or 10^6 positions are currently in development [11]. Current technology therefore allows for the identification of all major types of genetic damage including; deletions, duplications, rearrangements and single nucleotide alterations. The recent development of a *C. elegans* specific aCGH platform for identification of novel single gene deletions has reliably identified deletion ranging in size from several million basepairs (bp) to as small as 200 bp [11, 10]. Given the success of this approach in *C. elegans* aCGH technology lends itself to the high-resolution analysis of genomes from animals returning from space experimentation. We have explored this possibility using two approaches.

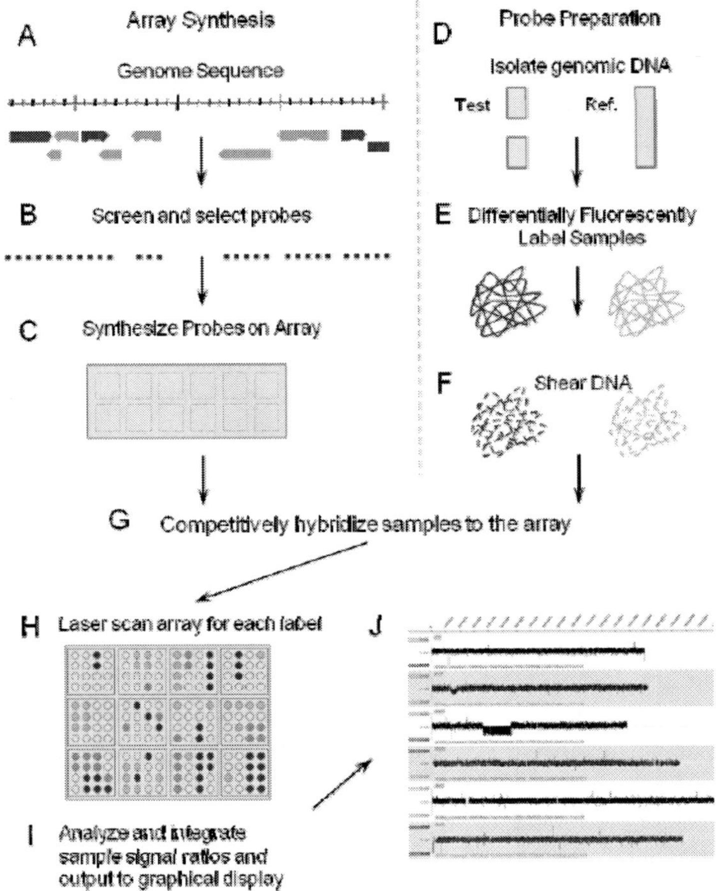

Figure 2. Overview of aCGH analysis. A-C, Oligo-array synthesis. A. Target genome sequence (A) is used to select suitable oligo sequences (B), which are synthesised on an array (C). D-F, Probe preparation. Genomic DNA for test and reference samples (D) is isolated and differentially labelled with fluorescent dyes (E). DNA is sheared to facilitate hybridisation to the array (F). Samples are hybridised to the array together (G) and the array processed and scanned (H). Signal output is processed to create a ratio of fluorescence, which is proportional to the ratio of reference to test samples (I). This data visualized using graphical user interface applications (J).

Confirmation of Characterised Mutations

Even though the *eT1*-system showed no statistically significant differences among the mutation rates of ground control and test populations in the ICE-First mission two mutational events predicted to be multiple gene deletions of Chromosome V were identified: *hDf36* and *hDf37* [39]. As proof of principal to ascertain if aCGH could be used to rapidly and efficiently characterise these predicted deletions we employed a whole genome aCGH array which specifically targets exons to high resolution to identify all genes deleted and to position deletion breakpoints with a high degree of accuracy. This analysis revealed that *hDf36* is an end-terminal deletion of between 184 and 194 kbp while *hDf37* is an internal deletion of approximately 150 kbp (unpublished results) (figure 3). Provisional sequence analysis at the known breakpoints of *hDf37* reveal the presence of a transposon at the right end of the deleted region indicating that in this case the deletion is likely to be due to transposon activity and therefore not due to damage from high energy irradiation. The result of this analysis highlights the benefit of using aCGH to characterise large genetic disruptions captured in space flight.

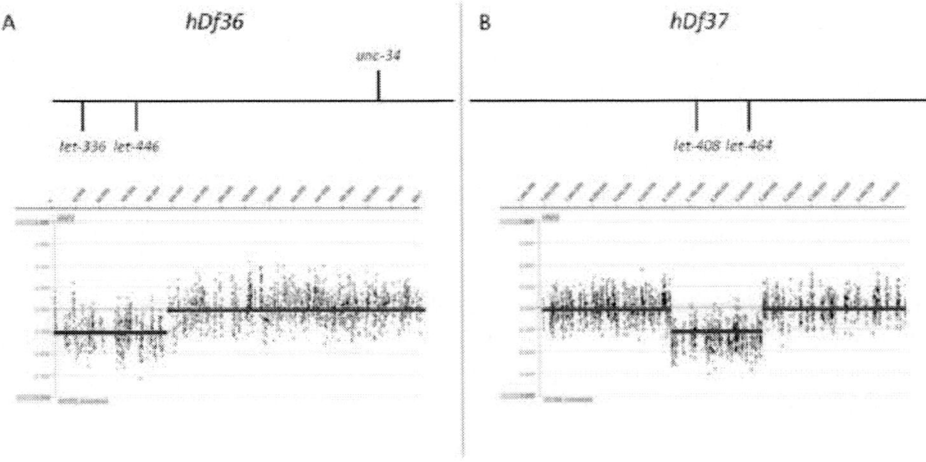

Figure 3. aCGH analysis of multiple gene deletions. A. *hDf36*, an end-terminal deletion of 184-194 kbp on Chromosome V. B. *hDf37*, an internal deletion of 150 kbp on Chromosome V. A schematic of the Genetic mapping data for each multiple gene deletion is shown above the relevant array data.

Non-Biased Analysis of Genomic Regions

A further advantage of high resolution aCGH analysis is to identify genetic changes not associated with a visible phenotype. Identification of damage in both coding and intergenic regions gives a true indication of the amount of damage occurring within a defined region of the chromosome. To this end we developed a high resolution tiling array covering the captured region of *eT1*. By limiting the probe range to the region balanced by *eT1* we have both eliminated analysis of DNA not 'captured' in the experiment and increased the probe density attainable with this array. The overlapping probes give a theoretical resolution of a few base pairs allowing the possible detection of very small genetic damage. We have tested a captured mutant strain identified from the most recent experiment performed aboard the ISS. In this case a small duplication of almost 47kb associated with the *eT1* balanced region was detected. No other significantly detectable damage in this experiment suggests that in this case at least, the variation was the only one of this type present in this region of the genome.

The Future of aCGH in Space Analysis

So far our studies have been limited to detection of CNVs, the next obvious step for this analysis is to be able to directly assess the level of damage reliably at the level if individual nucleotides. Several strategies to identify single nucleotide changes using aCGH have been successfully demonstrated. Both take advantage of the detrimental change in probe hybridization efficiency in the presence of a polymorphism. The most comprehensive method is to generate probes to represent all possible single base-pair mismatches at each position of the region being tested, though the high number of probes required for this type of analysis significantly reduces the region of the genome which can be tested on a single array. This approach has been successfully applied to detect heterozygous variations in BRCA1 [56] and for tracking the evolution of the relatively small SARS virus genome [57]. A further advantage of this approach is that at this level of resolution, sequence variation can be inferred directly from the hybridization data. Additionally, positional information with regard to genomic rearrangements can also be elucidated enabling in essence, all major types of genetic damage to be rapidly identified. The second approach to identify single nucleotide variation relies on generating probes at a high enough density that every nucleotide position is represented multiple times by the overlap of multiple probes. This approach has been successful used to investigate larger genomes including *A. thaliana* [58] and mosquitoes [59]. The application of this approach to the *C. elegans* genome is however currently only reliable for detecting around 60% of polymorphisms tested [60]. While this approach does not directly give information on the genetic change at the nucleotide level it does isolate the change to a small enough region to be able to systematically identify the change using standard sequencing analysis.

Conclusion

The Committee on the Evaluation of Radiation Shielding for Space Exploration report to NASA [1] states that the lack of knowledge of the biological effects of space radiation is the most important factor limiting prediction of radiation risks for human for extended space exploration. Based on our current knowledge and technology, a NASA based Mars mission should not be made because the astronauts' radiation exposure would exceed NASA's permissible exposure levels, indeed this would also limit long-term Moon activity. Fortunately, astonishing progress is being made in the biological sciences which could lead to the development of the necessary countermeasures for long-duration manned space-flights. In addition, the Chinese are planning to create a space weather forecast system with the Kuafu mission satellites placed at the L1 Lagrangian Point which will allow reliable predictions of solar storm activity thus allowing for effective use of the countermeasures. These could take the forms of more effective shielding and radiation damage-limiting medical procedures, both of which require much more complete understanding of the biological mechanisms by which radiation and micro-gravity affects human gene expression, health, and physical and mental performance than we currently possess. In addition to the risks from space radiation, micro-gravity has adverse biological effects on humans which included bone density loss, decreased muscle strength and endurance, postural instability, and reductions in aerobic capacity. We must increase our knowledge of the biological effects of space environments in order to preserve the health of space travelers.

The model organism, *C. elegans* is being used to help develop the necessary understanding of the biological effects of space radiation and micro-gravity. One type of analysis was to look at altered gene expression. Comparisons of *C. elegans* gene expression with that of the fruit fly *Drosophila melanogaster* showed no evidence for master genes regulating the organisms' response to spaceflight. Comparing space grown and Earth grown *C. elegans* gene expression shows reproducibly increased expression of nineteen genes and reproducibly decreased expression of ten genes. The altered gene response is likely specific adaptations to allow metabolism to function properly during spaceflight. Interestingly, both sets of gene changes could be the result of altered Insulin signalling as the identified genes are largely regulated by Insulin signalling. In addition, there was decreased expression of the muscle specific transcription factor MyoD and three pharyngeal muscle specific transcription factors suggesting decreased expression of muscle genes. This may result in increased difficulty in healing muscle damaged during spaceflight. The six month flight showed that there are no major gravity-dependent processes associated with spaceflight that preclude essentially normal *C. elegans* growth and development for at least ten generations.

C. elegans has only flown to LEO which is protected by the Earth's magnetosphere. Both the short flights and the multi-generational flight radiation studies showed relatively small increased mutation rates compared to ground controls although the samples sizes were small. The *C. elegans* eT1-system has proven robust and an excellent biological dosimeter for these types of studies and could be utilized to increase the samples sizes. State-of-the-art NimbleGen micro-array chips were used to analyze lethal mutations in worms returned from the multi-generation flight. These analyzes proved very effective for rapid and accurate identification of gene expression pattern changes and for determining the molecular basis of captured deficiency and duplication mutations. A future extension would be to use a massively parallel sequencing instrument, such as an Illumina Sequence Analyzer to do whole genome resequencing of space derived genomes. This would allow for a comprehensive look at the spectrum of changes following exposure to space radiation. It would also allow a comparison of the eT1 balanced region, in which all mutations are captured, with the reminder of the genome where mutations can be "lost" through segregation. *C. elegans* could also be used for mutational studies on deep space missions. For example, an in-flight detectable reporting system could be developed (such as a fluorescent coupled to a DNA damage repair system) that does not require the worms to be returned to Earth for analysis.

REFERENCES

[1] Committee on the Evaluation of Radiation Shielding for Space Exploration Aeronautics and Space Engineering Board. Managing Space Radiation Risk in the New Era of Space Exploration. *Division on Engineering and Physical Sciences*. The National Academies Press 2008 in preparation.

[2] Mole, RH; 1984 The LD50 for uniform low LET irradiation of man. *The British Journal of Radiology*. 57, Issue 677: 355-369.

[3] Szewczyk, NJ; Mancinelli, RL; McLamb, W; Reed D; Blumberg, BS; Conley CA. Caenorhabditis elegans survives atmospheric breakup of STS-107, space shuttle Columbia. Astrobiology. 2005 Dec;5(6):690-705

[4] Sulston, JE; Horvitz HR. Post-embryonic cell lineages of the nematode, Caenorhabditis elegans. *Dev. Biol.* 1977 56(1):110-56

[5] Sulston, JE; Schierenberg, E; White, JG; Thomson, JN. The embryonic cell lineage of the nematode Caenorhabditis elegans. *Dev. Biol.* 1983 100(1):64-119

[6] Rosenbluth, RE; Cuddeford, C; Baillie, DL. Mutagenesis in Caenorhabditis elegans: I. A Rapid eukaryotic mutagen test system using the reciprocal translocation, eT1(III;V). *Mutation Research.* 1983 110 39-48.

[7] Rosenbluth, RE; Rogalski, TM; Johnsen, RC; Addison, LM; Baillie, DL. Genomic organization in Caenorhabditis elegans: deficiency mapping on linkage group V (left). *Genetical Research.* 1988 52: 105-118.

[8] Johnsen, RC; Baillie, DL. Genetic analysis of a major segment [LGV(left)] of the genome of Caenorhabditis elegans. *Genetics.* 1991 129 735-752.

[9] Kim, S.K., et al., A gene expression map for Caenorhabditis elegans. *Science.* 2001. 293(5537): p. 2087-92.

[10] Jones, MR; Maydan, JS; Flibotte, S; Moerman, DG; Baillie DL. OligonucleotideArray Comparative Genomic Hybridization (oaCGH) based characterization of genetic deficiencies as an aid to gene mapping in Caenorhabditis elegans. *BMC Genomics.* 2007 8: 402.

[11] Maydan, JS; Flibotte, S; Edgley, ML; Lau, J; Selzer, RR; Richmond, TA; Pofahl, NJ; Thomas, JH; Moerman, DG. Efficient high-resolution deletion discovery in Caenorhabditis elegans by array comparative genomic hybridization. *Genome Res.* 2007 17: 337-347

[12] McKay, SJ; Johnsen, R; Khattra, J; Asano, J; Baillie, DL; Chan, S; Dube, N; Fang, L; Goszczynsk,i B; Ha, E; Halfnight, E; Hollebakken, R; Huang, P; Hung, K; Jensen, V; Jones, SJ; Kai, H; Li, D; Mah, A; Marra, M; McGhee, J; Newbury, R; Pouzyrev, A; Riddle, DL; Sonnhammer, E; Tian, H; Tu, D; Tyson, JR; Vatcher, G; Warner, A; Wong, K; Zhao, Z; Moerman DG. Gene expression profiling of cells, tissues, and developmental stages of the nematode C. elegans. *Cold Spring Harb. Symp. Quant. Biol.* 2003;68:159-69. Review.

[13] Hunt-Newbury, R; Viveiros, R; Johnsen, R; Mah, A; Anastas, D; Fang, L; Halfnight, E; Lee, D; Lin, J; Lorch, A; McKay, S; Okada, HM; Pan, J; Schulz, AK; Tu, D; Wong, K; Zhao, Z; Alexeyenko, A; Burglin, T; Sonnhammer, E; Schnabel, R; Jones, SJ; Marra, MA; Baillie, DL; Moerman, DG. High-throughput in vivo analysis of gene expression in Caenorhabditis elegans. *PLoS Biol.* 2007 Sep;5(9):e237.

[14] Hillier, LW; Marth, GT; Quinlan, AR; Dooling, D; Fewell, G; Barnett, D; Fox, P; Glasscock, JI; Hickenbotham, M; Huang, W; Magrini, VJ; Richt, RJ; Sander, SN; Stewart, DA; Stromberg, M; Tsung, EF; Wylie, T; Schedl, T; Wilson, RK; Mardis ER. Whole-genome sequencing and variant discovery in C. elegans. *Nature Methods.* 2008 5: 183-188.

[15] Rosenbluth, RE; and Baillie, DL. The genetic analysis of a reciprocal translocation, eT1(III;V), in Caenorhabditis elegans. *Genetics.* 1981 99 415-428.

[16] Szewczyk, N.J., et al., Description of International Ceonorhabditis elegans Experiment first flight (ICE-FIRST). *Adv. Space Res.* 2008 In Press.

[17] Selch, F., et al., Genomic response of the nematode Caenorhabditis elegans to spaceflight. *Adv. Space Res.* 2008 41(5): p. 807-815.

[18] Pinkston-Gosse, J. and C. Kenyon, DAF-16/FOXO targets genes that regulate tumor growth in Caenorhabditis elegans. *Nat. Genet.* 2007 39(11): p. 1403-9.

[19] Elshaikh, M., et al., Advances in radiation oncology. *Annu Rev. Med.* 2006 57: p. 19-31.

[20] Higashibata, A., et al., Decreased expression of myogenic transcription factors and myosin heavy chains in Caenorhabditis elegans muscles developed during spaceflight. *J. Exp. Biol.* 2006 209(Pt 16): p. 3209-18.

[21] Hill, A.A., et al., Genomic analysis of gene expression in C. elegans. *Science.* 2000 290(5492): p. 809-12.

[22] Leandro, L.J., et al., Comparative analysis of Drosophila melanogaster and Caenorhabditis elegans gene expression experiments in the European Soyuz flights to the International Space Station. *Adv. Space Res.* 2007. 40(4): p. 506-512.

[23] Oser, H. and B. Battrick, eds. Life Sciences Research in Space. Vol. ESA-SP-1105. 1989, European Space Agency Publications Division: Paris. 135.

[24] Nicogossian, A.E., C.L. Huntoon, and S.L. Pool, eds. Space Physiology and Medicine. 3rd ed. 1994, Lea and Fibiger: Philadelphia. 481

[25] Board, S.S., ed. A Strategy for Research in Space Biology and Medicine. 1998, National Academy Press: Washington, D.C. 296.

[26] Vandenburgh, H., et al., Space travel directly induces skeletal muscle atrophy. *FASEB J.* 1999. 13(9): p. 1031-8.

[27] Piano, F., et al., C. elegans network biology: a beginning. *WormBook.* 2006: p. 1-20.

[28] Roy, P.J., et al., Chromosomal clustering of muscle-expressed genes in Caenorhabditis elegans. *Nature.* 2002. 418(6901): p. 975-9.

[29] Fox, R.M., et al., The embryonic muscle transcriptome of Caenorhabditis elegans. *Genome Biol.* 2007. 8(9): p. R188.

[30] Adachi, R., et al., Spaceflight results in increase of thick filament but not thin filament proteins in the paramyosin mutant of Caenorhabditis elegans. *Adv. Space Res.* 2008. 41(5): p. 816-823.

[31] Szewczyk, N.J. and L.A. Jacobson, Signal-transduction networks and the regulation of muscle protein degradation. *Int. J. Biochem. Cell. Biol.* 2005. 37(10): p. 1997-2011.

[32] Tobin, B.W., P.N. Uchakin, and S.K. Leeper-Woodford, Insulin secretion and sensitivity in space flight: diabetogenic effects. *Nutrition.* 2002. 18(10): p. 842-8.

[33] Higashitani, A., et al., Checkpoint and physiological apoptosis in germ cells proceeds normally in spaceflown Caenorhabditis elegans. *Apoptosis.* 2005. 10(5): p. 949-54.

[34] Szewczyk, NJ; Kozak, E; Conley, CA. Chemically defined medium and Caenorhabditis elegans. *BMC Biotechnol.* 2003 Oct 27;3:19.

[35] Johnson, TE; Nelson, GA. Caenorhabditis elegans: a model system for space biology studies. *Exp. Gerontol.* 1991 26(2-3) 299-309.

[36] Rosenbluth, RE; Cuddeford, C; Baillie, DL. Mutagenesis in Caenorhabditis elegans. II. A spectrum of mutational events induced with 1500 r of gamma-radiation, *Genetics.* 109 1985 493-511.

[37] Nelson, GA; Schubert, WW; Marshall, TM; Benton ER; Benton EV. Radiation effects in Caenorhabditis elegans. Mutagenesis by high and low LET ionizing radiation. *Mutation Research.* 1989 212: 181-192.

[38] Zhao, Y; Johnsen, RC; Baillie, DL; Rose, AM. Worms in Space? A model biological dosimeter. *Gravitational and Space Biology Bulletin.* 2005 18(2):11-16.

[39] Zhao, Y., et al., A mutational analysis of Caenorhabditis elegans in space. *Mutation Research.* 2006 601(1-2): p. 19-29.

[40] Moerman, DG; Baillie, DL. Formaldehyde mutagenesis in Caenorhabditis elegans. *Mut. Res.* 1981 80: 273-279.

[41] Nelson, GA; Schubert, WW; Kazarians, GA; Richards, GF; Benton, EV; Benton ER; Henke, R. Radiation effects in nematodes: results from IML-1 experiments. *Adv. Space Res.* 1994a 14(10): 87-91.

[42] Nelson, GA; Schubert, WW; Kazarians, GA; Richards, GF; Benton, EV; Benton ER; Henke, R. Nematode radiobiology and development in space. Results from IML-1. *Proceedings of the Fifth European Symposium on Life Sciences Research in Space.* 1994b :187-191.

[43] Hartman, PS; Hlavacek, A; Wilde, H; Lewicki, D; Schubert, W; Kern, RG; Kazarians, GA; Benton, E V; Benton, ER; Nelson, GA. A comparison of mutations induced by accelerated iron particles versus those induced by low earth orbit space radiation in the FEM-3 gene of Caenorhabditis elegans. *Mutat. Res.* 2001 474, 47-55.

[44] Rosenbluth RE, Baillie DL: The genetic analysis of a reciprocal translocation, eT1(III;V), in Caenorhabditis elegans. *Genetics.* 1981, 99:415-428

[45] Johnsen RC, Baillie DL: Formaldehyde mutagenesis of the eT1 balanced region in Caenorhabditis elegans: dose-response curve and the analysis of mutational events. *Mutat. Res.* 1988, 201:137-147.

[46] Gresham D, Dunham MJ, Botstein D: Comparing whole genomes using DNA microarrays. *Nat. Rev. Genet.* 2008, 9:291-302.

[47] Schuster SC: Next-generation sequencing transforms today's biology. *Nat. Methods.* 2008, 5:16-18.

[48] Kallioniemi A: CGH microarrays and cancer. *Curr. Opin. Biotechnol.* 2008, 19:36-40.

[49] Dumas L, Kim YH, Karimpour-Fard A, Cox M, Hopkins J, Pollack JR, Sikela JM: Gene copy number variation spanning 60 million years of human and primate evolution. *Genome Res.* 2007, 17:1266-1277.

[50] Shen Y, Irons M, Miller DT, Cheung SW, Lip V, Sheng X, Tomaszewicz K, Shao H, Fang H, Tang HS, Irons M, Walsh CA, Platt O, Gusella JF, Wu BL: Development of a focused oligonucleotide-array comparative genomic hybridization chip for clinical diagnosis of genomic imbalance. *Clin. Chem.* 2007, 53:2051-2059.

[51] Stankiewicz P, Beaudet AL: Use of array CGH in the evaluation of dysmorphology, malformations, developmental delay, and idiopathic mental retardation. *Curr. Opin. Genet. Dev.* 2007, 17:182-192.

[52] Hughes TR, Mao M, Jones AR, Burchard J, Marton MJ, Shannon KW, Lefkowitz SM, Ziman M, Schelter JM, Meyer MR, Kobayashi S, Davis C, Dai H, He YD, Stephaniants SB, Cavet G, Walker WL, West A, Coffey E, Shoemaker DD, Stoughton R, Blanchard AP, Friend SH, Linsley PS: Expression profiling using microarrays fabricated by an ink-jet oligonucleotide synthesizer. *Nat. Biotechnol.* 2001, 19:342-347.

[53] Pease AC, Solas D, Sullivan EJ, Cronin MT, Holmes CP, Fodor SP: Light-generated oligonucleotide arrays for rapid DNA sequence analysis. *Proc. Natl. Acad. Sci. U. S. A.* 1994, 91:5022-5026.

[54] Hinds DA, Stuve LL, Nilsen GB, Halperin E, Eskin E, Ballinger DG, Frazer KA, Cox DR: Whole-genome patterns of common DNA variation in three human populations. *Science.* 2005, 307:1072-1079.

[55] NimbleGenSystemsInc.: [www.nimblegen.com].

[56] Hacia JG, Brody LC, Chee MS, Fodor SP, Collins FS: Detection of heterozygous mutations in BRCA1 using high density oligonucleotide arrays and two-colour fluorescence analysis. *Nat. Genet.* 1996, 14:441-447.

[57] Wong CW, Albert TJ, Vega VB, Norton JE, Cutler DJ, Richmond TA, Stanton LW, Liu ET, Miller LD: Tracking the evolution of the SARS coronavirus using high-throughput, high-density resequencing arrays. *Genome Res.* 2004, 14:398-405.

[58] Borevitz JO, Liang D, Plouffe D, Chang HS, Zhu T, Weigel D, Berry CC, Winzeler E, Chory J: Large-scale identification of single-feature polymorphisms in complex genomes. *Genome Res.* 2003, 13:513-523.

[59] Turner TL, Hahn MW, Nuzhdin SV: Genomic islands of speciation in Anopheles gambiae. *PLoS Biol.* 2005, 3:e285.

[60] Moerman DG, Barstead RJ: Towards a mutation in every gene in Caenorhabditis elegans. *Brief Funct. Genomic Proteomic.* 2008.

In: Space Exploration Research
Editors: J.H. Denis and P.D. Aldridge

ISBN: 978-1-60692-264-4

Chapter 5

STRUCTURE AND TRANSPORT OF THE MARTIAN SURFACE MATERIAL

J. P. Merrison
Mars Simulation Laboratory, Aarhus University,
Ny Munkegade, Aarhus 8000C, Denmark

ABSTRACT

The driving force for much of the present exploration of our solar system is the search for clues to the origin of life, specifically the search for habitable environments either past or present. In short it is the search for liquid water and of most current interest is the near-surface environment of Mars. After recent intense study, from both the surface of the planet and orbit, Mars reveals itself to be relatively rich in near surface water ice mixed and overlaid by sand and dust at least down to a few meters. Despite the present cold, dry conditions Mars appears to have undergone extremely wet periods at some locations up to geologically recent times. In order to understand the transport of heat and moisture through this near surface material details of its structure, composition and transport mechanisms must be known, which are intimately linked to the nature and history of the Martian climate. Specifically dust is actively transported through the atmosphere, though the process is not understood and wind produced sand features are abundant on Mars, though it is not clear whether sand transport is still active or points towards a previous climatic environment. In this chapter the most recent laboratory simulations and modeling techniques will be presented in the context of current observations from Mars. Advances in instrumentation will also be discussed for application to future Martian surface studies.

1. INTRODUCTION

One of the most important and fundamental goals in science is to establish the origin of life and thereby understand its nature. Despite the wide variety of life on earth it seems to share a common primitive, though extinct, ancestor. Central, therefore, to this scientific goal is the search for life elsewhere in the solar system.

As a neighbor to Earth, Mars has the possibility of being similarly temperate. However, NASAs Viking mission in 1976 showed the surface of Mars to be cold, dry and essentially devoid of organic material, though in fact the answer to the search for life on Mars was not answered by this mission [1]. Since that time Mars has shown itself to be a complex planet with a rich climatic history involving liquid water and a diversity of local environments with the possibility of habitability [2,3]. Mars has become the focus of increasing scientific investigation both on the surface and from orbit.

Bacteria are found inhabiting almost every environment on Earth where liquid water can exist, despite harsh environments or limited resources. This realization has allowed the search for life on other bodies in the solar system to broaden. The search for life has become the search for places where liquid water is (or has been) present. Specifically for Mars this search will be in the near sub-surface.

It is becoming clear that the study of Mars is not a trivial one and that to reveal the secrets of Martian habitability, past or present, requires detailed understanding of the near surface region. Specifically understanding its structure and evolution, that is to say describe and explain the transport and induced alteration of the surface material. In this way it becomes possible to probe the climatic history of Mars and study the role of water. This can and is being achieved by performing observations of Mars both from orbit and on the surface and combining these observations with concerted earth based simulation and modeling which in turn lead to the development of new investigative technologies. This will be the focus of this chapter.

Another, somewhat less scientific, aspect to the habitability of Mars is the question of manned missions to the planet. Here again for explorers on Mars to be safe, over long periods of exposure to the local environment, a far greater degree of understanding is required of the Martian surface and its hazards.

Much has been learned from the NASA Mars Exploration Rovers (MER) and ESA Mars Express missions specifically about the role of liquid water on the mineralogical and geological evolution on Mars, these issues will also be discussed.

The use of a unique wind tunnel facility at Aarhus University will be presented and its application in the interpretation of experimental data from Mars landers (e.g. the NASA MER mission) [4]. This is done by using a combination of computer/mathematical modeling and laboratory simulations, a specific example is the MER magnet properties experiments studying airborne dust. Such work is also complemented by the development of new miniaturized instrumentation which as well as performing in-situ measurements in the simulator are planned to perform the same task on the surface of Mars. Parameters such as wind speed, turbulence, dust concentration, dust electrification and deposition rate can be measured with an integrated laser based optoelectronic sensor system. These parameters, among others, are crucially important for building a detailed understanding of the transport and structure of the Martian surface (down to the micrometer scale).

Future missions to Mars are being planned and the search is beginning for landing sites where the possibility is highest of finding near surface environments suitable for inhabitation. This search would benefit greatly from a deeper understanding and better simulation of the Martian surface [5,6].

1.1. Physical Conditions on the Martian Surface

The atmosphere of Mars is significantly lower density than earths, with measured surface pressure in the range 6-9 mbar. The typically low average temperature on Mars (around -60°C), leads to low humidity, typically less than 1%. The composition of the atmosphere is 95% CO_2, with trace amounts of N_2 (2.7%), Ar (1.6%), CO (0.07%) and O_2 (0.13%). As a result of the thin atmosphere short wavelength solar UV light penetrates to the surface on Mars (down to 200nm) and has been blamed for the lack of organic material on the surface as will be discussed [1].

Significant variations in atmospheric pressure occur as CO_2 seasonally freezes/sublimates from the poles, though CO_2 ice is not observed to vanish from the south pole. Despite the low humidity, water as well as CO_2 clouds are occasionally observed and surface frost has been seen.

There are also large diurnal (night and day) variations in surface temperature, as in desert areas, and temperatures in excess of 20°C have been observed for short periods during mid day.

Heat absorption and light scattering within the atmosphere is dominated by the presence of (seasonally) varying amounts of dust. Global dust storms occasionally entrain so much dust that the atmosphere is essentially opaque. Dust grain diameters in the Martian atmosphere have been measured to be around 3 μm [7,8], though this is averaged over the atmospheric column and may mask a strong variation in grain size with height. The mass density of the dust grains in the Martian atmosphere is unknown. Both from theoretical considerations [9] and from experimental evidence [10] it seems likely that the atmospheric dust grains on Mars actually are aggregates of smaller grains. Electrification and subsequent aggregation dust has been observed in wind tunnel experiments [11]. In this case the size of the dust grains is a dynamically varying parameter as aggregates form and break up. The mass density of the aggregates has been seen to be significantly lower than the bulk density of the material.

Compositionally the Martian soil consists mainly of: SiO_2 (43%), Fe_2O_3 (18%), Al_2O_3 (7%), SO_3 (7%), MgO (6%), CaO (6%) [12], assuming the surface material is fully oxidised. Most of the Martian surface at low latitude is characterized by dark basaltic sand and stones over-layered by a (generally) thin actively deposited/removed layer of red dust which, although being of similar composition to the sand is enriched in highly oxidized iron(III). At higher latitudes and closer to the poles water and CO_2 ice are seasonally deposited and an apparently complex mixture of ice and sand/dust layers exists. Aeolean sand transport features such as sand dunes, ripples and drifts are seen ubiquitously across Mars, though evidence does not indicate that they are presently active. There is large scale transport of dust on Mars. The most striking demonstration of this has been the observation of families of dust devils by the MER rovers and also from orbit by Mars Express [13,14]. This presents a paradox to the current transport theory which predicts that fine grained material (dust) should be dominated by adhesive forces and far more difficult to remove by the wind than sand sized grains (above 63μm).

This seemingly contradictory evidence has inspired a series of experimental laboratory based simulations of Mars analogue dust and sand grain transport under Martian atmospheric conditions presented in this chapter, section 5.

Mars has a north south dichotomy, with the north typically consisting of low plains that are more recent and consisting of material deposited from a series of extremely large volcanoes (the largest in the solar system) and the southern highlands being a far older and heavily cratered surface possible dating back to the heavy bombardment period (3.8 billion years). Discussion continues as to whether the northern hemisphere was at an earlier climatic period covered by an ocean.

There is abundant observational evidence for the flow of liquid (presumably water) across the Martian surface, for example features that resemble rivers, deltas, lake beds etc. Some of these are on large scale (100-1000km). Other features appear more typical of glacial flows [15].

One of the most exciting observations from Mars Express has been what appears to be the dried remains of a geothermally produced sea/lake (section 4.2). Interestingly methane (and formaldehyde) have also been detected regionally concentrated around what might be geothermally active zones close to the Martian surface, by Mars Express [16,17]. Again this points towards recent/present volcanic activity close to the surface and therefore the potential for melting of subsurface ice.

2. Mars Simulation, Modeling and In-Situ Measurement

Much can, and has, been learned on earth which can be used to understand processes occurring on the surface of Mars. Geologically earth environments can be used as analogues to the Martian environments, specifically desert areas of various kinds. Another approach is to construct experimental simulation chambers and reproduce the environmental conditions on the Martian surface in a more detailed though down scaled way. In both cases comparison of observations/experiments from the Martian surface can be compared to the simulations, here the use of computer modeling allows quantitative feedback and the subsequent development of better simulations.

2.1. Laboratory Simulation

Experimental Martian simulators are becoming more common among the now many research groups dealing with the study of Mars. In most cases these simulators reproduce the Martian atmosphere and perform a variety of mineralogical, (micro) biological and chemical studies. With respect to studies involving the transport of surface material (dust/sand) Mars simulation wind tunnel experiments are of most relevance and here there are only two facilities namely: the NASA Ames MARSWIT (Mars Surface Wind Tunnel, California USA) [18] and the Aarhus Mars Wind Tunnel [4]. They complement each other in many ways, with MARSWIT dealing mainly with higher (than normal on Mars) wind speeds of greater than 20m/s and up to 180m/s (at low pressure) whereas the Aarhus wind tunnel has generally concentrated on wind speed studies lower than 10m/s, though recent investigations of sand (and dust) detachment/entrainment have extended to wind speeds of up to 40m/s. In the next section description of the Aarhus wind tunnel facilities' capabilities and research strategies will be presented. In comparison the MARSWIT facility is larger: it is an open-circuit wind tunnel powered by a high pressure nozzle ejector system, the total length is 13m with a main test section of 1.2 by 0.9 m and is housed in a 4000 m^3 low-pressure chamber which can operate at pressures down to ~3.8 mbar [19,20]. Though it cannot be cooled, CO_2 gas can be used as the fluid medium.

An important aspect to simulation is the use of material which is analogous to that on Mars. Though mineralogical and chemical studies have been performed on Mars, it is still at present necessary to use granular material with only limited similarity to that seen on Mars. There are several such Mars analogue granular (dust/sand) materials each with their own areas of similarity, but also other properties which match the Martian material poorly.

A complex mixture of crushed minerals has been able to reproduce the composition (observed by the Viking missions) of the Martian regolith and a reasonable color match can be obtained by regulating the hematite fraction (which is a few %) [21]. The size and magnetic properties are however not reproduced.

A commercially available fine granular material (measured to be dominantly around 2-3 microns diameter [22]) called Carbondale clay has also been used as a Mars analogue material, it consists of mostly quartz grains and has a pale pinkish appearance. It has the advantage of being available in large quantities.

Palagonite material is the weathered products of (basaltic) pyroclastic deposits, samples from volcanoes in Hawaii have been widely used as a Mars analogue. Specifically the material JSC-1 has been used by NASA groups [23]. The color (also in the infrared) has been shown to be a reasonable analogue to Martian material. It can be fine grained (<20 microns) though typically little is entrained on injection during wind tunnel investigations, indicating a large fraction of coarse grains. Some of these Palagonite samples have been seen to be similarly magnetic to the dust collected in the Mars Pathfinder magnetic properties experiments [24].

The iron oxide rich Salten Skov dust simulant is thought to be a chemical sediment and is found naturally, it can therefore be readily obtained in large amounts. It has a high iron content (60% by weight) with mineralogical fractions: goethite (73%), hematite (14%) and maghemite (13%). It was originally chosen for its color, its fine granular nature and its magnetic properties. This analogue material has been well studied using X-ray diffraction (composition), Mössbauer spectroscopy (iron mineralogy), laser diffraction (size distribution) and magnetisation [4,25]. Though its optical/IR color is not spectrally as good as some palagonites, magnetically it has been shown through extensive experimental simulation and computational modeling to be a good analogue to the Martian dust [26, 27]. The dust grains injected into the Aarhus wind tunnel have a measured mean diameter of 2 microns [4] in reasonable agreement with the observations from Mars (3μm), little of this analogue is coarse grained. Though the material has a bulk mass density of 3000 kg/m^3(ref) the powder is generally loosely packed with a mass density around 1000 kg/m^3. As can be seen in figure 1 this dust analogue is found as aggregates of sub-micron single mineral grains, which in turn form larger aggregates giving a broad distribution in grain size, also in suspension. Interestingly dust aggregates have also been observed on Mars possibly indicating similar electrostatic/adhesive properties [ref.28]. This aspect of the material may have great importance to its transport, allowing it to be entrained and also for the transport of heat and moisture through the material.

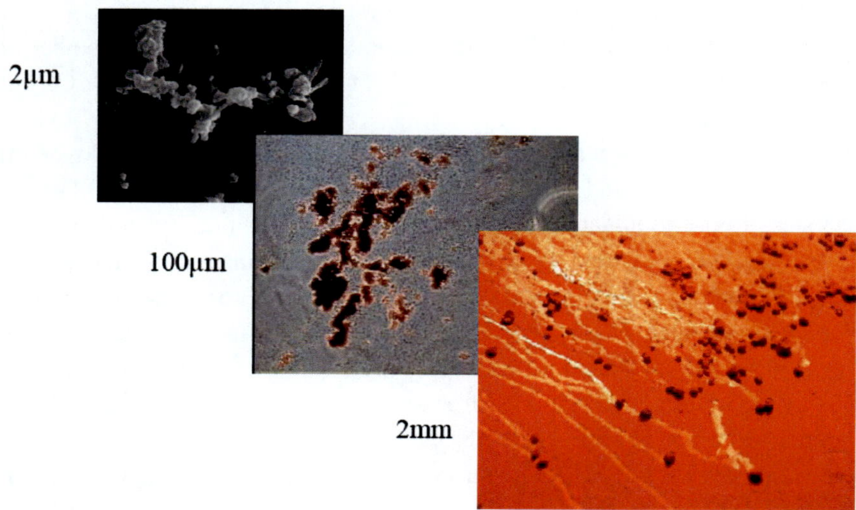

Figure 1. Photographs showing the step wise aggregation of the material to larger and larger size, the scale shows the approximate size of the aggregates: mm scale taken with a digital camera, 100micrometer scale taken with an optical microscope and on the micrometer scale using a scanning electron microscope.

Important recent findings about the Martian atmospheric/surface dust is that, despite its fine granular form, it contains a mixture of minerals including iron-II (of igneous origin) as well as oxidized iron-III. This shows that this dust has not had prolonged exposure to liquid water (this would lead to complete oxidation) and is probably an erosion product of the local sand. Simulation experiments performed in the Aarhus wind tunnel have shown that grinding basaltic rock (olivine basalt from Reykjavik, Iceland) produces the same phenomenology as observed on Mars, that the finer grained material has a higher content of iron-III (more oxidized). Here the same experimental technique: Mössbauer spectroscopy was used for the mineralogical analysis [29]. Also capture of this material on MER magnets in the simulation experiments again showed (section 3.4) a two component distribution consistent with the observation of Martian dust. Two possible processes are suggested for the oxidation of the fine grains. One being that the iron-III in the material is weaker and therefore, if locally enriched, erodes to finer material. Another possibility is that the grinding (erosion) process causes mineral alteration. This work may imply that such ground basaltic material would provide a good Mars analogue material and importantly would have implications for the structure of the finest atmospheric Martian dust, which possibly is not aggregated, but merely a mineral mixture or having an iron-III rich coating.

2.2. Mars Simulation Wind Tunnel

The wind tunnel facility at Aarhus University is intended to reproduce the environmental conditions observed at the surface of Mars, specifically the atmospheric pressure and composition, the temperature, wind conditions and the transport of airborne dust. It consists of a re-circulating wind tunnel housed inside an environmental chamber. This chamber can be evacuated to around 0.01mbar and re-pressurized and held at Mars like pressures (typically 6-

10mbar). The central wind tunnel is cylindrical, 40cm in diameter and 1.5m long. An axially mounted electric motor driven fan draws gas down the central wind tunnel and returns it in an outer cylinder. In this wind tunnel turbulence levels are typically between 3-16% (depending on the wind speed) and wind speeds in excess of 30m/s have been achieved.

Injection into the wind tunnel of a Mars analogue dust (Salten Skov) allows simulation of the Martian dust aerosol. Other Mars analogue materials or other granular minerals can and have been used. Injection involves passing a dust air mixture through a nozzle at high velocity, this is sufficient to disperse aggregated grains and suspend them.

It is a similar technique to that used in other Mars wind tunnel testing in the MARSWIT facility (section) though it is questionable as to whether it is a good representation of the entrainment processes on Mars, for example in a dust devil. Though some small scale vortex generators have been constructed [30] none have been combined with wind tunnel testing which might be expected to better reproduce the suspended dust characteristics on Mars especially with respect to electrification and dust grain (aggregate) size.

Although temperature and gas mixture can be controlled in the wind tunnel, for convenience low pressure air and room temperature are normally used for simulations. Typically 9.0mbar air pressure is used, this corresponds to the same fluid density as 6mbar of CO_2 at a temperature of -60°C.

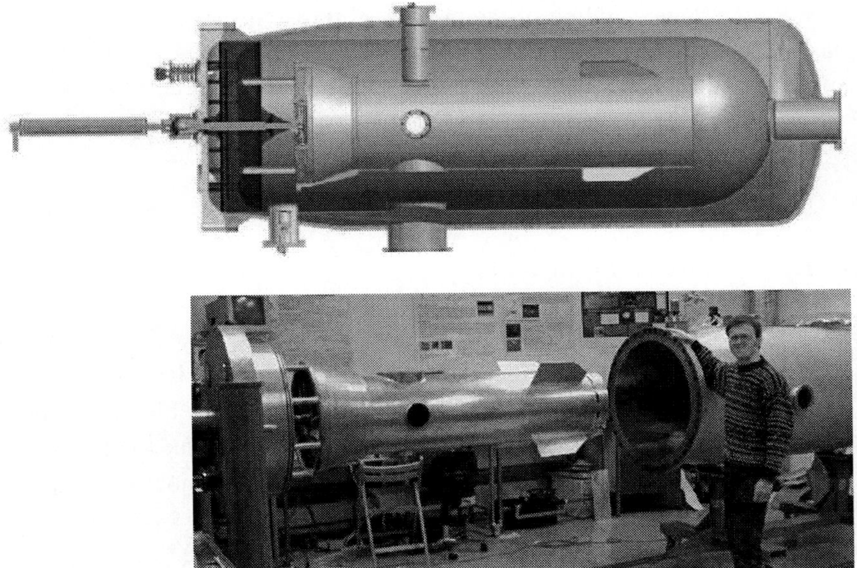

Figure 2. Schematic technical drawing of the Aarhus Wind Tunnel (top) and photograph of the (open) wind tunnel with moveable vacuum housing chamber.

Typically reproduction of the same Reynolds number would normally be used to allow for simulations to be performed under slightly different conditions. This however becomes problematic here since in order to reproduce the same fluid flow within the wind tunnel a Reynolds length scale of the order of the wind tunnel dimensions should be used (10cm), conversely if conditions for the transport of suspended fine grains is to be maintained the Reynolds length scale of the grain size should be used (μm). To summarize it must be decided whether to reproduce the micro or macro scale fluid behavior and in the case of most simulations in the Aarhus Mars wind tunnel the macro scale has been chosen. For very fine grained material the added complication of the 'slip' factor sets yet another constraint onto the effectiveness of simulations using different gas mixture/pressure/temperature.

Dust grains injected into the wind tunnel can interact with a surface and become adhered or adhere to each other and settle due to gravity. These processes mean that injected dust is lost out of suspension and the time taken varies inversely with wind speed (at higher wind speeds dust is lost faster to collisional processes) and is typically of the order of one to several minutes (time constant). Generally in wind tunnel tests Mars analogue dust grains are seen to cohere into aggregates, often forming lumps greater than 1mm in diameter [4], see figure 1.

The wind speed in the wind tunnel is measured using a commercial (Dantec) laser Doppler anemometer system (LDA), here the velocity and turbulent rms (root mean square) of individual dust grains is measured. The suspended dust concentration can also be monitored by counting the number of particles detected by the anemometer as a function of time and knowing the detection area of the LDA [4,31]. Though the effectiveness of this dust detector falls, at low wind speed most grains are assumed to be detected and this has been used to estimate the grain size. The value obtained (of 2μm diameter) is in agreement with measurements made with the dust suspended in water using a laser scattering instrument.

Ideally though a direct method of dust grain sizing (area or dimension) and/or mass determination is needed and would have great importance to modeling (where this parameter is crucial) and in the understanding of the physics of dust transport. Similar direct dust grain size and mass measurements are also desirable, but not (yet) available from Mars.

Presently the wind tunnels at Aarhus and MARSWIT are the only available facilities of this kind, this is surprising given the present interest in Mars research and presumably the future will see more and larger Mars simulation wind tunnels.

2.3. Mars Environmental Chamber

Having established that liquid water has played a major role in the mineralogy (and geology) of areas of the Martian surface in the past, the forthcoming missions (NASA and ESAs AURORA program) will begin to concentrate on more direct 'biological' investigations of the Martian (sub)surface. Many research groups world wide are tackling various aspects of this most difficult problem of remote identification of Martian biology/biological activity. Many groups concentrate on techniques for detection of organics whereas others study the effects of the Martian environments on biological samples and earth bacteria. Under this second category of research a broad range of earth and space based (the international space station) simulation projects are taking place broadly with the aim of quantifying the survivability of (known) bacteria to Martian conditions [32], typically the bacteria Bacillus Subtilis. In a prototype simulation facility at Aarhus University the activity and stability of a complex bacterial soil community has been studied under simulated Martian conditions [33]. At three different depths (down to 5cm) in a soil core the effects of Martian conditions with and without ultraviolet (UV) exposure corresponding to 8 Martian Sol (days) were compared. Community metabolic activities and functional diversity, measured as glucose respiration and versatility in substrate utilization, respectively, decreased after UV exposure, whereas they remained unaffected by Martian conditions without UV exposure. In contrast, the numbers of culturable bacteria and the genetic diversity were unaffected by the simulated Martian conditions both with and without UV exposure. The natural dominance of endospores and Gram-positive (spore forming) bacteria in the investigated Mars-analogue soil may explain the limited effect of the Mars incubations on the survival and community structure. These results suggest that UV radiation and desiccation are major selecting factors on bacterial functional diversity in terrestrial bacterial communities incubated under simulated Martian conditions. Importantly also these results suggest that forward contamination of Mars should be a matter of great concern in future space missions, a subject which has been the focus of much of the recent simulation work on bacterial survivability.

Following this pilot study the construction of the Mars Bio-chamber at Aarhus was carried out which is an experimental facility focusing on these kinds of biological exposure experiments. It consists of a vacuum chamber, in which the Martian atmosphere can be recreated, with the capability of being cryogenically cooled and allows samples to be irradiated by a 'solar simulator' i.e. an optical / ultra-violet (UV) light source resembling that of the sun as seen on the surface of Mars.

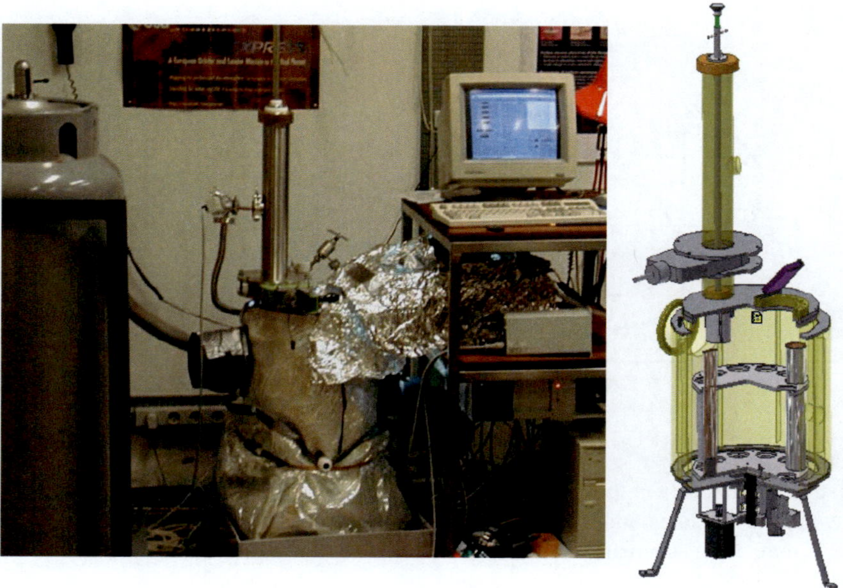

Figure 3. Schematic diagram of the Aarhus Mars Bio-Chamber (Right) and a photograph taken during testing (Left).

Several aspects of the environment on Mars are probably of importance; the low, but importantly variable temperatures of the Martian surface, the direct effect of hard solar UV (around 200-300nm) reaching the Martian surface and also the effect of oxidizing agents presumed to be present which may be created by the this UV light.

The Aarhus Bio-chamber houses a carousel where ten biological soil (mineral grain) samples can be placed and rotated. A load-lock system allows sample placement/replacement without interruption of an experimental series (at low temperature and under low pressure). Evacuation of the system down to 10^{-3}mbar is achievable. Injection of CO_2 gas and four other minor components: Ar, CO, O_2 and N_2 can be made. A rest gas analyzer is used to analyze/monitor the gas composition and test for leaks or out gassing from the samples.

Special attention is paid to the water content in the simulation atmosphere which should be maintained at the similarly low levels found on Mars, typically around a few hundred ppm by volume [12]. Interestingly, though this corresponds to around 30% relative humidity due to the low atmospheric temperature (-60°C). Similarly low humidity in the simulator is aided by the low temperatures (-40 to -80 °C). Relative humidity in the simulator is typically measured to be around 0-10%.

Cooling is performed by flowing cold nitrogen gas (from a liquid nitrogen flask) through a second outer compartment surrounding the vacuum chamber. Outside this is a wrapping of super insulating material. The cold gas flow is controlled by a computer system with thermometer feedback which ensures a stable and adjustable temperature. Using a quartz glass window light from a 200W Xenon-Mercury (Xe-Hg) discharge lamp can be passed onto the surface of one of the samples. Though it could be argued that the spectrum given by this choice of discharge lamp is not a good simulation of the solar flux, it does give a high intensity within the critical region 200-300nm such that simulation of many solar days (sols) can be performed in a shorter period (a few hrs) within the simulator.

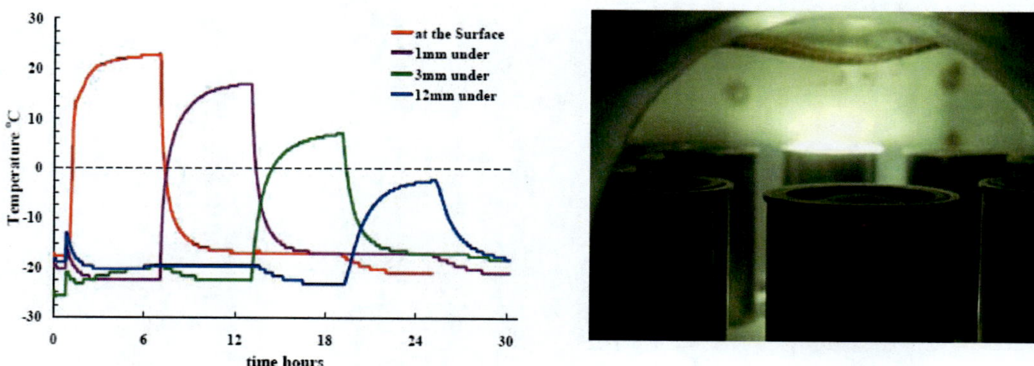

Figure 4. Photograph (right) of soil (granular analogue material) samples being irradiated by a UV generator (solar simulator). On the left temperature measurements from within the samples are shown as each new sample is rotated under the lamp, the temperature measured was at different depths under the surface: 12mm, 3mm, 1mm and at the surface.

Extensive testing of this simulator has been made and as with other such facilities a series of technical problems tackled, mostly related to the somewhat special desire to have both 'high vacuum' style gas purities, low pressure, cryogenics and moving parts which puts stringent demands upon the construction materials. The first long term bacterial-colony exposure has been performed of around 30 days, corresponding to around 60sol/sample effective exposure. Analysis will be forthcoming [34].

Despite the fact that the samples were freeze dried, out-gassing/sublimation of water from the samples was observed with the formation of ice on the chamber wall and increase in humidity (to a few percent). Humidity is an important [35], though often difficult aspect of the Martian surface to reproduce.

In addition to biological exposure testing related studies are planned in this facility, specifically investigation of the production and transport (through the surface layers) of UV induced reactive compounds such as ozone and hydrogen peroxide. Also, as will be discussed in section 7 the transport of humidity and temperature through the surface material can be studied, giving information on the stability of liquid water.

2.4. Mars Analogue Environments on Earth

There are many regions on earth where the environment or materials resemble aspects of Mars and have been used as analogues in interpreting observations from the planet. Some are cold (polar) others hot (equatorial), but generally all relatively dry. A selection will be discussed here with relevance to the Martian surface structure, transport and habitability.

On the subject of dry environments the Atacama desert in Chile is one of the most arid on earth and, although rarely at subzero temperatures, studies of the soil using techniques similar to the Viking pyrolitic experiments have shown depletion of organic material (analogous to the observations of Viking) and low levels of culturable bacteria [35]. This has subsequently made this region a (Mars analogue) testing ground for biological/organic instrumentation for Martian (sub)surface studies [36].

Bacterial resilience to UV irradiation has been studied in various polar latitudes, including sites such as the Antarctic and Devon Island in Canada (the arctic), it was found that many microbial communities were protected by physical and biological coverings and have improved tolerance to atmospheric ozone depletion in these regions [37] and the associated increase in solar UV exposure. Clearly the extreme cold (approaching the average temperatures found on Mars of -60°C), the arid conditions and increased UV doses found in these regions make them good Mars analogues. Specifically Devon Island has been adopted by the Mars Society as a location for Mars analogue tests for manned exploration [38].

Although many geological features on Mars remain strong evidence for fluid (liquid water) flow others resemble ice-rich glacial viscous flows (and sublimation) and can be compared with similar geological features found in polar regions on earth (such as Iceland or Antarctica [15]). Some of this glacial activity on Mars appears to be recent and is indicative of a variable climate [39].

Dust devils can be found in most geographic regions on earth, even at high latitudes though with varying frequency and are often and widely seen on Mars [40]. There are several research groups which have made extensive study of terrestrial dust devils, often in arid areas in Arizona or Navada USA, many of these researchers have an interest in the study of Martian dust devils [41,42]. Generally terrestrial dust devils are seasonal, the warm core vortices require dusty/sandy (preferably flat) areas for visibility and a lot of solar heating (for the creation of a thermal inversion) and a degree of gusty wind for initiation. Despite the differences in scale (Martian dust devils are substantially larger) and composition (on Mars they are visible due to dust, on earth it is commonly sand) the phenomenon is undoubtedly the same [13]. There is now a body of experimental and theoretical (modeling) work on dust devils [43,44,45], despite this a collected physical description of the phenomenon (especially on Mars) is still lacking. Possibly of importance, especially on Mars, is the study of the electrical (electromagnetic) properties of dust devils. Such studies are still in their infancy and should be repeated on Mars [46].

Figure 5. Photographs of a dust devil on earth, in Navada, USA, 2004. Similar dust devil phenomena have passed close by the NASA MER Spirit in Gusev Crater, 2005 [13].

3. Theoretical and Computational Modeling of Grain Transport on Mars

The aim of experimental simulation and earth analogue studies is to interpret the experiments and observation from Mars and thereby make conclusions about the planets environment and history. There are always aspects of the simulations/analogues which do not ideally reproduce the conditions on Mars, for grain transport examples could be gravity, scale of the simulation, material properties, surface structure and so on. It is here that modeling, both theoretical and computational, are of importance by allowing a deeper, more quantitative understanding of the processes being simulated and enabling parameters to be varied in the model which may be difficult/impractical by simulation. For the transport of material such modeling can be performed on many scales, from the microscopic (dust/sand grains) to the global scale encompassing the entire atmosphere. Some of the relevant modeling of transport mechanisms will be discussed here and an example will be presented of the application of computer modeling for interpretation of Mars data specifically concerning the MER magnetic properties experiments.

3.1. Modeling on the Micro Scale

Spherical particles encountering a fluid in relative motion will experience a drag force. This force is given by the equation: $F=C_dU^2\pi d^2\rho/8$, where ρ is the fluid density, d is the particle diameter, U is the relative velocity of the particle with respect to the fluid and C_d is the drag coefficient. C_d is dependent on the Reynolds number; $R_e=\rho Ud/\sigma$, where σ is the molecular viscosity of the gas [20]. For large values of Reynolds number C_d is constant (Newtonian flow). In the case of the Martian (analogue) dust the correspondingly low Reynolds numbers allows fluid drag to be well described by Stokes flow. In this case once entrained the dust is well suspended (even at low turbulence levels).

For the finest dust an additional correction must be applied to the Stokes model due to that the thin Martian atmosphere can be seen to be non continuous, this occurs when the size of the dust grains becomes comparable with that of the scattering length between atmospheric molecules, in the case of 7mbar CO_2 around 6μm. The correction has been empirically determined [47] and termed the slip factor S. The drag force on a spherical dust grain is given by a modified Stokes law as:

$$F_d = \frac{3\pi\sigma dU}{S}$$

Sand grains (and dust aggregates) are significantly larger in size than dust and together with the relatively large wind (friction) velocities required for entrainment these factors significantly increase the Reynolds number, in the case of sand entrainment to values close to 1. This is not large enough to allow use of simple Newtonian flow and not small enough for Stokes flow, but lies in the region referred to as Allen flow where the drag coefficient is a complicated function of Reynolds number (C_d = constant/R_e^n, where n=0 to 1) [48]. This means that the motion of sand sized grains passing through a fluid (for example in saltation) is not simple to describe. It is also different from the process of detachment from a surface which becomes further complicated by the proximity of a wall which produces a wind shear.

The motion of a dust grain in a gas is governed by a balance between the drag force from the gas and other forces such as gravity, electrical or magnetic forces. The solution to this equation of motion for a forced grain in a fluid is of the form:

$$U(t) = U_T + U_0 e^{(-t/\tau)}$$

where U_0 is the initial condition, U_T is the terminal (drift) velocity in the force direction and τ is given by (assuming the force is proportional to the mass), where ρ_g is the grain mass density:

$$\tau = \frac{S\rho_g d^2}{18\sigma}$$

τ is a characteristic time for a grain to accelerate from rest to the terminal velocity. The characteristic time is dependent on grain size, grain density and viscosity, but not on the strength of the force being applied, which only appears in the expression for the terminal velocity.

This characteristic time is all-important in determining the motion of dust grains and their response to outside forces. A large characteristic time corresponds to grains not overly affected by drag and as τ increases we leave the regime where Stokes law is valid and eventually recover the equations of conventional drag-free Newtonian mechanics. In the limit of small characteristic times grains respond rapidly to changes in the outside force and reach their (small) terminal velocities rapidly. Dust grains will thus always be moving at, or close to, the terminal velocity, which is proportional to the applied force.

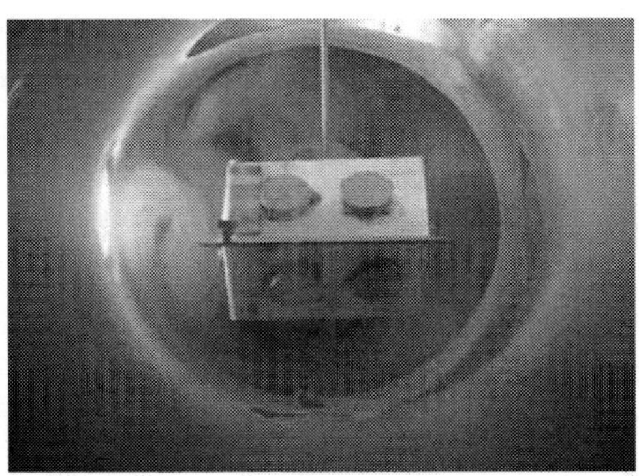

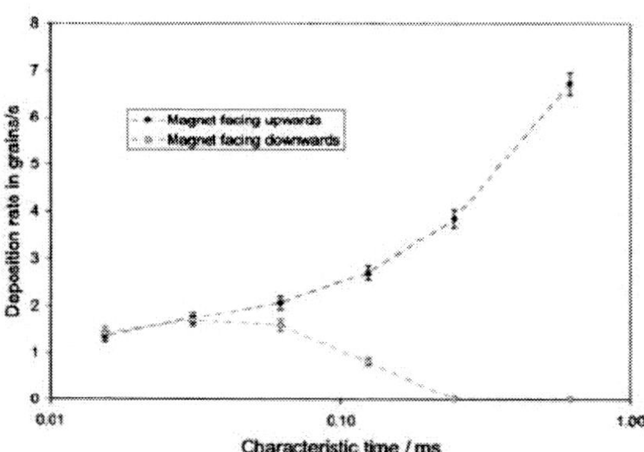

Figure 6. On the Right the calculated dust deposition rate as function of characteristic times for magnets facing upwards or downwards. The capture rate denotes the number of grains per second captured on a 1 cm-wide area across the surface of the magnet given a number density in the air of 1 grain/cm^3, wind speed of 1 m/s, Salten Skov dust and Terrestrial gravity. Left a picture of an analogous wind tunnel experiment where MER magnets face upward and downwards and capture dust.

For dust grains in the wind tunnel simulator we calculate characteristic times of a few milliseconds and conclude that the behavior of such grains is close to that described above for the limit of small characteristic times.

For computational purposes it is not necessary to vary grain density and grain size as separate parameters. Instead we vary a single parameter, the characteristic time, which expresses the strength of the drag force relative to the inertia of the grain. Interestingly this means that in this treatment the motion of a large low-density grain will be identical to the motion of a smaller grain of higher density. If possible it would be desirable to measure this parameter experimentally on Mars. One possible way of doing this, as illustrated in figure 6, might be to use gravity in combination with another force (electrostatics or magnetism) to vary the effective applied force and therefore the drift velocity (and capture height). Physically this would merely involve performing the same experiments also inverted with respect to gravity [26].

Experimentally the behavior of fine suspended grains allows a great deal of simplification since, as discussed, under the influence of a force fine grains reach a (terminal) velocity in the force direction. This means that grains subjected to an attractive force towards a surface placed in a wind stream will accumulate at a rate directly proportional to the force applied. Measurement of the accumulation of dust grains attracted to the surface of a magnet/electrode placed in the wind stream can thus be used directly to quantify the degree of magnetization/electrification of the dust.

3.2. Modeling on the Macro Scale

The treatment described for the micro scale can be used to describe the behavior of fine grains suspended in a continuous flow. On the larger scale (millimeters to meters) of experimental instruments/planetary landers, fluid dynamical treatment must be applied to determine the local flow. Assume a continuous fluid the standard Navier-Stokes equations can be applied and it is possible to numerically model the flow of the gas/dust grains and thus for example the process of dust capture. Making a discrete (finite element) computer model of the experimental geometry, including for example the wind tunnel structure (or lander structure) the fluid flow can be predicted given the properties of the fluid and wind conditions. Having determined the flow computationally grains can be injected into the fluid.

Such a computational fluid mechanics (CFD) model has been applied to wind tunnel investigations utilizing a commercially available STAR-CD code, which is a finite volume code based on the FORTRAN programming language. In this work the model was 2-dimensional with a geometry chosen to approach the geometry of the wind tunnel facility used for simulation experiments. The geometrical set-up, shown in figure 7, was used to model the magnetic capture on the MER magnets.

The gas density and dynamical viscosity were set at typical Martian values of 0.015 kg/m^3 and $1.5 \cdot 10^{-5}$ kg/m·s respectively [49], while the mean free path used in the slip factor calculation was set to 8 microns (mean free path of CO_2 gas at 273 K, 700 Pa). Most calculations were performed in a steady state mode without coupling between gas and dust grains. This means that a steady solution to the gas flow field was obtained first while ignoring the presence of dust grains, the paths of dust grains are determined based on this calculated flow field.

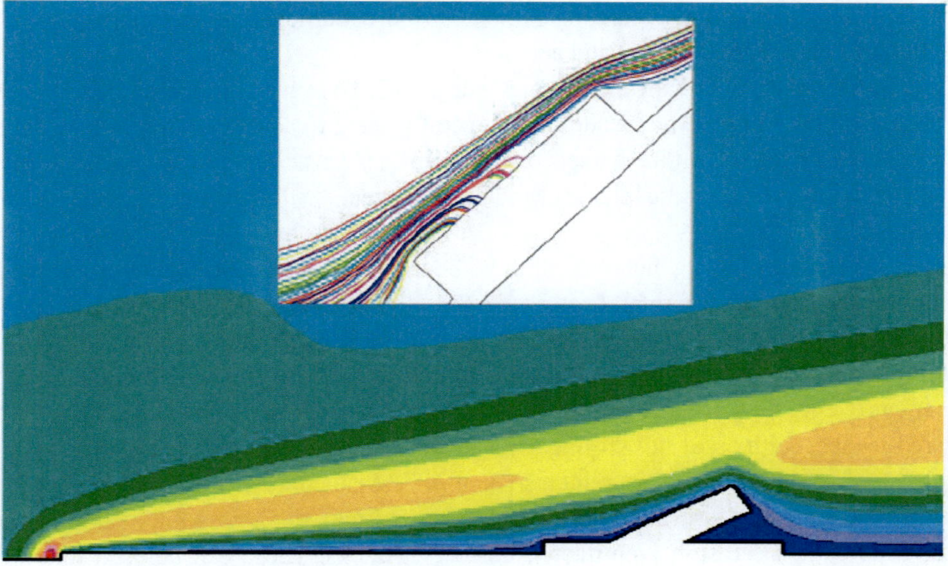

Figure 7. CFD calculation of MER magnets in the Aarhus wind tunnel, lower the fluid flow and upper (box) shows the magnet capturing grains from the wind stream.

In the standard case wind speed was set to 1 m/s. Turbulence was handled by a k-ε model [50], and for each step of the calculation of a dust grain path a random perturbation was added based on the calculated local intensity of turbulent motion in the gas. The main calculation was therefore essentially a Monte Carlo simulation of a large number of dust grains moving through a previously generated flow field under the influence of drag, gravity and in this case also a magnetic force.

As the standard template for dust grains the measured mean values for Salten Skov Mars analogue dust was used. For numerical modeling all grains were assumed to have a diameter of 2 microns and a mass density of 3000 kg/m^3. From these values the characteristic time was calculated to be 0.61 ms. This model was applied to the data obtained from the MER magnetic properties experiments as described in section 3.4.

Another aspect to such computer modeling is that variations in experimental design, for example in this case the magnet design, can be relatively easily made and the results tested. This can allow the development of new instruments/experiments which have certain desirable properties. Specifically designs could be developed which were sensitive to the grain time constant as discussed, thus having the capability of giving information about grain size and mass.

3.3. Modeling on the Global Scale

Modeling/measurements of local fluid flow and turbulence as discussed here has also global implications for transport (of heat and material) which can be based on/compared to local functions of wind speed and turbulence levels (related to vertical transport). Different global circulation models can be found for Mars, they are of great importance (as they are on earth) for understanding and predicting regional wind conditions and describing the (global and regional) transport of volatiles and aerosols.

Based on terrestrial Global Circulation Models (GCM's) similar models adapted for Mars have been available for several decades and generally seem to adequately reproduce the large scale circulation over the entire planet, over several seasons. Mentioning a couple of such models: Mars Regional Atmospheric Modeling System (MRAMS) [51], Mars General Circulation Model [52].

One of the challenges to such modeling is the importance of dust loading to the weather and climate by dominating solar heating of the atmosphere. This dust loading is highly variable and at present is not well understood, specifically the dynamics of dust devil events. Modeling therefore uses statistical approaches to reproducing the observed likelihood of dust entrainment events [53]. From these models the (near-surface) wind stress can be found and compared to observations of dust lifting rates. Models of the surface boundary layer can be compared to the wind speed measurements taken [49]. Also comparison of the models can be made to observations of surface features (aeolean forms) on Mars, for example: dunes, ripples, wind streaks, erosion (ventifacts), dust devil tracks. These can be used to map and study changes in local prevalent wind. Interestingly such models have shown indications of climatic change, and have been interpreted as a result of variations in the Martian orbit [54].

3.4. Analysis of the Dust Magnetic Properties Experiments on the MER Rovers

Detailed study of atmospheric dust on Mars was carried out by capturing magnetic dust grains using permanent magnets [55]. The captured dust has been observed optically (digitally using optical filters) as well as with chemical (Alpha Proton X-ray Spectroscopy [56]) and mineralogical (Mössbauer Spectroscopy [57]) analysis.

In the following section the extensive magnetic properties experiments which have been performed studying Martian dust will be presented. Analysis of some of this data will be made using a combination of wind tunnel simulations and computational fluid dynamics (CFD). Generally good agreement has been found, revealing interesting and useful information and showing that this approach of joint laboratory simulation and modeling can be a powerful tool for interpreting planetary data. Especially where new phenomenology must be characterized. It is however time consuming and requires careful and close collaboration during the experimental and modeling processes.

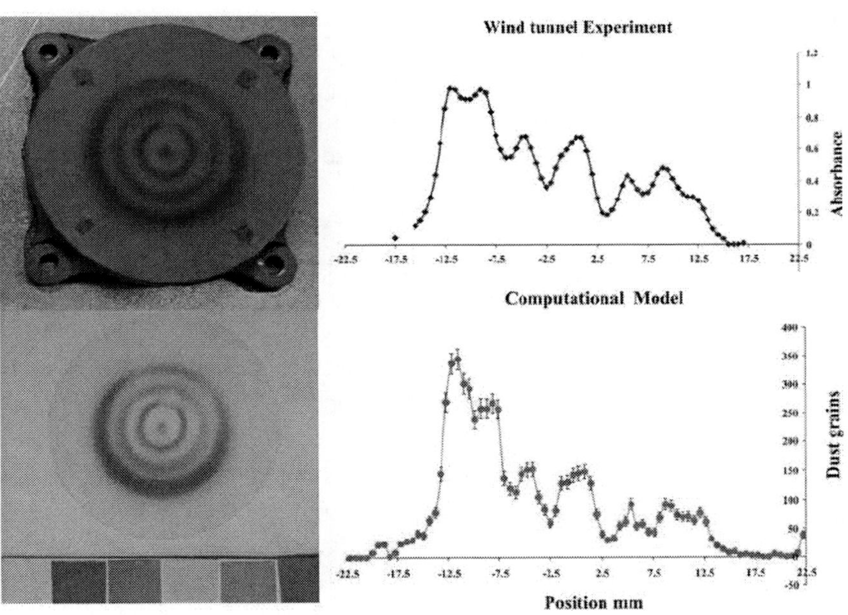

Figure 8. Top left shows a photograph of a MER capture magnet having accumulated dust in a wind tunnel test, a thin plastic film covering the magnet is removed while retaining the dust (bottom left). Top right shows the light absorbance measured from this film taken with a spectrometer and bottom right shows a CFD model result from the same magnet capturing ‘virtual’ dust grains.

Since the Viking missions, permanent magnets have been used to investigate the magnetic properties of the Martian dust, both Mars Pathfinder and the recent Mars Exploration Rover (MER) mission [55,58] also employed permanent magnets to capture magnetic dust carried by the wind. Experiments in a simulated Martian environment [4,59] have made it clear that the amount of dust deposited on such a magnet is strongly dependent on aerodynamic factors such as wind speed and pressure and should be performed under the most realistic conditions in order to be used for interpretation of data from magnets on Mars.

Furthermore it is necessary to understand how dust deposition depends on parameters such as wind velocity, dust grain size, magnetic properties of dust grains and the force of gravity. It was hoped to gain such an understanding by closely comparing experimental wind tunnel simulation results with computational fluid dynamics (CFD) modeling. Modeling offers the possibility of separating the different parameters and varying them in a highly controlled way, as well as monitoring the tracks of single dust grains.

Both NASA MER rovers carried a suite of permanent magnets designed to attract airborne dust for investigation by instruments on the spacecraft. Among these magnets were the capture magnet and the filter magnet, both placed directly in front of the camera mast, inclined at 45 degrees with respect to the horizontal. The capture magnet is designed to have a strong magnetic force close to the surface of the magnet, so that even weakly magnetic grains are captured. The filter magnet on the other hand is designed so as to attract highly magnetic grains more effectively than less magnetic grains, i.e. it 'filters' the magnetic dust.

During the mission multispectral images of these magnets were taken regularly using the panoramic camera (pancam) on the rover. Estimation of absolute dust deposition rates from such optical measurements is difficult due to, for example, varying camera angle, position of the sun or light scattering. For simplicity we concentrate on relative parameters such as the deposition on one magnet relative to the other and the deposition on the center of the capture magnet relative to total deposition on this magnet.

In the wind tunnel simulation Magnets, identical to those sent to Mars, were exposed to a dust-filled wind stream in a geometry reasonably close to the geometry used in the numerical model.

A thin (~0.2 mm) plastic film was placed on the surface of the magnet and after deposition of dust the transmission of the film was measured at 400 nm in an optical transmission spectrograph (Merrison et al. 2004a). From the measured transmission fraction T of the film we calculate the absorbance A as: $A = -\log(T)$. Subtracting the absorbance of the clean film we have the absorbance of the dust deposited at a given position, which is directly proportional to the amount of dust. Curves of dust deposition as a function of position on the magnet can thus be generated.

From the CFD calculations it is seen that three main factors determine dust pattern generation and deposition rates when capturing airborne dust on permanent magnets. These factors are wind speed, dust magnetization and characteristic time. In a qualitative sense we may moreover state that increasing grain magnetization has the same general effect as increasing grain size or mass density (i.e. the same effect as increasing characteristic time). This is understandable when we consider that stronger magnetic force and larger characteristic time both manifest themselves as a larger terminal velocity. We may also state that lowering the wind speed has the same effect on the pattern as raising the dust magnetization, at least in a qualitative sense by changing the velocity vector of the grains in the same way.

From the data for the dust pattern on Spirits capture magnet a central fraction of 0.23 was calculated. Clearly many different combinations of the three variables κ, v and τ can give the observed central fraction, however all of these variables can be restricted to some degree based on data from previous Mars missions. For the wind speed parameter 3 m/s was taken as a lower limit based on data from Viking showing wind speeds varying around 5 m/s [60]. Though at times the wind speed can be lower than 3 m/s the results will be dominated by the higher wind speeds since dust deposition is highest here. For the magnetization the value $\kappa = 1.03 \cdot 10^{-4}$ m^3/kg can be seen to be close to the upper limit based on the Magnetic properties experiments on Mars Pathfinder [58]. Taking v = 3 m/s and $\kappa = 1.03 \cdot 10^{-4}$ m^3/kg we calculate $\tau = 0.57$ ms, remarkably close to the value of 0.61 ms used in our standard case. Raising the wind speed above 3 m/s or lowering κ below $1.03 \cdot 10^{-4}$ m^3/kg would both result in a larger value for τ, so 0.57 ms becomes a lower limit. Taking dust grain diameters to be 3 microns [8], this results in a grain mass density of 1800 kg / m^3, consistent with the expectation that the grains are aggregates of lower density than the bulk density.

If the grains in the atmosphere have a wide distribution of sizes, the magnets will preferably capture the larger grains giving a mean diameter greater than 3 microns for grains captured on the magnets. In this case our estimate of grain mass density will need to be correspondingly lower, giving further support for dust aggregation. This set of values ($\kappa = 1.03 \cdot 10^{-4}$ m^3/kg, v = 3 m/s, $\tau = 0.57$ ms) is also consistent with twice as much deposition on the capture magnet as on the filter magnet as was observed on the MER experiments on Mars. Though wind speed and grain magnetization were not independently measured on the MER mission, these values are consistent with Viking wind speed measurements, and Mars Pathfinder magnetic properties results. Based on dust grain size estimates from both Viking, Pathfinder and MER this model also provides an indication that dust grains on Mars, as is the case for wind tunnel simulations, are aggregates with somewhat lower density than that of the bulk. While we cannot be sure that this set of parameters precisely matches conditions at the MER Spirit landing site, we have seen that this model can reproduce the observed dust deposition and it allows constraints to be placed on the physical parameters of interest (κ,v and τ).

4. In-situ Measurements and New Instrumentation

During the operation of the Mars simulation wind tunnel many parameters affect the experimental conditions such as: wind speed, turbulence, dust concentration, deposition rate, dust electrification. Knowledge of such parameters are crucial to the performance of quantitative and reproducible measurements and it has been necessary to develop new measurement techniques and instrumentation in order to determine them. Measurement of the same parameters is also crucial for understanding, modeling and quantifying the Martian surface conditions and for studying dust and sand transport. It has therefore become natural to develop miniaturized instruments which can both operate within the wind tunnel and have the possibility for transportation to Mars.

Three separate instrument development programs have (and are) being carried out at Aarhus Mars Laboratory: 1) the study of wind flow and dust suspension [59], 2) quantifying the accumulation (deposition/removal) of dust [64] and 3) quantification of the electrical properties of suspended dust [11]. All three prototype instrument designs have utilized optoelectronics and have now been integrated into a single instrument design prototype called a Laser Anemometer and Martian Dust Accumulator (LAMDA). The instrument and supporting research work will be presented in this section.

Ideally such dust and wind instrumentation would perform simultaneous measurements at various locations on Mars over a seasonal cycle and complemented by other dust and wind flow analysis instruments, specifically there can be found prototype Mars measurement technology capable of determining: dust size/mass using impact sensors [61], sand impact sensors [62], UV sensor for dust/dust devil detection [63] and on the coming NASA Phoenix mission a LIDAR for determining dust loading profiles in the lower atmosphere.

4.1. Anemometry on Mars

Only a single type of anemometer has been tested on Mars, one which is commonly used on earth and is generally reasonably accurate and reliable. It is called a 'hot wire' anemometer and utilises the cooling effect of the wind on an electrically heated wire or foil to determine wind speed, specifically the change in electrical resistance is measured. This parameter is however made greatly less sensitive due to the low pressure on Mars and factors such as the environmental temperature, gas pressure and composition will affect the measurements. Though measurements have been made on Viking and Pathfinder missions, uncertainties in calibration and stability have made interpretation problematic. Mechanical anemometers again have the problem of reduced signal due to the low pressure. Sonic anemometers are becoming common on earth and are being developed for use on Mars, though here again they become insensitive at low pressure and again rely on direct contact with the atmosphere. Laser anemometers are widely used in fluid flow applications on earth and their application to Mars is discussed in the following section.

4.2. Laser Anemometer

Laser Anemometry has the benefit of being a non-contact technique, which is insensitive to properties of the fluid medium such as temperature, pressure or composition and has the great benefit, for use on Mars, of being capable of measuring the suspended dust concentration.

Laser anemometers measure the velocity of suspended grains flowing with the fluid, fine grains as discussed in section 2 will have the fluid velocity. The most common type of instrument relies on the Doppler effect which is the shift in frequency of the reflected light due to the motion. Specifically this shift in frequency is measured by projecting an interference pattern onto the grain surface, as it passes, and measuring the frequency of time varying light scattered back to a detector. A commercial instrument of this kind; a Laser Doppler Anemometer (LDA), was used in wind tunnel testing.

With a starting point in the desire to make measurements of similar quality to those performed in the wind tunnel with the LDA system, efforts were made to develop a simplified and miniature system which could eventually perform such measurements on the surface of Mars.

The use of solid state components allow such a system to become practical since they are light weight and have low power consumption. Generally such solid state laser and optoelectronic technology is extremely well suited for planetary studies especially of the transport of fine suspended material. Examples of the use of such technology can already be found on earth where the detailed study of aerosols (air borne grains) or colloids (water suspensions) can be made in a direct manner, accurately and often also cheaply. Specific applications range from air/water pollution, climatic modelling/weather and medicine (the ingestion of fine grained material).

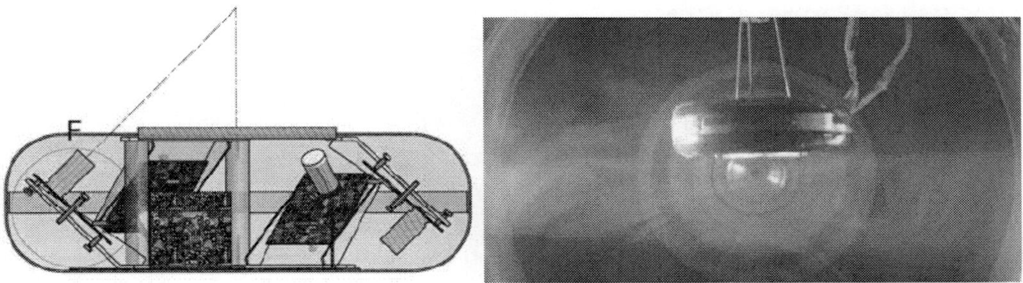

Figure 9. Technical drawing of the integrated Laser Anemometer and Martian Dust Accumulator (LAMDA) instrument (Left), a side cross section, and a photograph of the instrument under testing in the Aarhus wind tunnel under Martian simulated conditions (Right).

A time of flight system was chosen for the Mars laser anemometer, such a system has fewer demands on precision optics than the LDA and, with the (relatively recent) commercialization of holographic laser pattern generators, the optics become simple in design. Also the output signal from a time of flight system is easily interpreted, producing only a small data volume.

The design principle is the creation of a patterned laser light source, when a dust grain crosses this beam the pattern produces a time varying scattered light signal which is then collected using a Fresnel lens and detected using a photodiode. In the present design the light source consists of three separate laser diodes each with a collimating lens and a specially designed/etched holographic pattern generator. The pattern generator creates a cross sectional pattern to the beam of three lines approximately 0.2mm separation and 1mm broad. The three light beams are angled at 60° to one another such that the relative time separation between signals gives both the wind speed and angle (0-180°).

Tests have been performed in the Aarhus Mars simulation wind tunnel at wind speeds of around 1-20m/s and at low pressure air (9 mbar).

The measurement region is around 30mm above the surface of the Fresnel lens and the lasers are alternately (at present around 300ms each) illuminated (pulsed) so as to alternate measurement angle. Results from the CFD calculations described earlier have demonstrated that over the wind speed range required this measurement volume is outside the self induced turbulent boundary layer created by the instrument.

Calibration of the wind sensor has been verified with the use of the commercial Laser Doppler Anemometer (LDA). A major benefit to this wind sensor is the ability to determine dust grain concentration simply by counting the trigger signals and knowing the geometry of the laser beam/detection volume. Such direct measurement of dust concentration close to the surface would be invaluable in the study of the transport of surface material on Mars. They will also be unique since the current knowledge of Martian atmospheric dust is averaged over the entire atmospheric column.

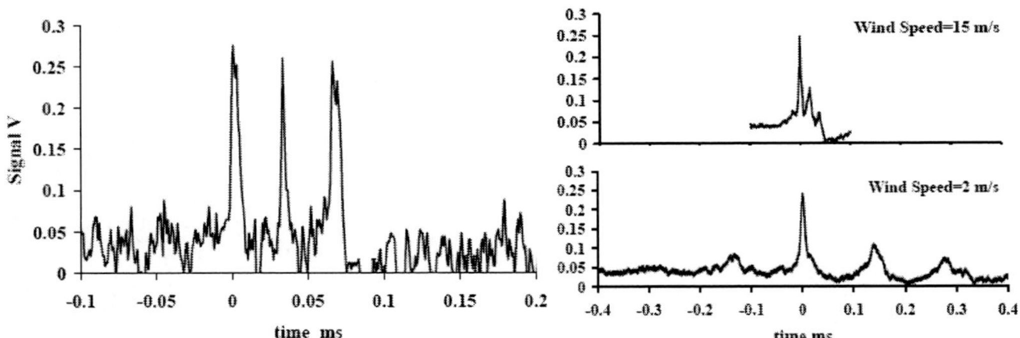

Figure 10. Testing of the Laser Anemometer showing on the left the raw signal of the detection of a dust grain and the three laser light reflection signals, on the right are two integrated signals (over 30 measurements) at high wind speed (upper right 15m/s) and lower wind speed (lower right 2m/s), the line spacing quantifies the wind velocity.

This instrument has proved to be more sensitive to the fine Martian dust analogue (Salten Skov dust) than the LDA especially at higher wind speeds.

The instrument presently collects data using an oscilloscope which is gated by the laser pulse and triggered by a dust grain detection, wind speed can be determined by integrating/averaging over many dust grain events thus gathering good statistics.

4.3. An Integrated: Laser Anemometer and Martian Dust Accumulator

Wind tunnel testing of the latest prototype of the laser anemometer discussed in the previous section will be described here. In this prototype the anemometer has been integrated with dust analysis instrumentation [64] which will be described with experimental test results in the following section.

The instrument consists of six laser light sources in total. Three of the laser systems on one side of the instrument contain holographic beam pattern generators which convert the beam into a three lined light intensity pattern and as described were used for wind speed measurement. The three non patterned laser beams are directed through the other side of the instrument.

All six laser beams have a photodiode mounted beside them which collects light scattered by the laser beams as they pass through the instrument surface. These surfaces are coated with a transparent conducting electrode (Indium oxide) which can be electrically biased by a high voltage supply in order to attract/repel electrically charged dust and thus quantify the amount and extent of dust electrification. The output from these six photo detectors are collected into a single 'dust accumulation' signal. The six lasers are pulsed in sequence in order to periodically sample each.

Below is collected a list of the parameters measured by this instrument with a short explanation and some experimental simulation data.

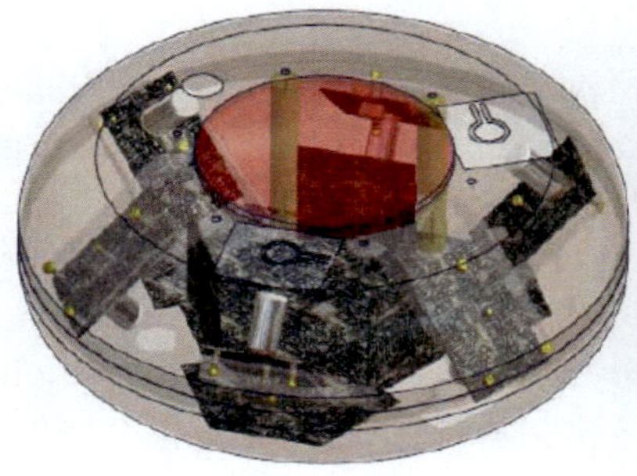

Figure 11. Technical drawing (Left) of the LAMDA instrument showing the outer transparent electrodes and inner laser and detector for determining dust accumulation. Also seen is a photograph of the instrument after wind tunnel testing showing the greatly enhanced (red/orange) dust accumulation on the electrodes (positive and negative).

Wind Speed

A wind carried dust grain crossing the one of the three patterned lasers will return a signal consisting of three peaks, the spacing in time of these pulses is a measure of that component of the wind speed (dust grain velocity). Wind speed has been measured as the fan rotation frequency was increased in the wind tunnel and was compared to the wind speed measured using a commercial LDA instrument. The wind speed measured by the LAMDA instrument shows good linearity and the calibration of the laser line spacing 0.27mm is close to that expected from the manufacture of the laser pattern generator of 0.2mm.

Wind Direction

The three patterned lasers are oriented at different angles such that comparing the relative wind speed measured during each laser pulse gives a measure of the wind angle (0-180°), though not strictly the direction since the sign of the angle is not determined. Figure 13 shows measurements of wind speed as the instrument was rotated with respect to the wind direction, the data has been fitted to the expected cosine behavior.

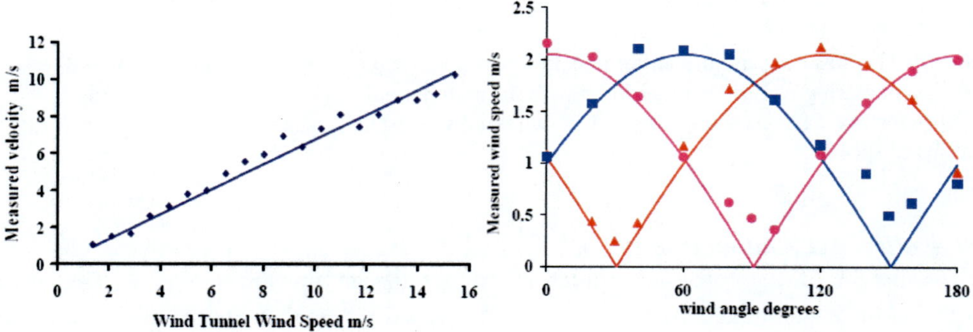

Figure 12. On the left the wind speed is measured by LAMDA at 30° to the wind (assuming a line spacing of 0.2mm) compared to the measured value with the commercial LDA, the line fit gives a calibrated line spacing of 0.27mm. Varying the wind angle alters the measured wind speed determined from each of the three lasers as shown on the plot on the left and fitting to a set of cosine curves (with offset of 60°). This demonstrates that the wind angle can be determined from the three laser measurements.

Wind Turbulence

Importantly performing a series of wind speed measurements gives turbulent induced variations in the wind speed, the rms (root mean square) variation in the measured wind speeds of an individual wind component will give a measure of the wind turbulence. This degree of turbulence is an important parameter for modeling and/or for comparison of grain transport in different environments, for example between simulators or between simulations and field measurements (on Mars). Quantification of the turbulent RMS has been previously measured using the wind sensor prototype and compared well with values measured using the LDA [31]

Dust Concentration in the Atmosphere

The trigger rate of dust grains crossing the detection area of the wind sensor ('wind speed' signal rate) will give a measure of the dust flux and thus the dust concentration in the wind flow. In the first laser anemometer prototype it was shown that this instrument had better detection efficiency than the commercial LDA system, meaning that lower dust concentrations could still be quantified [31]. The latest (integrated) design has improved laser pattern design and extra detector amplification and has given even greater efficiency, more than an order of magnitude greater than the LDA.

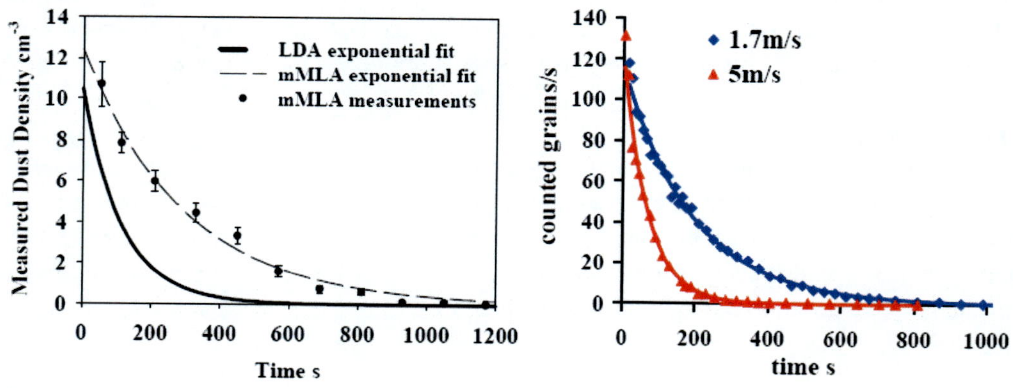

Figure 13. The dust concentration after injection was measured with a prototype laser anemometer (mMLA) comparing well to that measured with the commercial LDA (left) On the Right the concentration is measured using LAMDA at two wind speeds, as expected the dust decys faster at higher wind speed.

Dust Settling/Accumulation Rate

The height of the 'dust accumulation' signal for an upward facing surface gives a measure of the accumulated dust (gravitationally) deposited on that surface, the time variation of this signal gives the deposition/removal rate of dust on that surface. Dust deposited on the downward facing surface can quantify the amount of wind deposited dust. These parameters are of great technical and scientific interest to understanding/modeling dust transport, dust properties and protecting against dust build-up. See figures 11 and figure 14.

Dust Electrification

By comparing the dust accumulation rate of positively and negatively biased electrode surfaces with an unbiased surface the percentage, sign and amount of electrification of the dust can be quantified. This physical parameter is completely unknown on Mars though of tremendous importance to the behavior of the dust for example: adhesive effects, aggregation, electrical transport in the atmosphere, possible dust removal systems, etc. As shown in figure 15 (and figure 12) this dust accumulator prototype can attract electrified dust and show increased dust accumulation. Quantification of this is presented in the next section.

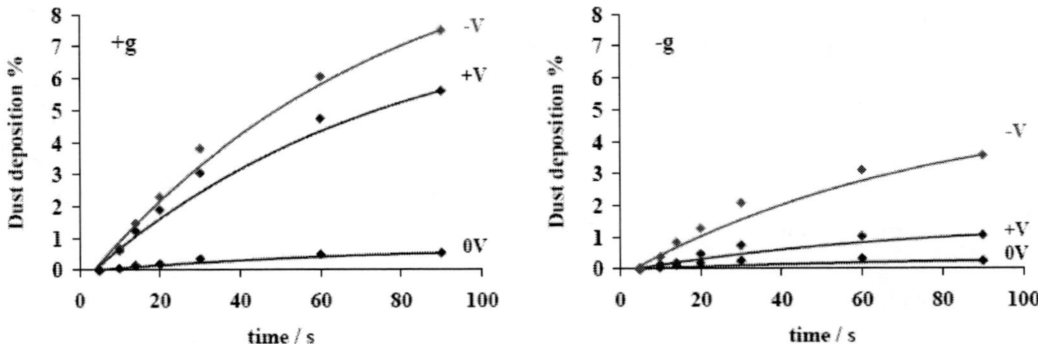

Figure 14. The dust accumulation is measured using LAMDA as a function of time after injecting dust, again the exponential decline in dust concentration (accumulation rate) is seen. Clearly more dust is accumulated on the upward facing surfaces due to gravity (dust settlement), and a large enhancement in dust accumulation is seen on the electrodes (+/- 392V) compared to that kept at 0V.

Dust Grain Size (Optical Area)

This parameter is not within the original scope of this instrument, but the amount of returned light to the wind sensor photodiode i.e. the pulse height of the 'wind speed' signal should be correlated with the size of the dust grain, specifically its surface area as seen by the instrument. This property has yet to be verified. The size (and distribution) of the dust grains is a fundamental physical parameter which is only known from indirect measurement of light scattering integrated over the atmospheric column and averaged over variations from grain to grain. An instrument has been proposed for dust grain sizing on Mars for the ESA ExoMars mission [61]

Although most of the results presented here have, for convenience been performed at room temperature and (low pressure) air, this prototype has been tested at low temperature (-30°C) and in a (7mbar) CO_2 atmosphere.

Though the operational parameters of this prototype have been demonstrated under simulated Martian environmental conditions, it remains to be integrated into a mission design including flight qualification and software/firmware development for interfacing and data treatment. It is hoped that this instrument may complement other dust analysis instrumentation and wind sensors on ESAs planned ExoMars lander (2011).

4.4. Dust Accumulator/Dust Electrification

Following observations consistent with dust electrification on Mars and in wind tunnel investigations, both at Aarhus [11] and MARSWIT [10], a series of investigations were made of Martian dust analogue electrification. A similar strategy was employed to that of the magnetic properties experiments, here dust was attracted onto the surface of an electrode placed in the wind stream. Optical analysis of the surfaces could then reveal the dust accumulation (rate) and therefore an estimate could be made of the electrification of the dust. Furthermore by quantifying the reduction in dust accumulation at low applied electric fields the percentage of electrified dust (positive and negative) could be obtained. In the first series of studies removable plastic films were used which were then optically analyzed by absorption of light using a spectrometer. It was found that a major fraction (90±16%) of the dust had a net electric charge, with almost equal quantities of negatively and positively charged grains. These grains were estimated to carry sufficient electrical charge to dominate the processes of adhesion and cohesion (around 10^5e per grain). There was also a great deal of evidence from this study for the formation of larger, more neutral dust aggregates while in suspension.

This work stimulated the hypothesis that wind displacement and break-up of such dust aggregates could be responsible for entraining dust on Mars.

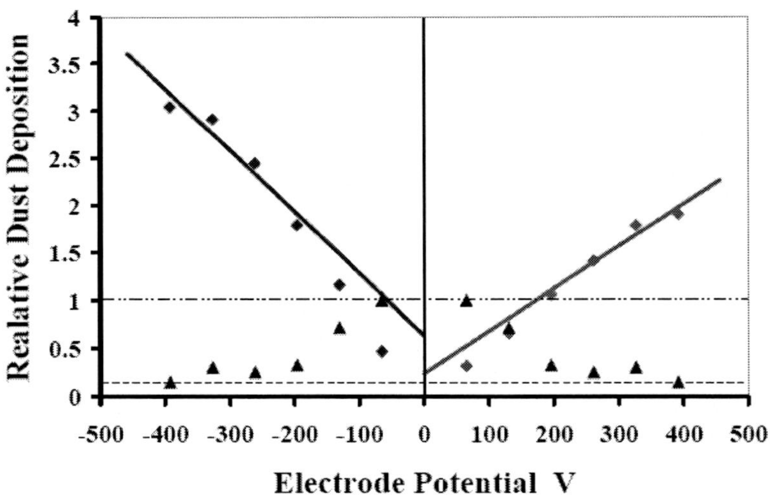

Figure 15. Plotting the relative dust deposition measured by LAMDA as the electrode voltage is varied qualitatively agrees well with previous studies. Crudely extrapolating the high voltage measurements gives the amount of repelled dust fraction and shows that 37(±13)% of the dust is negatively charged and 76(±7)% positively charged grains, i.e. that >93% is electrified. Note that the dust accumulated by the (downwind) zero volt electrode (black triangles) falls as the other two electrode potentials are raised, this indicates that around 85% of the dust was swept away at the highest applied electric field.

The results obtained from the LAMDA prototype instrument, which uses a laser/optoelectronic system for dust accumulation measurement, are shown in figure 16. They agree well with these previous electrification studies, showing a similarly high degree of electrification (also using electric fields of up to 10kV/m) with both positive and negative charging. Also a distinct difference between the dust accumulation observed for upward and downward pointing surfaces was seen. As discussed in section 3.4 modeling of dust accumulation with respect to the direction of gravity (+/- g) could reveal information about the dust physical properties (size and mass).

Extrapolation to low applied voltages showed that the deposition rate was lower than the unbiased electrode due to the deflection/repulsion of dust with the same charge as the electrode, also in agreement with the previous study. From this (see figure 15) the negatively charged dust was found to make up 37(±13) % of the dust and 76±7 % was found to be positive, in reasonable agreement with previous results of (44±15) % and (46±6) % respectively, though there is a pronounced bias towards there being more positive grains. Interestingly though the negative electrode shows a steeper gradient of dust accumulation with electric field, possibly indicating either larger negative grains or a higher degree of electrification. In this study over 93% of the dust was seen to be electrified. It was also observed that the unbiased electrode, which was downwind of the biased electrodes, collected decreasing amounts of dust as the applied voltage was increased (see figure 15), falling to a value of less than 15% of its original value. This (new observation) is consistent with a dust sweeping effect due to electrostatics and shows that a large fraction (over 85%) of the dust can be swept/retarded away. This could have great relevance to electrostatic systems for dust removal/prevention being developed, as will be discussed in section 8.

5. Transport of Martian Surface Material

Although wind is at present the most active process for modification of the Martian surface, in some cases causing changes on the time scale of days and weeks, it appears to be due to dust entrainment and deposition and that active sand transport has ceased. Observations of aeolian (wind related) processes include wind abraded rocks, ripples and drifts. Dunes, wind blown deposits and erosion features are also seen from orbit. However none of these features described have been observed to change (form, shape or position), also they appear to consist of a mixture of grain sizes and many even have a (mm thick) encrusted layer [65,83]. These observations argue against active saltation. This is somewhat paradoxical since it is expected from earth based studies that sand grains (of around 100 microns diameter) are entrained with the lowest wind speed (wind turbulence), significantly lower than the fine dust.

The motivation for the work described here was to help explain this seeming paradox. The observations of high electrification of Martian dust and subsequent aggregation of dust on deposition has led to the suggestion that movement/break-up of dust aggregates could be a possible mechanism to explain dust transport on Mars, specifically the formation of sand-sized dust aggregates with low mass density and therefore low threshold for detachment and subsequent transport. Additionally the study should complement earlier simulation work, performed in the MARSWIT wind tunnel, of sand entrainment which might enable further limits to be placed on the winds/pressures required for sand transport. As well as benefiting understanding of the present climate on Mars this could help to investigate climate change.

The experimental technique employed in the work described here involved the deposition of granular material, in a specific pattern, and observation of removal of this material by imaging using a digital camera. The wind speed/pressure were increased until the threshold for removal was exceeded. This differed from previous studies of the entrainment threshold wind speed in which the onset of saltation was observed, which would not include creeping/rolling of grains.

Note that saltation is the process responsible for ripples, dunes and the active sorting of grain sizes. Removal or detachment of grains is a pre-requisite for saltation, hence the detachment process described here should be considered a lower threshold value for saltation. However, once active saltation is initiated grain impacts 'seed' the process, leading to lowering of the threshold wind speeds (by around 20%) [66].

As well as using Mars dust analogue material (Salten Skov) and sand, the removal of glass spheres was studied some of which were hollow, allowing the weight of the grains to be lowered while maintaining the same surface properties. In this way effects of changing mass density or gravity could be studied.

Comparison of results between different wind tunnels and the surface of Mars can be difficult due to differences in wind flow and importantly turbulence. Parameters such as friction wind speed or shear stress can be difficult to measure directly (in simulation and in the field) and knowledge of the geometry/surface roughness is required for comparison using 'free' wind speed measurements [42].

5.1. Grain Transport: The Application of Detachment Theory

Grain transport theory on earth is a relatively mature and advanced field of study both experimentally and theoretically. However in most specific cases of studying the transport of grains a semi-empirical treatment is adopted since many of the necessary physical parameters are not specifically known or are dependent upon poorly controlled environmental factors for example surface cleanliness, surface roughness, humidity and so on. Applying this earth based work to the surface of Mars is therefore a non trivial exercise which should best be performed with a combination of experimental simulation/observation and theory.

In the detachment of fine grained materials (micro to mm scale) due to fluid flow a method developed by Cleaver and Yates [67], called the force balance approach, is often used in which the forces acting on individual grains are equated in order to quantify the threshold condition. In the case of spherical grains resting on a flat sheet or spheres resting on other spheres, four (effective) forces can be identified when describing the onset of movement. The forces of gravity (F_g) and adhesion (F_{adh}) resist movement, while the fluid drag force can induce movement in several ways which can be broken down into the actions of an effective lift force (F_L) and moment of torque (F_T) i.e. a rolling action due to shear stress. Shear stress may also cause sliding in some geometries though this is unlikely unless smooth surfaces are used. When these forces balance the onset for possible detachment is reached: i.e. $F_L+F_T=F_g+F_{adh}$, such that detachment threshold occurs when wind induced forces equal adhesive/gravitational forces.

With the exception of gravitation these forces are only known in an empirical or at best semi-empirical way such that it is not possible to quantify them for this study (or on the surface of Mars) without performing further experimentation. However of most importance to this investigation is the dependence of these forces on grain size, wind friction speed and fluid density and based on previous experimentation and/or theory it is possible to extract relevant expressions which can be applied. These forces will be discussed individually in this section, briefly summarizing the relevant conclusions:

Gravitational Force

Although the force of gravity is well described theoretically, $F_g=\pi d^3 g\rho_g/6$ since the mass density (ρ_g) is not known precisely and that the gravitational field is weaker on Mars ($g=3.8m/s^2$), it is difficult for direct comparison of experimental simulation and observation. For earth simulations this would require changing, for example, the material density without effecting any of its other properties specifically surface properties (composition, morphology, roughness etc..). Also there is lacking a detailed knowledge of the mass density of (all) surface materials on Mars.

As discussed in the introduction hollow glass spheres have been used in this study in order to vary the mass density while maintaining similar (if not identical) other physical conditions. Three different micro-glass spheres were used with bulk densities of: 2.7, 0.6 and 0.15 g/cm^3. The first corresponds to common quartz sand grains on earth, while the second would correspond to 1.6g/cm^3 in the lower Martian gravity, (somewhat low in mass density, though possibly still indicative of soil material). The latter glass sphere would correspond to 0.4g/cm^3 under Martian gravity which, as will be described in section 5.5, corresponds well to measurements of dust aggregate mass density determined from extraction from the wind tunnel.

Adhesive Force

Despite the large body of work which has been performed in studying adhesion, even on earth many questions remain and most applications involving adhesion become semi-empirical at best and measurements of specific adhesive properties generally show a great deal of variation probably reflecting that the process is not simple and dependent upon parameters which are not usually well controlled for example environmental factors such as humidity and surface cleanliness or of the detailed physical nature of the materials on a micro/nano meter scale.

Two electrically neutral surfaces in close proximity over a limited close range produce a self induced attractive force (Van de Waals interaction). Most theories of adhesion consider only this type of force and are derived from Hertzian theory. However other adhesive behaviour involving the exchange of electrons, discussed in relation to electrification, might be expected to have a similar dependence on grain size as it will also be related to an effective area of interaction/contact. Humidity (surface water) can also affect adhesion, though again it can in most cases be treated as a modification of the interaction area.

Numerous studies have been conducted looking at the adhesive force of spherical bodies. Bradley (1932) measured the force required to separate two rigid spherical bodies of radius d_1 and d_2 and found it to be [68]:

$$F_{adh} = \pi w d$$

Where $d=d_1d_2/(d_1+d_2)$ is the equivalent radius of curvature and w is the work of adhesion derived from the surface energies of the two bodies and the interface energy, in this case this parameter is not known.

Derjaguin, Muller and Toporov (DMT theory) considered also adhesive force outside the contact area and were in agreement with the form of Bradley's relation.

It is sufficiently useful for this analysis to apply an adhesive force of the form given i.e. $F_{adh} = C_{adh}.d$ where C_{adh} may be an adjustable fitting parameter.

Lift Force

As discussed in section 3.1 the expected form of the drag force and possibly therefore the lift force, is a (weak) power law dependence (n) on the Reynolds number ($R=u_*d\rho/\sigma$), u_* here is the friction velocity, specifically:

$$F_L = (\sigma^2/\rho)\ C\ R^{2-n}$$

Where C is a constant,σ is the molecular viscosity and ρ is the atmospheric mass density.

Experimentally Hall (1988) has measured the lift force for mm sized spherical grains and found a dependence [69]:

$$F_L = (\sigma^2/\rho)\ 20.9\pm1.57\ R^{2.31} \qquad 1.8 < R < 70$$

of more relevance to Martian sand transport was the study carried out by Mollinger and Nieustadt (1996) using 120 μm diameter particles glass spheres and 218 μm plastic spheres to measure the mean and fluctuating lift force, they obtained and empirical expression [70]:

$$F_L = (\sigma^2/\rho)\ 56.9\pm1.1\ R^{1.87},\ F_L = (\sigma^2/\rho)\ 33.4\pm3.0\ R^{1.98} \qquad 0.3 < R < 2$$

Comparison of this experimental work with the expected dependence of reveals that the power law factor n is measured to be close to zero, giving an expression for the lift force of, where C_L is a fitting factor:

$$F_L = C_L\, \rho^{1-n}\, u_*^{2-n}\, d^{2-n}$$

where n≈0

Rolling Force (Torque)

The motion of a sphere immersed in a fluid flow has been studied extensively and is well described theoretically, also for the case of flow close to a boundary wall. In this case as well as the translational force of drag the sphere experiences a torque due to gradients in the flow field. The simple expression derived for this case is also assumed to be valid for the case of grains in contact (ref) where it is this torque which is responsible for causing the grains to roll. The expression for this force will take the form of the expression for a drag force multiplied by a length scale which can be related to the diameter of the sphere, hence, one obtains:

$$F_T = C_T\, \rho^{1-n}\, u_*^{2-n}\, d^{3-n}$$

As before there is also be a (n) dependence of the drag coefficient on diameter (Reynolds number).

Lacking relevant experimental investigation of this force and the poor quality of calculated values for the torque-moment coefficient [20] the above expression may be taken as a guide for the expected grain size/wind speed dependence.

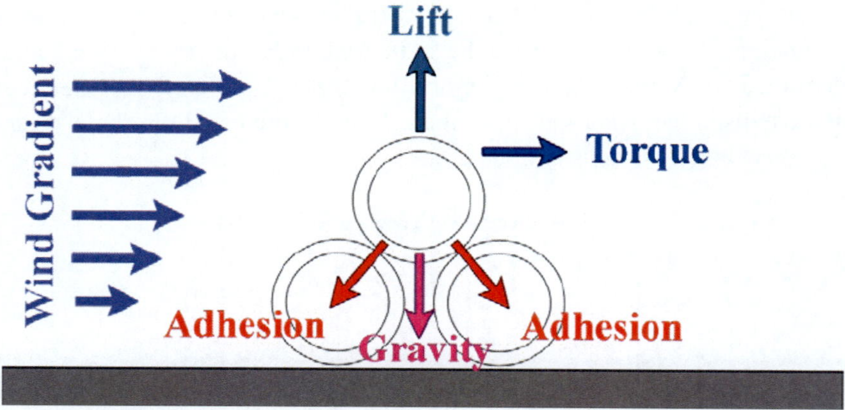

Figure 16. Schematic showing the forces applied to a spherical object (resting on similar spheres) on a flat plane and being subjected to the effects of wind flow which increases in speed (from zero) away from the surface.

In the case of both the lift and torque forces the power law dependence (parameter n) will be taken as the same (assumed) and will be obtained through fitting to experimental data.

Having obtained the form for these force factors the relation to the threshold condition can be formulated. Unfortunately in these wind tunnel investigations it has not been possible to achieve detachment for all grains within the limited range of wind speeds presently achievable and it has therefore been necessary to also increase pressure. In this case instead of determining the friction velocity, the threshold shear stress has been used as the measurement parameter even though it also has an unknown dependence on Reynolds number:

$$ShearStress \equiv \rho^{1-n} u_*^{\,2-n} \approx \frac{\frac{\pi}{6} g \rho_g d^3 + C_{adh} \cdot d}{C_L \cdot d^{2-n} + C_T \cdot d^{3-n}} \qquad \text{Equation 1}$$

This will be the form of the fitting functions.

5.2. Sand Transport

As discussed there is observational evidence for aeolean sand transport on Mars, though at present such transport does not seem to be active or occurring only at a low rate/frequency. It would be useful to be able to have simulation experiments and modeling upon which threshold surface shear stress could be predicted and compared to measurements from Mars to evaluate whether (and how much) sand transport should occur or under what change in climatic conditions sand transport would be possible.

A previous study of entrainment threshold (friction) wind speed was made in the MARSWIT simulator, with a semi-empirical fitting procedure. As seen in figure 17 the results predict a minimum threshold shear stress for solid sand grains of around of 115μm at a wind friction velocity of around 2m/s. This should be achievable with wind speeds of >25m/s assuming a surface roughness of z_0=10cm and where the wind speed (on Mars) was measured at around 1m height. This range overlaps the, at present, rather poorly confined range of wind speeds determined for Martian dust devils of 20-35m/s and indicates that sand could (or is close) to being transported. Also shown in figure 17 is the predicted threshold shear stress for dust aggregates (with a mass density of 0.4g/cm^3).

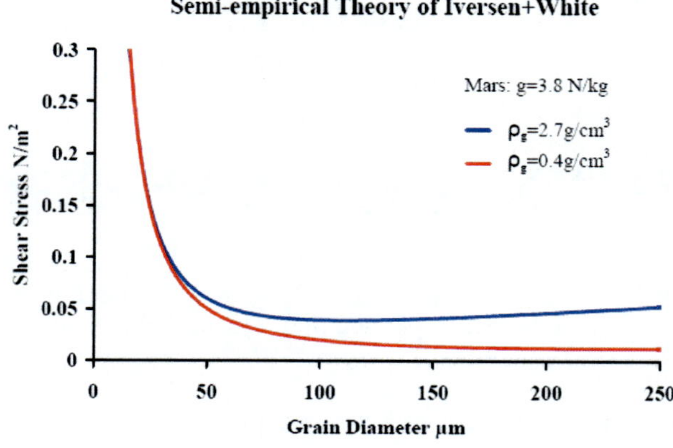

Figure 17. Using the semi-empirical equation [20]:Shear stress= $A^2/\rho g d$, where threshold parameter $A=0.129(1+0.006/\rho g d^{2.5})^{0.5}/(1.928\ R^{0.092}-1)^{0.5}$, here R is the Reynolds number at threshold, note that this formula is only valid for R between 0.3 and 10.

It was decided to undergo a systematic investigation of the effect of reducing the mass density and varying the grain size on the detachment threshold wind conditions by using glass bubbles (i.e. hollow spheres) of different effective mass density. Here the physical surface properties (hardness, smoothness uniformity etc.) of the glass will hopefully be similar while the mass density is reduced due to the presence of a hollow interior. It was in this way hoped to improve on previous studies where different materials, and therefore also different adhesive and morphological properties, were studied [20]

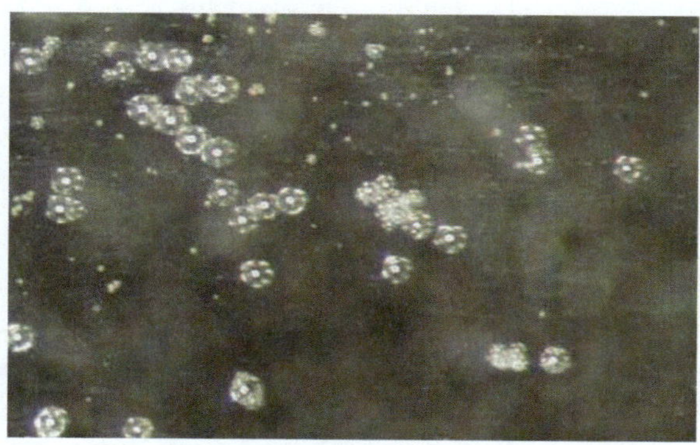

Figure 18. A microscope picture of hollow glass spheres (provided by the company 3M) of around 70μm diameter.

Ideally the friction velocity should be determined from the gradient of the wind velocity measured (very) close to the surface. This however is difficult to measure (locally) in the wind tunnel and measurements are not available from Mars. As the viscous boundary layer ends, the turbulent layer begins and in this region measurement of the turbulent component of the wind can be used as a reasonable approximation to twice the friction velocity [71,72]. Importantly also the turbulent wind component (RMS) is also (with most anemometers) a measurable quantity on Mars and so if having no other role it can be used to compare wind tunnel tests to measurements from Mars. It is known that, due to boundary effects, the turbulence in wind tunnel testing should be higher than on Mars, though this depends on the local geometry (for example surface roughness) at the site.

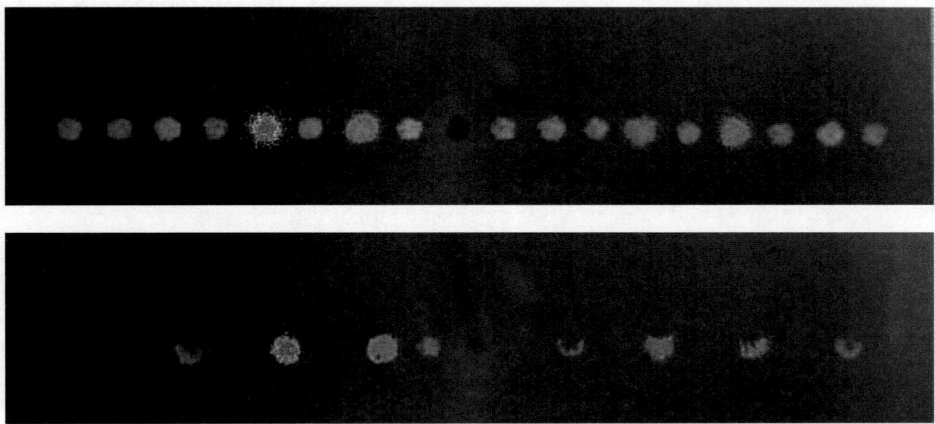

Figure 19. Photographs showing glass spheres (and a dust analogue sample) in the wind tunnel before testing (top) and during wind removal (bottom). The substrate was emery paper (aluminium oxide 35micron grain size). Of relevance to the next section on dust transport note that the Salten skov dust sample has been partially removed leaving a distinctive 'streak' of presumably smaller grained material which was not removed at this wind shear stress.

The experimental procedure consisted of forming discrete circular deposits of grains (glass spheres and bubbles) and photographing them (digitally) from above while the wind speed/pressure were increased. The picture frames were analyzed using a modified IDL based software and determining the relative reflected light intensity of the circular form relative to the background and in this way determine the removed fraction of grains. The threshold shear stress value was taken for which 50% of the grain material was seen to be removed.

As mentioned the shear stress has been used as a measurement parameter. As seen from equation 1 the shear stress also has a non linear dependency (n). A series of detachment threshold shear stress experiments were carried out to determine this dependency by varying wind (friction) speed while keeping pressure constant as well as varying the pressure for constant wind speed. Plotting curves one can vary the power law dependence of the shear stress (n) such that the data sets coincide. As expected a small value for n gave the best agreement (around 0.1) in fact within the experimental uncertainties it was not possible to distinguish from n=0. This has been used as justification for the simplification of taking n=0.

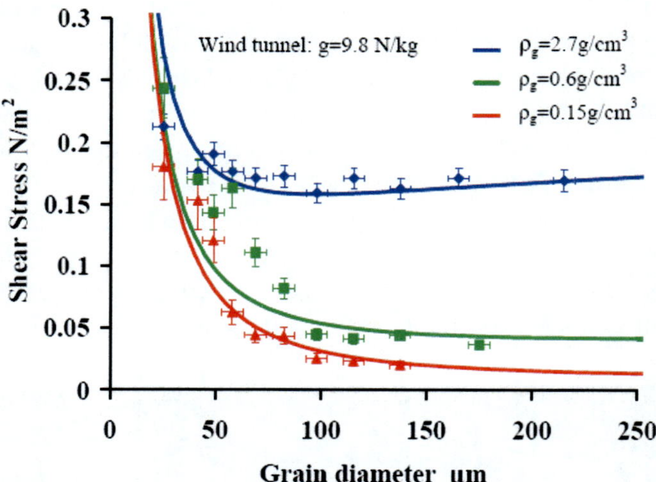

Figure 20. Shows a compilation of measured detachment threshold shear stress values for the three different glass sphere mass densities. This data has then been fitted to equation 1 to find the various force parameters. Again a small value of n was obtained, optimal value of n=0.07.

As seen in figure 20 the measured threshold shear stress for the different mass-density glass spheres were fitted to equation 2 and gave the following coefficients for the various forces:

Adhesion coefficient $C_{adh} = 2.5\times10^{-5}\ N/m$
Lift force coefficient $C_L = 1.3$
Roll force coefficient $C_T = 4.1\times10^{4}\ m^{-1}$

$$ShearStress \equiv \rho u_*^{\ 2} \approx \frac{\frac{\pi}{6} g\rho_g d^3 + C_{adh}\cdot d}{C_L\cdot d^2 + C_T\cdot d^3} \qquad \text{Equation 2}$$

Where ρ_g is the grain mass density, d is the grain diameter and u_* is the friction velocity which is approximately twice the turbulent wind speed close to the viscous boundary layer.

Careful comparison of the form of these detachment threshold results with the dependence of saltation threshold (figures 20 and 17) it can be seen that there is a less pronounced minimum at larger grain sizes and for the lowest mass density the minimum threshold occurs for grains larger than 250μm. This can be understood by the significance of the torque factor which increases greatly with grain size ($\alpha\ d^3$) thus dominating at large d where it will give an essentially constant threshold shear stress, canceling gravity (also $\alpha\ d^3$). Observations from the wind tunnel tests support this interpretation, where large grains were not seen to saltate but roll/bounce during removal. This process may be analogous to creep. Quantitative comparison of the measured shear stress can be made by conversion into the effective wind friction velocity thus expected for these values on Mars (assuming 9mbar CO_2 pressure and -60°C temperature). These are shown in table 1 for grains of around 100μm, the values agree rather well with the previous saltation studies. This is slightly unexpected given that detachment should occur below the threshold for saltation and given the different techniques/conditions/materials used for these studies.

It would be useful to compare these results also to wind speeds observed on Mars, this is shown in table 1. The shear stress has been translated into a free wind speed assuming a surface roughness of (z_0=10cm), giving u_*/U=8%, where wind speed (on Mars) was measured at around 1m height. This gives values of U=23m/s for the sand transport threshold and U=15m/s for dust (aggregate) transport.

Table 1. Using the semi-empirical formula derived in equation 2 the threshold shear stress for detachment is calculated for 100µm diameter grains under differing gravity and mass density. In the case of Mars (lower two) this corresponds to (g=3.8 m/s^2) a pressure of 9 mbar, temperature -60ºC (atmospheric mass density of CO_2 = 0,0225 kg/m^3) for sand and dust aggregates

Threshold for transport of:	d=100µm	ρu_*^2 N/m^2	u_* m/s	U m/s
Sand in the wind tunnel	g=9.8 m/s^2, ρ_g=2.7g/cm^3	0.158	2.6	33
Dust in the wind tunnel	g=9.8 m/s^2, ρ_g=0.4g/cm^3	0.044	1.4	17
Sand on Mars	g=3.8 m/s^2, ρ_g=2.7g/cm^3	0.076	1.8	23
Dust on Mars	g=3.8 m/s^2, ρ_g=0.4g/cm^3	0.032	1.2	15

A series of investigations were performed into the effect of turbulence around obscuring objects, which could for example on Mars be rocks or stones. In these wind tunnel simulations cylindrical metal tubes were used of various diameter. The altered wind flow was studied using the LDA which measured wind speeds and turbulence around the obstruction. Compared to the empty wind tunnel generally decreased wind speeds and increased turbulence was seen close to the obstruction. Figure 21 shows a cross section of wind speeds and turbulence taken 40mm behind the cylinder.

Glass bubbles (0.15g/cm^3) were seen to be removed close to the obstruction at (free) wind speeds significantly lower than in the empty wind tunnel, typically by as much as 40%. Interestingly heightened dust deposition is also seen close to the cylinder again probably due to the increased turbulence transporting dust to the surface.

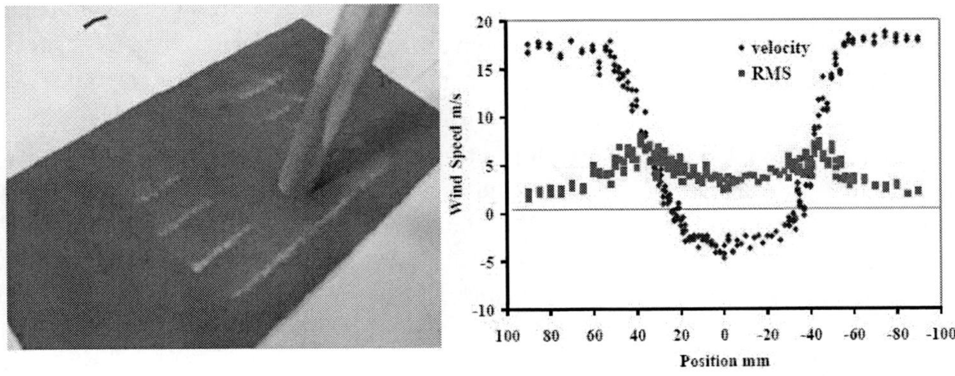

Figure 21. Shows a photograph (left) of a cylindrical obstruction after being in the wind tunnel showing removal of glass bubbles in a zone close to the object and interestingly increased Martian dust analogue deposition in the same area, also probably due to increased turbulence as shown in the measurements taken by an LDA of wind speed and turbulence around/behind such an obstruction (right) while in a wind stream under Martian conditions.

From this work it becomes possible to establish under what climatic conditions (pressure, wind) the observed aeolean surface features on Mars were made. Crudely extrapolating the semi-empirical formula to conditions on Mars (9mbar pressure) wind speeds in excess of 23m/s would be required. This wind speed is higher than the observed daily wind speeds (<15m/s [60]), but would allow for dust devils (with winds observed as high as 35 m/s [73]) to detach sand. Without knowing what wind conditions may occur during global dust storms it may be possible that here too such winds could be generated and sand transported. Comparison of the measured surface shear stress with predictions from global circulation models should make more detailed (quantitative) analysis possible, though still detailed measurements from Mars are lacking.

5.3. Martian Dust Transport

The majority of dust entrainment appears to occur in warm core vortices called dust devils. As discussed in section 2.2 and 2.4 both experimental laboratory simulations and field observations of dust devils have been made by several groups and some theoretical modeling also performed, although this work has given a consistent picture of the physical structure and occurrence of dust devils (at least terrestrially) for example: the pressure, wind flow and temperature gradients, there still lacks a detailed understanding of the physical process by which dust is entrained such that more accurate prediction/description of dust transport (rates) can be made.

Suggestions for entrainment processes have included saltation impact by sand [74], pressure gradient [30] and simply that sufficient wind speed (turbulent velocity) is created in such vortices. Generally, though sand transport would also be expected in these cases. Occasionally dust storms occur on a global scale, the origin and nature of these extreme climatic conditions are even more poorly understood.

Observations of many dust devil events were made by the Mars Pathfinder, of great importance here are wind speed measurements which were seen to peak up to 35m/s [73]. The observed pressure increase could also be used to calculate wind speeds (gradient) though found values closer to 15m/s [75]. As discussed earlier there were problems with calibration of the pathfinder hot wire anemometer, in comparison Viking measured typically wind speeds of 0-15m/s. These values are consistent with the threshold dust aggregate transport wind speeds determined in table 1 (section 5.2) and may also allow for gusting winds as well as dust devils to remove dust especially close to turbulence generating obstructions. One must conclude generally that more and better wind speed and turbulence measurements should be made on Mars, preferably a systematic study should be made.

Current robotic missions on Mars utilize solar panels and the effect of dust accumulation has been pronounced with significant reduction in power generation (around 0.3% per day). Dust devils were directly observed by the Mars Pathfinder NASA 1997 and associated dust removal was observed from the Magnetic Properties Experiment onboard the Pathfinder lander [76]. Unfortunately no anemometers were placed on the MER missions, though they have observed an abundance of dust devils and an increased power generation of one of the rovers, possibly indicating the passage of a dust devil.

5.4. Sand Electrification

Electrical charging of granular material can affect the way the grains stick to surfaces and to each other, affect their transport (detachment, saltation/suspension) as well as influencing their surface chemistry by the flow of ions and electrons. Environmentally electrification could cause electrical storms [77] and may affect the electrical behavior of the soil [78,79].

Even on Earth much has still to be learned about how much electrification of dust and sand occurs and what role this has on the physical properties and transport of the materials. When two materials/surfaces are removed from being in contact the transfer of electrical charge can occur, it can be referred to as contact electrification or tribo-electric charging (or even frictional charging), but it is essentially the same type of process. Though some studies of collision induced (tribo-) electrification have been performed and wind tunnel investigations have observed electrification of sand, these remain isolated empirical studies. Investigations under Martian simulation conditions led to the suggestion of dust electrification and subsequent aggregate formation.

Laboratory investigations have also shown that significant electrical charging of Mars analogue material (JSC-Mars-1, of around 100μm grain size) can be produced by collisional (triboelectric) charging [80] with surface charge densities of the order $2x10^{-6}C/m^2$ having been obtained. In a detailed study of small quartz grains impacting upon surfaces it was concluded that charging by collision (tribo-electric) is limited to the production of surface charge densities of $<1\times10^{-4}$ C/m^2. This appears to be in agreement with both theirs and other experimental results [81]. For dust this value is in reasonable agreement with the electrical charging studies presented of Salten Skov Martian dust analogue which gave around 10^5 per grain [11].

Even higher surface charge densities have been observed (in air) by chemically activating surfaces before performing contact-electrification. Here surface densities of $2\text{-}3mC/m^2$ have been observed [82]. In this study and possibly also generally, the electrification was limited by gas breakdown as the grains separated, in this case it will be highly dependent upon atmospheric conditions, for example humidity, pressure, composition, etc.

Despite this body of observational evidence for electrification of fine grained material the treatment of adhesion is still generally limited to Van de Waals effects and does not consider long range forces such as electrical attraction.

Observations of electrical charging of sand sized grains are easy to perform and there is little doubt that sand can become electrified, whether this has a significant effect on its transport may depend on whether net charging can occur due to variations in material or size. Experiments are planned at Aarhus to measure sand transport under applied electric field.

5.5. Dust Aggregation and Transport

As discussed in many of the previous sections there is a body of evidence from wind tunnel studies that electrification of dust leads to the formation of aggregates. In this section an investigation of the wind driven transport properties of dust aggregates will be made in a set of two complementary studies, one looking at the relative removal threshold wind speeds of sand and dust and the other simulating dust aggregate transport using low density glass bubbles.

Large dust aggregates (1-3mm diameter), as described in section 2 see figure 1, were collected after formation in the simulation wind tunnel. Their volume was determined by measurement under an optical microscope (2.37×10^{-3} cm^3). Their mass was then determined by weighing a collection of 20 such aggregates (0.9±0.3 mg). This results in an aggregate mass density=0.38±0.13g/cm^3. This is much lower than the bulk mass density of Salten Skov (around 3.5 g/cm^3) or even the mass density of the loose powder (around 1g/cm^3).

In the study presented earlier hollow glass spheres were used to simulate dust aggregates and the effect of the lower gravity on Mars (section 5.2). Using the semi-empirical expression (equation 2) derived from this work it can be seen that these 'simulated' dust aggregates were removed at significantly lower threshold shear stress than solid (glass/sand) grains. Taking grains of similar size (100μm) a reduction by around a factor of 3.6 (ρu_*^2(2.7g/cm^3)/ρu_*^2(0.4g/cm^3)) was seen (under wind tunnel conditions), see table 1.

In a separate study dust entrainment was studied by depositing a thick (around 60% coverage) of dust at a low wind speed of 1.4m/s upon a dust accumulator instrument similar to that described earlier in section 4.3, [64]. The wind speed was then increased while measuring the dust accumulation/removal rate. The wind speed and suspended dust concentration was also measured using the LDA. The threshold conditions were seen to occur for the Salten Skov dust analogue material at a wind speed of around 20m/s, measured turbulence level of 13% and pressure of 9.3mbar.

Using the same wind tunnel geometry the removal of sand (irregular grains of around 100μm diameter and mass density of 2.7g/cm^3) was performed allowing direct comparison to be made between the two tests. Assuming the sand grains and dust aggregates were of the same size the relative threshold shear stress for the dust aggregates compared to the sand grains was obtained: ρu_*^2(sand)/ρu_*^2(dust) = 8.9. This value can be compared to the ratio of 3.6 found above for glass spheres with densities corresponding to sand and dust aggregates. It seems that the real dust aggregates are significantly easier to remove than the glass 'simulated' dust aggregates (low density glass bubbles). One possible explanation for this could be the break-up of the dust aggregates, a process not available to the glass spheres. In this case it may have the effect of reducing the effective adhesive force by allowing fracture of the structure at the weakest point.

If adhesion is ignored and the difference in detachment threshold between sand and dust aggregates is due solely to the difference in mass density, given the mass density of the sand (around 2.6 g/cm^3) the mass density of the dust aggregates can therefore be predicted to be around 0.3g/cm^3. This value is in reasonable agreement with the measured value for dust aggregate mass density of 0.38±0.13g/cm^3. This is a crude analysis, but does seem to support this argument and these experiments clearly show that the dust transport threshold is substantially lower than that of sand and that the process is quantitatively consistent with what would be expected by removal of dust aggregates.

The breakup of dust aggregates is further supported by observational evidence from the wind tunnel of residual 'streaks' left by dust piles (not removed by high wind speeds) and the detection by the LDA of entrained (re-suspended) dust.

6. Structure and Composition of the Martian Surface Material

Two successful and scientifically important missions to Mars went into operation in 2004, one a team of two robotic landers the NASA Mars Exploration Rovers (MER) the other an ESA orbiter Mars Express. They have complemented each other in a series of studies of surface mineralogy, structure and composition and have both contributed to new understanding of the transport of Martian surface material. Their main goals however have been to pave the way for further study of the habitability of the planet and specifically the presence of water.

The following discussions will not be a review of these missions findings, instead a presentation of data with relevance to the topics of surface material and transport.

NASA Mars Exploration Rovers

The NASA Mars Exploration Rover mission consists of two identical robotic rovers designed to investigate Martian geology and geochemistry with a primary focus on the history of water on the planet. The first rover, Spirit, touched down in Gusev Crater on January 4th, 2004 UTC [2,3] while the second rover, Opportunity, landed in Meridiani Planum on January 25th, 2004, UTC. This NASA mission is the subject of numerous articles and books with the results having impacted most aspects of Martian research. Here a brief discussion will be made of the findings of this extremely successful mission with respect to the near surface structure, composition and transport.

The central goal for the MER mission was to gather mineralogical evidence for the presence (in the geological past) of liquid water on the Martian surface. This was to be achieved using a suite of instruments/techniques including Mössbauer Spectroscopy: analyses iron bearing mineralogy, APXS: Alpha-Proton X-ray fluorescence Spectroscopy for elemental composition and mini-TES: Thermal Emission Spectroscopy from which chemical composition can be extracted. Other instruments include a variety of cameras, a microscope and a drill.

Interesting and scientifically important soil analysis was performed by studying the tracks left (imprint depth) and performing soil excavation using the rovers wheels. From this (load-sinkage) information the bearing strength of the soil could be established. The soil was found to have a (few mm thick) encrusted layer, apparently cemented by salts [83]. These salts were presumably deposited by the passage of water, though not necessarily in the form of substantial amounts of liquid water.

Though not having revealed the results which had been anticipated the mission was successful in its primary goal in most part due to, for the first time, having encountered ancient Martian bedrock which had been mineralogically modified by the action of liquid water.

Despite the observation of fluid flows in the geology of the Spirit landing site the mineralogy was typical of volcanic (basaltic) origin with only minor modification due to humidity. The explanation here being that the surface material was relatively recent and overlaying the bedrock. This is supported by the observations of bedrock at the higher slopes Spirit has analyzed.

Generally the mission has suggested that Mars has experienced an ancient 'wet' past followed by volcanic activity and periods of aeolean transport.

The observed surface dust deposits were rather thin, easily disturbed by the airbags revealing the (dark) basaltic sand beneath.

ESA Mars Express

The ESA (European space agency) mission Mars Express went into orbit on January 2004, it contained a variety of scientific remote sensing instruments of direct relevance to studying the surface composition, transport of surface material and distribution of volatiles/aerosols. A list below is made of some of the important and relevant observations which have already been made:

Complementing the MER mineralogy studies Mars Expresses infrared mineralogical mapping spectrometer (OMEGA) has provided details of ice and dust/sand mixing maps.

The SPICAM UV spectrometer during stellar occulations of Mars should be capable of providing CO_2, but importantly also aerosol (including dust) vertical distributions through the atmosphere. This would be of great importance to transport models and though probably not of high enough resolution to be compared to surface measurements would be an improvement in the current level knowledge [84].

The MARSIS ground penetrating radar instrument should be capable of providing depth profiles of near surface ice, clearly invaluable for search for mission landing sites (search for habitable near surface environments) and accounting for the global water inventory (climatic modeling), though no data is available at this time.

Both observational (high resolution stereo camera) and spectroscopic/mineralogical evidence from Mars Express have pointed towards an ancient stable 'wet' period, with later periodic violent liquid water outbreaks followed by freezing and sublimation. These may be ongoing due to geo-thermal eruptions [16]. Also both MER and Mars Express have made detailed observations of dust transport events: 'dust devils', though have still failed to resolve the physical processes and issue of sand transport [14].

7. Water and Heat Transport through the Upper Surface

The central issue in the following discussion is: Can solar heating produce water liquification below the surface of Mars?. This is based on the observational evidence from Mars that at (mid latitude) regions there can be found water ice overlaid by (fine) granular material (sand/dust).

A situation which could lead to heating and the entrapment of humidity. To answer this question requires knowledge of heat and moisture transport properties of this granular material under these conditions. This has stimulated an experimental investigation involving the measurement of thermal conductivity and the diffusivity of water through Mars analogue dust. This investigation used an environment simulator where heat and moisture transport could be studied through layers of Mars analogue dust.

At temperatures and pressures common on Mars the presence of liquid water can be stable given high humidity (certainly given elevated salinity). Evidence for recent flow of liquid by NASA's Mars Global Surveyor and the probable detection of buried water by neutron scattering on board NASA's Mars Odyssey also support the speculation as to the possibility of stable liquid water close to the surface of Mars.

Experiments were performed in order to measure and compare the time taken for fine deposited dust (Salten Mars analogue) to become heated using a light source and the time taken for humidity to be transmitted. This dust contains extremely fine (<1μm) grains in the form of aggregates, this is fine in comparison with previous studies [85]. These simulations were performed in a small ($0.1m^3$) environmental chamber, gas mixture could be controlled and monitored using a rest gas analyser, temperature could be reduced using a liquid nitrogen flow through system. Humidity was measured using an array of three Honeywell HIH-3602-C polymer-dielectric capacitors which measure relative humidity and temperature. Two thermocouples were also used to measure temperature (see figure 22). In these experiments dry dust was obtained by baking to 110°C for several hours. The temperature and humidity was measured at various depths below the surface of the dust. The dry dust was then depressurized to 10mbar in the environmental chamber and exposed to a humidifying atmosphere. The relative humidity measured below the surface was observed to rise exponentially. By fitting the measured curves to an exponential a time constant was obtained.

A Xe/Hg discharge UV lamp (section) was used as a heat source at the surface of a dry dust layer at low pressure in a dry atmosphere. The temperature was measured at various depths as a function of time. The light intensity was 522 W/m^2. Similarly the temperature was observed to rise exponentially below the surface with a definite time constant.

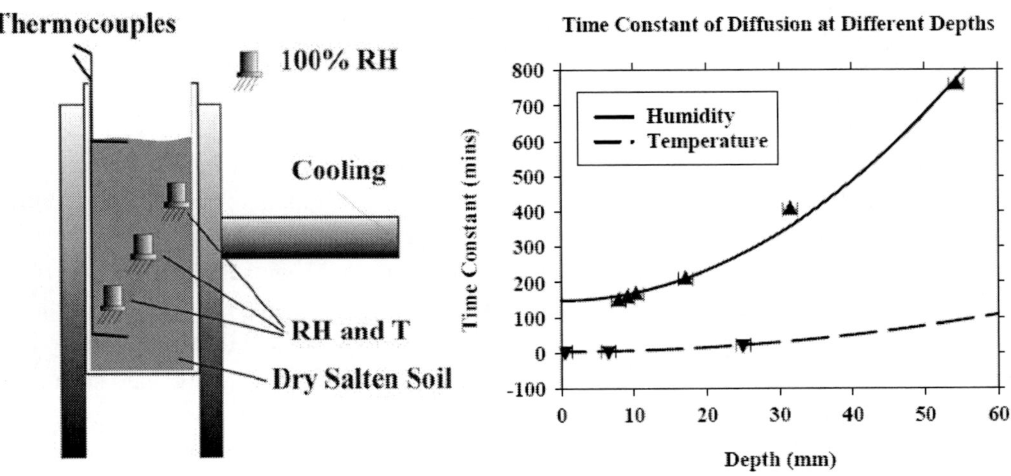

Figure 22. Schematic of a (dry) Mars analogue soil sample in which temperature and humidity has been measured at different depths while being held in a high humid environment (at low pressure). These measurements allow the diffusion time for heat and humidity to pass through the fine granular material (right).

At the low pressures used in these studies the gas atmosphere does not act as a continuous fluid when dealing with fine grains (pores). The water molecule diffusion process becomes so-called Knudsen diffusion and is well described by a simple one dimensional diffusion equation. In this case the time constant has a well known dependence upon depth x: $\tau_{humidity} = x^2/4D_e + T_{ad}$, where D_e is the diffusivity and T_{ad} is an additional time constant due probably in most part to adsorption of water on the extremely dry dust. A similar expression describes the thermal diffusion, $\tau_{thermal} = x^2/4K + T$, where K is the thermal diffusivity, the offset T in this case may be due to several effects, though probably primarily the thermal inertia of the detector (thermocouple) which will add a time constant to the diffusion, in this case around 4minutes. From these results a values for the water diffusivity was obtained: De=2.0 (±0.1) 10^{-8} m^2/s with an adsorption time constant of T_{ad}=149(±17) minutes. The thermal diffusivity was determined to be: K=1.5(±0.1)x10^{-7} m^2/s.

This means that generally the time taken for the dust to become heated (by a lamp) is at least a factor of ten shorter than the time taken for water (humidity) to pass through the layer. In fact this time becomes so long that it is not possible for humidity to be significantly lost through a thin soil layer in one Martian daily heating cycle (around 8-10 hours). Specifically a 5cm thick layer of dust would allow saturation humidity to be exceeded and allow the possibility for liquid water to be stable. More simulation/modeling is required to better quantify these observations.

8. Dust Adhesion and Hazards to Robotic/Manned Missions to Mars

Martian dust poses a current hazard to optical/optically power generating equipment on Mars and would pose a hazard to humans on the surface of the planet. It is therefore important for example to quantify processes of dust transport via EVA material into a Mars lander habitat. Wind tunnel testing is underway attempting to quantify the amount of dust mass entering a habitat on a mission to Mars based on Russian and American contemporary space suit samples [86].

Humans are most sensitive to small dust aerosols by ingestion into the lungs. Even at relatively small doses (mg) dust can cause side effects ranging from lung irritation to pneumonia or fibrosis [87].

All types of dust aerosol have a minimum Occupational Exposure Standard of 10mg/m^3 total inhalable dust concentration for 8 Hours (Time Weighted Average). Taking this least conservative exposure limit and given a habitat volume of 200m^3, the recommended health limit will be exceeded for 2g of dust transported and entrained into the habitat (for a duration of 8hrs). For silica grains this falls to 0.3g or for Iron oxide (the major component of Mars analogue dust) the limit for entrainment into the habitat falls to 1g. Clearly when dealing with such small amounts of material it will be important to quantify accurately the sources of dust contamination. As well as manned missions, current robotic instrumentation can and do suffer from exposure to dust which can affect optical, mechanical and possibly electrical equipment.

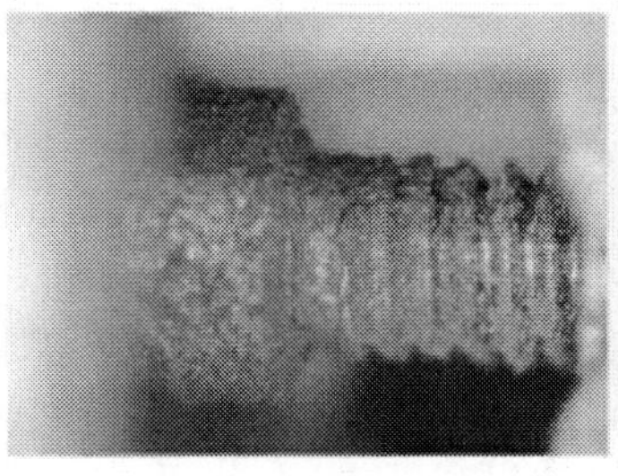

Figure 23. Dust can cause mechanical, electrical and optical damage/failure as shown in these wind tunnel tests of Beagle2 components: sample taking mechanism, serial connection and solar cell section.

As part of the MER magnetic properties experiments a magnet called the sweep magnet was used, its purpose to attract/repel dust away from a central region which should be then be 'swept' clear of magnetic dust. The resulting analysis has shown that within the accuracy of this experiment all the dust on Mars is magnetic (to some extent). It is however impractical (in weight or volume) to construct magnetic sweep systems for large areas and can therefore not be usefully used for keeping for example windows or solar panels clean. It was seen that dust repulsion and dust sweeping could be achieved by the application of dc electric fields (section 4.3). If as with the Salten Skov analogue dust studied, a large fraction of the Martian dust is electrified, electrical sweeping such as that observed in these experiments could be used. This would be simple to implement, cheap and with little weight/volume, involving simply thin conducting electrodes to be applied and a single low power high voltage source.

Electrostatic (and mechanical) dust removal/repulsion devices are being investigated for solar panels. An ESA funded program called wp210 SAMM: 'solar array for mars missions study' is being carried out by EADS Astrium in collaboration with the wind tunnel simulator at Aarhus.

Other options for dust removal include pressurised (with respect to Martian pressure; 7mbar) gas flow. Pressurized air is effective and is used in wind tunnel testing for dust removal, the required gas flow could be calculated from the detachment/entrainment wind threshold data presented in section 5.2. However such a procedure could cause the suspension of fine dust which was otherwise 'dormant' (break-up of dust aggregates) and has been seen to cause electrification.

9. Conclusion

It has become clear from the combined geological, mineralogical, biological and physical studies of the Martian surface and atmosphere that this is a complex, diverse and historically varied planet for which answers to its past/present habitation and habitability will not come easily but only from a concerted scientific investigation.

As most of the work here illustrates, it is necessary to apply a combination of realistic Mars simulations, theoretical modeling and to develop new instrumental techniques. It must also encompass a broad range of scientific disciplines which study similar phenomenology on earth.

In the wind tunnel work and associated modeling and observational data presented here a consistent picture of dust circulation has emerged in which dust electrification leads to dust aggregate formation. These aggregates form larger sand sized grains which have a mass density significantly lower than the bulk material. The threshold shear stress required for detachment has been observed, under Martian simulated conditions, to agree quantitatively with wind speeds observed in dust devil events on Mars. The threshold wind/pressure conditions for sand transport are significantly higher and have been quantified in a semi-empirical expression. More detailed measurements of wind conditions and dust/sand grain properties are required to make interpretations from this, though it should be possible to use this information to predict the climatic conditions under which the observed aeolean features (e.g. dunes and ripples) on Mars formed.

Specifically detailed wind speed, turbulence and direction should be measured accurately on Mars and over a broad seasonal and geographic basis. Also measurement of dust (and sand) grain physical properties such as size/mass, electrification, concentration and composition/mineralogy/structure are lacking. Much has been learned, but progress will come from the development of new instrumentation and it is crucial here to have realistic Mars simulation facilities for testing, development and calibration. This has been illustrated here by the development and testing of an integrated wind sensor and dust analysis instrument based on solid state laser and optoelectronic technology.

The electrical properties of the Martian surface material is a recurring theme when looking at the structure and transport of dust and sand. Nothing has been measured in this regard on Mars and even though studies have been performed on earth and some under Mars simulation conditions much of it is semi empirical and points to the importance of grain electrification, though there is lacking systematic studies capable of making predictions and allowing detailed modeling.

As an example of the application of combined modeling and Mars experimental simulation the magnetic properties experiments on the NASA MER mission have been investigated in the laboratory. As well as being able to reproduce the observed results for a set of parameters (wind speed, grain magnetization and characteristic time) constraints on these parameters and importantly on the dust grain size and mass have been made. Several independent findings in this work have again indicated that the dust grains have lower mass density than the bulk mass density of the material and may be aggregates.

Of direct importance to the present interest in the habitation of Mars is the liquification of water near the surface and whether it can occur due to solar heating or if geothermal sources are required. To answer this question detailed knowledge is needed of the surface material and structure. This may come partly from simulation experiments, but also requires direct study, meaning the subsurface exploration of Mars. Such exploration programs are planned for the coming missions of both NASA and ESA.

Acknowledgements

I would like to greatly thank the many colleagues and friends for their hard work and enthusiasm, especially those connected to the Mars Simulation Laboratory in Aarhus: Haraldur Palle Gunnlaugsson, Per Nørnberg, Kai Finster, Bente Aagaard Lomstein, Anna Eske Jensen, Kjartan Kinch, Torben Lau Jacobsen, Lars Liengård Jensen, Aviaja Hansen,

Keld Rømer Rasmussen, Kristian Von Bengtson, Søren Larsen, Morten Bo Madsen, Helge Wahlgreen, Henrik Bechtold, and to the memory of Jens Martin Knudsen.

I would also like to thank the Danish Science Research Agency and the faculty of Science, Aarhus University for their support.

REFERENCES

[1] Levin, G.V.; and Straat, P.A. *J. Geophys. Res.* 1977, 82, 4663-4668

[2] Squyres, S.W. et.al. *Science,* 2004, 305, 794-799

[3] Squyres, S.W. et.al. *Science,* 2004, 306, 1698-1703

[4] Merrison, J.P.; Bertelsen, P.; Frandsen, C.; Gunnlaugsson, H.P.; Knudsen, J.M.; Lunt, S.; Madsen, M.B., Mossin, L.A.; Nielsen, J.; Nørnberg, P.; Rasmussen, K.R. and Uggerhøj, E. *J. Geophys. Res. Planets,* 2002, 107, 5133-5141

[5] http://phoenix.lpl.arizona.edu/

[6] http://www.esa.int/SPECIALS/Aurora/SEM1NVZKQAD_0.html

[7] Pollack, et al. 1995, *J. Geophys. Res.* 100, 5235

[8] Lemmon, M.T.; Wolff, M.J.; Smith, M.D.; Clancy, R.T; Banfield, D.; Landis, G.A.; Ghosh, A.; Smith, P.H.; Spanovich, N.; Whitney, B.; Whelley, P.; Greeley, R.; Thompson, S.; Bell III, J.F.; Squyres, S.W. *Science.* 2004, 306, 1753-1756

[9] Farrell, W.M., Kaiser, M.L., Desch, M.D., Houser, J.G., Cummer, S.A., Wilt, D.M., and Landis, G.A., 1999, *J. Geophys. Res.* 104, 3795-3801

[10] Greeley, R. J. Geophys. Res. 1979, 84, 6248-6254

[11] Merrison, J.; Jensen, J.; Kinch, K.; Mugford, R.; Nørnberg, P.: *Planetary and Space Science. 2004*; 52: 279-290

[12] Klieffer, H.H.; Jakosky, B.M.; Snyder, C.W.; Matthews, M.S. Mars; Space Science Series, Arizona Press, 1992, p.30

[13] http://marsrovers.jpl.nasa.gov/home/

[14] Stanzel, C; Pätzold, M.; Wennmacher, A.; Hauber, E.; Neukum, G.; 1st Mars Express Science Conference, ESA Abstract Book, ESTEC, Holand 2005, p.63

[15] Head, J.W, Marchant, D.R. *Geology.* 2003, 31, 641-644

[16] Murray, J.B.; et al. *Nature.* 2005, 434, 352-355

[17] Formisano, V.; 1st Mars Express Science Conference, ESA Abstract Book, ESTEC, Holand 2005, p.113

[18] White, B.R.; *J. Fluids Engineering.* 1981, 103, 624-630

[19] Greeley, R; Leach, R.; White, B.R.; Iversen, J.D.; Pollack, J.B. *Geophysical Research Letters.* 1980, 7, 121-124

[20] Greeley, R.; Iversen, J.D.; Wind as a Geological Process on Earth, Mars and Venus, Cambridge Planetary Science Series, Cambridge University Press 1985, p.78-79

[21] Vaniman D.T.; Heiken, G.; Wohletz, K.; Blacic, J. LPSC XXIII, *NASA Astrophysics Data System,* p.1463-1464

[22] Nørnberg, P:; Private Communication of unpublished data.

[23] Gross, F.B.; Grek, S.B.; Calle, C.I.; Lee, R.U.; *Journal of Electrostatics.* 2001, 53, 257-266

[24] Morris, R.V.; Golden, D.C.; Ming, D.W.; Shelfer, T.D.; Jørgensen, L.C.; Bell II, J.F.; Graff, T.G.; Mertzman, S.A.*; J. Geophys. Res.* 2001, 106, 5057-5083

[25] Nørnberg, P.; Schwertmann, U.; Stanjek, H.; Andersen, T.; Gunnlaugsson, H.P.; *Clay Minerals*. 2004, 39, 85-98

[26] Kinch, K. M.; Merrison, J. P.; Gunnlaugsson, H. P.; Bertelsen, P. ; Madsen, M. B.; Nørnberg, P. *Planet. Space. Sci.* 2005, *Accepted for publication.*

[27] Madsen, M.B.; Bertelsen, P.; Goetz, W.; Binau, C.S.; Olsen, M.; Folkmann, F.; Gunnlaugsson, H.P.; Kinch, K.M.; Knudsen, J.M.; Merrison, J.; Nørnberg, P.; Squyres, S.W.; Yen, A.S.; Rademacher, J.D.; Gorevan, S.; Myrick, T.; Bartlett, P. *Journal of Geophysical research. 2003,* 108(E12), 8069

[28] Bell, J.F., III; McSween, H.Y., Jr.; Crisp, J.A.; Morris, R.V.; Murchie, S.L.; Bridges, N.T.; Johnson, J.R.; Britt, D.T.; Golombek, M.P.; Moore, H.J.; Ghosh, A.; Bishop, J.L.; Anderson, R.C.; Brückner, J.; Economou, T.; Greenwood, J.P.; Gunnlaugsson, H.P.; Hargraves, R.B.; Hviid, S.; Knudsen, J.M.; Madsen, M.B.; Reid, R.; Rieder, R.; Soderblom, L. *J. Geophys. Res.* 2000, 105, 1721.

[29] Gunnlaugsson, H.P.; et al. *Hyp. Int.* 2005, *submitted*

[30] Greeley, R.; Balme, M.R.; Iversen, J.D.; Metzger, S.; Mickelson, R.; Phoreman, J.; White, B. *J. Geophys. Res. Planets.* 2003, 108, 5041

[31] Merrison, J.P.; Gunnlaugsson, H.P.; Jensen, J.; Kinch, K.; Nørnberg, P.; Rasmussen, K.R. *Planetary and Space Science, 2004,* 52, 1177-1186.

[32] Mancinelli, R.L., Klovstad, M.; *Planetary and Space Science*. 2000, 48, 1093-1097

[33] Hansen, A.A.;Merrison, J.; Nørnberg, P.;Lomstein, B.Aa.; Finster, K. *International Journal of Astrobiology*. 2005, in press, 4 (2)

[34] Jensen, L. L.; Private communication, 2005

[35] Navarro-Gonzalez, R.; et al., *Science.* 2003, 302, 1018-1021

[36] Skelley, A.M.; et al. Proceedings of the National Academy of Sciences of the United States of America, 2005, 102, 1041-1046

[37] Cockell, C.; Rettberg, P.; Horneck, G.; Scherer, K.; Stokes, M.D. *Polar Biol.* 2003, 26, 62-69

[38] Cockell, C.S.; et al. *Photochemistry and Photobiology,* 2001, 74, 570-578

[39] Head, J.W.; et al. *Nature*. 2005, 434, 346-351

[40] Metzger, S.M.; Carr, J.R.; Johnson, J.R.; Parker, T.J.; Lemmon, M.T. *J. Geophys. Res. Lett.* 1999, 26, 2781-2784

[41] Renno, N.O.; et al. *J. Geophys. Res. Planets.* 2004, 109, 7001

[42] Balme, M.; Metzger, S.; Towner, M.; Ringrose, T.; Greeley, R.; Iversen, *J. Geophys. Res. Lett.* 2003, 30, 1830

[43] Toigo, A.D.; Richardson, M.I.; Ewald, S.P.; Giersch, P. J. *Geophys. Res. Planets.* 2003, 108, 5047

[44] Renno, N.O.; Nash, A.A.; Lunine, J.; Murphy, J. *Geophys. Res. Planets.* 2000, 105, 1859-1865

[45] Metzger, S.; Dentener, F.; Pandis, S.; Lelieveld, J. *Geophys. Res. Planets.* 2002, 107, 4312

[46] Farrell, W.M.; et al. *J. Geophys. Res. Planets.* 2004, 109, 3004

[47] Davies, C.N. Proceedings of the physical society. 1945, 57, 259-270

[48] Douglas, J.F., Gasiorek, J.M.; Swaffield, J.A.; Fluid Mechanics, Pearson Education Ltd., England 2001, p. 422

[49] Larsen, S.E.; Jørgensen, H.E.; Landberg, L.; Tillman, J.E.; *Boundary Layer Meteorology.* 2002, 105, 451-470
[50] Peric, M.; Ferziger, J.H. *Computational Methods for Fluid Dynamics,* Springer, 2 Ed., 1999
[51] Rafkin, S.C.R.; Haberle, R.M.; Michaels, T.I.; Icarus 2001, 151, 228-256
[52] Forget, F.; et al. *J. Geophys. Res.* 1999, 104, 24155-24175
[53] Newman, C.E.; Lewis, S.R.; Read, P.L. Icarus 2005, 174, 135-160
[54] Golombek, M.P.; Bridges, N.T.; *J. Geophys. Res.* 2000, 105, 1841-1853
[55] Bertelsen, P.; Goetz, W.; Madsen, M.B.; Kinch, K.M.; Hviid, S.F.; Knudsen, J.M.; Gunnlaugsson, H.P.; Merrison, J.; Nornberg, P.; Squyres, S.W.; Bell, J.F.; Herkenhoff, K.E.; Gorevan, S.; Yen, A.S.; Myrick, T.; Klingelhofer, G.; Rieder, R.; Gellert, R. *Science.* 2004, 305: 827
[56] Jensen, J. ; Folkmann, F.; Gunnlaugsson, H. P.; Merrision, J. P.; Kinch, K; Knudsen, J. M.; Nørnberg, P.; Madsen, M. B. *X-Ray Spectrometry.* 2005, 34, 359-362
[57] Goetz, W.; et al. *Nature.* 2005, 436, 62-65
[58] Madsen, M.B., Hargraves, R.B., Hviid, S.F., Gunnlausson, H.P., Knudsen, J.M., Goetz, W., Pedersen, C.T., Dinesen, A.R., Mogensen C.T. and Olsen, M. J. *Geophys. Res.* 1999, 104, 8761-8779
[59] Merrison, J.P.; Gunnlaugsson, H.P.; Mossin, L.A.; Nielsen, J.; Nørnberg, P.; Rasmussen K.R.; and Uggerhøj, E. *Planetary and Space Science.* 2002, 50, 371-374
[60] Tillman, J.E.; Landberg, L.; Larsen, S.E.; *J. Atmos. Sci.* 1994, 51, 1709-1727
[61] Palumbo, P.; et al. *Advances in Space Research.* 2004, 33, 2252-2257
[62] Towner, M.C.; et al. Plan*etary and Space Science.* 2004, 52, 1141-1156
[63] Patel, M.R.; Apostolos, A.C.; Cockell, C.S.; Ringrose, T.J.; Zarnecki, J.C.; Icarus 2004, 168, 98-115
[64] Gunnlaugsson, H.P.; Kinch, K.M.; Madsen, M.B.; Merrison, J.P.; Nørnberg, P.; Wahlgreen, H.: *Planetary and Space Science.. 2004,* 52, 693
[65] Greeley, R.; et. al. *Science.* 2004, 305, 810-821
[66] Rasmussen, K.R.; Private Comunication
[67] Cleaver, J.W.; Yates, B. *Journal of Colloid and Interface Science.* 1973, 44 464-474
[68] Bradley, R.S.; *Phil. Mag.* 1932, 13, 853
[69] Hall, D. *J. Fluid. Mech.* 1988, 187, 451-466
[70] Mollinger, A.M.; Nieuwstadt, F.T.M. *J: Fluid. Mech.* 1996, 316, 285-306
[71] Pye, K.; Tsoar, H. Aeolean Sand and Sand Dunes, Unwin Hymann Ltd, 1990, p.27-43
[72] Monin, A.S.; Yaglom, A.M.; Statistical Fluid Mechanics: *Mechanics of Turbulence* Volume 1, MIT Press, 1973, p.257-364
[73] Jørgensen, H.E.; et al. *Vejret,* 2004, 98,1
[74] Greeley, R. *Planetary and Space Science.* 2002, 50, 151-155
[75] Schofield J.T., Barnes J.R., Crisp D., Haberle R.M., Larsen S., Magalhaes J.A., Murphy J.R., Seiff S. and Wilson G. *Science.* 1997, 278, 1752-1758.
[76] Gunnlaugsson, H.P. Planet. Space Sci. 2000,48, 1491-1504
[77] Farrell, W.M., Kaiser, M.L., Desch, M.D., Houser, J.G., Cummer, S.A., Wilt, D.M., and Landis, G.A. *J. Geophys. Res.* 1999, 104, 3795-3801
[78] Berthelier, J.J., Grard, R., Laakso, H., and Parrot, M. *Planetary and Space Science.* 2000, 48, 1193-1200

[79] Calle, C.I.; Mantovani, J.G.; Buhler, C.R.; Groop, E.E.; Buehler, M.G., Nowicki, A.W. *Journal of Electrostatics*. 2004, 61, 245-257
[80] Sickafoose, A.A.; Colwell; J.E., Horányi; M., Robertson, S. J. *Geophys. Res.* 2001, 106, 8343-8356
[81] Poppe, T.; Blum, J.; and Henning, T. *Astropyhsical Journal.* 2000, 533, 472-480
[82] Horn, R.G.; Smith, D.T.; and Grabbe, A. *Nature.* 1993, 366, 442-443
[83] Arvidson, R.E.; et. Al. *Science*. 2004, 305, 821-824
[84] Bertaux, J.L.; et al. *Advances in Space Research*. 2005, 35, 31-36
[85] Titov, D.V.. Adv. *Space Res.* 2002, 29, 183
[86] Bengtson, K von; Skoog, Å.I.; Merrison, J.P. to be published 2006
[87] 'Occupational Exposure Limits', HSE Books 1998, Suffolk, U.K., Booklet EH.40/98, Guidance Note EH.59 'Crystalline Silica'.

In: Space Exploration Research
Editors: J.H. Denis and P.D. Aldridge

ISBN: 978-1-60692-264-4

Chapter 6

Supernovae as Probes for Dark Energy

Vasiliki A. Mitsou
Instituto de Física Corpuscular (IFIC), CSIC – Universitat de València,
Edificio Institutos de Paterna, P.O. Box 22085,
E-46071 Valencia, Spain
Nikolaos E. Mavromatos
King's College London, Department of Physics, Theoretical Physics,
Strand, London WC2R 2LS, UK

Abstract

Type Ia supernovae, acting as standard candles, play a leading rôle in the exploration of the Universe evolution. Initiated by similar stellar explosions whose physics is known in detail, they provide simultaneous measurements of the (luminosity) distance versus the redshift. Observations of this type of supernovae at high redshifts, being sensitive to the Hubble expansion rate, provide the most direct evidence for the accelerating expansion of the Universe and they are consistent with cosmological models proposing a dark energy component dominating the Universe energy budget. These findings have been corroborated by several independent sources, such as measurements of the cosmic microwave background, gravitational lensing, and the large scale structure of the Cosmos.

This chapter focuses on the importance of supernova observations for exploring the nature of dark energy. It briefly outlines the procedure followed in order to extract information relevant to cosmology from measurements of supernova luminosity and spectra and addresses the statistical and systematic errors involved. A complete review is given on supernova observational evidence starting from the first observations presented in 1995 by the Supernova Cosmology Project and the High-z Supernova Search Team and reaching the recent developments by the Hubble Space Telescope, the Supernovae Legacy Survey and the ESSENCE project.

The cosmological implications of supernova observations in conjunction with evidence collected from other astrophysical probes are discussed. A survey of the theoretical approaches devised to address the dark energy problem and their relation to observational questions is given.

The prospects for future improved measurements by facilities such as the Supernova Acceleration Probe are also discussed, together with the possibility of exploiting observations of other types of supernovae to construct a Hubble diagram and determine the cosmological parameters.

1. Introduction

The announcement that distant type Ia supernovae (SNe Ia) indicate an accelerated expansion of the Universe initiated an avalanche of theoretical scenarios and experimental proposals attempting to explain and confirm this finding. The accelerated rate stems from the observation that the distant SNe appear fainter than expected in a freely coasting Universe. These results were announced independently by two different research collaborations in 1998 and have since given rise to a burst of scientific investigations aiming at confirming, complementing and explaining them. These interpretations range from introducing new cosmological approaches, e.g. the existence of a "dark energy" component to the Cosmos energy budget, to alternative observational explanations, such as the absorption due to dust at large redshifts.

The central rôle of SNe Ia [1] in the exploration of the Universe history lies with the fact that they serve as *standard candles,* i.e. they share an identical peak luminosity and a characteristic light-curve shape, which may lead to the determination of the distance to their host galaxies. The significance of supernovae as cosmological probes has been extensively reviewed in the past [2].

Besides the supernovae measurements, there is a plethora of astrophysical evidence today, such as the spectrum of fluctuations in the Cosmic Microwave Background (CMB), galaxy cluster observations and other cosmological data, indicating that the expansion of the Universe is currently accelerating. The energy budget of the Universe seems to be dominated at the present epoch by a mysterious dark energy component, but the precise nature of this energy is still unknown. Nevertheless, current astrophysical data are capable of placing severe constraints on the nature of the dark energy, whose equation of state may be determined by means of an appropriate global fit.

Since this discovery was announced, astrophysicists have been attempting to thoroughly explore the systematics of the measurements, while cosmologists and particle physics theorists have been investigating possible alternative explanations. Among the most exciting theoretical proposals is the postulate of a new form of dark energy, i.e. energy with a negative pressure, like the cosmological constant [3], time-varying "quintessence" scenarios [4] —with a decaying particle field providing the acceleration—, or other exotic scenarios. Among the proposed observational explanations, on the other hand, the faintness of the distant SNe may be attributed to dust absorption [5] or to luminosity evolution of the SNe [6]. Further possibilities include changes in the properties of the observed SN ensemble from the nearby sample to the distant dataset.

The structure of this chapter is as follows. Section 2. describes the general features of SNe Ia, addresses the systematic errors involved and outlines the method applied to extract cosmological constraints from measurements of supernova luminosity and spectra. In Section 3., we give a detailed review on the supernova observational evidence supporting the accelerated expansion of the Universe, starting from the first observations presented in 1995 and reaching up to the recent developments. Further observations of dark energy by other cosmological probes are briefly presented in Section 4.. In Section 5., various theoretical approaches proposed to explain the existence of dark energy are discussed. The future facilities designed to further explore supernovae and their observational prospects are reviewed in Section 6.. Finally, the conclusions and an outlook are presented in Section 7..

2. From Type-Ia Supernovae to the Evolution of the Universe

As astronomical objects, supernovae are classified according to the presence or absence of specific absorption lines in their spectrum near maximum light [7]. Type-I supernovae are distinct from type II in the sense that they lack hydrogen absorption lines in their peak light spectrum. In addition, type-Ia SNe exhibit a strong absorption line near 6100 Å, which comes from a doublet of singly ionized silicon at wavelengths of 6347 Å and 6371 Å [8] (for further information on spectra of SNe Ia, see Section 2.2.). The image of a typical SN Ia is shown in Figure 1; brightness of the supernova alone rivals that of the entire host galaxy.

Figure 1. Type Ia supernova 1994D in galaxy NGC 4526 (credit: NASA/ESA, HST, HZT).

A supernova of type I may form through several ways, however they all share a common underlying mechanism. If a carbon-oxygen white dwarf accreted enough matter to reach the Chandrasekhar limit of about 1.4 solar masses (for a non-rotating star), it would no longer be able to support the bulk of its plasma through electron degeneracy pressure and would begin to collapse [9]. However, the current view is that this limit is not normally attained; increasing temperature and density inside the core ignite carbon fusion as the star approaches the limit (to within about 1%), before collapse is initiated. Within a few seconds, a substantial fraction of the matter in the white dwarf undergoes nuclear fusion, releasing enough energy ($\sim 10^{44}$ J) to unbind the star in a supernova explosion. An outwardly expanding shock wave is generated, with matter reaching velocities of roughly $0.03\ c$. There is also a significant increase in luminosity, reaching an absolute magnitude of around -20 with little variation, an interesting feature that renders type-I SNe reliable distance indicators, as we shall see below.

A supernova of this category may originate from a close binary star system. The more massive of the two stars is the first to evolve off the main sequence, and it expands to form a red giant. The two stars share a common envelope, causing their mutual orbit to shrink. The giant star then sheds most of its envelope, losing mass until it can no longer continue nuclear fusion. At this point it becomes a white dwarf star, composed primarily of carbon and oxygen. Eventually the secondary star also evolves off the main sequence to become a red giant. Matter from the giant is accreted by the white dwarf, causing the latter to increase in mass [9].

Another model for the formation of a type-Ia explosion involves the merging of two white dwarf stars, with the combined mass temporarily exceeding the Chandrasekhar limit. Alternatively, a white dwarf could accrete matter from other types of companions, including a main sequence star, if it orbits sufficiently close to it. The rise time and decay time of their light curve (magnitude as a function of time) are, respectively, $15-20$ days and ~ 2 months, in the SN rest frame.

2.1. Standardized Candles

Type-Ia supernovae follow a characteristic "light curve," as the graph of luminosity as a function of time after the explosion is called. This luminosity is generated by the radioactive decay of ^{56}Ni to ^{56}Co, which subsequently β-decays into ^{56}Fe. The peak luminosity of the light curve is considered to be universal across type-Ia supernovae (the vast majority of which are initiated with a uniform mass via the accretion mechanism), allowing them to be used as secondary standard candles to measure the distance to their host galaxies [10]. Nonetheless, the peak luminosity exhibits a narrow, albeit considerable variation among individual SNe.

This issue was tackled when it was discovered in 1993 by Mark Phillips that there is a correlation between the absolute magnitudes at maximum light of SNe Ia and the rate of decline of the light curves [11]. This correlation basically states that both the time scale and the overall energy of the supernovae explosion depend on the amount of Ni present in the progenitor. He introduced the decline rate parameter, $\Delta m_{15}(B)$, which is the number of magnitudes that a SN Ia declines in its B-band light curve in the first 15 days after maximum light.

There are four basic methods of analyzing and describing the decline rate relation. In addition to the $\Delta m_{15}(B)$ method [12], there is the Multi-color Light Curve Shape (MLCS) method [13], the "stretch method" [14], and the recently developed "C-magic" method [15]. For the B-band and V-band light curves, the SNe Ia which are intrinsically brighter at maximum light have wider light curves. Different methods for the light-curve shape corrections, however, do not compare well with each other; significant differences in the implementations of the corrections are found [1, 16].

It is therefore possible to normalize the peak flux and also "stretch" the time axis so that all type-Ia SNe fit a universal light curve as shown in Figure 2 for nearby SNe. This calibration yields an approximately universal intrinsic luminosity $\mathcal{L}$ equivalent to an absolute blue sensitive magnitude of -19.6. Thus, if we observe a SN Ia in a distant galaxy and measure the peak light output $\mathcal{F}$, we can use the inverse square law to infer its luminosity distance d_{L} (in Euclidean space):

$$\mathcal{F} = \frac{\mathcal{L}}{4\pi d_{\mathrm{L}}^2}. \tag{1}$$

Subsequently, since a difference of five magnitudes corresponds to a factor 100 in brightness, we have the following relation between d_{L}, the apparent magnitude m, and

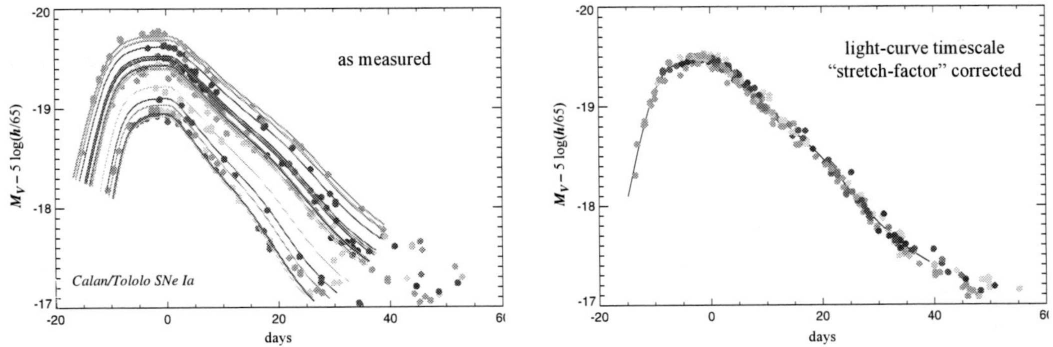

Figure 2. *Left:* Type-Ia SNe observed nearby showing a correlation between their peak absolute luminosity and the timescale of their light curve: the brighter SNe are slower and the fainter ones are faster. *Right:* the same SNe after fitting and removing the stretch factor and correcting the peak magnitude with a simple calibration relation. (Diagrams taken from Ref. [17].)

the absolute magnitude $\mathcal{M}$:

$$\mathcal{M} - m = 2.5 \log \frac{d_{\mathrm{L},0}^2}{d_{\mathrm{L}}^2} = 5 \log \frac{d_{\mathrm{L},0}}{d_{\mathrm{L}}}, \tag{2}$$

where $\mathcal{M} \simeq -19.6$ is the intrinsic magnitude of the standard candles at some nearby distance $d_{\mathrm{L},0}$ usually chosen equal to 10^{-5} Mpc. Thus one obtains for the "distance modulus" μ:

$$\boxed{\mu \equiv m - \mathcal{M} = 5 \log d_{\mathrm{L}} + 25.} \tag{3}$$

This way, the luminosity distance can be measured with an accuracy of $0.1 - 0.18\,\mathrm{mag}$ [12].

In addition, utilizing the spectral information of the host galaxy, one can measure the redshift of the SN and hence draw a distance modulus – redshift diagram, so-called "Hubble diagram". Such a diagram is shown in Figure 3 for a sample compiled out of SNe observed by various ground-based telescopes. Predictions for various values of the cosmological constant component (to be discussed in Section 2.4.) are superimposed. These observations and their implications for cosmology will be thoroughly discussed in Section 3.1.. The measurement of the relation between distance and redshift permits the use of SNe as extragalactic distance indicators, thereby facilitating the exploration of the evolution of the Cosmos.

2.2. Systematic Uncertainties

The results of the cosmological parameters that will presented in the following section are based on the comparison of nearby and distant SNe, under the assumption that the brightness measurements correspond to the same standard candle. A variety of systematic effects, however, may introduce a bias in the brightness versus redshift relation, ranging from differences in the measurement techniques to differing SN sample selections reaching up to fundamental variations in the properties of the standard candles or their environment. Should these effects are large and not well-understood and corrected, they will hamper

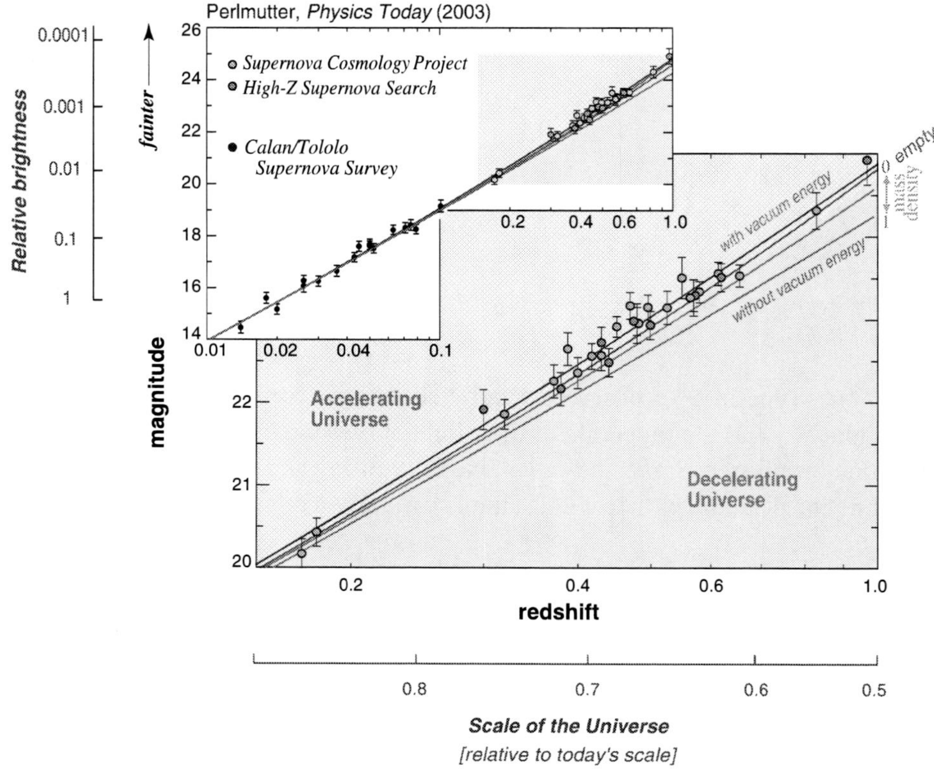

Figure 3. Observed magnitude versus redshift plotted for well-measured distant [18–20] and nearby [21] SN Ia. Curves represent cosmological model predictions (see text). (Diagrams reprinted with permission from [22]. ©2003 American Institute of Physics)

the accuracy of the cosmological parameters determination. A brief description of these potential sources of systematic uncertainties is given in the following.

Extinction by dust Intervening dust particles scatter and absorb light and can decrease the brightness of an object behind the dust screen. Extinction by dust particles along the line of sight may provide an alternative explanation to dark energy for the observed dimming of SNe Ia [23]. There are three possible contributions to extinction: dust in the host galaxy, intergalactic dust and dust in the Milky Way. Although the latter has been measured with sufficient accuracy, the estimation of the effect from the two former ones is more challenging.

Evolution Since the SN observations involve the observation of objects that exploded several billion years ago, it is possible that evolution has changed the explosions or their observable outcome [24]. There are two different processes that must be considered: the objects themselves may have changed, e.g. specific galaxies, or the average properties of the overall sample could have evolved. For the cosmological interpretation, a change of the SN Ia peak luminosity is the most important parameter to investigate.

Gravitational lensing The phenomenon of gravitational lensing, i.e. the deflection of SN light by massive objects while it propagates through the Universe towards the observer, has been studied thoroughly in the past [25]. Any considerably distant object experiences to some degree gravitational lensing. It appears that most distant sources are dimmed as light is scattered out of the line of sight, with only very few objects near deep potential wells being amplified [26]. In effect, the modal brightness is decreased, the dispersion is increased and the mean remains unchanged. For a limited sample of SNe, a bias may occur depending on the fraction of compact objects acting as lenses. Such a brightness bias may affect the cosmological interpretation of the SN measurements, especially those samples with high-redshift SNe.

Type contamination The measurements should be performed on type-Ia SNe, for which template light curves are available. Nevertheless, the light curve is not sufficient to distinguish the different types of SNe, especially between those of type Ia and type Ib/c. Spectroscopic data are needed to secure the classification, based on the most prominent spectral feature —the Si absorption line at 6100 Å—, as shown in Figure 4 for three different types of supernovae. Spectroscopy is less reliable for distant SNe due to higher noise or host-galaxy contamination in the spectra, as well as because the Si absorption line is red-shifted out of the optical regime for $z > 0.5$. In this case, the identification of SN Ia is based on other weaker spectral features on the blue part of the spectrum.

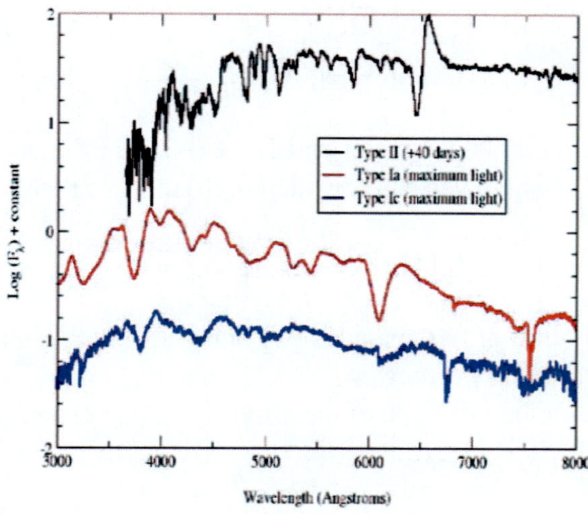

Figure 4. Measured spectra of three supernovae, from top to bottom: type II, type Ia, type Ic. The Si-II feature identifying a SNe Ia is clearly visible in the middle spectrum at about 600 nm. (From Ref. [27].)

Other possible sources of bias in the interpretation of SNe Ia observations include the Malmquist bias, i.e. an increase in average luminosity with distance due to the flux-limited nature of the distant SNe. Additionally, in the photometry and spectroscopy of an observed SN, other factors come into play such as the calculation of K-corrections [28].

2.3. The Redshift – Luminosity Distance Relation

A direct way to trace the evolution of the Universe observationally is to examine the redshift – luminosity distance relation. Redshift z, a reliably measured quantity by the spectroscopic study of a distant object, is directly related to the expansion of the Universe through the scale factor $a(t)$ by $1 + z = a^{-1}(t)$, if $a(0) \equiv 1$ is assumed for the present era.

The expanding Universe is described by the Friedmann-Lemaître-Robertson-Walker (FLRW) metric

$$\mathrm{d}s^2 = -\mathrm{d}t^2 + a^2(t)\left[\frac{\mathrm{d}r^2}{1 - kr^2} + r^2\left(\mathrm{d}\theta^2 + \sin^2\theta\,\mathrm{d}\phi^2\right)\right], \tag{4}$$

where k is a constant related to the spatial curvature: $k = 0$ for a flat Universe, $k > 0$ for a closed, and $k < 0$ for an open one. In this geometry, the measured flux $\mathcal{F}$ at the origin from the object of luminosity $\mathcal{L}$ located at a distance $r = r_1$ is given by

$$\mathcal{F} = \frac{\mathcal{L}}{4\pi r_1^2(1+z)^2}. \tag{5}$$

From the comparison of (1) and (5), the luminosity distance to the object is derived as $d_{\mathrm{L}} = r_1(1 + z)$.

Taking into account that photons propagate along a null geodesic ($\mathrm{d}s^2 = 0$), the general expression for the luminosity distance can be derived:

$$d_L(z) = \frac{1+z}{\sqrt{|\Omega_k|}} \times \begin{cases} \sin\left(\sqrt{|\Omega_k|}\int_0^z \frac{\mathrm{d}z'}{H(z')}\right), & \text{for a closed Universe;} \\ \sqrt{|\Omega_k|}\int_0^z \frac{\mathrm{d}z'}{H(z')}, & \text{for a flat Universe;} \\ \sinh\left(\sqrt{|\Omega_k|}\int_0^z \frac{\mathrm{d}z'}{H(z')}\right), & \text{for an open Universe,} \end{cases} \tag{6}$$

where $H \equiv \dot{a}/a$ is the Hubble parameter and $\Omega_k \equiv -k/H_0^2$ the density parameter of the curvature. If a spatially flat Universe is considered, (6) reduces to the simpler formula:

$$d_L(z) = (1 + z)\int_0^z \frac{\mathrm{d}z'}{H(z')}. \tag{7}$$

Other observables relevant to the analysis of supernova data are the *angular diameter distance* and the *deceleration parameter*. The former, d_A, represents the co-moving distance perpendicular to the line-of-sight of the observer and is related to d_L by

$$d_A(z) = \frac{d_L}{(1+z)^2}. \tag{8}$$

The deceleration parameter $q(t)$, on the other hand, reads

$$q(t) = -\frac{\ddot{a}}{aH^2}. \tag{9}$$

It has to be stressed that the relations (6)–(9) for the luminosity distance, the angular diameter distance and the deceleration parameter are derived by taking into account purely geometrical considerations, i.e. the FLRW metric (4), and hence they are independent of the dynamics of underlying cosmological model, which is concealed in the Hubble expansion rate as a function of the redshift.

2.4. The Standard Cosmological Model

Within the standard framework of general relativity —according to which the dynamics of the gravitational field is described by the Einstein-Hilbert action— the gravitational (Einstein) equations in a Universe with cosmological constant Λ read

$$R_{\mu\nu} - \frac{1}{2} g_{\mu\nu} R + g_{\mu\nu}\Lambda = 8\pi G_{\rm N} T_{\mu\nu}, \tag{10}$$

where $G_{\rm N}$ is the gravitational constant, $T_{00} = \rho$ is the energy density of matter, and $T_{ii} = a^2(t)p$ with p the pressure. From the FLRW metric (4), we arrive at the Friedmann equation

$$\left(\frac{\dot{a}}{a}\right)^2 = \frac{8\pi G_{\rm N}}{3}\rho + \frac{\Lambda}{3} - \frac{k}{a^2}. \tag{11}$$

From this equation one obtains the expression for the critical density, defined as the total density required for a flat ($k = 0$) and cosmological-constant-free ($\Lambda = 0$) Universe

$$\rho_c = \frac{3H^2}{8\pi G_{\rm N}}. \tag{12}$$

If we assume that the Universe components behave like ideal fluids in a co-moving cosmological frame, where all cosmological measurements are assumed to take place, they obey the equations of state $p_i = w_i \rho_i$, with time-independent w_i: for radiation, $w_R = 1/3$; for matter $w_M = 0$; and for the cosmological constant $w_\Lambda = -1$. These parameters determine the evolution of the energy densities at various epochs, related to the Hubble parameter as follows:

$$H(z) = H_0 \left(\Omega_k (1+z)^2 + \sum_i \Omega_i (1+z)^{3(1+w_i)} \right)^{1/2}, \tag{13}$$

with the notation $\Omega_i \equiv \rho_{i,0}/\rho_c$, $i = R, M, \Lambda$. In the present era, the densities Ω_i are linked by the relation $\Omega_k + \Omega_R + \Omega_M + \Omega_\Lambda = 1$.

As we shall see in the following, in some observational analyses, $w = w_\Lambda$ is treated as a free parameter —time-dependent in the general case—, with the intention to be determined by fitting the astrophysical data under study. The aforementioned cosmological scenario constitutes what is widely known as cosmological-constant model or, if the presence of a cold-dark-matter component is assumed, the ΛCDM concordance model.

In this context and in the case of a spatially flat Universe, the present-era value of the deceleration parameter, q_0, takes the form

$$q_0 = \frac{1}{2}\Omega_M - \Omega_\Lambda. \tag{14}$$

Thus, it becomes evident that Λ acts as "repulsive" gravity, tending to accelerate the Universe currently, and eventually dominates, leading to an eternally accelerating de-Sitter-type Universe, with a future cosmic horizon. This is schematically depicted in Figure 5. As we shall see in the following, the data favor a Universe that decelerated in the past, while it expands at an accelerated rate today!

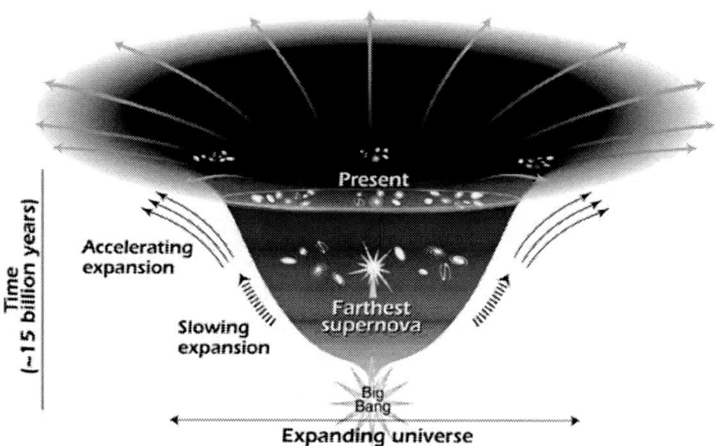

Figure 5. This diagram reveals changes in the rate of expansion since the Big Bang, which occurred ∼15 billion years ago. The more shallow the curve, the faster the rate of expansion. The curve changes noticeably about 7.5 billion years ago, when objects in the Universe began flying apart at a faster rate. Scientists theorize that the accelerated expansion rate is due to a mysterious repulsive force called dark energy. (Credit: NASA/STScI/Ann Feild.)

3. Observations of Dark Energy by Supernovae

In the past decade, significant improvements in terrestrial and extraterrestrial instrumentation have led to spectacular progress in precision measurements in astrophysics. From the point of view of interest to particle physics, the most spectacular claims from astrophysics came in 1998 from the study of distant (redshift of $z \sim 1$) supernovae of type Ia by two independent groups [29]. These observations pointed towards a current-era acceleration of our Universe, something that could be explained either by a non-zero *cosmological constant* in a Friedmann-Robertson-Walker-Einstein Universe, or in general by a non-zero *dark energy* component, which could be even relaxing to zero (the data are consistent with this possibility). In the past five years, many more distant ($z > 1$) supernovae have been discovered, exhibiting similar features as the previous measurements, thereby supporting the geometric interpretation of the acceleration of the Universe today, and arguing against the nuclear physics or intergalactic-dust effects. The time-line of these findings is presented in this section.

3.1. Accelerated Expansion of the Universe: First Evidence

The cosmological use of SNe Ia can be divided in two regimes. The first allows the determination of the Hubble constant at low redshifts ($z \lesssim 0.2$), where the effects of curvature are negligible. At these distances, SNe Ia can actually test the linearity of the expansion to a high degree [30], which in turn justifies their use as reliable distance indicators. In the second regime, at larger redshifts, the combination of the distinct cosmological models and evolution of SNe Ia peak luminosity can no longer be separated cleanly [16], and indirect evidence has to be deployed to tackle the lack of evolution of the SNe. A further difference between the determination of the current expansion rate, i.e. the Hubble constant H_0, and the measurement of the change of this parameter in the past, i.e. the deceleration parameter q_0, is the requirement to measure the absolute luminosity of the objects for H_0. In contrast to H_0, which requires an absolute measurement of the peak luminosity of SNe Ia, the determination of q_0 is independent of the absolute luminosity of SNe Ia, which, however, is assumed to be constant. In this chapter, emphasis is given on the high-redshift regime, where exciting discoveries for Cosmology arose.

Supernovae have a long history of employment initially in the measurement of the Hubble constant [30] and afterwards in the exploration of the expansion rate evolution. The first attempt to observe distant SNe was undertaken during the late 1980s in a ground-based effort by a Danish-British group, which discovered two distant SNe: SN 1988U, a SN Ia at $z = 0.31$ [31] and SN 1988T at $z = 0.28$, which according to the limited photometric information available was most probably a type-II SN [32]. This team employed modern image processing techniques to scale the brightness and resolution of images of distant clusters to match previous images and looked for supernovae in the difference frames.

Great progress was subsequently made by the Supernova Cosmology Project (SCP) [33], led by Saul Perlmutter, in the detection rate of high-redshift SNe Ia by employing large-format charged-coupled devices (CCDs), large-aperture telescopes, and more sophisticated image-analysis techniques [34]. These advances led to the detection of seven SNe Ia at $z \simeq 0.4$ between 1992 and 1994, yielding a confidence region that suggested a flat Universe without cosmological constant, but with a large range of uncertainty [35].

Brian Schmidt and Adam Riess, leading the High-z Supernova Search Team (HZT) [36], joined the hunt for high-redshift SNe Ia with their discovery of SN 1995K at $z = 0.48$ [37]. Both teams made rapid improvements in their ability to discover more type-Ia SNe at even larger redshifts. Before the samples of observed high-redshift SNe Ia become large enough to detect the acceleration signal, both teams found the data to be inconsistent with a Universe closed by matter [38, 39].

These first astrophysical findings were later followed by a thorough understanding of the SN data analysis, allowing thus the use of SN Ia observations to constrain the cosmological parameters. As mentioned in Section 2.1., empirical correlations between SN Ia light-curve shapes and peak luminosity improved the precision of distance estimates beyond the standard candle model. Degeneracies between Ω_M and Ω_Λ may be removed by means of SNe Ia measurements at different redshift bins [40]. Additional work on cross-filter K-corrections provided the ability to accurately transform the observations of high-redshift SNe Ia to the rest frame [28].

Nearby SNe Ia provided both the measure of the Hubble flow and the means to calibrate the relationship between light-curve shape and luminosity. The SCP used 20 nearby SNe Ia in the Hubble flow from the Calán/Tololo Survey [41], while the HZT adds to this set an equal number of SNe from the CfA sample [42]. The Hubble diagram compiled by these nearby SNe plus the distant ones observed by HZT [18, 19] and SCP [20] is shown in Figure 3 (measurements at the same redshift are combined) together with theoretical predictions (indicated by the curves). At redshifts beyond 0.1, the cosmological predictions begin to diverge, depending on the assumed cosmic densities of mass and dark ("vacuum") energy. The red (light grey) curves represent models with zero vacuum energy and mass densities ranging from the critical density ρ_c down to zero (an empty Universe). The best fit (blue/dark grey line) assumes a mass density of about $\rho_c/3$ plus a vacuum energy density twice that large —implying thus an accelerating cosmic expansion.

The determination of the cosmological parameters emerges from the fitting of these SN data to various models. For instance, a dark energy component Ω_X with an equation of state $w = p_X/\rho_X$ and a spatially-flat Universe ($\Omega_{\text{tot}} \equiv \Omega_M + \Omega_X = 1$) may be assumed. In this case, we obtain the confidence intervals in the (Ω_{M}, w) plane for the SN sample detected by the SCP team shown in Figure 6 (left). In such a diagram, Einstein's cosmological constant, Ω_Λ, corresponds to the equation of state $w = p_\Lambda/\rho_\Lambda = -1$. It is clear that the latter is (also) favored by the SN data at 68% confidence level, in conjunction with a matter density of $0.2 < \Omega_M < 0.4$.

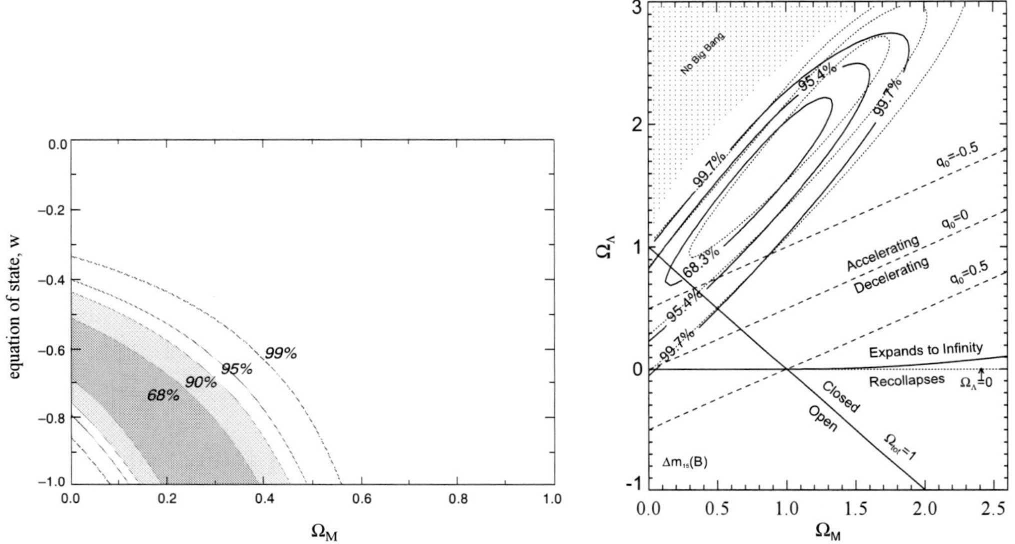

Figure 6. *Left:* Best-fit 68%, 90%, 95%, and 99% confidence regions in the (Ω_M, w) plane for an additional vacuum energy density component, characterized by an equation of state $w = p/\rho$, constrained to a flat cosmology (from SCP [20]). *Right:* Confidence intervals for $(\Omega_M, \Omega_\Lambda)$ from all SNe Ia of Ref. [18] (solid contours). The dotted contours are for the same objects excluding the unclassified SN 1997ck (from HZT [18]).

If we relax the requirement on flatness and we restrict the analysis to a cosmological constant component Ω_Λ, with $w = -1$, we acquire confidence contours for $(\Omega_M, \Omega_\Lambda)$, similar to those in Figure 6 (right), published by the HZT team. It is evident from this diagram, that the presence of a cosmological constant is favored by the SN indeed. Nevertheless, the possibility of a matter-only, open Universe is not entirely excluded. Regions representing specific cosmological scenarios are also illustrated, supporting the case for an accelerating expansion of the Universe today ($q_0 < 0$).

As we shall see in the following, these confidence intervals shrink considerably, when new SN data are added (c.f. Section 3.2.) and/or additional restrictions are posed by other astrophysical surveys (c.f. Section 4.). The reader should bear in mind that the statistical analysis followed to extract cosmological parameters from the absolute magnitude of observed SNe, does not require any assumption on priors, rendering it a direct observational method for probing the history of the Universe. Other astrophysical probes, on the other hand, do need such hypotheses for yielding estimations on observables, as discussed in Section 4..

3.2. Energy Budget of the Cosmos: Today's Picture

Following the traces of the first collaborations dedicated on the SNe Ia measurements, several teams are currently collect and analyze supernovae data to determine the cosmological parameters. During the Institute for Astronomy Deep Survey (Hawaii, US), which used the Canada-France-Hawaii 3.6 m Telescope (CFHT), 23 SNe were detected and monitored [43] leading to the confirmation of the findings of HZT and SCP made five years earlier. However, since only nine of these SNe were unambiguously classified as type Ia, the cosmological parameters were not constrained substantially.

The situation in the detection of SNe Ia drastically changed when the Hubble Space Telescope (HST) [44], as part of its scientific research program, was deployed to detect and monitor supernovae with very high redshift up to $z \sim 2$. The combination of its precision optics, location above the atmosphere, state-of-the-art instrumentation, and unprecedented pointing stability and control, allows HST to achieve the most detailed look at the farthest known objects in the Universe. This observational facility led, at a first stage, to the relatively precise measurement of the present value of the Hubble constant [45]: $H_0 = (72 \pm 8)\ \mathrm{km\,s^{-1}Mpc^{-1}}$. Although measurements of the cosmic microwave background provided a more accurate figure for H_0 (c.f. Section 4.1.), the SN-based determination does not strongly depend on the choice of assumptions and priors. The first results confirming the accelerated expansion of the Universe involving HST-detected SNe were released in 2003 by the SCP [46], which used 11 HST SNe of $z \lesssim 0.9$, but the cosmological parameters were severely constrained in 2004 when Riess *et al.* [47] analyzed a SN sample including 16 HST SNe, among which six were detected at very high redshift $(1.2 < z < 1.8)$.

As discussed in Section 2.2., the dimming of the distant SNe Ia may be attributed to the astrophysics of this type of supernovae or in the propagation of their light to us, instead of purely cosmological explanations. An analysis of this kind is attempted in Ref. [47] for three such alternative models: (i) a gray dust scenario representing a smooth background of dust present at high-z ($z > 2$); (ii) a model of "replenishing" dust continually replenished at the same rate in which is diluted by the expanding Universe; and (iii) a simple evolution model scaling as z in percent dimming. A comparison between the predictions of these models and SN data is shown in Figure 7. Only the replenishing-dust model fits the data sufficiently well and appears indistinguishable from the Ω_Λ model. However, the fine tuning required in the dust model renders it unattractive as a rival to the dark energy [47].

The same SN sample is also confronted with various purely kinematic scenarios, as shown in Figure 8. The jerk parameter is defined proportional to the third derivative of the scale factor, as $j(t) \equiv +(\dddot{a}/a)(\dot{a}/a)^{-3}$. From Figure 8, it is evident that neither a constantly accelerating nor a constantly decelerating expansion are supported by the SN observations. A recently ($z < 0.46$) accelerating and previously decelerating expansion is favored instead.

This effort was joined by the Supernova Legacy Survey (SNLS) [48], an experi-

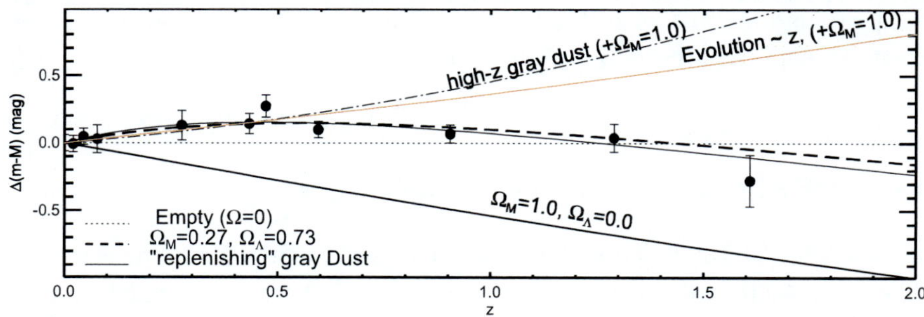

Figure 7. Residual Hubble diagram for weighted averages of SN Ia in fixed redshift bins. Comparison between cosmological models and astrophysical dimming predictions. Data and models are shown relative to a Milne Universe ($\Omega_{\rm tot} = 0, \Omega_\Lambda = 0$). (From Ref. [47].)

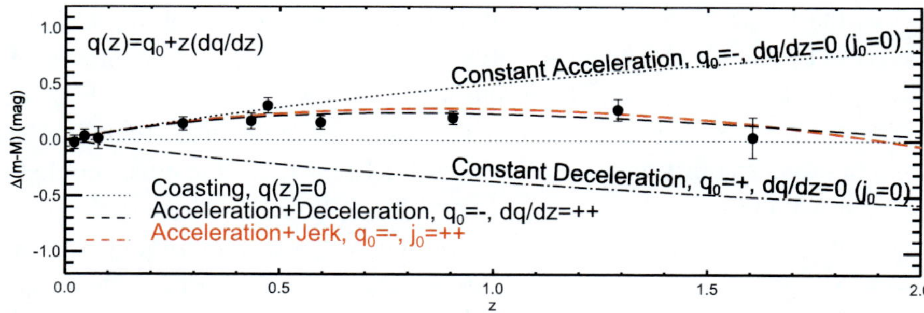

Figure 8. Residual Hubble diagram for weighted averages of SN Ia in fixed redshift bins. Comparison between specific kinematic models of the expansion history. Data and models are shown relative to an eternally coasting model, $q(z) = 0$. (From Ref. [47].)

ment/collaboration forming part of the CFHT Legacy Survey. A significant portion of its observation time is dedicated to the search and observation of distant supernovae, using the large field camera MegaPrime. The first year of operation yielded in 2005 71 SNe Ia of redshift up to $z \simeq 1$, which set further constraints on Ω_M, Ω_Λ and w [49] in agreement with earlier studies.

Early 2007 two new independent samples of distant supernovae became available. The first one released by Riess *et al,* [50] was enriched by 21 newly discovered SN Ia by HST, out of which 13 are of $z > 1$. This discovery, combined with nearby SNe and SNe detected by SNLS, narrowed even further the constraints on the early behavior of dark energy and is fully consistent with the existence of a cosmological constant.

In addition, the "Equation of State: SupErNovae trace Cosmic Expansion" (ESSENCE) [51], a NOAO (US) survey program, discovered and monitored 60 SNe Ia from 2002 through 2005 with redshifts up to $z = 0.78$ [52]. For comparison, two light-curve fitting methods were employed, namely the MLCS2k2 [53] and the SALT [54] fitters, to yield cosmological parameters constraints combined with HST [47] and SNLS [49] data.

A dataset of 192 SNe Ia has been compiled by all recently discovered type-Ia supernovae by HST [50], SNLS [49], ESSENCE [52] plus nearby SNe [42, 55] to yield the most restrictive cosmological parameters for the ΛCDM model, together with other astrophysical surveys [56,57]. Many studies make use of the SN sample compiled in Refs. [56,57] to test other possible theoretical proposals for the nature of the dark energy [58, 59] or to apply alternative statistical methods in the data analysis [60].

The cosmological constraints in the $(\Omega_M, \Omega_\Lambda)$ plane as imposed by the currently available SN sample is shown in Figure 9 (left). The enrichment of the SN sample with more and of higher redshift SNe Ia leads to a substantial narrowing of the confidence intervals as compared to the corresponding obtained after the first SN observations, shown in Figure 6 (right). The case for a spatially flat Universe continues to be favored at 68% confidence level, also supported by the measurements of baryon acoustic oscillations, which will be discussed in Section 4.2.. For a generally non-flat Universe, the favored matter and dark energy densities are $(\Omega_M, \Omega_\Lambda) \simeq (0.33, 0.85)$, while is spatial flatness is assumed, $\Omega_M = 0.259 \pm 0.019$ [59].

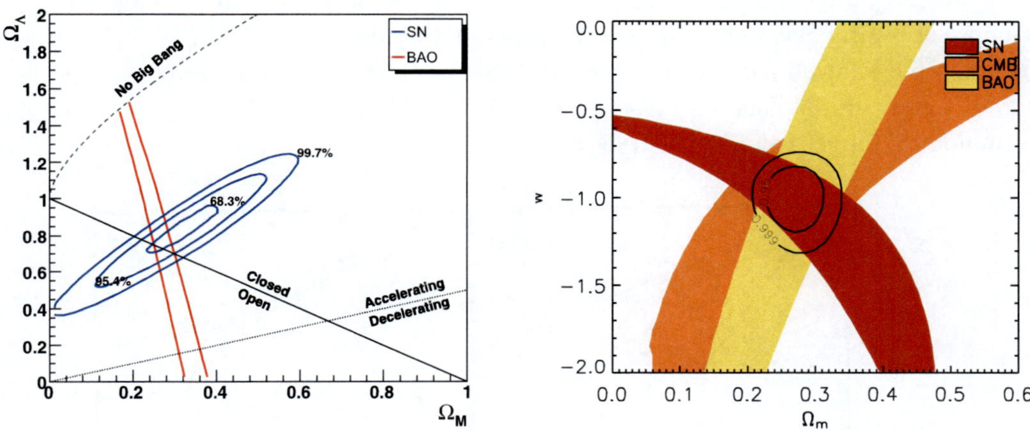

Figure 9. Cosmological constant model confidence intervals for the sample of 192 supernovae [56] from HST [50], SNLS [49], ESSENCE [52] plus nearby SNe [42, 55]. *Left:* 68.3%, 95.4% and 99.7% contours assuming $w = -1$. The result from BAO (1σ) is also superimposed. (From Ref. [59].) *Right:* Flat dark-energy model with constant w: 95% regions from each of the observational probes as shown in the legend. The combined contours (95% and 99.9% confidence) are overlayed in black. (From Ref. [56].)

Furthermore, if a flat dark-energy model with constant w is assumed, the confidence intervals are also more restricted as shown in Figure 9 (right). The SN constraints are depicted together with baryonic oscillations and CMB constraints (c.f. Section 4.1.). The complementarity of the different observational probes is clearly demonstrated in the differing angles of the overlapping contours. The combined data form a clear preference around the cosmological constant model ($w = -1$), however other scenarios for the equation of state of the dark energy may be assumed and tested.

In the case, for instance, of a time-dependent equation of state for the dark energy, the expansion history and geometry of the Universe can be expressed by the formula

$$\dot{a} = H_0\sqrt{\Omega_M/a + \Omega_R/a^2 + \Omega_k + \Omega_X[\rho_X(z)/\rho_X(0)]a^2}, \tag{15}$$

where $\rho_X(z)$ is the density of dark energy. If the dark-energy equation of state as a function of redshift is computed using the Chevallier-Polarski-Linder parametrization [61]

$$w = w_0 + 2w'(1 - a), \tag{16}$$

the dark energy density scales as

$$\frac{\rho_X(z)}{\rho_X(0)} = (1+z)^{3+3w_0+6w'} \exp\left(\frac{-6w'z}{1+z}\right). \tag{17}$$

This expansion history model has been tested with a combination of various astrophysical probes in many studies [62]. If a global fit is applied to SNe Ia data combined with observations of gamma-ray bursts, acoustic oscillations, nucleosynthesis, and large-scale structure [57], the contours in the (w, w') plane shown in Figure 10 are obtained. The bounds set by Big Bang nucleosynthesis, expressed by the stretch factor S^{BBNS}, severely constrain the allowed parameter space. In general, constraining a time-dependent equation of state with SN data only would be challenging, however a combination of various cosmological probes would make it feasible.

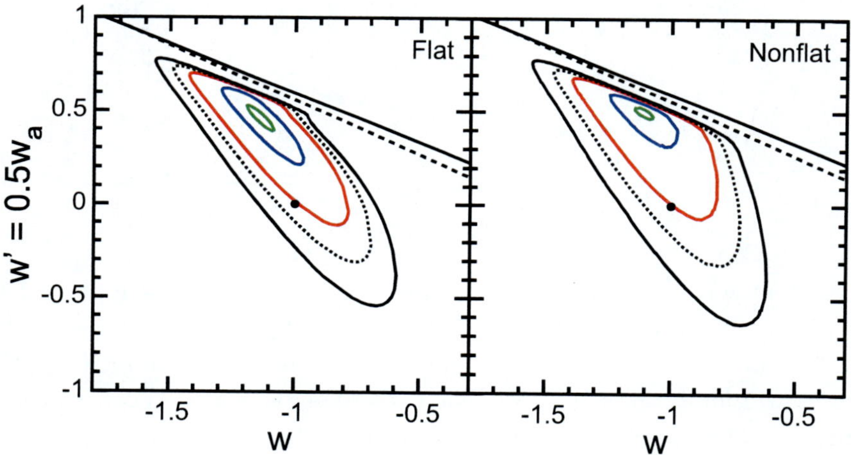

Figure 10. Confidence intervals in the (w, w') plane. The dotted contour shows the 95.4% C.L., the black dot shows the cosmological constant $w = -1$, and the dashed diagonal line shows where the dark energy is equal to the matter plus radiation density at last scattering. The solid diagonal line shows the 3σ limit on the stretch parameter S^{BBNS}. *Left:* The curvature Ω_k is fixed at zero and Ω_M is adjusted to minimize χ^2 at each point. *Right:* Both Ω_M and Ω_k are adjusted to minimize χ^2 at each point. (From Ref. [57].)

4. Complementary Constraints by Other Cosmological Probes

Although the supernova measurements provided the first hint for the accelerating expansion of the Universe, various other cosmological sources corroborated this finding. Such compelling evidence is discussed in the following.

4.1. CMB Anisotropy Measurements

After three years of running, the Wilkinson Microwave Anisotropy Probe (WMAP) [63] provided a much more detailed picture of the temperature fluctuations than its COBE predecessor [64], which can be analyzed to provide CMB-favored models for cosmology, leading to severe constraints on the energy content of various theoretical models, useful for particle physics, and in particular supersymmetric searches. Theoretically [65], the temperature fluctuations in the CMB radiation (shown in Figure 11, left) are attributed to: (i) our velocity with respect to the cosmic rest frame; (ii) gravitational potential fluctuations on the last scattering surface (Sachs-Wolf effect); (iii) radiation field fluctuations on the last scattering surface; (iv) velocity of the last scattering surface; and (v) dampening of anisotropies if Universe re-ionizes after decoupling.

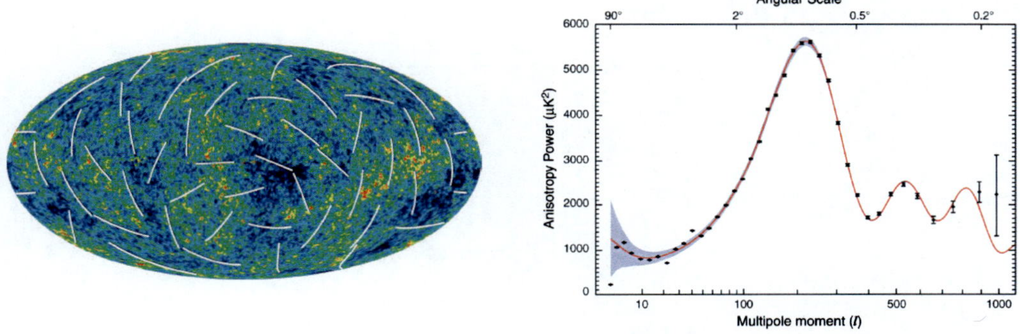

Figure 11. *Left:* CMB map of the Universe measured by WMAP, indicating the "warmer" (red) and "cooler" (blue) spots. The white bars show the "polarization" direction of the photons [66]. (Credit: NASA/WMAP Science Team.) *Right:* Angular power spectrum (black points), best-fit ΛCDM model (red curve), fit to WMAP data only [67], and 1σ cosmic variance error (grey band). (From Ref. [68].)

A Gaussian model of fluctuations [65], favored by inflation, is in very good agreement with the recent WMAP data [67], as shown in Figure 11 (right). The perfect fit of the first few peaks to the data allows a precise determination of the total density of the Universe, which implies its spatial flatness. The various peaks in the spectrum of Figure 11 (right) contain interesting physical signatures:

(i) The angular scale of the first acoustic peak (limited by the cosmic variance) determines the curvature (but not the topology) of the Universe.

(ii) The robust measurement of the second acoustic peak —truly the ratio of the odd peaks to the even peaks— determines the reduced baryon density.

(iii) The third acoustic peak can be used to extract information about the dark matter density. This is a model-dependent result, though; for instance, it requires the assumption of standard local Lorentz invariance.

The WMAP results constrain severely the equation of state $p = w\rho$ (with p the pressure), pointing towards $w = -1.08 \pm 0.12$, when the CMB data are combined with measurements of SNe Ia and large-scale structure. For comparison, we note that in the scenarios advocating the existence of a cosmological *constant* one has $w = -1$. The CMB results are in agreement with the type-Ia supernovae observations elaborated in Section 3.. The combination of WMAP three-year data plus the HST Key Project constraint on H_0 [45] implies $\Omega_k = -0.014 \pm 0.017$ and $\Omega_\Lambda = 0.716 \pm 0.055$, if $w = -1$ is assumed, due to high precision measurements of two secondary acoustic peaks as compared with previous CMB measurements (c.f. Figure 11, right). Essentially the value of Ω is determined by the position of the first acoustic peak in a Gaussian model, whose reliability increases significantly by the discovery of secondary peaks and their excellent fit with the Gaussian model [67].

The degeneracy of the cosmological parameter Ω_M and Ω_Λ from angular-size distances as measured by the cosmic microwave background is orthogonal to the one measured through luminosity distances, as shown in Figure 12.

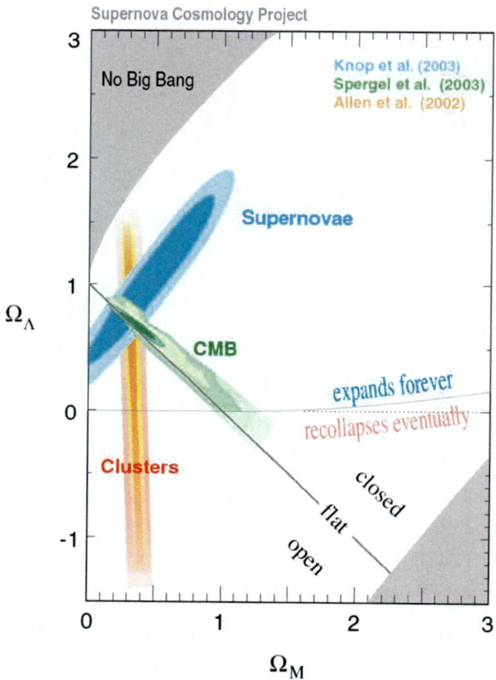

Figure 12. Confidence regions in the $(\Omega_M, \Omega_\Lambda)$ plane for supernovae [46], galaxy cluster [69] and CMB [70] data. The consistent overlap provides compelling evidence for a geometrically flat, dark-energy-dominated Universe (from [33]).

4.2. Baryon Acoustic Oscillations

Further evidence for the energy budget of the Universe is obtained by the detection of the baryon acoustic peak in the large-scale correlation function of luminous red galaxies (LRG) [71, 72] measured by the Sloan Digital Sky Survey (SDSS) [73] and the 2dF Galaxy Redshift Survey (2dFGRS) [74]. The underlying physics of baryon acoustic oscillations (BAO), used to constrain the mass density of the Universe, Ω_M, can be understood as follows. Since the Universe has a significant fraction of baryons, cosmological theory predicts that the acoustic oscillations (CMB) in the plasma will also be imprinted onto the late-time power spectrum of the non-relativistic matter. From an initial point perturbation common to the dark matter and the baryons, the dark matter perturbation grows in place while the baryonic perturbation is carried outward in an expanding spherical wave. At recombination, this shell is roughly 150 Mpc in radius. Afterwards, the combined dark matter and baryon perturbation seeds the formation of large-scale structure. Since the central perturbation in the dark matter is dominant compared to the baryonic shell, the acoustic feature is manifested as a small single spike in the correlation function at 150 Mpc separation [71]. The resulting redshift-space correlation function is shown in Figure 13.

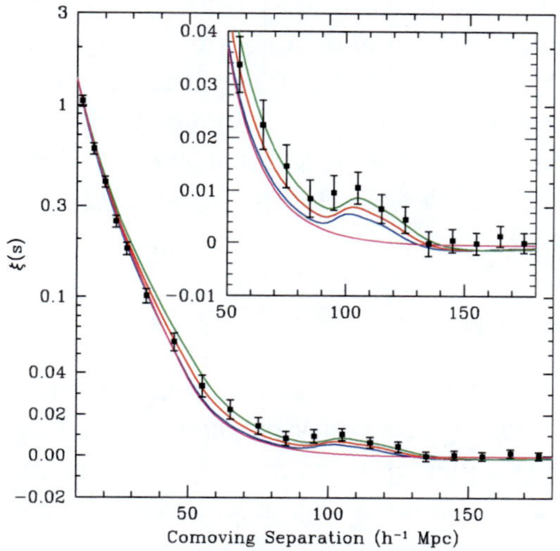

Figure 13. The large-scale redshift-space correlation function of the SDSS LRG sample. The inset shows an expanded view with a linear vertical axis. The models are $\Omega_M h^2 = 0.12$ (top, green), 0.13 (red), and 0.14 (blue). The magenta line (bottom) shows a pure CDM model ($\Omega_M h^2 = 0.105$), which lacks the acoustic peak (from [71]).

The acoustic signatures in the large-scale clustering of galaxies yield three more opportunities to test the cosmological paradigm with the Early-Universe acoustic phenomenon [75]:

1. They provide smoking-gun evidence for the theory of gravitational clustering, notably the idea that large-scale fluctuations grow by linear perturbation theory from $z \sim 1000$ to the present.

2. They give another confirmation of the existence of dark matter at $z \sim 1000$, since a fully baryonic model produces an effect much larger than observed.

3. They provide a characteristic and reasonably sharp length scale that can be measured at a wide range of redshifts, thereby determining purely by geometry the angular-diameter-distance-redshift relation and the evolution of the Hubble parameter.

In the current status of affairs of the BAO measurements, it seems that there is an underlying theoretical-model dependence of the interpretation of the results, as far as the predicted energy budget for the Universe is concerned. This stems from the fact that for small deviations from $\Omega_M = 0.3$, $\Omega_\Lambda = 0.7$, the change in the Hubble parameter at $z = 0.35$ is about half of that of the angular diameter distance. Eisenstein *et al.* [71] modeled this by treating the dilation scale as the cubic root of the product of the radial dilation times the square of the transverse dilation. In other words, they defined

$$D_V(z) = \left[D_M(z)^2 \frac{cz}{H(z)} \right]^{1/3}, \tag{18}$$

where $D_M(z)$ is the co-moving angular diameter distance. As the typical redshift of the sample is $z = 0.35$, we quote the result for the dilation scale as $D_V(0.35) = 1370 \pm 64$ Mpc [71]. The BAO measurements from large galactic surveys and their results for the dark sector of the Universe are consistent with the WMAP data, as far as the energy budget of the Universe is concerned, but the reader should bear in mind that they based their parametrization on standard FLRW cosmologies, so the consistency should be interpreted within that theory framework.

4.3. Large-Scale Structure

Accelerated expansion can be indirectly measured through its effect on the growth of large scale structures of matter, such as stars, quasars, galaxies and galaxy clusters. The recent accelerated expansion of space suppresses the growth due to gravitational attraction, forcing structure to have formed earlier to match the current pattern. This early growth and later slowdown should be measurable through the abundance of clusters of various masses at different epochs. This has been measured by two large galaxy surveys: the SDSS [76] and the 2dFGRS [77]. Both surveys converge to a value for matter density of $\Omega_M \simeq 0.3$ implying the existence of some form of dark energy. This constraint has been applied as a prior in SN analyses in the past [78] and has been combined with various astrophysical sources to probe dark energy models [57].

4.4. Other Astrophysical Sources

By combining the precision measurements of the CMB by WMAP with radio, optical and X-ray probes of the large-scale distribution of matter, further evidence pointing toward acceleration of the expansion rate are provided [79]. It appears that the cluster gravitational potential wells in the Universe have been stretched and made shallower over time, as if under the influence of repulsive gravity. This phenomenon, known as the integrated Sachs-Wolfe (ISW) effect, leads to a correlation between the temperature anisotropies in the CMB and the large-scale structure of the universe [79].

X-ray images of multimillion degree Celsius gas in galaxy clusters taken by the Chandra [80] telescope may provide a powerful method to probe the mass and energy content of the Universe. A recent study [81] of 26 clusters of galaxies confirms that the expansion of the Universe stopped slowing down about 6 billion years ago, and began to accelerate. The value of matter density inferred from such studies, which indicate $\Omega_M = 0.28^{+0.05}_{-0.04}$ when the $f_{\rm gas}$ and CMB data are combined, are consistent with that inferred by combining CMB and supernova data.

Weak gravitational lensing is the statistically detectable distortion in the shapes of distant galaxies by the intervening dark matter. The shear correlations due to weak lensing are sensitive to the growth rate of clustering as well as to angular diameter distances. Recent analyses [82] on LSS measurements set an upper limit of $\Omega_M \lesssim 0.4$ on matter density in agreement with SN and CMB data.

Additional bounds to the cosmological model parameters, albeit weak, may be set by considering robust and model-independent measurements of the age of the Universe and the Hubble parameter. Limits to the former are imposed by measurements of the age of the elements, of the age of distant astronomical objects and the temperature of the coolest white dwarfs [83]. The latter can be determined with a limited accuracy by the differential ages of passively evolving galaxies [59, 84]. Furthermore, high-redshift ($z \lesssim 5$) gamma ray bursts (GRBs), being the most powerful astrophysical events in the Universe, hold great potential to bridge up the gap between the relatively "recent" SN Ia $z \lesssim 2$ and the much earlier CMB ($z \sim 1100$) [57, 85].

To recapitulate, the combination of the SN measurements with results from the CMB fluctuations and determinations of the mass density from galaxy clusters and flow fields has turned out to be a powerful tool to constrain Ω_M and Ω_Λ [57, 86]. These measurements, are completely independent, use different astrophysical objects, are applied at largely different scales, and constrain the cosmological parameters in different ways, as clearly depicted in Figure 12.

5. Theoretical Interpretations of Dark Energy

All these measurements point towards the fact that around 74% of the Universe vacuum energy consists of a dark (unknown) energy substance, as shown in Figure 14, in agreement with the supernovae observations. This claim, if true, could revolutionize our understanding of the basic physics governing fundamental interactions in Nature. Indeed, only a few years ago, particle theorists were trying to identify an exact symmetry of nature that could set the cosmological constant (or more generally the vacuum energy) to zero. Now, astrophysical observations point to the contrary.

Theoretically, there may be several possible explanations regarding the dark energy part of the energy budget of the Universe. A numerous, yet not exhaustive, list of such theoretical approaches is outlined here.

Cosmological constant The dark energy is a cosmological constant $\Lambda \sim 10^{-122} M_P^4$, which does not change through space and time [3]. This has been the working hypothesis of many of the best fits so far, but it should be stressed that it is *not* the only explanation consistent with the data.

Quintessence The cosmological constant is dynamical, mimicked by a slowly-varying field ϕ, whose time until it reaches its potential minimum is (much) longer than the age of the Universe. Simple quintessence models [4] assume exponential potentials

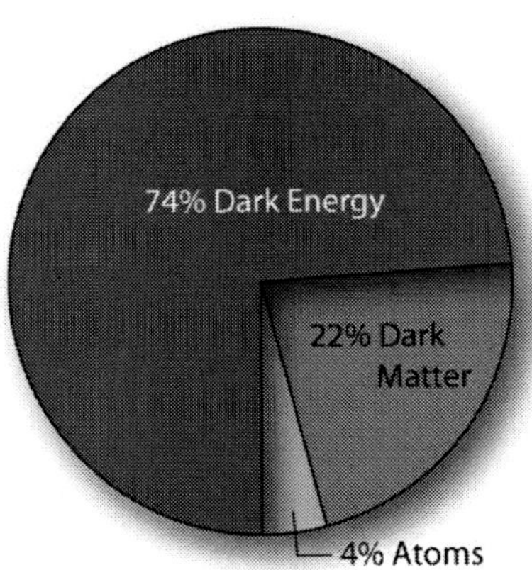

Figure 14. The energy content of our Universe as obtained by fitting data of WMAP satellite. The chart is in perfect agreement with earlier claims made by direct measurements of a current era acceleration of the Universe from distant supernovae of type Ia (credit: NASA).

$V(\phi) \sim e^{\phi}$, although more complicated combinations are most likely to characterize realistic quintessence cosmologies. In such a case the pertinent equation of state reads:

$$p_{\rm vac} = w_{\rm vac}\rho_{\rm vac}, \qquad w = \frac{\frac{(\dot{\phi})^2}{2} - V(\phi)}{\frac{(\dot{\phi})^2}{2} + V(\phi)}, \tag{19}$$

and one has a relaxing-to-zero vacuum energy, with a relaxation rate such that $V(\phi)$ has the right order of magnitude today to be compatible with observations. Such a situation could be met in some models [87] of string theory, where the rôle of the quintessence field could be played by the dilaton [88, 89], i.e. the scalar field of the string gravitational multiplet. The basic theoretical problem in such string-inspired scenarios, where the dilaton plays the rôle of the quintessence field, is to explain the late-era rôle of the dilaton as driving the acceleration of the Universe, while in the past its rôle was subdominant, as compared with radiation and matter fields, so as not to disturb the delicate balance between the Universe expansion and particle production and annihilation rates characterizing the nucleosynthesis era.

Modified gravity The Einstein-Friedmann model is incorrect or insufficient to describe the Universe in its entirety, and one could have modifications in the gravitational law at galactic or supergalactic scales. Models of this kind have been proposed as alternatives to dark matter, for instance Modified Newtonian Dynamics (MOND) by Milgrom [90], and its field theory version by Bekenstein [91], known as Tensor-Vector-Scalar (TeVeS) theory, which is Lorentz violating, in the sense of involving a preferred frame.

Brane cosmologies Other deviations from Einstein theory, which however maintain Lorentz invariance of the four-dimensional world, could be brane models for the Universe, which are characterized by a non-trivial —and in most cases time dependent—

vacuum energy [92]. The above-mentioned stringy dilaton quintessence could be accommodated [88] in such a framework for the Universe. It should be noted that such alternative models may lead to completely different energy budget.

Gauss-Bonnet In string/brane-inspired cosmologies, in general, there are higher-curvature corrections to the Einstein term in the effective action, the most studied form of which is the so-called Gauss-Bonnet gravitational ghost-free form,

$$\int \sqrt{-g}R + f(\Phi)\left(R_{\mu\nu\rho\sigma}R^{\mu\nu\rho\sigma} - 4R_{\mu\nu}R^{\mu\nu} + R^2\right) + \ldots, \qquad (20)$$

where Φ indicates scalar fields, e.g. the dilaton, from the gravitational multiplet of the string/brane-inspired theory, and the other terms $(\ldots)$ involve derivatives of this field [93]. Such higher-curvature corrections may give non trivial contributions to the dark energy sector of the model, and also can lead to singularity-free models [94].

$f(R)$ **models** In addition to the above modifications to standard General Relativity (GR), more *ad hoc* models, not necessarily derived from a microscopic string or brane theory, have been proposed in an attempt to account for the dark-energy sector of the Universe. As an example, one simple way to modify GR is to replace the Einstein-Hilbert Lagrangian density by a general function $f(R)$ of the Ricci scalar R. For appropriate choices of the function $f(R)$ it is then possible to obtain late-time cosmic acceleration without the need for dark energy [95]. However, evading bounds from precision solar-system tests of gravity turns out to be a much trickier matter, since such simple models are equivalent to a Brans-Dicke theory with $\omega = 0$ in the approximation in which one may neglect the potential, and are therefore inconsistent with experiment. To construct a realistic $f(R)$ model requires at the very least a rather complicated function, with more than one adjustable parameter in order to fit the cosmological data and satisfy solar system bounds.

It is natural to consider generalizing such an action to include other curvature invariants, such as the above-mentioned Gauss-Bonnet combination together with the $f(R)$ modifications [96], and it is straightforward to show these generically admit a maximally-symmetric solution: de Sitter space. Further, for a large number of such models (see e.g. [97]), solar system constraints, of the type I have described for $f(R)$ models, can be evaded. However, in these cases another problem arises, namely that the extra degrees of freedom that arise are generically ghost-like.

DGP braneworlds An alternative, and particularly successful approach, is that employed by Dvali and collaborators [98], in which an interesting modification to gravity arises from extra-dimensional models with both five and four dimensional Einstein-Hilbert terms. These Dvali-Gabadadze-Porrati (DGP) braneworlds allow one to obtain cosmic acceleration from the gravitational sector because gravity deviates from the usual four-dimensional form at large distances. One may also ask whether ghosts plague these models. However, it is our understanding that Dvali has claimed that this theory reaches the strong coupling regime before a propagating ghost appears. In fact, Dvali has shown that theories that modify gravity at cosmological distances must exhibit strong coupling phenomena, or else either possess ghosts or are ruled out by solar system constraints.

Liouville string We now remark that Cosmology, especially in the context of string and brane theory, may *not* necessarily be an *equilibrium* phenomenon, described entirely by means of on-shell fields, satisfying classical equations of motion. For instance, Early Universe cosmically catastrophic phenomena, such as the collision of two brane worlds, may result in significant departures from equilibrium, which from the point of view of string excitations on the brane world may be described [88] not by conformal theories on the world-sheet of the string, as is standard in the above-mentioned approaches to string cosmology, but by the so-called Liouville (non-critical) string [99]. The departure from equilibrium is described by specific terms, dictated by the requirement of restoration of world-sheet conformal invariance by means of the Liouville mode, which in such cosmologies plays the rôle of (an irreversible) cosmic time variable. For instance, the off-shell variations with respect to the graviton field $g^{\mu\nu}$, which replace the standard Friedmann equation for equilibrium string cosmologies, are of the form [59, 88, 100]:

$$0 = -\frac{\delta S^{G+\text{matter}}}{\delta g^{\mu\nu}} = \ddot{g}_{\mu\nu} + Q(t)\dot{g}_{\mu\nu} + \ldots, \tag{21}$$

where $Q(t)$ denotes the world-sheet conformal anomaly (central charge deficit of the Liouville string [99]). $S^{G+\text{matter}}$ is the total gravitational plus matter string-inspired action, including higher-curvature modifications, if appropriate, to the standard Einstein gravitational term. The dot indicates derivative with respect to the Liouville mode, which is identified dynamically in these models with (a function of) the cosmic time [88]. The dots indicate corrections that may describe Early epochs of the Universe. Theoretically, the above form, as given in (21), is actually valid for late eras, corresponding to redshifts $z \lesssim \mathcal{O}(10)$.

One such case of a non-critical string inspired cosmology (termed Q-cosmology) has been studied in detail in the context of brane models in Ref. [59, 100], entailing a relaxation dark energy contribution to the Universe energy budget. The latter is due to the (time dependent) dilaton field, $\Phi(t)$, that is required to be non trivial in this kind of non-equilibrium string models [88]. As it will be discussed below, such models still fit the astrophysical data with, however, exotic forms of "dark matter," not scaling like dust with the redshift at late epochs, and different percentages of dark (dilaton quintessence) energy. Most importantly, one cannot disentangle dark energy from dark matter contributions in such models. For instance, there are negative-dust contributions from the dark energy sector of the theory, which are crucial for consistency with the data, and may be attributed either to higher-string-loop contributions [59, 100], which are important due to the non-trivial dilaton $\Phi(t)$ configuration in this kind of models, given that the string coupling is $g_s \sim e^{\Phi(t)}$, or to the existence of bulk Kaluza-Klein graviton modes exerting pressure on the brane world pushing it outwards, thereby appearing effectively as negative energy density on the brane world for an observer on the brane [101].

Back-reaction models It has been suggested [102] that the currently observed acceleration of the Universe is caused not be the existence of a dark energy component, but it may appear as a result of back-reaction effects on the geometry of the Cosmos due to cosmological perturbations. The energy budget of these models is therefore characterized by $\Omega_M = 1$, if flatness is assumed. Unfortunately such models seemed to be ruled out when one combines SN Ia observations [100] with measurements of the Hubble parameter by differential ages of galaxies [59].

The aforementioned theoretical scenarios —among others— can provide alternative interpretations of the dark energy other than the cosmological constant. For instance, the behavior of a dust-dominated inhomogeneous Lemaître-Tolman-Bondi Universe model [103] may be directly confronted with supernova observations. It is found that such a model can easily explain the observed luminosity distance-redshift relation of supernovae without the need for dark energy, when the inhomogeneity is in the form of an under-dense bubble centered near the observer [104]. This is evident in the Hubble diagram of Figure 15, where the predictions of the aforementioned model, the standard ΛCDM model, and an Einstein-de Sitter model ($\Omega_M = 1$, $\Omega_\Lambda = 0$) are compared. It turns out that the statistics χ^2 is slightly better (lower) for the inhomogeneous-Universe model than the concordance one, rendering it as a possible interpretation of the observed accelerated expansion of the Universe [104]. Similar conclusions are extracted for other cosmological scenarios, such as a back-reaction and a Liouville string model, tested against SN data [59, 100].

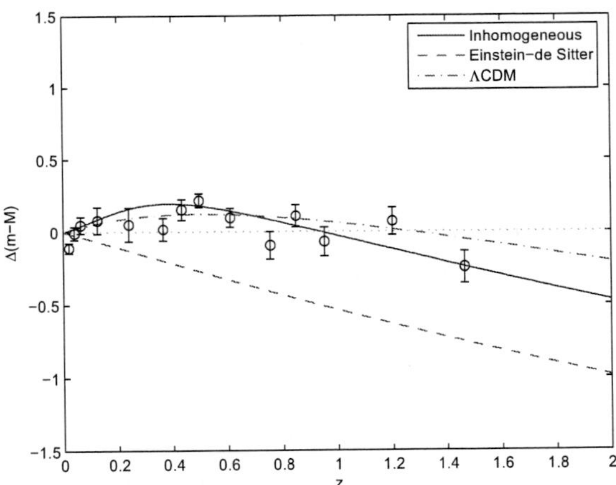

Figure 15. Distance modulus versus redshift for the ΛCDM model, the Einstein-de Sitter model and an inhomogeneous-Universe model (see text) together with type-Ia supernova observations. (Reprinted figure with permission from [104]. ©2006 by the American Physical Society)

We next remark that, since from most of the standard best fits for the Universe up to now, it follows that the energy budget of our Cosmos today is characterized by 73–74% vacuum energy, i.e. an energy density of order $\rho_{\text{vac}} \simeq (10^{-3}\ \text{eV})^4 = 10^{-8}\ \text{erg/cm}^3$, and about 27–26% matter (mostly dark), this implies the *Coincidence Problem*: "*The vacuum energy density today is approximately equal (in order of magnitude) to the current matter density.*" As the Universe expands, this relative balance is lost in models with a cosmological constant, such as the standard ΛCDM model, since the matter density scales with the scale factor as $\frac{\Omega_\Lambda}{\Omega_M} = \frac{\rho_\Lambda}{\rho_M} \propto a^3$. In this framework, at early times we have a vacuum energy much more suppressed as compared with that of matter and radiation, while at late times it dominates. There is only one brief epoch for which the transition from domination of one component to the other can be witnessed, and this epoch, according to the ΛCDM model, happened to be the present one! This calls for a microscopic explanation, an issue which is still lacking and which the theoretical models ought to address.

The smallness of the value of the dark energy today is another big mystery of particle physics. For several years the particle physics community thought that the vacuum energy was exactly zero, and in fact they were trying to devise microscopic explanations for such a vanishing by means of some symmetry. One of the most appealing, but eventually failed in this respect, symmetry justifications for the vanishing of the vacuum energy was that of *supersymmetry* (SUSY): if unbroken, supersymmetry implies strictly a vanishing vacuum energy, as a result of the cancelation among boson and fermion vacuum-energy contributions, due to opposite signs in the respective quantum loops. However, this cannot be the correct explanation, given that SUSY, if it is to describe Nature, must be broken below some energy scale $M_{\rm susy}$, which should be higher than a few TeV, as partners have not been observed as yet. In broken SUSY theories, in four dimensional space times, there are contributions to vacuum energy $\rho_{\rm vac} \propto \hbar M_{\rm susy}^4 \sim (\text{few TeV})^4$, which is by far greater than the observed value today of the dark energy $\Lambda \sim 10^{-122}\, M_P^4$, with $M_P \sim 10^{19}$ GeV. Thus, SUSY does not solve the *Cosmological Constant Problem*, which at present remains one of the greatest mysteries in Physics.

In this respect, the smallness of the value of the "vacuum" energy density today might point towards a relaxation problem. Our world may have not yet reached equilibrium, from which it departed during an early-epoch cosmically catastrophic event, such as a Big Bang, or —in the modern version of string/brane theory —a collision between two brane worlds. This non-equilibrium situation might be expressed today by a quintessence-like exponential potential e^ϕ, where ϕ could be the dilaton field, which in some models [87, 88] behave at late cosmic times as $\phi \sim -2\ln t$. This would predict a vacuum energy today of order $1/t^2$, which has the right order of magnitude, if t is of order of the age of the Universe, i.e. $t \sim 10^{60}$ Planck times. Supersymmetry in such a picture may indeed be a symmetry of the vacuum, reached asymptotically, hence the asymptotic vanishing of the dark energy. SUSY breaking may not be a spontaneous breaking but an *obstruction*, in the sense that only the excitation particle spectrum has mass differences between fermions and bosons. To achieve phenomenologically realistic situations, one may exploit [105] the string/brane framework, by compactifying the extra dimensions into manifolds with non-trivial "fluxes" (these are not gauge fields associated with electromagnetic interactions, but pertain to extra-dimensional unbroken gauge symmetries characterizing the string models). In such cases, fermions and bosons couple differently, due to their spin, to these flux gauge fields (a sort of generalized "Zeeman" effects). Thus, they exhibit mass splittings [106] proportional to the square of the "magnetic field," which could then be tuned to yield phenomenologically acceptable SUSY-splittings, while the relaxation dark energy has the cosmologically observed small value today.

In such a picture, SUSY is needed for stability of the vacuum, although today, in view of the landscape scenarios for string theory, one might not even have supersymmetric vacua at all. However, there may be another reason why SUSY could play an important physical rôle, that of providing candidates (e.g. neutralinos in Minimal Supersymmetric Standard Model extensions or gravitinos in other models) for (cold) dark matter. We shall not discuss this important issue here. Instead we only remark that stringent constraints on such models can be provided by combining cosmological and collider searches (Tevatron, Large Hadron Collider, etc.) of dark matter [107].

6. Future of Supernova Cosmology

The evidence for the existence of dark energy discovered so far call for a continuation in the exploration of supenovae with new, powerful apparatus. Such experiments under study are discussed in this section, as well as the use of type II supernovae.

6.1. The SNAP Satellite

A space observatory called SNAP (SuperNova Acceleration Probe) [108, 109] is proposed, designed to probe further into the expansion of the Universe and the nature of the mysterious dark energy that is accelerating this expansion. SNAP is being proposed as part of the Joint Dark Energy Mission (JDEM) [110] and, if selected, it will be launched before 2020. Other projects proposed in the frame of JDEM are Destiny [111], an infrared survey telescope and ADEPT, which will provide measurements on supernovae and the distribution of galaxies.

The SNAP satellite and mission design [112] has been optimized for efficient supernova detection and high quality follow-up measurements. The combination of a three-mirror, two-meter telescope and a $\sim$600-million-pixel optical to near-infrared imaging camera with a 0.7-square-degree field of view will allow simultaneous discovery and recording of multiple supernovae. The imaging system comprises 36 large format (3512×3512 pixels) CCDs and the same number of 2048×2048 HgCdTe infrared sensors. Both the CCDs and the near-infrared (NIR) detectors are placed in four symmetric 3×3 arrangements. Both the imager and a low resolution ($R \sim 100$) high-throughput spectrograph cover the waveband from 350 to 1700 nm, allowing detailed characterization of supernovae up to $z = 1.7$. This deep reach in redshift is essential to the mission as it will allow to resolve degeneracies in cosmological parameters and to discriminate between models of dark energy. Nine special filters fixed above the imaging sensors will provide overlapping red-shifted B-band coverage in the range $350 - 1700$ nm. As SNAP repeatedly steps across its target fields in the north and south ecliptic poles, every supernova will be seen in every filter in both the visible and NIR. Because of their larger linear size, each NIR filter will be visited with twice the exposure time of the visible filters. This, combined with the time-dilated light curve, will ensure that type-Ia supernovae out to redshift 1.7 will be detected with a signal-to-noise ratio higher than six at least two magnitudes below peak brightness. A schematic view of the SNAP satellite components is shown in Figure 16.

SNAP will conduct two primary surveys, a $\sim$15-square-degree ultra-deep supernova survey, and a $\sim$300-square-degree-deep weak-lensing survey. With this wealth of detailed data, SNAP will construct a Hubble diagram with unprecedented control over systematic

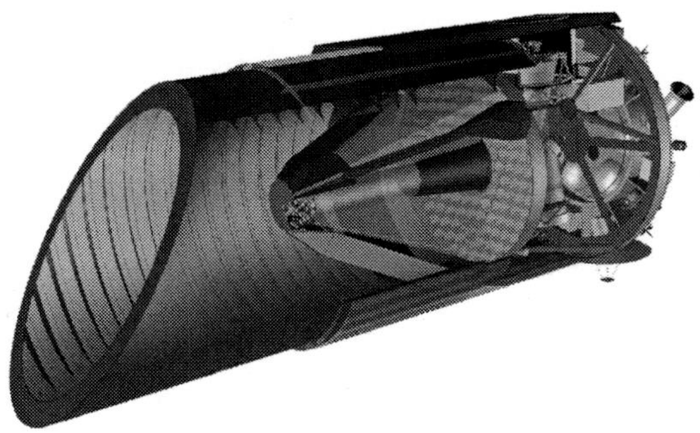

Figure 16. Cross-sectional view of the SNAP satellite. The principal assembly components are the telescope, optical bench, instruments, propulsion deck, bus, stray light baffles, thermal shielding and entrance door (from [108]).

uncertainties, addressing all known and proposed sources of error. The first goal is to provide precision measurements of the cosmological parameters: the matter density Ω_M, will be measured to ± 0.02, while Ω_Λ, and the curvature parameter Ω_k, will both be determined to an accuracy of ± 0.04. The SNAP measurements will be largely orthogonal to the CMB measurements in the $(\Omega_M, \Omega_\Lambda)$ plane, and the curvature measurement at $z \sim 1$ will test cosmological models by comparison with the CMB determination at $z \sim 1000$. The scientific reach of SNAP will then extend to an exploration of the nature of the dark energy, measuring the present equation of state, w, with an uncertainty of 5%. Of even more interest is a determination of w as a function of redshift. SNAP will maintain a tight control over systematics and the high statistics in each redshift bin will allow the determination of the dynamical variation of w, as shown in Figure 17.

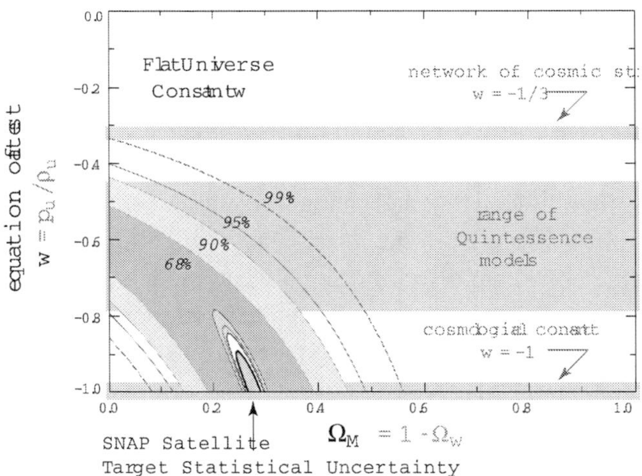

Figure 17. Best-fit 68%, 90%, 95%, and 99% confidence regions in the (Ω_M, w) plane for an additional energy density component w, characterized by an equation of state $w = p/\rho$ (from SCP [20]). For Einstein's cosmological constant Λ, $w = -1$. The fit is constrained to a flat cosmology ($\Omega_M + \Omega_W = 1$). Also shown is the expected confidence region allowed by SNAP assuming $w = -1$ and $\Omega_M = 0.28$ (from SNAP [108]).

To complement its supernova cosmology observations, SNAP will conduct a wide-area weak lensing survey. These weak lensing observations provide important independent measurements and complementary determinations of the dark matter and dark energy content of the Universe. They will substantially enhance ability of SNAP to constrain the nature of dark energy [113]. SNAP weak lensing observations benefit enormously from the high spatial resolution, the accurate photometric redshifts, and the very high surface density of resolved galaxies available in these deep observations.

6.2. Other Future Missions

Besides the JDEM/SNAP mission, the DUNE mission (Dark UNiverse Explorer) [114], due to be launched around 2012, should provide measurements of $\sim$10 000 SNe up to a redshift of $z \sim 1$, within a 18-month period. In addition, the Large Synoptic Survey Telescope (LSST) [115] is a proposed ground-based 8.4-meter, 10-square-degree-field telescope that will provide digital imaging of faint astronomical objects across the entire sky. The Dark Energy Survey (DES) [116] collaboration, finally, proposes to build an extremely red-sensitive 500-Megapixel camera and a one-meter-diameter, 2.2-degree field-of-view prime focus corrector, with a data acquisition system fast enough to take images in 17 seconds, used to conduct a large scale sky survey. It aims at extracting information on the dark energy from: (i) cluster counting and spatial distribution of clusters at $0.1 < z < 1.3$; (ii) the shifting of the galaxy spatial angular power spectra with redshift; (iii) weak lensing measurements on several redshift shells to $z \sim 1$; and (iv) about 2000 supernovae at $0.3 < z < 0.8$.

6.3. Type II Supernovae

Although the detailed study of type Ia SNe is the main focus of the future SN projects, the use of type II supernovae as cosmological probes is also feasible. In recent studies [117], the construction of a Hubble diagram for type II-P SNe at cosmologically significant redshifts has been demonstrated. The root-mean-square scatter of this method, 13% in distance, compares favorably to the 7–10% scatter typically seen in the SN Ia measurements. Further improvements may be sought with a view to reducing the scatter and increasing the cosmological power of the high-redshift data.

Exploring the utility of measuring distances to SNe II-P has potential benefits well beyond simply verifying, independently, the acceleration seen at redshifts $z < 1$. Several plausible models for the time evolution of the dark energy require distance measures to $z \gtrsim 2$. At such high redshifts, other cosmological probes may become less effective than at $z < 1$. However, current models for the cosmic star-formation history predict an abundant source of SNe II at these epochs and future facilities, such as the proposed JDEM telescope, SNAP, could potentially use SNe II-P to determine distances at these very high redshifts.

7. Conclusion

Type-Ia supernovae have provided so far "smoking gun" evidence for the accelerated expansion of the Universe and the existence of the dark energy. It constitutes a well-understood and relatively simple technique, allowing the direct probing of the history of the Universe. More precise measurements are expected to follow, should further improvements on the control of systematic errors, such as redshift evolution of SN properties and dust extinction corrections, are achieved. There is a vigorous current and future program of SN surveys, ranging from medium-z SNe from the ground, to high-z surveys from space.

On the interpretation of the SN findings front, a wide spectrum of theoretical models have been proposed from the existence of Einstein's cosmological constant, though models predicting a modified theory of gravity, to scenarios involving out-of-equilibrium strings. We should expect more insight on the nature of dark energy from current and future studies of type-Ia supernova samples.

Une affaire à suivre...

References

[1] For a comprehensive review, see Leibundgut, B. *Astron. Astrophys. Rev.* 2000, 10, 179–209.

[2] Gibson, B. K.; et al. *Astrophys. J.* 1999 529, 723–744.
Parodi, B. R.; Saha, A.; Sandage, A.; Tammann, G. A. *Astrophys. J.* 2000, 540, 634–651.
Leibundgut, B. *Ann. Rev. Astron. Astrophys.* 2001, 39, 67–98.

[3] Einstein, A. Sitzungsber. *Preuss. Akad. Wiss. Berlin (Math. Phys.)* 1917, 1917, 142–152.
Weinberg, S. *Rev. Mod. Phys.* 1989, 61, 1–23.
Carroll, S. M.; Press, W. H.; Turner, E. L. *Ann. Rev. Astron. Astrophys.* 1992, 30, 499–542.
Carroll, S. M. eConf C0307282 2003, *TTH09 & AIP Conf. Proc.* 2005, 743, 16–32, and references therein.

[4] Steinhardt, P. J. *Phil. Trans. Roy. Soc. Lond. A* 2003, 361, 2497–2513.

[5] Totani T.; Kobayashi C. *Astrophys. J. Lett.* 1999, 526, L65–L68.
Rowan-Robinson, M. *Mon. Not. Roy. Astron. Soc.* 2002, 332, 352–360.

[6] Umeda, H.; Nomoto, K.; Kobayashi, C.; Hachisu I.; Kato, M. *Astrophys. J. Lett.* 1999, 522, L43–L47.
Nomoto, K.; Umeda, H.; Kobayashi, C.; Hachisu, I.; Kato, M.; Tsujimoto, T. In *Cosmic Explosions;* Holt, S. S. and Zhang, W. W.; Ed.; American Institute of Physics: New York, NY, 2000; pp 35–52.

[7] Cappellaro, E.; Turatto, M. In *The influence of binaries on stellar population studies;* Vanbeveren, D.; Astrophysics and space science library (ASSL), xix, Kluwer Academic Publishers: Dordrecht, Netherlands, 2001; Vol. 264, pp 199–213.

[8] Minkowski, R. *Ann. Rev. Astron. Astrophys.* 1964, 2, 247–266.
Filippenko, A. V. *Ann. Rev. Astron. Astrophys.* 1997, 35, 309–355.

[9] Hillebrandt, W.; Niemeyer, J. C. *Ann. Rev. Astron. Astrophys.* 2000, 38, 191–230.

[10] Elias, J. H.; Matthews, K.; Neugebauer, G.; Persson, S. E. *Astrophys. J.* 1985, 296, 379–389.
Meikle, W. P. S. *Mon. Not. Roy. Astron. Soc.* 2000, 314, 782–792.

[11] Phillips, M. M. *Astrophys. J.* 1993, 413, L105–L108.

[12] Phillips, M. M.; Lira, P.; Suntzeff, N. B. a.; Schommer, R. A.; Hamuy, M.; Maza, J. *Astron. J.* 1999, 118, 1766–1776.

[13] Riess, A. G.; Press, W. H.; Kirshner, R. P. *Astrophys. J.* 1996, 473, 88–109.

[14] Perlmutter, S.; et al. *Astrophys. J.* 1997, 483, 565–581.
Goldhaber, G.; et al. *Astrophys. J.* 2001, 558, 359–368.

[15] Wang, X. F.; Wang, L. F.; Zhou, X.; Lou, Y. Q.; Li, Z. W. *Astrophys. J.* 2005, 620, L87–L90.

[16] Drell, P. S.; Loredo, T. J.; Wasserman, I. *Astrophys. J.* 2000, 530, 593–617.

[17] Perlmutter, S.; et al. *Bull. Am. Astron. Soc.* 1997, 29, 1351–1359.

[18] Riess, A. G.; et al. *Astron. J.* 1998, 116, 1009–1038.

[19] Garnavich, P. M.; et al. *Astrophys. J.* 1998, 509, 74–79.
Riess, A. G.; et al. *Astrophys. J.* 2001, 560, 49–71.

[20] Perlmutter, S.; et al. *Astrophys. J.* 1999, 517, 565–586.

[21] Hamuy, M.; et al. *Astron. J.* 1993, 106, 2392–2407.
Hamuy, M.; Phillips, M. M.; Maza, J.; Suntzeff, N. B.; Schommer, R. A.; Aviles, R. *Astron. J.* 1995, 109, 1–13.

[22] Perlmutter, S. *Physics Today* 2003, 56, 4, 53–60.

[23] Cardelli, J. A.; Clayton, G. C.; Mathis, J. S. *Astrophys. J.* 1989, 345, 245–246.

[24] Howell, D. A.; Sullivan, M.; Conley, A.; Carlberg, R. *Astrophys. J. Lett.* 2007 667, L37–L40.

[25] Blandford, R. D.; Narayan, R. *Ann. Rev. Astron. Astrophys.* 1992, 30, 311–358.
Mellier, Y. *Ann. Rev. Astron. Astrophys.* 1999 37, 127–189.

[26] Wambsganss, J.; Cen, R. y.; Xu, G. h.; Ostriker, J. P. *Astrophys. J.* 1997, 475, L81–L84.

[27] Miquel, R. *J. Phys. A* 2007, 40, 6743–6755.

[28] Kim, A.; Goobar, A.; Perlmutter, S. *Publ. Astron. Soc. Pac.* 1996, 108, 190–201.

[29] For a detailed review on the discovery, see: Riess, A. G. *Publ. Astron. Soc. Pac.* 2000, 112, 1284–1299.

[30] For a review, see: Branch, D. Ann. Rev. Astron. Astrophys. 1998, 36, 17–55.

[31] Norgaard-Nielsen, H. U.; Hansen, L.; Jorgensen, H. E.; Aragon Salamanca, A.; Ellis, R. S. *Nature* 1989, 339, 523–525.

[32] Hansen, L.; Jorgensen, H. E.; Norgaard-Nielsen, H. U.; Ellis, R. S.; Couch, W. J. *Astron. Astrophys.* 1989, 211, L9–L11.

[33] Supernova Cosmology Project (SCP): `http://panisse.lbl.gov`

[34] Perlmutter, S.; et al. *Astrophys. J.* 1995, 440, L41–L44.

[35] Perlmutter, S.; et al. *Astrophys. J.* 1997, 483, 565–581.

[36] High-Z Supernova Search Team (HZT): `http://cfa-www.harvard.edu/supernova//HighZ.html`

[37] Schmidt, B. P.; et al. *Astrophys. J.* 1998, 507, 46–63.

[38] Garnavich, P. M.; et al. *Astrophys. J.* 1998, 493, L53–L57.

[39] Perlmutter, S.; et al. *Nature* 1998, 391, 51–54.

[40] Goobar, A.; Perlmutter, S. *Astrophys. J.* 1995, 450, 14–18.

[41] Hamuy, M.; Phillips, M. M.; Suntzeff, N. B.; Schommer, R. A.; Maza, J. *Astron. J.* 1996, 112, 2408–2437.
Hamuy, M.; et al. *Astron. J.* 1996, 112, 2438–2447.

[42] Riess, A. G.; et al. *Astron. J.* 1999, 117, 707–724.

[43] Barris, B. J.; et al. *Astrophys. J.* 2004, 602, 571–594.

[44] Hubble Space Telescope (HST): `http://hubble.nasa.gov`

[45] Freedman, W. L.; et al. *Astrophys. J.* 2001, 553, 47–72.

[46] Knop, R. A.; et al. *Astrophys. J.* 2003, 598, 102–137.

[47] Riess, A. G.; et al. *Astrophys. J.* 2004, 607, 665–687.

[48] Supernova Legacy Survey (SNLS): `http://cfht.hawaii.edu/SNLS`

[49] Astier, P.; et al. *Astron. Astrophys.* 2006, 447, 31–48.

[50] Riess, A. G.; et al. *Astrophys. J.* 2007, 659, 98–121.

[51] Equation of State: SupErNovae trace Cosmic Expansion (ESSENCE): `http://www.ctio.noao.edu/essence`

[52] Wood-Vasey, W. M.; et al. *Astrophys. J.* 2007, 666, 694–715.

[53] Jha, S.; Riess, A. G.; Kirshner, R. P. *Astrophys. J.* 2007, 659, 122–148.

[54] Guy, J.; Astier, P.; Nobili, S.; Regnault, N.; Pain, R. *Astron. Astrophys.* 2006, 443, 781–792.

[55] Hamuy, M.; Phillips, M. M.; Schommer, R. A.; Suntzeff, N. B.; Maza, J.; Aviles, R. *Astron. J.* 1996, 112, 2391–2397.
Jha, S.; et al. *Astron. J.* 2006, 131, 527–554.

[56] Davis, T. M.; et al. *Astrophys. J.* 2007, 666, 716–725.

[57] Wright, E. L. *Astrophys. J.* 2007, 664, 633–639.

[58] Sullivan, S.; Cooray, A.; Holz, D. E. *JCAP* 2007, 0709, 004.
Chongchitnan, S.; Efstathiou, G. *Phys. Rev. D* 2007, 76, 043508.
Wu, Q.; Gong, Y.; Wang, A.; Alcaniz, J. S. *Phys. Lett. B* 2008, 659, 34–39.
Lazkoz, R.; Majerotto, E. *JCAP* 2007, 0707, 015.
Tartaglia, A.; Capone, M.; Cardone, V.; Radicella, N. *AIP Conf. Proc.* 2008, 1059, 39–47.

[59] Mavromatos, N. E.; Mitsou, V. A. *Astropart. Phys.* 2008, 29, 442–452.
Mavromatos, N. E.; Mitsou, V. A. In *The Identification of Dark Matter: Proceedings of the Sixth International Workshop;* Axenides, M.; Fanourakis, G.; and Vergados, J.; Ed.; World Scientific: Singapore, 2007; pp 623-634.

[60] Gong, Y.; Wu, Q.; Wang, A. *Astrophys. J.* 2008 681, 27–39.
Kurek, A.; Szydlowski, M. *Astrophys. J.* 2008 675, 1–7
Szydlowski, M.; Kurek, A. Preprint arXiv:0801.0638 [astro-ph] 2008.

[61] Chevallier, M.; Polarski, D. *Int. J. Mod. Phys. D* 2001, 10, 213–224.
Linder, E. V. *Phys. Rev. Lett.* 2003, 90, 091301.

[62] For some recent studies, see:
Johri, V. B.; Rath, P. K. *Phys. Rev. D* 2006, 74, 123516.
Wu, P.; Yu, H. *JCAP* 2007, 0710, 014.
Xu, L. X.; Zhang, C. W.; Liu, H. Y. *Chin. Phys. Lett.* 2007, 24, 2459–2462.
Nesseris, S.; Perivolaropoulos, L. *Phys. Rev. D* 2008, 77, 023504.

[63] Wilkinson Microwave Anisotropy Probe (WMAP):
`http://map.gsfc.nasa.gov`

[64] Bennett, C. L.; et al. *Astrophys. J.* 1996, 464, L1–L4.
Jaffe, A. H.; et al. *Phys. Rev. Lett.* 2001, 86, 3475–3479.

[65] Kolb, E. W.; Turner, M. S. *The Early Universe;* Frontiers in Physics, Addison-Wesley: Reading, MA, 1988; 719 pages.

[66] Page, L.; et al. *Astrophys. J. Suppl.* 2007, 170, 335–376.

[67] Spergel, D. N.; et al. *Astrophys. J. Suppl.* 2007, 170, 377–408.

[68] Hinshaw, G.; et al. *Astrophys. J. Suppl.* 2007, 170, 288–334.

[69] Bahcall, N. A.; Ostriker, J. P.; Perlmutter, S.; Steinhardt, P. J. *Science* 1999, 284, 1481–1488.

[70] Spergel, D. N.; et al. *Astrophys. J. Suppl.* 2003, 148, 175–194.

[71] Eisenstein, D. J.; et al. *Astrophys. J.* 2005, 633, 560–574.

[72] Percival, W. J.; Cole, S.; Eisenstein, D. J.; Nichol, R. C.; Peacock, J. A.; Pope, A. C.; Szalay, A. S. *Mon. Not. Roy. Astron. Soc.* 2007, 381, 1053–1066.

[73] Sloan Digital Sky Survey (SDSS): `http://www.sdss.org`

[74] 2dF Galaxy Redshift Survey (2dFGRS):
`http://www.mso.anu.edu.au/2dFGRS`

[75] Tegmark, M.; et al. Phys. Rev. D 2006, 74, 123507.

[76] Tegmark, M.; et al. *Astrophys. J.* 2004 606, 702–740.

[77] Cole, S.; et al. *Mon. Not. Roy. Astron. Soc.* 2005, 362, 505–534.

[78] Perlmutter, S.; Turner, M. S.; White, M. J. *Phys. Rev. Lett.* 1999, 83, 670–673.
Tonry, J. L.; et al. *Astrophys. J.* 2003, 594, 1–24.

[79] Boughn, S.; Crittenden, R. *Nature* 427, 2004, 45–47.
Giannantonio, T.; et al. *Phys. Rev. D* 2006, 74, 063520.

[80] Chandra X-ray Observatory: `http://chandra.harvard.edu`

[81] Allen, S. W.; Schmidt, R. W.; Ebeling, H.; Fabian, A. C.; van Speybroeck, L. *Mon. Not. Roy. Astron. Soc.* 2004, 353, 457–467.

[82] Benjamin, J.; et al., *Mon. Not. Roy. Astron. Soc.* 2007, 381, 702–712.

[83] Chaboyer, B.; Demarque, P.; Kernan, P. J.; Krauss, L. M. *Astrophys. J.* 1998, 494, 96–110.
Hansen, B. M. S.; et al. *Astrophys. J. Suppl.* 2004, 155, 551–576.

[84] Simon, J.; Verde, L.; Jimenez, R. *Phys. Rev. D* 2005, 71, 123001.

[85] Li, H.; Su, M.; Fan, Z.; Dai, Z.; Zhang, X. *Phys. Lett. B* 2008, 658, 95–100.

[86] The following works consist an indicative yet not exhaustive list:
Nesseris, S.; Perivolaropoulos, L. *JCAP* 2007, 0701, 018.
Nesseris, S.; Perivolaropoulos, L. *Phys. Rev. D* 2005, 72, 123519.
Melchiorri, A.; Paciello, B.; Serra, P.; Slosar, A. *New J. Phys.* 2006, 8, 325.
Huterer, D.; H. V.Peiris, H. V. *Phys. Rev. D* 2007, 75, 083503.
Wang, Y.; Mukherjee, P. *Astrophys. J.* 2006, 650, 1–6.
da Conceicao Bento, M.; Bertolami, O.; Santos, N. M. C.; Sen, A. A. *J. Phys. Conf. Ser.* 2006, 33, 197–202.
Capozziello, S.; Cardone, V. F.; Elizalde, E.; Nojiri, S.; Odintsov, S. D. *Phys. Rev. D*

2006, 73, 043512.
Jassal, H. K.; Bagla, J. S.; Padmanabhan, T. *Phys. Rev. D* 2005, 72, 103503.
Zhang, X.; Wu, F. Q. *Phys. Rev. D* 2005, 72, 043524.
Rapetti, D.; Allen, S. W.; Weller, J. *Mon. Not. Roy. Astron. Soc.* 2005, 360, 555–564.

[87] Antoniadis, I.; Bachas, C.; Ellis, J. R.; Nanopoulos, D. V. *Phys. Lett. B* 1988, 211, 393–399.
Antoniadis, I.; Bachas, C.; Ellis, J. R.; Nanopoulos, D. V. *Nucl. Phys. B* 1989, 328, 117–139.
Antoniadis, I.; Bachas, C.; Ellis, J. R.; Nanopoulos, D. V. *Phys. Lett. B* 1991, 257, 278–284.
Ellis, J. R.; Mavromatos, N. E.; Nanopoulos, D. V. *Phys. Lett. B* 2005, 619, 17–25.
Ellis, J. R.; Mavromatos, N. E.; Nanopoulos, D. V. Preprint arXiv:hep-th/0105206 2001.

[88] Diamandis, G. A.; Georgalas, B. C.; Mavromatos, N. E.; Papantonopoulos, E. *Int. J. Mod. Phys. A* 2002, 17, 4567–4589.
Diamandis, G. A.; Georgalas, B. C.; Mavromatos, N. E.; Papantonopoulos, E.; Pappa, I. *Int. J. Mod. Phys. A* 2002, 17, 2241–2266.
Diamandis, G. A.; Georgalas, B. C.; Lahanas, A. B.; Mavromatos, N. E.; Nanopoulos, D. V. *Phys. Lett. B* 2006, 642, 179–186.
Ellis, J. R.; Mavromatos, N. E.; Nanopoulos, D. V.; Westmuckett, M. *Int. J. Mod. Phys. A* 2006, 21, 1379–1444, and references therein.

[89] Gasperini, M. *Phys. Rev. D* 2001, 64, 043510.
Gasperini, M.; Piazza, F.; Veneziano, G. *Phys. Rev. D* 2002, 65, 023508.
Bean, R.; Magueijo, J. *Phys. Lett. B* 2001, 517, 177–183.
Gasperini, M.; Veneziano, G. *Phys. Rept.* 2003, 373, 1–212.

[90] Milgrom, M. *Astrophys. J.* 1983, 270, 365–370.

[91] Bekenstein, J. D. *Phys. Rev. D* 2004, 70, 083509 [Erratum-ibid. D 2005, 71, 069901].

[92] Maartens, R. *Living Rev. Rel.* 2004, 7, 7, and references therein.

[93] For a partial list of references, see:
Mavromatos, N. E.; Rizos, J. *Phys. Rev. D* 2000, 62, 124004.
Mavromatos, N. E.; Rizos, J. *Int. J. Mod. Phys. A* 2003, 18, 57–84.
Neupane, I. P. In *Dark Matter in Astroparticle and Particle Physics: Proceedings of the 6th International Heidelberg Conference;* Klapdor-Kleingrothaus, H. V.; and Lewis, G. F.; Ed.; World Scientific: Singapore, 2008; pp 228-242.
Leith, B. M.; Neupane, I. P. *JCAP* 2007, 0705, 019.
Nojiri, S.; Odintsov, S. D.; Sami, M. *Phys. Rev. D* 2006, 74, 046004.
Copeland, E. J.; Sami, M.; Tsujikawa, S. *Int. J. Mod. Phys. D* 2006, 15, 1753–1936.
Neupane, I. P. *Class. Quant. Grav.* 2006, 23, 7493–7520.
Nojiri, S.; Odintsov, S. D. *Int. J. Geom. Meth. Mod. Phys.* 2007, 4, 115–146.
Kofinas, G.; Maartens, R.; Papantonopoulos, E. *JHEP* 2003, 0310, 066.
Mavromatos, N. E.; Papantonopoulos, E. *Phys. Rev. D* 2006, 73, 026001.

Nojiri, S.; Odintsov, S. D.; Sasaki, M. *Phys. Rev. D* 2005, 71, 123509, and references therein.

[94] Antoniadis, I.; Rizos, J.; Tamvakis, K. *Nucl. Phys. B* 1994, 415, 497–514.
Binetruy, P.; Charmousis, C.; Davis, S. C.; Dufaux, J. F. *Phys. Lett. B* 2002, 544, 183–191.
Sami, M.; Singh, P.; Tsujikawa, S. *Phys. Rev. D* 2006, 74, 043514.

[95] Carroll, S. M.; Duvvuri, V.; Trodden, M.; Turner, M. S. *Phys. Rev. D* 2004, 70, 043528.
Capozziello, S.; Carloni, S.; Troisi, A. In *Recent Research Developments in Astronomy & Astrophysics;* Pandalai, S. G.; Ed.; Research Signpost: Kerala, India, 2003; Vol. 1, Part II, pp 625–670.
Bertolami, O.; Paramos, J. *Phys. Rev. D* 2008 77, 084018.
Bertolami, O.; Boehmer, C. G.; Harko, T.; Lobo, F. S. N. *Phys. Rev. D* 2007, 75, 104016.

[96] Carroll, S. M.; De Felice, A.; Duvvuri, V.; Easson, D. A.; Trodden, M.; Turner, M. S. *Phys. Rev. D* 2005, 71, 063513.
Nojiri, S.; Odintsov, S. D.; Tretyakov, P. V. *Phys. Lett. B* 2007, 651, 224–231.

[97] Navarro, I.; Van Acoleyen, K. *Phys. Lett. B* 2005, 622, 1–5.

[98] Dvali, G. R.; Gabadadze, G.; Porrati, M. *Phys. Lett. B* 2000, 485, 208–214.
Deffayet, C. *Phys. Lett. B* 2001, 502, 199–208.
Deffayet, C.; Dvali, G. R.; Gabadadze, G. *Phys. Rev. D* 2002, 65, 044023.

[99] David, F. *Mod. Phys. Lett. A* 1988, 3, 1651–1656.
Distler, J.; Kawai, H. *Nucl. Phys. B* 1989, 321, 509–527.
Mavromatos, N. E.; Miramontes, J. L. *Mod. Phys. Lett. A* 1989, 4, 1847–1853.
D'Hoker, E.; Kurzepa, P. S. *Mod. Phys. Lett. A* 1990, 5, 1411–1422.

[100] Ellis, J. R.; Mavromatos, N. E.; Mitsou, V. A.; Nanopoulos, D. V. *Astropart. Phys.* 2007, 27, 185–198.
Mitsou, V. A. In *Fundamental Interactions: Proceedings of the 22nd Lake Louise Winter Institute;* Astbury, A.; Khanna, A. F.; and Moore, R.; Ed.; World Scientific: Singapore, 2008; pp 363-367.

[101] Minamitsuji, M.; Sasaki, M.; Langlois, D. *Phys. Rev. D* 2005, 71, 084019.

[102] Kolb, E. W.; Matarrese, S.; Riotto, A. *New J. Phys.* 2006, 8, 322.
Kolb, E. W.; Matarrese, S.; Notari, A.; Riotto, A. Preprint arXiv:hep-th/0503117 2005.

[103] Lemaitre, G. *Gen. Rel. Grav.* 1997, 29, 641–680 & *Annales Soc. Sci. Brux. Ser. I Sci. Math. Astron. Phys. A* 1933, 53, 51–85.
Tolman, R. C. *Proc. Nat. Acad. Sci.* 1934 20, 169–176.
Bondi, H. *Mon. Not. Roy. Astron. Soc.* 1947, 107, 410–425.

[104] Alnes, H.; Amarzguioui, M.; Gron, O. Phys. Rev. D 2006, 73, 083519.

[105] Gravanis, E.; Mavromatos, N. E. *Phys. Lett. B* 2002, 547, 117–127.

[106] Bachas, C. Preprint arXiv:hep-th/9503030 1995.

[107] Lahanas, A. B.; Mavromatos, N. E.; Nanopoulos, D. V. *Int. J. Mod. Phys. D* 2003, 12, 1529–1591, and references therein.
Mavromatos, N. E. In *Fundamental Interactions: Proceedings of the 22nd Lake Louise Winter Institute;* Astbury, A.; Khanna, A. F.; and Moore, R.; Ed.; World Scientific: Singapore, 2008; pp 80-127, and references therein.

[108] SuperNova Acceleration Probe (SNAP): `http://snap.lbl.gov`

[109] Albert, J.; et al. Preprint arXiv:astro-ph/0507458 2005, white paper to Dark Energy Task Force.

[110] Joint Dark Energy Mission (JDEM):
`http://universe.nasa.gov/program/probes/jdem.html`

[111] Dark Energy Space Telescope (Destiny): `http://destiny.asu.edu`

[112] For a recent review, see: Levi, M. E. *Nucl. Instrum. Meth. A* 2007, 572, 521–525.
Lampton, M.; et al. *Proc. SPIE Int. Soc. Opt. Eng.* 2002, 4849, 215–226.
Lampton, M.; et al. *Proc. SPIE Int. Soc. Opt. Eng.* 2003, 4854, 632–639.

[113] Albert, J.; et al. *Astropart. Phys.* 2004, 20, 377–389.
Albert, J.; et al. Preprint arXiv:astro-ph/0507460 2005, white paper to Dark Energy Task Force.

[114] Dark Universe Explorer (DUNE): `http://www.dune-mission.net`

[115] Large Synoptic Survey Telescope (LSST):
`http://www.lsst.org/lsst_home.shtml`

[116] Dark Energy Survey (DES): `https://www.darkenergysurvey.org`

[117] Baron, E. A.; Nugent, P. E.; Branch, D.; Hauschildt, P. H. *Astrophys. J.* 2004, 616, L91–L94.
Nugent, P.; et al. *Astrophys. J.* 2006, 645, 841–850.

In: Space Exploration Research
Editors: J. H. Denis and P. D. Aldridge

ISBN: 978-1-60692-264-4

Chapter 7

FINGERPRINTS OF A LOCAL SUPERNOVA

Oliver Manuel[1] and Hilton Ratcliffe[2]

[1] Nuclear Chemistry, University of Missouri, Rolla, MO 65401 USA
[2] Astronomical Society Southern Africa, PO Box 354, Kloof 3640 SOUTH AFRICA

ABSTRACT

The results of precise analysis of elements and isotopes in meteorites, comets, the Earth, the Moon, Mars, Jupiter, the solar wind, solar flares, and the solar photosphere since 1960 reveal the fingerprints of a local supernova (SN)—undiluted by interstellar material. Heterogeneous SN debris formed the planets. The Sun formed on the neutron (n) rich SN core. The ground-state masses of nuclei reveal repulsive n-n interactions that can trigger axial n-emission and a series of nuclear reactions that generate solar luminosity, the solar wind, and the measured flux of solar neutrinos. The location of the Sun's high-density core shifts relative to the solar surface as gravitational forces exerted by the major planets cause the Sun to experience abrupt acceleration and deceleration, like a yoyo on a string, in its orbit about the ever-changing centre-of-mass of the solar system. Solar cycles (surface magnetic activity, solar eruptions, and sunspots) and major climate changes arise from changes in the depth of the energetic SN core remnant in the interior of the Sun.

Keywords: supernovae; stellar systems; supernova debris; supernova remnants; origin, formation, abundances of elements; planetary nebulae; solar system; planetology; solar nebula; origin and evolution of solar system; origin and evolution of planets; solar system objects; comparative planetology; solar wind; solar physics; solar composition; solar interior; photosphere; solar emissions; star formation, origin, evolution, age; stellar dynamics and kinematics; pulsars; neutron stars; late stages of stellar evolution; normal stars; stellar characteristics and properties; star formation; stellar structure, interiors, evolution, ages; accretion and accretion disks; chemical composition; luminosity and mass functions; binding energies and masses; nucleosynthesis in supernovae; nuclear physics in supernovae; nuclear aspects of neutron stars; solar neutrinos; nucleon-nucleon interactions; nuclear physics; nuclear structure; nuclear forces; nuclear matter;

fundamental astronomy and astrophysics; nuclear astrophysics; nuclear matter in neutron stars; transport processes; mass spectrometers.

PACS: 97.60.Bw; 98.; 98.38.Mz; 98.58.Mj; 98.80.Ft; 98.38.Ly; 98.58.Li; 98.35.Pr; 96.; 96.10.+i; 96.10.Bc; 96.12.-a; 96.12.Bc; 96.15.-g; 96.30.-t; 96.30.Bc; 96.50.Ci; 96.60.-j; 96.60.Fs; 96.60.Jw; 96.60.Mz; 96.60.Vg; 98.35.Ac; 98.10.+z; 97.60.Gb; 97.60.Jd; 97.60.-s; 97.20.-w; 97.; 97.10.-q; 97.10.Bt; 97.10.Cv; 97.10.Gz; 97.10.Tk; 97.10.Xq; 21.10.Dr; 26.30.+k; 26.50.+x; 26.60.+c; 26.65+t; 13.75.Cs; 20.; 21.; 21.30.-x; 21.65.+f; 95.; 95.10.-a; 95.30.-k; 26.; 26.30.+k; 26.60.+c; 05.60.-k; 07.75.+h.

INTRODUCTION

Fingerprints of a local supernova are not necessarily obscure and may be associated with such current issues as climate change, although that is not widely recognized in the debate over global warming. Repulsive interactions between neutrons—in the tiny nucleus that occupies a negligible fraction of the volume of individual atoms and stars—is the energy source that lights the Sun and warms planet Earth [1, 2], causes heavy nuclei to fission and massive stars to explode [3], and as will be discussed below, may drive the solar cycle of sunspots and the axial emission of material in the solar wind.

Fifty years ago, in October of 1957, Burbidge, Burbidge, Fowler and Hoyle [4, hereafter B2FH] published their classical paper on nuclear synthesis (nucleosynthesis) of elements heavier than hydrogen in stars. B2FH [4] were able to show that eight different types of nuclear reactions and a reasonable model of stellar evolution—from a hydrogen-rich first generation star to the final explosion of an evolved star as a supernova—could account for all of the isotopes of all elements heavier than hydrogen in the solar system today and also explain the correlation observed between nuclear properties and isotope abundances.

One year earlier, in 1956, two scientists who would later analyze meteorites and experimentally confirm important features of the B2FH scenario of element synthesis—especially r-products from the final supernova explosion—were actively pursuing related research projects. A young nuclear chemist who had studied nuclear fission at the University of Tokyo during Word War II, Dr. Kazuo Kuroda, correctly predicted that self-sustaining nuclear chain reactions (fission) occurred naturally in the Earth's geologic history [5, 6]. And a young physicist who had studied rare modes of nuclear decay at the University of Chicago, Dr. John H. Reynolds, developed the high sensitivity mass spectrometer [7] that would confirm—still preserved in the solar system today—decay products from short-lived isotopes and unmixed isotope anomalies from the stellar nuclear reactions that B2FH [4] would propose in 1957.

The B2FH paper [4] was the culmination of the nuclear era—a stage of science that had been feverishly pursued by competing nations during World War II and that held the entire world as hostage with the threat of mutual nuclear annihilation during the Cold War. The B2FH paper was published in October of 1957, and the Cold War competition ushered in the space age that same month when the Soviet Union successfully launched Sputnik I on October 4, 1957. The Cold War and the "space race" continued, and twelve years later precise isotope analysis on samples returned from the 1969 Apollo Mission to the Moon [8] provided the first hint of an unimaginable fingerprint—partially grasped at the time [9] but not fully recognized until 14 years later [10]—indicating that our Sun sorts atoms by mass and covers its surface with lightweight elements, but its interior consists almost entirely of even-Z elements of high nuclear stability from the deep interior of a supernova [10].

In 1960 Reynolds [11] reported the first hint of a supernova fingerprint in the material that formed the solar system—radiogenic ^{129}Xe from the decay of extinct ^{129}I ($t_{1/2}$ = 16 My) in a meteorite. Fowler, Greenstein and Hoyle [12] quickly noted that Reynolds' discovery was difficult to reconcile with the idea that the solar system formed from a typical interstellar cloud and suggested that ^{129}I and other short-lived radioactive nuclei might have been produced locally, by less violent nuclear reactions than those that occur in a supernova. However, four years later Rowe and Kuroda [13] reported a less ambiguous supernova fingerprint in another meteorite—the decay products of extinct ^{244}Pu ($t_{1/2}$ = 82 My). Unlike ^{129}I, the radioactivity of ^{244}Pu could not be explained by ordinary neutron capture on known nuclides in the solar system. The presence of ^{244}Pu in meteorites required rapid neutron capture, the r-process described by B2FH [4], in a supernova at the birth of the solar system.

The most recent evidence for an energized, high-density supernova core lurking inside the Sun came from long-term geological studies of periodic changes in Earth's climate that are related to periodic changes in solar activity and to acceleration and deceleration of the Sun as it orbits the centre-of-mass of the solar system [14, 15].

The identification of each supernova fingerprint in the solar system initially seemed far-fetched and encountered stiff opposition. For example, this interpretation of the empirical link between primordial He in meteorites with excess ^{136}Xe from the r-process of nucleosynthesis [16] was described in 1977 as *". . . too extreme to merit discussion"* [ref. 17, p. 209]. A decade later other critics stated that this idea is *". . . based on the use of an over-simplified model and a careless interpretation of the existing rare gas isotope data."* [ref. 18, p. 315].

Nevertheless, the empirical link of primordial He with excess ^{136}Xe from the r-process of nucleosynthesis formed the basis for the successful 1983 prediction [10] that the Galileo Mission would find excess ^{136}Xe from the r-process when the mass spectrometer on the Galileo probe entered Jupiter's He-rich atmosphere in 1996 [19].

Below is a summary of ten probable supernova fingerprints that have been recognized in the solar system since 1960, in approximately reverse chronological order—beginning with long-term, scholarly studies of periodicities in climate change and sunspots that might arise from a dense, energetic SN core inside the Sun and ending with isotope measurements that first revealed the decay products of short-lived radioactive nuclides from a supernova in meteorites and planets. These fingerprints are consistent with the scenario shown in figure 1: The Sun exploded as a supernova five billion years (5 Gy) ago, giving birth to the planetary system [16]. After ejecting the material that now orbits it as planets, moons, comets, and asteroids, the Sun reformed as an iron-rich plasma diffuser around the neutron-rich supernova core [20].

The most obvious fingerprint of a local supernova—a striking similarity in the chemical layers of elements inside the Earth, across the solar system, and in the onion-skin model of a pre-supernova star—was not recognized until quantitative and precise experimental data showed that meteorites contain the decay products of short-lived nuclides and isotopic anomalies from stellar nucleosynthesis that are still linked today with elements from different layers of a heterogeneous, evolved star [16].

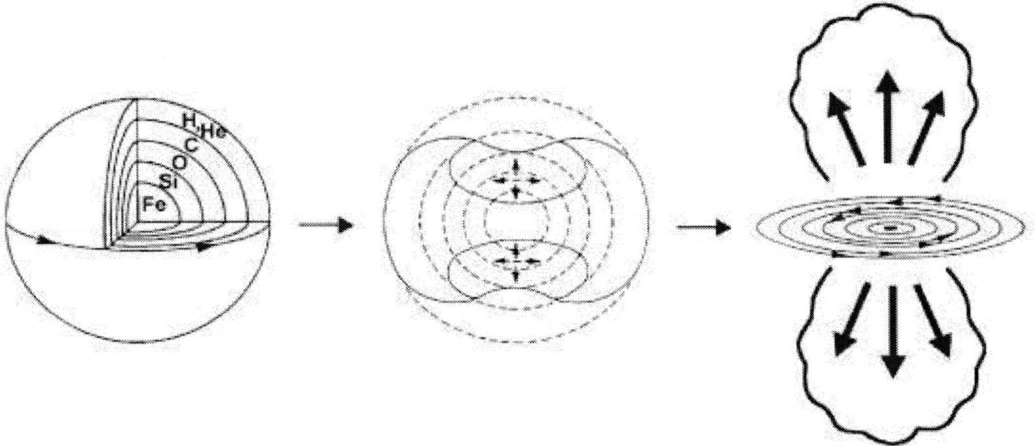

A massive, spinning precursor Sun was highly evolved and chemically layered when asymmetric collapse occurred to conserve angular momentum .

The infall of low-Z elements caused an axially directed SN explosion, producing a rapidly expanding bipolar nebula with an equatorial accretion disk.

The current Sun re-formed on the SN core; cores of inner planets formed in the central Fe-rich region; Jovian planets formed in the outer SN layers.

Figure 1. This is a schematic drawing of the birth of the solar system, first revealed by the decay products of short-lived nuclides, isotopic anomalies from stellar nucleosynthesis reactions, and primordial He in meteorites that is tightly linked with excess Xe-136 from the r-process of nucleosynthesis [16].

SUPERNOVA FINGERPRINTS

I. Sunspots, Solar Cycles, Solar Inertial Motion, Climate Changes

Empirically links between sunspots, solar cycles, solar inertial motion, and changes in Earth's climate exposed the latest and most controversial fingerprint of a local supernova. As this paper goes to press, there is a great deal of debate and political wrangling over the possibility that the release of carbon dioxide into the atmosphere from the burning of fossil fuels has been the primary cause of global warming since about 1950.

Before entering the political arena, many scholarly studies [e.g., 14, 15, 21-24 and references therein] had shown that past climate changes were linked with solar activity and sunspots as the Sun moved in an irregular orbit about the centre-of-mass of the solar system.

For example, Alexander *et al.* [15] state that, "*Sunspot production is a direct function of the sun's galactic acceleration and deceleration . . .*" [ref. 15, p. 42] as the Sun moves like a yoyo on a string, in orbit about the constantly changing centre-of-mass (barycentre) of the solar system.

Figure 2 illustrates this uneven, jerky motion that the Sun experiences because of sudden acceleration and deceleration in its orbit about the barycentre of the solar system [24]. The gravitational force exerted by the planets, especially the more massive ones like Jupiter and Saturn, determines the position of the centre-of-mass of the solar system. The empirical link between planetary motions, sunspots, solar activity and climate change [14, 15, 21-24] contains valuable information on the internal structure of the Sun.

Figure 2 shows three complete orbits of the Sun, each of which takes about 179 years. Each solar orbit consists of about eight, 22-year solar cycles [24]. The total time span shown in figure 2 is therefore three 179-year solar cycles [22], about 600 years.

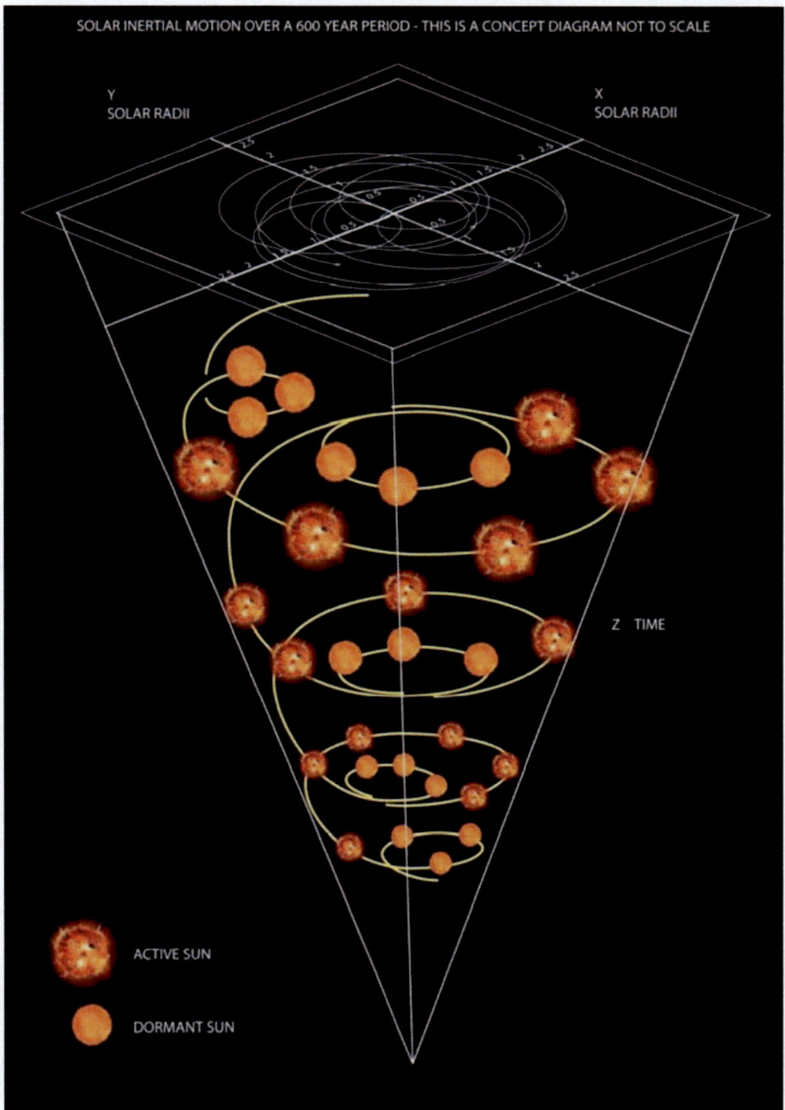

Figure 2. This schematic drawing illustrates sudden shifts in the solar inertial motion (SIM) as the Sun travels in an epitrochiod-shaped orbit about the centre-of-mass of the solar system. This 2006 drawing by Daniel Brunato, University of Canberra, was reproduced by permission from Richard Mackey [24].

The standard model of the Sun as a homogeneous ball of hydrogen remains very popular. That is perhaps why *"the mechanism for sunspot production as a result of galactic velocity changes in the sun has yet to be determined"* [15, p.42]. The observed link between sunspot production and changes in the sun's velocity would not be possible if the Sun were homogeneous [23, 24].

However, if the solar magnetic fields arise from iron-rich material in the deep interior of the Sun or from the compact neutron object at the solar core [25], then the location of this high-density material would likely shift relative to the visible solar surface as gravitational forces exerted by the major planets cause the Sun to experience abrupt acceleration and deceleration in its orbit about the ever-changing barycentre of the solar system [15].

We suggest that the empirical link between sunspots, solar cycles, solar inertial motion, and climate changes is the latest fingerprint of a local supernova—one that left a high abundance of iron and an extremely dense neutron core that moves about the solar interior as the Sun orbits the centre-of-mass of the solar system [14, 15, 21-24].

II. Bipolar Outflows and Axially Directed Jets

Repulsive interactions between neutrons in the solar core generate solar luminosity, solar neutrinos, and the solar wind [e.g., 1, 2, 20]. If the solar cycle of sunspots is produced by changes in the location of that dense object in the interior of the Sun, as suggested above, then other puzzling features of the solar surface may arise from the emission of high energy neutrons in the solar core and be telltale signs of the collapsed supernova remnant there [3].

The anisotropic outflow of the solar wind, illustrated in figure 3, is one such feature. This figure is from a 2005 NASA report on the primary mission results [http://ulysses.jpl. nasa.gov/science/mission_primary.html] of the Ulysses spacecraft. The top panel is an X-ray image of the Sun from the Soft X-ray Telescope on the Japanese Yohkoh spacecraft. The bottom panel shows the solar wind speed and density that the Ulysses spacecraft observed in measurements made at various latitudes, from near the South Pole to near the North Pole.

We suggest that the high-speed solar wind, 700-800 km/s, coming from the polar regions of the Sun is another fingerprint—faintly visible at the solar surface—of the compact object at the solar core. The solar-wind hydrogen is the decay product of neutrons emitted from the Sun's core [1, 2, 20], and the Spitzer telescope [26] has recently seen bipolar jets *"at a very early stage in the life of an embryonic sun-like star"* [http://www.spitzer.caltech.edu/Media/ releases/ssc2007-19/release.shtml]. Bi-polar jets are common in compact stellar objects, and Adam Frank notes in a recent review, *"Bipolar outflows and highly collimated jets are nearly ubiquitous features associated with stellar mass loss"* [ref. 27, p. 241].

If asymmetry in the solar wind (figure 3) begins by the same process that causes material to emanate axially from compact stellar objects, then this process may start with anisotropic emission from the solar core—as has been seen for $\alpha-$ [28] and n-emission [29] from nuclei. Neutron-emission requires penetration of the gravitational barrier around a neutron star [1, 2, 20], just as a-emission from a nucleus requires penetration of the Coulomb barrier. We suggest that neutrons are emitted axially from the solar core, perhaps for the same reason a-particles are emitted in an axial direction from nuclei [28]: The shortest path through the barrier and thus the highest probability of barrier penetration may occur in an axial direction, because that is perpendicular to both the surface and to the spin direction. Higher angular momentum of particles at the equator may reduce their probability of escape by sending the particle on a longer path through the barrier.

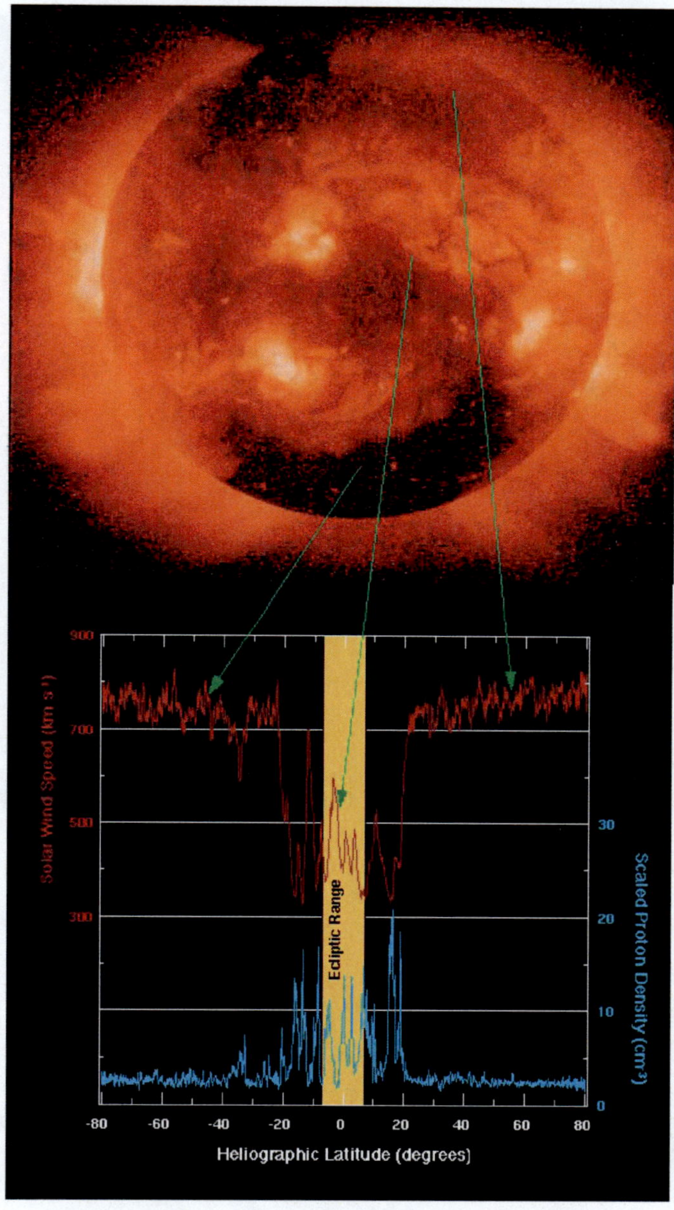

Figure 3. This is a summary of the primary findings from the Ulysses spacecraft, a combined NASA/ESA mission that orbited the Sun and measured properties of the solar wind as a function of solar latitude. This figure is from their report [http://ulysses.jpl.nasa.gov/science/mission_primary.html] dated 25 August 2005.

III. Rigid, Iron-Rich Structures Below the Solar Photosphere

Figure 4 shows the remarkable fingerprint of a local supernova that was first noticed as rigid, iron-rich structures beneath the fluid photosphere in 2005 [30, 31], almost three decades after it was first suggested that a supernova gave birth to the solar system and the Sun formed on the collapsed SN core [32].

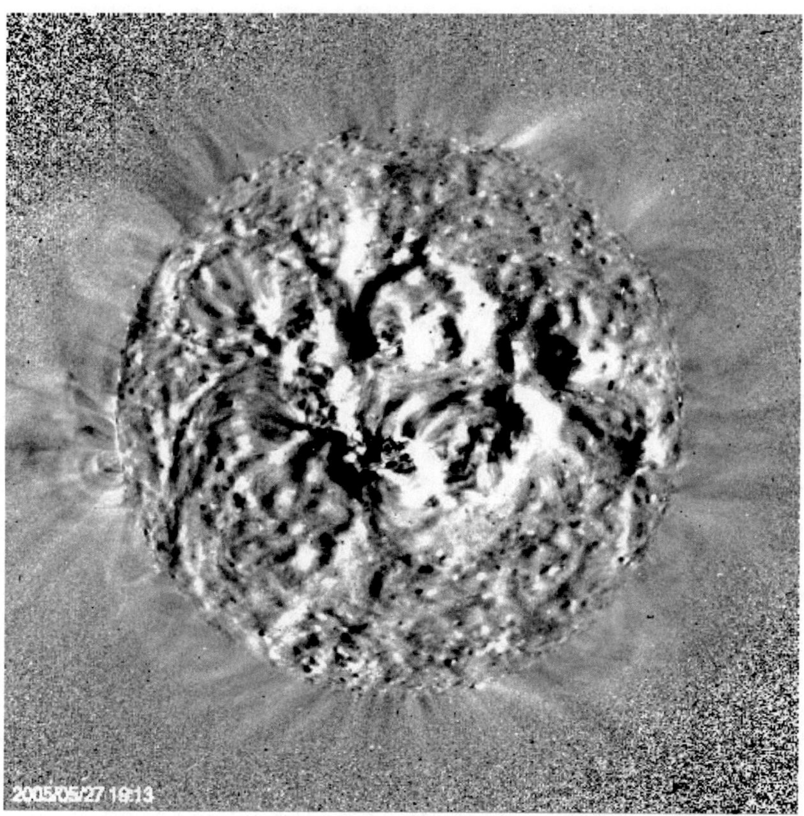

Figure 4. This "running difference" image of the Sun was taken by the SOHO spacecraft on May 27th 2005 at 19:13, using a 195 Å filter to enhance light from emissions by Fe (IX) and Fe (X). These rigid, iron-rich structures are visible for several days and even weeks later, rotating from left to right [30].

Michael Mozina, a software engineer in Mt. Shasta, CA, explained the impact of these solar images from the SOHO spacecraft on his opinion about the Sun [30]: *"While viewing images from SOHO's EIT program, I finally stumbled across the raw (unprocessed EIT images) marked "DIT" images that are stored in SOHO's daily archives. After downloading a number of these larger "DIT" (grey) files, including several "running difference" images, it became quite apparent that many of the finer details revealed in the raw EIT images are simply lost during the computer enhancement process that is used to create the more familiar EIT colorized images that are displayed on SOHO's website. That evening in April of 2005, all my beliefs about the sun changed*" [30, 31].

Mozina also noticed rigid, iron-rich structures in images that were taken with the TRACE satellite, using filters to enhance light emissions from iron ions [30].

Figure 5 shows active region AR 9143, using the 171Å filter that is specifically sensitive to the iron ion (Fe IX/X) emissions. On 28 August 2000, an eruption and mass ejection occurred from this small region of the visible solar surface. A video recording of the mass ejection event can be viewed here: http://trace.lmsal.com/POD/movies/T171_000828.avi

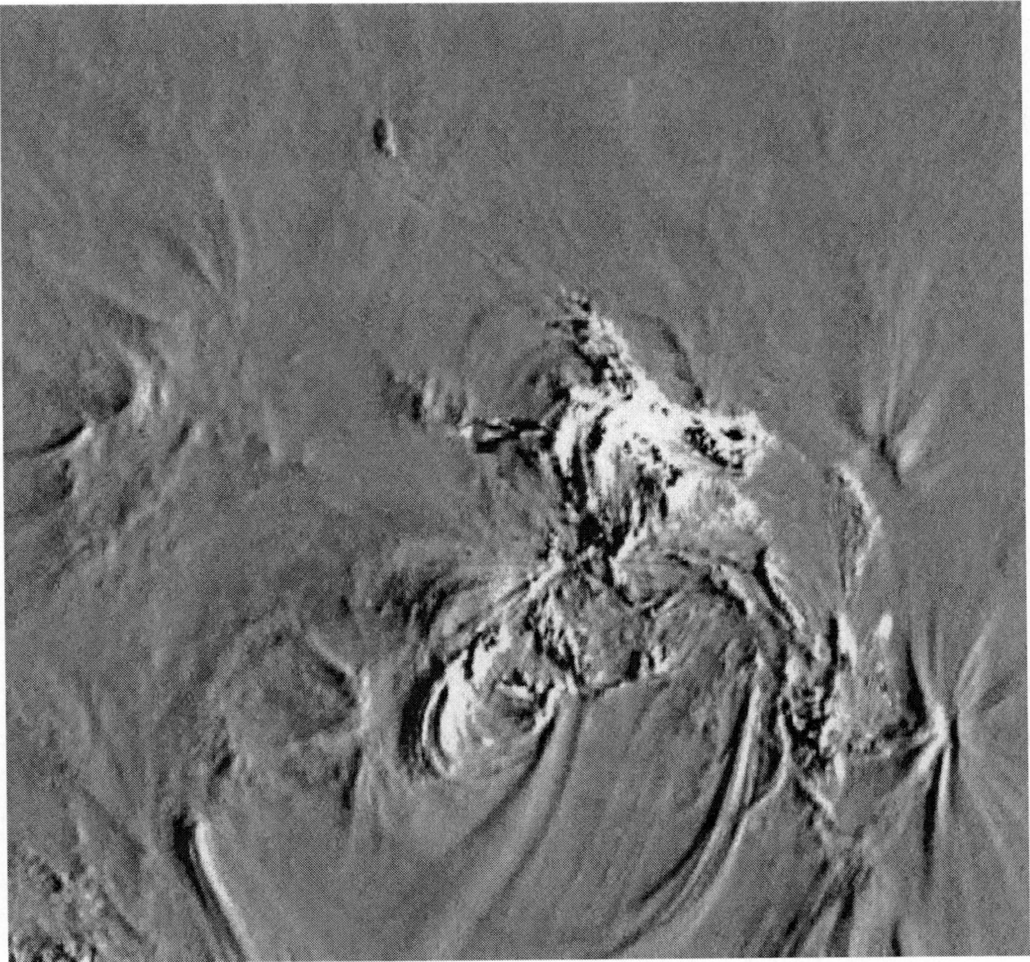

Figure 5. This is a "running difference" image of a small region of the Sun that the TRACE satellite recorded on 28 August 2000 using a 171 Å filter. This filter is specifically sensitive to light emissions from iron ions, Fe IX and FeX. The TRACE satellite later recorded an eruption and mass ejection from this same region. To view a video recording of the flare event, go to: http://trace.lmsal.com/POD/movies/T171_000828.avi.

IV. Solar Luminosity, Solar Neutrinos, and Solar Hydrogen

As noted in the introduction, a systematic enrichment of lightweight isotopes in samples from the 1969 Apollo Mission to the Moon [8] provided an early hint [9] that the Sun sorts atoms by mass and selectively moves lightweight ones to its surface. Subsequent measurements showed a common mass fractionation pattern, extending from 3 amu to 136 amu (atomic mass units) [10]. Correcting the photosphere for mass fractionation yielded Fe, Ni, O, Si, S, Mg and Ca as the main constituents of the solar interior [10]—the same even-Z elements that constitute ≈ 99% of the material in ordinary meteorites [33] and in rocky planets close to the Sun. The probability of this remarkable agreement being a coincidence is essentially zero, $P < 2 \times 10^{-33}$ [34].

These seven elements all have high nuclear stability. Therefore, the source of solar luminosity, solar neutrinos and solar-wind hydrogen pouring from the surface of an iron-rich Sun remained a mystery for almost two decades, from 1983 until 2000.

That year five graduate students in an advanced nuclear science class at the University of Missouri-Rolla (Chem. 471)–Cynthia Bolon, Shelonda Finch, Daniel Ragland, Matthew Seelke and Bing Zhang–helped the first author of this paper construct a 3-D plot of reduced nuclear variables, M/A (mass per nucleon, or potential energy per nucleon) and Z/A (charge density, or charge per nucleon), showing each of the 3,000 known nuclides in the ground state [35]. The results, published on the cover of the book, "*Origin of Elements in the Solar System: Implications of Post 1957 Observations*" [27] and elsewhere [1, 2], are shown on the left side of figure 6 as the "Cradle of the Nuclides."

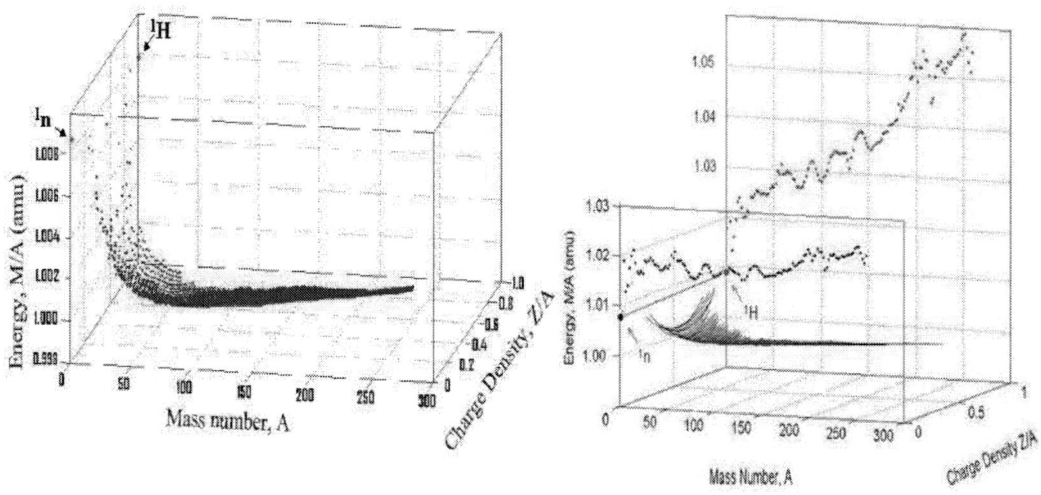

Figure 6. The "Cradle of the Nuclides" on the left shows the potential energy per nucleon for all stable and radioactive nuclides that were known in 2000 [34]. The more stable nuclides have lower values of M/A and occupy lower positions in the cradle. Nuclei that are radioactive or readily consumed by fusion or fission occupy higher positions. In the figure on the right, mass parabolas through data points at each value of A>1 intersect the front plane at $\{Z/A = 0, M/A = (M/A)_{neutron} + \sim 10 \text{ MeV}\}$.

Data points on the left side of figure 6 represent each atomic mass in the ground state [35]. The right side of figure 6 shows the mass parabolas defined by the data at each value of A>1 [35]. Intersections of mass parabolas with the front plane at Z/A = 0 show the potential energy per nucleon, M/A, for assemblages of pure neutrons. Intersections of mass parabolas with the back plane at Z/A = 1 show the potential energy per nucleon, M/A, for assemblages of pure protons. Repulsion between positive charges causes the value of M/A at Z/A = 1 to be larger than that at Z/A =0 [1]. The intercepts at Z/A = 0 and Z/A = 1.0 display peaks and valleys at the same mass numbers because of tight and loose packing of nucleons, respectively [2].

Systematic properties of nuclei in figure 6 reveal strong attractive interactions between unlike nucleons (n-p) and repulsive but symmetric interactions between like nucleons (n-n or p-p), that is further increased by Coulomb repulsion between positive charges on protons [1, 2]. Thus, empirical nuclear data indicate that a neutron is in an excited state in a neutron star, with about +10-22 MeV more energy than a free neutron [36]. On the other hand, theoretical studies suggest that neutron stars are "dead" nuclear matter, with each neutron having about -93 MeV less energy than a free neutron [37].

We prefer the conclusion based on empirical nuclear mass data [1, 2, 34, 36], and a recent review [38] agrees that useful information on neutron stars can be obtained by extrapolating atomic mass data out to *"homogeneous or infinite nuclear matter (INM)"* [ref. 38, p. 1042]. Thus solar luminosity, solar neutrinos and solar-wind hydrogen coming from the surface of an iron-rich object that formed on a collapsed SN remnant [16] are fingerprints of energetic neutrons in the solar core. These processes generate luminosity, neutrinos, and an outflow of 3×10^{43} H^+ per year in the solar wind by the following series of reactions [1, 2, 34, 36]:

1. Neutron emission from the solar core $<{}_0^1n> \rightarrow {}_0^1n + \sim 10\text{-}22$ MeV	Generates >57% of solar luminosity
2. Neutron decay ${}_0^1n \rightarrow {}_1^1H^+ + e^- + \text{anti-}\nu + 0.78$ MeV	Generates < 5% of solar luminosity
3. Fusion and upward migration of H^+ $4\ {}_1^1H^+ + 2\ e^- \rightarrow {}_2^4He^{++} + 2\ \nu + 27$ MeV	Generates <38% of solar luminosity
4. Escape of excess H^+ in the solar wind $3 \times 10^{43}\ H^+/yr \rightarrow$ Departs in solar wind	Generates 100% of solar-wind H

V. The Iron-Rich Plasma Diffuser with a Hydrogen-Rich Veneer

Abundant lightweight elements at the solar surface, identified with large diamonds in Figure 7, were for several decades too familiar to be seen as strange fingerprints. Atomic abundances at the solar surface are ≈91% H, ≈9% He and ≈ 0.2% for all eighty-one heavier elements [39]. Hydrogen (H) is the lightest of all elements; He is the next-lightest one. Growing empirical evidence of severely mass fractionated isotopes [9, 11, 40-47] in the solar system pointed clearly to a solar site of the mass fractionation by 1983 [10], and two papers report recent observational evidence that elements are sorted by mass in other stars [48, 49].

Excess lightweight isotopes of He, Ne, Ar, Kr and Xe implanted in lunar soils by the solar wind revealed a common mass-dependent fractionation power law in 1983 [10], as shown in figure 8. The abundance of a lighter isotope of mass (L) in the solar wind has been increased relative to that of a heavier isotope of mass (H) by a fractionation factor (f), where

$$\log (f) = 4.56 \log (H/L) \tag{1}$$

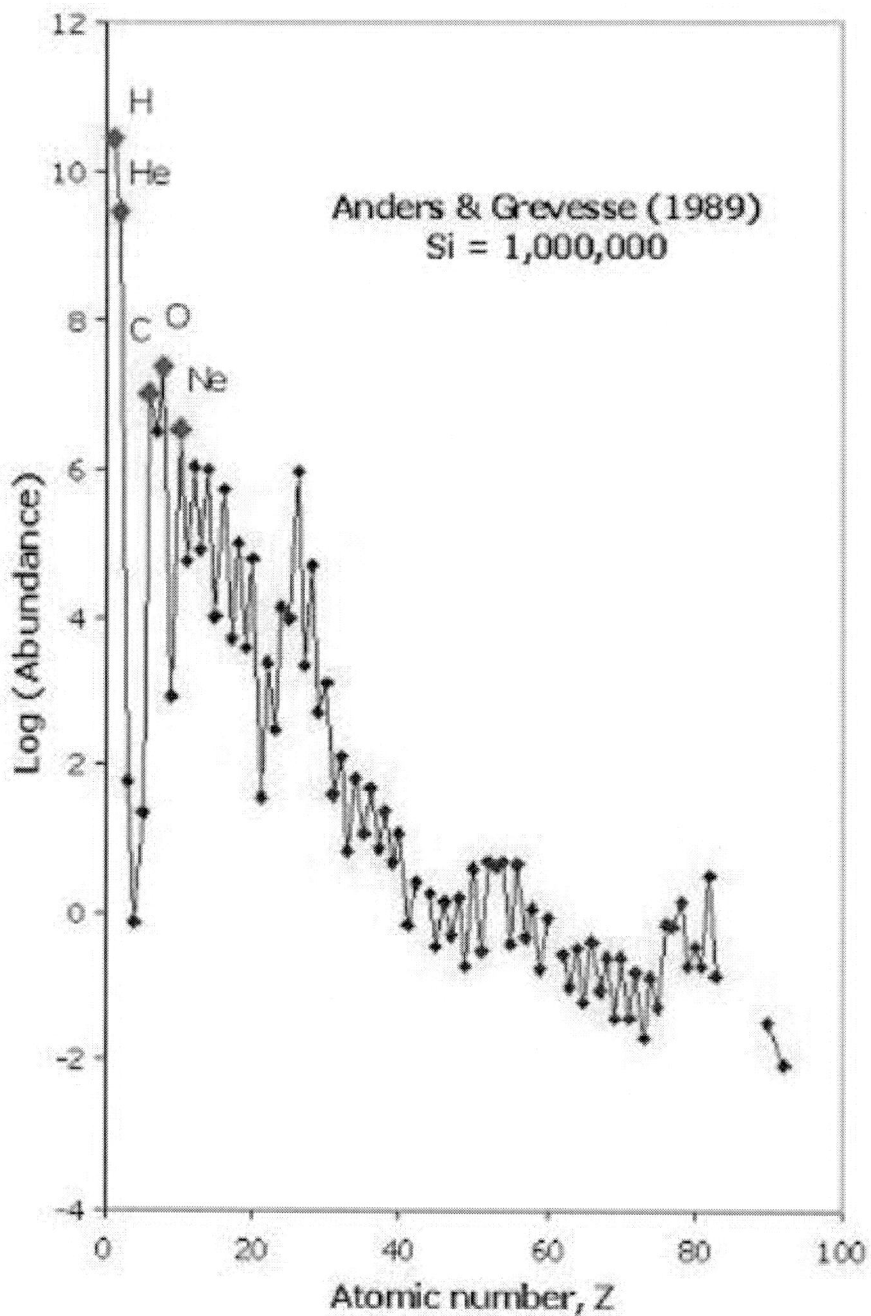

Figure 7. The abundance pattern of elements in the photosphere [39] generally declines with atomic (Z) number, except for a deficit of elements with unusually low nuclear stability (Li, Be and B) and an excess of those with unusually high nuclear stability (Fe and Ni).

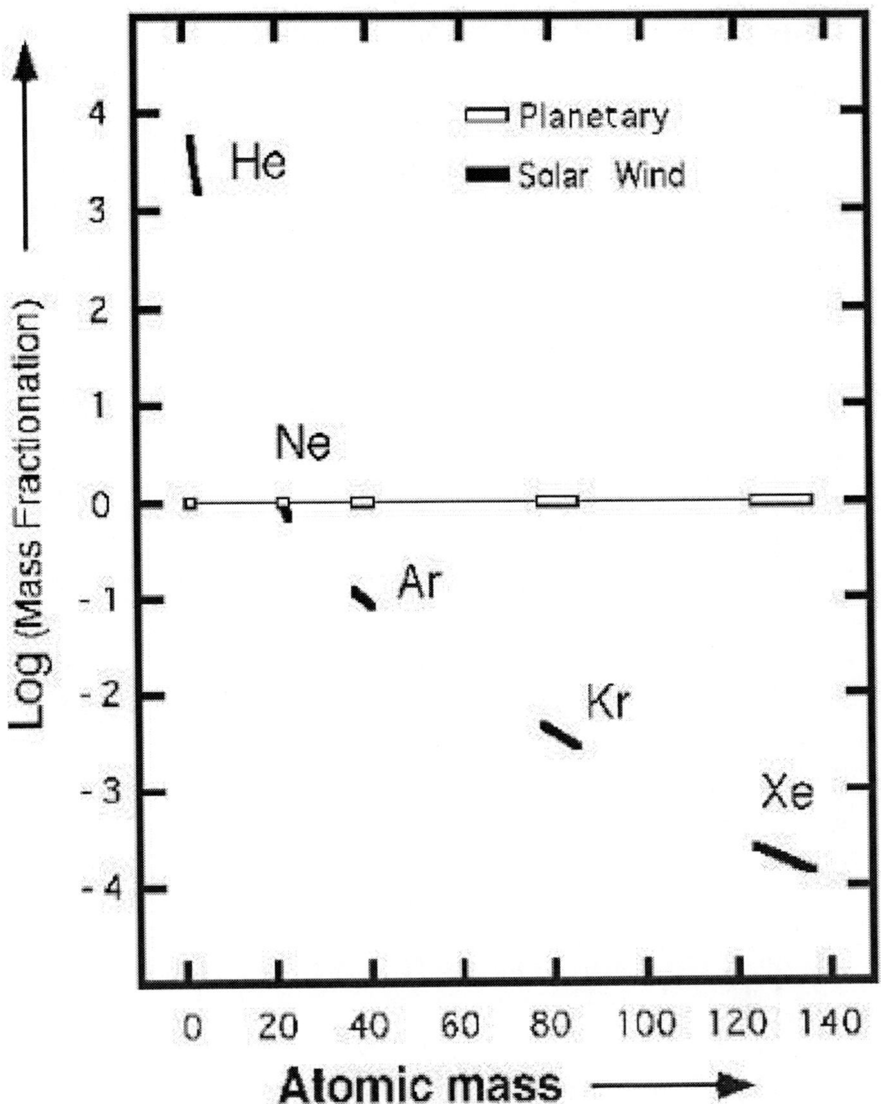

Figure 8. Isotope ratios of noble gases implanted in lunar soils exhibit a smooth, mass-dependent fractionation [10]. The percent enrichment per mass unit decreases as mass increases, from a value >200% per amu (atomic mass unit) for He to a value of 3.5% per amu across the nine stable isotopes of Xe. Mass fractionation is shown relative to ^{20}Ne.

Figure 9 shows the abundance of elements in the bulk Sun, after correcting the abundance of elements in the solar photosphere [39] for the mass fractionation seen across the isotopes of noble gases in the solar wind [10]. The most abundant elements in the interior of the Sun are also the most abundant elements in ordinary meteorites— Fe, O, Ni, Si and S [33]. In 2005 it was shown that the fractionation pattern of s-products in the photosphere [50] also yields Fe, O, Ni, Si and S as the most abundant elements in the bulk Sun.

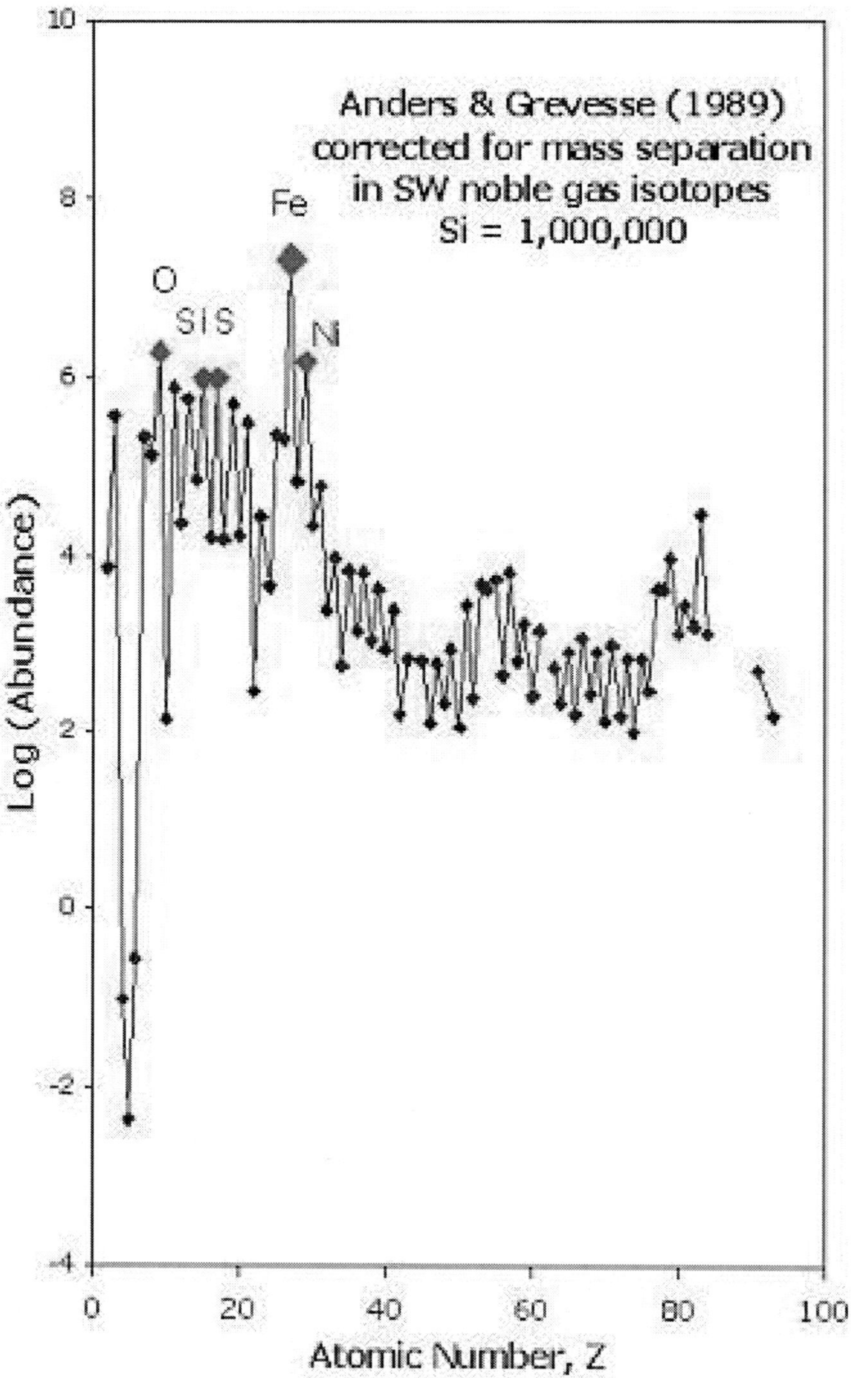

Figure 9. Large diamonds identify abundant elements inside in the interior of the Sun, calculated by correcting the abundance of elements in the photosphere [39] for the mass fractionation observed across the isotopes of elements in the solar wind [10].

VI. Mirror Image Isotope Anomalies: Excesses and Deficits:

Reynolds encountered two SN fingerprints in xenon isotopes from meteorites in 1960: Radiogenic ^{129}Xe from the *in situ* decay of SN-produced ^{129}I [11] and isotope anomalies [51] from mass fractionation and stellar nuclear reactions. This was later called "strange xenon" [17] or Xe-2 [49]—xenon that is enriched in ^{124}Xe and ^{136}Xe from the p- and r-processes of nucleosynthesis in a supernova [4].

A mirror image of "strange" xenon was discovered in silicon carbide (SiC) mineral separates of another meteorite in 1978 [53]. This mirror image component is enriched in the intermediate-mass xenon isotopes, ^{128,130}Xe, which were made by the slow, s-process of nucleosynthesis [4], before the star reached the terminal supernova stage.

Precise isotope analysis of other heavy elements in meteorites and mineral separates of meteorites over the next fifteen years also indicated that "normal" isotope abundances of heavy elements may be a mixture of heterogeneous stellar debris that had still retained the signatures of distinct stellar nuclear reactions when refractory solids started to form in the solar system: Diamonds and graphite trapped heavy elements that were enriched in isotopes made by rapid r- and p-nucleosynthesis reactions in a supernova [4]; silicon carbide minerals trapped heavy elements that were enriched in the intermediate-mass isotopes that were made earlier by the slow s-process of nucleosynthesis [54-56].

Such nucleogenetic isotopic anomalies had been reported in six elements—krypton (Z = 36), tellurium (Z = 52), xenon (Z = 54), barium (Z = 56), neodynium (Z = 60) and samarium (Z = 62) by 1993, when Professor Begemann [56] brought attention to the striking "mirror-image" (+ and -) isotopic anomaly patterns in the isotopes of barium, neodynium, and samarium in mineral separates of meteorites.

Figure 10 is an illustration of the "mirror-image" isotopic anomaly patterns that Begemann noted in Ba, Nd and Sm from mineral inclusions of the Allende and Murchison meteorites [56]. Elements in inclusion EK-1-4-1 of the Allende meteorite (top) are enriched in isotopes that were made by rapid nuclear reactions in a supernova—the r- and p-processes of element synthesis [4]. These same elements in silicon carbide (SiC) inclusions of the Murchison meteorite (bottom) are enriched in intermediate mass isotopes that were made earlier by the s-process.

The stellar debris that produced the isotopic anomaly patterns shown in figure 10 was heterogeneous in both chemical and isotopic compositions. That is why the isotope record of nucleosynthesis by the r-, p- and s-processes still survives today in chemically distinct minerals and inclusions of meteorites; isotopes made by s-process were trapped in different minerals than isotopes made by the r- and p-processes.

Major elements formed silicon carbide (SiC) in a region of the heterogeneous stellar debris where s-products were prominent in the isotopes of heavy trace elements like Kr, Te, Xe, Ba, Nd, and Sm [53-56]. Carbon condensed into graphite or diamond (C) in a region of the heterogeneous stellar debris where r- and p-products were prominent in the isotopes of heavy trace elements like Kr, Te, Xe, Ba, Nd, and Sm [53-56].

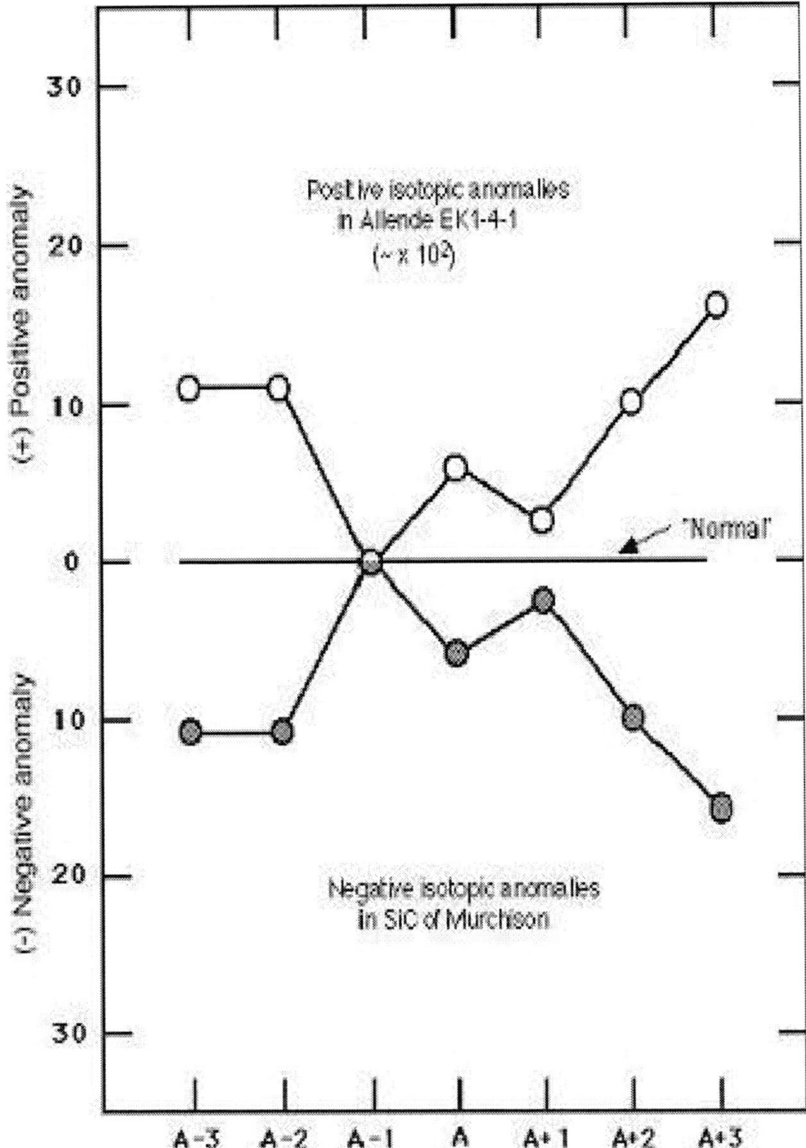

Figure 10. "Mirror-image" isotopic anomalies [56] observed in the isotopes of Ba, Nd and Sm from inclusion EK-1-4-1 of the Allende meteorite (top) and silicon carbide (SiC) inclusions of the Murchison meteorite (bottom). Isotope anomalies in these two meteorite inclusions are complementary, but the scale of s-products in the silicon carbide inclusion of the Murchison meteorite is about two orders-of-magnitude larger than that of the r- and p-products in inclusion EK 1-4-1 of the Allende meteorite [56].

VII. Linked Chemical and Isotopic Heterogeneities in Planets

As mentioned above, the nucleogenetic isotopic anomaly pattern of a trace element may be linked with the presence of a specific major element. The link of ^{136}Xe from the r-process with primordial He in meteorites [16-18] was the first example. Linked elemental and isotopic variations in elements of meteorites were pivotal to the search for variations in isotope abundances that are linked with major differences in the chemical compositions of planets and to the scenario shown in figure 1 for the birth of the solar system from a supernova [16, 32].

Aside from detailed isotopic information on the material in planet Earth, only limited data are available on a few elements in two extra-terrestrial planets, Jupiter and Mars. Fortunately these two planets lie on opposite sides of the asteroid belt—the great chemical divide that separates the small, dense, rocky planets made mostly of Fe, O, Si, Ni, and S from the large, gaseous planets that are made mostly of the lightweight elements like H, He, C, and N.

Below is a list of the linked chemical and isotopic anomalies that have been confirmed to date in large objects in the solar system:

- The Galileo probe into the He-rich atmosphere of Jupiter in 1996 recorded excess ^{136}Xe from the r-process of nucleosynthesis [19], like the excess ^{136}Xe that that was trapped with "strange xenon" [57] and primordial helium in carbonaceous grains of meteorites at the birth of the solar system [16].
- The abundance pattern of xenon isotopes in Earth and Mars [58]— planets that are rich in iron (Fe) and sulfur (S)—matches the xenon isotope pattern seen in iron-sulfide (FeS) minerals of metallic [59] and stony [60] meteorites.
- Xenon in the Sun is mostly like that in Earth and Mars [58], but light isotopes are enriched at the solar surface by 3.5% per amu from solar mass fractionation. Only about 7-8% of the ^{136}Xe in the Sun is the "strange xenon" that is seen in Jupiter and in carbonaceous grains of meteorites [61, 62].
- The level of mono-isotopic ^{16}O in oxygen from the Earth is like that seen in differentiated meteorites, but less than that in carbonaceous meteorites [63].
- The latest measurements indicate that the level of mono-isotopic ^{16}O (from He-burning) is lower in the Sun's oxygen than in the Earth or in carbonaceous meteorites [64], despite an earlier report that ^{16}O is enriched in the Sun [65].

VIII. Linked Chemical and Isotopic Heterogeneities in Meteorites

Figure 11 is a summary of a 1976 University of Chicago survey [63] on the level of mono-isotopic ^{16}O in different classes of meteorites and their mineral separates. Oxygen-16 is a He-burning product—a nucleosynthesis fingerprint that distinguishes six categories of meteorites and mineral separates with distinctive chemical characteristics. This empirical evidence of linked chemical and isotopic variations in meteorites may be explained by a local supernova (figure 1) [16] or perhaps by "mass-independent fractionation" at the solar system's birth [66, 67]. However, "mass-independent fractionation" [66, 67] does not explain the latest results that show a deficiency of ^{16}O in the Sun [64].

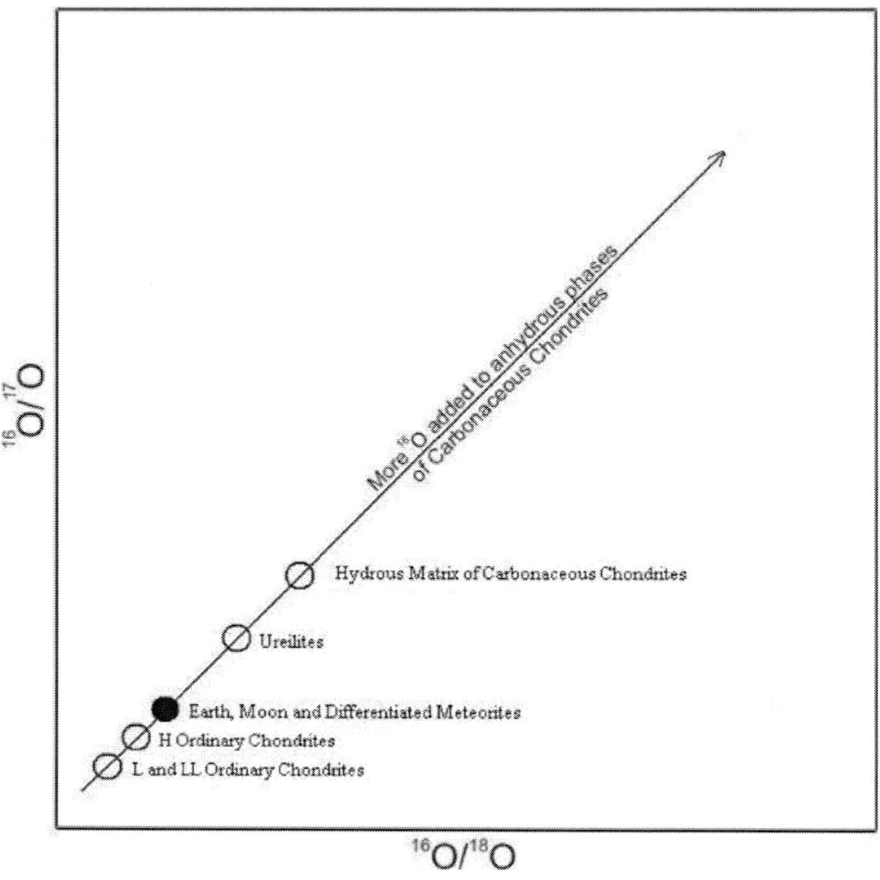

Figure 11. The abundance of ^{16}O, a product of He-burning, is characteristic of six types of meteorites and mineral separates with other distinguishing chemical qualities [63].

Scientists at the University of Chicago also provided the first evidence that the abundance of another major element, helium, was linked with an isotopic anomaly from stellar nucleosynthesis when meteorites started to form [57]. Figure 12 shows the link of primordial He with excess ^{136}Xe from the r-process of nucleosynthesis in the Allende meteorite [32]. Subsequent analysis confirmed that the link of primordial helium with excess ^{136}Xe still remains as a SN fingerprint imprinted in the noble gases of meteorites [68, 69] and planets [19].

Ordinary elements that form the matrix of meteorite minerals are also linked with these characteristic isotope signatures of nucleosynthesis in heavy elements that they incorporated:

- Carbonaceous (C) inclusions contain excess r- and p-products [54, 55].
- Silicon carbide (SiC) inclusions contain excess s-products and severely mass fractionated neon, Ne-E [47, 53-56].
- Iron sulfide (FeS) minerals contain little or no isotopic anomalies [59, 60].
- Iron nickel (Fe, Ni) contains small but significant isotopic anomalies [70].

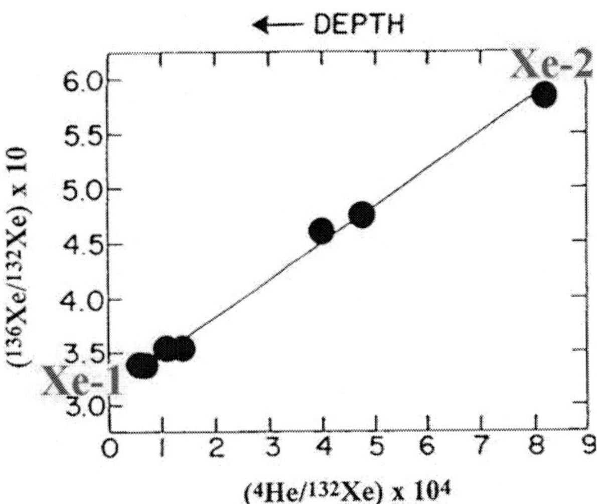

Figure 12. Analysis of the Allende meteorite in 1975 [57] first exposed a link [32] of primordial helium with excess ^{136}Xe from the r-process. Xe-1 is xenon of "normal" isotopic composition; Xe-2 is "strange" xenon." The correlation suggests that ^{136}Xe was made by the r-process in the outer layers of a supernova, where light elements like He were abundant. The arrow at the top of the figure indicates stellar depth [32].

IX. Ubiquitous Isotope Anomalies from Nucleosynthesis

Anomalous isotope abundances from stellar nucleosynthesis were an unexpected SN fingerprint when first seen [52], but they now seem to be ubiquitous. They have now been identified in over twenty elements in meteorites [71, 72]—C, N, O, Ne, Mg, Si, Ar, K, Ca, Ti, Cr, Kr, Sr, Zr, Mo, Te, Xe, Ba, Nd, Sm, Dy, and Os. Meteorite-like isotope abundance patterns have also been recognized in Mars [58-60], in Jupiter [19], and in the Sun [61, 62, 64, 65]. Even the "normal" isotope abundances here on Earth appear in some cases to be mixtures of two anomalous components that were trapped in meteorites [56]. Only xenon isotope anomalies will be discussed in detail here.

Ubiquitous isotope anomalies across the solar system are likely fingerprints of the supernova (figure 1) that gave birth to the solar system [16]. They are difficult to explain if 99.99% of the starting material of the solar system had "normal" isotopic composition and only 0.01% of the starting material had exotic nucleogenetic composition [73, 74].

The combined results of stellar nuclear reactions and mass fractionation were fortunately both noticeable in the unusual abundance pattern of xenon isotopes in meteorites when a nucleogenetic origin was first proposed in 1972 [52]. This is illustrated in figure 13, where the solid, 45° line shows the correlation of r- and p-products of stellar nucleosynthesis. The approximately perpendicular dashed line shows the effects of physical mass fractionation in xenon isotopes [52].

Primordial isotope anomalies from nucleosynthesis were later found in many other elements, frequently embedded in an isotope spectrum that has also been altered by physical mass fractionation. Reasons for the association of nuclear and physical fractionation remained unknown and mysterious [45, 46] until mass fractionation was recognized as a common stellar process [10, 20]. For example, the term FUN isotope anomalies was coined in 1977 to identify the combined roles of Fractionation plus Unknown Nuclear processes that generated correlated isotope anomalies of oxygen and magnesium in inclusions of the Allende meteorite [45, 46].

As shown by the dashed line in figure 13, major xenon reservoirs in the inner part of the solar system—the Earth and the Sun—are related by mass fractionation of normal xenon (Xe-1). "Strange" xenon (Xe-2), enriched in r- and p-products, occurs in carbonaceous inclusions of meteorites and in Jupiter [19]. Bulk xenon in carbonaceous chondrites [51], AVCC Xe, is a mixture of Xe-1 and Xe-2.

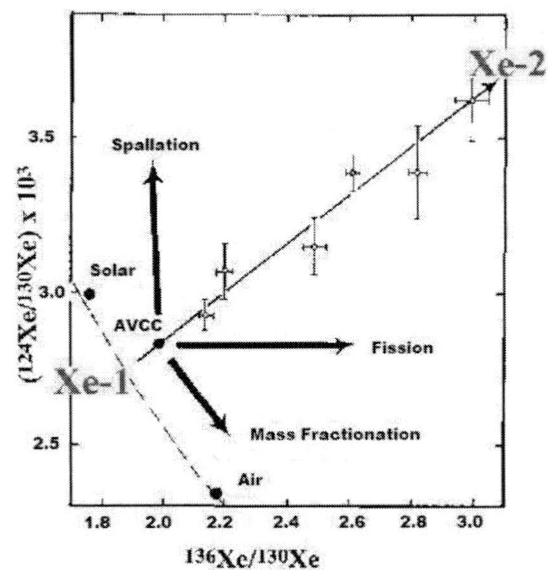

Figure 13. Xenon in meteorites is a mixture of normal xenon (Xe-1) with “strange” xenon (Xe-2) [49]. Mass fractionation (dashed line) of "normal" xenon (Xe-1) relates solar xenon with the xenon in air. "Strange" xenon (Xe-2) is normal xenon (Xe-1) plus extra ^{136}Xe from the r-process and ^{124}Xe from the p-process [4].

Figure 14 compares the "strange" abundance pattern of xenon isotopes in mineral separate 3CS4 of the Allende meteorite [57] with that in the Earth's atmosphere. Xenon in 3CS4 contains the smallest fractional amount of normal xenon, Xe-1. This measurement best represents the isotopic composition of Xe-2, which has also been referred to as Xe-X [52] and as "strange" xenon [57] in meteorites and in Jupiter [19].

Isotopic anomalies in xenon from the r- and p-process in Allende mineral separate 3CS4 (figure 14) are mirrored as excesses of the intermediate mass xenon isotopes from the s-process of nucleosynthesis in silicon carbide grains from the Murchison meteorite [53]. The mass spectrometer that Reynolds built in 1956 [7] made possible the detection of more SN fingerprints in the nine isotopes of xenon than in any other element—as r-, p- and s-products of nucleosynthesis, as stellar mass fractionation, and as the decay of extinct ^{129}I and ^{244}Pu.

Anomalous isotope abundances in about twenty other elements recorded fingerprints of a supernova [71, 72]—a remarkably high fraction of those elements with the requisite three or more stable isotopes so that alterations from nucleosynthesis reactions, mass fractionation, and radioactive decay can be separated and identified.

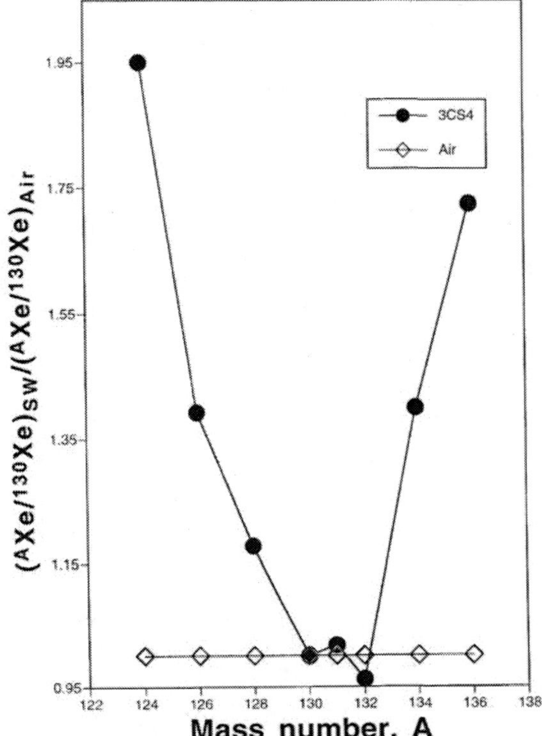

Figure 14. The p- and r-processes of nucleosynthesis [4] almost doubled the natural abundance of the lightest and the heaviest stable isotopes, ^{124}Xe and ^{136}Xe, of "strange" xenon trapped in Allende mineral separate 3CS4 [57]. This "strange" xenon was closely associated with the primordial helium incorporated into meteorites [68] and planets [19] at the birth of the solar system.

Problems may still arise when mass fractionation is more severe than anticipated. The isotopic anomalies discovered in neon illustrate the problem. In the 1960s and 1970s, ten distinct isotopic forms of neon were identified in meteorites and labelled as Ne-A, Ne-B, Ne-C, Ne-D, Ne-E, etc. [53, 75, 76].

The mysterious origin of so many distinct forms of neon was finally solved in 1980, when it was shown that all of the isotopic forms of neon that had been reported in meteorites could be explained by severe mass fractionation and the addition of products of cosmic-ray-induced spallation reactions [47].

X. Short-Lived Nuclides at the Birth of the Solar System

Below is a chronological listing of short-lived nuclides whose decay products were reported in meteorites during the period of 1960 [11] to 2006 [13, 78-83].

The 1960 discovery of radiogenic ^{129}Xe in a meteorite [11] from the decay of extinct ^{129}I ($t_{1/2}$ = 16 My) was the first hint of a supernova fingerprint in the solar system. In 1961 Fowler, Greenstein and Hoyle [12] suggested that local element synthesis might explain the unexpectedly high level of ^{129}I radioactivity found in meteorites. The finding of ^{244}Pu ($t_{1/2}$ = 82

My) decay products in a meteorite in 1965 [13] demonstrated the need for local element synthesis to include the r-process of nucleosynthesis [4].

Short-lived Nuclide	Half-life	Reference
1. Iodine-129	16 My	J. H. Reynolds [11]
2. Plutonium-244	82 My.	M. W. Rowe and P. K. Kuroda [13]
3. Aluminum-26	0.7 My	C. M. Gray and W. Compston [78]
4. Palladium-107	6.5 My.	W. R. Kelly and G. J. Wasserburg [79
5. Manganese-53	3.7 My.	J. L. Birck and C. J. Allègre [80]
6. Iron-60	1.5 My.	A. Shukolyukov and G. W. Lugmair [81]
7. Calcium-41	0.1 My.	G. Srinivasan, et al. [82]
8. Sodium-22 (???)	2.6 y.	S. Amari, et al. [83]

All of the above short-lived nuclides may be products of a supernova [77], including the decay products of ^{129}I and ^{244}Pu that were found in the interior of the Earth in 1971 [84]. The decay products of shorter-lived nuclides in meteorites, and the absence of convincing evidence that any meteorite grains were exposed to cosmic rays during an interstellar journey to the solar system, favour the arguments that Fowler, Greenstein and Hoyle [12] made in 1961 for production of these short-lived radioactive nuclides by local element synthesis.

The supernova event shown in figure 1 is compatible with the suggestion by Amari *et al.* [83] that ^{22}Ne in graphite grains extracted from the Murchison meteorite *". . . is from the in situ decay of ^{22}Na ($T_{1/2}$=2.6a) produced in the C-burning zone in pre-supernova stars"*, and we hope that this research group has finally succeeded in finding excess ^{22}Ne in a meteorite that is not accompanied by a proportionate excess of ^{21}Ne from mass fractionation [47].

Kuroda and Myers [85] combined the U-Pb and ^{244}Pu-^{136}Xe age dating chronometers to show that ^{244}Pu and other actinide elements in the solar system were produced in a supernova explosion 5000 My ago.

Kuroda and Myers [86] also noticed other convincing fingerprints of a local supernova when they compared physical properties and the amounts of radioactivity trapped in fallout grains from nuclear explosions with those in five special silicon carbide (SiC) grains that Amari *et al.* [87] recovered from the Murchison meteorite.

In both cases, the meteorite grains that formed first, grew larger and trapped more radioactivity than the grains that started to form later. The five special silicon carbide grains from the Murchison meteorite were physically much larger and trapped more radioactive ^{26}Al and more exotic isotopic anomalies than other silicon carbide grains. Mixing apparently occurred as short-lived nuclides continued to decay away in the highly radioactive supernova debris that formed the solar system: Amari *et al.* [87] state that the five special silicon carbide grains, *". . . exhibit extremely exotic isotopic compositions, distinct from the majority of the SiC grains, . . ."*.

Conclusion

The Sun exploded as a supernova five billion years ago and ejected the material that now orbits it as planets, moons, asteroids, comets and other planetary rubble. Fingerprints of the supernova that gave birth to the solar system are visible in the planetary system as linked chemical and isotopic heterogeneities and as decay products of short-lived nuclides. Fingerprints of the supernova are visible in the Sun as overabundances of lightweight isotopes in the solar wind and overabundances of lightweight s-products in the photosphere, as solar emissions of energy, neutrinos, and hydrogen that match the products expected from a collapsed SN core—defined by systematic properties of the 3,000 known types of nuclei, as the axially directed solar wind, and as solar cycles (surface magnetic activity, solar eruptions, and sunspots) and changes in Earth's climate that are linked with gravitational forces that torque the Sun in its ever-changing orbit about the centre-of-mass of the solar system. The vibratory motion that planets induce in the star that they orbit may limit elemental segregation and bring enhanced metallicity closer to the surface of stars that host planets [88], like that ejected from the interior of the Sun by impulsive flares [89].

ACKNOWLEDGMENTS

We are grateful to Professor D. Vincent Roach for reviewing this manuscript, to Richard Mackey for permission to reproduce the drawing by Daniel Brunato as figure 2, to NASA and ESA for the summary of Ulysses spacecraft measurements shown in figure 3 and the solar images shown in figures 4 and 5 from the SOHO and TRACE programs, and to the Foundation for Chemical Research, Inc. (FCR) for permission to reproduce the other figures from reports to FCR, Inc. Students that were enrolled in Advanced Nuclear Chemistry (Chem. 471) in the spring semester of 2000 – Cynthia Bolon, Shelonda Finch, Daniel Ragland, Matthew Seelke and Bing Zhang – helped develop the “Cradle of the Nuclides” (figure 6) that exposed repulsive interactions between neutrons as an important source of energy [1, 2, 35]. The moral support of former UMR Chancellors, Dr. Raymond L. Bisplinghoff (1975-1976) and Dr. Gary Thomas (2000-2005); the late Professors John H. Reynolds and Nobel Laureate Glenn T. Seaborg of UC-Berkeley; and former UMR Distinguished Curators' Professor and Chair of Chemistry, Professor Stig E. Friberg (1976-1979) is gratefully acknowledged. This paper is dedicated to the memory of the late Professor Paul Kazuo Kuroda (1917-2001) and his former student, Dr. Dwarka Das Sabu (1941-1991) for their contributions [9, 13, 16, 18, 31, 40, 46, 49, 58, 66, 82, 83] to the conclusions reached here and for their deep personal commitment to the basic precepts of science as a sacred path to truth. Finally, we remember with gratitude our colleague, the late Robert Bennett Blore, who quietly continued to take daily measurements on the Sun while we were arguing about theory.

REFERENCES

[1] O. Manuel, C. Bolon, A. Katragada and M. Insall, "Attraction and repulsion of nucleons: Sources of stellar energy", *J. Fusion Energy*. 19, 93-98 (2001).

[2] O. Manuel, E. Miller, and A. Katragada, "Neutron repulsion confirmed as energy source", *J. Fusion Energy*. 20, 197-201 (2002).

[3] O. Manuel, M. Mozina, H. Ratcliffe, "On the cosmic nuclear cycle and the similarity of nuclei and stars," *J. Fusion Energy*. 25, 107-114 (2006). http://arxiv.org/abs/nucl-th/0511051

[4] E. M. Burbidge, G. R. Burbidge, W. A. Fowler and F. Hoyle [B2FH], "Synthesis of elements in stars," *Rev. Mod. Phys.* 29, 547-650 (1957).

[5] P. K. Kuroda, "On the nuclear physical stability of the uranium minerals," *J. Chem. Phys.* 25, 781-782 (1956).

[6] P. K. Kuroda, "On the infinite multiplication constant and the age of the uranium minerals," *J. Chem. Phys.* 25, 1295-1296 (1956).

[7] J. H. Reynolds, "High sensitivity mass spectrometer for noble gas analysis," *Rev. Sci. Instruments.* 27, 928-934 (1956).

[8] LSPET (The Lunar Sample Preliminary Examination Team), "Preliminary examination of lunar samples from Apollo 11", *Science.* 165, 1211-1227 (1969).

[9] P. K. Kuroda and O. K. Manuel, "Mass fractionation and isotope anomalies in neon and xenon," *Nature.* 227, 1113-1116 (1970).

[10] O. K. Manuel and G. Hwaung, "Solar abundance of the elements", *Meteoritics.* 18, 209-222 (1983).

[11] J. H. Reynolds, "Determination of the age of the elements," *Phys. Rev. Lett.* 4, 8-10 (1960).

[12] W. A. Fowler, J. L. Greenstein, and F. Hoyle, "Deuteronomy. Synthesis of deuterons and light nuclei during the early history of the Solar System," *Am. J. Phys.* 29, 393-403 (1961).

[13] M. W. Rowe and P. K. Kuroda, "Fissiogenic xenon from the Pasamonte meteorite," *J. Geophys. Res.* 70, 709-714 (1965).

[14] Theodor Landscheidt, "Extrema in sunspot cycle linked to Sun's motion, "*Solar Physics.* 189, 413-424 (1999). http://bourabai.narod.ru/landscheidt/extrema.htm. For other papers over the past quarter century by Theodor Landscheidt on solar-induced climate changes, see: http://bourabai.narod.ru/landscheidt/publications.htm

[15] W. J. R. Alexander, F. Bailey, D. B. Bredenkamp, A. vander Merwe and N. Willemse, "Linkages between solar activity, climate predictability and water resource development," *J. South African Institut. Civil Eng.* 49, 32-44 (2007). http://www.lavoisier.com.au/papers/Conf2007/Alexander-etal-2007.pdf

[16] O. K. Manuel and D. D. Sabu, "Strange xenon, extinct super-heavy elements, and the solar neutrino puzzle," *Science.* 195, 208-209 (1977).

[17] R. S. Lewis, B. Srinvasan and E. Anders, Reply to "Strange xenon, extinct super-heavy elements, and the solar neutrino puzzle," *Science.* 195, 209-210 (1977).

[18] P. K. Kuroda and Z. Z. Sheng, "Elemental and isotopic heterogeneities of rare gases in the solar system," *Geochemical J.* 21, 315-317 (1987). See also the reply by O. K. Manuel and D. D. Sabu, *ibid.,* 21, 319-322 (1987).

[19] O. Manuel, K. Windler, A. Nolte, L. Johannes, J. Zirbel and D. Ragland, "Strange xenon in Jupiter", *J. Radioanalytical Nucl. Chem.* 238, 119-121 (1998) or http://www.omatumr.com/abstracts2001/windleranalysis.pdf

[20] O. Manuel, "The Sun is a plasma diffuser that sorts atoms by mass," *Physics of Atomic Nuclei* 69, 1847-1856 (2006). http://arxiv.org/abs/astro-ph/0609509

[21] P. D. Jose, "Sun's motion and sunspots", *Astron. J.* 70, 193-200 (1965).

[22] R. W. Fairbridge and J. H. Shirley, "Prolonged minima and the 179-yr cycle of the solar inertial motion," *Solar Physics.* 110, 191-220 (1987).

[23] J. Shirley, "Axial rotation, orbital revolution and solar spin-orbit coupling", *Monthly Notices of the Royal Astronomical Society.* 368, 280-282 (2006).

[24] Richard Mackey, "Rhodes Fairbridge and the idea that the solar system regulates the Earth's climate," *J. Coastal Research* SI 50 (Proceedings of the 9th International Coastal Symposium, Gold Coast, Australia, 2007) pp. 955-968. http://www.griffith.edu.au/conference/ics2007/pdf/ICS176.pdf

[25] O. K. Manuel, Barry W. Ninham and Stig E. Friberg, "Superfluidity in the solar interior: Implications for solar eruptions and climate," *J. Fusion Energy* 21, 193-198 (2002). http://arxiv.org/abs/astro-ph/0501441

[26] L. W. Looney, J. J. Tobin and W. Kwon, "A flattened protostellar envelope in absorption around L1157", *Ap. J. Letters* 670. L131–L134 (1 December 2007) http://www.spitzer.caltech.edu/Media/releases/ssc2007-19/release.shtml

[27] Adam Frank, "Bipolar outflows in stellar astrophysics," in *Origin of Elements in the Solar System: Implications of Post 1957 Observations,* O. K. Manuel, editor, Kluwer Academic/Plenum Publishers, New York, NY, pp. 241-249 (2000).

[28] J. Wouters, D. Vandeplassche, E. van Walle, N. Severijns, and L. Vanneste, "Anisotropic alpha-emission from on-line-separated isotopes," *Phys. Rev. Lett.* 56, 1901 - 1904 (1986) http://prola.aps.org/abstract/PRL/v56/i18/p1901_1

[29] M. Castillo Mejia, M. Milanese, R. Moroso, J. Poouzo, "Some experimental research on anisotropic effects in the neutron emission of dense plasma-focus devices," *J. Physics D: Applied Physics.* 30,1499-1506 (1997). http://adsabs.harvard.edu/abs/1997JPhD...30.1499C

[30] M. Mozina, http://www.thesurfaceofthesun.com/index.html, "The Surface of the Sun." This was also quoted in reference #30, below.

[31] O. Manuel, S. A. Kamat, and M. Mozina, "Isotopes tell Sun's origin and operation," *AIP Conference Proceedings.* 822, 206-205 (2006).

[32] O. K. Manuel and D. D. Sabu, "Elemental and isotopic inhomogeneities in noble gases: The case for local synthesis of the chemical elements", *Trans. Missouri Acad. Sci.* 9, 104 122 (1975).

[33] W. D. Harkins, "The evolution of the elements and the stability of complex atoms," *J. Am. Chem. Soc.* 39, 856-879 (1917).

[34] O. Manuel and Stig Friberg, "Composition of the solar interior: Information from isotope ratios," Proceedings of the SOHO/ GONG Conference on Local and Global Helioseismology, ESA SP-517 (editor: Huguette Lacoste) 345-348 (2003). http://arxiv.org/abs/astro-ph/0410717

[35] J. K. Tuli, *Nuclear Wallet Cards,* 6th ed., Upton, NY, Brookhaven National Laboratory, National Nuclear Data Center, 74 pp. (2000).

[36] O. Manuel, Cynthia Bolon and Max Zhong, "Nuclear systematics: III. The source of solar luminosity", *J. Radioanal. Nucl. Chem.* 252, 3-7 (2002).

[37] H. Heiselberg, "Neutron star masses, radii and equation of state", in *Proceedings of the Conference on Compact Stars in the QCD Phase Diagram,* eConf C010815, edited by R. Ouyed and F. Sannion, Copenhagen, Denmark, Nordic Institute for Theoretical Physics, pp. 3-16 (2002). http://arxiv.org/abs/astro-ph/?0201465

[38] D. Lunney, J. M. Pearson and C. Thibault, "Recent trends in the determination of nuclear masses," *Rev. Mod. Phys.* 75, 1021-1082 (2003).

[39] E. Anders and N. Grevesse, "Abundances of the elements: Meteoritic and solar," *Geochim. Cosmochim. Acta* 53, 197-214 (1989).

[40] O. K. Manuel, "Noble gases in the Fayetteville meteorite," *Geochim. Cosmochim. Acta* 31, 2413-2431 (1967).

[41] O. K. Manuel, R. J. Wright, D. K. Miller and P. K. Kuroda, "Heavy noble gases in Leoville: The case for mass-fractionated xenon in carbonaceous chondrites," *J. Geophys. Res.* 75, 5693-5701 (1970).

[42] E. W. Hennecke and O. K. Manuel, "Mass fractionation and the isotopic anomalies of xenon and krypton in ordinary chondrites," *Z. Naturforsch.* 20a, 1980-1986 (1971).

[43] B. Srinivasan and O. K. Manuel, "On the isotopic composition of trapped helium and neon in carbonaceous chondrites," *Earth Planet. Sci. Lett.* 12, 282-286 (1971).

[44] B. Srinivasan, E. W. Hennecke, D. E. Sinclair and O. K. Manuel, "A comparison of noble gases released from lunar fines (#15601164) with noble gases in meteorites and in the Earth," *Proceedings of the Third Lunar and Planetary Science Conference*, vol. 2, 1927-1945 (1972).

[45] R. N. Clayton and T. K. Mayeda, "Correlated oxygen and magnesium isotope anomalies in Allende inclusions, I: Oxygen", *Geophys. Res. Lett.* 4, 295-298 (1977).

[46] G. J. Wasserburg, T. Lee and D. A. Papanastassiou, "Correlated O and Mg isotope anomalies in Allende inclusions, II: Magnesium", *Geophys. Res. Lett.* 4, 299-302 (1977).

[47] D. D. Sabu and O. K. Manuel, "The neon alphabet game", *Lunar Planet. Sci.* XI, 878-899 (1980). http://www.omatumr.com/abstracts2005/Neon_alphabet_game.pdf

[48] P. Dufour, J. Liebert, G. Fontaine and N. Behara, "White dwarf stars with carbon atmospheres, " *Nature.* 450, 522-524 (22 Nov 2007). http://www.nature.com/nature/journal/v450/ n7169/abs/nature06318. html

[49] P. Dufour, J. Liebert, G. Fontaine and N. Behara,, "Hot DQ white dwarf stars: A new challenge to stellar evolution," *Hydrogen Deficient Stars, ASP Conference Series*, in press (2008). http://arxiv.org/PS_cache/arxiv/pdf/0711/0711.3458v1.pdf

[50] O. Manuel, M. Pleess, Y. Singh and W. A. Myers, "Nuclear Systematics: Part IV. Neutron-capture cross sections and solar abundance", *J. Radioanalytical Nucl. Chem.* 266, 159–163 (2005). http://www.omatumr.com/abstracts2005/Fk01.pdf

[51] J. H. Reynolds, "Isotopic composition of primordial xenon," *Phys Rev. Lett.* 4, 351-354 (1960).

[52] O. K. Manuel, E. W. Hennecke and D. D. Sabu, " Xenon in carbonaceous chondrites,", *Nature.* 240, 99-101 (1972).

[53] B. Srinivasan and E. Anders, "Noble gases in the Murchison meteorite: Possible relics of s-process nucleosynthesis," *Science.* 201, 51-56 (1978).

[54] R. V. Ballad, L. L. Oliver, R. G. Downing and O. K. Manuel, "Isotopes of tellurium, krypton and xenon in Allende meteorite retain record of nucleosynthesis," *Nature.* 277, 615-620 (1979).

[55] L. L. Oliver, R. V. Ballad, J. F. Richardson, and O. K. Manuel, "Isotopically anomalous tellurium in Allende: Another relic of local element synthesis," *J. Inorg. Nucl. Chem.* 43, 2207-2216 (1981).

[56] F. Begemann, "Isotopic abundance anomalies and the early solar system" in *Origin and Evolution of the Elements*, edited by N. Prantos, E. Vangioni-Flam, and M. Cassé, Cambridge: Cambridge University Press, pp. 518-527 (1993).

[57] R. S. Lewis, B. Srinivasan, and E. Anders, "Host phase of a strange xenon component in Allende," *Science.* 190, 1251-1262 (1975).

[58] T. D. Swindle, J. A. Grier and M. K. Burkland, "Noble gases in ortho pyroxenite AlH84001: A different kind of Martian meteorite with an atmospheric signature," *Geochim. Cosmochim. Acta.* 59, 793-801 (1995).

[59] Golden Hwaung and O. K. Manuel, "Terrestrial-type xenon in meteoritic troilite, " *Nature* 299, 807 - 810 (1982). http://tinyurl.com/392tka http://www.nature.com/nature/journal/v299/n5886/abs/299807a0.html

[60] J. T. Lee, Li Bin and O, K, Manuel, "Terrestrial-type xenon in sulfides of the Allende meteorite", *Geochem. J.* 30, 17-30 (1996). http://www.terrapub.co.jp/journals/GJ/pdf/3001/30010017.PDF

[61] D. D. Sabu and O. K. Manuel, "Xenon record of the early solar system," *Nature.* 262, 28-32 (1976).

[62] R. O. Pepin, R. H. Becker and P. E. Rider, "Xenon and krypton isotopes in extraterrestrial regolith soild and in the solar wind," *Geochim. Cosmochim. Acta.* 59, 4997-5022 (1995).

[63] R. N. Clayton, N. Onuma and T. K. Mayeda, "A classification of meteorites based on oxygen isotopes," *Earth Planet. Sci. Lett.* 30, 10-18.

[64] T. R. Ireland, P. Holden, M. D. Norman and J. Clarke, "Isotopic enhancements of ^{17}O and ^{18}O from solar wind particles in the lunar regolith," *Nature* 440, 776-778 (2006). http://www.nature.com/nature/journal/v440/n7085/abs/nature04611.html

[65] K. Hashizume and M. Chaussidon, "A non-terrestrial ^{16}O-rich isotopic composition for the protosolar nebula," *Nature.* 434, 619-622 (2005). http://www.nature.com/nature/journal/ v434/n7033/abs/nature03432.html

[66] M. H. Thiemans and J. E. Heidenreich III, "The mass-independent fractionation of oxygen: A novel isotope effect and its possible cosmochemical implications," *Science.* 219, 1073-1075 (1983).

[67] R. N. Clayton, "Self-shielding in the solar nebula," *Nature.* 415, 860-861 (2002).

[68] O. K. Manuel, "The enigma of helium and anomalous xenon," *Icarus.* 41, 312-315 (1980).

[69] O. K. Manuel and D .D. Sabu, "Noble gas anomalies and synthesis of the chemical elements", *Meteoritics.* 15, 117-138 (1980).

[70] Qi Lu and A. Masuda, "Variation of molybdenum isotopic composition in iron meteorites in meteorites," in *Origin of Elements in the Solar System: Implications of Post 1957 Observations,* O. K. Manuel, editor, Kluwer Academic/Plenum Publishers, New York, NY, pp. 385-400 (2000).

[71] U. Ott, "Isotope abundance anomalies in meteorites: Clues to yields of individual nucleosynthesis processes," in *Origin of Elements in the Solar System: Implications of Post 1957 Observations,* O. K. Manuel, editor, Kluwer Academic/Plenum Publishers, New York, NY, pp. 369-384 (2000).

[72] L. C. Reisberg, N. Dauphas, A. Luguet, D. G. Pearson and R. Gallino, "Large s-process and mirror osmium isotopic anomalies within the Murchison meteorite," *Lunar Planet. Sci.* XXXVIII, abstract 1177 (2007).

[73] W. A. Fowler, "The quest for the origin of the elements," *Science.* 226, 922-935 (1984).

[74] G. J. Wasserburg, "Isotopic abundances: inferences on solar system and planetary evolution", *Earth Planet. Sci. Lett.* 86, 129-173 (1987).

[75] D. C. Black and R. O. Pepin, "Trapped neon in meteorites II," *Earth Planet. Sci. Lett.* 6, 395-405 (1969).

[76] D. C. Black, "On the origins of trapped helium, neon and argon isotopic variations in meteorites-1I. Carbonaceous chondrites", *Geochim. Cosmochim. Acta.* 36, 377-394 (1972).

[77] J. N. Goswami, "Chronology of early solar system events: Dating with short lived nuclides," in *Origin of Elements in the Solar System: Implications of Post 1957 Observations,* O. K. Manuel, editor, Kluwer Academic/Plenum Publishers, New York, NY, pp. 407-430 (2000).

[78] C. M. Gray and W. Compston, "Excess ^{26}Mg in Allende meteorite," *Nature.* 251, 495-497 (1974).

[79] W. R. Kelly and G. J. Wasserburg, "Evidence for the existence of 107 Pd in the early solar system," *Geophys. Res. Letters.* 5, 1079-1082 (1978).

[80] J. L. Birck and C. J. Allègre, "Evidence for the presence of ^{53}Mn in the early solar system," *Geophys. Res. Letters.* 12, 745-748 (1985).

[81] A. Shukolyukov and G. W. Lugmair, "Live iron-60 in the early solar system," *Science.* 259, 1138-1142 (1993).

[82] G. Srinivasan, A. A. Ulyanov and J. N. Goswami, "41 Ca in the early solar system," *Ap. J. Letters.* 431, L67-L-70 (1994).

[83] S. Amari, R. Gallino, M. Limongi and A. Chieffi, "Presolar graphite from the Murchison meteorite: Imprint of nucleosynthesis and grain formation," AIP Conference Proceedings 847, 311-318 (2006). http://scitation.aip.org/getabs/servlet/Getabs Servlet? prog=normal and id=APCPCS 000847000001000311000001 and idtype=cvips and gifs=yes

[84] M. S. Boulos and O. K. Manuel, "The xenon record of extinct radioactivities in the Earth," *Science.* 174, 1334-1336 (1971).

[85] P. K. Kuroda and W. A. Myers, "Iodine-129 and plutonium-244 in the early solar system," *Radiochim. Acta.* 77, 15-20 (1996)

[86] P. K. Kuroda and W. A. Myers, "Aluminum-26 in the early solar system," *J. Radioanal. Nucl. Chem.* 211, 539-555 (1997).

[87] S. Amari, P. Hoppe, E. Zinner and R. S. Lewis, "Interstellar SiC with unusual isotopic compositions: Grains from a supernova?" *Ap. J.* 394, L43-L46 (1992).

[88] L. Pasquini, M. P. Doellinger, A. Weiss, L. Girardi, C. Chavero, A. P. Hatzes, L. da Silva, and J. Setiawan, "Evolved stars hint to an external origin of enhanced metallicity in planet-hosting stars," *Astronomy and Astrophysics,* in press (2007). http://arxiv.org/pdf/0707.0788.pdf

[89] D. V. Reames, "Abundances of trans-iron elements in solar energetic particle events," *Ap. J.* 540, L111-L114 (2000). http://epact2.gsfc.nasa.gov/don/00HiZ.pdf

In: Space Exploration Research
Editors: J.H. Denis and P.D. Aldridge

ISBN: 978-1-60692-264-4

Chapter 8

THE STUDY FOR BLACK HOLE IN M-THEORY

Makoto Tanabe
Kanagawa Institute of Technology, Atsugi, Kanagawa, Japan

Abstract

The black hole solutions in the M-theory are provided for the intersecting M-brane solutions. To solve the bosonic part of field equations of M-theory, the Einstein equation and the Maxwell equations, the metric is assumed by the stationary flat direction with the additional spaces occupied by M-branes, which have a time-like Killing vector or null Killing vector, and the time-like or null directed M-branes for the souse of three-form fields. In null case the solutions include the BMPV solution, otherwise in time-like Killing case, the supersymmetric black ring solution is included. The BMPV solution and the other null Killing solutions are characterized by the Harmonic functions, which given by the Laplace equation, thus the solutions are easy to add the other solutions to get the new solutions. Though the black ring solution are essentially contributed by the Charn-Simons terms, which makes the Laplace equation to the Poisson equation. Therefore in general the superposition on the solutions is not allowed. These black holes are the solution of the Einstein and Maxwell equations naturally, however, taking the Killing spinor equation, which related to the gravitinofs degree of freedom, into account, the integrated constants, which related to the black hole mass or charges and angular momentums, are limited and the solutions remained supersymmetries. Finally analysis of the low-dimensional black hole solutions with torus compactification, the conserved charges in lower dimension are possible to calculation in general.

1. Introduction

Black holes are one of the most exciting subjects in string theory. The Beckenstein-Hawking black hole entropy is proportional to the area surface of the black hole. An extreme black hole has no surface gravity at event horizon and it is obtain in string theory by intersecting D-brane with the equilibrium of force between the branes. The extreme black hole entropy is obtained by statistical counting of the microscopic state of a superstring between intersecting D-branes [1].

However, it is very difficult to construct a black hole solution in string theory because of its strong coupling. So far, we know several interesting black hole solutions in supergravity theories [2–7], which are obtained as an effective theory of a superstring model in a low energy limit. We also have black hole solutions in a higher-dimensional spacetime [8, 9], and there is no uniqueness theorem of black holes [10–12]. In fact, we have a variety of black objects such as a black brane [13–16]. One of the most remarkable solutions is a black ring, which horizon has a topology of $S^1 \times S^2$ [17], and a varieties of black ring solution are reviewed in [18].

Among such black objects, supersymmetric ones are very important. The black hole solutions in a supergravity include the higher-order effects of a string coupling constant, although these are solutions in a low energy limit. On the other hand, the counting of states of corresponding branes is performed at the lowest order of a string coupling. The results of these two calculations need not coincide each other. However, if there is supersymmetry, these should be the same because the numbers of dynamical freedom cannot be different in these BPS representations. Therefore, supersymmetric black hole solutions are often discussed in many literature [19–23].

The classification of supersymmetric solutions in minimal $\mathcal{N} = 2$ supergravity in four dimensional spacetime was first performed by a time-like or null Killing spinor [24]. Recently, solutions in minimal $\mathcal{N} = 1$ supergravity in five dimensions have been classified into two classes by use of G-structures analysis [25–29].

However, the fundamental unified heory is constructed in either ten or eleven dimensions. When we discuss the entropy of black holes, we have to show the relation between those supersymmetric black holes and more fundamental black branes either in ten or eleven dimensions, from which we obtain black holes by compactification. Not only a supersymmetric rotating black hole solution [30, 31] but also a supersymmetric rotating black ring solution [32, 33]. are obtained by compactification from black brane solutions in M or type II supergravity Such an object is btained also in lower dimensional supergravity theories. These solutions are in fact new classes of rotating solutions in four- or five-dimensional supergravity. The entropy of such an object is microscopically described by the quantized charges of branes [34]. The existence of such solutions suggests that the uniqueness theorem of black holes is no longer valid even in supersymmetric spacetime if the dimension is five or higher [35]. Thus we may need to construct more generic black brane solutions in the fundamental theory and the black holes by some compactification. M-theory is the best candidate for such a unified theory. Since its low energy limit coincides with the eleven-dimensional supergravity, it provides a natural framework to study black brane or BPS brane solutions.

We are also interested in non-BPS brane solutions, because the Hawking radiation have occurred in only non-BPS black hole. These new higher dimensional black hole solutions generate the new group of charged solutions in string theory and M-theory by using the U-duality in the string theory [36, 37]. Thus we must know how to construct the new solutions with arbitrary topology of event horizon, and key to finding these solutions is directly using the M-theory or superstring theory. Since the gravitational force and centrifugal force depend on equal power of the radius, the new topological black hole solutions appear, thus we must find a BPS state, which satisfies the equilibrium of whole force. In M-theory, we can easily find the BPS state to solve a Killing spinor equation, thus we give the rule for constructing BPS state for M-theory.

2. Basic Equation and Killing Spinor Formalism

We first provide the basic equation which must be satisfied the BPS or the non-BPS black hole solutions. Now we consider the action of $\mathcal{N} = 1$ eleven-dimensional supergravity (M-theory) [38];

$$S = \frac{1}{16\pi G_{11}} \int \sqrt{-g_{11}} \left[\mathcal{L}_{\text{boson}} + \mathcal{L}_{\text{CS}} + \mathcal{L}_{\text{fermion}} + \mathcal{L}_{\text{int}}\right] , \tag{1}$$

and the each component of Lagrangian can be written in

$$\begin{aligned}
\mathcal{L}_{\text{boson}} &= R_{11} - \frac{1}{2 \cdot 4!} F_{\mu\nu\rho\sigma} F^{\mu\nu\rho\sigma} , \\
\mathcal{L}_{\text{CS}} &= \frac{1}{\sqrt{-g}} \frac{1}{3!\, 4!\, 4!} C_{\mu_1\mu_2\mu_3} F_{\mu_4\cdots\mu_7} F_{\mu_8\cdots\mu_{11}} \epsilon^{\mu_1\cdots\mu_{11}} , \\
\mathcal{L}_{\text{fermion}} &= -\frac{1}{2} \psi_\mu \gamma^{\mu\nu\rho} D_\nu \psi_\rho , \\
\mathcal{L}_{\text{int}} &= -\frac{1}{8 \cdot 4!} \psi_\mu \gamma^{\mu\nu\rho_1\cdots\rho_4} F_{\rho_1\cdots\rho_4} \psi_\nu ,
\end{aligned} \tag{2}$$

where R_{11} is the Ricci scalar given by eleven-dimensional metric $g_{\mu\nu}$, $F_{\mu\nu\rho\sigma} = \partial_\mu C_{\nu\rho\sigma}$ is the fields strength for the three form field $C_{\nu\rho\sigma}$ and ψ_μ is gravitino, which is the superpartner of graviton $g_{\mu\nu}$ and it has a spin $3/2$.

First we assume the background metric $g_{\mu\nu}$ and the background gauge field $C_{\mu\nu\rho}$, then we find the basic equations of bosonic part as

$$\begin{aligned}
R_{\mu\nu} &= \frac{1}{2 \cdot 4!} \left[4F_\mu{}^{\rho_1\rho_2\rho_3} F_{\nu\rho_1\rho_2\rho_3} - \frac{1}{3} g_{\mu\nu} F_{\rho_1\rho_2\rho_3\rho_4} F^{\rho_1\rho_2\rho_3\rho_4} \right] , \\
\frac{1}{\sqrt{-g_{11}}} \partial_\mu \left(\sqrt{-g_{11}} F^{\mu\nu_1\nu_2\nu_3} \right) &= \frac{1}{4!\, 4!} \epsilon^{\nu_1\nu_2\nu_3\rho_1\cdots\rho_8} F_{\rho_1\cdots\rho_4} F_{\rho_5\cdots\rho_8} .
\end{aligned} \tag{3}$$

The Riemann tensor and the field strength are satisfy the Bianchi identity, $R^\mu{}_{[\nu\rho\sigma]} = 0$, $\nabla_{[mu} R^\nu{}_{|\rho|\sigma\tau]} = 0$ and $\partial_{[\mu} F_{\rho_1\cdots\rho_4]} = 0$. The fermionic part of basic equation named Killing spinor equation [39] is the equivalent to the supersymmetric transformation for fermionic infinitesimal parameter ζ :

$$\delta\psi_\mu = \left[\partial_\mu + \frac{1}{4} \omega_\mu{}^{ab} \gamma_{ab} + \frac{1}{144} \left(\gamma_\mu{}^{\nu\rho\sigma\tau} - 8\delta^\nu_\mu \gamma^{\rho\sigma\tau} \right) F_{\nu\rho\sigma\tau} \right] \zeta = 0 . \tag{4}$$

where $\gamma_{\mu\cdots\nu} \equiv \gamma^{[\mu} \cdots \gamma^{\nu]}$ is the gamma matrix in eleven dimension, and indices a is the local Lorentzian indices which related to the ordinary metric as $e^a{}_\mu e^b{}_\nu \eta_{ab} = g_{\mu\nu}$, where η_{ab} is the Minkowski metric, and $e^a{}_\mu$ is the vielbein. The spin connection $\omega_{\mu ab}$ are determined by

$$\omega_{\mu ab} = \frac{1}{2} e^\nu{}_a (\partial_\mu e_{b\nu} - \partial_\nu e_{b\mu}) - \frac{1}{2} e^\nu{}_b (\partial_\mu e_{a\nu} - \partial_\nu e_{a\mu}) - \frac{1}{2} e^\rho{}_a e^\sigma{}_b e^c{}_\mu (\partial_\rho e_{c\sigma} - \partial_\sigma e_{e\rho}) . \tag{5}$$

In general we will not solve these equation simultaneously, and then we assume some symmetries exist in the background metric and gauge fields. In this time we are interested in a lower dimensional black hole solution from intersecting M-branes, thus we assume the time-like Killing or null Killing vector with extra compact dimension. In the case of time-like Killing we assume lower-dimensional black hole space and compactifed space are orthogonal like a $ds_{11}^2 = ds_{\rm BH}^2 + ds_{\rm comp}^2$, and the black hole spacetime are independent for compact spaces. Otherwise the metric with null Killing vector we consider the one compactified direction are related to the black hole spacetime by the pp-wave for null direction. This description are consistent to fined the M-theory from the type IIA super string via strong coupling limit for the null wave direction.

The most easiest way to find solutions of black holes in lower dimension is solve the Killing spinor equation with some adequate assumption. For example we assume M2-brane which is the flat three dimension object with constant electric density, which means that $ds^2 = H_2^a(-dt^2 + dy_1^2 + dy_2^2) + H_2^b(dr^2 + r^2 d\Omega_7^2)$ and $C_{ty_1y_2} = E_2$. Substituting these background to the Killing spinor equation we find $E = 1/H_2$ and $a = -2/3$ and $b = 1/3$. We also have an additional condition for the gamma matrix as $(1 + \gamma_{0y_1y_2})\zeta = 0$, and this show M2-brane has a $1/2$ supersymmetry with $\gamma_{0y_1y_2} = 1$. If H_A is a harmonic function, the solutions must be satisfied the Einstein and Maxwell equations, thus we find the solution in M-theory. In general we can classify the solution generated by the globally covariant constant spinor ζ. With the torus extra dimension we can categorize the solution by the Hyper Köhler manifold [35].

The globally covariant spinor, however, must be a BPS state, because of the definition of it ($\delta\psi_\mu = 0$). Most of the black hole solutions in lower dimension are of cause non-BPS, i.e., Schwarzshild black hole, Kerr black hole, and so on. Therefore we would like to construct the non-BPS black hole solution in M-theory. There are two possible way to construct the non-BPS black hole solution, one is using the U-duality and charge up the lower dimensional vacuum black hole solution. The other one is solving the Einstein equation directory. U-duality method is using the vacuum solution with the flat extra dimension must be satisfy the vacuum Einstein equation in ten or eleven dimension. However this method is only possible to vacuum solution and some special configuration which satisfy the ten dimensional Einstein equation. The key role to add the new charge is using the boost for the other dimension, which gives new kk-wave solution in ten dimension. After that you can only use T-duality and S-duality, you will find the intersecting D-brane or M-brane solution, which gives non-BPS charged lower-dimensional black hole solution via torus compactification. However this method has a technically difficulty to apply for the stationary spacetime [40].

By the way, the most general way to get the black hole solution is just solve the bosonic part of basic equation with some specific metric and gauge ansatz. These black hole are always the solutions of higher dimensional basic equation, thus we can easily treat the non-BPS solution and BPS solution simultaneously. Thus we try to solve the Einstein equation and the Maxwell equation directly under some ansatz suitable for treat the stationary black hole [41], because only the stationary black hole contain the richest structure for the black hole phenomenology.

3. Null Killing Spacetime

In this below we consider the stationary black hole solution with compact space, thus we assume there exist the time-like Killing vector or null-like Killing vector for the background metric with compact extra dimension. First we consider null Killing case, and we assume that

$$ds^2 = e^{2\zeta_0(x)}du(dv + f(x)du) + e^{2\eta(x)}h_{ij}(x)dx^i dx^j + \sum_{\alpha=2}^{p} e^{2\zeta_\alpha(x)}dy_\alpha^2 ,$$

$$C_A = E^A(x)du \wedge dv \wedge dy_2 + \frac{1}{\sqrt{2}}B_j^A(x)dx^j \wedge du \wedge dy_2 , \tag{6}$$

where we use the light-cone coordinates; $\sqrt{2}u = -(t-y_1)$ and $\sqrt{2}v = t+y_1$ with the total dimension is given by $11 = d+p$. . We use the notation $\zeta_0 = \zeta_1$ for the simple expression.

We set the new functions for the both case, and these functions related to the $M2$-brane electric potential H_2 and $M5$-brane magnetic potential H_5 and volume of the spacetime V determined as

$$H_2 = \exp\left[-\sum_{\alpha=0}^{2}\zeta_\alpha\right] , \quad H_5 = \exp\left[-\sum_{\alpha=0}^{5}\zeta_\alpha\right] , \quad V = \exp\left[(d-3)\eta + \sum_{\alpha=0}^{p}\zeta_\alpha\right] , \tag{7}$$

We also set the new function related to the angular momentum and the new sign δ_α, A to distinguish between M-brane occupies or not as

$$\mathcal{F}_{ij} \equiv \partial_i \mathcal{A}_j - \partial_j \mathcal{A}_i , \quad \mathcal{F}_{ij}^{(A)} \equiv 2H_A\left(\mathcal{A}_{[i}\partial_{j]}H_A - \partial_{[i}B_{j]}^A\right)$$

$$\delta_{\alpha A} \equiv \begin{cases} 6 & \alpha \in \mathrm{M-brane} \\ -3 & \text{otherwise} \end{cases} \quad .. \tag{8}$$

In the null Killing background (6), the Chern-Simons terms are automatically vanishing. To solve under the arbitrary metric for the base space h_{ij} is impossible in general, thus in this section we assume $h_{ij} = \delta_{ij}$, then we find the Einstein equations as follows:

$$\partial^2 f + \partial_j f \partial^j \ln V = \frac{1}{8}e^{2(\xi-\eta)}\left[\mathcal{F}_{ij}^2 - \frac{1}{2}\sum_A\left(\mathcal{F}_{ij}^{(A)}\right)^2\right] , \tag{9}$$

$$\partial^j \mathcal{F}_{ij} + \mathcal{F}_{ij}\partial^j\left[2(\xi-\eta) + \ln V\right] = \sum_A H_A \mathcal{F}_{ij}^{(A)}\partial^j E_A, \tag{10}$$

$$\begin{aligned} &\left(\partial^2\eta + \partial_l\eta\partial^l \ln V\right)\delta_i^j + (d-3)\partial_i\eta\partial^j\eta + \sum_{\alpha=0}^{p}\partial_i\zeta_\alpha\partial^j\zeta_\alpha \\ &\quad + \partial_i\partial^j \ln V - \left(\partial_i\eta\partial^j \ln V + \partial^j\eta\partial_i \ln V\right) \\ &= \frac{1}{2}\sum_A H_A^2\left[\partial_i E_A \partial^j E_A - \frac{1}{3}(\partial E_A)^2\delta_i^j\right] , \end{aligned} \tag{11}$$

$$\partial^2\zeta_\alpha + \partial_j\zeta_\alpha\partial^j \ln V = \frac{1}{18}\sum_A \delta_{\alpha A}H_A^2(\partial E_A)^2, \tag{12}$$

and the Maxwell equations becomes $\partial_j(H_A^2 V\partial^j E_A) = 0$ and $\partial^j\left(V\mathcal{F}_{ij}^{(A)}\right) = 0$, where ∂_i is a partial derivative $\partial/\partial x^i$ in a flat $(d-1)$-space, $\partial^2 \equiv \partial_i\partial^i$. The square bracket denotes the anti-symmetrization of indices, i.e., $X_{[i}Y_{j]} \equiv \frac{1}{2}(X_iY_j - X_jY_i)$

Since E_A appears just with a spatial derivative ∂_i, we can replace it with $\tilde{E}_A = E_A - E_A^{(0)}$, where $E_A^{(0)}$ is a constant, which is fixed by a boundary condition. Using the Maxwell equation we obtain from Eqs. (12),

$$\partial^j \left[V \left(\partial_j \zeta_\alpha - \sum_A \frac{\delta_{\alpha A}}{18} H_A^2 \tilde{E}_A \partial_j \tilde{E}_A \right) \right] = 0. \tag{13}$$

This equation is a coupled system of elliptic-type differential equations, for which it is very difficult to find general solutions. Hence, we assume the following special relations:

$$\partial_j \zeta_\alpha = \sum_A \frac{\delta_{\alpha A}}{18} H_A^2 \tilde{E}_A \partial_j \tilde{E}_A\,. \tag{14}$$

These equations are relations between the first-order derivatives of variables just as the BPS conditions. Hence, these relations may be related to a BPS state, or an extremal black brane solution in supergravity. The metric function η is obtained from ζ_α, and this gives the equation

$$\frac{1}{V}\partial_j (V \partial^j \eta) = -\sum_A \frac{1}{6} H_A^2 (\partial \tilde{E}_A)^2 + \frac{\partial^2 V}{(d-3)V}\,. \tag{15}$$

We have, however, another equation for η, i.e., Eq. (11), which should be satisfied as well. We have to find a solution which satisfies both equations. This consistency gives two conditions for $\tilde{E}_A$. In order to derive them, we first take a trace of Eq. (11), and substituting Eqs. (14) and (15) into Eq. (11), we find the first condition:

$$\frac{1}{2}\sum_{A,B} M_{AB} H_A^2 H_B^2 \tilde{E}_A \tilde{E}_B (\partial \tilde{E}_A)(\partial \tilde{E}_B) - \sum_A H_A^2 (\partial \tilde{E}_A)^2 + 4\left(\frac{d-2}{d-3}\right) V^{-1/2} \partial^2 V^{1/2} = 0, \tag{16}$$

$$M_{AB} = \frac{1}{9^2}\left[2(8-q_A)(8-q_B) + (d-3)(q_A+1)(q_B+1) + \sum_{\alpha=2}^{p} \delta_{\alpha A}\delta_{\alpha B} \right]. \tag{17}$$

We have also a traceless part of Eq. (11) gives the second condition:

$$\begin{aligned}
&\frac{1}{2}\sum_{A,B} M_{AB} H_A^2 H_B^2 \tilde{E}_A \tilde{E}_B (\partial_i \tilde{E}_A)(\partial^j \tilde{E}_B) - \sum_A H_A^2 \partial_i \tilde{E}_A \partial^j \tilde{E}_A \\
&-\frac{1}{d-1}\delta_i^j \left[\frac{1}{2}\sum_{A,B} M_{AB} H_A^2 H_B^2 \tilde{E}_A \tilde{E}_B (\partial \tilde{E}_A)(\partial \tilde{E}_B) - \sum_A H_A^2 (\partial \tilde{E}_A)^2 \right] \\
&-2(d-3) V^{\frac{1}{d-3}} \left[\partial_i \partial^j \left(V^{-\frac{1}{d-3}} \right) - \frac{1}{d-1}\delta_i^j \partial^2 \left(V^{-\frac{1}{d-3}} \right) \right] = 0\,,
\end{aligned} \tag{18}$$

We have to find a solution for two conditions (16) and (18). Here we shall assume V= constant. We shall also impose the condition $M_{AB} = 0$ for $A \neq B$, which is called the intersection rule [?]. This rule is derived in the case of spherically symmetric spacetime from the condition that each E_A is independent. Our case is just an ansatz.

Suppose that the q_A-brane and q_B-brane are filled in different spatial dimensions, but those branes are crossing on $\bar{q}_{AB}$ dimensions ($\bar{q}_{AB} < q_A, q_B$). Calculating (17), we obtain

$$M_{AB} = \bar{q}_{AB} + 1 - \frac{(q_A+1)(q_B+1)}{9} = 0 \,. \tag{19}$$

Since we assume that it vanishes for $A \neq B$, we obtain the crossing dimensions $\bar{q}_{AB} + 1 = (q_A+1)(q_B+1)/9$.

Eqs. (16) and (18) are then reduced to

$$\sum_A \left[\frac{1}{2} M_{AA} H_A^2 \tilde{E}_A^2 - 1\right] H_A^2 (\partial \tilde{E}_A)^2 = 0 \,, \tag{20}$$

$$\sum_A \left[\frac{1}{2} M_{AA} H_A^2 \tilde{E}_A^2 - 1\right] H_A^2 \left[\partial_i \tilde{E}_A \partial^j \tilde{E}_A - \frac{1}{d-1} (\partial \tilde{E}_A)^2 \delta_i^j\right] = 0 \,, \tag{21}$$

thus we find $M_{AA} H_A^2 \tilde{E}_A^2 = 2$ or $\tilde{E}_A = \text{const}$. Since $M_{AA} = 2$ from Eq. (19), we find $\tilde{E}_A$ as $\tilde{E}_A = 1/H_A$ or $\tilde{E}_A = \text{const}$). If we impose that a spacetime is asymptotically flat (i.e., $H_A \to 1$ as $r \to \infty$) and the potential E_A vanishes at infinity, we find that $E_A = -(1 - 1/H_A)$ or $E_A = 0$. Inserting this relation into Maxwell equation we obtain the Laplace equation for H_A as $\partial^2 H_A = 0$ which means that H_A is a harmonic function on $\{x^i\} \in \mathbb{E}^{d-1}$. Thus we then obtain the solutions for metric functions in terms of the harmonic functions H_A.

We have two remaining equations (10) and Maxwell equation for $\mathcal{A}_i$ ($\mathcal{F}_{ij}$) and one Poisson equation (9) for f. We expect that each brane A has a charge $\mathcal{Q}_H^{(A)}$ (either electric or magnetic type), and then E_A becomes non-trivial, i.e., $H_A \neq 1$. In this case, if we set $B_i^A = -\tilde{E}_A \mathcal{A}_i = -\mathcal{A}_i / H_A$, we have $\mathcal{F}_{ij}^{(A)} = \mathcal{F}_{ij}$. Inserting the metric functions, we can show that two equations (10) and Maxwell equation are reduced to the following one Laplace equation: $\partial^j \mathcal{F}_{ij} = 0$. B_i^A describes a magnetic-type field produced by a current appearing through rotation of a charged brane.

Finally we discuss the last equation (9) for f. Here we assume we have N charged branes. Then, as for the metric f, we find

$$\partial^2 f = \frac{2-N}{4} \prod_A H_A^{-1} \left(\partial_{[j} \mathcal{A}_{i]}\right)^2 . \tag{22}$$

A describes charged branes which provide non-trivial potentials E_A.

If the number of M-brane is two then f is given by an arbitrary harmonic function on $\mathbb{E}^{d-1}$. In general, since the number of M-brane are determined by d and intersection rule $M_{AB} = 0$, thus we have to solve the Poisson equation (22).

The solution obtained in this section is summarized as follows:

$$ds^2 = \prod_A H_A^{\frac{q_A+1}{9}} \left[2 \prod_B H_B^{-1} du \left(dv + f du + \frac{\mathcal{A}}{\sqrt{2}}\right) + \sum_{\alpha=2}^{p} \prod_B H_B^{-\frac{\gamma_{\alpha B}}{9}} dy_\alpha^2 + \sum_{i=1}^{d-1} dx_i^2\right] \tag{23}$$

where $\gamma_{\alpha A} = \delta_{\alpha A} + q_A + 1$. H_A for each q_A-brane and $\mathcal{A} = \mathcal{A}_i dx^i$ are arbitrary harmonic functions, while the vector potential B_i^A can be chosen either as $B_i^A \propto \mathcal{A}_i / H_A$

(when $H_A \neq 1$), or an arbitrary harmonic function (when $H_A = 1$). The wave metric f usually satisfies the Poisson equation (22) with some source term originated by the rotation-induced metric $\mathcal{A}_i$, although it can be also an arbitrary harmonic function for some specific configuration of branes.

The M2-brane solution in this case is written as

$$\begin{aligned} ds_{11}^2 &= H_2^{1/3}\left[2H_2^{-1}du\left(dv+fdu+\frac{\mathcal{A}}{\sqrt{2}}\right)+H_2^{-1}dy_6^2+\sum_{i=1}^{8}dx_i^2\right], \\ F_4 &= d(1/H_2)\wedge du\wedge dv\wedge dy_6+\frac{1}{\sqrt{2}}dB_2\wedge du\wedge dy_6\,, \end{aligned} \tag{24}$$

where H_2 is a harmonic function on $\mathbb{E}^8$. Similarly, the field with a magnetic charge is related to the M5-brane because $^*q_A = {}^*n_A - 2 = D - n_A - 2 = 5$. The solution is described by

$$\begin{aligned} ds_{11}^2 &= H_5^{2/3}\left[2H_5^{-1}du\left(dv+fdu+\frac{\mathcal{A}}{\sqrt{2}}\right)+H_5^{-1}\sum_{\alpha=2}^{5}dy_\alpha^2+\sum_{i=1}^{5}dx_i^2\right], \\ *F_4 &= d(1/H_5)\wedge du\wedge dv\wedge dy_2\wedge dy_3\wedge dy_4\wedge dy_5 \\ &\quad +\frac{1}{\sqrt{2}}dB_5\wedge du\wedge dy_2\wedge dy_3\wedge dy_4\wedge dy_5\,, \end{aligned} \tag{25}$$

where H_5 is a harmonic function on $\mathbb{E}^5$. In both cases, $\mathcal{A}_i$ is also a vector harmonic function, while f is given by the Poisson equation (22) with $1/2$ coefficient because the number of brane is one.

These two branes (M2 and M5) can intersect if and only if

$$M2\cap M2 \to \bar{q}_{22}=0, \quad M2\cap M5 \to \bar{q}_{25}=1, \quad M5\cap M5 \to \bar{q}_{55}=3\,. \tag{26}$$

The crossing rule leads that there are only a possible configuration in four or five dimensional black hole [42,43]. In five dimension with null Killing vector $(f \neq 0)$, we find M2$\perp$ M5-brane solutions with traveling wave, and in four dimension we find M5 $\perp$ M5$\perp$ M5-brane with traveling wave [44]. We will consider M2 $\perp$ M5-brane solution in next section, because five dimensional solution is possible to take a diversity topology, charge and so on. Before consider the particular metric, we give the general method for compactification on torus. We also present the physical properties in lower dimension, mass, charge and angular momentum, in general null type metric with asymptotically flatness.

3.1. The Compactification of a Black Brane

The critical dimension for M-theory is eleven, thus we have to compactify extra dimensions to obtain an effective d dimensional spacetime. Rewriting the wave part of the metric as

$$2du\left(dv+fdu+\frac{\mathcal{A}}{\sqrt{2}}\right) = (1+f)\left[dy_1-\frac{1}{1+f}\left(fdt-\frac{\mathcal{A}}{2}\right)\right]^2-\frac{1}{1+f}\left(dt+\frac{\mathcal{A}}{2}\right)^2.$$

Introducing the conformal factors Ω_1 and Ω_α $(\alpha = 2,\ldots,p)$ by

$$\Omega_1^2=(1+f)\prod_A H_A^{-(8-q_A)/9},\ \Omega_\alpha^2=\prod_A H_A^{-\delta_{\alpha A}/9}, \tag{27}$$

and we also defined $\Omega^2 = \prod_{\alpha=1}^{p} \Omega_\alpha^2$. We perform a conformal transformation as

$$ds_D^2 = \Omega^{-\frac{2}{d-2}} d\bar{s}_d^2 + \Omega_1^2 \left[dy_1 - \frac{1}{1+f}\left(fdt - \frac{\mathcal{A}}{2}\right)\right]^2 + \sum_{\alpha=2}^{p} \Omega_\alpha^2 dy_\alpha^2 . \qquad (28)$$

With this conformal transformation, we obtain the Einstein gravity in d-dimensions;

$$\begin{aligned} d\bar{s}_d^2 &\equiv \bar{g}_{\bar{\mu}\bar{\nu}} dx^{\bar{\mu}} dx^{\bar{\nu}} = -\Xi^{d-3}\left(dt + \frac{\mathcal{A}}{2}\right)^2 + \Xi^{-1} \sum_{i=1}^{d-1} dx_i^2 , \\ \Xi &\equiv (1+f)^{-1/(d-2)} \prod H_A^{-1/(d-2)} , \end{aligned} \qquad (29)$$

where $\bar{\mu}, \bar{\nu}, \cdots$ are coordinate indices for d-dimensional spacetime. If the compactified space is sufficiently small, we find the effective d-dimensional world with the metric (29).

If this spacetime is asymptotically flat, which we impose, it may describe a black object in d-dimensions. From the asymptotic form of the metric, we can define the ADM mass $M_{\rm ADM}$ as

$$\bar{g}_{00} \sim -1 + \frac{16\pi G_d}{(d-2)\omega_{d-2}} \frac{M_{\rm ADM}}{r^{d-3}} , \qquad (30)$$

where $\omega_{d-2} \equiv 2\pi^{\frac{d-1}{2}} / \Gamma\left(\frac{d-1}{2}\right)$, and $r^2 \equiv \sum_{i=1}^{d-1} x_i^2$. Assuming $H_A \to 1 + \mathcal{Q}_H^{(A)}/r^{d-3}$ and $f \to \mathcal{Q}_0/r^{d-3}$, we obtain

$$M_{\rm ADM} = \frac{(d-3)\pi^{(d-3)/2}}{8G_d\Gamma\left((d-1)/2\right)} \left[\mathcal{Q}_0 + \sum_A \mathcal{Q}_H^{(A)}\right] , \qquad (31)$$

where A denotes a kind of charged branes. Once we find solutions described by the above set of equations, we have to study a spacetime structure. In particular, the horizon and the singularity of a spacetime are important geometrical objects. We then have to evaluate the curvature invariant $R_{\mu\nu\rho\sigma}^2$ of the metric (29).

3.2. Black Hole Solutions with M2⊥M5 Branes

The metric in five-dimensions compactified M2⊥M5-brane is written by

$$d\bar{s}_5^2 = -\Xi^2\left(dt + \frac{\mathcal{A}}{2}\right)^2 + \Xi^{-1} ds_{\mathbb{E}^4}^2 , \qquad (32)$$

where $\Xi = [H_2 H_5 (1+f)]^{-1/3}$. The unknown functions $H_A (A = 2, 5)$, $\mathcal{A}_i$ (or $\mathcal{F}_{ij} = \partial_i \mathcal{A}_j - \partial_j \mathcal{A}_i$) and f satisfy the following equations:

$$\partial^2 H_A = 0 , \;\; \partial_j \mathcal{F}^{ij} = 0 , \;\; \partial^2 f = 0 . \qquad (33)$$

In order to find the exact solutions, we assume that the 4-dimensional x-space as the hyperspherical coordinates, and we show explicitly how to construct the exact solutions. We adopt the hyperspherical coordinates: $x_1 + ix_2 = r\cos\theta e^{i\phi}, \quad x_3 + ix_4 = -r\sin\theta e^{i\psi}$

where $0 \leq \phi, \psi < 2\pi$ and $0 \leq \theta \leq \pi/2$. The line element of four dimensional flat space is $ds^2_{\mathbb{E}^4} = dr^2 + r^2 \left(d\theta^2 + \cos^2\theta d\phi^2 + \sin^2\theta d\psi^2\right)$. The symmetric axis is described by $\theta = 0$ and $\pi/2$, and the infinity corresponds to $r = \infty$.

The equation for H_A in Eq. (33) in this coordinate system is

$$\frac{1}{r}\partial_r \left(r^3 \partial_r H_A\right) + \frac{1}{\sin\theta\cos\theta}\partial_\theta \left(\sin\theta\cos\theta\partial_\theta H_A\right) = 0\,. \tag{34}$$

Setting $H_A = h_A(r) j_A(\theta)$, we separate the variables and obtain two equations:

$$\frac{1}{r}\frac{d}{dr}\left(r^3 \frac{dh_A}{dr}\right) - M h_A = 0\,, \tag{35}$$

$$\frac{1}{\sin\theta\cos\theta}\frac{d}{d\theta}\left(\sin\theta\cos\theta\frac{dj_A}{d\theta}\right) + M j_A = 0\,, \tag{36}$$

where M is a separation constant. Eq. (36) with $\mu = \cos 2\theta$ is just the Legendre equation. From regularity conditions on the symmetric axis ($\theta = 0, \pi/2$), we obtain $j_A = P_\ell(\cos 2\theta)$ by setting $M = 4\ell(\ell+1)$ ($\ell = 0, 1, \cdots$). Eq. (35) is easily solved as $h_A = r^{2\ell}$ or $r^{-2(\ell+1)}$. The general solution for H_A is then

$$H_A = \sum_{\ell=0}^{\infty} \left[g_\ell^{(A)} r^{2\ell} + h_\ell^{(A)} r^{-2(\ell+1)}\right] P_\ell(\cos 2\theta)\,, \tag{37}$$

where $g_\ell^{(A)}$ and $h_\ell^{(A)}$ are arbitrary constants. From the asymptotically flatness condition;

$$H_A = 1 + \sum_{\ell=0}^{\infty} h_\ell^{(A)} r^{-2(\ell+1)} P_\ell(\cos 2\theta)\,. \tag{38}$$

The spherically symmetric solution ($\ell = 0$) is given by $H_A = 1 + \mathcal{Q}_H^{(A)}/r^2$, where $\mathcal{Q}_H^{(A)}$ is a constant, which corresponds to a conserved charge.

Next, we discuss the equations for $\mathcal{A}_i$, which are written as

$$r\partial_r \left(r\partial_r \mathcal{A}_\phi\right) + \cot\theta\partial_\theta \left(\tan\theta\partial_\theta \mathcal{A}_\phi\right) = 0\,, \tag{39}$$

$$r\partial_r \left(r\partial_r \mathcal{A}_\psi\right) + \tan\theta\partial_\theta \left(\cot\theta\partial_\theta \mathcal{A}_\psi\right) = 0\,. \tag{40}$$

Setting $\mathcal{A}_\phi = a_\phi(r) b_\phi(\theta)$ and $\mathcal{A}_\psi = a_\psi(r) b_\psi(\theta)$, we have ordinary differential equations:

$$r\frac{d}{dr}\left(r\frac{d}{dr}a_\phi\right) - K a_\phi = 0 \tag{41}$$

$$\frac{d^2 b_\phi}{d\mu^2} - \frac{1}{1-\mu}\frac{db_\phi}{d\mu} + \frac{K}{4(1-\mu^2)} b_\phi = 0\,, \tag{42}$$

$$r\frac{d}{dr}\left(r\frac{d}{dr}a_\psi\right) - L a_\psi = 0 \tag{43}$$

$$\frac{d^2 b_\psi}{d\mu^2} + \frac{1}{1+\mu}\frac{db_\psi}{d\mu} + \frac{L}{4(1-\mu^2)} b_\psi = 0\,, \tag{44}$$

where $\mu = \cos 2\theta$, and K and L are separation constants. The solutions of Eqs. (42) and (44) are described by Gauss's hypergeometric functions as $b_\phi(\mu) = F(-\sqrt{K}/2, \sqrt{K}/2, 1, (1-\mu)/2)$ and $b_\psi(\mu) = F(-\sqrt{L}/2, \sqrt{L}/2, 1, (1+\mu)/2)$. The Gauss's hyper geometrical function $F(\alpha, \beta, \gamma, z)$ is defined by

$$F(\alpha, \beta, \gamma, z) = \frac{\Gamma(\gamma)}{\Gamma(\alpha)\Gamma(\beta)} \sum_{n=0}^{\infty} \frac{\Gamma(\alpha+n)\Gamma(\beta+n)}{\Gamma(\gamma+n)} \frac{z^n}{n!} \,. \tag{45}$$

From regularity conditions, we have to impose that $K = 4m^2$ and $L = 4n^2$, where $m, n = 1, 2, \cdots$. We then have the angular solutions as $b_\phi = F(-m, m, 1, \sin^2\theta)$ and $b_\psi = F(-n, n, 1, \cos^2\theta)$.

The equations for a_ϕ and a_ψ are easily solved under the asymptotically flatness;

$$\mathcal{A}_\phi = \sum_{m=1}^{\infty} \frac{b_m^{(\phi)}}{r^{2m}} F(-m, m, 1, \sin^2\theta), \tag{46}$$

$$\mathcal{A}_\psi = \sum_{n=1}^{\infty} \frac{b_n^{(\psi)}}{r^{2n}} F(-n, n, 1, \cos^2\theta)\,, \tag{47}$$

If we take the first two terms in the general solution, we obtain a simple solution as

$$\mathcal{A}_\phi = \frac{\cos^2\theta}{r^2}\left[J_1^{(\phi)} + \frac{J_2^{(\phi)}}{r^2}(1 - 3\sin^2\theta)\right] \tag{48}$$

$$\mathcal{A}_\psi = \frac{\sin^2\theta}{r^2}\left[J_1^{(\psi)} + \frac{J_2^{(\psi)}}{r^2}(1 - 3\cos^2\theta)\right], \tag{49}$$

where $J_1^{(\phi)}$, $J_1^{(\psi)}$, $J_2^{(\phi)}$ and $J_2^{(\psi)}$ are constants. The first two constants describe angular momenta of a black object. As we show in next section, if $\mathcal{F}_{ij}$ is self-dual, the spacetime is supersymmetric. This condition implies $J_1^{(\phi)} = -J_1^{(\psi)}$ and $J_2^{(\phi)} = J_2^{(\psi)}$.

Finally we discuss equation for f, but this is the almost same as H_A, but the asymptotically condition is different as $f \to 0$. Therefore we find the Laplace equation for f, which gives us a simple solution as $f = \sum_{\ell=0}^{\infty} \mathcal{Q}_\ell r^{-2(\ell+1)} P_\ell(\cos 2\theta)$, where $\mathcal{Q}_\ell$'s are constants. The solution with the lowest multipole moment is given by

$$H_A = 1 + \frac{\mathcal{Q}_H^{(A)}}{r^2}, \quad f = \frac{\mathcal{Q}_0}{r^2}, \quad \mathcal{A}_\phi = \frac{J_\phi \cos^2\theta}{r^2}, \quad \mathcal{A}_\psi = \frac{J_\psi \sin^2\theta}{r^2}\,. \tag{50}$$

The mass and the entropy of this spacetime are

$$M_{\mathrm{ADM}} = \frac{\pi}{4G_5}(\mathcal{Q}_0 + \mathcal{Q}_H^{(2)} + \mathcal{Q}_H^{(5)}), \tag{51}$$

$$S = \frac{A_h}{4G_5} = \frac{\pi^2}{3G_5} \frac{\Lambda_+^2 + \Lambda_+\Lambda_- + \Lambda_-^2}{\Lambda_+^{3/2} + \Lambda_-^{3/2}}, \tag{52}$$

where $\Lambda_{\pm} = \mathcal{Q}_0 \mathcal{Q}_H^{(2)} \mathcal{Q}_H^{(5)} - J^2/8 \pm \Delta J^2/16$, where J^2 and ΔJ^2 are defined by $J^2 \equiv (J_\phi^2 + J_\psi^2)/2$ and $\Delta J^2 \equiv J_\phi^2 - J_\psi^2$, respectively.

Fixing J^2, if we maximize entropy S, we find the maximum entropy with

$$S = S_{\max} = \frac{\pi^2}{2G_5}\sqrt{\mathcal{Q}_0 \mathcal{Q}_H^{(2)} \mathcal{Q}_H^{(5)} - \frac{J^2}{8}}\,, \tag{53}$$

if $\Delta J^2 = 0$, i.e., $J_\phi^2 = J_\psi^2 = J^2$. Note that supersymmetry implies $J_\phi = -J_\psi = J$, which corresponds to the BMPV solution [30, 31]. If $J_\phi \neq -J_\psi$, the above solution describes a regular rotating non-BPS black hole spacetime in five dimensions. At the event horizon $r = 0$, this solution obviously satisfy the Kretchmann invariance is finite.

In this section we adopt the hyperspherical coordinate, but in general we can we assume that the 4-dimensional x-space has two rotation symmetries which Killing vectors ($\xi^i_{(\phi)}$ and $\xi^i_{(\psi)}$) commute each other. In this case, $\partial_j \mathcal{F}^{ij} = 0$ is reduced to two uncoupled equations for two scalar fields, $\mathcal{A}_\phi = \mathcal{A}_i \xi^i_{(\phi)}$ and $\mathcal{A}_\psi = \mathcal{A}_i \xi^i_{(\psi)}$, as

$$\partial^2 \mathcal{A}_\phi - \partial_i \ln\left(\xi_{(\phi)} \cdot \xi_{(\phi)}\right) \partial^i \mathcal{A}_\phi = 0\,, \tag{54}$$

$$\partial^2 \mathcal{A}_\psi - \partial_i \ln\left(\xi_{(\psi)} \cdot \xi_{(\psi)}\right) \partial^i \mathcal{A}_\psi = 0\,. \tag{55}$$

Here we have assumed that the other components of $\mathcal{A}_i$ vanish.

We now have the Laplace equations or similar equations (the Poisson equation or Eqs. (54) and (55)) for several scalar functions (H_A, $\mathcal{A}_\phi$, $\mathcal{A}_\psi$, and f). Each equation is linear and uncoupled, thus it is very easy to find general solutions because the Laplace-Beltrami operator is defined on the flat Euclidian space. Once we obtain a complete set of solutions in an appropriate curvilinear coordinate system, we can construct any solutions by superposing them [45].

3.3. Supersymmetry in Eleven-dimensional Black Branes

We obtain the Killing equations (4) to get together the same metric function of M2 $\perp$M5 branes solution, using the properties of gamma metric as

$$\delta\psi_u = \frac{1}{6}H_2^{-1/6}H_5^{-1/3}\left[\frac{\partial_j H_2}{H_2}\gamma^{jv}\gamma_{M2} + \frac{1}{2}\frac{\partial_j H_5}{H_5}\gamma^{jv}\gamma_{M5} - 3\partial^j f\gamma^{ju}\right]\epsilon$$
$$-\frac{1}{4\sqrt{2}}H_2^{-2/3}H_5^{-5/6}\left[\left\{\frac{1}{2}\mathcal{F}_{ij} + \frac{1}{3}{}^*\mathcal{F}_{ij} - \frac{1}{6}\mathcal{F}^*_{ij}\right\}\gamma^{ij} - \frac{1}{3}{}^*\mathcal{F}_{ij}\gamma^{ij}\gamma_{M5} + \frac{1}{6}\mathcal{F}^*_{ij}\gamma^{ij}\gamma_{M2}\right]\epsilon\,,$$
$$\delta\psi_v = \frac{1}{6}H_2^{-1/6}H_5^{-1/3}\left[\frac{\partial_j H_2}{H_2}\gamma^{ju}\gamma_{M2} + \frac{\partial_j H_5}{2H_5}\gamma^{ju}\gamma_{M5}\right]\epsilon\,,$$
$$\delta\psi_i = H_2^{-1/6}H_5^{-1/3}\left(\partial_i + \frac{\partial_i H_2}{6H_2} + \frac{\partial_i H_5}{12H_5}\right)\epsilon$$
$$+\frac{1}{\sqrt{2}}H_2^{-2/3}H_5^{-5/6}\left[\left(\frac{1}{4}\mathcal{F}_{ij} - \frac{1}{6}\mathcal{F}^*_{ij} - \frac{1}{12}{}^*\mathcal{F}_{ij}\right)\gamma^{ju} + \frac{1}{6}\mathcal{F}^*_{ij}\gamma^{ju}\gamma_{M2} + \frac{1}{12}{}^*\mathcal{F}_{ij}\gamma^{ju}\gamma_{M5}\right]\epsilon$$
$$+\frac{1}{6}H_2^{-1/6}H_5^{-1/3}\left[\left(\frac{\partial_j H_2}{2H_2}\gamma^{ij} - \frac{\partial_i H_2}{H_2}\right)\gamma_{M2} + \left(\frac{\partial_j H_5}{H_5}\gamma^{ij} - \frac{\partial_i H_5}{2H_5}\right)\gamma_{M5}\right]\epsilon\,,$$
$$\delta\psi_{y_2(\cdots 5)} = -\frac{1}{12}H_2^{-1/6}H_5^{-1/3}\left[\frac{\partial_j H_2}{H_2}\gamma^{jy_2(\cdots 5)}\gamma_{M2} - \frac{\partial_j H_5}{H_5}\gamma^{jy_2(\cdots 5)}\gamma_{M5}\right]\epsilon$$
$$+\frac{1}{24\sqrt{2}}H_2^{-2/3}H_5^{-5/6}\mathcal{F}^*_{ij}\gamma^{ijuy_2(\cdots 5)y_6}\epsilon\,,$$
$$\delta\psi_{y_6} = \frac{1}{6}H_2^{-1/6}H_5^{-1/3}\left[\frac{\partial_j H_2}{H_2}\gamma^{jy_6}\gamma_{M2} - \frac{\partial_j H_5}{H_5}\gamma^{jy_6}\gamma_{M5} - \frac{1}{2\sqrt{2}}(H_2H_5)^{-1/2}\mathcal{F}^*_{ij}\gamma^{iju}\right]\epsilon\,.$$

where we defined the combination of magnetic gauge fields strength as ${}^*\mathcal{F}_{ij} = {}^*\mathcal{F}^{(2)}_{ij} + \mathcal{F}^{(5)}_{ij}$ and $\mathcal{F}^*_{ij} = \mathcal{F}^{(2)}_{ij} + {}^*\mathcal{F}^{(5)}_{ij}$, and typical combinations for the gamma matrix as $\gamma_{M2} = 1 - \gamma^{uvy_6}$ and $\gamma_{M5} = 1 - \gamma^{uvy_2\cdots y_5}$. Most parts of the above equations vanish if we impose the following condition for the Killing spinor ϵ: $\gamma_{M2}\epsilon = 0$, $\gamma_{M5}\epsilon = 0$ and $\gamma^{\hat{u}}\epsilon = 0$. These conditions can be rewritten as $(1 + \gamma^{0y_1y_6})\epsilon = 0$, $(1 + \gamma^{0y_1\cdots y_5})\epsilon = 0$ and $(1 + \gamma^{0y_1})\epsilon = 0$ However, two terms remain. One term is $\left(\partial_i + \partial_i \ln H_2^{1/6} + \partial_i \ln H_5^{1/12}\right)\epsilon$ and the other term is $\left(\mathcal{F}_{ij}/2 + {}^*\mathcal{F}_{ij}/3 - \mathcal{F}^*_{ij}/6\right)\gamma^{ij}\epsilon$. The former term vanishes if ϵ is described as $\epsilon = H_2^{-1/6}H_5^{-1/12}\epsilon_0$, where ϵ_0 is a constant spinor. The latter term also vanishes if $\mathcal{F}^{(A)}_{ij} \propto \mathcal{F}_{ij}$ $(A = 2, 5)$ and $\mathcal{F}_{ij}$ is self-dual (${}^*\mathcal{F}_{ij} = \mathcal{F}_{ij}$). In fact, this term is proportional to $\mathcal{F}_{ij}\gamma^{ij}\epsilon$, which vanishes for the self-dual field $\mathcal{F}_{ij}$.

In the case of the BMPV type solution discussed, this self-dual condition gives the relation between J_ϕ and J_ψ, that is, $J_\phi = -J_\psi = J$ [30]. Thus the only the difference of the non-BPS solution and the BPS solution is the angular momentum. Under the near horizon limit, the angular momentum of BMPV black hole on the event horizon are cancelled.

4. Time-like Killing Spacetime

We can also solve the time-like Killing spacetime almost the same way in null Killing case. We consider the ansatz with the time-like Killing vector as

$$ds^2 = -e^{2\zeta_0(x)}(dt + \omega)^2 + e^{2\eta(x)}h_{ij}(x)dx^i dx^j + \sum_{\alpha=1}^{p} e^{2\zeta_\alpha(x)}dy_\alpha^2\,,$$
$$C_A = E^A(x)dt \wedge dy_1 \wedge dy_2 + \frac{1}{\sqrt{2}}B^A_j(x)dx^j \wedge dy_1 \wedge dy_2\,. \qquad (56)$$

We must pay attention to remaining the Chern-Simons terms, because of the gauge ansatz, then we find the basic equations as follows:

$$\partial^2\zeta_0 + \partial_j\zeta_0\partial^j \ln V + \frac{1}{4}e^{2(\zeta_0-\eta)}(\mathcal{F}_{ij})^2 = \frac{1}{3}\sum_A \left[H_A^2(\partial E_A)^2 + \frac{1}{4}\left(\mathcal{F}_{ij}^{(A)}\right)^2 e^{2(\zeta_0-\eta)}\right], \tag{57}$$

$$\partial^j\mathcal{F}_{ij} + \mathcal{F}_{ij}\partial^j\left[2(\zeta_0-\eta)+\ln V\right] = \sum_A H_A\mathcal{F}_{ij}^{(A)}\partial^j E_A, \tag{58}$$

$$\begin{aligned}&\left(\partial^2\eta + \partial_l\eta\partial^l \ln V\right)\delta_i^j + (d-3)\partial_i\eta\partial^j\eta + \sum_{\alpha=0}^{p}\partial_i\zeta_\alpha\partial^j\zeta_\alpha + \partial_i\partial^j\ln V - \left(\partial_i\eta\partial^j\ln V + \partial^j\eta\partial_i\ln V\right)\\ &\quad - \frac{1}{2}e^{2(\zeta_0-\eta)}\mathcal{F}_{i\ell}\mathcal{F}^{j\ell} + \frac{1}{2}e^{2(\zeta_0-\eta)}\sum_A\left[\mathcal{F}_{i\ell}^{(A)}\mathcal{F}^{(A)j\ell} - \frac{1}{6}\delta_i^j(\mathcal{F}_{k\ell}^{(A)})^2\right]\\ &= \frac{1}{2}\sum_A H_A^2\left[\partial_i E_A\partial^j E_A - \frac{1}{3}(\partial E_A)^2\delta_i^j\right],\end{aligned} \tag{59}$$

$$\partial^2\zeta_\alpha + \partial_j\zeta_\alpha\partial^j\ln V = \frac{1}{18}\sum_A\delta_{\alpha A}\left[H_A^2(\partial E_A)^2 - \frac{1}{2}e^{2(\zeta_0-\eta)}\left(\mathcal{F}_{ij}^{(A)}\right)^2\right], \tag{60}$$

$$\partial_j(H_A^2V\partial^j E_A) + \frac{1}{2}VH_Ae^{2(\zeta_0-\eta)}\mathcal{F}_{ij}^{(A)}\mathcal{F}^{ij} = \frac{1}{4}\epsilon^{ijkl}\frac{1}{H_BH_C}\mathcal{F}_{ij}^{(B)}\mathcal{F}_{kl}^{(C)}, \tag{61}$$

$$\partial^j\left(VH_Ae^{2(\zeta_0-\eta)}\mathcal{F}_{ij}^{(A)}\right) = \frac{1}{2}\epsilon^{ijkl}\left[\partial_j E_B\frac{1}{H_C}\mathcal{F}_{kl}^{(C)} + \partial_j E_C\frac{1}{H_B}\mathcal{F}_{kl}^{(B)}\right], \tag{62}$$

We can also replace it with $\tilde{E}_A = E_A - E_A^{(0)}$, where $E_A^{(0)}$ is a constant, which is fixed by a boundary condition. Setting $\mathcal{F}_{ij}^{(A)} = \mathcal{F}_{ij} + H_Aq_{ij}^{(A)}$, where $\mathcal{F}_{ij}$ is self-dual, and $q_{ij}^{(A)}$ is anti self-dual, we find Eqs. (58) and (62) as

$$\partial^j\left(Ve^{2(\zeta_0-\eta)}\mathcal{F}_{ij}\right) = Ve^{2(\zeta_0-\eta)}\left[\left(\sum_A H_A\partial^j\tilde{E}_A\right)\mathcal{F}_{ij} + \sum_A H_A^2q_{ij}^{(A)}\partial^j\tilde{E}_A\right] \tag{63}$$

$$\begin{aligned}&\partial^j\left(VH_Ae^{2(\zeta_0-\eta)}\mathcal{F}_{ij}\right) + \partial^j\left(VH_A^2e^{2(\zeta_0-\eta)}q_{ij}^{(A)}\right)\\ &\quad = \left[\partial^j\tilde{E}_B\frac{1}{H_C}\left(\mathcal{F}_{ij} - H_Cq_{ij}^{(C)}\right) + \partial^j\tilde{E}_C\frac{1}{H_B}\left(\mathcal{F}_{ij} - H_Bq_{ij}^{(B)}\right)\right],\end{aligned} \tag{64}$$

Inserting Eq. (63) into Eq. (64), we obtain

$$\begin{aligned}&Ve^{2(\zeta_0-\eta)}\left[\left(\sum_X H_X\partial_j\tilde{E}_X\right)\mathcal{F}_{ij} + \sum_X H_X^2\partial_j\tilde{E}_Xq_{ij}^{(X)} + \frac{\partial_jH_A}{H_A}\mathcal{F}_{ij}\right]\\ &- \frac{H_B\partial_j\tilde{E}_B + H_C\partial_j\tilde{E}_C}{H_AH_BH_C}\mathcal{F}_{ij} + \frac{1}{H_A}\partial_j\left(VH_A^2e^{2(\zeta_0-\eta)}q_{ij}^{(A)}\right) + \left(\partial_j\tilde{E}_Bq_{ij}^{(C)} + \partial_j\tilde{E}_Cq_{ij}^{(B)}\right) = 0,\end{aligned}$$

Here we assume that $Ve^{2(\zeta_0-\eta)}\prod_X H_X = 1$. We then find

$$\begin{aligned}&H_A\left(\mathcal{F}_{ij} + H_Aq_{ij}^{(A)}\right)\partial_j\left(\tilde{E}_A - \frac{1}{H_A}\right) + H_AH_BH_C\partial_j\left(\frac{1}{H_BH_C}q_{ij}^{(A)}\right)\\ &+ \left(H_B\partial_j\tilde{E}_B + H_C\partial_j\tilde{E}_C\right)\left(H_Bq_{ij}^{(B)} + H_Cq_{ij}^{(C)}\right) = 0,\end{aligned} \tag{65}$$

Now we assume that $\tilde{E}_A = 1/H_A$, then we obtain the equation for $q_{ij}^{(A)}$ as

$$H_A\partial_jq_{ij}^{(A)} = \left(\frac{1}{H_B}\partial_jH_B + \frac{1}{H_C}\partial_jH_C\right)Q_{ij} \tag{66}$$

where $Q_{ij} = \sum_X H_X q_{ij}^{(X)}$. Eqs. (63) (61) are now

$$\partial_j \mathcal{F}_{ij} = -\sum_X \partial_j H_X q_{ij}^{(X)} \tag{67}$$

$$\partial_j \left(V \partial_j H_A\right) = \frac{1}{2} q_{ij}^{(B)} q_{ij}^{(C)} \tag{68}$$

where we have used the formula $S_{ij}A^{ij} = 0$ if S_{ij} is self-dual and A^{ij} is anti self-dual.

Using Eq. (68),Eqs. (57) and (60) can be written as

$$\partial \left(\partial \zeta_\alpha + \frac{1}{18} \sum_X \delta_{\alpha X} \frac{1}{H_X} \partial H_X \right) = \frac{1}{36 \prod_Y H_Y} \sum_X \delta_{\alpha X} H_X q_{ij}^{(X)} Q^{ij} \,, \tag{69}$$

where we set $V = 1$ just for simplicity. Hence if we assume that $Q_{ij} = 0$, we can set

$$\partial \zeta_\alpha + \frac{1}{18} \sum_X \delta_{\alpha X} \frac{1}{H_X} \partial H_X = 0 \tag{70}$$

From the definition of V with Eqs. (70), we find $\partial \eta - \frac{1}{6} \sum_X \partial H_X / H_X = 0$, and inserting it into Eq. (59), we find

$$\mathcal{F}_{i\ell} \mathcal{F}_{j\ell} - \frac{1}{4} \delta_{ij} \left(\mathcal{F}_{k\ell}\right)^2 = 0 \tag{71}$$

$$q_{i\ell}^{(A)} q_{j\ell}^{(A)} - \frac{1}{4} \delta_{ij} \left(q_{k\ell}^{(A)}\right)^2 = 0 \tag{72}$$

These equations are satisfied for some self-dual or anti self-dual solutions. The anti self-dual variables $q_{ij}^{(A)}$ are given by vector potential $h_j^{(A)}$ as $q_{ij}^{(A)} = 2\partial_{[i} h_{j]}^{(A)}$, where $h_j^{(A)} = -\left(\mathcal{A}_j / H_A + B_j^{(A)}\right)$. We will show the manifest representation about this solution in the next section.

4.1. Black Ring Solutions with M2⊥M2⊥M2 Branes

In order to find the exact solution having a regular event horizon, we assume the hyperpolorical coordinates (ξ, η, ϕ, ψ), which are defined by the transformation

$$x_1 + i x_2 = \frac{R \sinh \xi}{\cosh \xi - \cos \eta} e^{i\psi}, \quad x_3 + i x_4 = \frac{R \sin \eta}{\cosh \xi - \cos \eta} e^{i\phi} \,, \tag{73}$$

where $\xi \geq 0$, $0 \leq \eta \leq \pi$, and $0 \leq \phi, \psi \leq 2\pi$. This coordinates could be used to describe a ring topology. In this case, the infinity corresponds to $\xi = 0$, which also describes one of the symmetric axis. The line element is given by

$$ds_{\mathbb{E}^4}^2 = \frac{R^2}{(\cosh \xi - \cos \eta)^2} (d\xi^2 + \sinh^2 \xi d\psi^2 + d\eta^2 + \sin^2 \eta d\phi^2) \,. \tag{74}$$

Now we discuss Eqs. (68) using the vector potential $h_j^{(A)}$, then we find the equations as

$$\frac{1}{\sinh\xi}\partial_\xi(\sinh\xi\partial_\xi h_\phi^{(A)}) + \sin\eta\partial_\eta\left(\frac{1}{\sin\eta}\partial_\eta h_\phi^{(A)}\right) = 0 \tag{75}$$

$$\sinh\xi\partial_\xi\left(\frac{1}{\sinh\xi}\partial_\xi h_\psi^{(A)}\right) + \frac{1}{\sin\eta}\partial_\eta(\sin\eta\partial_\eta h_\psi^{(A)}) = 0\,. \tag{76}$$

The equations of h_ϕ and h_ψ can be solved under the regularity conditions on the symmetric axis and the asymptotically flatness condition, and we find

$$h_\phi^{(A)} = \sum_{m=1}^{\infty}\frac{b_m^{(\phi)}}{m+1}P_m(\cosh\xi)\left[\cos\eta\, P_m(\cos\eta) - P_{m-1}(\cos\eta)\right] \tag{77}$$

$$h_\psi^{(A)} = \sum_{n=1}^{\infty}\frac{b_n^{(\psi)}}{n+1}\left[\cosh\xi\, P_n(\cosh\xi) - P_{n-1}(\cosh\xi)\right]P_n(\cos\eta)\,, \tag{78}$$

where $b_m^{(\phi)}$ and $b_n^{(\psi)}$ are arbitrary constants.

To simplify we consider the $m = n = 0$ case, then

$$h_\phi^{(A)} = -\frac{q_\phi^A}{2R}(1-\cos\eta)\,,\quad h_\psi^{(A)} = \frac{q_\psi^A}{2R}(\cosh\xi - 1)\,, \tag{79}$$

and we find Eq. (68) as

$$\frac{1}{\sinh\xi}\partial_\xi\left(\frac{\sinh\xi\partial_\xi H_A}{(\cosh\xi-\cos\eta)^2}\right) + \frac{1}{\sin\eta}\partial_\eta\left(\frac{\sin\eta\partial_\eta H_A}{(\cosh\xi-\cos\eta)^2}\right) = \frac{q_\phi^B q_\phi^C + q_\psi^B q_\psi^C}{8R^2}\,.$$

Changing the variable by using $H_A(\xi,\eta) = 1 + (\cosh\xi - \cos\eta)\tilde{H}_A(\xi,\eta)$, we find

$$\partial_\xi^2\tilde{H}_A + \coth\xi\partial_\xi\tilde{H}_A + \partial_\eta^2\tilde{H}_A + \cot\eta\partial_\eta\tilde{H}_A = \frac{q_\phi^B q_\phi^C + q_\psi^B q_\psi^C}{8R^2}(\cosh\xi - \cos\eta)\,, \tag{80}$$

Setting $\tilde{H}_A = \tilde{h}_A(\xi)\tilde{j}_A(\eta)$, we can separate the variables and find the following two ordinary differential equations:

$$(\rho^2-1)\frac{d^2\tilde{h}_A}{d\rho^2} + 2\rho\frac{d\tilde{h}_A}{d\rho} - M\tilde{h}_A = -\frac{q_\phi^B q_\phi^C + q_\psi^B q_\psi^C}{8R^2}\rho \tag{81}$$

$$(1-\mu^2)\frac{d^2\tilde{j}_A}{d\mu^2} - 2\mu\frac{d\tilde{j}_A}{d\mu} + M\tilde{j}_A = \frac{q_\phi^B q_\phi^C + q_\psi^B q_\psi^C}{8R^2}\mu\,, \tag{82}$$

where $\rho = -\cosh\xi$ and $\mu = -\cos\eta$, and M is a separation constant.

We can find that the general solution by using the hypergeometric function, but it is too complicated. Thus we show typical case about $M = 0$ with the BPS sate $-q_\phi^A = q_\psi^A = q^A$. The case are already given by [32] named supersymmetric black ring solution as

$$H_A = 1 + (\cosh\xi - \cos\eta)\left[\frac{Q_A - q^B q^C}{2R^2} + \frac{q^B q^C}{4R^2}(\cosh\xi + \cos\eta)\right]\,, \tag{83}$$

where we assume the regularity on the symmetric axis. Q_A is an arbitrary constant.

The Eqs (67) can be written by

$$\frac{1}{\sinh\xi}\partial_\xi(\sinh\xi\partial_\xi\mathcal{A}_\phi)+\sin\eta\partial_\eta\left(\frac{1}{\sin\eta}\partial_\eta\mathcal{A}_\phi\right)$$
$$=-\frac{\sin^2\eta}{4R}\left[\sum_X Q_X q^X-3q^Aq^Bq^C(1-\cos\eta)\right] \quad (84)$$
$$\sinh\xi\partial_\xi\left(\frac{1}{\sinh\xi}\partial_\xi\mathcal{A}_\psi\right)+\frac{1}{\sin\eta}\partial_\eta(\sin\eta\partial_\eta\mathcal{A}_\psi)$$
$$=\frac{\sinh^2\xi}{4R}\left[\sum_X Q_X q^X+3q^Aq^Bq^C(\cosh\xi-1)\right]. \quad (85)$$

and the spacial solutions for this equation is

$$\mathcal{A}_\phi=-\frac{\sin^2\eta}{8R^2}\left[\sum_X Q_X q^X-q^Aq^Bq^C(3-\cosh\xi-\cos\eta)\right]$$
$$\mathcal{A}_\psi=-\sum_A\frac{q^A}{2}(\cosh\xi-1)-\frac{\sinh^2\xi}{8R^2}\left[\sum_X Q_X q^X-q^Aq^Bq^C(3-\cosh\xi-\cos\eta)\right].$$

This is a solutions of supersymmetric black ring solutions which introduce by Elvang et. al. ([33]), and satisfies the Killing spinor equation and preserve $1/8$ supersyymetries [46].

We can also applies the hyperspherical coordinate for this time-like brane configuration easily. Using the condition for the smooth event horizon, only the trivial solution ($q_j^{(A)}=0$) is possible, and this solutions are the exactly the same as BMPV black hole solution, we have already shown.

5. Conclusion

We study a stationary black hole solution written in intersecting M-branes. Assuming a BPS type relation between the first-order derivatives of metric function, we have shown the solutions contain solutions given by solving the Killing spinor equation. The solution with null Killing vector space gives an unique M-brane configuration M2 $\perp$M5-brane with pp-wave solution. Then the null solution gives five dimensional BMPV solution compactified six dimensional torus. We also have an unique configuration of M2 $\perp$M2$\perp$M2-brane with time-like Killing vector. The supersymmetric black ring solution are given by the time-like solution with specific physical parameters on torus compactification.

These solutions are given by the base space Laplace equation or Poisson equation, thus we can easily construct general solutions by superposition of harmonic functions. The each solution has the infinity series of the indexes of the harmonic function H_A and $\mathcal{A}_i$. We show the lowest order of the solution is the exactly the same in the previous solution, BMPV and supersymmetric black ring, but the higher order of the solution are related to the quantum collection or the interaction term in ordinary solution.

Using the hyperspherical coordinate system for our conformally flat base space (r, θ, ϕ, ψ), we present exact solutions in M theory with nulii Killing vector.Compactifying these solutions into five dimensions, we show that these solutions include the BMPV black hole and the Brinkmann wave solution [47], and those extension to non-BPS ones. We have proved that the solutions preserve the $1/8$ supersymmetry if $\mathcal{F}_{ij}$ is self-dual. All solutions found in the hyperspherical coordinates preserve the 1/8 supersymmetry if the angular momenta satisfy some relation (e.g., $J_\phi = -J_\psi$).

We also present exact solutions in time-like Killing case under the hyperbipolor coordinate system (ξ, η, ϕ, ψ). We show the solution include the supersymmetric solution with the BPS condition $-q_\phi = q_\psi$. We also have non-BPS solutions with infinity series of the indexes. We have proved that the black ring solution also preserve the $1/8$ supersymemtry.

The charges of branes of the BMPV black hole correspond to the numbers of D-brane tension. While SO(4) rotational symmetries, which describe angular momenta of the black hole, corresponds to endmorphisms in the graded algebra that rotate the fermionic generators G^i_m [31]. By this correspondence (AdS/CFT correspondence [48]), we can discuss the properties of our solutions in the SCFT side.

Although we assume the BPS type relations for the metric, we have to solve the elliptic type differential equations if we want to find most general solutions, especially non-BPS spacetimes. For this purpose, we need a completely different approach such as a soliton technique to generate new solutions [49, 50].

We have found that the BPS and non-BPS rotating asymptotically flat stringy black holes, from which we may learn more about connections between microscopic and macroscopic states of gravitating objects. In our framework, we consider a toroidally compactified string theory, but one may embed the BMPV type geometry in M-theory compactified on generic Calabi-Yau spaces, or black ring solution on Einstein manifold, which would be more interesting.

References

[1] A. Strominger and C. Vafa, *Phys. Lett. B* **379** (1996) 99.

[2] G.W. Gibbons, *Nucl. Phys. B* **207** (1982) 337.

[3] R.C. Myers, *Nucl. Phys. B* **289** (1987) 701.

[4] G. W. Gibbons, and K. Maeda, *Nucl. Phys. B* **298** (1988) 741.

[5] C.G. Callan, R.C. Myers, and M.J. Perry, *Nucl. Phys. B* **311** (1988) 673.

[6] D. Garfinkle, G.T. Horowitz and A. Strominger, *Phys. Rev. D* **43** (1991) 3140, *Erratum-ibid. D* **45** (1992) 3888.

[7] G.T. Horowitz and A. Strominger, *Nucl. Phys. B* **360** (1991)197.

[8] F.R. Tangherlini, *Nuovo Cim.* **27** (1963) 636.

[9] R.C. Myers and M.J. Perry, *Ann. Phys.* **172** (1986) 304.

[10] G. W. Gibbons, D. Ida and T. Shiromizu, *Phys. Rev. Lett.* **89** (2002) 041101.

[11] H. S. Reall, *Phys. Rev. D* **68** (2003) 024024.

[12] H. Elvang, R. Emparan *JHEP* **0311** (2003) 035.

[13] R.R. Khuri, and R.C. Myers, *Fields Inst. Comm.* **15** (1997) 273.

[14] M.J. Duff and J.X. Lu, *Nucl. Phys. B* **416** (1994) 301.

[15] M.J. Duff, J.X. Lu and C.N. Pope, *Phys. Lett. B* **382** (1996) 73.

[16] C. G. Callan, J. M. Maldacena, *Nucl. Phys.* **B472** (1996) 591.

[17] R. Emparan and H.R. Reall, *Phys. Rev. Lett.* **88** (2000) 101101.

[18] H. S. Reall, *Phys.Rev.* **D68** (2003) 024024.

[19] R. Kallosh, *Phys. Lett. B* **282** (1992) 80.

[20] S. Ferrara, R. Kallosh, A. Strominger, *Phys. Rev. D* **52** (1995) 5412.

[21] M. Cvetic, and D. Youm, *Nucl. Phys. B* **453** (1995) 259.

[22] M. Cvetic, and A. A. Tseytlin, *Phys. Lett. B* **366** (1996) 95.

[23] K. Behrndt, G. L. Cardoso, B. de Wit, R. Kallosh, D. Lüst, and T. Mohaupt, *Nucl. Phys. B* **488** (1997) 236.

[24] K. P. Tod, *Phys. Lett. B* **121** (1983) 241.

[25] W.A. Sabra, *Mod. Phys. Lett. A* **13** (1997) 239.

[26] J.P. Gauntlett, R.C. Myers, and P.K. Townsend, *Class. Quantum Grav.* **16** (1999) 1.

[27] J.P. Gauntlett, J.B. Gutowski, C.M. Hull, S. Pakis and S. Reall, *Class. Quant. Grav.* **20** (2003) 4587.

[28] J.P. Gauntlett, *Fortsch. Phys.* **53** (2005) 468.

[29] J.P. Gauntlett and S. Pakis, *Commun. Math. Phys.* **247** (2004) 421.

[30] J.C. Breckenridge, R.C. Myers, A.W. Peet and C. Vafa, *Phys. Lett. B* **391** (1993) 93.

[31] C.A.R. Herdeiro, *Nucl. Phys. B* **582** (2000) 363.

[32] H. Elvang, R. Emparan, D. Mateos, H.S. Reall, *Phys. Rev. Lett.* **93** (2004) 211302.

[33] H. Elvang, R.Emparan, D. Mateos, H.S. Reall, *Phys. Rev. D* **71** (2005) 024033.

[34] M. Cvetic, and C.M. Hull, *Nucl. Phys. B* **519** (1988) 141.

[35] J.B. Gutowski, H.S. Reall, *JHEP* **0402** (2004) 006.

[36] M. Cvetic and C. M. Hull, *Nucl. Phys. B* **480**, 296 (1996).

[37] H. Elvang, R Emparan and P. Figueras, *JHEP* **0502** (2005) 031.

[38] M. J. Duff, R. R. Khuri and J. X. Lu, *Phys. Rep.* **259** (1995) 213.

[39] E. Cremmer, B. Julia and J. Sherk, *Phys. Lett. B* **76** (1978) 409.

[40] M. Cvetic and D Youm, *Nucl.Phys.* **B499** (1997) 253.

[41] K. Maeda and M Tanabe, *Nucl. Phys.* **B738** (2006) 184.

[42] R. Argurio, F. Englert, and L. Houart, *Phys. Lett. B* **398** (1997) 61.

[43] N. Ohta, *Phys. Lett. B* **403** (1997) 218.

[44] J. G. Russo and A. A. Tseytlin, *Nucl. Phys. B* **490**, 121 (1997).

[45] K. I. Maeda, N. Ohta and M. Tanabe, *Phys.Rev.* **D74** (2006) 104002.

[46] M. Bertolini, P. Frè and M. Trigiante, *Class. Quant. Grav.* **16** (1999) 1519.

[47] H.W. Brinkmann, *Proc. Natl. Acad. Sci. U.S.* **9** (1923) 1.

[48] O. Aharony, S. S. Gubser, J. M. Maldacena, H. Ooguri and Y. Oz, *Phys. Rept.* **323**, 183 (2000).

[49] V. A. Belinski and V. E. Zakharov, *Sov. Phys. JETP* **48** (1978) 985.

[50] V. A. Belinski and V. E. Zakharov, *Sov. Phys. JETP* **50** (1979) 1.

In: Space Exploration Research
Editors: J.H. Denis and P.D. Aldridge
ISBN: 978-1-60692-264-4

Chapter 9

On The 5D Extra-Force According to Basini-Capozziello-Ponce De Leon Formalism and Four Important Features: Strong Gravitational Fields, Kar-Sinha Gravitational Bending of Light in Extra Dimensions, Gravitational Red Shift Affected by Extra Dimensions and the Experimental Research of Extra Dimensions On-Board International Space Station(ISS) Using Laser Beams

Fernando Loup*
Residencia de Estudantes Universitas Lisboa, Portugal

Abstract

We use the $5D$ Extra Dimensional Force according to Basini-Capozziello-Ponce De Leon,Overduin-Wesson and Mashoon-Wesson-Liu to demonstrate that in flat $5D$ Minkowsky Spacetime or weak Gravitational Fields we cannot tell if we live in a $5D$ or a $4D$ Universe. But, in the extreme conditions of Strong Gravitational Fields we demonstrate that the effects of the $5D$ Extra Dimension becomes visible and perhaps the study of the extreme conditions in Black Holes can tell if we live in a Higher Dimensional Universe. We also analyze the possibility of Experimental Research of Extra Dimensions On-Board International Space Station (ISS) by using a Satellite carrying a Laser device(optical Laser) on the other side of Earth Orbit targeted towards ISS.The Sun will be between the Satellite and the ISS so the Laser will pass the neighborhoods of the Sun at a distance R in order to reach ISS. The Laser beam will be Gravitationally Bent according to Classical General Relativity and the Extra Terms

*E-mail address: spacetimeshortcut@yahoo.com

predicted by Kar-Sinha in the Gravitational Bending Of Light due to the presence of Extra Dimensions can perhaps be measured with precision equipment. By computing the Gravitational Bending according to Einstein we know the exact position where the Laser will reach the target on-board ISS. However, if the Laser arrives at ISS with a Bending different than the one predicted by Einstein and if this difference is equal to the Extra Terms predicted by Kar-Sinha then this experience would proof that we live in a Universe of more than 4 Dimensions.We demonstrate in this work that ISS have the needed precision to detect these Extra Terms(see eq 137 in this work). Such experience would resemble the measures of the Gravitational Bending Of Light by Sir Arthur Stanley Eddington in the Sun Eclipse of 1919 that helped to proof the correctness of General Relativity although in ISS case would have more degres of accuracy because we would be free from the interference of Earth Atmosphere. The Laser Satellite could also test the Gravitational Red Shift affected by the presence of the Extra Dimensions. We also outline the fact that the huge number of Elementary Particles seen in $4D$ are as a matter of fact a small number of particles seen in $5D$ and an experimental proof of the Existence of Extra Dimensions can leads towards a major breakthrough in the theories of Physics unification.

1. Introduction

Much has been said about the so-called Extra Dimentional nature of the Universe.It was first proposed by Theodore Kaluza and Oskar Klein in 1918[1] in an attempt to unify Gravity and Electromagnetism. However the physical nature of the Extra Dimension was not well defined in a clear way. They used a so-called Compactification Mechanism to explain why we cannot see the $5D$ Extra Dimension but this mechanism was not clearly understood. Later on and with more advanced scientific knowledge other authors appeared with the same idea under the exotic concept of the so-called BraneWorld. In the BraneWorld concept our visible Universe is a $3 + 1$ Dimensional sheet of Spacetime :a Brane involved by a Spacetime of Higher Dimensional nature. This idea came mainly from Strings Theory where Gravitational Forces are being represented by an Elementary Particle called Graviton while other interactions are represented by other sets of Elementary Particles:Electromagnetic Interaction is being represented by an Elementary Particle called Photon. According to Strings Theory Gravitons are Closed Loops and can leave easily our $3 + 1$ Dimensional Spacetime and escape into the Extra Dimensions while Photons are Open Strings and are "trapped" in our $3 + 1$ Spacetime. To resume: In 1918 Kaluza-Klein tried to unify Gravity and Electromagnetism and Einstein among other scientists were interested in the same thing. But however inside the framework of the so-called Strings Theory how can a Closed Loop be unified with a Open String??? How can an Interaction that with some degrees of freedom is allowed to probe the Extra Dimensional Spacetime be unified with Interactions confined to our $3 + 1$ Spacetime??? A puzzle to solve.So the so-called Strings Theory is trying to unify Gravity with Electromagnetism and other Interactions but the framework is not completed or not well understood. On the other hand the Compactification Mechanism in the original Kaluza-Klein theory explains why we cannot see beyond the $3 + 1$ Spacetime because the Extra Dimensions are Compactified or Curled Up but it does not explains why we have $3+1$ Uncurled or Uncompactified Dimensions while the remaining Extra Dimensions

[1] see [21] for an excellent account on Kaluza-Klein History

are Curled and what generates this Compactification Mechanism in the first place??? We adopt in this work the so-called Basini-Capozziello Ponce De Leon formalism coupled to the formalisms of Mashoon-Wesson-Liu and Overduin-Wesson in which Extra Dimensions are not compactified but opened like the $3+1$ Spacetime Dimensions we can see.There are small differences between these formalisms but Basini-Capozziello Ponce De Leon admits a non-null ${}^5R_{AB}$ Ricci Tensor while the others make the Ricci Tensor ${}^5R_{AB} = 0$[2] but essentially these formalisms are mathematically equivalent.In the Basini-Capozziello Ponce de Leon the ordinary Spacetime of $3+1$ Dimensions is embedded into a large Higher Dimensional Spacetime,however in a flat or Minkowsly Spacetime the Spacetime Curvature eg Ricci and Einstein Tensors of the Higher Dimensional Spacetime reduces to the same Ricci and Einstein Tensors of a $3+1$ Spacetime. This explains without Compactification Mechanisms why we cannot see beyond the $3+1$ Spacetime:our everyday Spacetime is essentialy Minkowskian or flat [3]and a $5D$ Ricci Tensor reduces to a Ricci Tensor of a $3+1$ Spacetime. On the other hand in this formalism all masses,electric charges and spins of all the Elementary Particles seen in $4D$ are function of a $5D$ rest-mass coupled with Spacetime Geometry. We can observe in $4D$ Spacetime a multitude of Elementary Particles with different masses,electric charges or spins but according to the Basini-Capozziello Ponce De Leon formalism eg the $5D$ to $4D$ Dimensional Reduction all these different Elementary Particles with all these $4D$ rest-masses,electric charges or spins are as a matter of fact a small group of $5D$ Elementary Particles with a $5D$ rest-mass and is the geometry of the $5D$ Spacetime coupled with the Dimensional Reduction that generates these apparent differences. Hence two particles with the same $5D$ rest-mass M_5 can be seen in $4D$ with two different rest-masses m_0 making ourselves think that the particles are different but the difference is apparent and is generated by the Dimensional Reduction from $5D$ to $4D$.Look to the set of equations below:We will explore in this work these equations with details but two particles with the same $5D$ rest mass M_5 can be seen in the $4D$ with two different rest masses m_0 if the Dimensional Reduction from $5D$ to $4D$ or the Spacetime Geometric Coupling $\sqrt{1 - \Phi^2(\frac{dy}{ds})^2}$ is different for each particle. All the particles in the Table of Elementary Particles given below[4] with non-zero rest-mass m_0 seen in $4D$ can as a matter of fact have the same rest-mass M_5 in $5D$ and the Dimensional Reduction term $\sqrt{1 - \Phi^2(\frac{dy}{ds})^2}$ generates the apparent different $4D$ rest-masses.This is very attractive from the point of view of a Unified Physics theory.There exists a small set of particles in $5D$ and all the huge number of Elementary Particles in $4D$ is a geometric projection from the $5D$ Spacetime into a $4D$ one([2] eq 20,[11] eq 21 and [20] eq 8)([2] eq 14,[20] eq 1 and eq 2).

$$m_0 = \frac{M_5}{\sqrt{1 - \Phi^2(\frac{dy}{ds})^2}} \tag{1}$$

$$dS^2 = g_{uv}dx^u dx^v - \Phi^2 dy^2 \tag{2}$$

[2]see [21] pg 31 after eq 48 and see [2] eq 20 [11] eq 21 and [20] eq 8. We prefer to assume that exists matter in the $5D$ due to the last section of [20] about the particle Z

[3]we consider our Spacetime as a Schwarzschild Spacetime however at a large distance from the Gravitational Source it reduces to a Minkowsky SR Spacetime due to a large R and the ratio $\frac{M}{R}$ tends to zero

[4]extracted from the Formulary Of Physics by J.C.A. Wevers available on Internet

$$dS^2 = ds^2 - \Phi^2 dy^2 \tag{3}$$

Particle	spin ($\hbar$)	B	L	T	T_3	S	C	B^*	charge (e)	m_0 (MeV)	antipart.
u	1/2	1/3	0	1/2	1/2	0	0	0	$+2/3$	5	$\overline{u}$
d	1/2	1/3	0	1/2	$-1/2$	0	0	0	$-1/3$	9	$\overline{d}$
s	1/2	1/3	0	0	0	-1	0	0	$-1/3$	175	$\overline{s}$
c	1/2	1/3	0	0	0	0	1	0	$+2/3$	1350	$\overline{c}$
b	1/2	1/3	0	0	0	0	0	-1	$-1/3$	4500	$\overline{b}$
t	1/2	1/3	0	0	0	0	0	0	$+2/3$	173000	$\overline{t}$
e^-	1/2	0	1	0	0	0	0	0	-1	0.511	e^+
μ^-	1/2	0	1	0	0	0	0	0	-1	105.658	μ^+
τ^-	1/2	0	1	0	0	0	0	0	-1	1777.1	τ^+
ν_e	1/2	0	1	0	0	0	0	0	0	0(?)	$\overline{\nu}_e$
ν_μ	1/2	0	1	0	0	0	0	0	0	0(?)	$\overline{\nu}_\mu$
ν_τ	1/2	0	1	0	0	0	0	0	0	0(?)	$\overline{\nu}_\tau$
γ	1	0	0	0	0	0	0	0	0	0	γ
gluon	1	0	0	0	0	0	0	0	0	0	$\overline{\text{gluon}}$
W^+	1	0	0	0	0	0	0	0	$+1$	80220	W^-
Z	1	0	0	0	0	0	0	0	0	91187	Z
graviton	2	0	0	0	0	0	0	0	0	0	graviton

We employ in this work the Basini-Capozziello Ponce De Leon Formalism to demonstrate that while in flat or Minkowsky Spacetime the curvature in $5D$ reduces to a one in $4D$ due to the Dimensional Reduction suffered by the Ricci and Einstein Tensors and we cannot tell if we live in a $5D$ or in a $4D$ Universe due to the absence of Strong Gravitational Fields but in an environment of Strong Gravity the $5D$ Ricci and Einstein Tensors cannot be reduced to similar $4D$ ones and the Curvature of a $5D$ Spacetime is different than the one of a $4D$ because the $5D$ extra terms in the Ricci and Einstein Tensors have the terms of the Strong Gravitational Field and cannot be reduced to $4D$.Perhaps the study of the conditions of extreme Gravitational Fields in large Black Holes will tell if we live in a $5D$ Universe or in a $4D$ one.We also demonstrate that the International Space Station (ISS) can perhaps be used to study the Experimental Detection Of Extra Dimensions using the Gravitational Bending of Light of the Sun or the similar for large Black Holes.Higher Dimensional Spacetimes affects the Gravitational Bending Of Light adding Extra Terms as predicted by Kar-Sinha(see abstact of [3]).(see also pg 73 in [20]).International Space Station is intended to be a laboratory designated to test the forefront theories of Physics and ISS can provide a better environment for Physical experiences without the interference of Earth secondary effects that will disturb careful measures specially for gravity-related experiments.(see pg 602-603 for the advantage of the free-fall conditions for experiments in [8]) We argue that the Gravitational Bending Of Light first measured by Sir Arthur Stanley Eddington in a Sun Eclipse in 1919 is widely known as the episode that made Einstein famous but had more than 30 percent of error margin. In order to find out if the Extra Dimension exists(or not) we need to measure the factor C^2 of Kar-Sinha coupled to a Higher Dimensional definition of m_0 with a accurate precision of more than $\triangle\omega < 2.8 \times 10^{-4}$.(see pg 1783 in [3]).ISS can use the orbit of the Moon to create a "artificial eclipse"[5] to get precise measurements of the Gravitational Bending Of Light and according to Kar-Sinha detect the existence of the $5D$ Extra Dimension although in this work we will propose a better idea.ISS will be used to probe the foundations of General Relativity and Gravitational Bending of Light certainly will figure out in the experiences(see pg 626 in [6]).ISS Gravitational shifts are capable to detect measures of $\frac{\triangle\omega}{\omega}$ with Expected Uncertainly of 12×10^{-6}(see pg 629 Table I in [6]) smaller than the one predicted by Kar-Sinha for the Extra Dimensions although the ISS shifts are red-shifts and not Bending of Light similar precision can be achieved.Also red-shifts are due to a time delay in signals and the $5D$ can also affect this measure as proposed by Kar-Sinha.(see pg 1782 in [3]).The goal of ISS is to achieve a Gravitational Shift Precision of 2.4×10^{-7}(see pg 631 Table II in [6]) by far more than enough to detect the existence of the $5D$ Extra Dimension.We believe that Gravitational Bending Of Light can clarify the question if we live in a Higher Dimensional Universe or not.A small deviation in a photon path different than the one predicted by Einstein can solve the quest for Higher Dimensional Spacetimes. We propose here the use of a Satellite with a Laser beam in the other side of the Earth Orbit targeted towards ISS. The Laser would pass the neighborhoods of the Sun at a distance R in order to reach ISS and would be Gravitational Bent according to General Relativity and affected by the Kar-Sinha Extra Terms due to the presence of the Higher Dimensional Spacetime.Hence the Gravitational Bending Of Light can be measured with precision equipment of ISS.We demonstrate in this work that ISS have the needed precision to detect these Extra Terms(see eq 137 in this work).By computing the Gravitational Bending according to Einstein we know where the photons would reach ISS.However if the photons arrives at ISS with a Bending angle different than the one predicted by Einstein and if this difference is equal to the Kar-Sinha Extra Terms then we would have a proof that we live in a Universe of more than 4 Dimensions.We also examine Gravitational Red Shifts affected by the presence of the Extra Dimension.This experience made on-board ISS would have the same impact of the Sir Arthur Stanley Eddington measures of Gravitational Bending Of Light in the Sun Eclipse of 1919 and if the result is "positive" then the International Space Station ISS would change forever our way to see the Universe.

2. The Basini-Capozziello Ponce De Leon Formalism and Resemblances with Mashoon-Wesson-Liu and Overduin-Wesson Formalisms

Basini-Capozziello Ponce de Leon argues that our $3+1$ Dimensional Spacetime we can see is a Dimensional Reduction from a larger $5D$ one and according to a given Spacetime Geometry we can see(or not) the $5D$ Extra Dimension .This is also advocated in the almost similar Formalisms of Mashoon-Wesson-Liu and Overduin-Wesson. A $5D$ Spacetime metric is defined as([1] eq 32,[5] eq 18,[20] eq 62 and [9] pg 556 Section 2) and contains all the $3+1$ Spacetime Dimensions of our observable Universe plus the $5D$ Extra Dimension.Then $A, B = 0, 1, 2, 3, 4$ where $0, 1, 2, 3$ are the Dimensons of the $4D$ Spacetime and 4 is the script of the $5D$ Extra Dimension(see [5] pg 2225 after eq 18 and again [9] pg 556).

[5] we agree that idea is weird but is better than to wait for a Sun Eclipse in the proper conditions

$$dS^2 = g_{AB}dx^A dx^B \tag{4}$$

Note that this equation is common not only to Basini-Capozziello-Ponce De Leon but also to Mashoon-Wesson-Liu and Overduin-Wesson Formalisms.These formalisms advocates the Dimensional Reduction from $5D$ to $4D$[6] and we need to separate in this Spacetime Metric both the $3+1$ Components of our visible Universe and the components of the $5D$ Extra Dimension.The resulting equation would then be([1] eq 56,[2] eq 12 and 14,[5] eq 42,[9] eq 32 and 33 without vector potential,[11] eq 10, [12] pg 308,[20] eq 109 and [13] pg 1346)[7][8][9]

$$dS^2 = g_{AB}dx^A dx^B = g_{\alpha\beta}dx^\alpha dx^\beta - \Phi^2 dy^2 \tag{5}$$

Note that when the Warp Field[10] $\Phi = 1$ the Spacetime Metric becomes:

$$dS^2 = g_{AB}dx^A dx^B = g_{\alpha\beta}dx^\alpha dx^\beta - dy^2 \tag{6}$$

Writing the ${}^5R_{\alpha\beta}$ Ricci Tensor and the ${}^5R$ Ricci Scalar according to Basini-Capozziello using these equations:($\alpha, \beta = 0, 1, 2, 3$)([1] eq 58, [5] eq 44 and [20] eq 111.See also [21] eq 48 for ${}^5R_{\alpha\beta}$)[11]

$${}^5R_{\alpha\beta} = R_{\alpha\beta} - \frac{\Phi_{,a;b}}{\Phi} - \frac{1}{2\Phi^2}(\frac{\Phi_{,4}g_{\alpha\beta,4}}{\Phi} - g_{\alpha\beta,44} + g^{\lambda\mu}g_{\alpha\lambda,4}g_{\beta\mu,4} - \frac{g^{\mu\nu}g_{\mu\nu,4}g_{\alpha\beta,4}}{2}) \tag{7}$$

$${}^5R = R - \frac{\Phi_{,a;b}}{\Phi}g^{\alpha\beta} - \frac{1}{2\Phi^2}g^{\alpha\beta}(\frac{\Phi_{,4}g_{\alpha\beta,4}}{\Phi} - g_{\alpha\beta,44} + g^{\lambda\mu}g_{\alpha\lambda,4}g_{\beta\mu,4} - \frac{g^{\mu\nu}g_{\mu\nu,4}g_{\alpha\beta,4}}{2}) \tag{8}$$

Simplifying for diagonalized metrics we should expect for:

$${}^5R_{\alpha\beta} = R_{\alpha\beta} - \frac{\Phi_{,a;b}}{\Phi} - \frac{1}{2\Phi^2}(\frac{\Phi_{,4}g_{\alpha\beta,4}}{\Phi} - g_{\alpha\beta,44} + \frac{g^{\mu\nu}g_{\mu\nu,4}g_{\alpha\beta,4}}{2}) \tag{9}$$

$${}^5R = R - \frac{\Phi_{,a;b}}{\Phi}g^{\alpha\beta} - \frac{1}{2\Phi^2}g^{\alpha\beta}(\frac{\Phi_{,4}g_{\alpha\beta,4}}{\Phi} - g_{\alpha\beta,44} + \frac{g^{\mu\nu}g_{\mu\nu,4}g_{\alpha\beta,4}}{2}) \tag{10}$$

Note that the term ${}^4\Box\Phi = \nabla_\alpha\Phi^\alpha = g^{\alpha\beta}(\Phi_\alpha)_{;\beta} = g^{\alpha\beta}[(\Phi_\alpha)_\beta - \Gamma^K_{\beta\alpha}\Phi_K]$ corresponds to the D'Alembertian in $4D$[12][13] so we can write for the Ricci Scalar the following expression:

[6] We will skip a tedious definition and concentrate on the Dimensional Reduction.A unfamiliar reader must study first [1] pg 122 Section 2.2 to pg 127,[5] pg 2225 Section 3 to pg 2229 and [20] pg 1434 Section 4 to pg 1441.see also [21] pg 29 Section 6 to pg 31

[7] [12] with spacelike signature

[8] see [13] pg 1341 the Campbell-Magaard Theorem

[9] see [1] eq 57,[5] eq 43 and [21] eq 47.

[10] the term "Warp" appears in pg 1340 in [2]

[11] Working with diagonalized metrics the terms α,λ,μ,β and ν are all equal

[12] see pg 129 in [1] and pg 2230 in [5]

[13] see also pg 311 in [12]

$$^5R = R - \frac{^4\Box\Phi}{\Phi} - \frac{1}{2\Phi^2}g^{\alpha\beta}(\frac{\Phi_{,4}g_{\alpha\beta,4}}{\Phi} - g_{\alpha\beta,44} + \frac{g^{\mu\nu}g_{\mu\nu,4}g_{\alpha\beta,4}}{2}) \tag{11}$$

If according to Basini-Capozziello the terms $g_{\alpha\beta}$ have no dependance with respect to to the Extra Coordinate y after the Reduction from $5D$ to $4D$ then all the derivatives with respect to y vanish and we are left out with the following expression for the Ricci Scalar:([1] eq 59,[5] eq 45 and [20] eq 116)).We will analyze this in details when studying the $5D$ to $4D$ Dimensional Reduction.

$$^5R = R - \frac{^4\Box\Phi}{\Phi} \tag{12}$$

Writing the remaining Ricci Tensors we should expect for([21] eq 48)[14]:

$$\begin{aligned}
\hat{R}_{\alpha\beta} &= R_{\alpha\beta} - \frac{\nabla_\beta(\partial_\alpha\Phi)}{\Phi} - \frac{1}{2\Phi^2}\left(\frac{\partial_4\Phi\,\partial_4 g_{\alpha\beta}}{\Phi} - \partial_4 g_{\alpha\beta}\right.\\
&\quad \left. + g^{\gamma\delta}\,\partial_4 g_{\alpha\gamma}\,\partial_4 g_{\beta\delta} - \frac{g^{\gamma\delta}\,\partial_4 g_{\gamma\delta}\,\partial_4 g_{\alpha\beta}}{2}\right) \quad ,\\
\hat{R}_{\alpha 4} &= \frac{g^{44}g^{\beta\gamma}}{4}(\partial_4 g_{\beta\gamma}\,\partial_\alpha g_{44} - \partial_\gamma g_{44}\,\partial_4 g_{\alpha\beta}) + \frac{\partial_\beta g^{\beta\gamma}\,\partial_4 g_{\gamma\alpha}}{2}\\
&\quad + \frac{g^{\beta\gamma}\,\partial_4(\partial_\beta g_{\gamma\alpha})}{2} - \frac{\partial_\alpha g^{\beta\gamma}\,\partial_4 g_{\beta\gamma}}{2} - \frac{g^{\beta\gamma}\,\partial_4(\partial_\alpha g_{\beta\gamma})}{2}\\
&\quad + \frac{g^{\beta\gamma}g^{\delta\epsilon}\,\partial_4 g_{\gamma\alpha}\,\partial_\beta g_{\delta\epsilon}}{4} + \frac{\partial_4 g^{\beta\gamma}\,\partial_\alpha g_{\beta\gamma}}{4} \quad ,\\
\hat{R}_{44} &= \Phi\Box\Phi - \frac{\partial_4 g^{\alpha\beta}\,\partial_4 g_{\alpha\beta}}{2} - \frac{g^{\alpha\beta}\,\partial_4(\partial_4 g_{\alpha\beta})}{2}\\
&\quad + \frac{\partial_4\Phi\, g^{\alpha\beta}\,\partial_4 g_{\alpha\beta}}{2\Phi} - \frac{g^{\alpha\beta}g^{\gamma\delta}\,\partial_4 g_{\gamma\beta}\,\partial_4 g_{\alpha\delta}}{4} \quad ,
\end{aligned} \tag{13}$$

where "$\Box$" is defined as usual (in four dimensions) by $\Box\Phi \equiv g^{\alpha\beta}\nabla_\beta(\partial_\alpha\Phi)$.

Note that the Overduin-Wesson definition is exactly equal to the one presented by Basini-Capozziello Ponce De Leon.Both Formalisms are equivalent except that Basini-Capozziello-Ponce De Leon admits a $^5R_{AB}$ not null.

Working with diagonalized Spacetime Metrics of signature (+,-,-,-,-)[15] the Ricci Tensors would be written as:

$$\begin{aligned}
\hat{R}_{\alpha\alpha} &= R_{\alpha\alpha} - \frac{\nabla_\alpha(\partial_\alpha\Phi)}{\Phi} - \frac{1}{2\Phi^2}\left(\frac{\partial_4\Phi\,\partial_4 g_{\alpha\alpha}}{\Phi} - \partial_4 g_{\alpha\alpha}\right.\\
&\quad \left. + g^{\alpha\alpha}\,\partial_4 g_{\alpha\alpha}\,\partial_4 g_{\alpha\alpha} - \frac{g^{\alpha\alpha}\,\partial_4 g_{\alpha\alpha}\,\partial_4 g_{\alpha\alpha}}{2}\right) \quad ,\\
\hat{R}_{\alpha 4} &= \frac{g^{44}g^{\alpha\alpha}}{4}(\partial_4 g_{\alpha\alpha}\,\partial_\alpha g_{44} - \partial_\alpha g_{44}\,\partial_4 g_{\alpha\alpha}) + \frac{\partial_\alpha g^{\alpha\alpha}\,\partial_4 g_{\alpha\alpha}}{2}
\end{aligned}$$

[14] adapted from the arXiv.org LaTeX file of [21] eq 48.Note the difference between the first term $\hat{R}_{\alpha\beta}$ between [21] eq 48 ,[1] eq 58,[5] eq 44 and [20] eq 111

[15] $\alpha = \beta = \gamma = \delta = \epsilon$

$$+\frac{g^{\alpha\alpha}\,\partial_4(\partial_\alpha g_{\alpha\alpha})}{2} - \frac{\partial_\alpha g^{\alpha\alpha}\,\partial_4 g_{\alpha\alpha}}{2} - \frac{g^{\alpha\alpha}\,\partial_4(\partial_\alpha g_{\alpha\alpha})}{2}$$
$$+\frac{g^{\alpha\alpha}g^{\alpha\alpha}\,\partial_4 g_{\alpha\alpha}\,\partial_\alpha g_{\alpha\alpha}}{4} + \frac{\partial_4 g^{\alpha\alpha}\,\partial_\alpha g_{\alpha\alpha}}{4},$$
$$\hat{R}_{44} = \Phi\Box\Phi - \frac{\partial_4 g^{\alpha\alpha}\,\partial_4 g_{\alpha\alpha}}{2} - \frac{g^{\alpha\alpha}\,\partial_4(\partial_4 g_{\alpha\alpha})}{2}$$
$$+\frac{\partial_4\Phi\, g^{\alpha\alpha}\,\partial_4 g_{\alpha\alpha}}{2\Phi} - \frac{g^{\alpha\alpha}g^{\alpha\alpha}\,\partial_4 g_{\alpha\alpha}\,\partial_4 g_{\alpha\alpha}}{4}, \tag{14}$$

Compare the first of the Ricci Tensors above with [20] eq 113. Note that the Mashoon-Wesson-Liu Formalism is exactly equal to the Basini-Capozziello-Ponce De Leon and Overduin-Wesson Formalisms. Look to the equations [9] eq 32 and 33 without vector potential.Compare with

$$dS^2 = g_{\alpha\beta}dx^\alpha dx^\beta - \Phi^2 dy^2 \tag{15}$$

One can see that we already presented this equation proving without shadows of doubt that the three formalisms are equivalent.Mashoon-Wesson-Liu in [9] pg 557 makes $g_{44} = -\Phi^2$. They also makes $g_{44} = -1$ (see pg 558)giving the equation below:

$$dS^2 = g_{\alpha\beta}dx^\alpha dx^\beta - dy^2 \tag{16}$$

We already presented this equation:is the $5D$ Spacetime Geometry without the Warp Field.

One thing advocated by Mashoon-Wesson-Liu and Basini-Capozziello Ponce De Leon is the fact that the $5D$ Extra Dimension generate a $5D$ Extra Force that can be detected in $4D$.(see [9] abstract and pgs 556,562 look to eq 24,pg 563 look to eq 31 and the comment below this equation,pg 565 eq 38,39 and the comments on the de-acceleration,pg 566 definition of β,pg 567 look to the comment of a small force but detectable).(see also [2] abstract and pgs 1336,1337,1341 eq 20, pg 1342 eq 25,pg 1343 eq 30).We will now prove that the $5D$ Extra Force in both formalisms is equivalent.

If we have a Spacetime Geometry defined as:

$$dS^2 = g_{\alpha\beta}dx^\alpha dx^\beta - \Phi^2 dy^2 \tag{17}$$

$$dS^2 = g_{\alpha\beta}dx^\alpha dx^\beta - [\phi(t,x)\chi(y)]^2 dy^2 \tag{18}$$

where we defined the Warp Field Φ according to Basini-Capoziello([1] eq 76,[5] eq 70 and [20] eq 132) we have two choices:

- M_5 the $5D$ Mass is not zero and we have matter in the $5D$ Extra Dimension according to one of the Ponce De Leon Options making also ${}^5R_{AB}$ the Ricci Tensor in $5D$ not null.

- M_5 The $5D$ Mass is zero and we have no matter in the $5D$ Extra Dimension according to another of the Ponce De Leon Options making also ${}^5R_{AB}$ the Ricci Tensor in $5D$ null.

Overduin-Wesson and Mashoon-Wesson-Liu formalisms agree with the second option of Ponce De Leon.(see [21] pg 31 after eq 48 and [9] pg 557 eq 2)

According to Ponce De Leon in option 1 if we have a rest-mass in $5D$ M_5 this rest-mass will be seen in $4D$ as a rest-mass m_0 as follows([2] eq 20,[11] eq 21 and [20] eq 8):

$$m_0 = \frac{M_5}{\sqrt{1 - \Phi^2(\frac{dy}{ds})^2}} \tag{19}$$

$$m_0 = \frac{M_5}{\sqrt{1 - [\phi(t,x)\chi(y)]^2(\frac{dy}{ds})^2}} \tag{20}$$

We have Quantum Chromodynamics for Quarks and a Quantum Electrodynamics for Leptons like Electron but as a matter of fact two particles with the same rest-mass in $5D$ M_5 can appear in our $4D$ Spacetime with different rest masses m_0 making one appear as a Quark and the other as a Lepton depending on the Dimensional Reduction from $5D$ to $4D$ or the Spacetime Coupling $\sqrt{1 - \Phi^2(\frac{dy}{ds})^2}$,$\sqrt{1 - [\phi(t,x)\chi(y)]^2(\frac{dy}{ds})^2}$ although in $5D$ both particles are the same.

This is a very interesting perspective of Modern Physics.Why Quantum Electrodynamics and Quantum Chromodynamics in $4D$ while as a matter of fact in $5D$ both are the same???.Look again to the table below[16] :

Particle	spin ($\hbar$)	B	L	T	T_3	S	C	B^*	charge (e)	m_0 (MeV)	antipart.
u	1/2	1/3	0	1/2	1/2	0	0	0	$+2/3$	5	$\overline{u}$
d	1/2	1/3	0	1/2	$-1/2$	0	0	0	$-1/3$	9	$\overline{d}$
s	1/2	1/3	0	0	0	-1	0	0	$-1/3$	175	$\overline{s}$
c	1/2	1/3	0	0	0	0	1	0	$+2/3$	1350	$\overline{c}$
b	1/2	1/3	0	0	0	0	0	-1	$-1/3$	4500	$\overline{b}$
t	1/2	1/3	0	0	0	0	0	0	$+2/3$	173000	$\overline{t}$
e^-	1/2	0	1	0	0	0	0	0	-1	0.511	e^+
μ^-	1/2	0	1	0	0	0	0	0	-1	105.658	μ^+
τ^-	1/2	0	1	0	0	0	0	0	-1	1777.1	τ^+
ν_e	1/2	0	1	0	0	0	0	0	0	0(?)	$\overline{\nu}_e$
ν_μ	1/2	0	1	0	0	0	0	0	0	0(?)	$\overline{\nu}_\mu$
ν_τ	1/2	0	1	0	0	0	0	0	0	0(?)	$\overline{\nu}_\tau$
γ	1	0	0	0	0	0	0	0	0	0	γ
gluon	1	0	0	0	0	0	0	0	0	0	gluon
W^+	1	0	0	0	0	0	0	0	$+1$	80220	W^-
Z	1	0	0	0	0	0	0	0	0	91187	Z
graviton	2	0	0	0	0	0	0	0	0	0	graviton

The Extra Force generated by the $5D$ seen in $4D$ for a massive $5D$ particle M_5 seen in $4D$ as m_0 according to Ponce De Leon is defined as follows([2] eq 25 and [20] eq 15):

$$\frac{1}{m_0}\frac{dm_0}{ds} = -\frac{1}{2}u^u u^v \frac{\partial g_{uv}}{\partial y}\frac{dy}{ds} - \Phi u^u \frac{\partial \Phi}{\partial x^u}(\frac{dy}{ds})^2 \tag{21}$$

[16] extracted from the Formulary Of Physics by J.C.A. Wevers available on Internet

We have here two choices:

- The Warp Field $\Phi = [\phi(t,x)\chi(y)]$([1] eq 76,[5] eq 70 and [20] eq 132) is not null and we have a Warp Field coupled to the $5D$ Extra Dimension.
- The Warp Field $\Phi = 1$ and we have no Warp Field at all.

For a $5D$ Extra Dimension coupled with a Warp Field according to Basini-Capozziello([1] eq 76,[5] eq 70 and [20] eq 132) the Extra Force is given by:

$$\frac{1}{m_0}\frac{dm_0}{ds} = -\frac{1}{2}u^u u^v \frac{\partial g_{uv}}{\partial y}\frac{dy}{ds} - \phi(t,x)\chi(y)u^u \frac{\partial \phi(t,x)\chi(y)}{\partial x^u}(\frac{dy}{ds})^2 \tag{22}$$

$$\frac{1}{m_0}\frac{dm_0}{ds} = -\frac{1}{2}u^u u^v \frac{\partial g_{uv}}{\partial y}\frac{dy}{ds} - \phi(t,x)\chi(y)^2 u^u \frac{\partial \phi(t,x)}{\partial x^u}(\frac{dy}{ds})^2 \tag{23}$$

If we have no Warp Field at all $\Phi = 1$ the equation is simply:

$$\frac{1}{m_0}\frac{dm_0}{ds} = -\frac{1}{2}u^u u^v \frac{\partial g_{uv}}{\partial y}\frac{dy}{ds} \tag{24}$$

Note that this equation is exactly equal to the $5D$ Extra Force equation as defined by Mashoon-Wesson-Liu([9] eq 24 and 38 because they used $g_{44} = -\Phi^2 = -1$ see [9] pg 558).Of course we expected this result because Basini-Capozziello Ponce De Leon and Mashoon-Wesson-Liu formalisms are equivalent.

For the case of a null $5D$ rest-mass M_5 the option 2 of Ponce De Leon the equation of the $5D$ Extra Force seen in $4D$ is given by([2] eq 30 and [20] eq 19):

$$\frac{1}{m_0}\frac{dm_0}{ds} = \mp \frac{1}{2\Phi}\frac{\partial g_{uv}}{\partial y}u^u u^v - \frac{u^u}{\Phi}\frac{\partial \Phi}{\partial x^u} \tag{25}$$

$$\frac{1}{m_0}\frac{dm_0}{ds} = \mp \frac{1}{2\phi(t,x)\chi(y)}\frac{\partial g_{uv}}{\partial y}u^u u^v - \frac{u^u}{\phi(t,x)}\frac{\partial \phi(t,x)}{\partial x^u} \tag{26}$$

If we have a no Warp Field at all the equation becomes:

$$\frac{1}{m_0}\frac{dm_0}{ds} = \mp \frac{1}{2}\frac{\partial g_{uv}}{\partial y}u^u u^v \tag{27}$$

This is equal to ([9] eq 24 and 38 with $dy = ds$ a Null-Like $5D$ Spacetime Geometry)

According to the following Spacetime Geometry as defined by Basini-Capozziello Ponce De Leon,Mashoon-Wesson-Liu and Overduin-Wesson formalisms([1] eq 56,[2] eq 12 and 14,[5] eq 42 and [20] eq 1)

$$dS^2 = g_{uv}dx^u dx^v - \Phi^2 dy^2 \tag{28}$$

$$dS^2 = ds^2 - \Phi^2 dy^2 \tag{29}$$

$$ds^2 = g_{uv}dx^u dx^v \tag{30}$$

We have three different types of Spacetime Geometries:

- Timelike $5D$ Geometry

$$dS^2 > 0 \dashrightarrow ds^2 - \Phi^2 dy^2 > 0 \dashrightarrow ds^2 > \Phi^2 dy^2 \dashrightarrow \frac{1}{\Phi^2} > (\frac{dy}{ds})^2 \dashrightarrow Timelike5D \quad (31)$$

- Null-Like $5D$ Geometry

$$dS^2 = 0 \dashrightarrow ds^2 - \Phi^2 dy^2 = 0 \dashrightarrow ds^2 = \Phi^2 dy^2 \dashrightarrow \frac{1}{\Phi^2} = (\frac{dy}{ds})^2 \dashrightarrow Nulllike5D \quad (32)$$

- Spacelike $5D$ Geometry

$$dS^2 < 0 \dashrightarrow ds^2 - \Phi^2 dy^2 < 0 \dashrightarrow ds^2 < \Phi^2 dy^2 \dashrightarrow \frac{1}{\Phi^2} < (\frac{dy}{ds})^2 \dashrightarrow Spacelike5D \quad (33)$$

Note that for a Null-Like $5D$ Geometry the equation of the $4D$ rest-mass m_0 in function of the $5D$ rest-mass M_5 is not valid.([2] eq 20,[11] eq 21 and [20] eq 8).

$$m_0 = \frac{M_5}{\sqrt{1 - \Phi^2 (\frac{dy}{ds})^2}} \quad (34)$$

Hence we suppose that for a Null-Like $5D$ Geodesics the Extra Dimension have no mass at all or all matter in the $5D$ Extra Dimension obeys Timelike $5D$ Geometries.

Then we can say that the Basini-Capozziello Ponce De Leon $5D$ formalism is for Timelike $5D$ Geometries bacause they admit a non-null $5D$ rest-mass while the formalisms of Mashoon-Wesson-Liu and Overduin-Wesson are valid for a Null-Like $5D$ Geometry where we have a $M_5 = 0$ a null $5D$ Ricci Tensor or a flat $5D$ Spacetime.

- $$\frac{1}{\Phi^2} > (\frac{dy}{ds})^2 \dashrightarrow 1 > \Phi^2 (\frac{dy}{ds})^2 \quad (35)$$

- $$\frac{1}{\Phi^2} = (\frac{dy}{ds})^2 \dashrightarrow 1 = \Phi^2 (\frac{dy}{ds})^2 \quad (36)$$

- $$\frac{1}{\Phi^2} < (\frac{dy}{ds})^2 \dashrightarrow 1 < \Phi^2 (\frac{dy}{ds})^2 \quad (37)$$

Note that a small Warp Field $0 < \Phi^2 < 1$ will generate a large $\frac{1}{\Phi^2}$ ideal for a $5D$ Timelike Geodesics.A small Warp Field will appear in the Pioneer Section due to the work of Bertolami-Paramos[17].

Although we can have a Null $5D$ rest-mass M_5 the Warp Field in the $5D$ Extra Dimension can still account for the generation of rest-masses in $4D$.

[17] the Yukawa Potential defined by Bertolami-Paramos eq 7 in [?] uses a rest-mass M in $4D$ that can be defined in function of the $5D$ rest-mass M_5 according to eq 20 in [2] where the Warp Field appears

See these Ponce De Leon Equations for the $4D$ rest-mass m_0 ([2] eq 27 and 28,[20] eq 16,17 and 18)

$$m_0 =^{+}_{-} \Phi \frac{dy}{d\lambda} \tag{38}$$

$$d\lambda = \frac{1}{m_0} ds \tag{39}$$

Combining eqs 45 and 46 we can clearly see that:[18]

$$\frac{dy}{ds} = \frac{1}{\Phi} \tag{40}$$

The $5D$ Extra Force seen in $4D$ for massless particles in $5D$ is given by:([2] eq 30,[20] eq 19)[19][20]

$$\frac{1}{m_0}\frac{dm_0}{ds} =^{-}_{+} \frac{1}{2\Phi}\frac{\partial g_{uv}}{\partial y} u^u u^v - \frac{u^u}{\Phi}\frac{\partial \Phi}{\partial x^u} \tag{41}$$

This equation although for massless $5D$ particles have many resemblance with its similar for massive $5D$ particles as pointed out by Ponce De Leon and can easily be obtained combining eqs 15 and 18 of [20](see pg 1343 in [2]).

According to the Table of Elementary Particles already presented in this work(two times and we think its enough) Photons or Gravitons have a $4D$ rest-mass $m_0 = 0$ corresponding to a $5D$ Null-Like Spacetime Geometry or in hence a stationary particle a particle that is at the rest in the $5D$ Spacetime,a particle with a $m_0 =^{+}_{-} \Phi \frac{dy}{d\lambda} \dashrightarrow m_0 = 0 \dashrightarrow \frac{dy}{d\lambda} = 0 \dashrightarrow \frac{dy}{ds} = 0$.

But of course we can have a $5D$ rest-mass $M_5 = 0$ giving a non-null $4D$ rest-mass $m_0 \neq 0$ even with a Warp Field $\Phi = 1$ if $\frac{dy}{d\lambda} \neq 0$ according to the following equations although we believe that non-null rest-masses m_0 in $4D$ comes from non-null rest-masses M_5 in $5D$(see sections 8 and 9 about particle Z in [20]) :

$$m_0 =^{+}_{-} \frac{dy}{d\lambda} \tag{42}$$

$$d\lambda = \frac{1}{m_0} ds \tag{43}$$

$$\frac{dy}{ds} = 1 \tag{44}$$

$$\frac{1}{m_0}\frac{dm_0}{ds} =^{-}_{+} \frac{1}{2}\frac{\partial g_{uv}}{\partial y} u^u u^v \tag{45}$$

[18] see pg 1343 in [2] but the result is obvious from [20] eq 10

[19] note that like for its analogous $5D$ massive counterpart the Warp Field function only of the Extra Coordinate makes the second term vanish(examine eqs 50 and 52 in [1])

[20] compare this equation with [9] eq 24 and look for the + signal in this equation while [9] eq 24 only have the - sign

Note that if the Warp Field $\Phi = 1$ with $\frac{dy}{ds} = 1$ and $dS^2 = 0$ the equation of the $5D$ Extra Force for a massless particle in $5D$ $M_5 = 0$ becomes equivalent to [9] eq 24 proving that the Ponce De Leon equations are equivalent to the Mashoon-Wesson-Liu ones.

$$\frac{1}{m_0}\frac{dm_0}{ds} =^{-}_{+} \frac{1}{2}\frac{\partial g_{uv}}{\partial y}u^u u^v \tag{46}$$

3. Dimensional Reduction from $5D$ to $4D$ According to Basini-Capozziello Ponce De Leon, Mashoon-Wesson-Liu and Overduin-Wesson. Possible Experimental Detection of Extra Dimensions in Strong Gravitational Fields or On-Board the International Space Station (ISS) Using the Gravitational Bending of Light in Extra Dimensions

The most important thing to keep in mind when we study models of BraneWorlds or Extra Dimensions is to explain why we cannot "see" directly the presence of the Extra Dimension although we can "feel" its effects in the $4D$ everyday Physics.We avoid here the models with compactification or "curling-up" of the Extra Dimension because these models don't explain why we have $3+1$ Large Dimensions while the remaining ones are small and "unseen" Extra Dimensions and also these models dont explain what generates the "Compactification" or "Curling" mechanism in the first place.Also some of these models develop "Unphysical" features.An excellent account of the difference between compactified and uncompactified models of Extra Dimensions is given by [21](see pgs 2 to 31).We prefer to adopt the fact that like the $3+1$ ordinary Large Spacetime Dimensions the Extra Dimensions are Large and uncompactified but due to a Dimensional Reduction from $5D$ to $4D$ we cannot "see" these Extra Dimensions although we can "feel" some of its effects. (see abs and pg 123 of [1] when Basini-Capozziello mentions the fact that we cannot perceive Time as the fourth Dimension and hence we cannot perceive the Spacelike Nature of the $5D$ Extra Dimension).(see also pg 1424 and pg 1434 begining of section 4 in [20]).(see also abs pg 2218 and 2219 of [5].Note the comment on Dimensional reduction and a $4D$ Spacetime embedded into a larger $5D$ one).We will now demonstrate how the Dimensional Reduction from $5D$ to $4D$ work and why in ordinary conditions we cannot "see" the $5D$ Extra Dimensions but we can "feel" some of its effects.Also we will see that changing the Spacetime Geometry and the shape of the Warp Field the $5D$ Extra Dimension will become visible.(Dimensional Reductions from $5D$ to $4D$ appears also in pg 2040 of [4]). We know that in ordinary $3+1$ Spacetime the curvature of the Einstein Tensor is neglectable and Spacetime can be considered as Minkowskian or flat where Special Relativity holds.A Minkowskian $5D$ Spacetime with a Warp Field can be given by(see eq 325 in [20]):

$$dS^2 = dt^2 - dX^2 - \Phi^2 dy^2 \tag{47}$$

The Warp Field considered here have small values between 0 and 1 nearly close to 0 and we recover the ordinary Special Relativity Ansatz.A Minkowskian $5D$ Spacetime with no Warp Field at all would be given by(see eq 326 in [20]):

$$dS^2 = dt^2 - dX^2 - dy^2 \tag{48}$$

The Ricci Tensors and Scalars for the Basini-Capozziello $5D$ Spacetime Fomalism and Ansatz given by $dS^2 = g_{\mu\nu}dx^\mu dx^\nu - \Phi^2 dy^2$ are shown below:(see pg 128 eq 58 in [1],pg 2230 eq 44 in [5] and pg 1442 eqs 111 to 115 in [20])

$$^5R_{\alpha\beta} = R_{\alpha\beta} - \frac{\Phi_{,a;b}}{\Phi} - \frac{1}{2\Phi^2}(\frac{\Phi_{,4}g_{\alpha\beta,4}}{\Phi} - g_{\alpha\beta,44} + \frac{g^{\mu\nu}g_{\mu\nu,4}g_{\alpha\beta,4}}{2}) \tag{49}$$

$$^5R = R - \frac{\Phi_{,a;b}}{\Phi}g^{\alpha\beta} - \frac{1}{2\Phi^2}g^{\alpha\beta}(\frac{\Phi_{,4}g_{\alpha\beta,4}}{\Phi} - g_{\alpha\beta,44} + \frac{g^{\mu\nu}g_{\mu\nu,4}g_{\alpha\beta,4}}{2}) \tag{50}$$

$$^5R = R - \frac{^4\Box\Phi}{\Phi} - \frac{1}{2\Phi^2}g^{\alpha\beta}(\frac{\Phi_{,4}g_{\alpha\beta,4}}{\Phi} - g_{\alpha\beta,44} + \frac{g^{\mu\nu}g_{\mu\nu,4}g_{\alpha\beta,4}}{2}) \tag{51}$$

For a $5D$ Spacetime Metric without Warp Field defined as $dS^2 = g_{\mu\nu}dx^\mu dx^\nu - dy^2$ the Ricci Tensor and Scalar would then be(see eqs 330 and 331 pg 1477 in [20]):

$$^5R_{\alpha\beta} = R_{\alpha\beta} - \frac{1}{2}(-g_{\alpha\beta,44} + \frac{g^{\mu\nu}g_{\mu\nu,4}g_{\alpha\beta,4}}{2}) \tag{52}$$

$$^5R = R - \frac{1}{2}g^{\alpha\beta}(-g_{\alpha\beta,44} + \frac{g^{\mu\nu}g_{\mu\nu,4}g_{\alpha\beta,4}}{2}) \tag{53}$$

But remember that a Minkowskian $5D$ Spacetime in which Special Relativity holds have all the $3+1$ Spacetime Metric Tensor Components defined by $g_{\mu\nu} = (+1, -1, -1, -1)$(see pg 1476 and 1477 in [20]) and the derivatives of the Metric Tensor vanishes and hence we are left with the following results(see eqs 332 and 333 in [20])

$$^5R_{\alpha\beta} = R_{\alpha\beta} \tag{54}$$

$$^5R = R \tag{55}$$

From the results above in a flat Minkowsky $5D$ Spacetime the Ricci Tensor in $5D$ is equal to its counterpart in $4D$ and since the Spacetime is flat then both are equal to zero.Then its impossible to tell if we live in a $4D$ Spacetime or in a larger $5D$ Extra Dimensional one.(see pg 1477 after eq 333 in [20]).If the Geometry of a flat $5D$ Extra Dimensional Spacetime is equivalent to the Geometry of a $3+1$ Spacetime we cannot distinguish if we live in a $4D$ or a $5D$ Universe.This is one of the most important things in the Dimensional Reduction from $5D$ to $4D$ as proposed by Basini-Capozziello.The $5D$ Extra Dimension is Large and Uncompactified but the physical reality we see is a Dimensional Reduction from $5D$ to $4D$ because we live in a nearly flat Minkowsky Spacetime[21] where Special Relativity holds and the $5D$ Ricci Tensor is equal to the $3+1$ counterpart.No Compactification mechanisms needed.The $5D$ Extra Dimension have a real physical meaning(see pg 2226 in [5] and pg 127 in [1]).(see also pg 2230 in [5] the part of the reduction of the Ricci Tensor

[21]The Systen Earth-Sun have a weak Gravitational Field so around Earth the Spacetime is cosidered flat

from $5D$ to $4D$ eqs 44 and 45.if the Warp Field $\Phi = 1$ both $5D$ and $4D$ Ricci Tensors from eq 45 are equal.the same can be seen in pg 128 to 129 eqs 58 to 59 in [1].see also pg 1442 eqs 115 to 116 in [20]). If in a flat $5D$ Minkowsky Spacetime we cannot "see" the Extra Dimension then we have three choices in order to tell if we live in a $5D$ Extra Dimensional Spacetime or a $3+1$ Ordinary Dimensional one.The choices are:

- Making the Warp Field $\Phi \neq 1$ in order to generate a difference between the $5D$ Extra Dimensional Ricci Tensor and the $3+1$ Spacetime counterpart according to eq 45 in [5],eq 59 in [1] and eq 116 in [20].This difference can tell the difference between a $5D$ Universe and a $3+1$ one.(see pg 1477 in [20])
- Making the $3+1$ Spacetime Metric Tensor components be a function of the $5D$ Extra Dimension in order to do not vanish the derivatives of the Metric Tensor with respect to the Extra Dimension generating a difference between the $5D$ Ricci Tensor and the $3+1$ counterpart according to eqs 330 and 331 pg 1477 in [20]. A Strong Gravitational Field of a Large Maartens-Clarkson $5D$ Schwarzschild Black String have the Spacetime Metric Tensor Components defined in function of the $4D$ rest-mass M but the $4D$ rest-mass is function of the $5D$ Extra Dimensional Spacetime Geometry according to eq 20 in [2].
- Making both conditions above hold true

We will examine all of the items above in this section. We live in a region of Spacetime where the Warp Field $\Phi = 1$ then we cannot see the $5D$ Extra Dimension.Or we can live in a region of Spacetime where the Warp Field $\Phi = 0$ and this cancels out the term $\Phi^2 dy^2$ in the $5D$ Spacetime Ansatz making the Extra Dimension invisible.Or perhaps we can live in a region of spacetime where $0 \leq \Phi \leq 1$ but near to 0 or 1 so its very difficult to detect the presence of the $5D$ Extra Dimension although we can "feel" some of its effects.Considering now a Warp Field $\Phi \neq 1$ the Minkowsky $5D$ Spacetime Ansatz would still have the terms of the $3+1$ Spacetime Metric Tensor given by $g_{\mu\nu} = (+1, -1, -1, -1)$.Hence the $5D$ Spacetime Ansatz would then be:

$$dS^2 = dt^2 - dX^2 - \Phi^2 dy^2 \tag{56}$$

The derivatives of the $3+1$ components of the Spacetime Metric Tensor vanishes but note that the $5D$ component do not vanish.The Ricci Tensor and Scalar would be given by the following expressions(see eqs 335,336 and 337 in [20]):

$${}^5R_{\alpha\beta} = R_{\alpha\beta} - \frac{\Phi_{,a;b}}{\Phi} \tag{57}$$

$${}^5R = R - \frac{\Phi_{,a;b}}{\Phi} g^{\alpha\beta} \tag{58}$$

$${}^5R = R - \frac{{}^4\Box\Phi}{\Phi} \tag{59}$$

Note that now the scenario is different:while with the Warp Field $\Phi = 1$ the $5D$ Ricci Tensor is equal to its $3+1$ counterpart and we cannot tell if we live in a $5D$ or in a $3+1$

Universe but when the Warp Field $\Phi \neq 1$ there exists a difference between the Ricci Tensor in $5D$ and the $3+1$ one.The Geometrical Properties of Spacetime of the $5D$ Spacetime are now different than the $3+1$ equivalent one and this makes the $5D$ Extra Dimension visible.(see also pg 1478 after eq 337 in [20])

According to Basini-Capozziello the Warp Field can be decomposed in two parts:one in $3+1$ ordinary Spacetime and another in the $5D$ Extra Dimension given by the following equation:([1] eq 76,[5] eq 70 and [20] eq 132 and 338)

$$\Phi = \phi(t,x)\chi(y) \tag{60}$$

Note that when we compute the covariant derivative of the Warp Field with respect to the $3+1$ Spacetime the terms of the $5D$ Extra Dimension are cancelled out and we are left with derivatives of the $3+1$ components of the Warp Field(see eq 339 in [20])

$$\Phi_{,a;b} = \chi(y)[\phi_{,a;b}] = \chi(y)[(\phi_\alpha)_\beta] - \Gamma^K_{\beta\alpha}\phi_K] \dashrightarrow \frac{\Phi_{,a;b}}{\Phi} = \frac{\chi(y)[\phi_{,a;b}]}{\phi(t,x)\chi(y)} = \frac{[\phi_{,a;b}]}{\phi(t,x)} = \frac{[(\phi_\alpha)_\beta - \Gamma^K_{\beta\alpha}\phi_K]}{\phi(t,x)} \tag{61}$$

The result shown below is very important.It demonstrates that only the $3+1$ component of the Warp Field $\phi(t,x)$ fortunately the component that lies in "our side of the wall" and its derivatives with(again fortunately) respect to our $3+1$ Spacetime coordinates can make the $5D$ Ricci Tensor be different than its $3+1$ counterpart and since we are considering in this case a flat Minkowsky Spacetime the $4D$ Ricci Tensor reduces to zero and this means to say that ${}^5R_{\alpha\beta} = -\frac{\phi_{,a;b}}{\phi}$ or better ${}^5R_{\alpha\beta} = -\frac{[(\phi_\alpha)_\beta - \Gamma^K_{\beta\alpha}\phi_K]}{\phi(t,x)}$

$${}^5R_{\alpha\beta} = R_{\alpha\beta} - \frac{\phi_{,a;b}}{\phi} \tag{62}$$

$${}^5R_{\alpha\beta} = R_{\alpha\beta} - \frac{[(\phi_\alpha)_\beta - \Gamma^K_{\beta\alpha}\phi_K]}{\phi(t,x)} \tag{63}$$

Note that if the $4D$ Ricci Tensor vanishes due to a flat Minkowsky Spacetime and we are left with derivatives of the $3+1$ Spacetime components of the Warp Field with respect to (again fortunately for the second time) $3+1$ Spacetime coordinates and we are left with a result in which the $5D$ Ricci Tensor and our capability to detect the existence of the $5D$ Extra Dimension depends on the shape of the $3+1$ component of the Warp Field.If we can detect the derivatives of the Warp Field we can detect the existence of the $5D$.

The other way to make the $5D$ Extra Dimension visible is to make the derivatives of the $3+1$ Spacetime Metric Tensor components $g_{\mu\nu} = (g_{00}, g_{11}, g_{22}, g_{33})$non-null with respect to the $5D$ Extra Dimension.

$$dS^2 = dt^2 - g_{\mu\mu}d(X^\mu)^2 - \Phi^2 dy^2 \tag{64}$$

For our special case of diagonalized metric:

$$dS^2 = dt^2 - g_{\mu\mu}d(X^\mu)^2 - \Phi^2 dy^2 \tag{65}$$

Considering the 3 + 1 Spacetime Metric Tensor Components g_{00} and g_{11}(see eqs 353 and 354 in [20]).

$$^5R_{00} = R_{00} - \frac{\Phi_{,0;0}}{\Phi} - \frac{1}{2\Phi^2}(\frac{\Phi_{,4}g_{00,4}}{\Phi} - g_{00,44} + \frac{g^{00}g_{00,4}g_{00,4}}{2}) \tag{66}$$

$$^5R_{11} = R_{11} - \frac{\Phi_{,1;1}}{\Phi} - \frac{1}{2\Phi^2}(\frac{\Phi_{,4}g_{11,4}}{\Phi} - g_{11,44} + \frac{g^{11}g_{11,4}g_{11,4}}{2}) \tag{67}$$

Now we can see that if the derivatives of g_{00} and g_{11} do not vanish with respect to the Extra Coordinate then the terms $\frac{\Phi_{,0;0}}{\Phi} - \frac{1}{2\Phi^2}(\frac{\Phi_{,4}g_{00,4}}{\Phi} - g_{00,44} + \frac{g^{00}g_{00,4}g_{00,4}}{2})$ and $\frac{\Phi_{,1;1}}{\Phi} - \frac{1}{2\Phi^2}(\frac{\Phi_{,4}g_{11,4}}{\Phi} - g_{11,44} + \frac{g^{11}g_{11,4}g_{11,4}}{2})$ will generates a difference between the $5D$ Ricci Tensor and its 3 + 1 Ordinary Spapcetime Dimensional counterpart.Remember also that g_{00} and g_{11} can be defined as the Spacetime Metric Tensor Components of the Maartens-Clarkson $5D$ Schwarzschild Black String centerd on a large Black Hole for example in which M is the $4D$ rest-mass of the Black Hole but M can be defined in function of the $5D$ Extra Dimensional rest-mass M_5 and also defined in function of the $5D$ Spacetime Geometry according to Ponce De Leon eq 20 in [2].This can make the $5D$ Extra Dimension becomes visible. Writing the Maartens-Clarkson $5D$ Schwarzschild Cosmic Black String as follows:([7] eq 1,[20] eq 380):

$$dS^2 = [(1 - \frac{2GM}{R})dt^2 - \frac{dR^2}{(1 - \frac{2GM}{R})} - R^2 d\eta^2] - \Phi dy^2 \tag{68}$$

Where the Spacetime Metric Tensor Components of the Black String are given by: $g_{00} = (1 - \frac{2GM}{R})$ and $g_{11} = -(1 - \frac{2GM}{R})^{-1}$.The derivatives with respect to the Extra Coordinate are then[22]:

$$\frac{\partial g_{00}}{\partial y} = \frac{\partial(1 - \frac{2GM}{R})}{\partial y} = -2G\frac{\partial \frac{M}{R}}{\partial y} = -2G[\frac{\partial M}{\partial y} \times R^{-1} + \frac{\partial R^{-1}}{\partial y} \times M] \tag{69}$$

We know that the $4D$ rest-mass M of the Maartens-Clarkson $5D$ Schwarzschild Cosmic Black String can be defined in function of the Ponce De Leon $5D$ rest-mass M_5 eq 20 in [2].The final result would then be:

$$\frac{\partial g_{00}}{\partial y} = -2G[\frac{1}{R}\frac{M_5}{[\sqrt{1 - \Phi^2(\frac{dy}{ds})^2}]^3}[\Phi^2\frac{dy}{ds}\frac{\partial \frac{dy}{ds}}{\partial y} + (\frac{dy}{ds})^2\Phi\frac{\partial \Phi}{\partial y}] - \frac{1}{R^2}(\frac{\partial R}{\partial y})\frac{M_5}{\sqrt{1 - \Phi^2(\frac{dy}{ds})^2}}] \tag{70}$$

$$\frac{\partial g_{11}}{\partial y} = \frac{\frac{\partial g_{00}}{\partial y}}{g_{00}^2} = \frac{-2G}{g_{00}^2}[\frac{1}{R}\frac{M_5}{[\sqrt{1 - \Phi^2(\frac{dy}{ds})^2}]^3}[\Phi^2\frac{dy}{ds}\frac{\partial \frac{dy}{ds}}{\partial y} + (\frac{dy}{ds})^2\Phi\frac{\partial \Phi}{\partial y}] - \frac{1}{R^2}(\frac{\partial R}{\partial y})\frac{M_5}{\sqrt{1 - \Phi^2(\frac{dy}{ds})^2}}] \tag{71}$$

[22] only time and radial components are considered here.

Note that in a Strong Gravitational Field these derivatives will have high values and this will make the $5D$ Ricci Tensor be highly different than its $3+1$ counterpart making the $5D$ Extra Dimension be visible but far away from the center of the Black String $\frac{M_5}{R} \simeq 0$ and the derivatives will vanish due to the Weak Gravitational Field however the term corresponding to the Warp Field will remain as shown below:

$$^5R_{00} = R_{00} - \frac{\Phi_{,0;0}}{\Phi} \tag{72}$$

$$^5R_{11} = R_{11} - \frac{\Phi_{,1;1}}{\Phi} \tag{73}$$

Then in a Weak or Null Gravitational Field[23] is the Warp Field that can make the $5D$ Ricci Tensor be different than its $3+1$ counterpart and can tell if we live in a $5D$ Extra Dimensional Spacetime or in a ordinary $3+1$ one.Remember that at faraway distances from Gravitational Field the Spacetime is flat or Minkowskian and the $3+1$ Ricci Tensor is zero or nearly zero.Then we could rewrite the two equations above as follows:

$$^5R_{00} = -\frac{\Phi_{,0;0}}{\Phi} \tag{74}$$

$$^5R_{11} = -\frac{\Phi_{,1;1}}{\Phi} \tag{75}$$

We already know that when computing derivatives of the Warp Field with respect to $3+1$ Coordinates the $5D$ Extra Dimensional terms are cancelled out and we will get these results:

$$^5R_{00} = -\frac{[(\phi_0)_0 - \Gamma^K_{00}\phi_K]}{\phi(t,x)} \tag{76}$$

$$^5R_{11} = -\frac{[(\phi_1)_1 - \Gamma^K_{11}\phi_K]}{\phi(t,x)} \tag{77}$$

Writing the Ricci Tensors with the derivatives of the $3+1$ Spacetime Metric Tensor Components of the Warp Field explicitly written we have:

$$^5R_{00} = -\frac{[\frac{\partial^2\phi(t,x)}{\partial t^2} - \Gamma^K_{00}\frac{\partial\phi(t,x)}{\partial x^K}]}{\phi(t,x)} \tag{78}$$

$$^5R_{11} = -\frac{[\frac{\partial^2\phi(t,x)}{\partial R^2} - \Gamma^K_{11}\frac{\partial\phi(t,x)}{\partial x^K}]}{\phi(t,x)} \tag{79}$$

$$\Gamma^K_{00}\frac{\partial\phi(t,x)}{\partial x^K} = \Gamma^0_{00}\frac{\partial\phi(t,x)}{\partial t} + \Gamma^1_{00}\frac{\partial\phi(t,x)}{\partial R} \tag{80}$$

$$\Gamma^K_{11}\frac{\partial\phi(t,x)}{\partial x^K} = \Gamma^0_{11}\frac{\partial\phi(t,x)}{\partial t} + \Gamma^1_{11}\frac{\partial\phi(t,x)}{\partial R} \tag{81}$$

[23] eg Earth-Sun System or a Spaceship far away from a Black Hole

We still don't know the shape of the Warp Field but remember that the $3+1$ component of the Warp Field can be coupled to Gravity as defined by Basini-Capozziello in [1] pg 119 and [5] pg 2235.

Considering only valid Christoffel Symbols we have [24]:

$$\Gamma^{K}_{00}\frac{\partial\phi(t,x)}{\partial x^{K}} = \Gamma^{0}_{00}\frac{\partial\phi(t,x)}{\partial t} = \frac{1}{2}\frac{1}{g_{00}}\frac{\partial g_{00}}{\partial t}\frac{\partial\phi(t,x)}{\partial t} = \frac{1}{2}\frac{1}{g_{00}}\frac{\partial g_{00}}{\partial y}\frac{\partial\phi(t,x)}{\partial t}\frac{\partial y}{\partial t} \tag{82}$$

$$\Gamma^{K}_{11}\frac{\partial\phi(t,x)}{\partial x^{K}} = \Gamma^{1}_{11}\frac{\partial\phi(t,x)}{\partial R} = \frac{1}{2}\frac{1}{g_{11}}\frac{\partial g_{11}}{\partial R}\frac{\partial\phi(t,x)}{\partial R} = \frac{1}{2}\frac{1}{g_{11}}\frac{\partial g_{11}}{\partial y}\frac{\partial\phi(t,x)}{\partial R}\frac{\partial y}{\partial R} \tag{83}$$

These expressions are valid for Strong or Weak Gravitational Fields .But we are considering here Weak Gravitational Fields where the Gravitational Force almost vanishes and the derivatives of the Spacetime Metric Tensor Components of the Maartens-Clarkson $5D$ Schwarzschild Cosmic Black String vanishes due to the term $\frac{M_5}{R}$[25] and the final expression for the $5D$ Ricci Tensors can be given by:

$$^{5}R_{00} = -\frac{[\frac{\partial^2\phi(t,x)}{\partial t^2}]}{\phi(t,x)} \tag{84}$$

$$^{5}R_{11} = -\frac{[\frac{\partial^2\phi(t,x)}{\partial R^2}]}{\phi(t,x)} \tag{85}$$

The $5D$ Ricci Scalar would be given by:

$$^{5}R = ^{5}R_{00} + ^{5}R_{11} = -\frac{[\frac{\partial^2\phi(t,x)}{\partial t^2}]}{\phi(t,x)} + -\frac{[\frac{\partial^2\phi(t,x)}{\partial R^2}]}{\phi(t,x)} = -\frac{1}{\phi(t,x)}{}^{4}\Box\phi(t,x)^2 \tag{86}$$

Remarkably we can extract the Ricci Scalar from the $5D$ to $4D$ Dimensional Reduction of Basini-Capozziello .

If the Warp Field Coupled to Gravity is defined by Basini-Capozziello then this can be regarded as a final proof that the $5D$ Extra Dimension really exists

Considering now the case of the $5D$ Spacetime Metric with no Warp Field at all $\Phi = 1$ the difference between the $5D$ and the $4D$ Ricci Tensors will depend on the derivatives of the Spacetime Metric Tensor Components with respect to the $5D$ Extra Dimension that will vanish far away from the center of the Maartens Clarkson $5D$ Schwarzschild Cosmic Black String making the $5D$ be invisible but in the regions of intense Gravitational Field the $5D$ Ricci Tensor will be different than its $4D$ counterpart making the $5D$ Extra Dimension becomes visible

$$^{5}R_{\alpha\beta} = R_{\alpha\beta} - \frac{1}{2}(-g_{\alpha\beta,44} + \frac{g^{\mu\nu}g_{\mu\nu,4}g_{\alpha\beta,4}}{2}) \tag{87}$$

$$^{5}R = R - \frac{1}{2}g^{\alpha\beta}(-g_{\alpha\beta,44} + \frac{g^{\mu\nu}g_{\mu\nu,4}g_{\alpha\beta,4}}{2}) \tag{88}$$

[24] diagonalized metrics

[25] making $g_{00} = 1$ and $g_{11} = 1$

Writing the Ricci Tensor for the time and radial components we have(see eqs 343 and 344 pg 1479 in [20]):

$$^{5}R_{00} = R_{00} - \frac{1}{2}(-g_{00,44} + \frac{g^{00}g_{00,4}g_{00,4}}{2}) \tag{89}$$

$$^{5}R_{11} = R_{11} - \frac{1}{2}(-g_{11,44} + \frac{g^{11}g_{11,4}g_{11,4}}{2}) \tag{90}$$

We we know that the derivatives of the Spacetime Metric Tensor for the $5D$ Maartens-Clarskon Schwarzschild Cosmic Black String are given by:

$$\frac{\partial g_{00}}{\partial y} = -2G[\frac{1}{R}\frac{M_5}{[\sqrt{1-(\frac{dy}{ds})^2}]^3}\frac{dy}{ds}\frac{\partial(\frac{dy}{ds})}{\partial y} - \frac{M_5}{\sqrt{1-(\frac{dy}{ds})^2}}\frac{1}{R^2}\frac{\partial R}{\partial y}] \tag{91}$$

$$\frac{\partial g_{11}}{\partial y} = \frac{\frac{\partial g_{00}}{\partial y}}{g_{00}^2} = -\frac{1}{g_{00}^2}2G[\frac{1}{R}\frac{M_5}{[\sqrt{1-(\frac{dy}{ds})^2}]^3}\frac{dy}{ds}\frac{\partial(\frac{dy}{ds})}{\partial y} - \frac{M_5}{\sqrt{1-(\frac{dy}{ds})^2}}\frac{1}{R^2}\frac{\partial R}{\partial y}] \tag{92}$$

Note that far away from the Black String the ratio $\frac{M_5}{R} \simeq 0$ and the derivatives of the Spacetime Metric Tensor Components will vanish making the $5D$ Ricci Tensor equal to its $3+1$ counterpart and the $5D$ Extra Dimension will become invisible.But in the neighborhoods of the Black String center Gravity becomes so high that the Extra Terms will make the $5D$ Ricci Tensor be different than its $3+1$ counterpart. Another way to measure the presence of the $5D$ Extra Dimension is to measure how the Extra Dimension affects the Gravitational Bending of Light in the vicinity of the Black String according to Kar-Sinha(see abstract of [3]).In one of our works([20] pg 1495 section 8) we proposed the use of the International Space Station ISS[26][27] to measure the Kar-Sinha Gravitational Bending of Light of the Sun to find out if it can be affected by the presence of the $5D$ Extra Dimension.(see also pg 1467 before eq 290 in [20]) While the Sun have a "weak" Gravitational Field a Black String is a Black Hole in $5D$ and in the vicinity of the Black String perhaps the Gravitational Bending Of Light affected by the presence of the $5D$ Extra Dimension according to Kar-Sinha would be better noticeable(ISS could still be used in this fashion to measure Gravitational Bending Of Light affected by the presence of Extra Dimensions by observing accretion disks of Black Holes free from the disturbances of Earth Atmosphere.We propose here to use beams of neutrons or photons to measure the Extra Terms On-Board ISS). Writing the Kar-Sinha Gravitational Bending Of Light affected by the presence of the $5D$ Extra Dimension in the neighborhoods of a Black String or in the neighborhoods of the Sun to be measured On-Board ISS with a non-null Warp Field as follows(see pg 1467 eq 288 to 291 and pg 1468 eq 295 to 296 in [20])(see also pg 1781 eq 18 in [3])[28][29]:

[26]more on General Relativity and ISS in [6],[8],[16],[17] and [18]

[27]ISS will also appear in the next section

[28]equations written without Warp Factors and with the Gravitational Constant

[29]see also eqs 156 to 158 pg 70 section 8.7 in [21].see also in the same reference the comment on the velocity along the $5D$ Extra Dimension in pg 71 after eq 159 $\frac{d\psi}{dt}$ similar to our $\frac{dy}{dt}$.see also between page 70 and 71 the

$$\Delta\omega = \frac{2GM}{c^2R}(2 + [\Phi\frac{dy}{cdt}]^2) \tag{93}$$

$$\Delta\omega = \frac{2GM}{c^2R}(2 + [\phi(t,x)\chi(y)\frac{dy}{cdt}]^2) \tag{94}$$

$$\Delta\omega = \frac{2G}{c^2R}\frac{M_5}{\sqrt{1 - \Phi^2(\frac{dy}{ds})^2}}(2 + [\Phi\frac{dy}{cdt}]^2) \tag{95}$$

$$\Delta\omega = \frac{2G}{c^2R}\frac{M_5}{\sqrt{1 - \phi(t,x)^2\chi(y)^2(\frac{dy}{ds})^2}}(2 + [\phi(t,x)\chi(y)\frac{dy}{cdt}]^2) \tag{96}$$

The same expression for a null Warp Field would be given by:

$$\Delta\omega = \frac{2GM}{c^2R}(2 + [\frac{dy}{cdt}]^2) \tag{97}$$

$$\Delta\omega = \frac{2G}{c^2R}\frac{M_5}{\sqrt{1 - (\frac{dy}{ds})^2}}(2 + [\frac{dy}{cdt}]^2) \tag{98}$$

In the above equations M_5 and M are the $5D$ and $4D$ rest-masses of the Sun or the Black Hole and R the distance between the accretion disk and the Black Hole or the distance between the photon beam and the Sun. We know that the Warp Field must have values between 0 and 1 so the shift in the Gravitational Bending Of Light must be very small making the value of the expression in $5D$ be close to its $4D$ counterpart.The presence of the Gravitational Constant $G = 6,67 \times 10^{-11}\frac{Newton\ timesm^2}{kg^2}$ divided by the square of the Light Speed would make the things even worst.This is the reason why we need a Black String of large rest mass M or M_5 to make the shift noticeable.Perhaps in the Sun we would never be able to measure the shift. Note also the comment on [21] pg 71 that the derivative $\frac{d\psi}{dt}$ analogous to our $\frac{dy}{dt}$ is null for photons and we know from Ponce De Leon that the $5D$ Spacetime Metric $dS^5 = ds^2 - \Phi^2dy^2$ is null for photons making $ds^2 = \Phi^2dy^2$ $M_5 = 0$ and $m_0 = 0$.Remember that in $4D$ SR $ds^2 = 0$ for photons and $\frac{dy}{ds} = 0$ making the shift in $5D$ be equal to its $4D$ counterpart.Kar-Sinha mentions in pg 1783 [3] the fact that if the photon propagates in $5D$ the value of $\frac{dy}{ds} < 2,8 \times 10^{-4}$ and the shift $\Delta\omega$ affected by the $5D$ Extra Dimension must lie between the error margins of the observed values of pgs 39 to 41 in [19][30].Remember also that the Spacetime at a distance R from the Sun where the photon passes by in order to be Bent or Deflected is described by the $5D$ Black String centered on the Sun according to [7] eq 1,[20] eq 380.Although $\frac{dy}{dt}$ is zero for photons it may be not for the Sun Mass M and then the Extra Terms in the Gravitational Bending Of Light for photons may still be measurable anyway.The Gravitational Bending can be

comment that the shift is physically measurable.we will examine photon paths in the $5D$ Maartens-Clarkson Schwarzschild Cosmic Black String in this section also but we will use the Ponce De Leon point of view of pg 1343 after eq 30 in [2].look to the Ponce De Leon comment of genuine manifestation of the $5D$ Extra Dimension before section 4

[30]this reference contains one of the best explanations for the Gravitational Bending Of Light Geometry and describes even the 30 percent margin of error in the 1919 measurements

observed for other particles with a non-null M_5 and a non-null m_0 and perhaps the study of the motion of high-speed relativistic particles from accretion disks of large Black Holes can tell the difference between the $5D$ $\Delta\omega$ and its $4D$ counterpart.For a non-null $\frac{dy}{ds}$ particle the Gravitational Bending formulas could be given by:

$$\Delta\omega = \frac{2GM}{c^2R}(2 + [\Phi\frac{dy}{ds}\frac{ds}{cdt}]^2) \tag{99}$$

$$\Delta\omega = \frac{2GM}{c^2R}(2 + [\phi(t,x)\chi(y)\frac{dy}{ds}\frac{ds}{cdt}]^2) \tag{100}$$

$$\Delta\omega = \frac{2G}{c^2R}\frac{M_5}{\sqrt{1 - \Phi^2(\frac{dy}{ds})^2}}(2 + [\Phi\frac{dy}{ds}\frac{ds}{cdt}]^2) \tag{101}$$

$$\Delta\omega = \frac{2G}{c^2R}\frac{M_5}{\sqrt{1 - \phi(t,x)^2\chi(y)^2(\frac{dy}{ds})^2}}(2 + [\phi(t,x)\chi(y)\frac{dy}{ds}\frac{ds}{cdt}]^2) \tag{102}$$

The same for a Warp Field $\Phi = 1$

$$\Delta\omega = \frac{2GM}{c^2R}(2 + [\frac{dy}{ds}\frac{ds}{cdt}]^2) \tag{103}$$

$$\Delta\omega = \frac{2G}{c^2R}\frac{M_5}{\sqrt{1 - (\frac{dy}{ds})^2}}(2 + [\frac{dy}{ds}\frac{ds}{cdt}]^2) \tag{104}$$

Note that a relativistic beam of neutrons would not suffer the deflection by electromagnetic fields and could be used to measure the Extra Terms in the Gravitational Bending of Light due to the presence of the Higher Dimensional Spacetime but a beam of photons is more easy to be obtained.Consider a Satellite carrying a small Laser device in the other side of the Earth Orbit targeting the beam towards a target in ISS.The beam in order to reach ISS must pass at a distance R from the Sun.The Extra Terms in $\Delta\omega$ could perhaps be measured in Outer Space On-Board International Space Station ISS free of the interference of the Earth Atmosphere.Computing the Classical Bending of Light as(see pg 1781 eq 18 in [3]):

$$\Delta\omega_{Classic} = \frac{4GM}{c^2R} \tag{105}$$

If we observe these deviations of the Bending Angle

$$\Delta\omega_{ExtraTerms} = \frac{2GM}{c^2R}[\Phi\frac{dy}{cdt}]^2 \tag{106}$$

$$\Delta\omega_{ExtraTerms} = \frac{2GM}{c^2R}[\frac{dy}{cdt}]^2 \tag{107}$$

different than the original Classical value then we can demonstrate that we live in a $5D$ Higher Dimensional Spacetime.

The Laser beam would be affected by the Sun Mass M while passing at a distance R from the Sun but would reach ISS.Computing the Classical Einstein Bending we know where the Laser will reach the target.If the observed Bending is equal to the Classical Einstein then this would mean that there are no Extra Dimensions in the Universe.But if the deviated photon arrives at ISS with an angle different than the one predicted by Einstein and if this difference in the angle matches the Extra-Terms predicted by Kar-Sinha then ISS would proof that we live in a Universe of more than 4 Dimensions.Then this experiment onboard ISS would have the same degree of importance of the measures of the Gravitational Bending of Light by Sir Arthur Stanley Eddington in the Sun Eclipse of 1919

Computing the Classical Einstein Bending Of Light for a Laser beam passing the Sun at distance $R = 150.000.000km$ we would have:

Mass of the Sun(in $4D$):

$$M = 1,9891 \times 10^{30} kg \tag{108}$$

Newton Gravitational Constant (in $4D$):

$$G = 6,67 \times 10^{-11} Newton \times m^2/kg^2 \tag{109}$$

There exists a common factor between the Classical Bending Of Light in $4D$ and the Kar-Sinha Extra Terms given by :

$$\triangle\omega_{CommonFactor} = \frac{2GM}{c^2 R} \tag{110}$$

The common factor would be given by:

$$\triangle\omega_{CommomFactor} = 2\times 6,67\times 10^{-11}\times 1,9\times 10^{30}/(1,5\times 10^{8}\times 9\times 10^{16}) = 1,877481481481\times 10^{-5} \tag{111}$$

$$2 \times 6,67 \times 10^{-11} \times 1,9 \times 10^{30} = 25,346 \times 10^{19} \tag{112}$$

$$1,5 \times 10^{8} \times 9 \times 10^{16} = 13,5 \times 10^{24} \tag{113}$$

$$\frac{25,346}{13,5} = 1,877481481481 \tag{114}$$

The Classical Einstein Bending Of Light would be given by:

$$\triangle\omega_{Classic} = 4 \times 6,67 \times 10^{-11} \times 1,9 \times 10^{30}/(1,5 \times 10^{10} \times 9 \times 10^{16}) = 3,754962962962 \times 10^{-5} \tag{115}$$

$$4 \times 6,67 \times 10^{-11} \times 1,9 \times 10^{30} = 50,692 \times 10^{19} \tag{116}$$

$$\frac{50,692}{13,5} = 3,754962962962 \tag{117}$$

We already outlined in the Introduction Section the fact that the goal of ISS is to achieve a Gravitational Shift Precision of 2.4×10^{-7}(see pg 631 Table II in [6]) and the faxt that ISS Gravitational shifts are capable to detect measures of $\frac{\triangle\omega}{\omega}$ with Expected Uncertainly of 12×10^{-6}(see pg 629 Table I in [6])

The difference between Kar-Sinha and Classical Gravitational Bending Of Light is given by:

$$\triangle\omega_{KarSinha} - \triangle\omega_{Classic} = \frac{2GM}{c^2R}(2 + [\Phi\frac{dy}{cdt}]^2) - \frac{4GM}{c^2R} \tag{118}$$

$$\triangle\omega_{KarSinha} - \triangle\omega_{Classic} = \frac{4GM}{c^2R} + \frac{2GM}{c^2R}[\Phi\frac{dy}{cdt}]^2 - \frac{4GM}{c^2R} \tag{119}$$

$$\triangle\omega_{KarSinha} - \triangle\omega_{Classic} = \frac{2GM}{c^2R}[\Phi\frac{dy}{cdt}]^2 \tag{120}$$

$$\triangle\omega_{KarSinha} - \triangle\omega_{Classic} = \triangle\omega_{CommonFactor}[\Phi\frac{dy}{cdt}]^2 \tag{121}$$

$$\triangle\omega_{KarSinha} - \triangle\omega_{Classic} = 1,877481481481 \times 10^{-5}[\Phi\frac{dy}{cdt}]^2 \tag{122}$$

$$\triangle\omega_{KarSinha} - \triangle\omega_{Classic} = \triangle\omega_{ExtraTerms} \tag{123}$$

And this difference is equal to the Common Factor multiplied by the Kar-Sinha Extra Terms due to the presence of the Higher Dimensional Spacetime

The Kar-Sinha additional terms depends on the derivative of the $5D$ Extra Coordinate or the value of the Warp Field Φ due to the Extra Dimensional factors $[\Phi\frac{dy}{cdt}]^2$.We know that for a Timelike $5D$ geodesics according to Ponce De Leon we have $0 < \Phi < 1$ and $0 < [\Phi\frac{dy}{ds}]^2 < 1$(see eq3 pg 1426 in [20] without the Warp Factors $\Omega = 1$)and for photons we would have a Null-Like Geodesics(see eq4 pg 1426 in [20] without the Warp Factors $\Omega = 1$) giving $0 < [\Phi\frac{dy}{ds}]^2 = 1$.Assuming also a small derivative of the Extra Coordinate with respect to time $\frac{dy}{cdt}$ then we would have very small values for the Extra Dimensional Term $[\Phi\frac{dy}{cdt}]^2$ at least for a Timelike Geodesics.

Examining now the case for photons in a Null-Like $5D$ Geodesics in the neighborhoods of the Sun as a Maartens-Clarkson Schwarzschild Black String[31] :

$$dS^2 = 0 \dashrightarrow ds^2 - \Phi^2 dy^2 = 0 \dashrightarrow ds^2 = \Phi^2 dy^2 \dashrightarrow \frac{1}{\Phi^2} = (\frac{dy}{ds})^2 \dashrightarrow 1 = \Phi^2(\frac{dy}{ds})^2 \tag{124}$$

$$1 = \Phi^2(\frac{dy}{ds})^2 \dashrightarrow 1 = \Phi^2(\frac{dy}{cdt})^2(\frac{cdt}{ds})^2 \dashrightarrow \Phi^2(\frac{dy}{cdt})^2 = (\frac{ds}{cdt})^2 \tag{125}$$

[31] We already outlined the fact that Kar-Sinha mentions in pg 1783 [3] the fact that if the photon propagates in $5D$ the value of $\frac{dy}{ds} < 2,8 \times 10^{-4}$ and the shift $\triangle\omega$ affected by the $5D$ Extra Dimension must lie between the error margins of the observed values of pgs 39 to 41 in [19]

$$\triangle\omega_{ExtraTerms} = 1,877481481481 \times 10^{-5}[\Phi\frac{dy}{cdt}]^2 = 1,877481481481 \times 10^{-5}(\frac{ds}{cdt})^2 \quad (126)$$

$$dS^2 = [(1 - \frac{2GM}{c^2R})(cdt)^2 - \frac{dR^2}{(1-\frac{2GM}{c^2R})} - R^2d\eta^2] - \Phi dy^2 \quad (127)$$

$$ds^2 = [(1 - \frac{2GM}{c^2R})(cdt)^2 - \frac{dR^2}{(1-\frac{2GM}{c^2R})} - R^2d\eta^2] \quad (128)$$

$$\frac{ds^2}{(cdt)^2} = [(1 - \frac{2GM}{c^2R}) - \frac{1}{(1-\frac{2GM}{c^2R})}\frac{dR^2}{(cdt)^2} - R^2\frac{d\eta^2}{(cdt)^2}] \quad (129)$$

$$\frac{ds^2}{(cdt)^2} = [(1 - \triangle\omega_{CommonFactor}) - \frac{1}{(1-\triangle\omega_{CommonFactor})}\frac{dR^2}{(cdt)^2} - R^2\frac{d\eta^2}{(cdt)^2}] \quad (130)$$

$$1-\frac{2GM}{c^2R} = 1-\triangle\omega_{CommonFactor} = 1-1,877481481481\times10^{-5} = 0,9998122518518519 \quad (131)$$

$$\frac{ds^2}{(cdt)^2} = (1 - \triangle\omega_{CommonFactor}) - \frac{1}{c^2}[\frac{1}{(1-\triangle\omega_{CommonFactor})}\frac{dR^2}{(dt)^2} - R^2\frac{d\eta^2}{(dt)^2}] \quad (132)$$

Note that the term $\frac{1}{c^2}$ will make the second right term neglectable due to the factor 10^{-16}.

$$\frac{ds^2}{(cdt)^2} = (1-\triangle\omega_{CommonFactor}) = 0,9998122518518519 = 9,998122518518519\times10^{-1} \quad (133)$$

$$\triangle\omega_{KarSinha} - \triangle\omega_{Classic} = \triangle\omega_{ExtraTerms} = 1,877481481481 \times 10^{-5}[\Phi\frac{dy}{cdt}]^2 \quad (134)$$

$$\triangle\omega_{KarSinha} - \triangle\omega_{Classic} = \triangle\omega_{ExtraTerms} = 1,877481481481 \times 10^{-5}(\frac{ds}{cdt})^2 \quad (135)$$

$$\triangle\omega_{KarSinha}-\triangle\omega_{Classic} = \triangle\omega_{ExtraTerms} = 1,877481481481\times10^{-5}\times9,998122518518519\times10^{-1} \quad (136)$$

$$\triangle\omega_{KarSinha}-\triangle\omega_{Classic} = \triangle\omega_{ExtraTerms} = 18,771289878096695909454046639\times10^{-6} \quad (137)$$

The result above is the most important of this work and means to say that the Kar-Sinha Extra Terms in the Gravitational Bending of Light due to presence of the Extra Dimensions are in the range of the detection capability of the International Space Station ISS enclosing the Gravitational Shift Precision of 2.4×10^{-7}(see pg 631 Table II in [6]) and the Gravitational Shifts of $\frac{\triangle\omega}{\omega}$ with Expected Uncertainly of 12×10^{-6}(see pg 629 Table I in [6]). If this Extra Term is detected with "positive" results then the International Space Station ISS can demonstrate for the first time that our Universe have more than $4D$ Dimensions making the Physics of Extra Dimensions an Experimental Branch of Modern Physics.

4. Experimental Detection of Extra Dimensions Using Gravitational Red-Shifts On-Board the International Space Station ISS

From the abstract of [6] we know that Gravitational Red Shifts are also considered to experiments On-Board the International Space Station ISS.We already outlined before the Gravitational Shift precision of ISS.We also know from Kar-Sinha that Extra Dimensions affects the Gravitational Red Shift(see pg 1782 in [3]) generating Extra Terms in a way similar to the ones for the Gravitational Bending Of Light. We will propose in this Section a way to detect these Extra Terms On-Board ISS as a second proof that Extra Dimensions exists(or not).
There are two Classical expressions for the Gravitational Red-Shift $\triangle\lambda$(The wavelength displacement in Spectral Lines due to Gravity as seen by a far away observer in free space).One approximate and one exact.
The approximate expression is given by:

$$\triangle\lambda_{approximate} = \triangle\lambda_{Classical} = \frac{GM}{c^2R} \tag{138}$$

And the exact one by:

$$\triangle\lambda_{exact} = \frac{1}{\sqrt{1-\frac{2GM}{c^2R}}} - 1 \tag{139}$$

These expressions considering the Kar-Sinha Extra Terms due to the presence of the Extra Dimensions would be given by:

$$\triangle\lambda_{approximate_{ks}} = \frac{GM}{c^2R}(1+[\Phi\frac{dy}{ds}\frac{ds}{cdt}]^2) \tag{140}$$

$$\triangle\lambda_{exact_{ks}} = \frac{1}{\sqrt{1-\frac{2GM}{c^2R}(2+[\Phi\frac{dy}{ds}\frac{ds}{cdt}]^2)}} - 1 \tag{141}$$

$$\triangle\lambda_{approximate_{ks}} = \frac{GM}{c^2R}(1+[\Phi\frac{dy}{cdt}]^2) \tag{142}$$

$$\triangle\lambda_{exact_{ks}} = \frac{1}{\sqrt{1 - \frac{2GM}{c^2R}(2 + [\Phi\frac{dy}{cdt}]^2)}} - 1 \quad (143)$$

The approximate expression is more than enough to illustrate our point of view.

$$\triangle\lambda_{KarSinha} = \frac{GM}{c^2R}(1 + [\Phi\frac{dy}{cdt}]^2) \quad (144)$$

$$\triangle\lambda_{KarSinha} = \triangle\lambda_{Classical}(1 + [\Phi\frac{dy}{cdt}]^2) = \triangle\lambda_{Classical} + \triangle\lambda_{Classical}[\Phi\frac{dy}{cdt}]^2] \quad (145)$$

$$\triangle\lambda_{KarSinha} - \triangle\lambda_{Classical} = \triangle\lambda_{Classical}[\Phi\frac{dy}{cdt}]^2 = \triangle\lambda_{ExtraTerms} \quad (146)$$

Our idea is to send a second Satellite with another Laser beam to the Venus Orbit but directed towards the ISS.The Satellite would send the Laser beam to ISS with a certain blue wavelength but when arriving at ISS due to the difference of Sun Gravitational Fields between Earth and Venus the beam would be Red-Shifted and the Kar-Sinha Extra Terms due to the presence of Extra Dimensions could be detected.

In Venus the Gravitational Red Shift would be given by:

$$\triangle\lambda_{KarSinhaVenus} = \frac{GM}{c^2D_{Venus}}(1 + [\Phi\frac{dy}{cdt}]^2_{Venus}) \quad (147)$$

From the Previous Section we know that

$$\triangle\lambda_{KarSinhaVenus} = \frac{GM}{c^2D_{Venus}}(1 + [\frac{ds}{cdt}]^2_{Venus}) \quad (148)$$

With D_{Venus} being the Sun Radius R plus the distance d_{Venus} from Sun to Venus

On Earth the Gravitational Red Shift would be given by:

$$\triangle\lambda_{KarSinhaEarth} = \frac{GM}{c^2D_{Earth}}(1 + [\Phi\frac{dy}{cdt}]^2_{Earth}) \quad (149)$$

$$\triangle\lambda_{KarSinhaEarth} = \frac{GM}{c^2D_{Earth}}(1 + [\frac{ds}{cdt}]^2_{Earth}) \quad (150)$$

With D_{Earth} being the Sun Radius R plus the distance d_{Earth} from Sun to Earth

Since in Venus the Laser is still Blue-Shifted when sent towards ISS the Red Shift detected by ISS would be generated in the Venus-Earth trip.

$$\frac{GM}{c^2D_{Earth}} - \frac{GM}{c^2D_{Venus}} = \frac{GM}{c^2}[\frac{1}{D_{Earth}} - \frac{1}{D_{Venus}}] \cong \frac{GM}{c^2}[\frac{1}{d_{Earth}} - \frac{1}{d_{Venus}}] \quad (151)$$

The Kar-Sinha Gravitational Red Shift in Extra Dimensions epresions from the Venus Earth trip would be given by the following expressions

$$\triangle\lambda_{KarSinhaEarthVenus} = \frac{GM}{c^2}[\frac{1}{d_{Earth}} - \frac{1}{d_{Venus}}](1 + [\Phi\frac{dy}{cdt}]^2) \tag{152}$$

$$\triangle\lambda_{KarSinhaEarthVenus} = \frac{GM}{c^2}[\frac{1}{d_{Earth}} - \frac{1}{d_{Venus}}](1 + [\frac{ds}{cdt}]^2) \tag{153}$$

And the Extra Terms due to the presence of the Extra Dimensions would be given by:

$$\triangle\lambda_{ExtraTermsEarthVenus} = \frac{GM}{c^2}[\frac{1}{d_{Earth}} - \frac{1}{d_{Venus}}][\Phi\frac{dy}{cdt}]^2 \tag{154}$$

$$\triangle\lambda_{ExtraTermsEarthVenus} = \frac{GM}{c^2}[\frac{1}{d_{Earth}} - \frac{1}{d_{Venus}}][\frac{ds}{cdt}]^2 \tag{155}$$

From the previous Section we know that ISS can measure these Extra Terms if the Extra Dimensions exists.

5. Conclusion-Physics of Extra Dimensions as an Experimental Branch of Physics for the First Time

Our approach to the study of Extra Dimensions was centered on the Basini-Capozziello-Ponce de Leon Formalism.While other formalisms of Extra Dimensions uses $3+1$ uncompactified ordinary spacetime dimensions while the Extra Dimensions are compactified bringing the question of why $3+1$ large ordinary dimensions and the rest of the Extra Dimensions "curled-up" over themselves and what causes or generates the "compactification mechanism"????.In the Basini-Capozziello-Ponce de Leon Formalism the Extra Dimensions are large the same size of the $3+1$ ordinary dimensions avoiding the need of "exotic" compactification mechanisms but we cannot "see" these dimensions in normal conditions due to the reasons presented in this work. Also it can explain the multitude of particles seen in $4D$ as Dimensional Reductions from a small group of particles in $5D$ allowing perhaps the "unification" of Physics from the point of view of the Extra Dimensional Spacetime. This is very attractive from the point of view of a Unified Physics theory.There exists a small set of particles in $5D$ and all the huge number of Elementary Particles in $4D$ is a geometric projection from the $5D$ Spacetime into a $4D$ one([2] eq 20,[11] eq 21 and [20] eq 8).[32]

$$m_0 = \frac{M_5}{\sqrt{1 - \Phi^2(\frac{dy}{ds})^2}} \tag{156}$$

[32] We know that we are repeating the table for the third time but the table coupled with the Ponce De Leon equation illustrates the beauty of this point of view

Particle	spin ($\hbar$)	B	L	T	T_3	S	C	B*	charge (e)	m_0 (MeV)	antipart.
u	1/2	1/3	0	1/2	1/2	0	0	0	+2/3	5	$\overline{\text{u}}$
d	1/2	1/3	0	1/2	−1/2	0	0	0	−1/3	9	$\overline{\text{d}}$
s	1/2	1/3	0	0	0	−1	0	0	−1/3	175	$\overline{\text{s}}$
c	1/2	1/3	0	0	0	0	1	0	+2/3	1350	$\overline{\text{c}}$
b	1/2	1/3	0	0	0	0	0	−1	−1/3	4500	$\overline{\text{b}}$
t	1/2	1/3	0	0	0	0	0	0	+2/3	173000	$\overline{\text{t}}$
e^-	1/2	0	1	0	0	0	0	0	−1	0.511	e^+
μ^-	1/2	0	1	0	0	0	0	0	−1	105.658	μ^+
τ^-	1/2	0	1	0	0	0	0	0	−1	1777.1	τ^+
ν_e	1/2	0	1	0	0	0	0	0	0	0(?)	$\overline{\nu}_e$
ν_μ	1/2	0	1	0	0	0	0	0	0	0(?)	$\overline{\nu}_\mu$
ν_τ	1/2	0	1	0	0	0	0	0	0	0(?)	$\overline{\nu}_\tau$
γ	1	0	0	0	0	0	0	0	0	0	γ
gluon	1	0	0	0	0	0	0	0	0	0	gluon
W^+	1	0	0	0	0	0	0	0	+1	80220	W^-
Z	1	0	0	0	0	0	0	0	0	91187	Z
graviton	2	0	0	0	0	0	0	0	0	0	graviton

Look to the Elementary Particles Table above:we have a multitude of Quarks,Leptons,Muons and Heavy particles with apparently different rest-masses m_0 in $4D$ but these particles can have the same rest-mass M_5 in the $5D$ Extra Dimension and the differences are being generated by the Dimensional Reduction from $5D$ to $4D$ according to the Basini-Capozziello-Ponce De Leon formalism.This can bring new perspectives for the desired dream of the Unification of Physics We proposed here the use of a Satellite[33] with a Laser device placed in the other side of Earth Orbit with the Sun between the Satellite and ISS.The satellite will send the Laser beam towards ISS and the beam must pass in the neighborhoods of the Sun in order to reach ISS. The bemm will be Gravitationally Bent according to Einstein and if Extra Dimensions exists then the Extra Terms in the Gravitational Bending Of Light affected by the presence of the Extra Dimensions predicted by Kar-Sinha will appear. We demonstrated here that ISS have the needed precision to spot the Extra Terms predicted by Kar-Sinha and ISS could answer for the first time the question if the Universe have or not more than 4 Dimensions predicted by many Physics Theories but never seen before. Such an experiment would have the same degree of importance of the measures of Gravitational Bending Of Light made by Sir Arthur Stanley Eddington in the Sun Eclipse of 1919 that proved valid the Einstein General Theory Of Relativity The implications of a "positive" result would be enormous making the Physics of Extra Dimensions an Experimental Branch of Physics for the first time [34] [35]. All the theories of Physics Unification that predicts the existence of Extra Dimensions would be regarded as valid Physical Descriptions of Nature and not only mere Mathematical Models adjusted to "normalize consistently " some calculations and this would pose major modifications in Particle Physics since it could be proved that the multitude of apparent

[33] Such a Satellite could perfectly be christened as "Eddington" as a Homage to Sir Arthur Stanley Eddington

[34] An excellent account of the progresses in this area is given by the link below:

[35] $http://arXiv.org/find/grp_physics/1/abs:+AND+dimensions+AND+experimental+extra/0/1/0/all/0/1$

different particles we see in 4 Dimensions are Dimensional Reductions or Dimensional Projections from a small group of perhaps the same particles in 5 Dimensions according to the Basini-Capozziello Ponce De Leon formalism If this experiment is performed with "positive" results then the International Space Station ISS could change drastically and forever our way to understand the Universe.Young Jedi Knight Padawan:Stay Away From The Dark Side And May The Force Be With You[36]

6. Epilogue

- "The only way of discovering the limits of the possible is to venture a little way past them into the impossible."-Arthur C.Clarke[37]
- "The supreme task of the physicist is to arrive at those universal elementary laws from which the cosmos can be built up by pure deduction. There is no logical path to these laws; only intuition, resting on sympathetic understanding of experience, can reach them"-Albert Einstein[38][39]

Acknowledgements

We would like to express the most profound and sincere gratitude towards Doctor Frank Columbus Editor-Chief of NOVA Scientific Publishers United States of America for the invitation to write a paper for his presentation conference meeting entitled "Space Stations: Crew, Experiments and Missions." This arXiv.org paper is our answer to the invitation. We also would like to Acknowledge Professor Doctor Martin Tajmar of University Of Viena, Austria – ESA (European Space Agency) – ESTEC-SV (European Space Technology and Engineering Center - Space Vehicles Division) and Seibesdorf Austria Aerospace Corporation GmBH-ASPS (Advanced Space Propulsion Systems), for his kindness and goodwill for being our arXiv.org sponsor and to Acknowledge Paulo Alexandre Santos and Dorabella Martins da Silva Santos from University of Aveiro Portugal for the access to the scientific publication General Relativity and Gravitation(GRG). In closing we would also like to Acknowledge the Administrators and Moderators of arXiv.org at the Cornell University, United States of America for their agreement in accepting this document.

7. Remarks

The bulk of the bibliographic sources used in our research came from the refereed scientific publication General Relativity and Gravitation (GRG) from Springer-

[36]Slightly modified from Frank Oz as Master Yoda in the George Lucas movie Star Wars Episode II The Attack Of The Clones

[37]special thanks to Maria Matreno from Residencia de Estudantes Universitas Lisboa Portugal for providing the Second Law Of Arthur C.Clarke

[38]"Ideas And Opinions" Einstein compilation, ISBN $0-517-88440-2$, on page 226."Principles of Research" ([Ideas and Opinions],pp.224-227), described as "Address delivered in celebration of Max Planck's sixtieth birthday (1918) before the Physical Society in Berlin"

[39]appears also in the Eric Baird book Relativity in Curved Spacetime ISBN $978-0-9557068-0-6$

Verlag GmBH (formerly Kluver/Plenum Academic Publishing Corp)(ISSN:0001-7701 paper)(ISSN:1572-9532 electronic) under the auspices of the International Comitee on General Relativity and Gravitation and quoted by Deutsche Zentralblatt Math of EMS(European Mathematical Society). The Volume 36 Issue 03 March 2004 under the title:"Fundamental Physics on the ISS" was totally dedicated to test experimentally General Relativity,Quantum Gravity,Extra Dimensions and other physics theories in Outer Space on-board International Space Station(ISS) under the auspices of ESA(European Space Agency) and NASA(National Aeronautics and Space Administration). All the mention to pages of the references in the main text and in the footnotes of this work are for GRG and Liv Rev Rel references originally from the published version since we have access to GRG and Liv Rel Rel although we provide the number of the arXiv.org available GRG and Liv Rev Rel papers but for PhysRpt the page numbers are originally from the arXiv.org version since we cannot access this journal and sometimes exists differences in page numbers between the arXiv.org version and the published version due to different editorial styles preferred by scientific journals.[40] We choose to adopt in our research mainly refereed published papers from these publications not only due to their prestige and reputation among the scientific community but also because we are advocating new points of view in this work but based on the solid ground of certifiable and credible references

8. Legacy

This work is dedicated to the memory of the British Astronomer Sir Arthur Stanley Eddington that measured for the first time the Gravitational Bending Of Light in the Sun Eclipse of 1919 proving valid the Einstein General Theory Of Relativity. This work is also dedicated with a feeling of gratitude to all the people of NASA(National Aeronautics and Space Administration),ESA(European Space Agency),to all the people of the Space Agencies of Canada,Russia,Japan,China,Brazil,Argentina.Mexico,Israel and India all of these people with major contributions and Manned Space Missions or responsible and involved in one way or another with the project of the International Space Station ISS

References

[1] Basini G. and Capozziello S. (2005).*Gen Rel Grav* **37** 115 .

[2] Ponce De Leon J. (2004).*Gen Rel Grav* **36** 1335 ,gr-qc/0310078.

[3] Kar S. and Sinha M. (2003).*Gen Rel Grav* **35** 1775 .

[4] Loup F. Santos P. and Santos D. (2003).*Gen Rel Grav* **35** 2035 .

[5] Basini G. and Capozziello S. (2003).*Gen Rel Grav* **35** 2217 .

[6] C.Lammerzahl;G. Ahlers N. Ashby, M. Barmatz,P. L. Biermann, H. Dittus, V. Dohm, R. Duncan, K. Gibble, J. Lipa, N. Lockerbie, N. Mulders and C. Salomon. (2004). *Gen Rel Grav* **36** 615

[40] readers that can access GRG can compare for example gr-qc/0310078 with [2] or gr-qc/0603106 with [20]

[7] Clarkson C. and Maartens R. (2005).*Gen Rel Grav* **37** 1681,astro-ph/0505277

[8] Dittus H. (2004). *Gen Rel Grav* **36** 601

[9] Mashhoon B., Wesson P. and Liu H. (1998). *Gen Rel Grav* **30** 555

[10] Loup F. Santos P. and Santos D. (2003).*Gen Rel Grav* **35** 1849

[11] Ponce De Leon J. (2003).*Gen Rel Grav* **35** 1365 ,gr-qc/0207108

[12] Wesson P. (2003). *Gen Rel Grav* **35** 307,gr-qc/0302092

[13] Seahra S. and Wesson P. (2005). *Gen Rel Grav* **37** 1339

[14] Seahra S. and Wesson P. (2001). *Gen Rel Grav* **33** 1731,gr-qc/0105041

[15] Billyard A. and Sajko W. (2001).*Gen Rel Grav* **33** 1929,gr-qc/0105074

[16] Paik H. , Moody M. and Strayer D. (2004).*Gen Rel Grav* **36** 523

[17] Dittus H. , Lammerzahl C. and Selig H. (2004).*Gen Rel Grav* **36** 571

[18] Walz J and Hansch T. (2004).*Gen Rel Grav* **36** 561

[19] Will C.M.,(2006).*Liv Rev Rel*,(9) lrr-2006-3,gr-qc/0103036

[20] Loup F (2006).*Gen Rel Grav* **38** 1423,gr-qc/0603106

[21] Overduin J.M. and Wesson P. (1997). *Phys.Rept.* **283** 303-380,gr-qc/9805018

In: Space Exploration Research
Editors: J.H. Denis and P.D. Aldridge

ISBN: 978-1-60692-264-4

Chapter 10

HISTORY OF THE DISCOVERY OF SATURN'S RINGS AND MOONS

Matthias Risch
University of Applied Sciences, Augsburg, Germany

ABSTRACT

The history of the discovery of Saturn's rings started right at the beginning of sky observations with a telescope after it had been invented in the Netherlands in about 1608. This discovery includes many great names of researchers in astronomy such as Galileo, Scheiner, Hevelius, Gassendi and Huygens, until Cassini finally settled the very nature of the rings.

The first person to make observations about Saturn's rings was Galileo during the time he made his sky observations that led to the acceptance of the heliocentric model. Galileo put his findings into a riddle as he was not sure enough about what he saw.

The French philosopher Gassendi described his observations in 1640 as "ansae" (Latin, English: handles), which became a scientific term for Saturn's appearance for a long time. It was not until 1656 when Huygens proved the existence of Saturn's moon Titan, using an improved telescope, that he conjectured the appearance he saw was actually a ring. This discovery was so unusual to accepted science at that time, thus again, like Galileo; he released his findings in an anagram, which he did not disclose earlier than 1659. His ideas were hardly accepted, however. While he conjectured the ring to be solid throughout, Cassini found the true nature of the ring formed by small ice particles in 1705 when he described the gap within the ring. Then finally the ring structure model was widely accepted.

The discovery of the moon Phoebe by Pickering in 1898 was both the first major discovery in astronomy made by photography as well as the first discovery of a moon from the southern hemisphere.

INTRODUCTION: OBSERVATIONS OF SATURN IN ANTIQUITY

The strong belief in the Ptolemaic system laid down in the Almagest throughout antiquity and the Middle Ages was due to the precision it allowed in the calculation of the eclipses and position of planets [Freeth 2006]. This Greek theory of the skies was based considerably on Babylonian and Chaldean measurements of planetary positions [Galter 1993, Kugler 1900]. For example, the Chaldeans determined the synodic revolution time of Saturn to 57 Saturnian's revolutions in 59 years, the so-called Goal Year, which was a deviation of only 0,0028 from the scientific number of today. Using these value extensive cuneiform tables of Saturn's visibility periods and velocity were recorded in Chaldean antiquity [Neugebauer 1955]. Ptolemy quoted in the Almagest Hipparchus to have determined that number more precisely to be 57 Saturnian's revolutions in 59 years plus one and three quarter days, improving accuracy to better than 0,0001. Prior to this improvement, the Chaldeans determined a so-called giant period of 256 Saturn's revolutions in 265 years (for 589 years) [Kugler 1907, Betzold 1911]. Some Assyrians wrote in cuneiform texts that Saturn is an extinct kind of sun [Boll 1950].

History of the Discovery of Saturn´s Rings

Throughout the early history of Science the heliocentric world model was refused.

In the time of Galileo, Science still thought about the world in Aristotle's terms of dichotomy meaning that the planet Earth is made up of the elements water, air, fire and earth while the sky, the other planets as well as the stars, are made of aether [Sambursky 1974]. Dichotomy was seriously shaken by Galileo's observations of the sky by telescope laid down in his famous script „Sidereus Nuncius", in English: News from New Stars 1610 / 1611, [Galileo 1967].

The telescope was invented in 1604 or 1608 in the Netherlands. The invention of the telescope has two stories: an official and an unofficial one.

It is well recorded that the lens grinder and spectacle maker Jan Lippershey, immigrant from Germany to Middleburg in Zeeland in the Netherlands (at this time on an island in the North Sea close to Antwerp) offered a telescope to Count Maurits of the Netherlands and the Generalstaat of the Netherlands, which was effectively the parliament, for military use asking for a patent on October 2, 1608 [Günther 1883]. Since one could use only one eye with that telescope, Lippershey was asked on October 6, 1608 to develop a binocular, which he did by December 1608 [Helden 1977]. For this Lippershey got an enormous reward of 900 Guilders but was denied the patent on the grounds that the telescope was already well known throughout the Netherlands in that year and nothing new [Wolf 1870]. This rejection is support for the unofficial story based on tales told in the Netherlands for a long time. Children playing with two flawed lenses close to the shop of Lippershey´s or Janssen´s shop, situated at opposite sides of the Nieuwe Kerk (church) in Middelburg in about 1604 saw the weather vane of the Nieuwe Kerk of Middleburg considerably enlarged holding the lenses far apart and looking through both of them in 1604. Zacharias Janssen, an eye glass maker in Middelburg, however, did not pay much attention to that invention since the picture of the church tower seen through two lenses was "het onderste boven gekeerd" (English: upside down), Lippershey, however, would have had second thoughts about this incident [Wolf 1890]. At the Frankfurt fair in 1609 telescopes were offered to the public by Belgian merchants who distributed them all over Europe within a short time. There are claims that Giambattista della Porta of Naples, Italy, observed magnification by two lenses in 1589, but there are no sound records and it is not clear whatever he saw by means of the two lenses. Also, there are stories that an Italian merchant offered that invention to either Lippershey or Jansen, but there are no reliable records. The French royal doctor Borel in Paris recorded the history of the invention of the telescope [Borel 1655].

The telescope meant nothing less than a revolution to astronomy. It enabled Galileo's revolutionary observations of mountains on the moon, of spots on the sun, phases of Venus and four moons encircling Jupiter made in 1609 and in 1610 all contradicting the Ptolemaic system from antiquity. These observations by a telescope paved the way for acceptance of the Copernican system [Galileo 1957]. Besides observation of Jupiter and seeing its moons for the first time, Galileo also observed Saturn and he saw an elongated structure rather than small disks which he saw when looking at the other planets (figure 1).

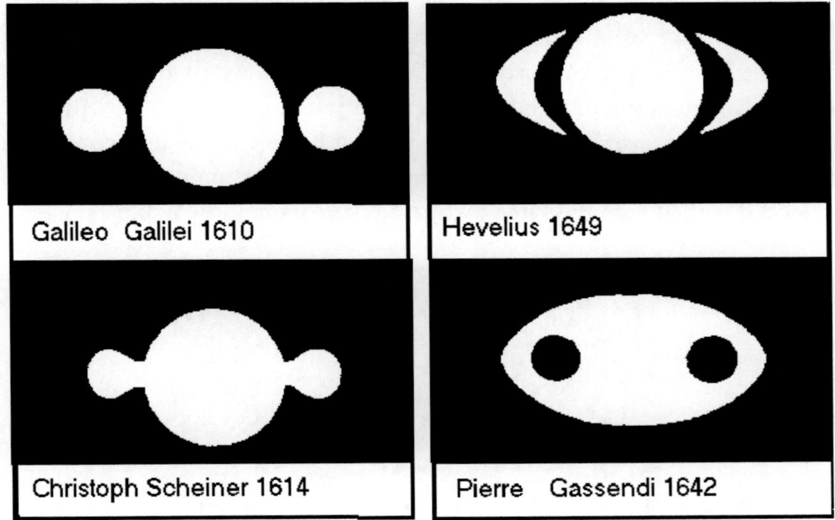

Figure 1. Several historic views of Saturn through a telescope from 1609-1655, an artist's design following the descriptions of the astronomers, *according to Ley [1963].*

He did not describe these observations in his famous Latin book „Sidereus Nuntius" which was released in 1610 [Galileo 1967], but he reported it in a letter to Johannes Kepler in the same year. In this letter he claimed to have made a further discovery beyond those described in the book, but he did not want to release it yet. Rather than release it Galileo wanted to keep the priority of the discovery by putting it in a riddle, an anagram, which was a method well used in the times of Galileo.

This anagram was:

smais mr milm epoeta lev mibun enug tta viras

Kepler tried to solve this unsuccessfully. He thought it could refer to moons of Mars, but this was not correct. A later letter of Galileo's to the Ambassador of the Duke of Tuscany to the Court in Prague, Guiliano de Medici, solved this extraordinary riddle. Galileo wrote that in the telescope he saw Saturn in changing forms. The sphere of Saturn looked as if two other little spheres like „two servants helping an old master" accompanied it. Galileo asked the ambassador to forward the solution of the riddle to the Emperor´s Mathematician, Kepler, in Latin which was the language of Science in those times: „Altissimus planetam tergeminum observavi“ meaning in English: „I saw the utmost (outermost) planet thrice“. First he thought the two little spheres to be moons rotating about Saturn. However, continuing these observations, he noticed that the small spheres stayed well in their places. He was even more astonished when both small spheres seemed to vanish simultaneously in 1612. About this, he noted, „What shall I say about such a strange metamorphosis? Did they disappear or escape? Did Saturn even dare to devour his very own children?“. Saturn devouring his own children indeed was a theme of the arts at the time of Galileo (figure 2).

Today we can explain this behaviour. Since Saturn's´ rings are very thin, much less than one kilometre in thickness, they can hardly be seen from earth for several months when the line of sight from earth falls onto the plane of the rings, which will happen twice during every sideric revolution of Saturn in 29 and a half years. Galileo had no explanation for the vanishing of the structure he saw, but still he continuously hoped to see it again. Indeed, after several years the small spheres accompanying Saturn showed up again but they were changing their form and seemed to be no spheres any more. Thus Galileo still did not know what he continued to observe regularly. He drew what he saw in the telescope and one sketch of 1618 was so precisely drawn that Abetti [Pannekoek 1961] conjectured that Galileo would have understood if he had seen the same picture eight years before (figure 1).

Figure 2. Saturn devouring his own children, woodcutting by Hans Sebald Beham, Nuremberg and Frankfurt, Germany, 1540 to 1550. Such a picture may have been the trigger for Galileo's remark about Saturn.

In the first half of the 17th century most astronomers observed Saturn by telescope after Galileo, next was Christopher Scheiner from the hamlet Wald in the episcopacy of Augsburg at Ingolstadt University in Bavaria, Germany. Since he invented the pantograph enabling mechanic magnifications of drawings in Dillingen in the episcopacy of Augsburg before, he was able to produce fine drawings and pictures of what he saw in his telescope [Braunmuehl 1891], for example exact reproductions of the surface of the moon and his observations of Saturn (figure 1).

Scheiner improved the telescope to attain a better magnification by using three lenses rather than two [Daxecker 1996]. He observed sunspots in January 1611 at the university from a church tower in Ingolstadt (Bavaria) behind a fog without knowing Galileo's findings [Braunmuehl 1891]. He wrote his findings in four letters to M. Welser, a rich merchant and mayor to the city republic of Augsburg who had conquered "little Venice" and took her gold at this time, this land was ceded to Spain later and known by the name Venezuela. Welser published these reports one year later using a pseudonym "Apelles" rather than Scheiner. Welser sent a copy to Galileo [Braunmühl 1891], Galileo replied that he saw sunspots in 1610 before and had reported this as private communications to several cardinals in Rome. Scheiner watched Saturn with his improved telescope and reported to have seen two companions (not moons) close to the planet, like three stars with the largest in the middle; he diligently made drawings of their changing positions [Daxecker 2006].

Gassendi too watched Saturn with his improved telescope. Pierre Gassendi was born in 1592 in Champtercier in Provence; he studied theology in Aix-en-Provence and Digne [Jones 1981], was consecrated as a priest (minister) in Digne and became a professor in Aix [Detel 1978, Egan 1964]. Gassendi had exchanged letters with Galileo since 1625 [Taussig 2004]. Travelling frequently to Paris he co-operated with the salon (scientific circle) of Mersenne. Since scientific academies were founded no earlier than 1657 in Florence, after the death of Galileo and Gassendi (there the pope banned them in 1667), then they were founded in 1662 in London. Therefore, this circle around Mersenne was the most important scientific audience at that time.

The discussions there caused Gassendi to perform several astronomical observations with his improved telescope, for example:

- He observed Mercury's eclipse of sun by telescope in 1630.
- He was the first to draw a map of the moon to scale; a crater south of the moon's equator is named „Gassendi" in memory of this [Detel 1978, Fisher 2005].
- He discovered the particular appearance of Saturn with rings, describing Saturn's view through the telescope as elongated and wrinkled with handles in the second „de motu"- letter in Article IX, folio 124, [Gassendi 1642] „...duos vetuli lululae... ...oblongam, quasque sejunge ab eo". This view he interpreted as caused by two moons of Saturn like the four moons of Jupiter discovered by Galileo just before. When he wrote his findings in 1640, the rings had maximum visibility and were invisible from earth in 1626 and again in 1656.

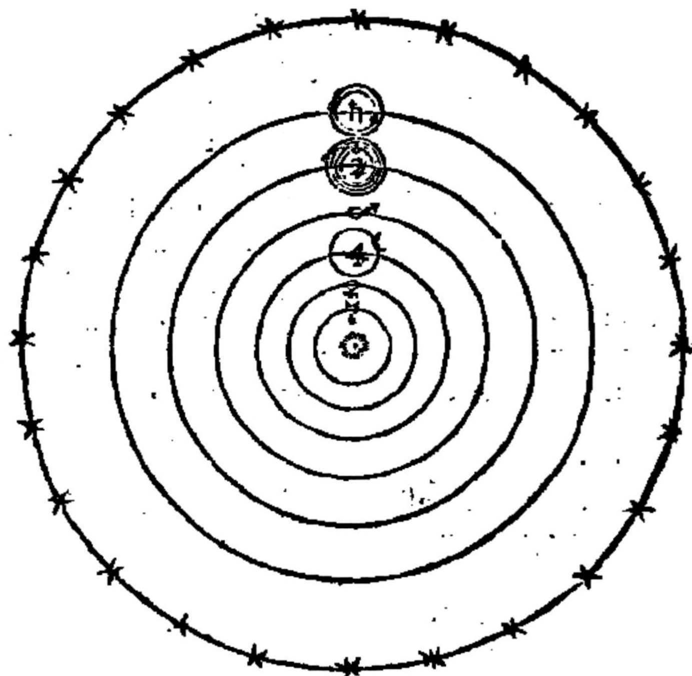

Figure 3. The first known drawing of the solar system with moons of Jupiter as well as Saturn in the second „de motu"- letter, by Gassendi, 1640, in reprint: Gassendi [1642].

This particular expression of Saturn having „ansae" coined by Gassendi was in use until 19th century in spite of the knowledge of the true cause of this appearance for the sake of lingual simplicity.

Helvenius saw Saturn as a single sphere, triple sphere and as sphere with ansae, bur he never doubted the reality of the „ansae"- structure [Prince 2007]. Helvenius studied in France in the circle of M. Mersenne and P. Gassendi in 1632. In 1641 and 1650 he built observatories at the Baltic Sea, which were the most elaborate at this time mainly for observations of the moon discovering her third liberation. Between 1650 and 1660 he published observations of Saturn. In 1663 at the age of 52 he married his second wife Catharina Elisabeth Koopmann aged only 16 who was very eager to help him with his astronomical observations (figure 3) since his eyes could not see as sharply any more. In a fire at the observatory in 1679, however, most of the papers and copper plates ready for print were destroyed [Kaempfert 1987]. What was left of the drawings after that incineration has been published in a dissertation kept at the Bavarian State Library [Wünsch 1986].

The first astronomer who speculated about the structure of Saturn's "ansae" was Peronne de Roberval from France who guessed it must be a vapour emanating from Saturn's equator being denser and visible in Saturn's summers and diluted faintly and invisible in winters.

Figure 4. Helvenius and his second wife Catharina Elisabeth observing with the big sextant. From: Machina coelestis I, 1673. Two persons usually operated such big instruments.

Sir Christopher Wren also had some ideas about these „ansae". He observed Saturn with his elaborate telescope in Oxford from 1654 onwards. He was appointed professor of Astronomy in Gresham College in London in1657 and later held the same position in Oxford. In his inaugural address he said „that a true Description of the Body of Saturn only, were enough for the Life of one Astronomer" [Helden 1968]. He thought Saturn had an elliptic corona and calculated rotation period, axis and inclination of this corona in such a way that it could explain all the observations of the „ansae" [Bennett 1982]. He made an early wax model of Saturn in 1654 or 1655 and in that year he observed with Ball „a certain zone, darker than the rest of the area of the disk and slightly narrower than Jupiter's belts" and he thought he could make out a series of spots. In 1656 he wrote that on the basis of many pictures of Saturn his „ansulae" and „his spots had attained to a theory of his Rotation and various inclination of his Body". As for the nature of the „Ansulae" he formulated a hypothesis in 1657 and worked it out in two pasteboard models (figure 5) in turn, with Saturn surrounded by an elliptical ring whose width varies from maximum at two points where it is furthest from the planet and zero at two other points where it touches the planet. The „fluid corona" of these „Ansae" is thus bound by two ellipses and so thin as to be invisible when it is viewed edge-on. A rotating or reciprocation motion around the corona could then explain Saturn's appearances about the major axis. Two ellipses in touch became an element in his later architecture such as St.-Mary-le- Bow in 1671. Wren's model was a considerable advance to previous attempts to explain Saturn's nature. He proposed but did not publish his results in a book nor did he built an even more elaborate model of Saturn with „ansae", since the true nature of Saturn, the ring structure, had been found meanwhile in 1659 by someone else [Alexander 1961]. Wren's drawings for the church St. Mary–le-Bow include ellipses touching each other which can be deduced to the astronomical experiences Wren had had as professor of astronomy in London and Oxford when he observed Saturn [Whinney 1971].

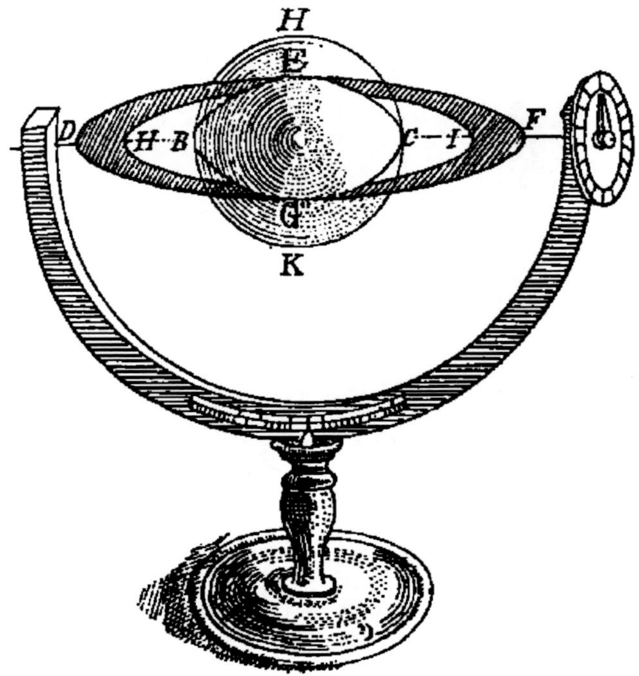

Figure 5. Wren's model of Saturn from his book „De corpore Saturni“, reproduced in *Huygens, Ouvres complètes, The Hague, 1888,* Volume 3, 424, [Bennett 1982].

The first astronomer who really understood what he saw in the telescope was Christiaan Huygens from the Netherlands who worked for Louis XIV in France and so he was granted the honour of the discovery of Saturn's rings. The precondition was that he first developed both an improved method to grind optical lenses as well as a much better ocular part of the telescope combined by two lenses rather than a single lens. With his improved telescope, he discovered moon Titan of Saturn in 1656. He also believed to have found the true nature of the „ansae“, but he was not quite sure and so he did not disclose his discovery except in an anagram just like Galileo did before him.

It was impossible for anyone else to solve this anagram:

aaaaaaa	ccccc	d	eeeee	g	h	
iiiiiii	llll	mm	nnnnnnnnn			
oooo	pp	q	rr	s	tttt	uuuuu

The solution of this anagram was published by Huygens in his book about Titan „Systemii Saturnis“ in 1660 and was the following sentence in Latin:

„Annulo cingitur, tenui plano, nusquam cohaerente, ad eclipticam inclinatu", translated in English, „surrounded by a ring, thin and flat, nowhere touching, inclined against the ecliptic".

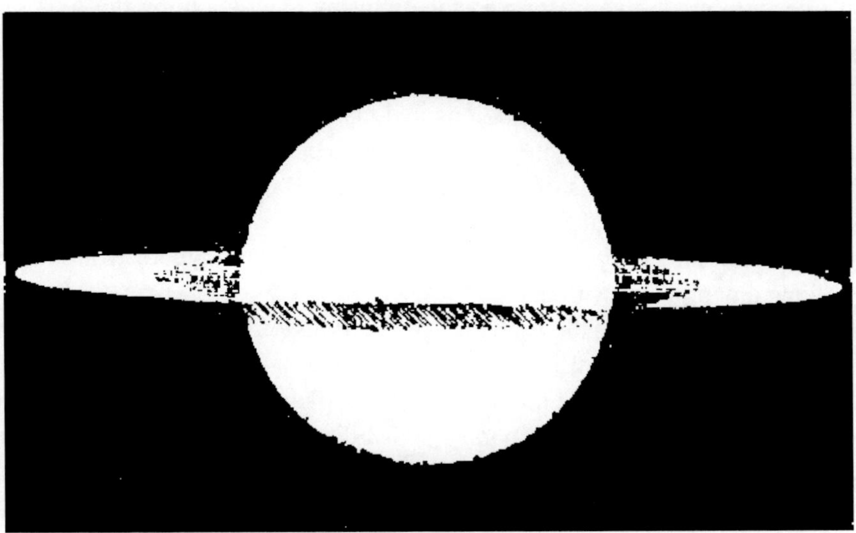

Figure 6. Saturn with ring as seen and drawn by Huygens in his book „Systemii Saturnis" in 1660.

Of course such a keen and new explanation evoked a lot of objections from astronomers of that time. Based on Aristotle's philosophy, people of that time thought only an ideal body, a sphere, can exist in space. Furthermore nobody ever saw anything like a ring in a telescope. But Huygens′ keen explanation could indeed provide a shrewd logic in accordance with all observations ever made about Saturn. Such a ring could very well rotate about its axis just like a sphere and is really visible unlike the epicycles claimed by the Ptolemaic system which was still applied by some astronomers of that time. Huygens could even predict that the ring would vanish again in July 1671, March 1685, and in December 1700 basing on his ideas [Huygens 1659].

Sir Christopher Wren was intrigued by the idea of a ring around Saturn and defended this model eloquently. However, many scientists objected to this idea, claiming a ring could never disappear. They could not imagine the ring to be as thin as Huygens claimed and as we measured nowadays to be only about 20 metres thick [Davis 1987]. The Italian astronomer Divini and Fabry, a French Jesuit, wrote a book objecting to all Huygens′ ideas triggering a lively discussion among astronomers of the 17th century: „Brevis Annotatio in Systema Saturnium", in English: A short note about the Systema Saturnium. They disclaimed the ring structure, they refused the improved telescope of Huygens claiming it was useless, and they refused even the idea of a planet earth moving in space. Divini claimed Huygens′ idea of a freely rotating ring would be a self-illusion and an erroneous conclusion. As a highly controversial alternative Fabri proposed a model of the Saturnine System they thought could explain the astronomical observations. Fabri claimed that the black areas seen within the „ansae" are black moons of Saturn, which are moons in addition to white moons causing the light parts of the „ansae". This model would require a total of four satellites, just like Jupiter, two white and two black ones. But it was in contradiction even to the knowledge of that time, because the moons had to rotate about Saturn and inevitably had to cover the visible disk of this planet regularly, which was not observed at all. Even since the discovery of moon Titan by Huygens was accepted by the scientific audience of the time, Fabri still challenged this achievement by claiming Titan would be continuously behind Saturn and thus could not be a true moon of Saturn's. He overly stressed this point by even claiming the four satellites of Jupiter were not true moons to that planet.

Huygens answered these challenges in his booklet „Brevis Assertio Systematis Saturnii sui“ in 1660, elaborating on the observations and claiming that four moons can not explain the „ansae“, and that Titan moves on a circular path seen from earth as an ellipse. Huygens further mentioned that the English astronomer William Ball in a letter written to Huygens confirmed his observations about Saturn in 1656.

As an answer Divini and Fabri increased the number of their hypothetical moons to six in 1661 not solving any problem with that.

As a result of this dispute, the existence of the rings of Saturn became more and more accepted, but the question of the true nature or structure of it remained to be discussed.

Huygens himself proposed a solid structure. To explain the solid structure of the ring, he argued that this could explain the temporary invisibility of the ring. He believed the opaque solid ring would temporarily turn its shadow side towards the line of sight to the earth and would thus become invisible. Today, we know that was wrong, but in those times this explanation seemed to be logical.

Meanwhile, some additional astronomical observations had been made. An Italian developer and maker of telescopes, Campani, observed that the outer part of the ring was indeed darker than the inner part. Several astronomers saw that the ring cast a shadow on the surface of Saturn, which has been suggested by Huygens before.

The Italian astronomer Cassini discovered two additional moons of Saturn, Japetus and Rhea, in 1671 and 1672. In his report in a letter printed in 1675 [Cassini 1676] he reported that „the breadth of the ring is separated by a dark, seemingly elliptic, but really circular line in two parts like in two concentric rings, the inner of which is lighter than the outer“ (figure 7). This was the discovery of Cassini's separation with the name honouring the discoverer and the coining of the expression „Saturn's rings“ in plural. Cassini did not say if one of the rings is broader than the other, but his drawing showed indeed the inner ring broader (figure 7).

Cassini diligently continued to observe Saturn with his elaborate telescope and discovered two additional moons of Saturn, Tethys and Dione, in 1684. Their paths of revolution around Saturn lay well within the path of the other moons discovered so far and are very close to the outer rim of the rings (table 1).

Today we know from the encounters of the space probes Voyager 1 and Voyager 2 with Saturn on November 13, 1980 and August 27, 1981, that moon Tethys is within the very faint E- ring which has been found by overexposing the images from the Voyager probes [Elliot 1984, Davis 1987].

Cassini very successfully argued about the true nature of the rings in 1705 saying that they consist of clusters of satellites too small to be observed directly: „un essaim des petites satellites“ (a swarm of small satellites). This theory was supported by his son in 1715 and has been confirmed by the Voyager missions.

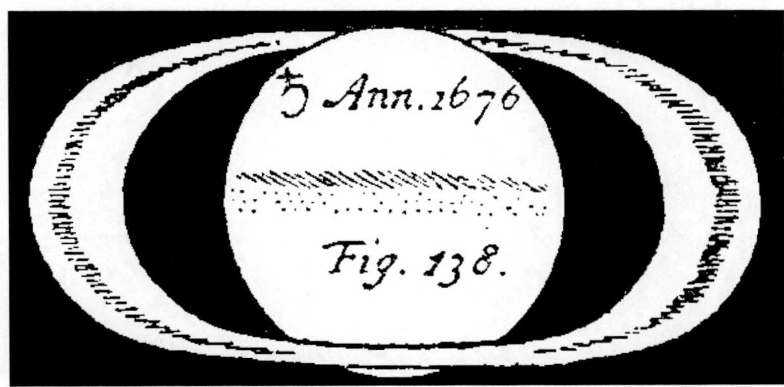

Figure 7. Cassini´s drawing of Saturn with ring, first picture of the „Cassini Separation“ , [Cassini 1676].

Before Herschel from Hanover in Germany discovered planet Uranus in 1781 and later his moons Titania and Oberon in 1787 with his elaborate mirror telescope with 1,25 m aperture and 13 m focus near Slough, England (figure 8), he also observed Saturn with his rings starting before 1778 in Bath in England where he held a position as an organist and operated a 2 metre focus mirror telescope constructed all by himself. Since he saw only the northern plane of the ring he hesitated to draw any conclusions at that time. He believed the ring to be a solid structure and observing Cassini´s separation by his telescope he conjectured it should be a dark surface structure on a solid ring. In May 1789 looking towards the edges of the rings and noticing they became invisible from earth he concluded that the rings have to be very thin. Comparing the diameter of the inner moons to the view of the ring before vanishing he could even estimate the maximum thickness to be in the order of a few hundred kilometres.

Furthermore he saw bright spots on the surface of the rings, he believed them to be mountains. But later, starting to watch the tilting from north to south face of the rings, he noticed that one of the bright spots seemed to separate away from the rings and became visible as a separate body while the rings were barely visible almost on their edges. This brought up the question if all bright spots on the ring were all small moons of Saturn, which eventually could be seen in front of one of the rings and thus were merged to the rings optically. Therefore, he drew the conclusion in 1789 that two of the bright spots observed in connotation with the rings must be moons of Saturn rotating very close to the outer boundary of the rings: Mimas and Enceladus. He reported this in the 80th volume of „Philosophical Transactions“ together with the observation of bright and dark belts on Saturn and a big blurred spot on Saturn, which he saw as an indication for an extended gaseous atmosphere on that planet. It is certain that his sister Lucretia Caroline assisted Herschel in his observations especially during his obligations in his main profession as an oboist. Since his eyes at the age of 51 were inferior to his sisters´, who was twelve years younger, it could very well be that one of the moons was detected by his younger sister first. That would be the first major discovery in astronomy made by a woman, besides the faint nebula and eight weak comets Lucretia is known to have discovered for sure (figure 8).

Figure 8. William Herschel, his sister Lucretia, and his son the later Sir John Herschel, all three of them famous astronomers, at their huge telescope with 1,25 m aperture and 13 m focus near Slough, England.

Today we know from the encounters of the two space probes Voyager 1 and Voyager 2 with Saturn on November 13, 1980 and August 27, 1981 respectively, that moon Enceladus (like Tethys) is within the very faint E-ring and Mimas is shepherding the outer rim of the very small twisted and braided ring G. Both E- and G-ring have been found by overexposing the images from the Voyager probes [Elliot 1984, Davis 1987]. An overview of the ring and inner moon system of Saturn as Voyager has detected it is sketched in figure 9 adapted from Morrison [1983]. By telescope only rings A to C can be seen, the other extremely faint rings were detected indirectly by the measurements from the Voyager spacecraft and might constitute merely debris from the main rings and or from the inner moons circling temporarily on a more or less stable orbit [Porco 2007].

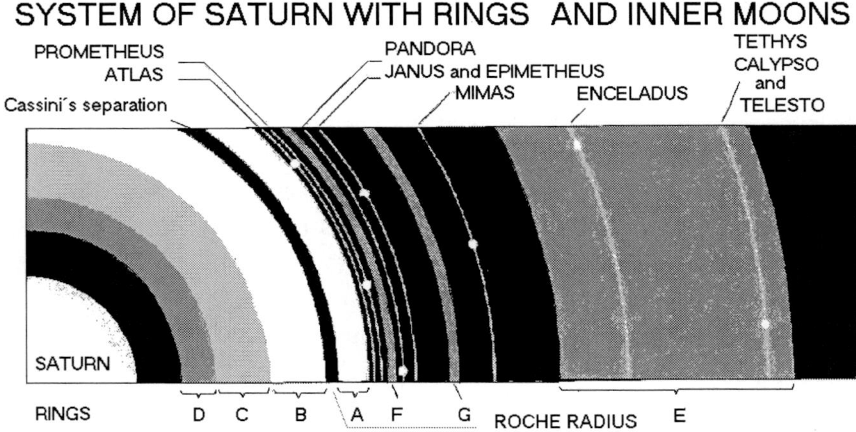

Figure 9. Overview of the system of rings and inner moons of Saturn as detected by Voyager, adapted from Morrison [1983].

Before looking on the edges of the ring and detecting the additional moons, Herschel proposed a final decision if the ring is separated by Cassini´s division into two rings or if they are dark areas on the ring: Observation of light of a star in transition by the ring structure would prove the reality of Cassini´s division if the light of this star could be seen penetrating in between the two ring parts in the division. Unfortunately, there was no transition of a star by Saturn in the lifetime of Herschel but afterwards such light from a star served as a final proof of Cassini´s division. He reported to the Philosophical Society in London in 1794 the measurement of the rotation period of the rings by observation of the movement of irregularities resulting in a period of ten and a half hours [Ley 1963, chapter 15].

In 1785 the French mathematician Pierre Simon de Laplace published a treatise about stability and structure of the solar system using Newton's law of gravity „Théorie des Attractions des Sphéroïdes et de la Figure des Planètes“ and concluded that one Saturnine ring without Cassini´s separation can not be a stable structure. Laplace proved in his treatise that the ring has to be in a swift rotation to be stable. If the ring were a solid structure, it would be torn apart by tidal forces. Laplace concluded the ring should be made up of separated areas rotating each with a speed of a satellite at the same distance. In a solid ring this would result in mechanical tensions and forces, which together with gravitational forces of Saturn would tear every material apart. It would have been logic to conclude from these arguments that the rings consist of a herd of small solid particles confirming Cassini´s conjecture, but rather than this Laplace thought the ring consists of two ring systems gaped by Cassini´s separation each comprising of a great number of very narrow rings. These calculations of Laplace and the additional observation of the southern face of the rings led Herschel to the conclusion in 1791 that the „black belt“ he had observed on the ring was indeed a separation between two rings: "I believe I can say that the planet Saturn has two concentric rings with different dimensions and widths laying in a plane which I guess is hardly tilted against the equator of the planet“.

Laplace´s treatise showed clearly that there could be many more than two ring sections and that the ring could contain more than one separation in order to be as stable as possible. This was the beginning of an extended search for more gaps in the ring as well as several theoretical analyses and model calculations about the ring.

In 1830 Captain Kater released a careful collection of all results of the hunt for additional ring gaps so far in volume IV of the „Memoirs of the Royal Astronomical Society“ including his own observations of December 1825: „I believe to have seen ... the outer ring to be separated by several dark lines... lying very closely together, one of these stronger than the others, dividing this ring in almost equal partitions. ... According to a careful observation lasting several hours I do not doubt that the ring considered to be the outmost ring of Saturn consists of several parts. The inner ring does not show this kind of structure...“. In January 1826 Kater, the vice-president of the Royal Astronomical Society, made similar observations but in January 1828 he was not able to see these partitions any more. Kater found other observers of additional partitions of the ring: Quetelet in Paris using a 26 cm aperture telescope seeing one additional partition of the outer ring and Lalande reporting that Mr. Smith saw three or four additional partitions of the outer ring. Following that written account, many astronomers tried to see additional separations, however, only a few of these could see any contradicting in both number and position of these partitions. Only one such observation of an additional partition seemed to persist. The partition J. F. Encke saw in Berlin using the giant telescope of the Urania observatory, which had been used later to discover Neptune, on April 25, 1837 and May 28, 1837 again dividing the inner third from the rest of the outmost ring. His observations were confirmed four times by H. Schwabe in Dessau in Germany in 1841. Both W. Lassell and Reverend W. R. Dawes saw in 1843 and 1850 a clear partition within the outer ring dividing a third of it from the remainder, but measuring from the outside rather than from inside as Encke did. Both observers pointed out that they were able to see this partition "in both ansae" using the expression coined by P. Gassendi in 1640. In spite of being seen at different positions this division got the name „Enckes ´s Partition" by German astronomers. At the turn of 1853 to 1854 Dawes saw the partition again and Padre A. Secci who described this partition „like a faint pencil's line" confirmed his discovery. These extended searches for the additional partitions were somehow inefficient because these observations were not clearly reproducible, but they resulted in the simultaneous discovery of a new ring, the C-ring inside the B-ring, called „Floss-ring" at these times of discovery, by W. C. and G. P. Bond (father and son) in the U. S. A., Rev. Dawes and Lassell in Britain, and C. W. Tuttle, almost simultaneously. To make the story even more complicated, it was claimed that W. Struve in Russia saw the C-ring in 1826 before, but he did not report it himself. The vice-president of the Royal Society Kater wrote that Struve saw the inner boundary of the inner ring of Saturn without any sharp limit. In 1838, Galle obviously could see the „Fleece-ring" in Berlin describing his observations „The inner rim of the first ring seem to be swallowed gradually by the dark space between ring and sphere of Saturn" ... „it seems that the inner ring stretches out almost half of the space down to the surface of the planet" [Galle 1838], but he was not aware that he saw an entirely new ring structure. The clear distinction of this C-ring from both background and inner rim of the B-ring was reported as late as 1850 by W. C. Bond (father) in „Astronomical Journal" II as well as by Dawes in „Monthly Notes of the Royal Astronomical Society" XI and a confirmation by Lassell in the same year. This discovery, however, was nothing more than an explanation of what others had seen before. Even that explanation had been attributed by C. W. Bond to someone else, the astronomer C. W. Tuttle: „Mr. Tuttle first had the idea that the half-dark light at the inner rim of the bright ring as well as the dark line running across the disk (of Saturn) ... could both be referred to the existence of an inner dark ring which is now discovered to be part of the ring system. ... This explanation needs just to be released to be accepted as a true and fully satisfactory solution for all extraordinary observations...". Lassell coined the name „Floss-ring" in a letter to the Royal Astronomical Society. The expression used today, the C-ring, was proposed by Otto Struve, son to W. Struve, counting with letters from outside to inside. Therefore, the C- ring has at least five discoverers.

After the C-ring was discovered theoretical analyses and model calculations about the ring were made. The French mathematician Roche proved in 1850, that a satellite to a star or planet within a radius called „Roche Radius" today would be torn apart by tidal forces and can not be stable. In the case of Saturn, this radius turned out to be exactly the radius of Cassini´s separation. In the year of the final discovery of the C-ring G. P. Bond (the son) published a treatise about the ring structure in 1850 claiming that observing additional gaps at different distances some times and some times not can be explained by a ring changing its structure spontaneously. This means the ring cannot be solid or composed of a finite number of small rings. Rather than that, the ring should be a liquid with the zones of the liquid moving against each other. Spontaneous instabilities of this structure would show up as temporary gaps in the rings, which will disappear after some time has elapsed. Up to then, structures in the sky always were thought to be eternal and not subjected to any irregular change. Bond's idea of a dynamic ring system in a constant change of shape was completely new and was a conceptual change just like the change from Ptolemaic to Copernican system before. Even if the idea of a liquid was the wrong explanation, the ever-changing small ring separations have been confirmed by the Voyager and Cassini missions [Cuzzi 1987, Porco 2007].

This liquid ring hypothesis was supported in1855 by a mathematical analysis of B. Peirce from America. He made calculations about the separate solid ringlets idea of Laplace. His result was that the number of solid ringlets proposed by Laplace was too small or the breadth of the ringlets too big to be stable. Therefore smaller and more ringlets were necessary to be stable mathematically. The increasing number of ringlets to attain stability would increase interaction between these above all limits resulting in a breakdown of the entire system. Therefore he supposed the whole ring system to be liquid rather than solid. In the year of Peirce´s publication, Cambridge University promised a reward for an essay solving the problem of stability and nature of the ring, the „Adam Price Essay". Two years later the price was rewarded to James C. Maxwell for his publication „Essay on the stability of Saturn's rings". Maxwell proved firstly that small solid ringlets as proposed by Laplace were very well possible within the limits of nature's laws, but they had to be of varying thickness, which contradicted observations. On the other hand liquid rings would be stable only in a system perfectly symmetric to rotation. With perturbation from Saturn's moons the liquid rings lacking any stiffness would coagulate to a limited number of satellites under the forces of gravitation from moons as well as by self-gravitation. Therefore the only ring system permanently stable would be a system of numerous small solid particles, which have different speeds depending on the distance from Saturn. This way Maxwell's mathematical analysis finally confirmed Cassini´s early hypothesis. Honouring this bright analysis, the separations: to a faint C-ring inside the B-ring, with the C-ring barely observable to telescopes, was called Maxwell's separation later.

While Maxwell's separation is rather an idea than a fact, the reason for the most prominent Cassini separation remained to be explained. In 1857 the American astronomer C. Kirkwood explained the gap in the belt of asteroids by the gravitational influence of Jupiter. Then it was straightforward for Kirkwood to explain Cassini`s separation in the ring by the gravitational influence of Saturn's moons one year later. A hypothetical particle rotating about Saturn just in the distance of Cassini´s separation would have a rotational period of exactly half of that of moon Mimas, a third of moon Enceladus and a quarter of that of Tethys. The periodic gravitational drags of these satellites would disturb that hypothetical particle periodically and would toss it out of its path soon. Kirkwood was even able to solve the mystery of Encke´s separation which was seen once separating the inner third of the A-ring and in other times separating the outer third of the A-ring from the rest. The period of rotation would be three fifths of Enceladus´ period in the first position and would be two fifths of Enceladus´ period in the latter. Therefore, instability due to moon Enceladus can sweep out particles out of either of these positions supporting both groups of former observations and proving Bond's hypothesis of an instability in the ring changing its structure periodically and may be even chaotically as a realistic possibility. Due to Kirkwood's calculations, the separation to a faint C-ring inside the B-ring, which was called Maxwell's separation later, can be explained by a rotational period of a third of that of moon Mimas sweeping out that area. With the C-ring barely observable to telescopes, that explanation has been doubted in those times, however.

In 1907 and 1908 the tilt of the ring was large enabling thorough observations by many an astronomer. The French astronomer E. Barnard and Georges Fournier both observed a weakly glowing area outside the A-ring „une zone lumineuse très pâle" an area with swiftly moving illuminated tiny points. In Switzerland E. Schaer observed similar faint lights outside the A-ring independent of Barnard and Fournier, which he called the „outer Floss ring". Later

Lasalle and English astronomers confirmed these observations. However, many other astronomers could not confirm to have seen these faint lights. Nevertheless, these findings can be weighed as the discovery of the faint G-ring detected by the Voyager missions [Cuzzi 1987].

The transition of a star through the Saturn system, which Herschel hoped to see in the 17th century, finally took place in 1917. It was observed incidentally by the British astronomers John Knight and Commander M. C. Ainslie independent of each other. On February 9th in 1917 a star 7th grade in sign Gemini (Twins) passed behind the Saturn's A-ring and Cassini´s partition. Observing this star behind the Saturn rings both astronomers found that the star remained visible all the time behind the rings with brightness diminished to about a quarter and it was at full brightness in the Cassini partition. At two instances within the A-ring, the brightness increased to about half for a few seconds, which was interpreted by Ainslie as indications for Encke´s partitions. This observation of the ring covering a star did finally confirm the existence of Cassini´s division as a true gap in the ring and not a dark belt. Furthermore, it allowed setting a boundary for both maximum ring thickness and maximum particle size. Due to observations of the ring being invisible in telescopes when seen edge-on, Huygens determined a maximum thickness of the ring to be 450 km. With improved telescopes, G. B. Bond in 1828 and Barnard set this limit to 80 km in the 19th century. Due to the observations of the ring covering a star Crommelin set a new limit to the maximum size of particles in the ring. The speed of Saturn's A-ring covering the star was 26 km/s and a particle size of 1 km or more would have caused a considerable flickering of the light of the star which is essentially a point source of light. This was observed by neither of the astronomers who watched the coverage of the star and thus Dr. Crommelin conjectured that the particles in the ring should be much less than 1 kilometre in diameter. On March 14th 1920 a coverage of a star 7th grade by the planet Saturn and his rings was observed by W. Reid in Rondebosch near Cape Town with the star visible in entire brightness through Cassini´s division and dimly even through the dense B-ring.

Observations with improved telescopes yielding spectroscopic analyses of light in the 20th century allowed additional discoveries of the nature of the ring like spectra with hints on water ice.

The structure of the rings became much clearer due to the observations from the Voyager space probes in 1980 and 1981, however, these measurements put a couple of new riddles and challenges for contemporary astronomers [Elliot 1984, Davis 1987]. The Cassini-Huygens mission on a close flyby observed the structure of the rings in 2004, which measured structure and rotation time of the partitions of the ring. Due to evaluations of these measurements, the A-, B-, and C- rings seem to be formed by different collision disintegrating from different progenitor bodies and therefore were of different ages [Porco et al 2007]. The moons at the rings seem to have grown to their present size by the accumulation of porous ring material throughout ages [Porco et al 2007]. Thus the investigations about Saturn's rings became a clue to the old miracle of how planetary systems form.

PARTICULARITIES ABOUT DISCOVERIES OF SATURN'S MOONS

Christian Huygens discovered the satellite Titan and his time of revolution in 1655, published in his book "De Saturni Luna observatio nova". G. D. Cassini discovered another satellite in 1671 and still another in 1672 in Paris published in the "Philosophical Transactions" XII, 831. In his publication he pointed out that he could see the first moon (now called Japetus) only on the West side of Saturn and not on the East side, explaining that strange observation with the moon having a dark and a bright side. The Voyager and Cassini Missions confirmed this shrewd explanation in the 20th and 21st centuries [Davis 1987, Porco et al. 2007]. In 1684 Cassini discovered two additional satellites. At that time satellites were not named but counted by numbers. Since William Herschel discovered two new innermost moons to Saturn in 1789 that caused some confusion because it is not obvious if number one is the first or the innermost.

In 1848 two astronomers independent of each other discovered the next moon, W. C. Bond in Harvard in America on October 16th and W. Lassell in England on October 18th making Bond the lucky official discoverer.

To stop confusion Sir John Herschel, son of William Herschel, proposed mythological names for the satellites in 1858, which are in use today: Mimas, Enceladus, and so on, see table 1.

Table 1. Names of moons of Saturn as given by J. Herschel

Name of moon	Year of discovery	Discoverer	Time of revolution , days	Mean distance from planet, 1000 km	Diameter, modern measurement, kilometres
Mimas	1789	W. Herschel	0,9	185	394
Enceladus	1789	W. Herschel	1,4	238	502
Tethys	1684	Cassini	1,9	295	1048
Dione	1684	Cassini	2,7	377	1120
Rhea	1672	Cassini	4,5	526	1530
Titan	1655	Huygens	15,9	1214	5150
Hyperion	1848	Bond	21,3	1480	270
Japetus	1671	Cassini	79,3	3556	1435
Photographic discovery: Phoebe	1898	Pickering	retrograde 550,5	12855	220

In 1898 the American astronomer Edward Charles Pickering from Harvard, Boston, announced the discovery of an additional satellite of Saturn by a new method different from methods used so far in astronomy. For this new moon being very much further apart from Saturn than the other moons, 13 million kilometres, he proposed the name „Phoebe" which is a sister of Saturn in mythology. This published discovery was extraordinary in two ways. First, the discovery was made by comparing photographic plates rather than by observation with the eye. Three plates, exposed in August 1898, showed the moon wandering slowly from one position to another constituting an inevitable distinction from a fixed star. Secondly, these observations were made from the southern hemisphere, from the Arequipa observatory in Peru, which was operated by Harvard University. As the manager of the Harvard Observatory, Pickering had started the first systematic sky surveillance by photography and this discovery of Phoebe was the first major fruit of this new photographic method. Also, Phoebe was the first major astronomical object discovered in modern times from the southern hemisphere. The discovery of this nineth moon of Saturn seemed straightforward, but it resulted in a fierce philosophical discussion just like about the rings before. In 1904, Pickering made additional photographic observations of Phoebe, and Barnard was able to observe Phoebe directly by the giant mirror telescope of Yerkes observatory getting an additional astronomic position of Phoebe. The long time elapsed between these observations was caused by the fact that Saturn was in the vicinity of bright stars around the turn of centuries, thus observation of a faint object like Phoebe was not possible against this background. The British mathematician Crommelin calculated the revolution of Phoebe to have a period of 550 ½ days and to be retrograde, in opposition to all planets and all other moons of the solar system [Crommelin 1905]. He suggested that Phoebe might be an asteroid from the asteroid belt captured by Saturn, as it was proposed later for the outer small moons of Jupiter. Furthermore, the plane of revolution was found to be inclined considerably against the ecliptic plane. Since Phoebe was the first body in the close solar system found to be retrograde, many an astronomer doubted these calculations and observations and additional observations were necessary. Therefore the British astronomer Melotte made additional photographic exposures of Phoebe in 1907 and 1908 confirming the calculations of Crommelin. An additional particularity of Phoebe showed up. Cassini guessed in 1705 [Ley 1963, ch.15] that all satellites in the solar system are bound to their planets meaning the same hemisphere of the moon keeps to face towards the mother planet like the earth's moon shows the same face to earth continuously. But in 1935 the astronomer Antoniadi published a paper claiming Phoebe is the only satellite in the solar system which spins about its axis independently from the rotation around the planet [Antoniadi 1935]. Antoniadi´s claims were proven by observations of Phoebe by the Cassini-Huygens mission on a close flyby in June 11th, 2004, which measured the rotation time to be nine and quarter hours.

In 1944 the American astronomer Kuiper was first to measure an atmosphere on moon Titan by spectroscopy and found this atmosphere to contain the gas methane [Kuiper 1944, Alexander 1962]. The Voyager and Cassini space missions confirmed this in modern times.

CONCLUSION

Throughout all times concept changes in Science have influenced the interpretation of observations about Saturn and these observations have induced or supported concept changes as well. As in other great theories in physics and astronomy, such as the Copernican model [Risch 2007], the discoveries and ideas about Saturn's rings and moons have been influenced by philosophical ideas as well as technological background.

One of the most beautiful objects in sky has also stories to tell about its discoveries triggering beautiful Science theories.

REFERENCES

Alexander, A., F., O`Donnel; *The Planet Saturn,* Faber & Faber, London, 1961, Macmillan, New York, NY, 1962, chap 4

Antoniadi, E., M.; *Journal of the British Astronomical Association*, vol. 45, 1935.

Bennett, J., A.; *The Mathematical Science of Christopher Wren*, Cambridge University Press, Cambridge, and New York, 1982, 26-44

Betzold, C.; *Himmelsschau und Astrallehre bei den Babyloniern, Sitzungsberichte Heidelberger Akademie der Wissenschaften, Philosophisch- Historische Klasse,* 1-2, Carl Wintwer′s Universitäts -Druckerei, 1910, 1911, 14-17

Boll, F.; *Kleine Schriften zur Sternkunde des Altertums, ed. Stegemann, V.*, Koehler & Amelang, Leipzig, 1950, 374-376

Borel, Pierre; (Castres in Languedoc 1620- Paris 1689), *De Vero telescopii inventore*, (in Latin), The Hague, in tom. 4

Braunmuehl, A. von; *Christoph von Scheiner als Mathematiker, Physiker und Astronom*, Buchnersche Verlagsbuchhandlung; Bamberg, Bavaria, Bayerische Bibliothek, Vol. 24, 1891, 56-68

Cassini, Giovanni Domenico, in: *Philosophical Transactions,* Volume XI, 1676, p. 689f

Crommelin, A., C., D.; in: *Journal of the British Astronomical Association,* 15, 1905

Cuzzi, J., N., Esposito, L., W.; The Rings of Uranus, (in comparison to Saturn's), Scientific American, July 1987, 42-48

Davis, J.; *Flyby: The Interplanetary Odyssey of Voyager 2,* Atheneum, 1987.

Daxecker, F.; *Das Hauptwerk des Astronomen P. Christoph Scheiner SJ "Rosa Ursina sive Sol",* Universitätsverlag Wagner. Innsbruck, Austria, 1996, 26-68

Daxecker, F.; English translation: *The Physicist and Astronomer C. Scheiner*, VUI 246, Innsbruck University, Austria, 2006, 49-101

Detel, W.; *Scientia Rerum Natura Occultarum, Methodologische Studien zur Physik Pierre Gassendis*, De Gruyter, Berlin, (1978) 123-195

Egan, H. T.; *Gassendi′s view of knowledge*, Lanha, New York (1964)

Elliot, J., Kerr, R.; *Rings: Discoveries s from Galileo to Voyager,* The MIT Press, 1984

Fisher, Saul (2005) *Pierre Gassendi′s Philosophy and Science - Atomism for Empiricists*, Brill, Leiden, Boston, 225-276

Freeth, T., Bitsakis, Y., Moussas, X., Seiradakis, J. H., Tselikas, A., Mangou, H. Zafeiropoulou, M., Hadland, R., Bate, D., Ramsay, A., Allen, M., Crawley, A., Hockley, P., Malzbender, T., Gelb. D., Ambrisco, W., Edmunds, M. G.; Decoding the ancient Greek astronomical calculator known as the Antikythera Mechanism, *Nature 444 (30),* 2006, 587-591

Galilei, Galileo, *Discoveries and Opinions of Galilei,* translation St. Drake, Anchor Books, NY, 1957

Galilei, Galileo, *Sidereus Nuncius*- (news of new stars) - *Dialogue Concerning the Two Chief World Systems*, 1638, translator St. Drake, Univ. of Cal. Press, Berkeley, CA, 1967

Galle, J., G., in: *Verhandlungen der Berliner Akademie der Wissenschaften,* 1838

Galter, H., D., ed. in: *Die Rolle der Astronomie in den Kulturen Mesopotamiens, Proc. 3. Morgenländisches Symposium Graz*, r m- Druck- und Verlagsgesellschaft Graz 1993, 61-94

Gassendi, P. (1642); de Motu Impresso a Motore translato…, Paris, 1640 / 1642, letter, Bibliotheka Windhaginna, copy of the Bavarian State Library, Munich

Günther, S.; *Lippersheim, Hans, in: Allgemeine Deutsche Biographie, 18.,* Leipzig 1883, 734-735

van Helden, A.; Christopher Wren's "De Corpore Saturni", Notes and Records of the Royal Society, London, 23, 1968, 213-229

van Helden, A.; The Invention of the Telescope, Transactions of the American Philosophical Society, 67, no.4, 1977.

Jones, H.; *Pierre Gassendi´s „Institutio Logica" (1658),* critical edition and transl., Van Gorcum, Assen, The Netherlands, 1981, 160-161

Kaempfert, H. J.; *Johannes Helvenius, Arbeitshilfe 49, Bund der Vertriebenen,* Bonn, Catalogue for exhibition Helvenius, Schloss Drostenhof, Münster, Germany, 1987, 4-13

Kugler, F. X..; *Die Babylonische Mondrechnung*, Herder, Freiburg, Germany, 1900, 192-211

Kugler, F., X ., SJ; *Sternenkunde und Sternendienst in babylonischen, assyriologischen, astronomischen und astralmythologischen Untersuchungen, Band I: Entwicklung der babylonischen Planetenkunde von ihren Anfängen bis auf Christus,* Münster, Germany, 1907, 41-53

Kuiper, G., P. ; Astrophysical Journal 100, 1944

Ley, W.; *Watchers of the Sky,* Viking Press, New York, N. Y., 1963

Morrison, D.; *Exploring Planetary Worlds*, Scientific American Library, New York, 1983

Neugebauer, O.; *Astronomical Cuneiform Texts, 3 vols*., Lund & Humphries, London 1955, nos. 801-802 and 811, reprint: Springer, Berlin, Heidelberg, New York, 1983

Pannekoek, A., Holland, *History of Astronomy*, Interscience Publishers, New York, NY, = Allen & Unwin, London, 1961 pp 108-109, 230, 254-256

Porco, C. C., Thomas, P. C., Weiss, J. W., Richardson, D. C., Thomas P. C.; Saturn's small inner Satellites, clues to their origin, *Science 318,* 2007, p. 1602.

Prince, C. Leeson, *The Illustrated Account given by Hevelius in his Machina Celestis, of the Method of Mounting His Telescopes and Erecting an Observatory,* American Paperback Reprint Series, 2007

Risch, M.; Das erste Großexperiment der Physik auf einer Galeere: Pierre Gassendi und die kopernikanische Zeitenwende, (The first big Science experiment in physics on a galley; Pierre Gassendi and the change to The Copernican system), *Physik In Unserer Zeit 38,* (5) 2007, 249-253 , DOI 10.1002/phiuz.200601148

Sambursky, S.; *Physical Thought from the Presocratics to the Quantum Physicists. An anthology*, Hutchinson & Co, 1974, London

Taussig, Sylvie; *Pierre Gassendi, Lettres latines*, 1211/ XVI; Aix en Provence, Turnhout editions, Paris, 2004, 231 (1) and 172-173 (2)

Whinney, M. ; *Wren,* Thames and Hudson, London, 1971, 41-71

Wolf, R., *Handbuch der Astronomie, ihrer Geschichte und Literatur, I,* Zürich, Switzerland , 1890

Wolf, R.; *Handbuch der Mathematik, Physik, Geodäsie und Astronomie, I,* Verl. Fr. Schulthess, Zürich, Switzerland, 1870, 403-404

Wünsch, J.; Die Auswertung der Sonnen– und Mondbeobachtungen des Danziger Astronomen Johannes Hevelius, Dissertationsdruck Frank, Munich, Germany, 1986, 147-161

In: Space Exploration Research
Editors: J. H. Denis, P. D. Aldridge

ISBN: 978-1-60692-264-4

Chapter 11

ORIGIN OF THE SATURN RINGS: ELECTROMAGNETIC MODEL OF THE SOMBRERO RINGS FORMATION

Vladimir V. Tchernyi [*] ***(Cherny)***
Modern Science Institute, Moscow, Russia

ABSTRACT

For the first time the role of superconductivity of the space objects within the Solar system located behind a belt of asteroids is considered. Observation of experimental data for the Saturn's rings shows that the rings particles may have superconductivity. Theoretical electromagnetic modeling demonstrates that superconductivity can be the physical reason of the origin of the rings of Saturn from the frozen particles of the protoplanetary cloud. The rings appear during some time after magnetic field of planet appears. It happened as a result of interaction of the superconducting iced particles of the protoplanetary cloud with the nonuniform magnetic field of Saturn. Finally, all the Kepler's orbits of the superconducting particles are localizing as a sombrero disk of rings in the magnetic equator plane, where the energy of particles in the magnetic field of Saturn has a minimum value. Within the sombrero disc all iced particles redistributing by the rings (strips) like it is happened for the iron particles nearby the magnet. Electromagnetism and superconductivity allow us to understand why planetary rings in the solar system appear only for the planet with the magnetic field after the belt of asteroids where the temperature is low enough and why there are no rings for the Earth, and many other phenomena.

1. INTRODUCTION

There is no yet clear picture of the origin of Saturn's rings. There are two versions of the rings origin. The most common version saying that the rings originated when an asteroid type body approached Saturn and was destroyed by the gravity and centrifugal forces and then

[*] Modern Science Institute, SAIBR. 20-2-702, Osennii blvd. Moscow 121614, Russia; Tel.: +7(926)592-6066; E-mail: chernyv@bk.ru

from the debris the rings were created. It looks like this version contains a mysterious fact. Another idea relates to the origin of rings from the particles of the protoplanetary cloud around Saturn, and this problem has not been resolved yet. This paper compensates the lack of this knowledge, demonstrating how rings could originate and form from the frozen particles of the protoplanetary cloud after the appearance of the magnetic field of Saturn due to electromagnetic interaction of icy particles with the planetary magnetic field.

The founder of the theory of electromagnetic waves J.K. Maxwell in his award winning paper on the subject "On the stability of the motion of Saturn's rings" (1859), deduces that the rings of Saturn cannot be solid and the rings could be stable only if they consist of "an indefinite number of unconnected particles orbiting Saturn in much the same way as our Moon orbits the Earth" [1]. Otherwise gravitational forces would destroy them. Ground-based experiments and the data of the Pioneer, Voyager 1, 2 and Cassini-Huygens space missions have revealed the rings to be composed of icy particles, and icy particles with impurities.

After G. Galileo (1610) many researchers have studied the nature of the rings [for example, 1-16]. From the consideration of the gravity and celestial movements it follows that different ring systems are morphologically quite distinct and are all shaped by a few common processes. This is the outward transport of angular momentum by rings particles and by gravitational interactions between satellites, moons and ring material. Orbital resonances between satellites, moons and ring particles play an important role in enhancing the influence of satellites and forming specific structure of the rings and gaps. At the same time, extensive experimental data confirmed the importance of magnetohydrodynamic plasma phenomena and, particularly, electromagnetism of the rings structure origin.

Despite the rich available database, there is no yet physically satisfactory model of the Saturn rings and the mystery of many experimental data has no explanation: origin, evolution and dynamics of the rings; why particles are separated and at the same time they could stick together; considerable flattening and the sharp edges of the rings; thin periodic structure of the rings; deformation of the magnetic field lines nearby ring *F;* formation of "spokes" in the ring *B*; high radio-wave reflectivity and low brightness of the rings; anomalous reflection of circularly polarized microwaves (like from magnetic mirror); strong pulse electromagnetic radiation of the rings in the 20.4 kHz - 40.2 MHz range; spectral anomalies of the thermal radiation of the rings; why substance of the rings does not mix, but preserves its small-scale color differences; existence of an atmosphere of unknown origin nearby the rings; why there is existence of the waves of density and bending waves within the rings; why the planetary rings in the solar system appear for the planets which are located only outside the belt of asteroids and why the Earth has no rings, etc.

The superconductivity of the ring particles may follow from the fact that the ring particles are relics of the early days of the Solar system and particles were never subject to coalescence and heating. Indeed, the Sun heats the rings weakly, because temperature in the area of the rings is about 70-110 K. It makes possible the existence of the superconducting substance in the space behind the belt of asteroids. The superconductive particles cannot stick together because the magnetic field emanates from them and pushes the particles apart. In 1933 W. Meissner and R. Ochsenfeld found that a superconducting material will repel a magnetic field. The high-temperature superconductivity was discovered by Bednortz and Muller in

1986 [17]. In 1986 superconductivity of ice was experimentally demonstrated by A.N. Babushkin[1] et al. [18].

It is assumed that superconducting matter of the particles of the rings of Saturn allows extending classical theories of the planetary rings (gravitational, mechanical, magnetohydrodynamic and plasma interactions) by non-conflicting superconducting model [20-44].

An interesting fact of this paper is that even though its subject concerns problem related to astronomy, the solution may come from the general electromagnetic theory.

2. An Experimental Data Observation

Thin Structure and Sharp Edges of the Rings

Similar to magnetic particles creating dense and rarefied areas in a nonuniform magnetic field, the superconductive ice particles also form their groups, which from outside look like a system of rings. Superconducting particles will collapse into a system of rings as the result of their replacement in the area with the less density of the magnetic flow, within plane of magnetic equator, with the force: F = - mdH/dz, where m – magnetic moment of a particle, dH/dz – gradient of intensity of a magnetic field along an axis z of the magnetic dipole. The force of diamagnetic push-out is forming the sharp edges of the ring: F= -mdH/dy, where dH/dy - gradient of intensity of a magnetic field along the radius of the ring. The casual break in the ring will be stabilized by the force of diamagnetic push-out F = - mdH/dx, where dH/dx - gradient of intensity of a magnetic field in tangential direction. Measurements of the magnetic field nearby the rings *F* by the Pioneer mission [5, 6, 12] have registered deformation (distortion) of the magnetic field lines like it is happened for the superconducting disc in laboratory under Meissner - Ochsenfeld state [17].

Planetcentrical Dust Flow

For superconducting particles there is London's depth λ_L of penetration of the magnetic field inside superconductor [17]. Influence of penetration of the magnetic flow becomes appreciable for particles with the size comparable with London's depth of penetration. Smaller particles are not cooperating with the planetary magnetic field, because they lost their superconductivity by size. Dynamics of these particles is different from dynamics of particles with bigger size which is $>2\lambda_L$. These particles will fall down to the planet due to the gravity. Thus, existence of the planetcentrical dust flows of submicron's size particles related to disappearance of superconductivity of the matter of the rings particles due to reducing their size. It is also possible for the particles to loose their superconductivity by influence of collisions and by fluctuations of magnetic field.

---

[1] Professor Aleksey N. Babushkin is a Dean of the Physics Department at the Ural State University, Yekaterinburg,

Change of the Azimuth Brightness of the *A* Ring of Saturn

There is a number of theories explaining this fact, based on assumptions of a synchronous rotation of the ring's particles with their asymmetrical form as extended ellipsoids directed under a small angle to the orbit, or with asymmetrical albedo of the surface [5, 6, 12]. Let's go to superconducting model. If superconductor is placed in the magnetic field, additional moment directed opposite to the external field arises. The matter is magnetized, not along the external magnetic field but in the opposite direction. The rod of superdiamagnetic substance of the ring particle tries to locate itself perpendicularly to the magnetic field lines. From the science that studies ice [19] it is a known fact that at the temperature below – 22 ^{0}C growing snowflakes take the form of prisms. Thus, the prism of the superconducting iced particle will be oriented perpendicularly to the field lines of the polhoidal and toroidal constituents of the magnetic fields of Saturn. So, it's clear that variable azimuth brightness of the Saturn's rings system *A* relates to orientation of the elongated ellipsoid of superconducting particles to the normal direction to the magnetic field of the planet.

Spokes in the Ring *B* of Saturn

As well as the spokes of any wheel, are located almost along radiuses. According to the laws of Kepler any radial formation should be distorted and washed out for a few tens of minutes. However, experimental data show that lifetime of separated spoke 10^3 - 10^4sec though Kepler washing outs in them are nevertheless proven. The size of spokes themselves is about 10^4km along the radius and about 10^3km along the orbit of the ring. The matter of the spokes consists of micron and submicron size particles [11]. There were many attempts to explain the nature of the spokes. Mostly all theories are based on the action of the force of gravity. At the same time, there were some ideas that registration of rotating spokes somehow related to electromagnetic interaction because of its rotation synchronously going along with the magnetosphere of the Saturn [5, 6, 12]. The analysis of spectral radiant power of spokes provides specific periodicity about 640.6±3.5 min, which is almost coincident to the period of rotation of the magnetic field of Saturn, which is 639.4 min. Moreover, the strong correlation of maxima and minima of activity of spokes with the spectral magnetic longitudes is connected to presence or absence of the radiation of Saturn's Kilometric Radiation (SKR). It confirms the assumption of the dependence of the spokes dynamics on the magnetic field of Saturn and testifies to the presence of large-scale anomalies in the magnetic field of Saturn. Superconducting iced particles of the rings matter are rotating in accordance with Kepler's law, and at the same time magnetic field is rotating along with the planet and has anomalies itself. Superconducting particles coming to anomalies of the magnetic field positions and the balance of the three forces acting with each particle will change due to change of the electromagnetic force. Then within the anomaly of the planetary magnetic field all particles will try to get another position and observer can see all these chaotic movements of the particles on the picture of the rings as a spokes. After passing position of the planetary magnetic field anomaly all particles will get positions in accordance with Kepler law.

Russia. Tel: +7(912)243-6892, +7(343)261-1885, +7(343)261-6058; E-mail: alexey.babushkin@usu.ru

High Reflection and Low Brightness of the Rings Particles in the Radiofrequency Range

We can explain also using superconducting model. The discovery of strong radar-tracking reflection from the rings of Saturn in 1973 was surprising [5, 6, 12]. It turned out that the rings of Saturn actually have the greatest radar-tracking section among all bodies of the Solar system. It was explained by the metallic nature of the particles. The data of the Voyager excludes this possibility. The disk of superconducting particles completely reflects radiation with frequencies below 10^{11}Hz and poorly reflects radiation with higher frequencies, as in the case of superconductor. The superconductors have practically no resistance up to frequencies of 100MHz. At frequencies about 100GHz there comes a limit, above which the frequent quantum phenomena cause a fast increase of resistance, as it is shown on figure 1. Hence a specific picture of the dependence of brightness.

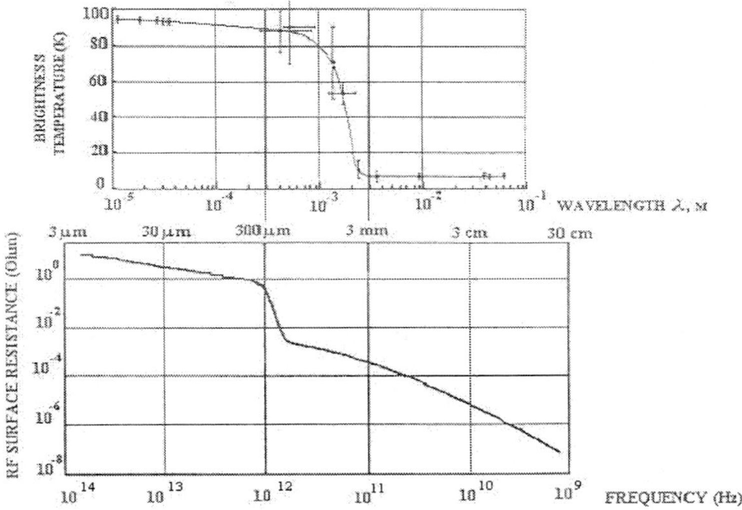

Figure 1. Top diagram is the dependence of the brightness temperature of the rings on the wavelength: transition from the radiation of the almost black body to practically complete reflection is observed [5, 6, 12]. Bottom diagram is the dependence of the surface resistance of the superconductor on frequency for Nb at T=4,2K [17].

Own Wide Band Pulse Radiation of the Rings

Data from the Voyager have shown that the rings radiate own wide band pulse radiation within the 20 KHz-40,2MHz [5, 6, 12]. These waves, probably, are a result of interaction of charged particles with the particles of ice or destruction and friction of iced particles when co-striking occur. These incidental radio discharges are named as Saturn's Electrostatic Discharges (SED). The average period SED is well determined and was established as 10 hour 10±5 min and 10 hour 11±5 min by Voyager-1, -2. If the ring has a source of SED, the area of this source can be located at the distance of 107,990 – 109,000km from the planet according to measured periodicity. Data of the experiments can help specify electrodynamics coupling between the planetary ring system and the magnetosphere, in which SKR, SED and activity of spokes are subordinated to longitude regulation. In accordance with superconducting model, approaching of superconducting rings' particles up to distance of about 10^{-8}m, or, the existence of narrowing or dot contact will result in the formation of a weak link (superconducting transition) through which superconducting electrons can be tunneled. When the difference of phases between superconductors under action of the electrical or magnetic field occurs, the weak link will generate electromagnetic radiation with frequency proportional to power failure on this transition (nonstationary Josephson phenomenon) [17, 20-36]. The radiation frequency is proportional to the voltage in the transition, ν=2eV/h, where 2e/h= 483,6 MHz/μV, e is a charge of electron, h is Plank constant.

Frequency Anomalies of Thermal Radiation of the Rings in the 100μm - 1cm Range

The measured brightness temperature on the short waves is less than true brightness temperature of the rings, and on the longer waves the rings look much colder than in the case when the radiation corresponds to their physical temperature [5, 6, 12]. On the range 100μm - 1mm brightness temperature of the ring (figure 2) sharply falls to the meanings smaller than those ones characteristic of an absolutely black body. On the wavelengths longer than 1cm a ring behaves as the diffusion screen, reflecting planetary and cold space radiation. The central part of the spectral range 100μm – 1cm is the most sensitive to the parameter of refraction, and may contain the important determining information of fundamental properties of the substance. In accordance with the model, under the superconducting condition the electrons do not interact with a crystal lattice and do not exchange energy with it, therefore they cannot transfer heat from one part of the body into another. Hence, when the substance passes into a superconducting condition, its heat conductivity is lowered. This effect can be obvious under temperatures much less critical, when there are very few conventional electrons capable of transferring heat [20-36].

Color Difference of Rings in a Small Scale

The balance of three forces determines the position of the superconducting particles in the gravitational and magnetic planetary fields: gravitational force, centrifugal one and magnetic levitation (diamagnetic push out), figure 1. Going along with our model let's consider the distribution of three particles (a, b, c) with equal weights on close orbits. Let a particle *a* be wholly superconducting, *b* – have an impurity clathrate-hydrates of ammonia or methane (NH_3; CH_4 H_2O), *c* - has an impurity of sulphur and Ferro containing silicates (H_2S). Each impurity will give the contribution to reduction of the volume of superconducting phase and will determine the color of the particle. Force of diamagnetic push out - F_L depends on the volume of the superconducting phase, therefore for each of considered particles the balance of three forces will be carried out in the orbits with different radiuses [20-36].

Anomalous Inversion Reflection of Microwaves with Circular Polarization Above 1 Cm

The research of reflection of radiowaves above 1 cm from the rings was carried out with the use of ground based radio-locators and by the space probes [5, 6, 12]. The reflection appeared rather large, and the geometrical albedo is equal approximately to 0.34 and has no strong functional dependence on the wavelength or on the angle of the inclination of ring's pitch. The rings are strong depolarizers. Therefore, in order to get any information from the reflection it is necessary to measure separately the intensity of two orthogonal polarized reflected signals. It provides information of the factor of the ring's polarization, which carries information about properties of particles. For the majority of single objects of the Solar system, for example the planets, factor of reflection unobserved polarization (orthogonal to observable) is rather small. As to the rings, the supervision in some range of wavelengths and angles of inclination give reflection factor of unobserved polarization between 0.4 – 1.0 . Let's go to our superconducting model [20-36]. The superconductors have an essential difference from ideal conductors; besides almost infinite conductivity they also demonstrate an ideal diamagnetism. The falling electromagnetic wave will induce circular currents in superconductor, which will completely compensate action of the magnetic field of the incident wave. So that the absence of the magnetic field in the volume of the superconductor should be carried out. Superconductor will be acting as a magnetic mirror. Thus, if the falling on superconductor electromagnetic wave has a determined direction of a circular polarization (the spirality), the direction of circular polarization (spirality) will be kept in the reflected wave.

An Atmosphere of "Unknown" Origin at the Rings

The atmosphere of Saturn's rings can exist as a result of thin balance of forces of gravitational attraction and diamagnetic push-out of gas molecules. The levitation of gas molecules comes true at the expense of forces of diamagnetic push-out induced in superconducting particles by molecular magnetic moment of gas [20-36]. The similar situation can be observed under laboratory conditions when atmospheric water steam is precipitated on substance as white-frost at the transition moment of substance from a superconducting into a conventional one.

Existence of Wave of Density and Bending Waves within the Rings

The existence of waves of density and bending waves in the Saturn's rings has no complete explanation based only on gravitation phenomena. Let's use the superconducting model. It is possible to note that the external magnetic field is directed along the free surface of diamagnetic liquid which represents a disk of the rings. In case of periodic deformation of a free surface a normal field under a hollow decreased, and it increased under chamber. Consequently, the ponderomotive force works on the side of restoration of the flat form of the free surface. Thus, the field increases a rigidity of a free surface. The change of rigidity of the free surface in a field gives a chance to excite its parametrical fluctuations. When the quantity of the amplitude of intensity of a variable field is larger than the critical one, the standing wave is occurring on a free surface of superconducting liquid. The constant phase lines of this wave are transverse to the vector of intensity of the field. The wavelength is determined by the condition of the parametric resonance. The constant phase lines begin to be bent when intensity of the field increases, and the excited ripple becomes casual [20-36].

3. Theoretical Solution of the Saturn's Rings Origin

Following from the solution of the electromagnetic problem we will demonstrate how rings of Saturn could be originated from the iced particles located within the protoplanetary cloud. Before appearance of the magnetic field of Saturn all particles within the protoplanetary cloud are located on such an orbit as Kepler's, where there is a balance of the force of gravity and the centrifugal force. With the occurrence of the magnetic field of the Saturn the superconducting particles of the protoplanetary cloud begin to demonstrate an ideal diamagnetism (Meissner-Ochsenfeld phenomenon). Particles start to interact with the magnetic field and all particles become to be involved in additional azimuth-orbital movement. Let's estimate the result of this movement [37-38].

If the magnetic field of the planet is equal *H*, and the planetary magnetic moment is equal $\vec{\mu}$, then the magnetic field at any particular point within the protoplanetary cloud, located on the distance $\vec{r}$, can be presented as:

$$\mathbf{H} = \frac{3 \cdot \mathbf{r} \cdot (\mathbf{r}, \boldsymbol{\mu})}{r^5} - \frac{\boldsymbol{\mu}}{r^3} \tag{3.1}$$

Then the superconducting ball with the radius *R*, which is located within the protoplanetary cloud, gets the magnetic moment equal:

$$\mathbf{M} = -R^3 \cdot \mathbf{H}. \tag{3.2}$$

That energy of a superconductor in a magnetic field gets the following value:

$$U_H = -(\mathbf{M}, \mathbf{H}) = R^3 \cdot H^2 \tag{3.3}$$

If the beginning of coordinates to place in the center of a planet, and an axis z to direct along the magnetic moment of a planet (orthogonal to equator) magnetic energy thus will be equal:

$$U_H = \frac{R^3 \mu^2}{r^6}(3\cos^2\theta + 1). \tag{3.4}$$

Here θ - an angle between a vector $\vec{r}$ and an axis Z. We can see from the expression (3.4), that magnetic energy of the superconducting particle has a minimum value when the radius-vector $\vec{r}$ (position of the superconducting particle) appears in a plane of magnetic equator, a perpendicular to the axis Z $(\cos\theta = 0)$. It is clear, that for one particle its trajectory (orbit) in the azimuth-orbital movement will be only disturbed by the magnetic field. However, in case of a huge amount of particles, like its happened within the protoplanetary cloud, after some time, collisions between particles will compensate their azimuth-orbital movements, and, as a result, eventually, during some time, all orbits of the particles of the protoplanetary cloud should come together to magnetic equator plane and create highly flattening disc around planet. Within the disc of rings all particles will be located on such an orbit as Kepler's, where there is a balance of the force of gravity, the centrifugal and electromagnetic forces. At the same time, orbital resonances (due to gravity force) between satellites, moons and the rings particles play an important role in forming specific structure of the rings and gaps.

4. Separation and Collision of the Particles within the Sombrero of Rings

We can define the energy of the interaction of two superconducting particles with magnetic moment μ_1 и μ_2 if they are located on the distance $\mathbf{r}_1$ и $\mathbf{r}_2$, respectively as:

$$U = -\boldsymbol{\mu}_1 \mathbf{H}_2, \tag{4.1}$$

Where magnetic field $\mathbf{H}_2$ is produced by the magnetic moment μ_2 it can be presented as

$$\mathbf{H}_2 = \frac{3(\mathbf{r}_1 - \mathbf{r}_2)(\boldsymbol{\mu}_2(\mathbf{r}_1 - \mathbf{r}_2))}{|\mathbf{r}_1 - \mathbf{r}_2|^5} - \frac{\boldsymbol{\mu}_2}{|\mathbf{r}_1 - \mathbf{r}_2|^3} \tag{4.2}$$

If we place the particles with magnetic moment μ_2 at the beginning of coordinate ($\mathbf{r}_2$=0) then the expression for the energy of the interaction of two particles (4.1) will be the following:

$$U = -\frac{3(\boldsymbol{\mu}_1\mathbf{r}_1)(\boldsymbol{\mu}_2\mathbf{r}_1)}{|\mathbf{r}_1|^5} + \frac{\boldsymbol{\mu}_1\boldsymbol{\mu}_2}{|\mathbf{r}_1|^3}\ . \tag{4.3}$$

In the plane of the rings of Saturn the magnetic field of the planet coincides with the rotation axis of the planet. If the axis Z is directed along with the rotation axis of the planet then magnetic moment of the particles also will be directed along the axis Z. Using the cylindrical system of coordinate (ρ, φ, z) we can represent expression (4.3) as:

$$U = -\left(\frac{3z^2}{\left(\rho^2 + z^2\right)^{5/2}} - \frac{1}{\left(\rho^2 + z^2\right)^{3/2}}\right)\mu_{1z}\mu_{2z} = \frac{\rho^2 - 2z^2}{\left(\rho^2 + z^2\right)^{5/2}}\mu_{1z}\mu_{2z}, \tag{4.4}$$

Let us use the expression (4.4) to estimate how two superconducting particles will be interacting in two different cases. The first one is when two particles are located in the same plane within the sombrero of the rings (Z=0), and the second situation is when two particles are located on the different planes but on the same axis $(\rho = 0)$.

From the expression (4.4) follows that for the particles with magnetic moment μ_{1z} and μ_{2z} which are located on the same plane, Z=0, we can get that energy of their interaction is equal:

$$U = \frac{\mu_{1z}\mu_{2z}}{\rho^3}, \tag{4.5}$$

From the expression (4.5) follows that in this situation both particles will be pushing each other and they will be holding separation distance in between them. This result has been confirmed by the data of Cassini mission: the particles are separated[2].

Then for another situation both particles are located on the same axis but on the different planes, and, as it follows from (4.4) the expression for the interaction energy is:

$$U = -\frac{\mu_{1z}\mu_{2z}}{|z|^3}, \tag{4.6}$$

We can see that in this case both particles will be attracting each other, they could even collide or stick together and form bigger pieces or lumps of ice. This fact has an experimental conformation by Cassini mission. As we know from the data of Cassini mission it was registered that particles within the sombrero of the rings can collide or even stick together and form bigger pieces of ice[3]. Then later the particles with 50 meters or more in diameter can be destroyed into smaller pieces by the common action of gravity and centrifugal force.

4. Conclusion

Theoretical electromagnetic model of the origin of the rings of Saturn from superconducting particles of the protoplanetary cloud, which is presented in this paper, is a direct continuation of the paper by J.K. Maxwell (1859). Unfortunately, at his time there was no knowledge about superconductivity (1911) and force of diamagnetic push-out of superconductor (1933). Superconductivity of ice and high temperature superconductivity was discovered just recently, in 1986 [17, 18]. As it follows from above consideration, observation of experimental data and electromagnetic modeling confirmed the suggested model.

---

[2] http://pds-rings.seti.org

[3] saturn.jpl.nasa.gov/multimedia/images/index.cfm

The model considered here makes possible a "magnetic coupling" between protosun and superconducting particles of its protoplanetary cloud, that in the process of formation of the solar system at an early stage of its development, when the temperature was low enough to have superconductivity, lead to the carrying of the moment of momentum from the Sun to other planets by the electromagnetic means of the superconducting substance of the space environment. Following this we can conclude that the idea of H. Alfen [10-11] that "solar system history as recorded in the Saturn rings structure" becomes to be a physical reality.

From the approach presented above, with the possibility of an electromagnetic origin for the rings, it follows as important for the space physics to take into account the natural space superconductivity for the space substance after the belt of asteroids. It may have a fundamental importance for analyzing the data of the Cassini-Huygens probe, and striking parallels to those that occur in more remote disc systems such as galactic discs and accretion discs around stars and black holes. The force of diamagnetic push-out of superconductor in the magnetic field may be a driving force for propagation of organic molecules within the interstellar space by electromagnetic means, and organic molecules can also be contained within the rings of Saturn, as it is presented in [45-47].

The author would like to express greatest thanks for valuable discussions to A.M. Prokhorov, V.N. Strakhov, V.V. Migulin, Yu.V. Gulyaev, V.I. Pustovoit, A.A. Rukhadze, B.I. Rabinovich, V.G. Kurt, A.Yu. Pospelov, A.N. Malov, O.N. Rzhiga, V.A. Miliaev, E.V. Chensky, V.N. Lugovenko, S.V. Girich, M.V. Belodedov, E.P. Bazhanov, A.V. Zaitsev from the Russian Academy of Sciences; S.V. Vasilyev (SVVTI, CA), V.P. Vasilyev (SOLERC, Ukraine), J.R. Whinnery and T.K. Gustafson (UC Berkeley); J.A. Kong (MIT), D. Osheroff (Stanford), E.A. Marouf (SJSU); J.N. Guzzi, R.B. Hoover, J.F. Spann, R. Sheldon, D. Gallagher, K. Mazuruk and A. Pakhomov (NASA); P. Goldreich (Caltech), L. Spilker (JPL), C.T. Russel and Y. Rahmat-Samii (UCLA), G. Shoemaker (CSUS), L.N. Vanderhoef, P. Rock, R.T. Shelton, R.R. Freeman, A. Albrecht, B. Backer, W. Ko and W. Pickett (UC Davis), R.P. Kudritzki and R.D. Joseph (IFA UH), J.A. Burns (Cornell University, NY), M. Pardavi-Horvath (GTU), P.M. Cincotta (IAFE, Buenos Aires), G. Gerlach (Oriongroup, CA), A. Mendis, L. Peterson, M. Fomenkova (UCSD), N. Castle, R.S. Henderson (San Diego, CA), K. Fischer (Los Gatos, CA), C.B. Vesecky (Palo Alto, CA), A. Pagliere (Sacramento, CA), A.V. Aliaev, V.V. Tsykalo, P.A. Razvin, E.N. Muraviev, O.I. Chernaya (Moscow, Russia) for support. The author is also grateful for fruitful discussions to all participants of the seminars and conferences at the NASA Marshall Space Flight Center and the Huntsville Space Physics Colloquium, the Institute for Astronomy at the University of Hawaii, Astrophysics and the Space Research Center at the University of California in San Diego, CA, the Electrical Engineering and Computer Sciences Department at the University of California in Berkeley, the University of California in Davis, the Institute of Astronomy and Physics at La Plata in Buenos Aires, the Progress In Electromagnetic Research Symposium (PIERS) organized by MIT, the 42nd – 50th SPIE Annual Meetings, the National Bureau of Standard in Washington D.C., the 30th Annual Meeting of the Division of Planetary Sciences of the American Astronomical Society.

REFERENCES

[1] Maxwell J.C., Brush S.G., Everitt C.W.F. , Garber E. (Eds.). Maxwell on Saturn's rings. Cambridge, MA. MIT Press, 1983.

[2] V.S. Safronov. Evolution of protoplanet cloud and formation of the Earth and planets. *Nauka.* (Russian), Moscow, 1969.

[3] Goldstein R.M., Morris G.A. Radar observations of the rings of Saturn // *Icarus.* 1973. V. 20. P. 249-283.

[4] Kaiser M.L., Desch V.D., Lecacheus A. Saturnian kilometric radiation: statistical properties and beam geometry // *Nature.* 1981. V. 292, P. 731-733.

[5] Brahic A. (Ed.). Planetary rings. Toulouse: Copadeus, 1984.

[6] Greeberg R., Brahic A. (Eds.). *Planetary rings.* Tucson: University of Arizona Press, 1984.

[7] A.N. Bliokh, V.V. Yaroshenko. "Spokes" in the rings of Saturn // *Nature.* (Russian), 1991, N4, P.19-25.

[8] Spilker L.J. (Ed.). The Cassini-Huygens mission to Saturn and Titan. Washington, D.C.: JPL, Caltech, NASA SP-533, October 1997.

[9] L. Rowan, A. Sanchez-Lavega, T.I. Gombosi and K.S. Hansen, C.C. Porco et al., F.M. Flasar., L.W. Esposito et al., D.A. Gurnett et al., J.H. Waite Jr. et al., D.T. Young et al., M.K. Dougherty et al., S.M. Krimigis et al., S. Kempf et al. Cassini at Saturn // *Science.* 25 February 2005. V. 307, №. 5713. P. 1222 – 1276.

[10] Alfven H. Cosmic plasmas. Dordrecht, 1981.

[11] Alfen H. Solar system hystory as recorded in the Saturnian rings structure // *Astrophysics Space Science.* 1983. V. 97. P. 79-94.

[12] Mendis D.A., Hill J.R., Ip W.H., Goertz C.K., and Grun E. Electrodynamics processes in the ring system of Saturn. Saturn / *Saturn.* T. Gehrels, M. Mathews (Eds.). Tucson: University of Arizona Press, 1984. P. 546-589.

[13] Esposito L.W., Cuzzi J.N., Holberg J.B., Marouf E.A., Tyler G.L., Porco C.C. Saturn's rings, structure, dynamics and particle properties / *Saturn.* Gehrels T., Matthews M.S. (Eds.). Tucson: University of Arizona Press, 1984. P. 463-545.

[14] N.N. Gor'kavyi, A.M. Fridman. Physics of the planetary rings: celestial mechanics of continuous medium, *Nauka.* (Russian), Moscow, 1994. 348 p.

[15] Rabinovich B.I. Magnetohydrodynamic of rotating vortex rings with magnetized plasma // *DAN* (Russian), 1996, V.351, N3, P.335-338.

[16] Rabinovich B.I. Rotating plasma ring in gravitational and magnetic fields. Stability problems. *DAN* (Russian), 1999, V.367, N.3, P.345-348.

[17] Ginzburg V.L., Andryushin E.A. Superconductivity. World Scientific, 2004. (Please, also see: www.superconductivity.org).

[18] G.V. Babushkina, L.Ya. Kobelev, E.N. Yakovlev, A.N. Babushkin. Superconductivity of ice under high pressure // *Physics of Solid State.* (Russian), 1986. V.28, N12. P.3732-3734.

[19] N. Maeno. Science about ice. *Mir.* (Russian), Moscow, 1988. 231 p.

[20] A.Yu. Pospelov, V.V. Tchernyi. Electromagnetic properties material forecast in the planet rings by the methods of functionally physical analysis // *Proc. of International Scientific-Methodical Conference on Innovative Design in Education, Techniques, and*

Technologies". Volgograd State Technical University, Volgograd, Russia, 1995. P. 75-77.

[21] Pospelov A. Yu., Tchernyi V. V., Girich S.V. Planet's rings: super-diamagnetic model and new course of investigations // *Proc. of SPIE 42nd Annual Meeting. San Diego*, CA. 27 July-1August 1997. *"Small spacecraft, Space Environments and Instrumentation Technologies"*. 1997. V. 3116. № 15. P. 117-128.

[22] Pospelov A.Yu.,Tchernyi V.V., Girich S.V., Korendovich V.V. Superdiamagnetic model of planetary rings // *Electrodynamics and Techniques of HF and SWF*. Moscow, 1997. T. 5. № 3. P. 248-262.

[23] Pospelov A.Yu., Tchernyi V.V., Girich S.V. Planetary rings: new mission concept // Kona, HW: *SPIE International Symposiumon Astronomical Telescopes and Instrumentation*. 20-28 March 1998. № 132.

[24] Pospelov A.Yu., Tchernyi V.V., Girich S.V. Possible explanation of the planet's rings behavior in the radio and mm-wave range via superdiamagnetic model // Kona, HW: *SPIE International Symposiumon Astronomical Telescopes and Instrumentation*. 20-28 March 1998. № 132. № 73.

[25] Pospelov A.Yu., Tchernyi V.V., Girich S.V. Possible explanation of the planet's rings behavior in the radio and mm-wave range via superdiamagnetic model *// International Aerospace Abstracts of American Institute of Aeronautics and Astronautics*, Inc. January 1999. № 1. P. A99-10781.

[26] Pospelov A.Yu., Tchernyi V.V., Girich S.V. Superdiamagnetic model of planetary rings behavior in the millimeter and submullimeter range // *Digest 3465 – 4th International Conference on MM and SMM Waves and Applications*. San Diego, CA: Proc. SPIE 43 Annual International Symposium. 20-23 July 1998. San Diego, CA, 1998. P. 172-173.

[27] Girich S.V., Pospelov A.Yu., Tchernyi V.V. Radar data explanation via superdiamagnetic model of the Saturn's rings // *Annual Report of the AAS. 30th Meeting Division of Planetary Science. Madison,* WI. 11-16 October 1998. Bulletin of the American Astronomical Society. 1998. V. 30. № 3. P. 1043.

[28] Pospelov A.Yu., Tchernyi V.V., Girich S.V. Anomalous inversion of polarization of icy satellites and Saturn's rings: superdiamagnetic model // *Proc. 44th SPIE Annual Meeting. Denver,* CO. 18-23 July 1999. "Polarization: measurements, analysis and remote sensing II". 1999. V. 3754. P. 329-333.

[29] Pospelov A.Yu., Tchernyi V.V., Girich S.V. Are Saturn's rings superconducting? // Progress In Electromagnetic Research Symposium (PIERS). 5-14 July 2000. Cambridge, MA: *MIT*. 2000. P. 1158.

[30] Pospelov A.Yu., Tchernyi V.V., Girich S.V. What data could confirm Saturn's rings superconductivity? // *SPIE conference on Astronomical Telescopes and Instrumentation*. Munich, Germany. 27-31 March 2000. N. 4015-67.

[31] Pospelov A.Yu., Tchernyi V.V. Magnetic levitation of Saturn's rings. // *Progress In Electromagnetic Research Symposium*. (PIERS). 5-14 July 2002. MIT. Cambridge, MA. 2002. C. 135.

[32] Tchernyi V.V., Pospelov A.Yu. Possible role of space electromagnetism for Saturn's rings existence // *Progress In Electromagnetic research Symposium. (PIERS)*. 13-16 October 2003. Honolulu, HW. 2003.

[33] Tchernyi V.V., Pospelov A.Yu. Possible magnetic levitation of Saturn's (planetary) rings. Pisa, Italy. // *Progress in Electromagnetic Research Symposium. (PIERS).* 28-31 March, 2004. N. P08.

[34] Pospelov A.Yu., Tchernyi V.V. Space electromagnetism: modeling of magnetic levitation of superconducting Saturn rings // *Progress In Electromagnetic Research Symposium (PIERS).* Hangzhou, China. 22-26 August 2005.

[35] Tchernyi V.V., Pospelov A.Yu. Possible electromagnetic nature of the Saturn's rings: superconductivity and magnetic levitation // *Progress in electromagnetic research (PIER).* Cambridge, MA: MIT Press. 2005. V. 52. P. 277-299.

[36] Tchernyi V.V., Pospelov A.Yu. About possible electromagnetic nature of the planetary rings: magnetic levitation of superconducting rings of Saturn // *Fizika volnovykh protzessov i radiotekhnicheskikh sistem.* (In Russian). 2005. T. 8. № 2. P. 4-16.

[37] Tchernyi V.V., Chensky E.V. Electromagnetic background for possible magnetic levitation of the superconducting rings of Saturn // *Journal of Electromagnetic Waves and Applications.* Cambridge, MA: MIT Press. 2005. V. 19. № 15. P. 1997-2006.

[38] Tchernyi V.V., Chensky E.V. Movements of the protoplanetary superconding particles in the magnetic field of Saturn lead to the origin of rings // *IEEE Geoscience and remote sensing letters.* 2005. V. 2. No. 4. P. 445-446. Corrections // *IEEE GRSL.* 2006. V. 3. No. 2.

[39] Tchernyi V.V. (Cherny). About possible role of electromagnetism and superconductivity for the origin of Saturn's rings. *Prikladnaya fizika.* (Applied Physics). 2006. N. 5. P. 10-16. (In Russian).

[40] Tchernyi V.V. (Cherny). Possible role of superconductivity and electromagnetism for the origin of the rings of Saturn. Proc. Intern. Conf. "Fundamental principles of engineering sciences". Devoted to 90-years birthday of Nobel prize winner A.M. Prokhorov. Moscow, Oct. 25-27, 2006. P.257-259. (In Russian).

[41] Tchernyi V.V. Responsibility of electromagnetism for the origin of the rings of Saturn from superconducting particles of the protoplanetary cloud // *Progress in electromagnetic research symposium (PIERS).* Tokyo, Japan. 2006.

[42] Tchernyi V.V., Pospelov A.Yu. About hypothesis of the superconducting origin of the Saturn's rings // *Astrophysics and space science.* Springer. 2007. V. 307. No. 4. P. 347-356.

[43] Tchernyi V.V. To the glory of G. Galileo and J.K. Maxwell: electromagnetic modelling of the origin of Saturn's rings from superconducting particles of the protoplanetary cloud // *The 23 rd Annual review of progress in applied computational electromagnetics.* Verona, Italy. March 19-23, 2007.

[44] Tchernyi V.V. Modeling of electromagnetic origin of the rings of Saturn from superconducting particles of the protoplanetary cloud // *Progress In Electromagnetic Research Symposium (PIERS).* Hangzhou, China. 2007.

[45] Tchernyi V.V., Kapranov S.V. Possible role of superconductivity for simplest life propagation within interstellar space by electromagnetic force of magnetic levitation // *Journal of Electromagnetic Waves and Applications.* Cambridge, MA: MIT Press. 2005. V. 19. № 15. P. 1997-2006.

[46] Tchernyi V.V., Kapranov S.V. Is electromagnetic force a possible means for life transmission in the universe? // *Progress In Electromagnetic Research Symposium (PIERS).* Hangzhou, China. 22-26 August 2005.

[47] Tchernyi V.V., Kapranov S.V. Contribution of superconductivity to possible interstellar propagation of organic molecules by electromagnetic way // *Proc. 50th SPIE Annual Meeting.* July 31-Aug. 4, 2005. San Diego, CA. Hoover R.B. et al. (Eds.). Astrobiology and planetary missions. *SPIE,* 2005. V. 5906.

In: Space Exploration Research
Editors: J.H. Denis and P.D. Aldridge
ISBN: 978-1-60692-264-4

Chapter 12

PERIODIC FINE-SCALE STRUCTURE IN SATURN'S RINGS: A THEORY OF SELF-GRAVITY DENSITY WAVES

Evgeny Griv*and Michael Gedalin
Department of Physics, Ben-Gurion University,
Beer-Sheva, Israel

Abstract

A highly flattened, rapidly and differentially rotating disk of primarily large $>$ cm size mutually gravitating and elastically colliding ice particles orbiting a central object is oftenly taken as an idealized model of Saturn's main A, B, and C rings. This article considers the problem of the stability of the Saturnian main ring system with special emphasis on its fine-scale of the order of 100 m density wave structure (almost regularly spaced, aligned cylindric density enhancements and rarefications). We attribute this periodic microstructure to the propagation of compression density waves in the ring plane. The wave propagation is a process of rotation as a solid about the center at a fixed phase velocity, despite the general differential rotation of the system; the structure consists of different material at different times. It seems likely that the key factor contributing to the generation of density waves is the classical Jeans instability of gravity perturbations (e.g., those produced by a spontaneous disturbance). This gravitational instability associated with small departures of macroscopic parameters from the dynamical equilibrium is hydrodynamical in nature and has nothing to do with any explicit resonant effects. We analyse Jeans' gravitational instability analytically through the use of hydrodynamic equations. It is shown that the instability in the rotating Saturnian ring layer may be stabilized by a peculiar particle motion, or "temperature" of a suitable magnitude. A stability criterion is given to suppress the instability of all perturbations including the most unstable spiral ones. We demonstrate that exclusively trailing spirals can be formed in Saturn's A nd B rings. The very existence and the value of the critical wavelength of the fine-scale structure is explained. Theoretical predictions are compared with numerical simulations. The stability analysis presented here would have to be regarded as an explanation of the almost regular periodic structure in the range of few tens to few hundreds meters in Saturn's A and

*E-mail address: griv@bgu.ac.il

B rings that has been recently revealed by Cassini spacecraft high-resolution measurements.

Keywords: planetary rings—Saturn, rings—planetary dynamics—instabilities and waves

1. Introduction

The dynamics of highly flattened gravitating systems has now been studied quite thoroughly. This research has aimed to explain the origin of various observed structures: spiral and ring formation in flat galaxies and protostellar/protoplanetary clouds, ring formation in disks around supermassive black holes, the fine ringlets around Saturn, the narrow and widely separated rings of Uranus, etc. One of the main trends has therefore been to analyze the perturbation dynamics in such systems, in both linear and nonlinear regimes (Binney & Tremaine 1987; Bertin 2000).

Saturn's ring disk is composed predominantly of water-ice particles ranging between about 1 cm and 10 m in radius. Above this size range, the number of particles drops sharply (Zebker et al. 1985; French & Nicholson 2000). Numerical simulations indicated that the larger particles are nearly in a monolayer, with the smaller particles filling in the spaces between the larger particles. Voyager flybys of Saturn have revealed that the disk about the planet is not simply divided into several main bands (the A, B, and C rings) but that in fact the entire disk assembly is subdivided into a huge number of fine thread-like rings (Smith et al. 1981, 1982; Lane et al. 1982; Stone & Miner 1982). The Voyager 2 spacecraft close-up view of Saturn's rings shows that even the so-called gaps demonstrate a complicated structure—the Cassini Division, for example, contains a large number ringlets. It is important that the Voyager's photopolarimeter PPS data revealed some *indirect* evidence for "finest" structuring in the densest central parts of the opaque Saturn's B ring down to the 100 m length scale (Showalter & Nicholson 1990). However, below a few kilometres scale, the PPS data is too noisy to extract information about the structure: the finest structure observed by PPS is well fit by models of statistical noise combined with stochastic variations resulting from large particles or clumps of particles (Showalter & Nicholson 1990). It was suggested that much more precise Cassini spacecraft observations would help to settle the question (Griv 1998; Griv et al. 2000a, 2003a, b; Griv & Gedalin 2003).

The almost regular finest of the order of 100 m or even less density structure of the Saturnian brightest A and B rings has recently discovered by Cassini science images (Porco et al. 2005, Figs. 5A and 5F therein), UVIS observations (Colwell et al. 2006, 2007), Visual and Infrared Mapping Spectrometer observations (Hedman et al. 2007), and diffraction of coherent radio waves (Thomson et al. 2007). Infrared observations of Saturn's rings by Cassini CIRS also indicate the presence of fine density structure in the A ring (Layrat et al. 2008). The revealed by Cassini periodic fine-scale structure was found only in limited regions of the rings, where particles are densely packed together, such as the B ring and the innermost part in the A ring; particle clumps, called the self-gravity wakes, separated by nearly empty gaps. Both nonaxisymmetric structures that have a characteristic trailing orientation of $\sim 20^\circ$ relative to local direction of orbital motion (Colwell et al. 2006, 2007; Hedman et al. 2007) and axisymmetric structures (Thomson et al. 2007) were indicated, characterised by a periodic radial variation in optical depth. The spacing of the microstructure vary from 30 m to 250 m, depending on the location in the rings. Both spacecraft missions have shown that these relatively large "irregular variations" in optical depth with the appearance of record-grooves (Brahic 2001, Figs. 7–9 therein; Cuzzi et al. 2002, Fig. 2b therein; Esposito 2002, Fig. 5 therein) are not associated with any resonances with known satellites. Surprisingly, on a small scale the rings have been observed to be undergo variation and oscillations with time and ring longitude (Smith et al. 1982). The latter indicates that probably the irregular variations are wave phenomena, and different instabilities of small-amplitude gravity perturbations (e.g., those produced by a spontaneous disturbance or, in rare cases, by a companion system) may play important roles in ring's dynamics. The wavelet analysis of the structure of A, B, and C rings has shown that these rings exhibit various wave perturbations, which weakly interact with each other (Postnikov & Loskutov 2007).

Almost regular density wave structure has been also detected by Salo (1992, 1995) and others (Richardson 1994; Osterbart & Willerding 1995; Griv 1998, 2005a; Daisaka & Ida 1999; Ohtsuki & Emori 2000; Griv & Gedalin 2005; Griv et al. 2004, 2006a) in simplified N-body simulations of an orbiting patch of Saturn's rings. It was shown that a wave-like spiral structure ("wake" structure) is formed by the self-gravitational instability and that in such situations coherent motion of particles in spiral arms is dominant rather than random motions, which leads to an increase in velocity dispersion of particles. [1] In the process, the disk self-gravity is essential, while particle's physical impacts play just a modest role (Daisaka & Ida 1999); density wave structure develops even in a collisionless system (Griv et al. 1999). Griv & Gedalin (2003, 2005), Griv et al. (2004, 2006a), and Griv (2005a) have already explained the computer-generated wave-like spiral structure in terms of the gravitational instability.

On the other hand, there are regular ringlet complexes in the A, B, and C rings connected to resonances with external satellites, including Lindblad horizontal resonances and vertical resonances (Holberg et al. 1982; Lissauer & Cuzzi 1982). The structures associated with this kind of resonances are directly observed as the so-called wave trains (Shu et al. 1983; Shu 1984; Rosen et al. 1991; Esposito 2002, p. 1752; Porco et al. 2005, Figs. 5G and 5H therein; Sicardy 2005, p. 463; Tiscareno et al. 2006, 2007). In this process the ring self-gravity is not essential. The trains are the density waves that decay as they propagate away from the resonances, e.g., a strongly damping wave from Mimas 5:3 resonance in the A ring (Esposito 2002, Fig. 6 therein). The waves can be used as diagnostic to obtain fundamental physical parameters that characterize the dynamical state of the ring such as mass, thickness, and collision velocities. Most of the structures in Saturn's main rings, however, are do not correspond to resonances with known satellities: wave trains associated with known resonances cover less than 1% of the radial extent of the A and B rings (Goldreich & Tremaine 1982; Horn & Cuzzi 1996). The study of these regular wave ringlets is beyond the scope of the present paper. Also, we do not consider few truly isolated ringlets with ad-

[1]In simplified simulations dynamics of particles in small regions of the disk are assumed to be statistically independent of dynamics of particles in other regions. One has to realize, however, the shortcomings of simplified N-body simulations are the neglection of gravitational forces from distant particles, the neglection of nonlinear higher order effects, and the use of the periodic boundary conditions (Griv et al. 1999; Huber & Pfenniger 2001). Nothing similar exists in nature. Furthermore, a critical role in the behavior of systems dominated by long range interactions is played by their exact geometry. Such simulations, therefore, provide us with results which can serve only as a convenient starting point for more complicated realistic simulations.

jacent empty gaps, located in the low-density C ring, in the inner B ring, in the A ring, and in the Cassini Division, resembling those of Uranus (Porco & Nicholson 1987; Porco 1990; Esposito 2002, p. 1756). Many of the narrow ringlets with typical widths of a few tens of kilometers and extremely sharp edges are found in the isolated resonance locations of different satellites. Examples of the isolated narrow ringlets in the empty gaps are the ringlet in the Prometheus 2:1 inner Lindblad resonance at the outer C ring (Rosen et al. 1991). The outer edges of Saturn's A and B rings coincide with strong isolated resonances. An adequate theoretical explanation for these isolated narrow ringlets is still missing (Willerding 1986; Hanninen & Salo 1995; Goldreich et al. 1995; Melita & Papaloizou 2005; Griv 2007a). An external satellite takes angular momentum from the particles of a disk, and the resultant angular momentum transfer can open gaps or terminate rings. One associates the occurrence of the Cassini Division with the action of a spiral wave which existed earlier from a 2:1 resonance with Mimas (Goldreich & Tremaine 1978a).

The number of mechanisms producing the ubiquitous fine-scale structure of Saturn's rings grows in the last three decades. In a review by Griv & Gedalin (2003) seven mechanisms are listed. However, along with this growth in the number of known mechanisms (a new one has been suggested by Tremaine 2003), there has been a growth in understanding of the fact that a *universal* mechanism may generate the density structure in all regions of Saturn's main rings. We shall try to present the subject of Saturn's rings instabilities in such a way as to emphasize such a universal mechanism as exists. We regard the periodic fine-scale structure of rings about Saturn as a wave pattern, which does not remain stationary in a frame of reference rotating around the planet at a proper speed, excited as a result of Jeans' gravitational instability. Accordingly, the nonaxisymmetric (spiral) structure rotates uniformly although the material rotates differentially; the spirals (and rings) consist of different material at different times.

A very popular model of the particles in Saturn's rings is a smooth ice sphere, whose restitution coefficient is quite high (exceeding 0.63) and decreases as the collision velocity increases (Goldreich & Tremaine 1978b; Bridges et al. 1984; Kerr 1985). In our model, Saturn's rings consist of primarily large $>$ cm size identical, almost elastically colliding, and gravitating particles. The model formation is thought to start with inelastically colliding particle settling to the central plane of a rotating protoring cloud to form a thin and relatively dense disk around the plane. Because of inelastic physical collisions between particles, the disk radiates heat from its surface, and, therefore, it cools down and becomes thinner and thinner. Subsequently, as a result of local instability, on attaining a certain critical thickness, small in comparison with the outer radius of the system R (and, correspondingly, very low temperature), the disk disintegrated spontaneously into a number of separate rings and spirals.[2] It is natural to assume that the growth rate of the gravity perturbations of an originally stable, slowly evolving protoring cloud was rather small. Therefore, we may investigate perturbations with values relatively small as compared to equilibrium (though

[2]Destabilizing self-gravity in much more "dangerous" in thin disks than in thick disks. If a rotating disk has a large vertical thickness owing to a high internal temperature, then it is stabilized against all gravitational instabilities. Instabilities arise as the thickness of the disk is reduced (Safronov 1980; Shu 1984). We are interested only in thin astrophysical disks with a ratio of the half-thickness h to the radius R much smaller than unity: $h/R \ll 1$. This ratio is characteristic for all spiral galaxies, the planet-forming disks of protostars, and planetary rings.

with finite amplitude).

This paper addresses the stability of unforced density waves propagating in Saturn's rings. We present a discussion of various aspects of the rings' dynamics, collective phenomena, WKB approximation, and proceed via linearized hydrodynamical treatment to derive both a marginal stability condition and a dynamical "heating." A connection between several plasma physics phenomena and the dynamics of Saturn's rings is established. Similarities between self-gravitating systems and ordinary plasmas arise from the common long-range nature of the basic forces, whereas differences arise from the opposite signs of these forces. It seems that such a connection deepens our understanding of the nature of this phenomenon and broadens the reader audience. To emphasize it again, the present work is intended to draw parallels between plasmas and Saturn's rings as well as disklike galaxies and protoplanetary disks. These systems share common dynamics, even though their physics and spatial scales are very different (Goldreich & Tremaine 1982; Tremaine 1989).

The difficulties for a satisfactory understanding of the dynamics of particulate systems are due to the well-known fact that in a system of N gravitationally interacting particles Debye screening, as distinct from plasma, is absent. This circumstance makes the statistical description of gravitational systems more complicated. The other basic difference with respect to ordinary plasma physics is that all gravitational systems are naturally inhomogeneous, because gravitational forces between gravitational "charges" (which are always "charges of the same sign") are always attractive. The density of the gravitating disk decreases towards its periphery. This inherent inhomogeneity character is the origin of some mathematical problems. For instance, in the case of quasi-stationary Saturn's rings one has to deal with complex boundary problems while in a plasma the model of a homogeneous infinitely extended background already allows one to obtain many familiar effects. Folowing plasma physics, however, the problem may be simplified by considering the so-called local WKB approximation.[3] Under the local consideration (in the vicinity of a given point) of perturbations with scales small compared to a characteristic linear dimention R of the system, one may assume parameters of the stationary state equal to its values in a given point.

The Saturnian ring disk of mutually gravitating particles, the behavior of which is governed by collective effects, is highly dynamic and is subject to various instabilities of gravity perturbations (Maxwell 1859; Goldreich & Tremaine 1982; Shu 1984; Griv et al. 2000a). This is because the evolution of the system is primarily driven by angular momentum redistribution (Goldreich & Tremaine 1982). The system may then fall toward the lower potential energy configuration and use the energy so gained to increase its coarse grained entropy (Griv & Gedalin 2004). As for the present study, it is the gravitational instability in the Saturnian rings which is responsible for the redistribution of angular momentum and hence the evolution of the disk as a whole.

In the dynamics of Saturn's rings the physics of the collective interactions should be supplemented by taking into additional account physical collisions between particles. In plasma physics, methods for investigating oscillations and the stability of a collisional system have been developed using the exact Boltzmann integral formulation or the model integral for elastic particle collisions. Reviews of plasma kinetic theory, taking into ac-

[3]From the initials of Wenzel, Kramer, and Brillouin who initially introduced this approximation in a different subject matter.

count collisions between particles, are given by Mikhailovskii (1974) and Alexandrov et al. (1984). Such calculations are very complex mathematically: the Boltzmann equation is nearly intractable because of the complicated collision integral. In this work, we make the study by using the greatly simplified theory based on hydrodynamical equations (Alexandrov et al. 1984, p. 47; Landau & Lifshitz 1987). That is, we assume that the motion of the particles is so strongly correlated that the system behaves like a neutral conducting fluid. The greatest defect of this approach is that it does not take into account detailed mechanisms of the inelastic interaction such as spin degrees of freedom, the particle size distribution, the finite size of the particles, etc. (Shukhman 1984; Araki & Tremaine 1986; Araki 1988). In particular, since collective self-gravity becomes most important in high densities, one needs carefully to include the nonlocal pressure terms arising via frequent physical impacts (Mosqueira 1996). Since the spin of a particle in Saturn's rings is comparable to the orbital frequency, a large ammount of energy may be stored in this degree of freedom. The latter increases the dissipativity of the system, and thus, accelerates the formation of waves. The coefficient of restitution in elastic collisions can be a function of the impact velocity and decreases as the collision velocity increases (Bridges et al. 1984), and this fact should be also taken into account by modelling in great detail the equilibrium distribution and dynamical evolution of Saturn's rings. All such minor effects can be included in the analysis if necessary. The present work has precedents in earlier studies of gravity disturbancies in galactic disks and protoplanetary disks (Lin & Shu 1966; Lin et al. 1969; Lau & Bertin 1978; Lin & Lau 1979; Bertin 1980, 2000; Morozov 1980; Bertin et al. 1989; Bertin & Lin 1996; Griv et al. 1999, 2008; Griv 2007b).

Like a high-temperature plasma, a system of particles of Saturn's rings exhibits collective modes of motions: because of its long-range Newtonian forces, a self-gravitating medium (a particulate "gas," say) would possess collective motions in which all the particles of the system participate. These properties would be manifested in the behavior of small gravity perturbations arising against the equilibrium background. Collective processes—modes of motion in which the particles in large regions move coherently, or in unison—are completely analogous to two-body collisions, except that one particle collides not with another one but with many which are collected together by some coherent process such as a wave. The collective processes are random, and usually much stronger than the ordinary two-body collisions and leads to a random walk of the particles that takes the complete system towards a thermal quasi-steady state. Thus, relaxation in particulate systems could occur without ordinary physical collisions through the influence of collective motions of the particulate gas upon the particle distribution (Kulsrud 1972; Lin & Bertin 1984; Bertin & Lin 1996, p. 72; Bertin 2000, p. 68; Griv et al. 2006b).

Since collisions tend to dissipate the highly ordered motions involved in wave propagation, collisions will lead to wave dissipation. Except for rings of a very small collision frequency, impacts are certain to dominate over individual gravitational encounters. Some of the properties of physical collisions in Saturn's rings are discussed in the article. The stabilizing influence of relatively rare and weak interparticle collisions on the development of Jeans' gravitational instability is examined in Appendix A below.

It is convenient to divide instabilities into two broad classes: (a) macroscopic and (b) microscopic. The macroscopic, or hydrodynamic instabilities imply the displacement of macroscopic portions of rings—all the particles in a given macroscopic volue execute the same average motion. The gasdynamic equations and the continuity equation are necessary for the theoretical analysis of this class of instabilities. The microscopic, or kinetic instabilities can be defined as those for which the differences in the motion of different particles in the same volume are important. The Boltzmann (Vlasov) equations are necessary for the analysis of these instabilities. The aim of this paper is to present the theory of only Jeans' macroscopic instabilities.

The present linearized theory assumes that the departure of the system from dynamical equilibrium is infinitesimal and then ask whether this infinitesimal departure grows or decays. We assume that the ring self-gravity plays a major role in the observed phenomenon. Arbitrary perturbations are expressed as a superposition of eigenmodes, with each eigenmode evolving independently. It is clear that the linear theories are incapable of answering many of the questions which are of major importance. Other problems, which can be treated within the framework of nonlinear theory (e.g., Griv et al. 2003a, b), are account of the reaction of the excited oscillations on the equilibrium parameters of the system and the determination of the amplitude of the oscillations that are produced. Our discussion of natural nonlinear effects is given below in this paper by extending the theory into a weakly nonlinear, or quasi-linear regime (§ 4.). The quasi-linear approach to nonlinear theory is usually referred to as the theory of weak turbulence, i.e., the case when the dynamics of the system can be described in the language of weakly interacting linear waves. There are many random collective oscillations present in the system and it is permissible to treat the phases of these oscillations as being random in some sense. It can be justified if the energy in the excited spectrum is small compared with the total mechanical energy in particles but large compared to thermal noise. The theory of strong turbulence is still far from complete. Notice that although the linear theory does not establish the amplitude of the perturbations, it does yield values of their dispersion properties and by means of stability criteria it can determine some of the equilibrium parameters of the system.

The Jeans instability (see, e.g., Bertin 2000, p. 137) is set in when the destabilizing effect of the self-gravity in the disk exceeds the combined restoring action of the pressure and Coriolis forces. The instability can be an efficient mechanism to generate turbulence in disks. The main force in this gravitational instability process is the self-gravitation force, hence the energy source for the Jeans instability is the gravitational potential energy of the matter involved in unstable oscillations. The typical growth time is of the order of several rotation periods. The wave propagation is a process of rotation as a solid about the center at a fixed phase velocity, despite the general differential rotation of the system. The classical Jeans instability of gravity disturbances in sufficiently flat, rapidly rotating systems is one of the most frequent and most important instabilities in the stellar and in the planetary cosmogony, and galactic dynamics; according to the standard cosmological model, gravitational instability is the main process responsible for the formation of observed structures in the universe. The term gravitational instability, as introduced by Jeans (1929), deals with the question of whether initial density fluctuations will be amplified or will die down. Jeans instability, which is algebraic in nature, identifies *nonresonant* instabilities of gravity fluctuations associated with almost aperiodically growing accumulations of mass, and the dynamics of Jeans perturbations can be characterized as a fluidlike wave–particle interaction. In other words, the instability associated with departures of macroscopic quantities from the dynamical equilibrium is hydrodynamical in nature and has nothing to do with any explicit resonant $\omega = \boldsymbol{k}\cdot\boldsymbol{v}$ effects, where ω is the oscillation frequency, $\boldsymbol{k}$ is the wavenumber, and $\boldsymbol{v}$ is the particle's velocity. Following Lin and Shu (Lin et al. 1969), a relatively simple hydrodynamical model can be used to investigate the instability (Lau & Bertin 1978; Lin & Lau 1979; Morozov 1985; Montenegro et al. 1999; Griv 2006); a kinetic description yields results almost no different frome those obtained hydrodynamically. In a general sense, the instability represents the ability of a gravitating system to relax from a nonthermal state by collective collisionless processes in much less time than the binary collision time. Apparently, Ginzburg et al. (1972) first examined the possible gravitational instability of Saturn's rings of the type discussed by Lin and Shu in context of the formation of spiral arms of normal galaxies.[4] Nonresonant, or algebraic instabilities are well known in plasmas, e.g., electrostatic bunching instabilities or a firehose instability (Sagdeev & Galeev 1969, p. 67; Ichimaru 1973; Galeev & Sagdeev 1983). In plasma physics an instability of the Jeans type is known as the negative-mass instability of a relativistic charged particle ring or the diocotron instability of a nonrelativistic ring that caused azimuthal clumping of beams in synchrotrons, betatrons, and mirror machines (Davidson 1974)

Usually, one chooses for the model of Saturn's rings an infinitesimally thin disk of particles with the surface density equal to the projection of the full mass density on the plane perpendicular to the rotating axis, i.e., a disk, the equilibrium half-thickness h of which is many times less than the perturbation radial wavelength $\lambda_r = 2\pi/k_r$, where k_r is the radial wavenumber. It has been stated that such an accuracy is sufficient for the discussion of oscillation modes in Saturn's rings of very small disk thickness.[5] The standard Lin–Shu dispersion relation for planar density waves (Lin et al. 1969; Rohlfs 1977, p. 100; Binney & Tremaine 1987, p. 360; Bertin & Lin 1996, p. 77; Bertin 2000, p. 192) has been slightly improved either in a heuristic manner (Lin & Shu 1968; see also Safronov 1980) or by introducing an approximate reduction factor providing the correction for finite thickness (Vandervoort 1970). See Shu (1984) and Bertin (2000) for a discussion.

Thus, a hydrodynamical model can be used to investigate the stability of Jeans perturbations. The relevant equations are the standard Euler and continuity gasdynamic equations supplemented by an equation of state and the Poisson equation. This approach is relatively simple and, far from resonances, results are found to be in good agreement with the more rigorous kinetic analysis explored by Griv & Gedalin (2003) and Griv et al. (2000, 2003a, b, 2006a).

We solve a self-consistent system of the gasdynamic equations and the Poisson equation describing the motion of a self-gravitating ensemble of particles, looking for time-dependent waves which propagate in a rapidly and nonuniformly rotating disk. In the analysis performed below, the thin disk approximation is adopted, and therefore one deals with

[4]Both optical and near-infrared observations of pre-main-sequence stars of intermediate mass have also revealed the structure of rings and spirals, and thus presumably the Jeans instability of axisymmetric and non-axisymmetric perturbations, in the circumstellar disk with structure more than 100 AU from the parent star (Grady et al. 2001; Clampin et al. 2003; Corder et al. 2005; Fukagawa et al. 2004, 2006). Apparently, Bodenheimer (1974) and Cassen et al. (1981) first explored the scenario of planet formation via gravitational instability using numerical simulations. See also Tomley et al. (1991, 1994), Laughlin & Bodenheimer (1994), Mayer et al. (2002, 2007), Pickett et al. (2003) Boss (2005, 2007), Durisen et al. (2007), and Griv (2007b) for a discussion.

[5]Based on recent Cassini images of Saturn's rings, Tiscareno et al. (2007) have placed upper limits on the Cassini Division thickness (3–4.5 m) and the inner A ring thickness (10–15 m).

vertically integrated quantities. Limiting ourselves to the case of infinitesimally thin disk simplifies the algebra without introducing any fundamental changes in the physical results. Self-gravitating evolution of a thick disk is generally very similar to that of a razor-thin disk, because the induced motions are almost planar (see, e.g., Fig. 9 below). One qualitative difference the disk's finite but small thickness makes is that it tends to be stabilizing by reducing self-gravity at the midplane (Toomre 1964; Vandervoort 1970; Safronov 1980; Shu 1984; Romeo 1992; Osterbart & Willerding 1995). N-body simulations have confirmed that the perturbed motion takes place predominantly in the plane of the disk and the primary effect of performing of simulation in three-dimensions is just a very slight reduction of the growth rate of gravitational instabilities (Hohl 1978; see also Griv et al. 2006a). The latter justifies the two-dimensional treatment of the main part of a rotating disk. In our study we restrict the analysis to a treatment of Jeans' "sausage-like" perturbations (Kulsrud et al. 1971; Bertin & Casertano 1982; Bertin & Lin 1996, p. 73; Bertin 2000, p. 104) which are symmetric with respect to the $z = 0$ equatorial plane of the disk (which do not cause it to bend). The perturbed pressure, density, gravitational potential, and horizontal velocity components are even functions of z, while the perpendicular velocity v_z is odd in z: $v_z(z) = -v_z(-z)$, in particular, $v_z(z = 0) = 0$. See Fig. 1 for an explanation. The even "sausage-like" perturbations, forming the basis of the Lin–Shu density wave theory, can release gravitational energy and are subject to Jeans' gravitational instability. These perturbations are associated with such phenomena as, for example, the appearance of the spiral structure of galaxies, protoplanetary clouds, and a protolunar disk (Lin et al. 1969; 1969; Shu 1970; Bertin & Lin 1996; Takeda & Ida 2001; Griv 2006, 2007b; Griv et al. 2008). Examples of even Jeans-unstable (that is, growing) perturbations are given in both two-dimensional and three-dimensional N-body simulations of astrophysical disks (Hohl 1972, 1978; Athanassoula & Sellwood 1986; Sellwood & Athanassoula 1986; Tomley et al. 1991, 1994; Salo 1992, 1995; Richardson 1994; Osterbart & Willerding 1995; Griv 1998, 2005a; Daisaka & Ida 1999; Ohtsuki & Emori 2000; Takeda & Ida 2001; Liverts et al. 2003, Fig. 1 therein) and hydrodynamical simulations (Laughlin & Bodenheimer 1994; Laughlin & Różyczka 1996; Gammie 2001; Mayer et al. 2002, 2007; Pickett et al. 2003; Boss 2005, 2007; Durisen et al. 2007).

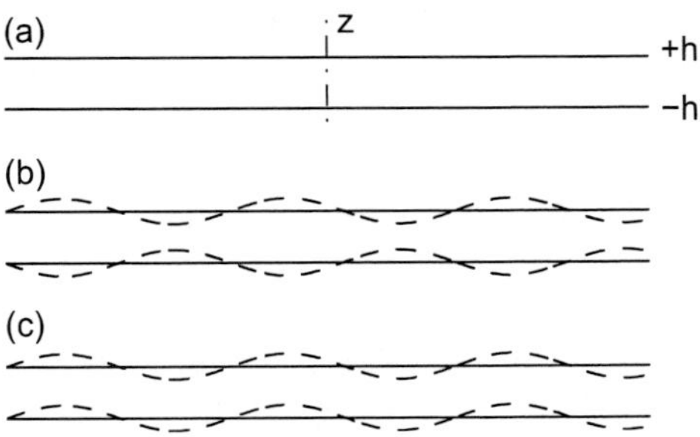

Figure 1. Sketch of perturbations of a three-dimensional disk. In (a) a section of the disk is shown edge-on. In (b) a mode of even symmetry with respect to the equatorial plane, or an even Jeans-type perturbation is shown (the dashed line). In (c) a mode of odd symmetry with respect to the equatorial plane, or an odd bending-type perturbation is illustrated (the dashed line).

The bending type of motions (Fig. 1c) can be either caused by tidal influence of a satellite (Shu et al. 1983; Shu 1984), or excited by the so-called bending firehose-type instability. The firehose instability is a collective phenomenon, so it is different from resonant excitation.[6] Contrary to the case of even perturbations, the perturbed pressure, density, gravitational potential, and horizontal velocity components are odd functions of z, while the vertical velocity of such motions is an even function of z: $v_z(-z) = v_z(z)$ and $v_z(z = 0) \neq 0$. The firehose-type instability of a sufficiently thin stellar disk has been predicted by Toomre (1966, 1983) by using the simplified theory based on moment equations. Toomre (1966) considered the collisionless analog of the Kelvin–Helmholtz instability in an infinite, two-dimensional, nonrotating sheet of stars. (See also Kulsrud et al. 1971 and Mark 1971, where the bending instability developing in nonrotating disks was investigated by using an energy principle.) This is the usual way to discuss the conditions of the firehose instability in plasma physics (Krall & Trivelpiece 1986). It has been demonstrated by Toomre that the instability is driven by the stellar "pressure" anisotropy: the source of free energy in the instability is the intrinsic anisotropy of a velocity dispersion ("temperature"). The bending perturbations do not release gravitational energy and, therefore, they are expected to be Jeans-stable (Bertin & Casertano 1982). The firehose instability is well known in plasma physics for transferring energy from one degree of freedom to another (perpendicular) degree of freedom (Ichimaru 1973). Raha et al. (1991), Griv & Chiueh (1998), Liverts et al. (2003, Fig. 2 therein), Snytnikov et al. (2004, Fig. 5 therein), and Sotnikova & Rodionov (2005) have presented nonresonant bending-unstable oscillations of three-dimensional rotating computer models. Bending waves caused by resonant particle–satellite interactions, also present in Saturn's rings, are vertical corrugations in the rings rather than compression waves (Shu et al. 1983; Shu 1984).

The organization of the paper is as follows. The basic equations of the theory and the equilibrium conditions for the particulate disk are given in § 2.. In § 3., the spectrum of oscillations is obtained. In § 4., the basic equations of the weakly nonlinear theory are presented. § 5. contains a discussion of turbulent viscosity. The predictions of the theory are verified by simplified N-body simulations in § 6.. A summary of the principal conclusions is given in § 7.. In Appendix A, the weak damping of Jeans-unstable waves by rare interparticle collisions is demonstrated. In Appendix B, the existence of solutions to gasdynamic equations of the form of normal modes is examined.

2. Dynamical Equilibrium

An extended flat disk of identical mutually gravitating particles orbiting the planet is studied, with M_p being Saturn's mass, M_d being the mass of the disk, and $M_\mathrm{d}/M_\mathrm{p} \ll 1$. The fact that h/R is small means that the disk is considered by us is rather cold and that the pressure gradient in it is much smaller than the two main forces—the gravitational and the centrifugal forces. Newton's equations of three-dimensional motion for any individual

[6]Actually, the instability has nothing whatsoever to do with a gyrating firehose, whose motion is driven by water gushing from the unsupported nozzle.

particle of unit mass in an inertial frame, with the origin at the disk center, are given by

$$\frac{\mathrm{d}^2 r}{\mathrm{d}t^2} = r\,(\dot{\varphi})^2 - \frac{\partial \Phi_\mathrm{p}}{\partial r} - \frac{\partial \Phi_\mathrm{d}}{\partial r} - \nu_\mathrm{coll} v_r \,, \tag{1}$$

$$\frac{\mathrm{d}}{\mathrm{d}t}\left(r^2 \dot{\varphi}\right) = -\frac{\partial \Phi_\mathrm{p}}{\partial \varphi} - \frac{\partial \Phi_\mathrm{d}}{\partial \varphi} - \nu_\mathrm{coll} v_\varphi \,, \tag{2}$$

$$\frac{\mathrm{d}^2 z}{\mathrm{d}t^2} = -\frac{\partial \Phi_\mathrm{p}}{\partial z} - \frac{\partial \Phi_\mathrm{d}}{\partial z} - \nu_\mathrm{coll} v_z \,, \tag{3}$$

where the dots indicate time derivatives, $\Phi_\mathrm{p}(\boldsymbol{r}, t)$ is the planetary gravitational potential, $\Phi_\mathrm{d}(\boldsymbol{r}, t)$ is the disk potential, the friction term $\boldsymbol{F} = -\nu_\mathrm{coll}\boldsymbol{v}$ approximates the force produced by physical impacts, $\nu_\mathrm{coll} = n\langle sv\rangle$ is the effective collision frequency, n is the number density of particles, s is the effective radius of a particle, $\langle\cdots\rangle$ denotes the average over particles of all random velocities v, (r, φ, z) are planetocentric cylindrical coordinates, and the axis of the disk rotation is along the z-axis. In a self-consistent problem, Eqs. (1)–(3) must be solved simultaneously with the Poisson equation and the continuity equation.

Let us assume that $M_\mathrm{d}/M_\mathrm{p} \to 0$. Consider a free particle orbiting in a circle of radius r with angular speed $\Omega(r)$ in the equatorial plane $z = 0$ of the non-spherical planet with an associated gravitational potential $\Phi_\mathrm{p}(r, z)$. The equilibrium of a system is expressed by the condition

$$r\Omega^2 = \left.\frac{\partial \Phi_\mathrm{p}}{\partial r}\right|_{z=0} . \tag{4}$$

If this test particle is displaced by an arbitrary small amount, it will oscillate freely in the horizontal and vertical directions about the reference circular orbit with epicyclic frequency $\kappa(r)$ and vertical frequency $\mu(r)$ given by Lindblad's theory of epicyclic motion (Shu 1984; Borderies & Longaretti 1994; Griv et al. 1999),

$$\kappa^2(r) = r^{-2}\frac{\mathrm{d}}{\mathrm{d}r}\left[\left(r^2\Omega\right)^2\right] , \tag{5}$$

$$\mu^2(r) = \left.\frac{\partial^2 \Phi_\mathrm{p}}{\partial z^2}\right|_{z=0} . \tag{6}$$

In the epicyclic approximation, the motion of a particle is represented as in epicyclic motion along the small ellipse (epicycle) with a simultaneous circulation of the epicenter about the planetary center (Rohlfs 1977, p. 52; Goldreich & Tremaine 1982; Binney & Tremaine 1987, p. 103; Bertin 2000, p. 153). Of course, the epicyclic approximation may be applied only when the actual particle motion is nearly circular, as in planetary rings, and the collision frequency is small in comparison with $\kappa \approx \Omega \approx \mu$. It follows from Eq. (5) that a nearly circular orbit is stable in the equatorial plane of the disk if $\kappa^2 > 0$; in other words, if Φ_p tends to $-\infty$ as $-\mathrm{const}/r^n$ with $n \leq 2$. It is clear, however, that the stability of the individual particle orbits will not guarantee the stability of an actual system ($M_\mathrm{d}/M_\mathrm{p} \neq 0$) against collective oscillations when in addition the time-dependent perturbation of the basic *total* equilibrium potential is taken into account: the problem may be resolved only by considering the self-consistent system.

Near the equatorial plane the planetary potential is

$$\Phi_\mathrm{p} \approx -\frac{GM_\mathrm{p}}{r}\left[1 + \frac{1}{2}J_2\left(\frac{R_\mathrm{p}}{r}\right)^2\right] , \tag{7}$$

where G is Newton's constant, R_p is the planet's radius, J_2 is its second multipole moment, and $J_2(R_\mathrm{p}/r)^2 \ll 1$. Using the potential given by Eq. (7), we have

$$\left\{ \begin{array}{c} \Omega^2(r) \\ \kappa^2(r) \\ \mu^2(r) \end{array} \right\} = \frac{GM_\mathrm{p}}{r^3} \left[1 + \left\{ \begin{array}{c} A \\ B \\ C \end{array} \right\} J_2 \left(\frac{R_\mathrm{p}}{r} \right)^2 \right] ,$$

with $A = 3/2$, $B = -3/2$, and $C = 9/2$ (Goldreich & Tremaine 1982; Shu 1984; Borderies & Longaretti 1994). The presence of non-spherical harmonics thus lead to having instead of a single characteristic frequency $\Omega = \kappa = \mu$ three slightly different ones: $\mu > \Omega > \kappa$. One calls this a multiplication of frequencies, and there will be different types of resonance in accordance with the existence of different harmonics. In the Saturnian system, $R_\mathrm{p} \sim 60,00$ km, $J_2 \sim 10^{-3}$ at $r = R_\mathrm{p}$, and $M_\mathrm{d}/M_\mathrm{p} \sim 10^{-7}$. In practical work, one can accept $\Omega = \kappa = \mu$.

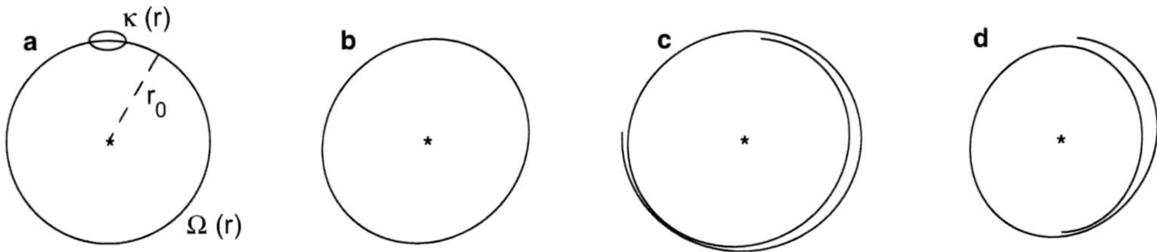

Figure 2. A schematic representation of epicyclic motion in the Saturnian ring disk with nearly circular particle motions. The disk angular velocity is $\Omega(r)$, the epicyclic frequency is $\kappa(r)$, and the direction of the disk rotation is clockwise. The current particle radius vector is $\boldsymbol{r}(t) = \boldsymbol{r}_0 + \boldsymbol{r}_1(t)$, where vector $\boldsymbol{r}_0$ uniformly rotates with angular velocity $\Omega = \Omega(r_0)$ and $|r_1/r_0| \ll 1$. In a coordinate system that rotates with velocity $\Omega(r_0)$, the particle moves along the epicycle in a retrograde sense. Particle orbit: (a) epicyclic orbit in a rotating frame ($\kappa = \Omega$), (b) closed orbit in an inertial frame ($\kappa = \Omega$), (c) rosette orbit in an inertial frame ($\Omega > \kappa$), and (d) rosette orbit in an inertial frame ($\Omega < \kappa$). The resulting motion in the inertial frame is a rosette orbit, generally not closed.

The problem of epicyclic motion in its most general form is equivalent to the problem of the motion of an electrically charged particle in a given electromagnetic field, in which the solution can be decomposed into two parts: the guiding center motion and the epicyclic motion. The role of the magnetic part of the Lorentz force is assumed by the Coriolis force, and the epicycle radius used in particle dynamics is analogous to the gyroradius in a plasma (Marochnik 1966). This is an important source of analogies between Saturn's rings dynamics and plasma physics. In the problem of a gravitational medium the analog of the plasma cyclotron frequency is the epicyclic frequency κ. A characteristic feature of the epicycles in Saturn's rings, in contrast to a plasma, is that they are elliptical: the ratio of the semiaxes of the ellipse in the r and φ directions is approximately $\kappa/2\Omega \approx 0.5$ (Fig. 2a). The motion along the epicycle proceeds in the opposite direction to the primary rotation (to conserve an angular momentum). In Saturn's rings, the resulting motion in the inertial frame is a rosette orbit, almost closed ($\kappa \approx \Omega$). In the Kepler's problem, $\kappa = \Omega$, after one complete revolution along the circular orbit and one complete revolution along the epicycle, the particle will occupy its original position.

Some simple relations exist between optical depth, ring thickness, particle density, and other quantities that define the physical state of a planetary ring. Thus the optical depth is approximately equal to the ratio of the collision frequency $\nu_{\rm coll}$ to the orbital frequency Ω:

$$\tau \approx \pi \frac{\nu_{\rm coll}}{\Omega} \tag{8}$$

(Jeffreys 1947; Cook & Franklin 1964; Goldreich & Tremaine 1982; Bridges et al. 1984; Stewart et al. 1984). This is fundamental result which was nicely rediscovered by Shu & Stewart (1985), employing a Krook approximation to the collision integral. In what follows we argue that fine-scale $\lesssim 100$ m structures could be primarily produced by the classical Jeans instability in low and moderately high optical depth regions of the system under study with $\tau < 3$, that is, in regions of the system with relatively rare particle's impacts, $\nu_{\rm coll} < \Omega$. Such regions which can be modeled as a rarefied gas can be found in the main parts of the rings A, B, and C.

The unperturbed disk is assumed to have no motion except for rotation. The present theory suggests some perturbed radial and azimuthal motions of the fluid element distributed in the form of a spiral-like flow field which is a small correction to the basic circular, equilibrium motion. In the plane, the equilibrium motion is described by the following equation:

$$r\Omega^2 = \frac{\partial \Phi_0}{\partial r} + \frac{c_{\rm s}^2}{\rho_0}\frac{\partial \rho_0}{\partial r}\,, \tag{9}$$

where $\Phi_0(r,z)$ is the mean total potential, $\rho_0(r)$ is the mean volume mass density, $c_{\rm s}$ is the speed of sound which is a measure of the thermal motion of particles (the random velocity spread), and the term $\propto c_{\rm s}^2$ is a small correction motion. As is seen, planar equilibrium is established in a simple manner in such a disk, i.e., it is governed mainly by the balance between the centrifugal and gravitational forces, $r\Omega^2 \approx \partial\Phi_0/\partial r$. According to observations, in the Saturnian rings $c_{\rm s}/r\Omega \sim 10^{-6}$. The unperturbed disk has velocity $\boldsymbol{v}_0 = (0, r\Omega, 0)$, where the angular rotational velocity $\Omega = \Omega(r)$ is taken to be a function of r alone (Binney & Tremaine 1987; Fridman & Gorkavyi 1999, p. 341).

The equation of hydrostatic equilibrium along the z coordinate (for $z \ll r$) is obviously

$$\frac{\partial \Phi_0}{\partial z} + \frac{c_{\rm s}^2}{\rho_0}\frac{\partial \rho_0}{\partial z} = 0\,. \tag{10}$$

By considering the thin disk, $2h \ll R$, where $2h$ is the effective thickness, in Eq. (10) one can expand $\partial\Phi_0/\partial z$ about the orbit plane as

$$\frac{\partial \Phi_0}{\partial z} = \mu^2 z \tag{11}$$

and $\mu^2 = (\partial^2\Phi_0/\partial z^2)|_{z=0}$ is the frequency of natural vertical oscillations. Equations (10) and (11) then imply

$$\rho_0(r,z) = \rho_0(r,0)\exp\left(-\frac{\mu^2 z^2}{2c_{\rm s}^2}\right)\,, \tag{12}$$

where $\rho_0(r,0)$ is the density on the equatorial plane. The disk is geometrically thin if $2h \ll r$, which from the equation $h \approx c_{\rm s}/\Omega$, is equivalent to the disk being dynamically cold, $c_{\rm s} \ll r\Omega$, i.e., the sound speed of the ring disk is much less than the orbital speed.

To reiterate, since realistic Saturn's rings are extremely difficult to treat, we shall consider a simple idealized model. First, we consider the simplified case of dilute rings whose particles are not tightly packed, $\nu_{\rm coll} < \Omega$. Second, we restrict ourselves to consideration of only Lin–Shu density waves, or "heavy sound" which are nothing but longitudinal compression waves in which the self-gravitation of the fluctuations in density is taken into account, Fig. 1b (cf. Chandrasekhar's 1955 singular modes; Goldreich & Lynden-Bell 1965a). These modes have wave vectors perpendicular to the axis of rotation. For such modes there are no variations in the z direction, the distribution of the basic gravitational potential is assumed to be symmetric with respect to the $z = 0$ plane, and the vertical velocity of the element is equal to zero. N-body experiments have already shown such longitudinal collective motions for a system of mutually gravitating particles in computer-generated Saturn's rings, that is, the particle motion is restricted to be almost parallel to the equatorial plane of the system (Griv 2005a, Fig. 5 therein; Griv et al. 2006a, Fig. 5 therein; Fig. 9 below). The modes of odd symmetry with respect to the equatorial plane as shown in Fig. 1c deserve a separate investigation. Third, the weak radial dependence of the surface mass density (and the basic gravitational potential) are assumed, thus the localized solutions are considered. It is this simple equilibrium model of the Saturnian ring disk that is to be examined for stability in the present investigation.

3. Oscillation Spectrum

We begin with the set of equations involving the equations of plane hydrodynamics and the Poisson equation:

$$\frac{\partial \sigma}{\partial t} + \mathrm{div}(\sigma \boldsymbol{v}_\perp) = 0\,, \tag{13}$$

$$\frac{\partial \boldsymbol{v}_\perp}{\partial t} + (\boldsymbol{v}_\perp \cdot \boldsymbol{\nabla})\boldsymbol{v}_\perp = -\boldsymbol{\nabla}\Phi - \frac{1}{\sigma}\boldsymbol{\nabla}P_\perp\,, \tag{14}$$

$$\Delta\Phi = 4\pi G\sigma\delta(z)\,, \tag{15}$$

where σ is the surface density of the "gas" in the disk, $\boldsymbol{v}_\perp$ is the velocity of the gas element in the (r,φ)-plane, Φ is the total gravitaional potential (including the planet), $P_\perp = \int P dz$ is the plane pressure, P is the ordinary gas pressure, and $\delta(z)$ is the Dirac delta-function with respect to the spatial coordinate z. In a more detailed representation, the Poisson equation in cylindrical coordinates is given by

$$\frac{1}{r}\frac{\partial}{\partial r}\left(r\frac{\partial \Phi}{\partial r}\right) + \frac{1}{r^2}\frac{\partial^2 \Phi}{\partial \varphi^2} + \frac{\partial^2 \Phi}{\partial z^2} = 4\pi G\sigma\delta(z)\,. \tag{16}$$

Notice that the Poisson equation for the electric potential differs from the Poisson gravitational equation (15) by the sign on the right-hand side. For the five unknown functions, namely v_r, v_φ, σ, $P_\perp$, and Φ, we have so far written down four equations. The fifth equation which is needed to close the set of equations is the equation of state, $P_\perp = P_\perp(\sigma, c_{\rm s}^2)$. In the following, we will assume that the matter in the disk satisfies the barotropic equation of state

$$P_\perp = P_\perp(\sigma)\,. \tag{17}$$

The original nonlinear equations (13)–(14) together with the Poisson equation (15), the equation of state (17), and appropriate boundary conditions give a complete description of the problem for disk modes of collective oscillations. This set of equations is called the set of equations with a self-consistent field, and is the counterpart of the system of hydrodynamic and Maxwell equations in electromagnetic plasmas. Hence the term "gravitational plasma" is appropriate for the description of mutually gravitating particles of Saturn's rings (Lin & Bertin 1984; Griv et al. 2006b). We proceed as follows. The system of Eqs. (13)–(15) may be simplified by considering particular limiting cases. In particular, we apply the standard procedure of the linear approach as already developed in plasma theory (Mikhailovskii 1974; Alexandrov et al. 1984; Krall & Trivelpiece 1986; Swanson 1989). In the following, using the WKB, or short-wavelength approximation, we will assume that the wavelength is shorter than the characteristic lengths of change of the stationary parameters of the disk. The latter assumption makes it possible to consider the disk essentially to be infinite and releases us from taking into account the boundary conditions.

3.1. Perturbation

The time-dependent surface density $\sigma(\boldsymbol{r}, t)$, the total gravitational potential $\Phi(\boldsymbol{r}, t)$, the pressure $P_\perp(\boldsymbol{r}, t)$, and the fluid velocity $\boldsymbol{v}_\perp(\boldsymbol{r}, t)$ of a spatially inhomogeneous along the r coordinate disk are splited up as

$$X(\boldsymbol{r}, t) = X_0(r) + X_1(\boldsymbol{r}, t) \,.$$

Here $X(\boldsymbol{r}, t)$ stands for any of the above mentioned physical variables, $X_0(r)$ describes the basic flow, and $|X_1/X_0| \ll 1$ represents the perturbations. These quantities σ, Φ, $P_\perp$, and $\boldsymbol{v}_\perp$ are then substituted into the equations of motion of the gas, the continuity equation, the Poisson equation, and the second order terms of the order of σ_1^2, Φ_1^2, P_1^2, $v_\perp^2$ are neglected with respect to the first order terms. The resultant equations of motion are cyclic in the variables t and φ, and hence by applying the widely used WKB method one may seek solutions in the form of normal modes (assuming that the perturbation scale is sufficiently small for the disk to be regarded as only weakly inhomogeneous) by expanding

$$X_1(\boldsymbol{r}, t) = \sum_{\boldsymbol{k}} \tilde{X}_{\boldsymbol{k}} \exp\left[\imath k_r(r) r + \imath m\varphi - \imath\omega_{\boldsymbol{k}} t\right] + \text{c.c.}\,, \tag{18}$$

where $\tilde{X}_{\boldsymbol{k}} = \text{const}$ is a real amplitude, m is the azimuthal (nonnegative) mode number, or the number of spiral arms, $\omega_{\boldsymbol{k}} = \Re\omega_{\boldsymbol{k}} + \imath\Im\omega_{\boldsymbol{k}}$ is the complex natural frequency (normal mode) of excited waves, suffixes $\boldsymbol{k}$ denote the $\boldsymbol{k}$th Fourier component, t is the elapsed time from the onset of the perturbation, $r \sim R$, $|k_r| r \gg |k_r L| \gg 1$, R is the outer radius of the disk, in Saturn's rings $R \approx 150,000$ km, $|L|$ is the scale of radial inhomogeneity, and c.c. means the complex conjugate. This is accurate for short wave perturbations only, $|k_r| r \gg 1$, but qualitatively correct even for perturbations with a longer wavelength, of the order of the disk radius. The coexistence of several spiral (and ring) waves is possible. Generally, the WKB method is a powerful approach that can be used in a large class of wave problems. The WKB treatment leads to a relatively simple phase-integral type relation as a dispersion relation from which in many cases stability boundaries can be obtained analytically. In the

WKB approximation, the radial wavenumber k_r is presumed to be of the form

$$k_r(r) = \mathcal{A}\Psi(r),$$

where $\mathcal{A}$ is a large parameter and $\Psi(r)$ is a smooth, slowly and monotonically varying function of the radial distance, i.e., $\mathrm{d}\ln k_r/\mathrm{d}\ln r = O(1)$. In the lowest, or local WKB theory we are interested in $k_r = \mathrm{const}$. In other words, one neglects all derivatives of $k_r(r)$ in the local approximation, the first-order derivatives are accounted for the next order approximation, etc. In the local WKB approximation it is assumed that the wave vector and the wavefrequency vary continuously. By utilizing the more accurate nonlocal approximation, it may be shown that in fact the characteristic oscillation frequencies of an inhomogeneous disk must be quantized, i.e., must pass through a discrete series of values (Alexandrov et al. 1986, p. 243). The existence of solutions to gasdynamic equations of the form $\exp(-\imath\omega t)$ adopted here is examined in Appendix B below. We assume that unstable perturbations develop rapidly on the instability timescale t_{inst}. See Appendix B for the definition of t_{inst}.

Evidently X_1 is a periodic function of φ, and hence m must be an integer. In the linear theory, one can select one of the Fourier harmonics:

$$X_1 = \tilde{X}\exp(\imath k_r r + \imath m\varphi - \imath\omega t) + \mathrm{c.c.}\,. \tag{19}$$

The meaning of localized solution has been discussed in plasma physics (Alexandrov et al. 1984, p. 245; Krall & Trivelpiece 1986, p. 424; Swanson 1989, p. 142). The solution in such a form represents in the plane a spiral wave with m arms or a ring ($m = 0$). With φ increasing in the rotation direction, we have $k_r > 0$ for trailing spiral patterns and $k_r < 0$ for leading ones. With $m = 0$, we have the density waves in the form of concentric rings that propagate away from the planet when $k_r > 0$, or toward the planet when $k_r < 0$. The imaginary part of ω corresponds to a growth ($\Im\omega > 0$) or decay ($\Im\omega < 0$) of the components in time, $X_1 \propto \exp(\Im\omega t)$, and the real part to a rotation with angular velocity $\Omega_{\mathrm{p}} = \Re\omega/m$. When $\Im\omega > 0$, the medium transfers its energy to the growing wave and oscillation buildup occurs. A perturbation is considered to be a superposition of different oscillation modes. This disturbance in the disk will grow until it is limited by some nonlinear effect.

The main shortcoming in our presentation is the short-wave assumption under which all the below considerations are only valid. Without this assumption, Fourier normal modes are no longer solutions of the linearized equations and one has to switch to much more complicated mathematics (e.g., Rüdiger & Kitchatinov 2000).

Schematic model of self-gravity Jeans-unstable density waves we are investigating is shown in Fig. 3. In the lowest approximation of the theory, the density waves are regularly spaced, aligned three-dimensional structures (Fig. 3a).

3.2. Perturbed Velocities

The equations of two-dimensional motion of the fluid element in the frame of reference rotating with angular velocity Ω at the reference position r_0 can be written in Hill's approx-

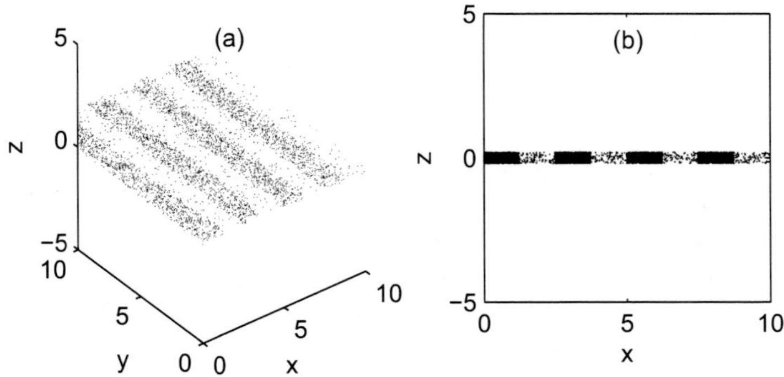

Figure 3. Schematic model of the fine-scale density wave structure in Saturn's rings. Self-gravity density waves, which were first studied by Lin and Shu on larger scales as they exist in galactic disks, manifest themselves as evenly spaced elongated clusters of ring particles. Shown are both a three-dimensional distribution of particles (a) and a distribution of particles in the (x, z)-plane (b).

imation as (Goldreich & Lynden-Bell 1965b)

$$\frac{\mathrm{d}v_r}{\mathrm{d}t} - 2\Omega v_\varphi + 2rr_1\Omega\frac{\mathrm{d}\Omega}{\mathrm{d}r} = -\frac{\partial\Phi_1}{\partial r} - \frac{c_\mathrm{s}^2}{\sigma_0}\frac{\partial\sigma_1}{\partial r}, \tag{20}$$

$$\frac{\mathrm{d}v_\varphi}{\mathrm{d}t} + 2\Omega v_r = -\frac{1}{r}\frac{\partial\Phi_1}{\partial\varphi} - \frac{c_\mathrm{s}^2}{r\sigma_0}\frac{\partial\sigma_1}{\partial\varphi}, \tag{21}$$

where v_r and v_φ are the radial and azimuthal velocities, the subscript of r_0 is dropped as we are considering linearized quantities, $\Omega \equiv \Omega(r_0)$, $c_\mathrm{s} = (\partial P_\perp/\partial\sigma)_0^{1/2}$, the influence of the planet enters through $\Omega(r)$, and we use the expansion

$$P_\perp = P_0 + \sigma_1 (\partial P_\perp/\partial\sigma)_0 \quad \text{and} \quad \big|(\sigma_1/P_0)(\partial P_\perp/\partial\sigma)\big| \ll 1\,.$$

In the absence of any perturbing gravity, $\Phi_1 = \sigma_1 = 0$, Eqs. (20)–(21) yield the ordinary epicyclic velocities

$$v_r = K\sin(\phi_0 - \kappa t)\,, \tag{22}$$

$$v_\varphi = K\frac{2\Omega}{\kappa}\cos(\phi_0 - \kappa t)\,, \tag{23}$$

where K and ϕ_0 are constants of integration. The values of r and φ coordinates of the element as functions of time are readily obtained by direct integration of Eqs. (22)–(23).

The particular solution of the system of Eqs. (20)–(21) is

$$v_r = \frac{\aleph}{\omega_*^2 - \kappa^2}\left(\omega_* k_r + \imath 2\Omega\frac{m}{r}\right), \tag{24}$$

$$v_\varphi = \frac{\aleph}{\omega_*^2 - \kappa^2}\left(\frac{4\Omega^2 - \kappa^2 + \omega_*^2}{\omega_*}\frac{m}{r} - \imath 2\Omega\omega_* k_r\right), \tag{25}$$

where $\aleph = \Phi_1 + c_s^2\sigma_1/\sigma_0$, $\omega_* = \omega - m\Omega$ is the Doppler-shifted (in a circular rotating frame) wavefrequency, $\omega_* \neq 0$, $\omega_*^2 - \kappa^2 \neq 0$, and

$$\kappa(r) = 2\Omega\left(1 + \frac{r}{2\Omega}\frac{\mathrm{d}\Omega}{\mathrm{d}r}\right)^{1/2}$$

is the epicyclic frequency. The solutions (24)–(25) describe the perturbed velocities of the element under the action of the small gravity perturbation, $|v_r|$ and $|v_\varphi| \ll r\Omega$. As is seen the present theory suggests some systematic motions of the element distributed in the form of a spiral-like flow field which is a correction to the basic circular, equilibrium motion described by Eq. (9) (cf. Yuan 1969).

A special analysis of the solution near corotation ($\Re\omega_* \equiv m(\Omega_\mathrm{p}-\Omega) = 0$) and Lindblad resonances is required. The Lindblad resonances occur when the mean motion of the wave and that the disk particles are in the ratio

$$\frac{\Omega - \Omega_\mathrm{p}}{\kappa} = \pm\frac{1}{m}\,. \tag{26}$$

The most important are strong low-order resonances with $m < 10$ (see, e.g., Goldreich & Tremaine 1982, Shu 1984, and Griv 2007a for an explanation). When conditions given by Eq. (26) hold for the upper (lower) choice of sign, we refer to r_ILR and r_OLR as the inner (outer) Lindblad horizontal resonance, respectively.

3.3. Perturbed Density

Equations (24) and (25) must be solved simultaneously with the continuity equation

$$\begin{aligned}\sigma_1 &= -\int_{-\infty}^{t}\left[\frac{1}{r}\frac{\partial}{\partial r}(\sigma_0 r v_r) + \frac{1}{r}\frac{\partial}{\partial\varphi}(\sigma_0 v_\varphi)\right]\mathrm{d}t' \\ &\approx -\int_{-\infty}^{t}\left(\sigma_0\frac{\partial v_r}{\partial r} + v_r\frac{\partial\sigma_0}{\partial r} + \frac{\sigma_0}{r}\frac{\partial v_\varphi}{\partial\varphi}\right)\mathrm{d}t'\,,\end{aligned} \tag{27}$$

where v_r and v_φ are the perturbed velocities and the relatively small term $\sigma_0 v_r/r$ is omitted, i.e., the curvature effect is neglected. This is a valid approximation if $|k_r|r$ is large (Lau & Bertin 1978; Lin & Lau 1979).

Using Eqs. (24)–(25), from Eq. (27) it is straightforward to show that (e.g., Griv 2006)

$$\sigma_1 \approx \frac{\sigma_0\aleph}{\omega_*^2 - \kappa^2}\left(k_r^2 + \frac{4\Omega^2 - \kappa^2 + \omega_*^2}{\omega_*^2}\frac{m^2}{r^2} + \frac{2\Omega}{\omega_*}\frac{m}{rL}\right) + \mathrm{c.c.}\,, \tag{28}$$

where

$$|L| = \left|\frac{\partial}{\partial r}\ln\left(\Omega\sigma_0\kappa^{-2}\right)\right|^{-1} \tag{29}$$

is the radial scale of inhomogeneity, $|L|/r \ll 1$, and the terms $\propto m$ are small corrections. Only nonresonant low-frequency ($\omega_* \neq 0$ and $|\omega_*|^2 \lesssim \kappa^2$) perturbations developing between the Lindblad resonances are considered (Griv et al. 1999). Equation (28) contains the second-order term in the bracket:

$$\frac{4\Omega^2 - \kappa^2 + \omega_*^2}{\omega_*^2}\frac{m^2}{r^2} \ll k_r^2$$

as well as the term $\propto mL^{-1}$, which play a crucial role in modifying Safronov–Toomre's stability criterion from $Q \equiv c_s/c_T \geq 1$ to $Q \gtrsim 2\Omega/\kappa \approx 2$ (§ 3.6.). Equation (28) describes the response of a differentially rotating ($\mathrm{d}\Omega/\mathrm{d}r \neq 0$, that is, $4\Omega^2 > \kappa^2$) and spatially inhomogeneous ($L^{-1} \neq 0$) disk to both radial ($m = 0$) and spiral ($m \neq 0$) perturbations. According to Lau & Bertin (1978) and Lin & Lau (1979), in Eq. (28) the term $\propto mL^{-1}$ corresponds to a wave–fluid resonance $\omega_* \to 0$, and sufficiently far from the resonance this term may be omitted. However, Griv & Gedalin (2004), Griv (2006), and Griv et al. (2008) have proved that the existence of spatial inhomogeneity is critically important for the exchange of angular momentum in the wave–particle system, and therefore the term $\propto mL^{-1}$ must be retained. The wave–fluid, or corotation resonance $\omega_* = 0$ has been studied by Lovelace & Hohlfeld (1978) and Morozov (1980). In contrast to Griv et al. (1999), in Eq. (28) only the most important low-frequency ($|\omega_*|^2 \lesssim \kappa^2$) perturbations developing in the equatorial $z = 0$ plane between the inner and outer Lindblad resonances are considered (Lin et al. 1969; Shu 1970; Griv et al. 1999). Thus, the distortion of the wave packet due to the disk spatial inhomogeneity is included through the term $\propto mL^{-1}$ in Eq. (28). By including this term one obtains also a retardation of the phase velocity of the azimuthal perturbations, the effect responsible for a trailing configuration in a disk (§ 3.7.).

Lovelace & Hohlfeld (1978) have argued that in Eq. (29) the quantity

$$f(r) = \Omega\sigma_0\kappa^{-2} \tag{30}$$

has the role of the distribution function for angular momentum. In Saturn's rings, $f(r) \approx \sigma_0\Omega^{-1} \propto r^{3/2}\sigma_0$; in A and B rings, $\partial f/\partial r \leq 0$.

3.4. Perturbed Potential

For such a form of perturbation, the Poisson equation (16) becomes

$$\left(\frac{\mathrm{d}^2}{\mathrm{d}z^2} - k^2\right)\Phi_1 = 4\pi G\sigma_1\delta(z)\,. \tag{31}$$

In the vacuum ($z > 0$ and $z < 0$), Eq. (31) is reduced to the Laplace equation

$$\Delta\Phi_2 = 0\,,$$

and, therefore, in these regions the solutions are

$$\Phi_+ \equiv \Phi_{1,z>0} = C_1 e^{-|k|z}\,, \quad \Phi_- \equiv \Phi_{1,z<0} = C_2 e^{|k|z}\,, \tag{32}$$

where C_1 and C_2 are constants. By integrating Eq. (31) over z, the boundary conditions relating Φ_1 to Φ_2 on the surface of the disk–vacuum partition ($z = 0$) are found:

$$\Phi_+ = \Phi_-\,|_{z=0}\,; \quad \left(\frac{\partial\Phi_+}{\partial z} - \frac{\partial\Phi_-}{\partial z}\right)_{z=0} = 4\pi G\sigma_1\,. \tag{33}$$

On substituting the solutions (32) into the boundary conditions (33), one obtains the required connection between the perturbed potential $\Phi_1(\boldsymbol{r}, t)$ and the perturbed surface density $\sigma_1(\boldsymbol{r}, t)$ of the infinitesimally thin disk

$$\Phi_1 = -\frac{2\pi G\sigma_1}{|k|}e^{-|k|z} + \text{c.c.} \tag{34}$$

(Bertin & Mark 1978; Lau & Bertin 1978; Lin & Lau 1979; Bertin 1980). Thus Eq. (16), an integral relation for arbitrary values of $|k_r|r$, becomes a local relation (34) in the short-wave, or WKB limit $|k_r|r \gg |k_r L| \gg 1$.

Solution (34) determines the perturbed surface density required to support the perturbed potential up to second order in the Lin–Shu asymptotic approximation (Bertin & Mark 1978; Lau & Bertin 1978; Lin & Lau 1979). In the first approximation ($m = 0$), from Eq. (34) it follows the old Lin–Shu result (Lin & Shu 1966; Lin et al. 1969; Shu 1970; Rohlfs 1977, p. 95; Binney & Tremaine 1987, p. 355): in the plane $z = 0$ maxima of the surface density correspond to minima of the potential, $\sigma_1 = -|k_r|\Phi_1/2\pi G + \text{c.c.}$. Note that the condition $|m/k_r r|^2 \ll 1$ used by Bertin & Mark (1978), Lau & Bertin (1978), and Lin & Lau (1979) to obtain the asymptotic second-order solution (34) of the Poisson equation is equivalent to the requirement

$$|\tan\psi| \lesssim 1\,, \tag{35}$$

where the pitch angle $\psi = \arctan(m/rk_r)$ is the angle between the direction of the wave front and the tangent to the circular orbit of the gas element, with "$\lesssim$" in place of "$\ll$" in contrast to the first-order approximation, $\Phi_1 = -(2\pi G\sigma_1/|k_r|)\exp(-|k_r|z) + \text{c.c.}$, of the Lin–Shu theory. The requirement (35) limits the consideration to moderately tightly-wound disturbances with pitch angle ψ less than (or equal to) 45°, which are appropriate for a large portion of Saturn's rings where $m \sim 1$. The asymptotic Lin–Shu approximation used here is actually the WKB approximation, which is tight-winding approximation for small m (Griv et al. 1999).

3.5. Dispersion Relation

Equating the density σ_1 (Eq. (28)) to the perturbed density given by Eq. (34), one obtains the generalized Lin–Shu dispersion relation $\omega_* = \omega_*(r, k)$, which determines the spectrum of collective oscillations developing in the plane $z = 0$,

$$1 = \frac{c_\mathrm{s}^2 - 2\pi G\sigma_0/|k|}{\omega_*^2 - \kappa^2}\left(k_r^2 + \frac{4\Omega^2 - \kappa^2 + \omega_*^2}{\omega_*^2}\frac{m^2}{r^2} + \frac{2\Omega}{\omega_*}\frac{m}{rL}\right), \tag{36}$$

where $2\pi G\sigma_0|k| \gg k^2 c_\mathrm{s}^2$, $k_r^2 \gg m^2/r^2$, and $|k_r|r \gg |k_r L| \gg 1$. It is valid for ring and relatively open spiral structures excluding the resonance zones. In Eq. (36), the second term in the bracket proportional to m^2/r^2 is a small correction (this is because $4\Omega^2 \sim \kappa^2$ and $m^2/k_r^2 r^2 \ll 1$) and $|\omega_*|^2 \lesssim \kappa^2$. The dispersion relation (36) is of the order of ω_*^5 and is consistent with the dispersion relation derived by Montenegro et al. (1999, their Eq. (20)). Neglecting the effects of spatial inhomogeneity ($|L| \to \infty$), in Griv et al. (1999), the generalized dispersion relation is given by Eq. (37), which is $\sim \omega_*^4$ (cf. Eq. (36) above) and is consistent with the dispersion relation obtained by Bertin et al. (1989, their Eq. (3.1)), which is also $\sim \omega_*^4$.

Equation (36) is complicated: it is nonlinear in ω_*. In order to deal with the most interesting oscillation types, let us consider various limiting cases of perturbations described by some simplified variations of Eq. (36). For instance, similar to the plasma physics method, one solves Eq. (36) by successive approximations. Indeed, in the low-frequency ($|\omega_*|^2 < \kappa^2$) and local WKB approximations we are explored in Eq. (36), the terms that describe azimuthal forces are assumed to be small in comparison with other terms. By neglecting the small terms $\propto m^2/r^2$ and m/rL, one obtains, in the first approximation, the familiar Lin–Shu dispersion relation for axisymmetric waves

$$\omega_*^2 = \kappa^2 - 2\pi G\sigma_0|k_r| + k_r^2 c_s^2 \tag{37}$$

(Lin et al. 1969; Rohlfs 1977, p. 100; Binney & Tremaine 1987, p. 352; Bertin & Lin 1996, p. 77; Bertin 2000, p. 192). For stable waves with real frequencies we must have $\omega_*^2 > 0$ while an unstable situation is met if $\omega_*^2 < 0$. Equation (37) describes the effects of main forces which act on each element of matter: the stabilizing non-inertial Coriolis force $\propto \kappa^2$, the destabilizing gravitational force $\propto G\sigma_0|k_r|$, and the stabilizing pressure gradient (thermal motion) $\propto k_r^2 c_s^2$. Long-wavelength ($|k_r| \to 0$) perturbations are stabilized by the Coriolis force (rotation) while short-wavelength ($|k_r| \to \infty$) ones are stabilized by the thermal motion, or turbulence. As a result, it turns out for a system only intermediate size perturbations are unstable. This leads to the appearance of structures in the system: rings and spirals. By a proper choice of κ^2 and c_s^2 provided σ_0 is given we can construct a stable disk (Rohlfs 1977, p. 60; Sicardy 2005, p. 461). It can be seen from Eq. (37) that the thermal motion has a stabilizing effect on the Jeans instability which therefore develops most strongly in a "cold" ($c_s^2 \to 0$) disk. Since $h \propto c_s$, we may say that the smaller the thickness of the disk, the more unstable it is—provided the other parameters are not changed. We conclude that the generalized dipersion relation tends to the ordinary Lin–Shu dispersion relation in the limit $\psi \to 0$. This establishes a continuity between the two formulae.

From Eq. (37), the disk is Jeans-unstable ($\omega_*^2 < 0$) to axisymmetric perturbations if

$$c_s < c_T\,, \tag{38}$$

where

$$c_T = \frac{\pi G\sigma_0}{\kappa} \tag{39}$$

is the Safronov–Toomre (Safronov 1960, 1980; Toomre 1964) critical sound speed to suppress the instability of only axisymmetric $\tan\psi \equiv m/rk_r = 0$ perturbations.

In the next approximation, in Eq. (36), considering the Safronov–Toomre stable disk ($c_s \geq c_T$), in the small term $\propto m^2/r^2$ one can replace ω_*^2 by κ^2. As a result, the simplified generalized dispersion relation is obtained in the form:

$$\omega_*^3 - \omega_*\omega_J^2 + 4\pi G\sigma_0\Omega\,(m/r|k|L) = 0\,, \tag{40}$$

where $|\omega_*| \lesssim \kappa$,

$$\omega_J^2 = \kappa^2 - 2\pi G\sigma_0(k_*^2/|k|) + k_*^2 c_s^2 \tag{41}$$

is the square of the Jeans frequency, $k = \sqrt{k_r^2 + m^2/r^2}$ is the total wavenumber, and

$$k_*^2 = k^2\left\{1 + [(2\Omega/\kappa)^2 - 1]\sin^2\psi\right\}$$

is the squared effective wavenumber. Equation (40) differs from the ordinary Lin–Shu dispersion relation (37) by the appearance of the total k and effective k_* wavenumbers, which originate from the consideration of the nonaxisymmetrical modes $\propto \psi$, and by the factor

$\propto mL^{-1}$, which originate from the consideration of the effects of inhomogeneity (see Lau & Bertin 1978, Lin & Lau 1979, Morozov 1980, 1985, Bertin & Lin 1996, p. 214, Bertin 2000, p. 193, and Griv et al. 2003b for a discussion). Lynden-Bell & Kalnajs (1972, their Eq. (A11)) first obtained the Lin–Shu type dispersion relation for open waves propagating in a homogeneous disk. Simplified dispersion relation (40) for low-frequency $|\omega_*|^2 \lesssim \kappa^2$ perturbations we are investigating can be easily obtained from Eqs. (D12) (actually, in Eq. (D12) it should be $T_1/(1-\nu^2)$ instead of T_1; Montenegro et al. 1989) and (D14) of Lin & Lau (1979) by ignoring the "out-of-phase" term ik_rA and using the expansion $2\Omega/\kappa \approx 1-(r/4\Omega)(\mathrm{d}\Omega/\mathrm{d}r)$.

When $\Omega \to 0$, $L \to \infty$, and $h \to R$, Eq. (40) reduces to the well-known Jeans dispersion relation for a sound wave propagating in an infinite homogeneous nonrotating medium

$$\omega^2 = k^2 c_{\mathrm{s}}^2 - 4\pi G\rho_0 \,,$$

where ρ_0 is the volume density and an instability takes place at $\omega^2 < 0$, or $\lambda > \lambda_{\mathrm{J}} = \pi^2 c_{\mathrm{s}}^2/G\rho_0$. For an infinite homogeneous medium uniformly rotating with an angular velocity Ω the Jeans-type dispersion relation for a singular set of modes with wave vectors exactly perpendicular to rotation axis was obtained by Chandrasekhar (1955):

$$\omega^2 = 4\Omega^2 - 4\pi G\rho_0 + k^2 c_{\mathrm{s}}^2 \,.$$

A more general equation for the waves propagating in a plane perpendicular to the axis of rotation when the rotation is differential ($\Omega = \Omega(r)$) was found by Bel & Schatzman (1958). Safronov (1960, 1980) modified the Bel & Schatzman (1958) dispersion relation by considering the radial wave propagation in a thin ($h/R \ll 1$) three-dimensional disk.

The dispersion relation (40) has three roots that describe three branches of oscillations: two ordinary Jeans branches modified by the spatial inhomogeneity and a gradient one modified by the Jeans mode. Let us investigate these branches of oscillations following closely Morozov (1980, 1985) and Griv et al. (2003b). From Eq. (40) in the most important high-frequency range

$$|\omega_*|^3 \sim |\omega_{\mathrm{J}}|^3 \gg 4\pi G\sigma_0\Omega \left(m/r|kL|\right) \,, \tag{42}$$

we determine the dispersion law for the Jeans branch:

$$\omega_{*1,2} \approx \pm p|\omega_{\mathrm{J}}| - 2\pi G\sigma_0 \frac{\Omega}{\omega_{\mathrm{J}}^2}\frac{m}{r|k|L} \,, \tag{43}$$

where $p = 1$ for gravity-stable perturbations with $\omega_*^2 \approx \omega_{\mathrm{J}}^2 > 0$, $p = \imath$ for gravity-unstable perturbations with $\omega_*^2 \approx \omega_{\mathrm{J}}^2 < 0$, and the term involving mL^{-1} is the small correction. Equation (43) determines the spectrum of oscillations. Accordingly, an inhomogeneity $\propto L^{-1}$ will not influence the stability condition of Jeans modes. As velocity dispersion increases, the system moves toward a more stable situation.

In the another, opposite to (42) frequency range,

$$|\omega_*|^3 \sim 4\pi G\sigma_0\Omega(m/r|kL|) \ll |\omega_{\mathrm{J}}|^3 \,, \tag{44}$$

that is, $|\omega_*| \ll \Omega$, Eq. (40) has another root equal to

$$\omega_{*3} \approx 4\pi G\sigma_0 \frac{\Omega}{\omega_J^2} \frac{m}{r|k|L} \,. \tag{45}$$

The root (45) describes the gradient ($L^{-1} \neq 0$) branch of oscillations. As is seen, the gradient perturbations are stable and are independent of the stability of Jeans modes. These low-frequency ($|\omega_{*3}| \ll \Omega$), stable ($\Im\omega_{*3} = 0$) oscillations are obviously not important in dynamics of Saturn's rings. According to Morozov (1985), analogous oscillation branches, with frequencies proportional to the gradients of the undisturbed equilibrium quantities, occur in the terrestrial atmosphere (internal gravity waves), in inhomogeneous plasmas (drift waves), in the terrestrial oceans (Rossby waves), and in other spatially inhomogeneous systems.

3.6. Stability Criterion

Saturn's rings can be subject to collective effects that induce both clumping and structure formation. From Eq. (41), at the limit of gravitational stability, the two conditions $\partial\omega_J^2/\partial k = 0$ and $\omega_J^2 \geq 0$ are fulfilled. The first condition determines the most unstable wavelength (the modified Jeans–Toomre wavelength)

$$\lambda_{\mathrm{crit}} = \lambda_{\mathrm{JT}} \left\{1 + \left[(2\Omega/\kappa)^2 - 1\right] \sin^2\psi\right\}^{1/2} \approx 2\lambda_{\mathrm{JT}} \,, \tag{46}$$

where $\lambda_{\mathrm{JT}} = c_s^2/G\sigma_0$ is the ordinary Jeans–Toomre wavelength, corresponding to the minimum on the the dispersion curve (41). It can be shown that λ_{JT} is the two-dimensional version of the classical Jeans wavelength. Indeed, if we use here the fact that for a thin disk $\sigma_0 \sim 2\rho_0 h$ and $k \sim 1/h$, we find $\lambda_J = \pi^2 c_s^2/G\rho_0 \sim c_s^2/G\sigma_0$. It follows from Eq. (46), by including the effects of the nonaxisymmetric forces, the limit of gravitational stability is shifted toward a longer wavelength than it follows from the ordinary Jeans–Toomre wavelength. The effect is relatively large, $\lambda_{\mathrm{crit}} \approx 2\lambda_{\mathrm{JT}}$. Use of the second condition determines the critical sound speed c_{crit} for the stability of arbitrary but not only axisymmetric perturbations.

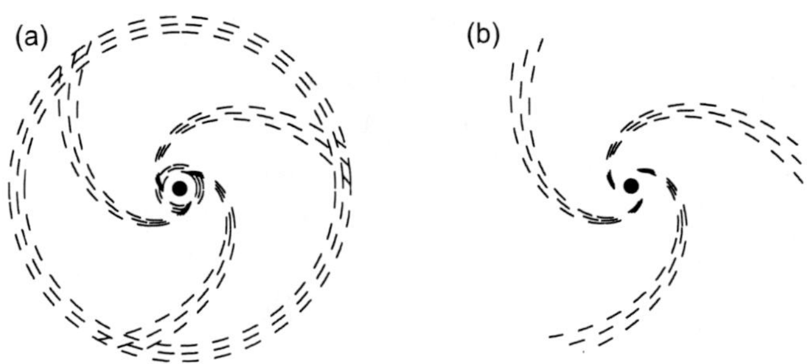

Figure 4. A schematic model of a Jeans-unstable disk; (a) the Safronov–Toomre unstable disk ($c_s < c_T$, or $Q < 1$, respectively) and (b) the Safronov–Toomre stable disk ($Q \geq 1$ but $Q \lesssim 2\Omega/\kappa$, and in Saturn's rings $2\Omega/\kappa \approx 2$).

One concludes that the disk is Jeans-unstable ($\omega_J^2 < 0$) to both axisymmetric (radial) and nonaxisymmetric (spiral) perturbations if $c_s < c_T$ (Fig. 4a). This result suggest that disks with Toomre's $Q = c_s/c_T$ parameter < 1 (with the local surface density $\sigma_0 > c_s\kappa/\pi G$) break up into fragments of preferred mass

$$M_{\rm frag} \sim \lambda_{\rm crit}^2 \sigma_0 \sim \frac{4c_s^4}{G^2\sigma_0}\,. \tag{47}$$

Similar results hold for a self-gravitating Safronov–Toomre unstable disk of finite but small thickness in hydrostatic equilibrium with the critical wavenumber $k_{\rm crit} = 2\pi/\lambda_{\rm crit}$ reduced by a factor of 2 (Morozov 1981). In Saturn's rings, the Safronov–Toomre sound speed $c_T \sim 0.05$ cm s^{-1} and

$$M_{\rm frag} \sim 10^7 \text{ g}\,.$$

These values for the ring disk is based on a surface density of 50 g cm^{-2} and an angular speed of 2×10^{-4} s^{-1} (Nicholson et al. 2005).

Nonaxisymmetric ($\psi \neq 0$) instabilities in a differentially rotating disk are more difficult to stabilize; stability is achieved only for sufficiently large sound speed (although still of the order of c_T)

$$c_s \gtrsim c_{\rm crit} = c_T \left\{1 + \left[(2\Omega/\kappa)^2 - 1\right]\sin^2\psi\right\}^{1/2} \approx \frac{2\Omega}{\kappa} c_T \tag{48}$$

(Lau & Bertin 1978; Lin & Lau 1979; Morozov 1980, 1985; Griv et al. 2003b). Note that for a rigidly rotating disk $2\Omega/\kappa = 1$. The parameter

$$\{1 + \left[(2\Omega/\kappa)^2 - 1\right]\sin^2\psi\}^{1/2}$$

is an additional stability parameter which depends on both the pitch angle ψ and the amount of differential rotation in the galaxy $d\Omega/dr$ (cf. the parameter $\mathcal{J}$ introduced by Bertin & Mark 1978, Lau & Bertin 1978, Lin & Lau 1979, Bertin 1980, Toomre 1981, Bertin & Lin 1996, p. 214, and Bertin 2000, p. 193). As one can see from Eq. (48), the modified critical sound speed grows with ψ in a differentially rotating system. Even though in the case of very open spirals with $\psi > 45°$ a special more accurate analysis is needed, it is clear that the generalized stability criterion for the local stability of a gaseous disk against arbitrary Jeans perturbations should be approximately of the form of Eq. (48).

The modified Safronov–Toomre stability criterion (48) means that if the sound speed (the velocity dispersion) drops below the critical value $c_{\rm crit} \approx 2c_T$ than the disk of mutually gravitating particles will develop exponentially growing density fluctuations. The modified stability criterion (48) is obviously only an approximation, as it assumes spiral arm pitch angles ψ that are relatively small ($|\tan\psi| \lesssim 1$) and neglects (among other things) the thickness of the disk. Nevertheless, numerical experiments suggest that the formulation is approximately correct (Khoperskov et al. 2003; Liverts et al. 2003; Griv 2005a; Griv et al. 2006a).

As is seen, nonaxisymmetric disturbances in a nonuniformly rotating system ($d\Omega/dr \neq 0$, or $2\Omega/\kappa > 1$, respectively) are more difficult to uppress than the axisymmetric ones, in general agreement with the work by Goldreich & Lynden-Bell (1965a) and Julian & Toomre (1966). Maxwell (1859) has considered just this kind of spiral instabilities with $m = 1$ in his study concerning the stability of the Saturnian uniform rings whose radial extent was considerably larger than the average interparticle distance. Maxwell made a correct that, in such a system, the azimuthal force resulting from azimuthal displacements was more important in determining the stability than was the radial force resulting from radial displacements. Goldreich & Lynden-Bell (1965a, p. 123) especially noted that "... in the galaxy the tangential modes may be most unstable." Lau & Bertin (1978, p. 509) have clarified the problem by considering the motion of a fluid element: the density response that is in phase with the potential minimum is found to exceed, by an amount proportional to both $d\Omega/dr$ and m, the corresponding response due to an axisymmetric field of equal strength. In Toomre (1981), this amplification was discussed in terms of "swing mechanism," very reminiscent of the way we reach the modified stability criterion (48). (Swing amplified mechanism was discovered by Goldreich Goldreich & Lynden-Bell (1965b). Swing works on leading waves and turns them into trailing waves giving strong amplification in the process.) The free kinetic energy associated with the differential rotation of the system is one possible source for the growth of the energy of these spiral Jeans-type perturbations, and appears to be released when angular momentum is transferred outward. Lin & Shu (1966), Lin et al. (1969), Shu (1970), as well as Ginzburg et al. (1972), Nakamura et al (1975), Safronov (1980), and Lovelace et al. (1997) allowed for a departure from axial symmetry of the perturbations only partially by introducing a wavefrequency $\omega_* \to \omega - m\Omega$ but omitting all other $m-$ (or φ-) dependent terms in the exponential factor of Eq. (18), that is, in fact they ignored the effect of the finite spiral inclination. However, as it has been pointed out in plasma theory, the modes propagating in the azimuthal direction can maintain communication between two poits separated in the φ direction, and all φ-dependent terms must therefore be fully taken into account. See also Lin & Lau (1979), Bertin (1980), Morozov (1980, 1985), and Griv et al. (1999, 2003b) for a discussion of the problem. N-body simulations in shearing box approximation have demonstrated that these spiral waves ("wakes") trail at an average angle of $\sim 20^\circ$ relative to the azimuthal direction in a Keplerian ring (Salo 1992, 1995, 2001; Richardson 1994; Osterbart & Willerding 1995; Daisaka & Ida 1999; Ohtsuki & Emori 2000; Griv et al. 2006a). Fourier analysis shows close correspondence of these waves to Jeans-type perturbations (Griv 2005a; Griv & Gedalin 2005).

Morozov (1980, 1981) took into account the additional weak destabilizing effect of a density inhomogeneity and a radial gradient of a velocity dispersion, and the weak stabilizing effect of a finite thickness. The result is that the stabilizing effect introduced by the finite thickness and the destabilizing effect introduced by the inhomogeneity and the velocity gradient practically cancel out each other, at least in the solar vicinity of our own Galaxy. In practical work, one can neglect all these corrections. Saturn's rings are likely to be marginally gravitationally unstable. We expect therefore that in Saturn's rings $c_s \sim c_T$, or Toomre's stability parameter $Q \sim 1$, in agreement with available observations (Lane et al. 1982, p. 543).

It is crucial to realize that the various dynamical properties of the perturbations with different ψ are peculiarities of the differentially rotating disks only. In a way of contrast, in the rigidly rotating disk $2\Omega/\kappa = 1$ (and/or for axisymmetric perturbations, $\psi = 0$) and the modified sound speed (48) is in fact equal to c_T. Similar stability criterion can also be derived from the Lynden-Bell & Kalnajs (1972, Eq. (A11) therein) dispersion relation for open spirals. Apparently, Hunter (1973) was the first who obtained the stability criterion generalizing the Safronov–Toomre criterion c_T for nonaxisymmetric perturbations. From Goldreich & Tremaine's (1978c) Eq. (34) for the surface density perturbation, Chiueh & Tseng (2000) derived a stability criterion for localized perturbations in Goldreich & Lynden-Bell's (1965a, b) and Julian & Toomre's (1966) shearing-sheet approximation. At least, for the disk with a flat rotation curve ($2\Omega/\kappa = \sqrt{2}$), the marginal stability condition that was found by Chiueh & Tseng (2000) ($Q \approx \sqrt{3}$) almost coincides with Eq. (48) in the case $\psi \to 90°$ ($Q \approx \sqrt{2}$). Griv et al. (1999, 2006a), Khoperskov et al. (2003), Liverts et al. (2003), Griv (2005a), and Griv & Gedalin (2005) used computer simulations to test the validity of the modified stability criterion c_{crit}. Observations have shown that both the gas density and the stellar density in the disks of most galaxies including the Milky Way are maintained at a level close to the threshold marginal stability of gaseous and stellar layers to local gravitational perturbations with $Q \approx 2$; the disks of spiral galaxies underwent no significant dynamical heating after they reached a quasi steady-state (Bottema 1993; Zasov et al. 2004; Zasov & Smirnova 2005).

A general impression of how the spectrum of nonaxisymmetric Jeans perturbations behaves in a homogeneous nonuniformly rotating disk can be gained from Fig. 5, which shows the dispersion curves in the cases of Jeans-unstable systems ((a) and (b)), a marginally Jeans-stable system (c), and a Jeans-stable one (d) (as determined on a computer from Eq. (40)). In this figure, the ordinate is the effective wavenumber k_* measured in terms of the inverse "epicyclic radius" $\rho = c_s/\kappa$ and the abscissa is $\nu = \omega_*/\kappa$, i.e., the dimensionless angular frequency at which the gas element meet with the pattern, measured in terms of the epicyclic frequency κ. In general, for fixed dimensionless wavefrequency ν there are two solutions in $k_*\rho$, comprising a long-wavelength wave, $k_*\rho \lesssim 1$, and a short-wavelength wave, $k_*\rho > 1$. A property of the solution (40) is that in a homogeneous system the Jeans-stable modes those with $Q > 2\Omega/\kappa$ are separated from each other by frequency intervals where there is no wave propagation: gaps ("forbidden zones") occur between each harmonic (cf. the longitudinal Bernstein modes in a hot magnetized plasma; Krall & Trivelpiece 1986, p. 409). Lin et al. (1969), Nakamura et al. (1975), Lovelace et al. (1997), and Griv et al. (2006a) have pointed out that the stable density waves in a thin stellar disk derived by Lin & Shu (1966), Lin et al. (1969), and Shu (1970) are no other than the representation of the Bernstein mode in the gravitational system.

Thus, if the disk is thin, $c_s \ll r\Omega$, and dynamically cold, $c_s < c_T$ (Toomre's stability parameter $Q < 1$), then such a model will be gravitationally unstable to both axisymmetric and nonaxisymmetric perturbations, and it should almost instanteneously (see below for a time estimate) taken on the form of a cartwheel, that is, a structure of spirals and rings (Fig. 4a). One understands, however, that in the nonlinear stage of evolution there will be some exchange of energy and angular momentum between the axisymmetric and nonaxisymmetric modes which will give rise to a pattern much more complex than the cartwheel shown in Fig. 4a.[7] Cassini observations have indicated that axisymmetric periodic structure in Saturn's A and B rings co-exists with nonaxisymmetric structure (Colwell 2006, 2007;

[7]The most striking example of the cartwheel-like galaxies is the classical ring galaxy A0035-324 (the "Cartwheel"), which earned its nickname from the prominent system of spiral spokes which connect the bright outer ring to the inner ring (Higdon 1996; Horellou et al. 1998). The ring of the Cartwheel can be considered as the crest of a high amplitude $m = 0$ Jeans-unstable density wave. In turn, the spokes are probably nothing but density maxima of spiral density waves (Griv 2005b).

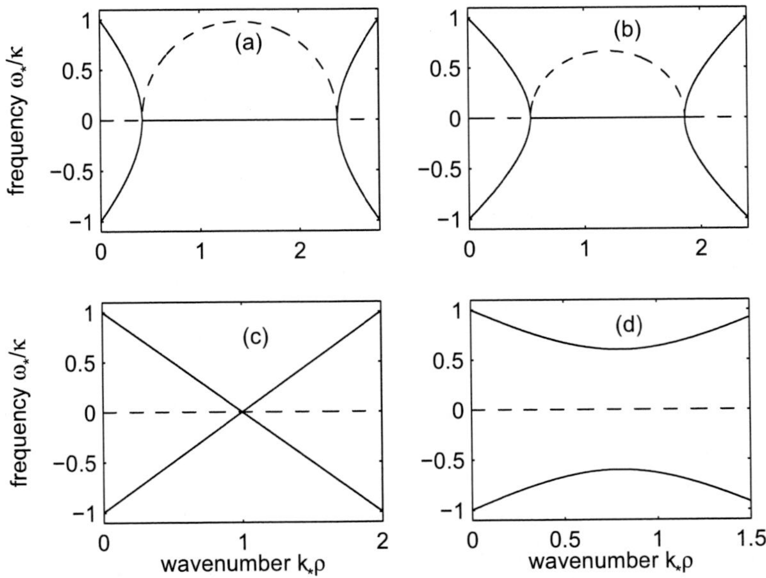

Figure 5. The generalized Lin–Shu dispersion relation of a homogeneous ($|L| \to \infty$) differentially rotating ($2\Omega/\kappa > 1$) gaseous disk in the case $2\Omega/\kappa = 2$ and $|\sin\psi| = 1$ for the different Toomre's Q-values: (a) $Q = 0.6 \times (2\Omega/\kappa)$, (b) $Q = 0.8 \times (2\Omega/\kappa)$, (c) $Q = 2\Omega/\kappa$, and (d) $Q = 1.2 \times (2\Omega/\kappa)$. The solid curves represent the real part of the dimensionless wavefrequency $\nu = \omega_*/\kappa$ of low-frequency ($|\omega_*| \lesssim \kappa$), long-wavelength ($k_*\rho \lesssim 1$, where $\rho = c_s/\kappa$) and short-wavelength ($k_*\rho > 1$) Jeans oscillations we are interested in. The dashed curves represent the imaginary part of ν.

Thomson et al. 2007). We suggest, therefore, that in Saturn's A and B rings $Q < 1$. Clearly, in this case of both radial and spiral excitation, the distribution of the surface density along the spiral arms is not uniform, but describes a sequence of maxima, that might be identified with forming embedded clusters of particles.[8] Such a disk should break up into discrete *porous* blobs of matter ("clumpy moons") of preferred mass $M_{\rm frag}$ distributed in spirals around the spin axis (Fig. 6). The Safronov–Toomre unstable modes might be proposed as potential clusters-forming mechanism (cf. Snytnikov et al. 2004, Figs. 2 and 4 therein; Griv 2005b, Fig. 2 therein). Interestingly, low values of thermal inertias of B and C ring particles derived from infrared observations of Saturn's rings might be characteristic of very porous particle aggregates (Ferrari et al. 2005).

Contrary, if the disk is thin and warm, $Q \geq 1$ but $Q \lesssim 2\Omega/\kappa \approx 2$, then such a model will be unstable only with respect to spiral perturbations (Fig. 4b) and cannot therefore fragment. An uncooled hot model with $Q \gtrsim 2$ is Jeans-stable to all small-amplitude perturbations, including the most unstable spiral ones. To emphasize it again, this is not an entirely new idea: Lau & Bertin (1978), Lin & Lau (1979), and Morozov (1980, 1985) first obtained the modified stability criterion $c_{\rm crit}$. Simulated disks that prove stable evolve to

[8]The latter is an important step towards an understanding of a main question of protoplanetary disk evolution, as well as the evolutionary processes in galactic disks: what kind of evolutionary processes lead to the formation of moons, planets, and stars in a different astrophysical disk system?

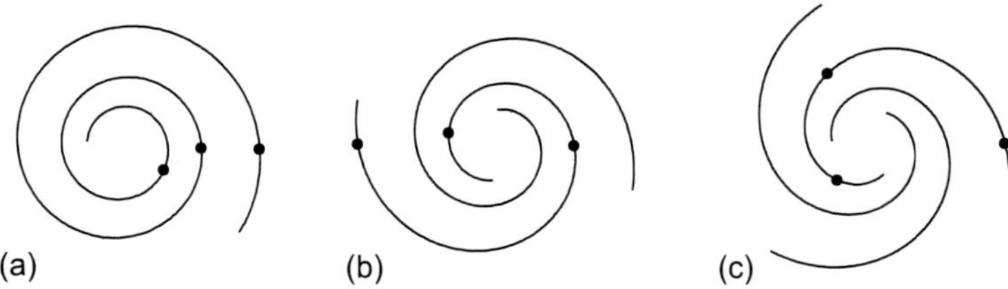

Figure 6. (a) Gravitationally unstable density waves with $m = 1$ arm in the (r, φ)-plane, (b) density waves with $m = 2$ arms, and (c) density waves with $m = 3$ arms. The filled circles represent the maxima of the perturbed density of Jeans waves, which are unstable to both axisymmetric and nonaxisymmetric gravity perturbations. The distribution of the surface density along the spiral arms is not uniform, but describes a sequence of maxima, that might be identified with forming embedded complexes of particles of a mass of $M_{\rm frag}$ each.

Q values in the range 1.5–2.5 (Hohl 1972; Sellwood & Carlberg 1984; Athanassoula & Sellwood 1986; Sellwood & Athanassoula 1986; Tomley et al. 1991; Griv 1998; Bottema 2003; Khoperskov et al. 2003; Liverts et al. 2003; Griv & Gedalin 2005). These values agree with modified stability criterion (48). The instability is driven by a strong nonresonant interaction of the gravity fluctuations with the bulk of the particle population, and the dynamics of Jeans perturbations can be characterized as a nonresonant wave–particle interaction: in Eq. (28), $\omega_* - l\kappa \neq 0$ and $l = 0, \pm 1$. The waves are created if self-gravity is included; only particle's collisions do not create the structure. Note that the spatially limited wave–particle resonances have been investigated by Lynden-Bell & Kalnajs (1972), Goldreich & Tremaine (1979, 1980), Meyer-Vernet & Sicardy (1987), Griv et al. (2000b), and Griv (2007a).

The growth rate of the instability is relatively high,

$$\Im\omega_* \sim \sqrt{2\pi G\sigma_0(k_*^2/|k|)} \sim \Omega\,,$$

and depends of the azimuthal mode number m. That is, the instability develops rapidly on a dynamical timescale. An important feature of the instability under consideration is the fact that in a rotating frame it is almost aperiodic ($|\Re\omega_*/\Im\omega_*| \ll 1$). At the boundary of instability ($Q \sim 1$), the wavelength of the most unstable mode $\lambda \approx \lambda_{\rm crit} = 2\pi^2 G\sigma_0/\kappa^2 \sim 4\pi h$; thus $\lambda_{\rm crit} \gg h$. It means that of all harmonics of initial gravity perturbation, one perturbation with $\lambda_{\rm crit} \sim 10h$, with the associated number of spiral arms $m_{\rm crit}$, and with the pitch angle $\psi_{\rm crit}$ will be formed in time of a single disk rotation. Because $\lambda_{\rm crit} \gg h$, the approximation of an infinitesimally thin disk used throughout the theory does not fail.

3.7. Prevalence of Trailing Spirals

In an inertial frame, a condition for constancy of phase of oscillations is $k_r r + m\varphi - \Re\omega t = \mathrm{const}$. We differentiate this equation with respect to time:

$$\frac{\mathrm{d}\varphi}{\mathrm{d}t} = -\frac{k_r}{m}V_r + \frac{\Re\omega}{m}\,, \tag{49}$$

where $\mathrm{d}\varphi/\mathrm{d}t$ represents the azimuthal velocity of the perturbation (the rate at which the nonaxisymmetric pattern rotates, the pattern speed), the term $\propto V_r$ describes mass transport in a strictly radial direction, and this term may be omitted, since we are interested only in the azimuthal motion of the perturbation. In the spirit of Mikhailovskii & Fridman (1973), from Eq. (49), the condition for the formation of trailing spirals (spirals trail the direction of rotation of a system) is obviously:

$$\Re\omega < m\Omega\,. \tag{50}$$

In the gravity-unstable case ($\omega_*^2 \approx \omega_\mathrm{J}^2 < 0$), the local equilibrium parameters of the disk determine the pattern speed of exponentially growing nonaxisymmetric perturbations (in a rotating frame):

$$\Omega_\mathrm{p} \equiv \Re\omega_*/m \approx 2\pi G\sigma_0(\Omega/|\omega_\mathrm{J}^2|)(1/r|k|L) \ll \Omega\,, \tag{51}$$

where $2\pi G\sigma_0|k| \sim \Omega^2$, $|\omega_\mathrm{J}^2| \sim \Omega^2$, $rk^2|L| \gg 1$ (Eq. (43)). The first conclusion is that because Ω_p does not depend on m, each Fourier component of a perturbation in an inhomogeneous system will rotate with the same constant angular velocity even while the perturbation (for instance the density disturbance) is otherwise growing. Spirals rotate rigidly with a single fixed pattern speed, Eq. (51). In other words, any "mixing" or wind up of the spirals are absent. The theory states that in homogeneous ($|L| \to \infty$) disks $\Omega_\mathrm{p} = 0$. Second, according to observations in Saturn's A and B rings $L \propto \left[(\partial/\partial r)(r^{3/2}\sigma_0)\right]^{-1} < 0$, and from Eq. (51) we find (in the frame of reference rotating with angular velocity Ω at the reference position r_0)

$$\Omega_\mathrm{p} \equiv \Re\omega_*/m < 0\,. \tag{52}$$

In accordance with Eq. (50), then, exclusively trailing spirals can develop in Saturn's A and B rings. Interestingly, only trailing spirals have been observed both in Saturn's rings (Colwell et al. 2006, 2007) and protoplanetary disks (Clampin et al. 2003; Fukagawa et al. 2004, 2006). Utter prevalence of trailing spirals in galaxies has been demonstrated by Pasha & Smirnov (1982).

4. Turbulent Heating

The Jeans instability of a disk will in fact be suppressed because of the efficient heating of the medium during the growth of that same instability (Morozov 1978; Fuchs 2001; Griv et al. 2001, 2002, 2003a, b). Following the physics of plasmas (e.g., Kadomtsev 1965), by turbulent heating of astrophysical disks we will understand the transfer of ordered energy of rotation or the energy of discrete oscillations, into energy of random motion and ultimately into heat, due to turbulence, i.e., the nonlinear interaction between oscillations. Gravitational instabilities convert gravitational potential energy (the orbital energy of the disk) into kinetic energy of random motions. This process—the self-suppression of instability by a rise in the amplitude of the unstable harmonics—recalls the case of nonresonant quasi-linear relaxation in plasmas, which can effectively heat the medium without raising the entropy (Sagdeev & Galeev 1969; Galeev & Sagdeev 1983; Alexandrov et al. 1984,

p. 420; Krall & Trivelpiece 1986, p. 531; Swanson 1989, p. 307). Indeed, the perturbed potential energy U is a negative quantity:

$$U = \left\langle \frac{\sigma_1 \Phi_1^*}{2} \right\rangle = -\frac{|k||\tilde{\Phi}|^2}{2\pi G} e^{2|\Im\omega_*|t} , \tag{53}$$

where $\langle \cdots \rangle$ denotes the time average over the fast oscillations, $\sigma_0 = \langle \sigma \rangle$, $\Phi_0 = \langle \Phi \rangle$, $\langle \sigma_1 \rangle = \langle \Phi_1 \rangle = 0$, and Φ_1^* is the complex conjugate potential. Clearly as gravitational instability develops the amplitudes $|U|$ of the unstable Fourier-harmonics of the perturbation will grow (via the mass and orbital momentum redistribution), setting energy free to heat the medium. This effect will in turn diminish the growth rate of the instability and ultimately cause it to become saturated. Fresh waves must be continually created to maintain the density wave pattern. Inelastic collisions are the only option for dissipating energy in a particle disk. Numerical experiments have already confirmed that uncooled gravitationally unstable disks are heated to the point of stability and cease to produce density waves in a time of a single rotation (Hohl 1972; Sellwood & Carlberg 1984; Tomley et al. 1991; Laughlin & Bodenheimer 1994; Salo 1995; Bottema 2003; Khoperskov et al. 2003; Liverts et al. 2003). Notice that random motions of gas elements of an unstable disk will grow under the action of both the radial and the azimuthal forces ($k \equiv \sqrt{k_r^2 + k_\varphi^2} \neq 0$) (see Eq. (53) above).

Following Morozov (1978), let us obtain mathematical expressions describing the heating process. For simplicity, the gas is supposed to obey a simple polytropic law

$$P = P_0 (\sigma/\sigma_0)^n , \tag{54}$$

where the two-dimensional adiabatic index n can be mapped to a three-dimensional adiabatic index in the low-frequency (static) limit; in galactic and protostellar disks $n = 1.5 - 2$. Equation (54) can be expanded as far as terms quadratic in the amplitude of the perturbed density

$$P = P_0 + n\frac{P_0}{\sigma_0}\delta\sigma + \frac{n(n-1)}{2}\frac{P_0}{\sigma_0^2}\delta\sigma^2 , \tag{55}$$

where $\delta\sigma = \epsilon\sigma_1 + \epsilon^2\sigma_2$ and $\epsilon \ll 1$. Next, the result can be averaged in the sense indicated above:

$$\langle P - P_0 \rangle = \delta P(t) = \frac{n(n-1)P_0}{\sigma_0^2} |\tilde{\sigma}|^2 e^{2\Im\omega_* t} , \tag{56}$$

where the growth rate $\Im\omega_*(t)$ is determined from Eq. (41) by the relation

$$\Im\omega_* = \left[-\kappa^2 + 2\pi G\sigma_0 (k_*^2/|k|) - n(P_0 + \delta P(t))k_*^2/\sigma_0\right]^{1/2} . \tag{57}$$

Equations (56) and (57) completely describes the heating process. Of course, the system of Eqs. (56)–(57) is correct only in the linear approximation used throughout the theory, $|\sigma_1/\sigma_0| \ll 1$ and $|\Phi_1/\Phi_0| \ll 1$. The linear wave considered in the present study is growing on the timescale of about one cycle rotation period ($\sim \Omega^{-1}$) and in one cycle rotation period it will reach nonlinear amplitudes. Strictly speaking, this theory of weak turbulence describes only the tendency of the disk to be heated by Jeans-unstable density waves and shows the direction of the disk's evolution. One understands that we still have to develop the theory of strong turbulence.

Taking into account the fact that $c_s^2 \propto \langle P - P_0 \rangle$, from Eq. (56) one obtains that during the quasi-linear stage of the instability the effective temperature c_s^2 grows with time according to the law

$$c_s^2 \propto t \,. \tag{58}$$

Of course, this is not really temperature, it is ordered motion, i.e., coherent mechanical oscillations of the element in response to the fluctuating fields. The steady growth in $c_r \propto \sqrt{t}$, where c_r is the radial velocity dispersion, has already been detected in an unstable N-body disk (Binney & Tremaine 1987, p. 480; Liverts et al. 2003). Observations indicate almost the same $\propto t^{1/2}$ heating of the Galactic disk of stars (Wielen 1977; Binney 2001; Nordström et al. 2004). The latter means that an unknown mechanism increases the velocity spread of stars in the Galaxy after they are born. Griv et al. (2001, 2002) have elaborated the idea of the collisionless collective-type relaxation; unstable perturbations of the equilibrium gravitational potential affect the averaged velocity distribution of stars.

It is obvious that the uncooled disk manages to keep its local stability parameter close to the critical value, $Q \approx 2$ (Griv et al. 1999, 2002; Griv & Gedalin 2004). In this case, once the differentially rotating disk has been heated to values $Q \approx 2$ (or $c_s \approx 2c_T$, respectively) by gravitationally unstable density waves, no further unstable waves can be sustained by virtue of the Jeans instability—unless some cooling mechanism is available leading to Toomre's Q-value under approximately 2, or to the value of c_s smaller than approximately $2c_T$, respectively. To stress it again, it has been found in simulations that the stability number Q of Toomre in relaxed equilibrium disks does not fall below a critical value, which lies about $Q_{crit} = 2 - 2.5$ (Hohl 1972; Sellwood & Carlberg 1984; Athanassoula & Sellwood 1986; Sellwood & Athanassoula 1986; Tomley et al. 1991; Laughlin & Bodenheimer 1994; Osterbart & Willerding 1995; Salo 1995; Daisaka & Ida 1999; Ohtsuki & Emori 2000; Bottema 2003; Durisen et al. 2007). However, no adequate explanation of the latter has been presented. Unlike Hohl, Sellwood, Carlberg, Athanassoula, Tomley, and other, we give the condition $Q \lesssim 2$ of the wave formation which follows directly from our theory. See also Khoperskov et al. (2003), Liverts et al. (2003), Griv (2005a), and Griv et al. (2006a) for a discussion.

The development of gravitational instabilities would lead to a self-regulation process. Namely, if the differentially rotating disk is initially cold ($Q < Q_{crit}$), then gravitational instabilities would repeadly heat it according to the law (58) on the dynamical timescale $\lesssim \Omega^{-1}$, bringing it toward stability. On the other hand, if the disk is initially hot enough ($Q \geq Q_{crit}$), then radiative cooling is going to bring the value of Toomre's stability parameter Q down toward an unstable configuration. A similar approach has been suggested in computer simulations (e.g., Gammie 2001). Just this approach is assumed in the modern version of the long-term spiral structure in numerical models of a stellar–gaseous galactic disk—suggesting the dissipation in the gas and formation of new dynamically cold stars (Sellwood & Carlberg 1984; Griv & Chiueh 1998). In Saturn's rings, inelastic (dissipative) particle's impacts are necessary for the reconstruction of the wavelike $\sim$ 100 m microstructure by reducing the random velocity spread of the entire particle component. Thus, the fine-scale density waves in Saturn's rings may be a long-term *recurrent* phenomenon, being created and destroyed on the timescale of an order of Keplerian period $\sim$ 20 hr (Griv et al. 2006). Cassini observations have shown ring variability in time and space; recycling could extend the microstructure lifetime essentially.

5. Turbulent Viscosity

As we have shown above, growing gravity perturbations can be self-excited in the main domain of the disk between the inner and outer Lindblad resonances via the dynamical Jeans instability (gravitational collapse) in a nonresonant wave–fluid interaction, and thus can transport the angular momentum and mass in the disk. The turbulent viscosity transfers energy from a large-scale shear flow into heat—the random motion of the particles. Generally speaking, in astrophysical self-consistent disk configurations it is likely that some, perhaps most, of the evolution is driven by diffusion of angular momentum within the disk. The growing gravity disturbances are expected to enhance angular momentum transport through the gravitational torque and particles' collective motion associated with the wave structure (Griv et al. 2008). The long-term evolution of astrophysical disks is controlled by the rate at which angular momentum is transported outwards so allowing the disk material to sink deeper in the gravitational potential well. There is observational evidence that redistribution of angular momentum does occur in astrophysical disks. Some kind of hydrodynamical turbulence is usually postulated to explain the observed phenomenon (e.g., Pringle 1981).

The gravity-driven turbulence in astrophysical disk configurations has been studied numerically (e.g., Wada et al. 2002). Numerical simulations of the gravitational instability have shown that the turbulence generated can enhance essentially the radial angular momentum transport (see, e.g., Griv et al. 2008 for a discussion of the problem). In the nonself-gravitating limit hydrodynamic turbulence cannot transport angular momentum effectively in astrophysical disks (Ji et al. 2008). The power spectra of optical emission in gas-rich galaxies have indicated that gravitational instabilities form flocculent spiral arms but also these instabilities generate much of the turbulence in the interstellar medium; young stars and clusters are distributed in a fractal pattern that is generated by a turbulent gas (Elmegreen et al. 2003; Elmegreen & Scalo 2004).

5.1. Molecular Viscosity

The dynamics of unperturbed viscous gas of particles has been considered by Lynden-Bell & Pringle (1974) and Goldreich & Tremaine (1978a, 1982). It was shown that the presence of microscopic viscosity (particle collisions) and differential rotation induces a viscous stress that leads to an outward transfer of angular momentum: an unperturbed disk experiences radial spreading due to collisional diffusion.[9] On the macroscopic level this follows from conservation of the disk angular momentum while the total disk energy decreases. The viscous torque which is exerted by the material inside radius r on the material outside this radius due to the viscous interaction is

$$\Gamma^{\rm vis} \approx 3\pi r^2 \Omega \sigma_0 \nu_{\rm vis} \,, \tag{59}$$

where $\nu_{\rm vis}$ is the molecular viscosity and σ_0 is the unperturbed local surface density (Lynden-Bell & Pringle 1974). In planetary rings, collisions between ring particles lead

[9]Following Pringle (1981), by viscosity we mean the mechanism, whatever it is, which enables angular momentum to be transferred and energy to be dissipated. The influence of viscosity on a rotating mass of gas is well known: "the dissipative processes act to spread the disk out, allowing the inner parts to move in and necessitating, through conservation of angular momentum, the outer parts to move out" (Pringle 1981).

to friction, which generally works to destroy structure in the ring system (Shu 1984). The level of shear friction in the rings is characterized by the kinematic viscosity:

$$\nu_{\rm kin} = \Omega r_{\rm part}^2 \tau \quad \text{for} \quad c_{\rm s} < 2\Omega r_{\rm part} \tag{60}$$

and

$$\nu_{\rm kin} = \left(\frac{c_{\rm s}^2}{2\Omega}\right)\left(\frac{\tau}{1+\tau^2}\right) \quad \text{for} \quad c_{\rm s} > 2\Omega r_{\rm part}\,, \tag{61}$$

where $r_{\rm part}$ stands for particle characteristic size, τ for normal optical depth, and $c_{\rm s}$ for random speed (Goldreich & Tremaine 1978a, 1982). Equations (60) and (61) were obtained (a) considering a simple case of a ring of smooth, uniformly sized spherical particles with a particular velocity dispersion and ignoring interparticle gravitation, (b) considering a simple case of a rarefied ring when the particle density is not high enough that particle size becomes important (Wisdom & Tremaine 1988), and (c) assuming that τ is not very different from the order of unity. Most researches treat Eqs. (60) and (61) as upper limits. An unconfined ring of width $W \ll r$ spreads on the viscous timescale $t_{\rm vis} \sim W^2/\nu_{\rm kin}$. In a collision disk the rate at which angular momentum flows outward accross the radius r is given by Eq. (59). The angular momentum is steadily concentrated onto a fraction of the mass which is spiraling away while the rest is accreted onto the central body (Lynden-Bell & Pringle 1974).

5.2. Anomalous Viscosity

Let us now show—in a speculative manner—that the hydrodynamic turbulence due to non-resonant gravitational instability transports angular momentum outward, giving the disk a source of internal viscosity (the angular momentum flux). Anomalous turbulent viscosity would regulate evolution and structure formation of Saturn's rings. Notice that a turbulent viscosity could be one agent for angular momentum transport in a differentially rotating medium has been postulated (e.g., Pringle 1981). This idea is inherent in the accretion-disk models of Shakura & Sunyaev (1973) and Lynden-Bell & Pringle (1974). (The theory of accretion disks has been reviewed by Papaloizou & Lin 1995 and Lin & Papaloizou 1996.) Another agent is stable self-gravitational waves (including those generated by resonant interactions), as considered by Goldreich & Tremaine (1979, 1980). Paczynski (1978) considered a model of a thin disk, rotating around a central compact object, that is marginally stable to self-gravity. His model followed the accretion-disk theory by Shakura & Sunyaev (1973) and Lynden-Bell & Pringle (1974). These standard models have "understood" angular momentum transport in that they assumed that it is due to turbulent stresses. Accordingly, anomalous "turbulent viscosity" driven by local disk instabilities was assumed to transfer angular momentum outwards and makes the accretion of matter onto the central object possible. There is one important difference between the model presented here and the models by Shakura & Sunyaev (1973), Lynden-Bell & Pringle (1974), and Paczynski (1978), namely in the present model the physical cause for the turbulence is included self-consistently by considering a self-consistent system of the gasdynamic and Poisson equations. Paczynski (1978) also considered the vertical structure, which is a less simplified approach than assuming an infinitesimally thin disk, but this is not required for an understanding of angular momentum transport.

One of the goals of the present work is to explain the result of simulations of planetary rings by Daisaka et al. (2001, their Eq. (33)): in the presence of the gravitationally unstable density waves, the effective viscosity $\nu_{\rm eff}$ is given as

$$\nu_{\rm eff} = \frac{CG^2\sigma^2}{\Omega^3}\,, \tag{62}$$

where G, σ, and Ω are the gravitational constant, the surface mass density of a ring, and the angular velocity, respectively, and the non-dimensional correction factor $C \approx 10$. We will also show that Eq. (62) can be obtained by a simple dimensional analysis without the formalism elaborated in the text.

The conservation of total angular momentum, combined with the dissipative loss of total energy through inelastic collisions between particles, implies a ring system as a whole must spread with time. The aim of this section is also to calculate the characteristic spreading time for the system due to the instability. For the whole of Saturn's rings, spreading times are not much shorter than the age of the solar system (see an explanation below).

As is well known, the accretion of material onto rotating astrophysical disks is inefficient if the viscosity in the disk is determined by the classical microscopic transport coefficients; the presence of developed turbulence is usually postulated to explain the observed features. Numerical experiments have confirmed that heating, turbulence, and enhanced angular momentum transport follow the linear growth of the gravitational instability. The heating and the angular momentum transfer times are of the order of 5–10 orbital periods. Following Morozov et al. (1985) and Morozov & Khoperskov (1990), in this section we show that in Saturn's rings the turbulent viscosity arising by the Jeans instability may exceed the molecular viscosity substantially. In contrast to Morozov et al. and Morozov & Khoperskov, who considered the viscous instability and the gradient instability in the nonself-gravitating limit, respectively, we study the gravitational one.

Let us now turn to the question of how to calculate the maximum turbulent viscosity $\nu_{\rm tur}$ of the ring disk expressed through the equilibrium parameters of the system. The gravitational instability is most likely to be the mechanism which does produce the disk turbulent viscosity. The turbulence that may arise as a result of gravitational instability is related to stochastic motions of fluid elements. By considering a similar system with the long-range interaction between particles, a plasma, in which the fully developed hydrodynamic turbulence is established, Kadomtsev (1965) has estimated $\nu_{\rm tur}$ from the exponential damping factor determining its instability:

$$\nu_{\rm tur} \sim \frac{\Im\omega_*}{k_{\rm min}^2}\,, \tag{63}$$

where $k_{\rm min}$ represents the minimum value of the wavenumber (the maximum value of the wavelength, or the "principal scale") for which one still has an instability and for which the coefficient of viscosity is a maximum, and $|\Im\omega_*/\Re\omega_*| \gg 1$ (Kadomtsev 1965, p. 106; see also Horton 1984). To repeat ourselves, this approach has been used by Morozov et al. (1985) and Morozov & Khoperskov (1990) to estimate $\nu_{\rm tur}$ due to the development of the viscous and gradient instabilities. It has been stated that the estimate of $\nu_{\rm tur}$ given by Eq. (63) is valid if $\Im\omega_* \gg |\Re\omega_*|$ (otherwise, the nature of the estimate changes). The expression (63) given in Kadomtsev's review for the turbulent viscosity resulting from the development of drift or dissipative instabilities is quite universal: the order-of-magnitude estimate for the classical kinematic coefficient ν_{kin} of microscopic viscosity is $\nu_{\text{kin}} \sim c_{\text{s}} l$, where l is the mean free path of the particles. The corresponding estimate for the turbulent viscosity coefficient is $\nu_{\text{tur}} \sim v_{\text{tur}} l_{\text{tur}}$, where v_{tur} is the turbulence velocity (to order of magnitude) and l_{tur} is the turbulence scale in the disk plane. The velocity v_{tur} and the scale l_{tur} can be evaluated as follows: $v_{\text{tur}} \sim \omega_* l_{\text{tur}}^2 \sim \Im\omega_* l_{\text{tur}}^2$ (for $\Im\omega_* \gg |\Re\omega_*|$) and $l_{\text{tur}} \sim 1/k_{\text{min}}$. Then we will have a good estimate (63).

From dimensional considerations, Branover et al. (1999) considered the turbulent "diffusivity" $D_\perp$ in unstable plasma across the magnetic field $\mathbf{B}$ of the form

$$D_\perp \sim \tau^{-1} \lambda_\perp^2 , \tag{64}$$

where τ is the time of the correlation's disappearance that is reasonable to choose $\sim \Im^{-1}\omega_*$, and $\lambda_\perp$ is the characteristic scale of turbulent fluctuations in the direction normal $\mathbf{B}$. As a result, Eq. (64) also gives Eq. (63).

According to the results of the stability analysis described above, in the unstable ring disk $\Im\omega_* \sim \Omega$, $k_{\text{min}} \sim 1/2h \sim \pi G\sigma_0/c_{\text{s}}^2$, and $c_{\text{s}} \sim \pi G\sigma_0/\kappa$. Substituting these quantities into Eq. (63), we obtain

$$\nu_{\text{tur}} \sim 4\Omega h^2 \sim \frac{\pi^2 G^2 \sigma_0^2}{\Omega^3} . \tag{65}$$

In Eq. (65), we have adopted the requirement of the hydrostatic equilibrium in the z direction for a self-gravitating slab model, which gives

$$h \sim \frac{c_{\text{s}}}{\sqrt{4\pi G\rho_0}} \sim \frac{c_{\text{s}}^2}{2\pi G\sigma_0} \tag{66}$$

and $\rho_0 \sim \sigma_0/2h$ is the volume mass density in the midplane. A similar result, $\nu_{\text{tur}} \sim G^2\sigma_0^2/\Omega^3$, was already discussed by Larson (1984, 1989), Lin & Pringle (1987), and Takeda & Ida (2001) by using a simple dimensional analysis. A similar expression has been confirmed numerically by Daisaka et al. (2001) (see Eq. (62) above).

Thus, we have obtained that $\nu_{\text{tur}} \sim 4\Omega h^2$. This is of the same form as the standard Shakura & Sunyaev (1973) α prescription of accretion disks, $\nu_{\text{tur}} \sim \alpha h c_{\text{s}}$ where $h \sim c_{\text{s}}/\Omega$ and $\alpha \equiv v_{\text{tur}}/c_{\text{s}}$, and it suggests that α should be of order of unity in a gravitationally unstable system (see also Papaloizou & Lin 1995, p. 533 and Lin & Papaloizou 1996, p. 720 for a discussion).

Voyager 2 photopolarimeter measurements of the damping rate of density waves in the inner B ring yield (Lane et al. 1982) $\nu_{\text{kin}} \sim 10\ \text{cm}^2\ \text{s}^{-1}$ although this value of ν_{kin} is uncertain and should be regarded as an upper limit (Goldreich & Tremaine 1982). Taking the typical angular speed in rings $\Omega \sim 3 \times 10^{-4}\ \text{s}^{-1}$ at $92,000$ km distance from the center of Saturn (the C–B ring boundary) and the typical thickness of rings $2h \sim 30$ m (Zebker & Tyler 1984; Esposito 2002; Nicholson et al. 2005), we obtain $\nu_{\text{tur}} \sim 5 \times 10^3\ \text{cm}^2\ \text{s}^{-1}$. As one can see, the turbulent viscosity driven by the instability exceeds the microscopic molecular viscosity substantially,

$$\frac{\nu_{\text{tur}}}{\nu_{\text{kin}}} \sim 5 \times 10^2 . \tag{67}$$

Shu et al. (1985) have derived a very general formula for the kinematic shear viscosity (with a special kinetic model) in self-gravitating disks

$$\nu_{\mathrm{kin}}^{\mathrm{gen}} = \frac{9}{7}\,\beta(\alpha)\,B_{\mathrm{R}}(\alpha)\,\frac{c_{\mathrm{s}}^2}{\Omega}\,, \tag{68}$$

where α is the ratio of collision frequency of particles to the orbital rotation frequency, $\beta(\alpha)$ is a special function describing the elastic coefficient of restitution as function of α, and $B_{\mathrm{R}}(\alpha)$ is the real part of a complex "main" damping function, respectively. The viscosity formula by Shu et al. (1985) can be rewritten as

$$\nu_{\mathrm{kin}}^{\mathrm{gen}} = \frac{9}{7}\,\pi^2 Q^2\,\beta(\alpha)\,B_{\mathrm{R}}(\alpha)\,\frac{G^2\,\sigma^2}{\Omega^3},$$

but one can see here that viscosity depends on the collision frequency α, a non-hydrodynamical quantity. As is seen, Shu et al. (1985) obtained a general formula for the microscopic viscosity, which is determined by ordinary interparticle collisions. In contrast, we study collective modes of motions. In the framework of our model the viscosity is determined by collective effects: one particle collides not with another one but with many which are collected together by a wave. One concludes that the analysis presented here has nothing whatsoever to do with Shu et al.'s (1985) study.

The main result of this section, Eq. (65)—the derivation of the expression for the effective turbulent viscosity—can be obtained by dimensional analysis without the formalism elaborated in the text. The turbulent viscosity is, in order of magnitude, simply the square of the modified Jeans–Toomre length scale $\lambda_{\mathrm{crit}} = 2c_{\mathrm{s}}^2/G\sigma_0$, where $c_{\mathrm{s}} \sim \pi G\sigma_0/\kappa$, for self-gravitating disks times the angular velocity Ω of local disk rotation.

5.3. Mass Inflow and Lifetime of Saturn's Rings

We have seen that the fine-scale structure may be created by the Jeans instability. However, spiral density perturbations arising from nonresonant Jeans instabilities exert torques which effectively redistribute both mass and angular momentum (Griv & Gedalin 2004; Griv 2006; Griv et al. 2008). The torques exerted on the rings by the growing spiral perturbations lead to interesting physical questions concerning the dynamical timescale of the system under study: we must estimate the orbital evolution of particles of Saturn's rings due to the instability.

By order-of-magnitude, an estimate for the mass inflow rate toward the Saturn gives $\mathrm{d}M/\mathrm{d}t \sim M_{\mathrm{ring}}/t_{\mathrm{tur}}$, where M_{ring} is the mass of Saturn's rings, $t_{\mathrm{tur}} \sim W^2/\nu_{\mathrm{tur}}$ is the characteristic time for the angular momentum transport (viscous timescale, or spreading time), and W is the rings' width. Taking $M_{\mathrm{ring}} \sim 5\times10^{22}$ g, $W \sim 10^5$ km, and $\nu_{\mathrm{tur}} \sim 5\times10^3\ \mathrm{cm}^2\ \mathrm{s}^{-1}$, one gets

$$t_{\mathrm{tur}} \sim 10^9\ \mathrm{yr} \tag{69}$$

and

$$\mathrm{d}M/\mathrm{d}t \sim 5\times10^{13}\ \mathrm{g\ yr}^{-1}\,. \tag{70}$$

Accordingly, for the whole of Saturn' rings, spreading times are not much shorter that the age of the solar system $\sim 4.5\times10^9$ yr. Thus, both Saturn's rings and their fine-scale structure are not likely much younger than the solar system.

The question of lifetime of Saturn's rings is, however, very "complex," one has to discuss the dynamical influence of "shepherd moons" in and near planetary rings and of "impact events" on the disk. To reduce the question of lifetime to hydrodynamical viscosity, arising both from the strong shear flow in Keplerian disks and self-gravity, means over-simplifying the problem. The expression (65) for the anomalous turbulent viscosity only indicates the tendency of transport of angular momentum and mass by growing gravity perturbations. Interestingly, James Clerk Maxwell has already attemped to study this process in the year 1859 (Maxwell 1859).

In closing, even though these estimates based on a linear theory of turbulent viscosity show that growing density waves provide two to three orders of magitude more angular momentum redistribution than the ordinary "collisional" redistribution, the total flux is still very small ($t_{\rm tur} \gtrsim 10^9$ yr).

6. Numerical Simulations

Most of our knowledge of the nonlinear development of the gravitational instability in Saturn's rings relies on the results of numerical simulations employing the shearing box approximation, or the so-called "local numerical studies." N-body experiments in Hill's shearing-sheet approximation have been pioneered by Wisdom & Tremaine (1988), Toomre (1990), and Toomre & Kalnajs (1991). In these simulations dynamics of particles in small regions of the disk are assumed to be statistically independent of dynamics of particles in other regions. This simple model is useful for clarifying the physics of the phenomenon. Let us assume that the radial extent of any region of interest is much smaller than its distance from the center of rotation and any relative motion is only a small fraction of the full rotation velocity. In such a model, the linearized Newton's equations of the three-dimensional motion in Hill's approximation are

$$\frac{\mathrm{d}^2 x}{\mathrm{d}t^2} - 4\Omega A_0 x - 2\Omega\frac{\mathrm{d}y}{\mathrm{d}t} = F_x\,, \tag{71}$$

$$\frac{\mathrm{d}^2 y}{\mathrm{d}t^2} + 2\Omega\frac{\mathrm{d}x}{\mathrm{d}t} = F_y\,, \tag{72}$$

$$\frac{\mathrm{d}^2 z}{\mathrm{d}t^2} + \Omega^2 z = F_z\,, \tag{73}$$

where $x = r - r_0$, $y = r_0(\varphi - \Omega t)$, r_0 is the reference radius, $\Omega = \Omega(r_0)$, $A_0 = -(r/2)(\mathrm{d}\Omega/\mathrm{d}r) = (3/4)\Omega$ is a measure of the shear strength, and F_x, F_y, F_z are the forces due to interactions with other particles. The gravitational forces are

$$\boldsymbol{F}_i = G\sum_{j=1}^{N} \frac{\boldsymbol{r}_i - \boldsymbol{r}_j}{\left[(\boldsymbol{r}_i - \boldsymbol{r}_j)^2 + r_{\rm cut}^2\right]^{3/2}}\,,$$

where $\boldsymbol{r}_i$ is the position of the ith particle and $\boldsymbol{r}_j$ is the position of the jth particle. The small cutoff radius $r_{\rm cut}$ of the potential was introduced in order to avoid numerical difficulties caused by rare close encounters between the model particles. Of course, the linearized equations of motion (71)–(73) are valid only if $|x| \ll r_0$.

The system of equations (71)–(73) for N identical particles was integrated by the Runge–Kutta method of the fourth order. A rotating Cartesian coordinate system with origin at the reference position r_0 was chosen, the x axis pointing radially outward and the y axis pointing in the direction of the rotation (Salo 1995, 2001). The particles were initially placed on nearly circular orbits with an anisotropic Maxwellian (Schwarzschild) distribution of small random velocities (Griv & Gedalin 2003). Following Wisdom & Tremaine (1988), Salo (1992, 1995, 2001), Richardson (1995), and Daisaka & Ida (1999), we adopted the standard hard-sphere collision model.

The initial distribution of particles was generated by means of pseudo-random number generator placing particles uniformly in the box in real space. The box should be thought of as being embedded in Saturn's ring disk which has a constant angular velocity gradient in the x direction. Our model has the ratio $2\Omega/\kappa = 2$ and the radial velocity dispersion $c_r = c_{\rm T}$, so we expect the model to be initially violently unstable to the growth of spiral disturbances (see § 3.6. above for an explanation); the initial vertical velocity dispersion $c_z = 0.15c_r$. To maintain the system under the shearing stress in a steady state, the cyclic boundary conditions were used in the form suggested by Wisdom & Tremaine (1988), Toomre (1990), and Salo (1995).[10] The direction of the disk rotation was taken to be clockwise. Time $t = 1$ corresponds to a single revolution of the disk, and the orbital period was $T_{\rm orb} = 2\pi/\Omega$. All the particles moved with the same fixed time step $\Delta T = 0.001T_{\rm orb}$. In Saturn's rings, a measure of the fundamental vibration period, or the dynamical time is of the order of $T_{\rm dyn} \sim (G\rho_{\rm S})^{-1/2} \sim 1$ h if we assume the volume density $\rho_{\rm S} \sim 1$ g cm^{-3}. This means that $T_{\rm orb} \sim 10T_{\rm dyn}$, so a choice of the stepsize $\Delta T = 0.001T_{\rm orb}$ gives about 100 steps per particle dynamical time, which should be sufficient to accurately resolve particle–particle interactions. The value of $r_{\rm cut}$ was chosen to be $0.001\lambda_{\rm JT}$.

The following physical parameters for a simulated patch of the ring were chosen: the orbital period $T_{\rm orb} \equiv 2\pi/\Omega = 7.027$ h which corresponds to a typical orbital period of the C ring's particle at the distance $r = 85,000$ km from the planet (Saturn's mass is about 5.68×10^{29} g), the surface density $\sigma_0 = 15$ g cm^{-2}, and the total number of particles $N = 12,000$. The particle radius $r_{\rm p} = 4.2$ cm in a $L_x \times L_y \times L_z = 4\lambda_{\rm JT} \times 4\lambda_{\rm JT} \times \lambda_{\rm JT}$ three-dimensional model. The corresponding Jeans–Toomre wavelength $\lambda_{\rm JT} \approx 6.4$ m and the optical depth $\tau \approx \pi r_{\rm p}^2 n \approx 0.1$, where n is the number density per unit area. The constant coefficient of restitution was $\epsilon = 0.8$. The simplest procedure for detecting collisions by considering the inelastic impacts at the end of each time step was explored. We recognize that this is not the best collision detection strategy. In our point of view, in low and moderately high optical depth regions of Saturn's rings with $\tau < 3$ we are investigating, collisions are to be relatively unimportant and hence the treatment of collisions can be grossly simplified. The only important forces of consequence are gravity of Saturn, which accounts for the orbital motion, and self-gravity of Saturn's rings, which accounts for the distinct microstructure.

A few runs for systems containing $N = 24,000$ model particles and for smaller systems containing $N = 6,000$ ones were performed. It was found that the results obtained for those systems are qualitatively indistinguishable. Moreover, tests indicate that the results were insensitive to changes in other parameters (the area of unit cell, etc.).

[10]The "sliding brick" technique of Wisdom & Tremaine (1988), Toomre (1990), and Salo (1995) has been used in the past to simulate numerically transported properties of simple fluids under action of strong shearing force (e.g., Lees & Edwards 1972).

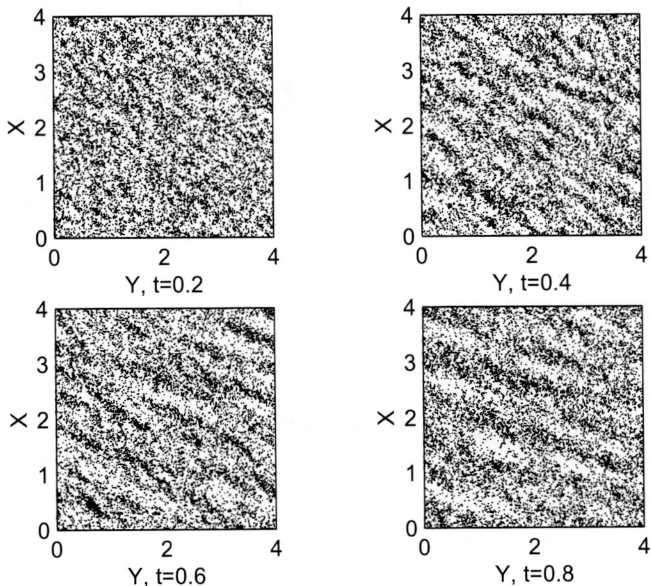

Figure 7. Time-development of the differentially rotating model (face-on view); $c_r = c_\mathrm{T}$. In good agreement with the theory, even though the initial radial velocity dispersion c_r is equal to the Safronov–Toomre one c_T, the model is still unstable against nonaxisymmetric, trailing ($\psi > 0$) perturbations.

6.1. Planar Structure

In Fig. 7 we show snapshots from a run with the warm model, i.e., the nonuniformly rotating model, in which initially particles all move along almost circular orbits and the radial dispersion of the random velocities is equal to the Safronov–Toomre one, $c_r = c_\mathrm{T}$, or Toomre's parameter $Q = 1$, respectively. As has been predicted in our theory, the instability develops quickly in the system. Figure 7 clearly shows that in a disk with the Keplerian shear profile a spiral pattern develops spontaneously in the initially featureless disk on a dynamical timescale $\lesssim \Omega^{-1}$. The spiral density waves are created if self-gravity is included; only collisions do not create the structure. The pattern speed Ω_p is zero, because the system is spatially homogeneous (Eq. (51)). The density wave structure is time dependent and transient.

The initial unstable growth is clear from Fig. 7. The structure consists of elongated trailing filaments with a definite pitch angle ψ with respect to the local shear flow. The pitch angle of spiral waves $\psi = 20° - 25°$ can be obtained by examining Fig. 7 directly, thus $\tan^2 \psi \ll 1$ and the asymptotic Lin–Shu approximation of moderately tightly-wound perturbations used throughout the theory does not fail. The translation symmetry in the direction ψ is clearly unbroken, while in the perpendicular direction a series of alternating peaks and dips are seen. The simulated wave structure rapidly at times $t \gtrsim 0.6$ reaches nonlinearity, and is thus beyond the scope of our theory (or any corresponding linear analysis of shearing modes). This fierce instability of the system with $Q = 1$ indicates that in a differentially rotating system the Jeans instability of nonaxisymmetric gravity perturbations cannot be suppressed by the ordinary Safronov–Toomre velocity spread c_T, in line with theoretical expectations (Griv et al., 2000a, 2003a, b; Griv & Gedalin 2003; § 3.6. above).

Figure 7 gives a feeling for the evolutionary process, but in order to trace and quantify the growth of instabilities in the disk, it is necessary to compute Fourier decompositions of the surface density distribution for various mode numbers. Shown in Fig. 8 are the sequences of density-spectrum evolution in the wavenumber space for the case shown in Fig. 7 at the calculation times $t = 0.2 - 0.8$. The power spectrum is constructed by employing a technique where one regards the particle, labeled by j, as a discrete δ-function to calculate the density Fourier component, i.e.,

$$A(k_x, k_y) = \frac{1}{N} \sum_{j=1}^{N} \exp(\boldsymbol{k} \cdot \boldsymbol{r}_j) . \tag{74}$$

In addition, k_y assumes discrete values compatible with the finite boundaries in the streamwise direction, that is, in the y-direction, whereas k_x assumes continuous values because of the background flow shear. The pitch angle of a spiral ψ is given by $\psi = \arctan(k_y/k_x)$, and positive k_x corresponds to trailing spirals and negative k_x to leading. The wavenumber is normalized to π/L_y. Thus, the quantity k_y gives the number of halfwaves of a spiral mode in the y-direction.

No particular set of k_x, k_y dominates at the beginning of calculations ($t = 0.2$), which corresponds to the initial noise. It is evident from Fig. 8 that the $c_r = c_{\rm T}$ case produces a rigorous instability. One can clearly see that at times $t \gtrsim 0.6$ a dominant k_y is equal about to $(2L_y/\lambda_{\rm JT}) \tan\psi \approx 4$, which is consistent with our linear stability analysis for the unstable mode. That is, a wavelength of the dominant mode in the streamwise $\lambda_y = \lambda_{\rm crit}/\tan\psi \approx 2\lambda_{\rm JT}$, where $\psi = 22° - 28°$. The latter fact convincingly indicates that we have dealt with a collective-type instability rather than with a random process (as advocated by Toomre 1990 and Toomre & Kalnajs 1991). It is natural to attribute the observed instability to the Jeans instability so far discussed in the paper. Also, we see that there are a number of $k_y = 2$, $k_y = 6$, and $k_y = 8$ discrete harmonics present.

6.2. Vertical Structure

Figure 7 shows the density distributions of an unstable model of Saturn's rings in the plane (face-on view). As has been mentioned by Salo (1992), Richardson (1994), and others, another feature of the simulations is the radical changes in vertical structures of unstable systems that result from the instabilities. Figure 9 shows distributions of particles and isodensity contours in the (z, η)-plane for the model at the simulations times $t = 0$ and $t = 0.4$. (We introduced a new coordinate system (z, η); η is perpendicular to the spiral waves, while z gives the normal to the plane position.) The disk surface generally evolves from a smooth height profile with η to complex structures with peaks and valleys that result from instabilities. The results presented in Fig. 9 are consistent with the hypothesis that we have dealt with the even Jeans perturbations, because the perturbed density is an even function of z (cf. Fig. 1b). As is seen, the vertical velocity of the fluid element $v_z(z) \approx -v_z(-z) \approx 0$. The latter indicates that we have dealt with Chandrasekhar's (1955) singular modes (Goldreich & Lynden-Bell 1965a).

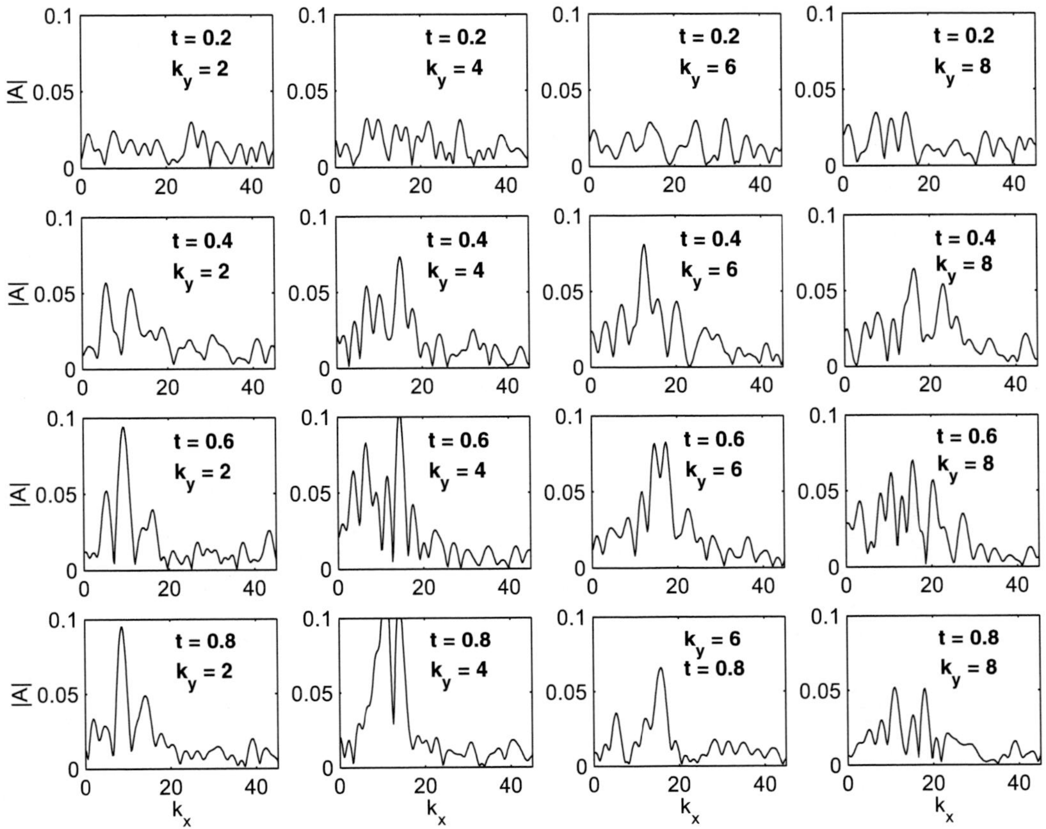

Figure 8. Density spectrum $|A(k_x, k_y)|$ of the particle distribution shown in Fig. 7 for different azimuthal k_y and radial k_x mode numbes at the calculation times $t = 0.2 - 0.8$.

7. Conclusion

We strongly believe that the gravitational instability is the main process responsible for the formation of resolved microstructures in Saturn's A and B rings. In the process, the behavior of the ring disk is governed by collective effects, while particle's physical impacts play a secondary role. In the preceding sections the theoretical model is described which explains the origin of this fine-scale stratification of the system under study, characterized by a periodic variation in optical depth. The idea of self-organising rings are developed from the formalism of the gravitational physics of collective processes which nearly fifty years earlier led to great achievements in the dynamics of magnetized plasmas. We show that the self-gravitating rings are able to form various three-dimensional structures—"sausage-like" moderately tightly-wound spirals and rings (Figs. 1b and 4), and clumpy moons (Fig. 6), continually forming and dispersing—without the interference of external forces such as embedded or external satellites. The model described here does include both azimuthally symmetric density waves and nonaxisymmetric waves, that is, the periodic radial density enhancements and rarefications may co-exist with the spiral ones. Thus, a "fluid" of ring particles might exhibit self-excited (i.e., intrinsic) Jeans' gravitational instability of sponta-

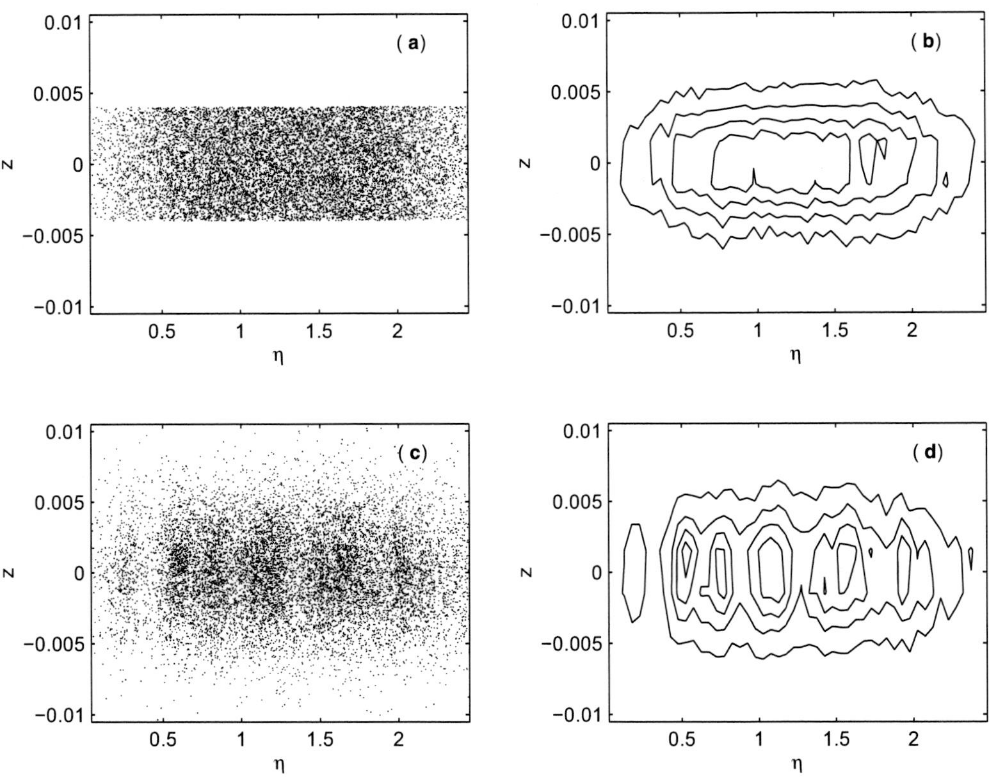

Figure 9. Comparison of vertical structure. Shown are both distributions of particles and isodensity contours in the (z, η)-plane for the three-dimensional model shown in Fig. 7 at the calculation times $t = 0$ and $t = 0.4$: (a) distribution of particles at $t = 0$, (b) isodensity contours at $t = 0$, (c) distribution of particles at $t = 0.4$, and (d) isodensity contours at $t = 0.4$. (c) and (d) give the end-on edge-on views of the positions and isocontours of the density of the disk particles, respectively. The unstable spiral disturbances significantly alter the disk's vertical surface.

neous gravity perturbations which takes place when the ring self-gravity enables the growth of density waves.

This paper examines analytically the stability of the Saturnian ring disk using a hydrodynamic approximation to particle dynamics. The hydrodynamic approximation is valid in the considered limit of highly interactive *self-gravitating* Saturn's rings. The simple model of the disk is studied: the disk is considered to be infinitesimally thin and its radial structure is considered in a horizontally local WKB, or short-wave approximation. As usual in the short-wave approximation, all perturbations are expressed by the modal representation. The linear analysis is restricted to odd-parity Lin–Shu type collective modes of oscillations for which the vertical velocity of the fluid element $v_z(r, \varphi, z = 0, t) = 0$; bending oscillations for which $v_z(r, \varphi, z = 0, t) \neq 0$ are not considered. We have interested in the dispersion relation for waves generated only in the region where there are no the corotation ($\omega_* = 0$) and inner (outer) Lindblad ($\omega_* \pm \kappa = 0$) resonances, or a resonance with a satellite. By assuming that equilibrium rotation and density vary over a much larger spatial scale than the modes wavelength, we found that the most important gravitational instability in Saturn's rings is a local instability, i.e., normal modes are driven unstable by the local value of the rotational flow shear.

An expression for the critical sound speed (the velocity dispersion) $c_{\mathrm{T}} = \pi G \sigma_0 / \kappa$ that the particles comprising a two-dimensional, rapidly rotating disk should have in order to suppress the instability of axisymmetric (ring) Jeans perturbations has already been derived (Safronov 1960, 1980; Toomre 1964; Lin et al. 1969; Rohlfs 1977, p. 100; Goldreich & Tremaine 1982; Binney & Tremaine 1987, p. 362; Sicardy 2005, p. 462). Following Lau & Bertin (1978), Lin & Lau (1979), Bertin (1980), and Morozov (1981, 1985), we extended the analysis by taking into additional account the nonaxisymmetric (spiral) gravity perturbations propagating in a nonuniformly rotating disk. Important conclusions were obtained about the enhanced amplification of spiral ($\psi \neq 0$) density waves in a differentially rotating ($2\Omega/\kappa > 1$, or $\mathrm{d}\Omega/\mathrm{d}r \neq 0$, respectively) system. The modified Safronov–Toomre type stability criterion for the disk to be Jeans-stable to both axisymmetric and nonaxisymmetric substructure formation by self-gravitation has been obtained, $c_{\mathrm{crit}} \gtrsim (2\Omega/\kappa) c_{\mathrm{T}} \approx 2 c_{\mathrm{T}}$.

The disk is Jeans-unstable to both axisymmetric and nonaxisymmetric perturbations if $c_{\mathrm{s}} < c_{\mathrm{T}}$ (Fig. 4a). This result suggests that disks with Toomre's $Q \equiv c_{\mathrm{s}}/c_{\mathrm{T}}$ parameter < 1 break up into fragments of preferred mass

$$M_{\mathrm{frag}} \sim \frac{4 c_{\mathrm{s}}^4}{G^2 \sigma_0} .$$

In Saturn's rings, the typical mass of such porous particle aggregates is predicted to be $M_{\mathrm{frag}} \sim 10^7$ g each. Interestingly, Cassini observations have indicated that axisymmetric periodic structure in Saturn's rings (Thomson et al. 2007) co-exists with nonaxisymmetric structure (Colwell 2006, 2007; Hedman et al. 2007). In the framework of our theory, the coexistence of several spatially correlated and preferentially oriented particle clusters (spirals) and axisymmetric clusters (rings) is possible, as long as $Q < 1$ and nonlinear effects are small; spirals rotate rigidly with a single fixed pattern speed. The Safronov–Toomre unstable Saturnian ring disk ($Q < 1$) evolves towards a quasi-stationary state where the disk is characterized by clumpy and filamentary structures.

Thus, the modified Safronov–Toomre stability criterion c_{crit} given by Eq. (48) is deduced, according to which the nonuniformly rotating system is violently Jeans-unstable against the growth of nonaxisymmetric perturbations, unless the sound speed is at least two-fold (Toomre's stability parameter $Q \approx 2$) as compared to that required by the usual Safronov–Toomre stability criteria for axisymmetric perturbations. A relationship exists between c_{crit} and what Toomre (1981) called "swing amplification." Exclusively trailing spirals can be formed in Saturn's A and B rings. The strongest growth occurs in the scale of $\lambda_{\mathrm{crit}} \sim 2c_{\mathrm{s}}^2/G\sigma_0$. In low and moderately high optical depth regions of the A and B rings $c_{\mathrm{s}} \sim 0.1$ cm s^{-1} and $\sigma_0 = 30 - 150$ g cm^{-2}, thus $\lambda_{\mathrm{crit}} = 30 - 150$ m. These values of $Q \sim 2$ and $\lambda_{\mathrm{crit}} \sim 100$ m predicted in our analysis are close to Salo (1992, 1995, 2001), Richardson (1994), Osterbart & Willerding (1995), Griv (1998, 2005a), Daisaka & Ida (1999), Ohtsuki & Emori (2000), and Griv et al. (2004, 2006a) numerical results.

Geometric model of self-gravity density waves in Saturn's rings is shown in Fig. 3. In the lowest approximation of the theory, the density waves are regularly spaced, aligned three-dimensional clusters of ring particles (Fig. 3a). At the limit of the instability ($c_{\mathrm{s}} \sim$

$\pi G\sigma_0/\kappa$), the physical scale of the waves in the $z = 0$ plane is predicted to be of the order of

$$\lambda_{\rm crit} \approx \frac{2c_{\rm s}^2}{G\sigma_0} \approx 4\pi h \,. \tag{75}$$

As one can see, the most unstable modes of the differentially rotating disk have wavelengths of some 2π the thickness $2h$ of the system. Observations have revealed a similar ~ 100 m or so fine stratification of Saturn's rings (Porco et al. 2005, p. 1230; Colwell et al. 2006, 2007; Hedman et al. 2007; Thomson et al. 2007). We conclude that the critical Jeans wavelength $\lambda_{\rm crit}$ *does* agree with the width of spirals and rings in Saturn's A and B rings. It is natural to attribute the observed microstructure to the Jeans instability of gravity perturbations so far discussed in this paper.

As for the present study, the Jeans-unstable spiral density waves also cause the ring A's quadrupole azimuthal brightness asymmetry observed first by Camichel (1958) in the ansae of Saturn's rings (see Griv & Gedalin 2003 for a discussion).[11] To emphasize it again, the physics underlying such self-gravity cylindric waves is essentially the same as the "density wave structure" which was studied in the context of galactic disks by Lin & Shu (1966), Lin et al. (1969), Shu (1970), Lau & Bertin (1978), Lin & Lau (1979), Bertin (1980, 2000), Morozov (1980), Bertin & Lin (1996), Griv et al. (1999, 2008), and Griv (2007b). These fully self-consistent density waves are not to be confused, of course, with Julian & Toomre's (1966) "forced" density wave proposal explored by Colombo et al. (1976) and Franklin & Colombo (1978). Actually, Julian & Toomre (1966) worked out the response of a disk when forced by an orbiting mass, like a giant molecular cloud in spiral galaxies. Julian & Toomre's (1966) calculations showed that even a stable stellar system (possessing a velocity dispersion more than sufficient for local axisymmetric stability) to be remarkably responsive in a spiral-like manner to localized forcing. In sharp contrast to Julian & Toomre (1966), we develope the theory of real instabilities of small spontaneous disturbances which propagate in the main part of the system. In other words, our model relies on normal modes of collective oscillations of a particulate disk to maintain spiral (and ring) coherence at all radii.

Even though the estimates based on a linear theory of turbulent viscosity (see § 5.3. above) show that growing density waves provide two to three orders of magnitude more angular momentum redistribution than the ordinary "collisional" redistribution, the total flux is still very small, $t_{\rm tur} \gtrsim 10^9$ yr. The energy source of the turbulence in this system originate in the shear driven by ring rotation and self-gravitational energy of the "gas." The rings were once 3–4 billion years ago a moon of Saturn whose orbit decayed until it came close enough to be ripped apart by tidal forces or are instead left over from the original nebular material from which Saturn formed, which never accreted to become one or more moons. In this regard, note that Saturn's rings have an integrated mass comparable to those of significant satellites like Mimas or Encelade (Sicardy 2005, p. 458). On the other hand, the surface density of particles of stratified Saturn's rings can be substantially larger than

[11]Nonaxisymmetric elongated particle conglomerations were first suggested to be the cause of this asymmetry by Colombo et al. (1976). Calculations by Franklin & Colombo (1978) and later by Lumme & Irvine (1979) showed that such periodic density structures with a pitch angle of about 20° would scatter light in a way that was consistent with the observed asymmetry (see, e.g., Dunn et al. 2007 and French et al. 2007 for a discussion).

one would infer from a uniform distribution of particles. The inferred total mass of Saturn's rings is therefore likely to be more massive than previously estimated, by a factor of ~ 3. Such a massive ring system is unlikely to have formed by the tidal disruption of a former satellite but instead one suggests that the rings could be of a primordial origin (Stewart et al. 2008).

Density waves are the way to form both axisymmetric and nonaxisymmetric structures (and also porous moonlets) in the Saturnian rings, but they are not steady-state modes. The nonlinear interaction of particles with almost aperiodically growing Jeans-unstable density waves increases random velocities of particles ("heating") in a short timescale of only 2–3 disk orbital revolutions. This growth in random velocities leads to an eventual stabilization, unless some effective cooling mechanism of the reconstruction of the wave structure exists. It is suggested that dissipative (inelastic) impacts between particles provide such a cooling mechanism, reducing the magnitude of the relative velocity of particles, and thus reducing the random velocity spread. We especially emphasize that the particles constituting Saturn's rings collide inelastically, with the subsequent loss of mechanical energy. This mechanical energy may be converted into thermal energy, and is essentially lost. The latter leads to some "cooling" of the system, and thus leads to a reccurent instability cycle (cf. Griv & Chiueh 1998, p. 209). We argue that both the Saturnian main rings and their axisymmetric and nonaxisymmetric particle clusters (being recurrently created and destroyed on the timescale of an order of Keplerian period $\sim$ 20 h) are not likely much younger the solar system; recycling could extend the microstructure lifetime essentially. Observations seem to back this suggestion: observations in the Saturnian ring system during approach and orbital insertion, with Cassini's visual and infrared mapping spectrometer, have shown that Saturn's rings have changed little in their radial structure since the Voyager flybys in the early 1980s (Brown et al. 2006).

Simplified numerical simulations employing the shearing box approximation are described which test the validities of both the modified Safronov–Toomre stability criterion $c_{\rm crit}$ and the modified Jeans–Toomre wavelength $\lambda_{\rm crit}$. The facts that the nonaxisymmetric perturbations are more unstable than the axisymmetric ones and in the nonaxisymmetric case the limit of stability is shifted toward a longer wavelength are taken into account in these quantities. It is shown that in agreement with a theoretical prediction at the limit of stability with respect to all gravity perturbations the critical wavelength becomes approximately equal to $\lambda_{\rm crit}$.

To summarize, one can attribute the periodic fine-scale structure already resolved in Saturn's A and B rings to the development of Lin–Shu type compression density waves developing in the plane of the system. The waves are described by the same physics that described the spiral arms of galaxies; both radial and spiral self-gravity density waves form and break apart on the orbital timescale. At the limit of gravitational stability, the physical scale of these three-dimensional longitudinal waves (the width and the separation of the waves) $\lambda_{\rm crit} \approx 4\pi h$ (Eq. (75)) and the disk's thickness is $2h$. In Saturn's rings, $\lambda_{\rm crit} = 30 - 150$ m. Thus, (a) the self-gravity recycling density enhancements as shown in Fig. 3 are flattened structures, with height/width ratio of about $1/\pi$. A separate investigation based on high-resolution of the order of 10 m observations of Saturn's A and B rings (and probably C ring as well) should be done to confirm (or deny!) this prediction. (b) Seen edge-on with the fine spirals seen end-on (with the line of sight along the spiral local major axis), Saturn's rings will show even with respect to the equatorial $z = 0$ plane "sausage-like" structures (Figs. 9c and 9d). (c) We expect also that observations will bring the value $Q \lesssim 1$ in large portions of Saturn's A and B rings. And (d) a number of porous moonlets of the order of 50 m in radius with the typical mass $M_{\rm frag} \sim 10^7$ g each are embedded in Saturn's rings, although this has yet to be directly measured.

Lastly, we add a brief comment on our work. We have investigated the collective instabilities of the self-gravitating Saturnian ring system of identical particles on the following fundamental assumption, i.e., the local WKB analysis. This assumption may has essential defects for the wave phenomena and instabilities of self-consistent systems though the physical mechanisms of the instabilities are well clarified (Alexandrov et al. 1984, p. 245; Krall & Trivelpiece 1986, p. 424). Therefore we have to investigate the effects of nonlocality in the next step (Alexandrov et al. 1984, p. 249; Rüdiger & Kitchatinov 2000). We speculate that by utilizing the more accurate nonlocal WKB approximation it will be shown that the characteristic oscillation frequencies of an inhomogeneous disk must be "quantized," i.e., must pass through a discrete series of values. According to the WKB method the spectrum of frequencies ω is determined by the quasiclassical rules of Bohr–Sommerfeld quantization:

$$\int_{r_1}^{r_2} k_r(r')\mathrm{d}r' = \left(n + \frac{1}{2}\right)\pi \,, \tag{76}$$

where r_1, r_2 are the "reflection" points (say, the inner and outer Lindblad resonances) and $n = 0, 1, 2, \cdots$. Equation (76) implies that ω is independent of r; that is, the spiral-wave front is not distorted by differential rotation. The pattern is composed of one or more modes, and discrete modes are stationary in a rotating frame and hence do not wind up. The weakly inhomogeneous approximation used throughout this paper has the meaning that the discrete spectrum will differ little from a continuous spectrum, and in the zero approximation may be regarded as continuous. Further, the nonlocal theory provides a formal basis for the main idea that formed the basis for the local description, namely, that of short-wavelength approximation $|k_r|r \gg |k_r L| \gg 1$.

Acknowledgements

It is a pleasure to acknowledge our collaborators and colleagues, especially Tzi-Hong Chiueh, David Eichler, Alexei Fridman, Edward Liverts, Michael Mond, Raphael Steinitz, and Chi Yuan for the various discussions in relation to this work, and Irena Zlatopolsky for technical assistance. Special thanks are to William Peter and Baruch Rosenstein for suggesting the improvement of the numerical calculations. This work was sponsored by the Israel Science Foundation and the U.S.–Israel Science Foundation. Evgeny Griv was supported in part by the Israeli Ministry of Immigrant Absorption in the framework of the programme "KAMEA."

Appendix A: The Effect of Collisions

Let us consider the influence of interparticle collisions on the dispersion law of Jeans oscillations using the simple method of single particle dynamics (Alexandrov et al. 1984, p. 46;

Griv 1998; Griv et al. 1999). The linearized Eqs. (1) and (2), $\Phi(\boldsymbol{r},t) = \Phi_0(r) + \Phi_1(\boldsymbol{r},t)$, $\boldsymbol{r}(t) = \boldsymbol{r}_0 + \boldsymbol{r}_1(t)$, and $|\Phi_1/\Phi_0| \ll 1$, $|\boldsymbol{r}_1/\boldsymbol{r}_0| \ll 1$, take the form:

$$\frac{\mathrm{d}^2 r_1}{\mathrm{d}t^2} \approx \frac{\left(r_0^2\dot{\varphi}_0\right)^2 - 2r_0^2\dot{\varphi}_0 \int_{-\infty}^{t}(\partial\Phi_1/\partial\varphi)\mathrm{d}t'}{(r_0+r_1)^3} - \frac{\partial\Phi_0}{\partial r} - \frac{\partial\Phi_1}{\partial r} - \nu_{\mathrm{coll}} v_r\,, \tag{77}$$

$$(r_0+r_1)^2(\dot{\varphi}_0+\dot{\varphi}_1) \approx r_0^2\dot{\varphi}_0 - \int_{-\infty}^{t}\frac{\partial\Phi_1}{\partial\varphi}\mathrm{d}t' - \nu_{\mathrm{coll}} r_0^2 \int_{-\infty}^{t}\frac{\partial\varphi_1}{\partial t}\mathrm{d}t'\,, \tag{78}$$

where $r_0^2\dot{\varphi}_0$ is the so-called area constant, $r_0\dot{\varphi}_0^2 = (\partial\Phi_0/\partial r)_0$, and the terms with ν_{coll} are small corrections (in the case of rare, $\nu_{\mathrm{coll}} < \Omega$, and weak collisions, $\nu_{\mathrm{coll}} < \omega_*$, in which we are especially interested). Equations (77) and (78) describe the small departure $\boldsymbol{r}_1(t)$ of the actual radius $\boldsymbol{r}(t)$ from $\boldsymbol{r}_0$, which is chosen so that the constant of areas for the circular orbit is equal to the angular momentum integral $J_z = r^2\dot{\varphi}$. From these equations we get

$$\frac{\mathrm{d}^2 r_1}{\mathrm{d}t^2} + \kappa^2 r_1 = -\frac{2\Omega}{r_0}\int_{-\infty}^{t}\frac{\partial\Phi_1}{\partial\varphi}\mathrm{d}t' - \frac{\partial\Phi_1}{\partial r} - \nu_{\mathrm{coll}}\frac{\mathrm{d}r_1}{\mathrm{d}t}\,, \tag{79}$$

$$(r_0+r_1)^2(\dot{\varphi}_0+\dot{\varphi}_1) - \Omega r_0^2 = -\int_{-\infty}^{t}\frac{\partial\Phi_1}{\partial\varphi}\mathrm{d}t' - \nu_{\mathrm{coll}} r_0^2 \int_{-\infty}^{t}\frac{\mathrm{d}\varphi_1}{\mathrm{d}t}\mathrm{d}t'\,, \tag{80}$$

where

$$\kappa^2 = \frac{3}{r_0}\left(\frac{\partial\Phi_0}{\partial r}\right)_0 + \left(\frac{\partial^2\Phi_0}{\partial r^2}\right)_0 \quad \text{and} \quad \Omega^2 = \frac{1}{r_0}\left(\frac{\partial\Phi_0}{\partial r}\right)_0 .$$

The homogeneous differential Eqs. (79) and (80) yield the ordinary Lindblad's elliptic-epicyclic orbits (cf. Eqs. (22)–(23)):

$$r = r_0 + \frac{v_\perp}{\kappa}\left[\cos(\phi_0 - \kappa t) - \cos\phi_0\right]\,, \tag{81}$$

$$\varphi = \Omega t - \frac{2\Omega}{\kappa}\frac{v_\perp}{r_0\kappa}\left[\sin(\phi_0 - \kappa t) - \sin\phi_0\right]\,, \tag{82}$$

where $v_\perp$ and ϕ_0 are constants of integration, $r_{\mathrm{epic}} = v_\perp/\kappa$ is the mean epicycle radius, and $r_{\mathrm{epic}}/r_0 \ll 1$.

The particular solutions yield the expressions for perturbed velocities (cf. Eqs. (24)–(25))

$$v_r = \frac{k_r\omega_*}{\omega_*^2 - \kappa^2 + \imath\omega_*\nu_{\mathrm{coll}}}\tilde{\Phi}e^{\imath k_r r + \imath m\varphi - \imath\omega_* t}\,, \tag{83}$$

$$v_\varphi = \frac{4\Omega^2 - \kappa^2 + \omega_*^2}{\omega_*\left(\omega_*^2 - \kappa^2 + \imath\omega_*\nu_{\mathrm{coll}}\right)}\frac{m}{r}\tilde{\Phi}e^{\imath k_r r + \imath m\varphi - \imath\omega_* t}\,, \tag{84}$$

where $|\omega_*| \sim \Omega > \nu_{\mathrm{coll}}$ and only the "in-phase" terms are included. As we can see from the equations above, in comparison with the collisionless disk in the collisional system one needs to replace the wavefrequency ω_* by $\omega_* + \imath\nu_{\mathrm{coll}}$; thus if $\nu_{\mathrm{coll}}/|\omega_*|$ is small enough we can ignore these collisions.

Paralleling the analysis leading to Eq. (40) and making use of Eqs. (83)–(84), it is straightforward to show that the simplified dispersion relation can now be expressed as

$$\omega_*^2 + \imath\omega_*\nu_{\mathrm{coll}} - \omega_{\mathrm{J}}^2 = 0\,, \tag{85}$$

where as usual $\omega_{\rm J}^2$ is the squared Jeans frequency. The dispersion relation (85) describes the spectrum of oscillations developing in a homogeneous, two-dimensional collisional disk. The solution of Eq. (85) is

$$\omega_{*1,2} \approx \pm p|\omega_{\rm J}| - \imath \frac{\nu_{\rm coll}}{2}\,, \tag{86}$$

where $p = \imath$ for Jeans-unstable perturbations ($\omega_{\rm J}^2 < 0$) and $p = 1$ for Jeans-stable ones ($\omega_{\rm J}^2 > 0$), $|\omega_{\rm J}| \sim \Omega$, and $\nu_{\rm coll} < |\omega_{\rm J}|$. According to Eq. (86), Jeans-stable perturbations will decay because of interparticle collisions. The effect is not large: the time necessary for the stable wave amplitude to fall to $1/e$ of its initial value is about the collision time, $\nu_{\rm coll}^{-1}$. This is longer than the characteristic time of a single revolution of a disk $\sim \Omega^{-1}$. Jeans-unstable perturbations will suffer weak stabilization. Clearly, however, these rare and weak collisions do not affect the local stability criterion $c_{\rm crit}$ and the critical wavelength $\lambda_{\rm crit}$ of the instability.

Appendix B: On the Solutions of Gasdynamical Equations in the Form of Normal Modes

Following Fridman (1989), let us consider a two-dimensional, spatially homogeneous ($\sigma = \sigma_0 + \sigma_1(r,\varphi,t)$ and $|\sigma_1/\sigma_0| \ll 1$) gaseous gravitating disk in equilibrium, which is rotating differentially with angular velocity $\Omega = \Omega(r)$ about the axis z. We have the following system of linearized dynamical equations (Landau & Lifshitz 1987):

$$\left(\frac{\partial}{\partial t} + \Omega\frac{\partial}{\partial\varphi}\right)\sigma_1 + \frac{1}{r}\frac{\partial}{\partial r}(r\sigma_0 v_r) + \frac{1}{r}\frac{\partial}{\partial\varphi}(\sigma_0 v_\varphi) = 0\,, \tag{87}$$

$$\left(\frac{\partial}{\partial t} + \Omega\frac{\partial}{\partial\varphi}\right)v_r - 2\Omega v_\varphi = -\frac{\partial\aleph}{\partial r}\,, \tag{88}$$

$$\left(\frac{\partial}{\partial t} + \Omega\frac{\partial}{\partial\varphi}\right)v_\varphi + \left(2\Omega + r\frac{\mathrm{d}\Omega}{\mathrm{d}r}\right)v_r = -\frac{1}{r}\frac{\partial\aleph}{\partial\varphi}\,, \tag{89}$$

$$\Delta\Phi_1 = 4\pi G\sigma_1\delta(z)\,, \tag{90}$$

where $\aleph = \Phi_1 + c_{\rm s}^2\sigma_1/\sigma_0$ and $c_{\rm s} = (\partial P_\perp/\partial\sigma)_0^{1/2}$. In the spirit of Goldreich & Lynden-Bell (1965b, p. 129), we now transform to co-moving (with the undisturbed flow) variables by writing

$$\varphi' = \varphi - \Omega(r)t, \quad t' = t, \quad \text{and} \quad r' = r\,. \tag{91}$$

Then

$$\frac{\partial}{\partial t} = \frac{\partial}{\partial t'} - \Omega\frac{\partial}{\partial\varphi'}\,,$$
$$\frac{\partial}{\partial\varphi} = \frac{\partial}{\partial\varphi'}\,,$$
$$\frac{\partial}{\partial r} = \frac{\partial}{\partial r'} - t'\frac{\mathrm{d}\Omega}{\mathrm{d}r}\frac{\partial}{\partial\varphi'}\,,$$

so

$$\frac{\partial}{\partial t} + \Omega\frac{\partial}{\partial\varphi} = \frac{\partial}{\partial t'}\,.$$

In these new variables, instead of Eqs. (87)–(90), we find (Goldreich & Lynden-Bell 1965b; Lominadze et al. 1988)

$$\frac{\partial \sigma_1}{\partial t'} + \sigma_0 \frac{\partial v_{r'}}{\partial r'} + \frac{\sigma_0 v_{r'}}{r'} + \sigma_0 t' \frac{\mathrm{d}\Omega}{\mathrm{d}r'} \frac{\partial v_{r'}}{\partial \varphi'} + \frac{\sigma_0}{r'} \frac{\partial v_{\varphi'}}{\partial \varphi'} = 0\,, \tag{92}$$

$$\frac{\partial v_{r'}}{\partial t'} - 2\Omega v_{\varphi'} = -\frac{\partial \aleph}{\partial r'} + t' \frac{\mathrm{d}\Omega}{\mathrm{d}r'} \frac{\partial \aleph}{\partial \varphi'}\,, \tag{93}$$

$$\frac{\partial v_{\varphi'}}{\partial t'} + \left(2\Omega + r' \frac{\mathrm{d}\Omega}{\mathrm{d}r'}\right) v_{r'} = -\frac{1}{r'} \frac{\partial \aleph}{\partial \varphi'}\,, \tag{94}$$

$$\Delta \Phi_1 = 4\pi G \sigma_1 \delta(z)\,. \tag{95}$$

In Eq. (95),

$$\Delta = \frac{\partial^2}{\partial r'^2} - t' \frac{\mathrm{d}^2\Omega}{\mathrm{d}r'^2} \frac{\partial}{\partial \varphi'} - 2t' \frac{\mathrm{d}\Omega}{\mathrm{d}r'} \frac{\partial}{\partial r'} \frac{\partial}{\partial \varphi'} + \left[1 + \left(t' r' \frac{\mathrm{d}\Omega}{\mathrm{d}r'}\right)^2\right] \frac{1}{r'^2} \frac{\partial^2}{\partial \varphi'^2} + \frac{\partial^2}{\partial z^2}\,.$$

We are interested in *localized* perturbations that are restricted to narrow radial ranges $\Delta r/r \ll 1$: in the local WKB approximation our attention is confined to the first terms in the expansions of coefficients in Eqs. (92)–(95). As a result, any perturbed function can be represented in the form

$$X_1\left(r', \varphi', t'\right) = \sum_{\boldsymbol{k}} \tilde{X}\left(t'\right) e^{\imath(k_{r'} r' + m\varphi')}\,, \quad k_{r'}^2 \gg k_{\varphi'}^2\,, \tag{96}$$

where $k_{r'}$ and $k_{\varphi'} \equiv m/r'$ are the radial and azimuthal wavenumbers, respectively (cf. Eq. (18)). Substituting Eq. (96) in Eqs. (92)–(95), one obtains the following gasdynamic equations for each harmonic:

$$\frac{\partial}{\partial t'}\left(\frac{\tilde{\sigma}}{\sigma_0}\right) + \imath\left(k_{r'} - t' k_{r'} r' \frac{\mathrm{d}\Omega}{\mathrm{d}r'}\right) \tilde{v}_{r'} + \imath k_{\varphi'} \tilde{v}_{\varphi'} = 0\,, \tag{97}$$

$$\frac{\partial \tilde{v}_{r'}}{\partial t'} - 2\Omega \tilde{v}_{\varphi'} = -\imath\left(k_{r'} - t' m \frac{\mathrm{d}\Omega}{\mathrm{d}r'}\right) \tilde{\aleph}\,, \tag{98}$$

$$\frac{\partial \tilde{v}_{\varphi'}}{\partial t'} + \left(2\Omega + r' \frac{\mathrm{d}\Omega}{\mathrm{d}r'}\right) \tilde{v}_{r'} = -\imath \frac{m}{r'} \tilde{\aleph}\,, \tag{99}$$

$$\left[\left(k_{r'} - t' m \frac{\mathrm{d}\Omega}{\mathrm{d}r'}\right)^2 + \frac{\partial^2}{\partial z^2}\right] \tilde{\aleph}_1 = 4\pi G \tilde{\sigma} \delta(z)\,. \tag{100}$$

As is seen, when

$$\frac{\mathrm{d}\Omega}{\mathrm{d}r'} \frac{\partial}{\partial \varphi'} \neq 0\,, \tag{101}$$

the coefficients in Eqs. (97)–(100) depend explicitly on time t'. Then from the definition of X_1, it follows that this system of equations may have solutions of the form $\exp(-\imath\omega t')$ only in three special cases (Fridman 1989): (1) the system is in rigid-body rotation, $\mathrm{d}\Omega/\mathrm{d}r' = 0$, (2) the perturbations are axially symmetric, $\partial/\partial\varphi' = 0$, and/or (3) the time interval t' which we are considering is much less than $t_{\mathrm{inst}} = \left|\left(k_{r'}/k_{\varphi'}\right)\left(r' \mathrm{d}\Omega/\mathrm{d}r'\right)^{-1}\right|$. Taking into account the fact that in astrophysical objects $r'\,(\mathrm{d}\Omega/\mathrm{d}r') \sim \Omega$, the latter condition may be rewritten in the form

$$t_{\text{inst}} = \left| \frac{k_{r'}}{k_{\varphi'}} \Omega^{-1} \right| \quad \text{and} \quad \left(k_{r'}/k_{\varphi'}\right)^2 \gg 1\,. \tag{102}$$

To emphasize it again, the growth of nonaxisymmetric normal modes $\propto \exp(-\imath\omega t')$ in the nonuniformly rotating medium may take place only on a timescale

$$t' \lesssim t_{\text{inst}}\,.$$

Goldreich & Lynden-Bell (1965b) first obtained that when the condition (101) is satisfied, the system (97)–(100) has the solution

$$\sigma_1 = \sigma_1(r', \varphi') + \left[\text{terms of period } 2\pi \left(\Omega + \frac{r'}{2}\frac{\mathrm{d}\Omega}{\mathrm{d}r'}\right) \text{ in } t'\right]$$
$$+t' \left[\text{terms of period } 2\pi \left(\Omega + \frac{r'}{2}\frac{\mathrm{d}\Omega}{\mathrm{d}r'}\right) \text{ in } t'\right]. \tag{103}$$

Thus there are secular terms—wave shearing terms—in $\partial\sigma_1/\partial t'$ (see also Julian & Toomre 1966). These terms are responsible for the non-normality of nonaxisymmetric ($\partial/\partial\varphi' \neq 0$) modes in a differentially rotating ($\mathrm{d}\Omega/\mathrm{d}r' \neq 0$) medium: nonaxisymmetric modes are sheared by the background shear. Goldreich & Lynden-Bell (1965b) have already explained the physical reasons for this interesting effect. This leads to transient algebraic growths, "shearing gravitational instabilities" (or decays depending on the sign of m) of waves (Goldreich & Lynden-Bell 1965b; Julian & Toomre 1966). The main consequence of this effect is that there is no trivial dispersion relation for nonaxisymmetric waves. More specifically, the dispersion relation of nonaxisymmetric waves involves a derivative in the Fourier space, which makes it difficult to use. Of course, there are ways to use the WKB approximation for such modes, but it is necessary to account for these transient growths (Julian & Toomre 1966; Chagelishvili et al. 2003).

In this work, however, we are interested in moderately tightly-wound, short-wavelength perturbations ($k_{r'}^2/k_{\varphi'}^2 \gg 1$), which develop rapidly on a dynamical timescale $\sim \Omega^{-1}$ (on a time of 2–3 disk rotations, or $\lesssim$ 50 h in Saturn's rings). In other words, we consider the development of violent (with a typical growth rate $\Im\omega \sim \Omega$) spiral perturbations on a time of "instability" t_{inst}, Eq. (102). Lominadze et al. (1988) and Fridman (1989) have defined the time interval between $t' = 0$ and $t' = t_{\text{inst}}$ as a time of instability. Clearly, for such "fierce," rapidly growing perturbations the solutions of gasdynamical equations (87)–(90) can be taken in the form of normal modes: on the time of instability t_{inst} the temporal growth of a perturbation can be approximated by an exponential proportional to $\exp(-\imath\omega t')$. We believe that the considerable growth in the amplitude of the gravity perturbation by the time $t' \sim t_{\text{inst}}$ may act as a source of local turbulence. The Saturnian ring disk accretion process is believed to result from an outward transfer of the angular momentum of accreting material, abetted by turbulent viscosity (Griv & Gedalin 2006; Griv et al. 2008).

References

[1] Alexandrov, A. F., Bogdankevich, L. S., & Rukhadze, A. A. (1984). *Principles of Plasma Electrodynamics*. New York: Springer

[2] Araki, S. (1988). The dynamics of particle disks. II. Effects of spin degrees of freedom. *Icarus* **76**, 182–198

[3] Araki, S., & Tremaine, S. (1986). The dynamics of dense particle disks. *Icarus* **65**, 83–109

[4] Athanassoula, E., & Sellwood, J. A. (1986). Bi-symmetric instabilities of the Kuz'min/Toomre disc. *Mon. Not. R. Astron. Soc.* **221**, 213–232

[5] Bel, N., & Schatzman, E. (1958). On the gravitational instability of a medium in nonuniform rotation. *Rev. Mod. Phys.* **30**, 1015–1016

[6] Bertin, G. (1980). On the density wave theory for normal spiral galaxies. *Phys. Rep.* **61**, 1–69

[7] Bertin, G. (2000). *Dynamics of Galaxies*. Cambridge: Cambridge Univ. Press

[8] Bertin, G., & Casertano, S. (1982). Excitation of warps in galaxies: fluid model of disk–halo interaction. *Astron. Astrophys.* **106**, 274–286

[9] Bertin, G., & Lin, C. C. (1996). *Spiral Structure in Galaxies. A Density Wave Theory.* Cambridge: MIT Press

[10] Bertin, G., Lin, C. C., Lowe, S. A., & Thurstans, R. P. (1989). Modal approach to the morphology of spiral galaxies. II. Dynamical mechanisms. *Astrophys. J.* **338**, 104–120.

[11] Bertin, G., & Mark, J. W.-K. (1978). Density wave theory for spiral galaxies: the regime of finite spiral arm inclination in stellar dynamics. *Astron. Astrophys.* **64**, 389–397

[12] Binney, J. (2001). Secular evolution of the Galactic disk. In: J. G. Funes, & E. M. Corsini (Eds.), *Galaxy Disks and Disk Galaxies* (63–70). San Francisco: ASP

[13] Binney, J., & Tremaine, S. (1987). *Galactic Dynamics*. Princeton: Princeton Univ. Press

[14] Bodenheimer, P. (1974). Calculations of the early evolution of Jupiter. *Icarus* **23**, 319–325

[15] Borderies, N., & Longaretti, P.-Y. (1994). Test particle motion around an oblate planet. *Icarus* **107**, 129–141

[16] Boss, A. (2005). Evolution of the Solar nebula. VII. Formation and survival of protoplanets formed by disk instability. *Astrophys. J.* **629**, 535–548

[17] Boss, A. (2007). Testing disk instability models for giant planet formation. *Astrophys. J.* **661**, L73–L76

[18] Bottema, R. (1993). The stellar kinematics of galactic disks. *Astron. Astrophys.* **275**, 16–36

[19] Bottema, R. (2003). Simulations of normal spiral galaxies. *Mon. Not. R. Astron. Soc.* **344**, 358–384

[20] Brahic, A. (2001). Dynamical evolution of viscous discs. Astrophysical application to the formation of planetary systems and to the confinement of planetary rings and arcs. In: T. Pöschel, & A. Luding (Eds.), *Granular Gases* (281–329). Berlin: Springer

[21] Branover, H., Eidelman, A., Golbraikh, E., & Moiseev, S. (1999). *Turbulence and Structures* (§ 7.2.2). San Diego: Academic Press

[22] Bridges, F. G., Hatzes, A. P., & Lin, D. N. C. (1984). Structure, stability, and evolution of Saturn's rings. *Nature* **309**, 333–335

[23] Brown, R. H., Baines, K. H., Bellucci, G., B. J. Buratti et al. (2006). Observations in the Saturn system during approach and orbital insertion, with Cassini's visual and infrared mapping spectrometer (VIMS). *Astron. Astrophys.* **446**, 707–716

[24] Camichel, H. (1958). Measures photometriques de Saturne et de son anneau. *Ann. d'Astr.* **21**, 231–242

[25] Cassen, P. M., Smith, B. F., Miller, R. H., & Reynolds, R. T. (1981). Numerical experiments on the stability of preplanetary disks. *Icarus* **48**, 377–392

[26] Chagelishvili, G. D., Zahn, J.-P., Tevzadze, A. G., & Lominadze, J. G. (2003). On hydrodynamic shear turbulence in Keplerian disks: via transient growth to bypass transition. *Astron. Astrophys.* **402**, 401–407

[27] Chandrasekhar, S. (1955). The gravitational instability of an infinite homogeneous medium when a coriolis acceleration is acting. *Vistas Astron.* **1**, 344–347

[28] Chiueh, T., & Tseng, Y.-H. (2000). Rotating halos and spirals in low surface brightness galaxies. *Astrophys. J.* **544**, 204–217

[29] Clampin, M., Krist, J. E., Ardila, D. R., D. A. Golimowski et al. (2003). Hubble Space Telescope ACS coronagraphic imaging of the circumstellar disk around HD 141569A. *Astron. J.* **126**, 385–392

[30] Colombo, G., Goldreich, P., & Harris, A. (1976). Spiral structure as an explanation for the asymmetric brightness of Saturn's A ring. *Nature* **264**, 344–347

[31] Colwell, J. E., Esposito, L. W., & Sremčević, M. (2006). Self-gravity wakes in Saturn's A ring measured by stellar occultations from Cassini. *Geophys. Research Lett.* **33**, L07201–07204

[32] Colwell, J. E., Esposito, L. W., Sremčević, M., Stewart, G. R., & McClintock, W. E. (2007). Self-gravity wakes and radial structure of Saturn's B ring. *Icarus* **190**, 127–144

[33] Cook, A. F., & Franklin, F. A. (1964). Rediscussion of Maxwell's Adams Prize Essay on the stability of Saturn's rings. *Astron. J.* **69**, 173–200

[34] Corder, S., Eisner, J., & Sargent, A. (2005). AB Aurigae resolved: evidence for spiral structure. *Astrophys. J.* **622**, L133–L136

[35] Cuzzi, J. N., Colwell, J. E., Esposito, L. W., C. C. Porco et al. (2002). Saturn's rings: pre-Cassini status and mission goals. *Space Sci. Rev.* **118**, 209–251

[36] Daisaka, H., & Ida, S. (1999). Spatial structure and coherent motion in dense planetary rings induced by self-gravitational instability. *Earth Planets Space* **51**, 1195–1213

[37] Daisaka, H., Tanaka, H., & Ida, S. (2001). Viscosity in a dense planetary ring with self-gravitating particles. *Icarus* **154**, 296–312

[38] Davidson, R. C. (1974). *Theory of Nonneutral Plasmas*. Reading: Benjamin

[40] Dunn, D. E., de Pater, I., & Molnar, L. A. (2007). Examining the wake structure in Saturn's rings from microwave observations over varying ring opening angles and wavelengths. *Icarus* **192**, 56–77

[40] Durisen, R. H., Boss, A. P., Mayer, L., Nelson, A. F., Quinn, T., & Rice, W. K. M. (2007). Gravitational instabilities in gaseous protoplanetary disks and implications for giant planet formation. In: B. Reipurth, D. Jewitt, & K. Keil (Eds.), *Protostars and Planets V* (607–622). Tucson: Univ. Arizona Press

[41] Elmegreen, B. G., Leitner, S. N., Elmegreen, D. M., & Cuillandre, J.-C. (2003). A turbulent origin for flocculent spiral structure in galaxies. II. Observations and models of M33. *Astrophys. J.* **590**, 271–283

[42] Elmegreen, B. G., & Scalo, J. (2004). Interstellar turbulence. I. Observations and processes. *Annu. Rev. Astron. Astrophys.* **42**, 211–273

[43] Esposito, L. W. (2002). Planetary rings. *Rep. Prog. Phys.* **65**, 1741–1783

[44] Ferrari, C., Galdemard, P., Lagage, P. O., Pantin, E., & Quoirin, C. (2005). Imaging Saturn's rings with CAMIRAS: thermal inertia of B and C rings. *Astron. Astrophys.* **441**, 379–389

[45] Franklin, F. A., & Colombo, G. (1978). On the azimuthal brightness variations of Saturn's rings. *Icarus* **33**, 279–287

[46] French, R. G., & Nicholson, P. D. (2000). Saturn's rings. II. Particle sizes inferred from stellar occultation data. *Icarus* **145**, 502–523

[47] French, R. G., Salo, H., McGhee, C. A., & Dones, L. (2007). HST observations of azimuthal asymmetry in Saturn's rings. *Icarus* **189**, 493–522

[48] Fridman, A. M. (1989). On the dynamics of a viscous differentially rotating gravitating medium. *Soviet Atron. Lett.* **15**, 487–491

[49] Fridman, A. M., & Gorkavyi, N. N. (1999). *Physics of Planetary Rings*. Berlin: Springer

[50] Fuchs, B. (2001). Density waves in the shearing sheet. III. Disc heating. *Mon. Not. R. Astron. Soc.* **325**, 1637–1642

[51] Fukagawa, M., Hayashi, M., Tamura, M., Y. Itoh et al. (2004). Spiral structure in the circumstellar disk around AB Aurigae. *Astrophys. J.* **605**, L53–L56

[52] Fukagawa, M., Tamura, M., Itoh, Y., T. Kudo et al. (2006). Near-infrared images of protoplanetary disk surrounding HD 142527. *Astrophys. J.* **636**, L153–L156

[53] Galeev, A. A., & Sagdeev, R. Z. (1983). Weak plasma turbulence approximation. In: M. N. Rosenbluth, & R. Z. Sagdeev (Eds.), *Basic Plasma Physics* (vol. 1, 679–722). Amsterdam: North-Holland

[54] Gammie, C. F. (2001). Nonlinear outcome of gravitational instability in cooling, gaseous disks. *Astrophys. J.* **553**, 174–183

[55] Ginzburg, I. F., Polyachenko, V. L., & Fridman, A. M. (1972). Ring stability of Saturn. *Soviet Astron.* **15**, 643–645

[56] Goldreich, P., & Lynden-Bell, D. (1965a). I. Gravitational stability of uniformly rotating disks. *Monthly Not. R. Astron. Soc.* **130**, 97–124

[57] Goldreich, P., & Lynden-Bell, D. (1965b). II. Spiral arms as sheared gravitational instabilities. *Monthly Not. R. Astron. Soc.* **130**, 125–158

[58] Goldreich, P., Rappaport, N., & Sicardy, B. (1995). Single-sided shepherding. *Icarus* **118**, 414–417

[59] Goldreich, P., & Tremaine, S. (1978a) The formation of the Cassini Division in Saturn's rings. *Icarus* **34**, 240–253

[60] Goldreich, P., & Tremaine, S. (1978b). The velocity dispersion in Saturn's rings. *Icarus* **34**, 227–239

[61] Goldreich, P., & Tremaine, S. (1978c). The excitation and evolution of density waves. *Astrophys. J.* **222**, 850–858

[62] Goldreich, P., & Tremaine, S. (1979). The excitation of density waves at the Lindblad and corotation resonances by an external potential. *Astrophys. J.* **233**, 857–871.

[63] Goldreich, P., & Tremaine, S. (1980). Disk–satellite interactions. *Astrophys. J.* **241**, 435–441

[64] Goldreich, P., & Tremaine, S. (1982). The dynamics of planetary rings. *Annu. Rev. Astron. Astrophys.* **20**, 249–283

[77] Grady, C. A., Polomski, E. F., Henning, Th., B. Stecklum et al. (2001). The disk and environment of the Herbig Be star HD 100546. *Astron. J.* **122**, 3396–3406

[66] Griv, E. (1998). Local stability criterion for the Saturnian ring system. *Planet. Space Sci.* **46**, 615–628

[67] Griv, E. (2005a). Fine-scale density wave structure in Saturn's A, B, and C rings. In: Z. Knežević, & A. Milani (Eds.), *Dynamics of Populations of Planetary Systems* (459–466). Cambridge: Cambridge Univ. Press

[68] Griv, E. (2005b). On the origin of the Cartwheel galaxy: gravitational instability? *Astrophys. Space Sci.* **299**, 371–385

[69] Griv, E. (2006). Gravitationally unstable gaseous disks of flat galaxies. *Astron. Astrophys.* **449**, 573–581

[70] Griv, E. (2007a). The torque at a Lindblad vertical resonance. *Astrophys. J.* **665**, 866–873

[71] Griv, E. (2007b). Formation of a star and planets around it through a gravitational instability in a disk of gas and dust. *Planet. Space Sci.* **55**, 547–568

[72] Griv, E., & Chiueh, T. (1998). Central NGC 2146: a firehose-type bending instability in the disk of newly formed stars? *Astrophys. J.* **503**, 186–211

[73] Griv, E., & Gedalin, M. (2003). The fine-scale spiral structure of low and moderately high optical depth regions of Saturn's main rings: a review. *Planet. Space Sci.* **51**, 899–927

[74] Griv, E., & Gedalin, M. (2004). Changes of angular momentum and entropy induced by Jeans-unstable density waves in stellar disks of flat galaxies. *Astron. J.* **128**, 1965–1973

[75] Griv, E., & Gedalin, M. (2005). Exploring local N-body simulations of Saturn's rings. *Planet. Space Sci.* **53**, 461–472

[76] Griv, E., & Gedalin, M. (2006). Turbulent viscosity and lifetime of Saturn's rings. *Planet. Space Sci.* **54**, 794–807

[77] Griv, E., Gedalin, M., & Eichler, D. (2001). On the Schwarzschild velocity distribution of the local stellar disk. *Astrophys. J.* **555**, L29–L32

[78] Griv, E., Gedalin, M., Eichler, D., & Yuan, C. (2000a). A gas-kinetic stability analysis of self-gravitating and collisional particulate disks with application to Saturn's rings. *Planet. Space Sci.* **48**, 679–698

[79] Griv, E., Gedalin, M., Eichler, D., & Yuan, C. (2000b). Landau excitation of spiral density waves in an inhomogeneous disk of stars. *Phys. Rev. Lett.* **84**, 4280–4283

[80] Griv, E., Gedalin, M., Livertz, E., & Yuan, C. (2004). Fine-scale irregular structure in Saturn's rings. In: G. G. Byrd, K. V. Kholshevnikov, A. A. Mylläri, I. I. Nikiforov, & V. V. Orlov (Eds.), *Order and Chaos in Stellar and Planetary Systems* (129–132). San Francisco: ASP

[81] Griv, E., Gedalin, M., & Lyubarsky, Yu. (2006a). Prediction of the fine-scale irregular structure in Saturn's A, B, and C rings. *Adv. Space Res.* **38**, 770–776

[82] Griv, E., Gedalin, M., & Yuan, C. (2006b). Spiral galaxies as gravitational plasmas. *Adv. Space Res.* **38**, 47–56

[83] Griv, E., Gedalin, M., & Yuan, C. (2002). Quasi-linear theory of the Jeans instability in disk-shaped galaxies. *Astron. Astrophys.* **383**, 338–351

[84] Griv, E., Gedalin, M., & Yuan, C. (2003a). On the stability of Saturn's rings to gravity disturbances. *Astron. Astrophys.* **400**, 375–383

[85] Griv, E., Gedalin, M., & Yuan, C. (2003b). On the stability of Saturn's rings: a quasi-linear kinetic theory. *Monthly Not. R. Astron. Soc.* **342**, 1102–1116

[86] Griv, E., Liverts, E., & Mond, M. (2008). Angular momentum transport in astrophysical disks. *Astrophys. J.* **672**, L127–L130

[87] Griv, E., Rosenstein, B., Gedalin, M., & Eichler, D. (1999). Local stability criterion for a gravitating disk of stars. *Astron. Astrophys.* **347**, 821–840

[88] Hanninen, J., & Salo, H. (1995). Formation of isolated narrow ringlets by a single satellite. *Icarus* **117**, 435–438

[89] Hedman, M. M., Nicholson, P. D., Salo, H., Wallis, B. D., Buratti, B. J., Baines, K. H., Brown, R. H., & Clark, R. N. (2007). Self-gravity wake structures in Saturn's A ring revealed by Cassini VIMS. *Astron. J.* **133**, 2624–2629

[90] Higdon, J. L. (1996). Wheels of fire. II. Neutral hydrogen in the Cartwheel ring galaxy. *Astrophys. J.* **467**, 241–260

[91] Hohl, F. (1972). Evolution of a stationary disk of stars. *J. Comput. Phys.* **9**, 10–25

[92] Hohl, F. (1978). Three-dimensional galaxy simulations. *Astron. J.* **83**, 768–778

[93] Holberg, J. B., Forrester, W. T., & Lissauer, J. J. (1982). Identification of resonance features within the rings of Saturn. *Nature* **297**, 115–120

[94] Horellou, C., Charmandaris, V., Combes, F., Appleton, P. N., Casoli, F., & Mirabel, I. F. (1998). Molecular gas in the Cartwheel galaxy. *Astron. Astrophys.* **340**, L51–L54.

[95] Horne, L. J., & Cuzzi, J. N. (1996). Characteristic wavelengths of irregular structure in Saturn's B ring. *Icarus* **119**, 285–310

[96] Horton, W. (1984). Drift wave turbulence and anomalous transport. In: M. N. Rosenbluth, & R. Z. Sagdeev (Eds.), *Handbook of Plasma Physics* (vol. 2, 383–448). Amsterdam: North-Holland

[97] Huber, D., & Pfenniger, D. (2001). Lumpy structures in self-gravitating disks. *Astron. Astrophys.* **374**, 465–493

[98] Hunter, C. (1973). Patterns of waves in galactic disks. *Astrophys. J.* **181**, 685–706

[99] Ichimaru, S. (1973). *Basic Principles of Plasma Physics.* Reading: Benjamin

[100] Jeans, J. H. (1929). *Astronomy and Cosmogony* (2nd Ed.). Cambridge

[101] Jeffreys, H. (1947). The effects of collisions on Saturn's rings. *Mon. Not. R. Astron. Soc.* **107**, 263–267

[102] Ji, H., Burin, M., Schartman, E., & Goodman, J. (2008) Hydrodynamic turbulence cannot transport angular momentum effectively in astrophysical disks. *Nature* **444**, 343–346

[103] Julian, W. H., & Toomre, A. (1966). Non-axisymmetric response of differentially rotating disks of stars. *Astrophys. J.* **146**, 810–830

[104] Kadomtsev, B. B. (1965). *Plasma Turbulence*. New York: Academic Press

[105] Kerr, R. A. (1985). Making better planetary rings. *Science* **220**, 1376–1377

[106] Khoperskov, A. V., Zasov, A. V., & Tyurina, N. V. (2003). Minimum velocity dispersion in stable stellar disks. Numerical simulations. *Astron. Rep.* **47**, 357–376

[107] Krall, N. A., & Trivelpiece, A. W. (1986). *Principles of Plasma Physics*. San Francisco: San Francisco Press

[108] Kulsrud, R. M. (1972). Enhancement of relaxation processes by collective effects. In: M. Lecar (Ed.), *Gravitational N-body Problem* (337–346). Dordrecht: Reidel

[109] Kulsrud, R. M., Mark, J. W.-K., & Caruso, A. (1971). The hose-pipe instability in stellar systems. *Astrophys. Space Sci.* **14**, 52–55

[110] Landau, L. D., & Lifshitz, E. M. (1987). *Fluid Mechanics*. Oxford: Pergamon

[111] Lane, A. L., Hord, C. W., West, R. A., L. W. Esposito et al. (1982). Photopolarimetry from Voyager 2: preliminary results on Saturn, Titan, and the rings. *Science* **215**, 537–543

[112] Larson, R. B. (1984). Gravitational torques and star formation. *Mon. Not. R. Astron. Soc.* **206**, 197–207

[113] Larson, R. B. (1989). The evolution of protostellar disks. In: H. A. Weaver, & L. Danly (Eds.), *The Formation and Evolution of Planetary Systems* (31–48). Cambridge: Cambridge Univ. Press

[114] Lau, Y. Y., & Bertin, G. (1978). Discrete spiral modes, spiral waves, and the local dispersion relationship. *Astrophys. J.* **226**, 508–520

[115] Laughlin, G., & Bodenheimer, P. (1994). Nonaxisymmetric evolution in protostellar disks. *Astrophys. J.* **436**, 335–354

[116] Laughlin, G., & Rózyczka, M. (1996). The effect of gravitational instabilities on protostellar disks. *Astrophys. J.* **456**, 279–291

[117] Layrat, C., Spilker, L. J., Altobelli, N., Pilorz, S., & Ferrari, C. (2008). Infrared observations of Saturn's rings by Cassini CIRS: phase angle and local time dependence. *Icarus* **56**, 117–133

[118] Lees, A. W., & Edwards, S. F. (1972). The computer study of transport processes under extreme conditions. *J. Phys. C.* **5**, 1921–1929

[119] Lin, C. C., & Bertin, G. (1984). Galactic dynamics and gravitational plasmas. *Adv. Appl. Mech.* **24**, 155–187

[120] Lin, C. C., & Lau, Y. Y. (1979). Density wave theory of spiral structure of galaxies. *SIAM Stud. Appl. Math.* **60**, 97–163

[121] Lin, C. C., & Shu, F. H. (1966). On the spiral structure of disk galaxies. II. Outline of a theory of density waves. *Proc. Natl. Acad. Sci.* **55**, 229–234

[122] Lin, C. C., & Shu, F. H. (1968). On the effects of finite disk thickness and gas content on spiral structure. In: M. Chretian, S. Deser, & J. Goldstein (Eds.), *Astrophysics and General Relativity* (vol. 2, 236–260). New York: Gordon and Breach

[123] Lin, C. C., Yuan, C., & Shu, F. H. (1969). On the spiral structure of disk galaxies. III. Comparison with observations. *Astrophys. J.* **155**, 721–746 (erratum, *156*, 797)

[124] Lin, D. N. C., & Papaloizou, J. C. B. (1996). Theory of accretion disks. II. Application to observed systems. *Annu. Rev. Astron. Astrophys.* **34**, 703–748

[125] Lin, D. N. C., & Pringle, J. E. (1987). A viscosity prescription for a self-gravitating accretion disc. *Mon. Not. R. Astron. Soc.* **225**, 607–613

[126] Lissauer, J. J., & Cuzzi, J. N. (1982). Resonances in Saturn's rings. *Icarus* **87**, 1051–1058

[127] Liverts, E., Griv, E., Gedalin, M., & Eichler, D. (2003). Dynamical evolution of galaxies: supercomputer N-body simulations. In: G. Contopoulos, & N. Voglis (Eds.), *Galaxies and Chaos* (340–347). Berlin: Springer

[128] Lominadze, D. G., Chagelishvili, G. D., & Chanishvili, R. G. (1988). The evolution of nonaxisymmetric shear perturbations in accretion disks. *Soviet Astron. Lett.* **14**, 364–367

[129] Lovelace, R. V. E., & Hohlfeld, R. G. (1978). Negative mass instability of flat galaxies. *Astrophys. J.* **221**, 51–61

[130] Lovelace, R. V. E., Jore, K. P., & Haynes, M. P. (1997). Two-stream instability of counterrotating galaxies. *Astrophys. J.* **475**, 83–96

[131] Lumme, K., & Irvine, W. M. (1979). A model for the azimuthal brightness variations in Saturn's rings. *Nature* **282**, 695-696

[132] Lynden-Bell, D., & Kalnajs, A. J. (1972). On the generating mechanism of spiral structure. *Mon. Not. R. Astron. Soc.* **157**, 1–30

[133] Lynden-Bell, D., & Pringle, J. E. (1974). The evolution of viscous discs and the origin of the nebular variables. *Mon. Not. R. Astron. Soc.* **168**, 603–637

[134] Mark, J. W.-K. (1971). Collective instabilities and waves for inhomogeneous stellar systems. II. The normal-modes problem of the self-consistent plane-parallel slab. *Astrophys. J.* **169**, 455–476

[135] Marochnik, L. S. (1966). A class of quasi-integrals in stellar dynamics. *Soviet Astron.* **10**, 442–447

[136] Maxwell, J. C. (1859). On the stability of the motion of Saturn's rings. In: W. D. Niven (Ed.), *Science Papers* (1960, vol. 1, 288–378). New York: Dover

[137] Mayer, L., Lufkin, G., Quinn, T., & Wadsley, J. (2007). Fragmentation of gravitationally unstable gaseous protoplanetary disks with radiative transfer. *Astrophys. J.* **661**, L77–L80

[138] Mayer, L., Quinn, T., Wadsley, J., & Stadel, J. (2002). Formation of giant planets by fragmentation of protoplanetary disks. *Science* **298**, 1756–1759

[139] Melita, M. D., & Papaloizou, J. C. B. (2005). Resonantly forced eccentric ringlets: relationships between surface density, resonance location, eccentricity and eccentricity-gradient. *Cel. Mech. Dyn. Astr.* **91**, 151–171

[140] Meyer-Vernet, N., & Sicardy, B. (1987). On the physics of resonant disk–satellite interaction. *Icarus* **69**, 157–175

[141] Mikhailovskii, A. B. (1974). *Theory of Plasma Instabilities* (vols. 1 and 2). New York: Consultants Bureau

[142] Mikhailovskii, A. B., & Fridman, A. M. (1973). "Fast" and "slow" density waves in spiral galaxies. *Soviet Astron.* **17**, 57–61

[143] Montenegro, L. E., Yuan, C., & Elmegreen, B. G. (1999). Curvature and acoustic instabilities in rotating fluid disks. *Astrophys. J.* **520**, 592–606

[144] Morozov, A. G. (1978). Self-suppression of Jeans instability in a rotating gravitating disk. *Soviet Astron. Lett.* **4**, 115–116

[145] Morozov, A. G. (1980). On the stability of an inhomogeneous disk of stars. *Soviet Astron.* **24**, 391–397

[146] Morozov, A. G. (1981). Constraints on the radial velocity dispersion of stars in the disk of a flat galaxy. *Soviet Astron. Lett.* **7**, 109–111

[147] Morozov, A. G. (1985). A local stability criterion for the gaseous subsystem of a flat galaxy. *Soviet Astron.* **29**, 120–124

[148] Morozov, A. G., & Khoperskov, A. V. (1990). The nature of turbulent viscosity in accretion disks. *Soviet Astron. Lett.* **16**, 244–246

[149] Morozov, A. G., Torgashin, Yu. M., & Fridman, A. M. (1985). Turbulent viscosity in a gravitating gaseous disk. *Soviet Astron. Lett.* **11**, 94–97

[150] Mosqueira, I. (1996). Local simulations of perturbed dense planetary rings. *Icarus* **122**, 128–152

[151] Nakamura, T., Takahara, F., & Ikeuchi, S. (1975). Collective instabilities of self-gravitating systems. II. Instabilities due to temperature anisotropy. *Prog. Theor. Phys.* **53**, 1348–1359

[152] Nicholson, P. D., French, R. G., Campbell, D. B., Margot, J., Nolan, M. C., Black, G. J., & Salo, H. J. (2005). Radar imaging of Saturn's rings. *Icarus* **177**, 32–62

[153] Nordström, B., Mayor, M., Andersen, J., & Holmberg, J. (2004). The Geneva–Copenhagen survey of the Solar neighbourhood. Ages, metallicities, and kinematic properties of $\sim$ 14 000 F and G dwarfs. *Astron. Astrophys.* **418**, 989–1019

[154] Ohtsuki, K., & Emori, H. (2000). Local N-body simulations for the distribution and evolution of particle velocities in planetary rings. *Astron. J.* **119**, 403–416

[155] Osterbart, R., & Willerding, E. (1995). Collective processes in planetary rings. *Planet. Space Sci.* **43**, 289–298

[156] Paczynski, B. (1978). A model of selfgravitating accretion disk. *Acta Astr.* **28**, 91–109

[157] Papaloizou, J. C. B., & Lin, D. N. C. (1995). 0n the dynamics of warped accretion disks. *Astrophys. J.* **438**, 841–851

[158] Pasha, I. I., & Smirnov, M. A. (1982). On the direction of rotation of the spirals in galaxies. *Astrophys. Space Sci.* **86**, 215–224

[159] Pickett, B. K., Mejía, A. C., Durisen, R. H., Cassen, P. M., Berry, D. K., & Link, R. P. (2003). The thermal regulation of gravitational instabilities in protoplanetary disks. *Astrophys. J.* **590**, 1060–1080

[160] Porco, C. C. (1990). Narrow rings: observations and theory. *Adv. Space Res.* **10**, 221–229

[161] Porco, C. C., Baker, E., Barbara, J., K. Beurle et al. (2005). Cassini imaging science: initial results on Saturn's rings and small satellites. *Science* **307**, 1226–1236

[162] Porco, C. C., & Nicholson, P. D. (1987). Eccentric features in Saturn's outer C ring. *Icarus* **72**, 437–467

[163] Postnikov, E. B., & Loskutov, A. Yu. (2007). Wavelet analysis of fine-scale structures in the Saturnian B and C rings using data from the Cassini spacecraft. *J. Exper. Theor. Phys.* **104**, 417–422

[164] Pringle, J. E. (1981). Accretion discs in astrophysics. *Annu. Rev. Astron. Astrophys.* **19**, 137–162

[165] Raha, N., Sellwood, J. A., James, R. A., & Kahn, F. D. (1991). A dynamical instability of bars in disk galaxies. *Nature* **352**, 411–412

[166] Richardson, D. C. (1994). Tree code simulations of planetary rings. *Monthly Not. R. Astron. Soc.* **269**, 493–511

[167] Rohlfs, K. (1977). *Lectures on density wave theory.* Berlin: Springer

[168] Romeo, A. B. (1992). Stability of thick two-component galactic discs. *Monthly Not. R. Astron. Soc.* **256**, 307–320

[169] Rosen, P. A., Tyler, G. L., Marouf, E. A., & Lissauer, J. J. (1991). Resonance structures in Saturn's rings probed by radio occultation. II. Results and interpretation. *Icarus* **93**, 25–44

[170] Rüdiger, G., & Kitchatinov, L. L. (2000). Nonlocal density wave theory for gravitational instability of protoplanetary disks without sharp boundaries. *Astron. Nachr.* **21**, 181–192

[171] Safronov, V. S. (1960). On the gravitational instability in flattened systems with axial symmetry and non-uniform rotation. *Ann. d'Astrophys.* **23**, 979–986

[172] Safronov, V. S. (1980). Some problems of evolution of the solar nebula and of the protoplanetary cloud. In: D. Lal (Ed.), *Early Solar System Processes* (73–81). Amsterdam: North-Holland

[173] Sagdeev, R. Z., & Galeev, A. A. (1969). In: T. M. O'Neil, & D. L. Book (Eds.), *Nonlinear Plasma Theory*. New York: Benjamin

[174] Salo, H. (1992). Gravitational wakes in Saturn's rings. *Nature* **359**, 619–621

[175] Salo, H. (1995). Simulations of dense planetary rings. III. Self-gravitating identical particles. *Icarus* **117**, 287–312

[176] Salo, H. (2001). Numerical simulations of the collisional dynamics of planetary rings. In: Pöschel, & A. Luding (Eds.), *Granular Gases* (330–349). Berlin: Springer

[177] Sellwood, J. A., & Carlberg, R. G. (1984). Spiral instabilities provoked by accretion and star formation. *Astrophys. J.* **282**, 61–74

[178] Sellwood, J. A., & Athanassoula, E. (1986). Unstable modes from galaxy simulations. *Monthly Not. R. Astron. Soc.* **221**, 195–212

[179] Shakura, N. I., & Sunyaev, R. A. (1973). Black holes in binary systems. Observational appearance. *Astron. Astrophys.* **24**, 337–355

[180] Showalter, M. R., & Nicholson, P. D. (1990). Saturn's rings through a microscope—particle size constraints from the Voyager PPS scan. *Icarus* **87**, 285–306

[181] Shu, F. H. (1970). On the density wave theory of galactic spirals. II. The propagation of the density of wave action. *Astrophys. J.* **160**, 99–112

[182] Shu, F. H. (1984). Waves in planetary rings. In: R. Greenberg, & A. Brahic (Eds.), *Planetary Rings* (513–561). Tucson: Univ. Arizona Press

[183] Shu, F. H., Cuzzi, J. N., & Lissauer, J. J. (1983). Bending waves in Saturn's rings. *Icarus* **53**, 185–206

[184] Shu, F. H., Dones, L., Lissauer, J. J., Yuan, C., & Cuzzi, J. N. (1985). Nonlinear spiral density waves—viscous damping. *Astrophys. J.* **299**, 542–573

[185] Shu, F. H., & Stewart, G. R. (1985). The collisional dynamics of particulate disks. *Icarus* **62**, 360–383

[186] Shukhman, I. G. (1984). Collisional dynamics of particles in Saturn's rings. *Soviet Astron.* **28**, 574–585

[187] Sicardy, B. (2005). Dynamics and composition of rings. *Space Sci. Rev.* **116**, 457–470

[188] Smith, B. A., Soderblom, L., Beebe, R. F., J. M. Boyce et al. (1981). Encounter with Saturn: Voyager 1 imaging science results. *Science* **212**, 163–191

[189] Smith, B. A., Soderblom, L., Batson, R., P. Bridges et al. (1982). A new look at the Saturn system: the Voyager 2 images. *Science* **215**, 504–537

[190] Snytnikov, V. N., Vshivkov, V. A., Kuksheva, E. A., Neupokoev, E. V., Nikitin, S. A., & Snytnikov, A. V. (2004). Three-dimensional numerical simulation of a nonstationary gravitating N-body system with gas. *Astron. Lett.* **30**, 124–137

[191] Sotnikova, N. Ya., & Rodionov, S. A. (2005). Bending instability of stellar disks: the stabilizing effect of a compact bulge. *Astron. Lett.* **31**, 15–29

[192] Stewart, G., Lin, D. N. C., & Bodenheimer, P. (1984). Collision-induced transport processes in planetary rings. In: R. Greenberg, & A. Brahic (Eds.), *Planetary Rings* (447–512). Tucson: Univ. Arizona Press

[193] Stewart, G., Robbins, S. J., & Colwell, J. E. (2008). Evidence for a primordial origin of Saturn's rings. In: American Astronomical Society, DPS meeting, #39, #7.06

[194] Stone, E. C., & Miner, E. D. (1982). Voyager 2 encounter with the Saturnian system. *Science* **215**, 499–504

[195] Swanson, D. G. (1989). *Plasma Waves*. Boston: Academic Press

[196] Takeda, T., & Ida, S. (2001). Angular momentum transfer in a protolunar disk. *Astrophys. J.* **560**, 514–533

[197] Thomson, F. S., Marouf, E. A., Tyler, L., French, R. G., & Rappaport, N. J. (2007). Periodic microstructure in Saturn's rings A and B. *Geophys. Res. Lett.* **34**, L24203–24206

[198] Tiscareno, M. S., Burns, J. A., Nicholson, P. D., Hedman, M. M., & Porco, C. C. (2007). Cassini imaging of Saturn's rings. II. A wavelet technique for analysis of density waves and other radial structure in the rings. *Icarus* **189**, 14–34

[199] Tiscareno, M. S., Nicholson, P. D., Burns, J. A., Hedman, M. M., & Porco, C. C. (2006). Unravelling temporal variability in Saturn's spiral density waves: results and predictions. *Astrophys. J.* **651**, L65–L68

[200] Tomley, L., Cassen, P., & Steiman-Cameron, T. (1991). On the evolution of gravitationally unstable protostellar disks. *Astrophys. J.* **382**, 530–543

[201] Tomley, L., Steiman-Cameron, T., & Cassen, P. (1994). Further studies of gravitationally unstable protostellar disks. *Astrophys. J.* **422**, 850–861

[202] Toomre, A. (1964). On the gravitational stability of a disk of stars. *Astrophys. J.* **139**, 1217–1238

[203] Toomre, A. (1966). A Kelvin–Helmholtz instability. In: *Geophysical Fluid Dynamics*, Notes on the Summer Study Programm in Geophysical Fluid Dynamics at the Woods Hole Oceanographic Institution, ref. no. 66–46, 111

[204] Toomre, A. (1981). What amplifies the spirals? In: S. F. Fall, & D. Lynden-Bell (Eds.), *The structure and evolution of normal galaxies* (111–136). Cambridge: Cambridge Univ. Press

[205] Toomre, A. (1983). Theories of warps. In: E. Athanassoula (Ed.), *Internal Kinetics and Dynamics of Galaxies* (177–186). Dordrecht: Reidel

[206] Toomre, A. (1990). Gas-hungry Sc spirals. In: R. Wielen (Ed.), *Dynamics and Interactions of Galaxies* (292–303). Berlin: Springer

[207] Toomre, A., & Kalnajs, A. J. (1991). Spiral chaos in an orbiting patch. In: B. Sundelius (Ed.), *Dynamics of Disc Galaxies* (341–358). Göteborg: Göteborg Univ. Press

[208] Tremaine, S. (1989). Common processes and problems in disc dynamics. In: J. A. Sellwood (Ed.), *Dynamics of Astrophysical Discs* (231–238). Cambridge: Cambridge Univ. Press

[209] Tremaine, S. (2003). On the origin of irregular structure in Saturn's rings. *Astron. J.* **125**, 894–901

[210] Vandervoort, P. O. (1970). Density waves in a highly flattened, rapidly rotating galaxy. *Astrophys. J.* **161**, 87–102

[211] Wada, K., Meurer, G., & Norman, C. A. (2002). Gravity-driven turbulence in galactic disks. *Astrophys. J.* **577**, 197–205

[212] Wielen, R. (1977). The diffusion of stellar orbits derived from the observed age–dependence of the velocity dispersion. *Astron. Astrophys.* **60**, 263–275

[213] Willerding, E. (1986). Theory of density waves in narrow planetary rings. *Astron. Astrophys.* **161**, 403–407

[214] Wisdom, J., & Tremaine, S. (1988). Local simulations of planetary rings. *Astron. J.* **95**, 925–940

[215] Yuan, C. (1969). Application of the density-wave theory to the spiral structure of the Milky Way system. I. Systematic motion of neutral hydrogen. *Astrophys. J.* **158,** 871–888

[216] Zasov, A. V., Khoperskov, A. V., & Tyurina, N. V. (2004). Stellar velocity dispersion and mass estimation for galactic disks. *Astron. Lett.* **30**, 593–602

[217] Zasov, A. V., & Smirnova, A. A. (2005). The gas content in galactic disks: correlation with kinematics. *Astron. Lett.* **31**, 160–170

[218] Zebker, H. A., Marouf, E. A., & Tyler, G. L. (1985). Saturn's rings: particle size distributions for thin layer models. *Icarus* **84**, 531–538

[219] Zebker, H. A., & Tyler, G. L. (1984). Thickness of Saturn's rings inferred from Voyager 1 observations of microwave scatter. *Science* **223**, 336–338

In: Space Exploration Research
Editors: J.H. Denis and P.D. Aldridge

ISBN: 978-1-60692-264-4

Chapter 13

CONTINUOUS WAVELET TRANSFORM AS AN EFFECTIVE TOOL FOR THE DETECTING OF SATURN RINGS' STRUCTURE

***Eugene B. Postnikov*[1] *and Alexander Loskutov*[2]**
[1]Humboldt Univ., Berlin and Kursk Univ., Russia
[2]Moscow State Univ., Russia

Abstract

The presented review is dedicated to the consideration of the Continuous Wavelet transform from the diffusion signal and image processing point of view. Such an approach is based on the consideration of diffusion smoothing via the solution of proper partial diferential equations.

Within this group of methods the real and complex wavelet transform with the wavelets of Gauss and Morlet families are considered. Especial attention is concentrated on the variety of numerical examples considering the processing of regular and irregular (random samples, chaotic ODE solutions etc.) signals. All of them are graphically illustrated.

1. Introduction: Various Structures in Saturn Rings

One of the modern problems of astrophysical data processing is development of the effective tools for the analysis of non-stationary data sequences. As an important example of such data processing one can mention the analysis of optical density of Saturn rings. These possess a rich variety of patterns, waves, wakes etc. discovered by the spacecrafts "Voyager–1" and "Voyager–2", which nature is not completely understood up to now (see e.g. the review [1]). Recently, new data become available due to the "Cassini" mission, see the primary report by the Cassini Imaging Team (Porco et al., [2]). In order to use these data and to verify various models of pattern formation, one needs a tool to resolve local spectra.

Thus, the local spectral properties could play a crucial role in waves classification and in analysis of the physical properties of the rings. The continuous complex wavelet–transform is the most powerful tool to achieve this goal. Its principal advantage with respect to other methods (see, e.g. [3, 4]) is a high localization of the basis functions in both spatial and frequency domains. Also, the size of a window and an instant period are correlated in this method: for high–frequency signals the window shrunks while for the low-frequency signals it is dilated. This keeps the effective number of oscillations in the window constant.

The complex wavelet analysis has been successfully applied to study the orbital period variations of asteroids in nearly–resonance zones [5] and for processing of the solutions obtained for Hamiltonian systems, in particular for the three–body problem (see [6]).

An application of the wavelet transform to analize the Saturn rings has been proposed by Bendjoya at al. [7], who studied the "Voyager–2"data for the Encke gap. The authors however used only real wavelets since their primary goal was to extract different–scale patterns from noisy images. At the same time, the problem of the local periodicity in the Saturn's rings requires applications of the complex wavelet transform. The relevance of such approach has been firstly shown by Postnikov and Loskutov, [8] and Porco et al. [2] and more detailed considered by these groups in the following papers [9–11].

1.1. The Structure of Saturn Rings

The fact that Saturn has a wide plane has been firstly detected in 1655 by Christian Hyigens. The further earth-based observations revealed his complicated structure in the form of concentric rings divided by empty gaps. New results concerning the fine ring structure have been obtained via developing of the telescopic technics up to the third quarter of the XX century.

The photographic observation of the last decades obtained by the spacecraft "Voyager" and recent researches by means of the "Cassini" mission lead to more precision understanding of the wide rings and ringlets structures. Also, it has been fount that in gaps of Saturn rings there exists a redial-nonuniform substance distribution.

The contemporary classification of main rings' structure following Planetary Data System [12] is shown in Table 1.

1.2. Physical Mechanisms of the Emerging of Wave Structures in the Rings

There exists two main approaches which, can presumably, describe the formation of such structures. The first one considers the gravitational interaction of the ring particle with Saturn's satellites (the basis of this approach is established by Shu et [13–15]. Within this approach, the appearance of a set of wave trains has been explained. Later, Spilker et al. [16] detected 40 resonance patterns in the ring A by applying the windowed Fourier transform. The search for the effects of high resonances in the ring B existing on a background of stochastic perturbations, is proposed in [17].

The second approach is based on the hydrodynamic description o the rings (see, e.g., [18, 19] and references therein). The corresponding theoretical analysis (see [18]) with the use of the Navier–Stokes equations for the viscous self-gravitating fluid reveals the radial structures as narrow peaks divided by interstice of about 80–100 meters. Horn and Cuzzi [20] described such a type of stable periodicity on the basis of the "Voyager" data.

In the paper [19], it has been shown that Jeans type instability in a self–gravitating continuum may be described by analogy with the plasma instability, where Coulomb forces are replaced by the unscreened gravitational interactions. It is found that characteristic wavelengths are about 30–200 m in the A ring, 7–30 m in the B ring and less than 7 m in the C ring.

Table 1. Classification of main rings' structure (from Planetary Data System [12]).

Feature	Inner boudary (km)	Outer boundary (km)	Description of structures
D Ring	66,900	74,658	Contains narrow ringlets at 67,580 and 71,710 km.
C Ring	74,658	91,975	Isolated "plateaus" among a surrounding, fainter ring.
Titan Ringlet	77,871	77,896	A narrow, eccentric ringlet inside a gap in the C Ring.
Maxwell Ringlet	87,491	87,555	A narrow, eccentric ringlet inside a gap in the C Ring.
1.470 Rs Ringlet	88,716	88,732	A narrow, eccentric ringlet inside a gap in the C Ring.
1.495 Rs Ringlet	90,171	90,232	A narrow, eccentric ringlet inside a gap in the C Ring.
B Ring	91,975	117,507	Contains fine structure on all scales. The most opaque of Saturn's rings.
Cassini Division	117,507	122,340	The prominent gap between the A and B Rings. It contains several features of low optical depth.
Huygens Ringlet	117,825	118,185	A narrow, eccentric ringlet near the inner edge of the Cassini Division.
A Ring	122,340	136,780	A fairly uniform ring with many density and bending waves near its outer edge.
Encke Gap	133,410	133,740	A gap in the A Ring "shepherded" open by the embedded moon Pan. One or more faint ringlets are also present.
Keeler Gap	136,510	136,550	An empty gap near the outer edge of the A Ring.
F Ring	140,194	140,244	An eccentric ringlet containing clumps and kinks.
G Ring	166,000	173,200	A very faint, isolated dust ring.
E Ring	180,000	480,000	A broad, faint dust ring encompassing the orbits of Mimas through Dione.

1.3. Recently Available Experimental Data

At the present, a large amount of raw and/or processed data concerning the Solar system's bodies was collected and published in Internet by NASA and Jet Propulsion Laboratory of California Institute of Technology (NASA/JPL-Caltech). Information is free for download via the project "Planetary Data System" (see Internet [21]).

Subsection of the database which accumulate information about rings of big planets in the Solar system is called "The Planetary Rings Node" [12]. Here, in particular, a total collection of images obtained by the missions "Voyager" in 80th. This collection includes both raw as well as calibrated and geometrically corrected versions recommended for the scientific purposes. The database obtained from "Cassini" contains images transmitted from this spacecraft beginning with the moment of arrival in the Saturn system up to the present. Now this database includes only raw images and and links to the software applicable to calibration.

2. Wavelet Methods for Extraction of Structures

This section provides a short review of the continuous wavelet transform (with the both, both real and complex wavelets) from the point its applicability to rings image data processing.

2.1. Theoretical Background: Real Wavelets

One of the image processing tasks is to reveal wave pattern from an initial data. Often, such a procedure may essentially simplify the interpretation of a general problem of the wave genesis due to the following reasons:

- The brightness change corresponding to density or bending waves of low intensity can be insufficiently contrasty in comparison with background of ring's matter. This task is closely related to the problem of the texture recognition from noisy images because the matter in highly inhomogeneous.

- The average brightness has a sharp gradient near the gap boundaries. This produces artifacts during the processing (for example, the Gibbs effect at the Fourier analysis generates extraneous lines; the intensive cone of influence at the wavelet transform suppresses the important patters; etc.).

- For some problem where we should find only the space dependence of the instantaneous period, it is possible to consider just the set of lines corresponding the density maxima.

- In case of narrow ringlets formed by satellites-shepherds it is necessary to use multiresolution of the close density maxima. For example, the authors of [7] have effectively studied the "Voyager–2"data for the Encke gap within this approach.

One of very effective methods for decision the above described aims is a real wavelet transform with the Gaussian wavelets [3, 22].

First of all we note that the rate of change of brightness in a certain point is determined by the gradient of the brightness function. It is obvious that its maximum will correspond to the sharp edge. It should be remember, however, that direct application of the gradient is possible for the case of smooth functions. Otherwise, even small-amplitude steps of noise lead to a quite large gradient modulus. This fact complicates the contour identification. That is the reason why it is necessary to use preliminary smoothing of the initial image by the Gaussian filter.

Consider as an example a one-dimensional case. Let $I_0(x)$ be an initial signal. Then the smoothed one has the form:

$$I(x) = \int_{+\infty}^{+\infty} I_0(x') \frac{e^{-\frac{(x-x')^2}{4t}}}{\sqrt{4\pi t}} dx'. \tag{1}$$

Here t is a smoothing parameter which gives the width of the average window. At $t = 0$ it transforms to the known Dirac delta-function:

$$\delta(x) = \begin{cases} \infty, & \text{if} \quad x = 0, \\ 0, & \text{otherwise} \end{cases}$$

and, in the correspondence with its properties we have:

$$I(x) = \int_{-\infty}^{+\infty} I_0(x)\delta(x - x')dx' = I_0(x).$$

In the theory of multiresolution methods, instead of parameter t the second independent variable $a = \sqrt{t}$ is used. In this case one can introduce self-similar combination $\xi = (x - x')/a$ and consider the following transform:

$$w(b, a) = C_n \int_{-\infty}^{+\infty} I_0(x)\psi_n\left(\frac{x - b}{a}\right)\frac{dx}{a}, \tag{2}$$

where C_n is a normalizing factor, and

$$\psi_n(\xi) = \frac{d^n}{dx^n}\psi_0(\xi), \tag{3}$$

$$\psi_0(\xi) = e^{-\frac{\xi^2}{2}}. \tag{4}$$

Application of the kernel (4) is equivalent to the simple Gaussian smoothing (1), and the kernel (3) with $n > 0$ produces smoothed derivatives. In particular, the use of

$$\psi_1(\xi) = -\xi e^{-\frac{\xi^2}{2}},$$

leads to the multiresolution version of the Canny Edge Detector, so that

$$\psi_2(\xi) = (1 - \xi^2)e^{-\frac{\xi^2}{2}}$$

relates to the generalized Marr-Hildreth approach. They may be formulated as follows: the signal has an edge at the scale a if the function modulus (2) with ψ_1 reaches in this point the maximal value (for Canny Edge Detector), or with ψ_2 is equal to zero (for Marr-Hildreth Edge Detector).

Apart from the pattern detection, as it has been shown in paper [22], continuous signal can be characterized if we take the double discrete sequence for the logarithmic set of scales. This means that we choose points of the edge localization (i.e. nulls of the second smoothed derivative) and the values of maxima (the first smoothed derivative in these points).

2.2. The Example Ring's Photo Processing

Let us consider decomposition of the image N1548116235 obtained 21.01.2007. The central part of the photo is shown in Fig. 1a. For more clearness this photo is vertically stretched. From the complete image we take a stripe of 10 pixels in this direction. This allows us to neglect by the ring curvature. The chosen image is a quite appropriate example because in the given part of the ring there are a slow radial brightness variation, extended bright plateau and developed fine wave structure.

Fig. 1b-d demonstrates contour lines extracted from Fig. 1a by means of Marr-Hildreth Edge Detector. In the given case we used two-dimensional wavelet-decomposition with the Marr wavelet:

$$w(a, b_x, b_y) = \int_{-\infty}^{+\infty} I(x, y) \left(1 - \frac{(x-b_x)^2 + (y-b_y)^2}{a^2}\right) e^{-\frac{(x-b_x)^2+(y-b_y)^2}{2a^2}} \frac{dxdy}{\pi a}.$$

Respectively, the contour lines are given as the solution of equations $w(a, b_x, b_y) = 0$ at $a = 1$ (Fig. 1b), $a = 2$ (Fig. 1c), $a = 4$ (Fig. 1d). One can clearly see that with the scale growth the number of detected edges decreases. If at $a = 1$ almost all ringlets may be revealed then at $a = 4$ we see only 5 contours bounding the large scale structures. In particular, 2nd and 3rd ones detect the most bright plateau.

Thus, the described method allows us to automatize the determination of coordinates of components with various characteristic length.

Consider now not only the edge location but also conception of the brightness curve change in different scales using the Mallat method [22]. Let us study a one-dimensional case of the radial distribution. In Fig. 2a a signal obtained by the average of Fig. 1a is shown. Wavelet-transform with the wavelet which is the first derivative of the Gaussian, calculated only in the points of the edge location. This allows to achieve us two goals: curtailment of computation and noise suppression in the initial image. The latter one is obtained by means of the absence of contribution in the restored signal the components of not leading hit. To reach more hight accuracy as localization points we introduce the following criterion: the edge is existing if in the lateral sample it appears for not less than half of pixels in the azimuthal direction. Fig. 2b-d show signals obtained as

$$I(x, a_i) = \sum_i w_1(a_i, b_i) \frac{x - b_i}{a_i} \frac{e^{\frac{(x-b_i)^2}{2a_i}}}{\sqrt{2\pi a_i}}$$

for the discrete sequence of the scales $a_i = 1, 2, 4$. Each of these graphs allows us to analyze not stationary patterns in the space at several levels of detailing. For example, the subsequent processing of the image 2b permits to carry out the analysis of the not stationary

Figure 1. A photo image of the rings part and contours detected with Marr wavelet.

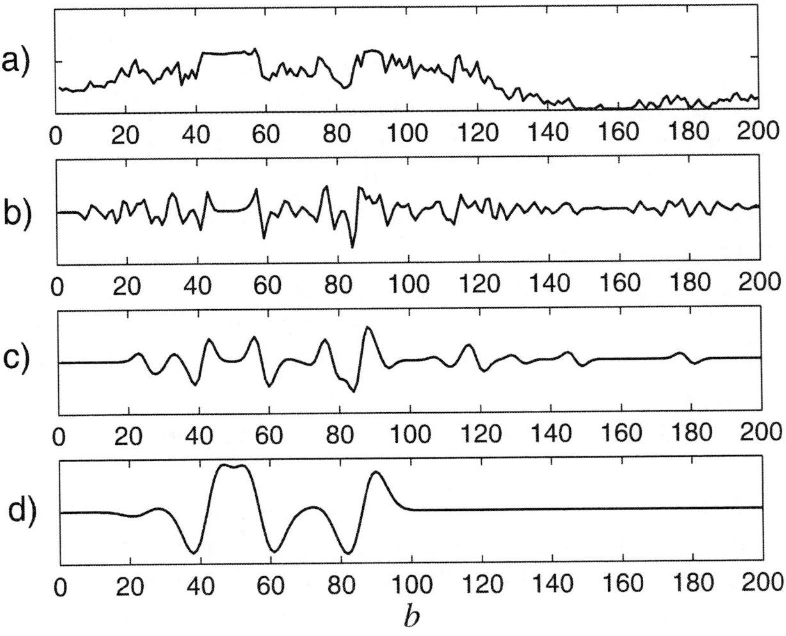

Figure 2. Decomposition of the brightness line change over different scales.

small-scale wave periodicity. In 2c it has been removed by increase of the cut-off threshold over the space frequencies. One can easily observe the structure of "swinging" of wave in the plateau edge (the part with the coordinate from 20th up to 40th pixels). The large scale shown in Fig. 2 keeps just "plateau".

In the conclusion, we note that wavelets of the Gaussian family have the number of the vanishing moments, which is equal to the order of derivative of the Gaussian. In particular, the considered transform zeros an average constant brightness level which is different in the left and right parts of the image. Therefore, the reconstructed distribution over the obtained contour wavelet transform completely reproduce the variable component of the original image without the background level difference.

3. Complex Continuous Wavelet Transform and Fields of Its Applications

Due to variety of wave structures in the Saturn rings, the local spectral properties could play a crucial role in waves classification and in analysis of the physical properties of the rings.

Based on the data obtained from the Voyager mission, Spilker et al [16] used the window Fourier transform that revealed about 40 resonance structures in Saturn's A ring. They were attributed to the influence of various Saturnian satellites. In the same time, authors of citied paper indicated a number of resonance regions, in which the achived resolution and the capacities of the precessing algorithm were unable to resolve any features in the ring matter density distribution. Authors of paper [20] have applied another windowed transform for the analysis of the characteristic wavelength in Saturn's B ring, namely the Burg tekhnique. This method solves for poles in the complex plane. In other words, the power spectrum is proportional $|1 - \sum_{j=1}^{M} g_j \exp(-i2\pi f_j)|^{-2}$, where f_j are the values of function sampled inside some sliding window and g_i are the values of filter of the same length M points.

The continuous complex wavelet–transform is the most powerful tool to achieve this goal. Its principal advantage with respect to other methods (see, e.g. [3], [4]) is a high localization of the basis functions in both spatial and frequency domains. Also, the size of a window and an instant period are correlated in this method: for high–frequency signals the window shrunks while for the low-frequency signals it is dilated. This keeps the effective number of oscillations in the window constant. In the connection with the celestial mechanics, the complex wavelet analysis firstly has been successfully applied to study the orbital period variations of asteroids in nearly–resonance zones [5] and for processing of the solutions obtained for Hamiltonian systems, in particular for the three–body problem (see [6]).

The wavelet approach to the problem of the local periodicity in the Saturn's rings has been started with the appearance of the high-resolution data from the Cassini spacecraft. Shortly after the arriving Cassini to the Saturn system, the relevance of such approach has been shown in [8] and [2]. The more detailed analysis has been evaluated in the articles [9] for the ring A and [10] for the rings B, C.

Recently, the most comprehensive review of results obtained by Cassini Imaging Team could be found in [11]. They have been analised by the complex wavelet transform with the Morlet wavelet the samples from all main rings including the inner parts of Cassini Division and the Enke gap.

3.1. The Complex Continuous Wavelet Transform with the Morlet Wavelet: How It Can Be Used for an Extraction of Ring's Periodic Features

The the continuous wavelet transform (CWT), according to its definition, reads [3]:

$$w(a,b) = \int_{-\infty}^{+\infty} f(t)\psi^*\left(\frac{t-b}{a}\right)\frac{dt}{a}, \tag{5}$$

where the asterisk denotes the complex conjugation. For the local spectral analysis it is convenient to use the following normalization:

$$\int_{-\infty}^{\infty} \left|\psi\left(\frac{t-b}{a}\right)\right| \frac{dt}{a} = C, \tag{6}$$

where C is a constant.

The wavelet exploited for the signal processing, is usually the complex Morlet wavelet:

$$\psi(\xi) = \frac{C}{\sqrt{2\pi}} e^{i\omega_0\xi} e^{-\frac{\xi^2}{2}}, \tag{7}$$

where ω_0 is large enough ($\omega_0 \geq \pi$) to make the application meaningful. The corresponding wavelet-transform $w(a,b)$ plays then a role of the local spectra in the neighborhood of the point b with period a. Let us illustrate this by two simple examples.

Consider first a harmonic $\exp(\mp i\omega t)$; the corresponding wavelet–transform and its modulus read:

$$w(a,b) \sim e^{\pm\omega\frac{b}{a}} e^{-\frac{1}{2}(\omega_0 \pm a\omega)^2}, \qquad |w(a,b)| \sim e^{-\frac{1}{2}(\omega_0 \pm a\omega)^2}.$$

The two–dimensional graph of the modulus distribution for the wavelet–transform clearly demonstrates the maxima line corresponding to the period $a = \pm\omega_0/\omega$, Fig.3a. The Gaussian factor reduces the noise by smoothing.

One can obtain an analytic expression for the wavelet–transform which refers to the matter distribution in the density wave due to resonance interactions with Saturn's satellites. The asymptotic expression in the area which are far from the perturbation source is [13]: $\sim \exp\left[i\left(\frac{x^2}{2\varepsilon} - s_\varepsilon\frac{\pi}{4}\right)\right]$. Here $s_\varepsilon = \pm 1$ and ε indicates the direction of the wave motion and ε is determined by the mass distribution in the ring:

$$\varepsilon = \frac{2\pi G\sigma}{r}\left[r\frac{d}{dr}\left(\mu^2 - (\omega - m\Omega)^2\right)\right]^{-1}.$$

Here G is a gravitational constant, σ is the particle surface density, ω and Ω are respectively the angular velocity of a satellite and ring particles, m is an integer, and $\mu = \left.\frac{\partial^2 V}{\partial z^2}\right|_{z=0}$, with $V(r,z)$ being the gravitational potential.

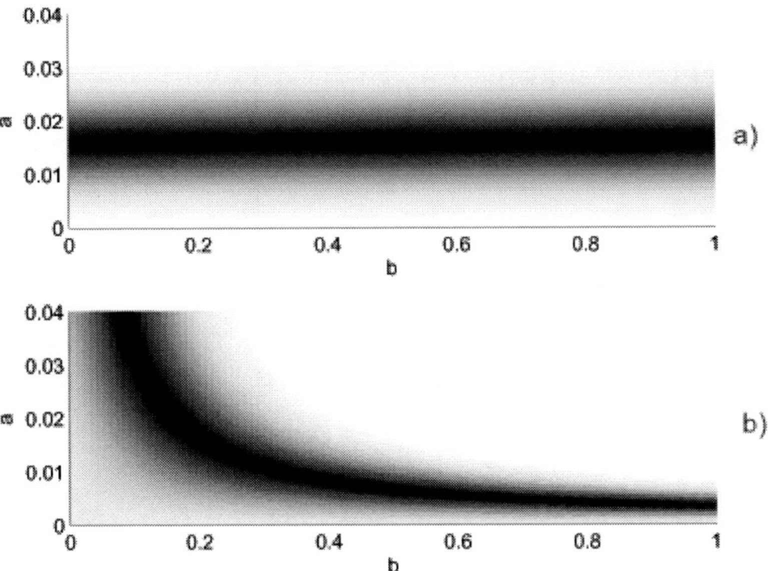

Figure 3. The wavelet–transform of the model examples.

The modulus of the corresponding wavelet–transform reads

$$|w(a,b)| \sim e^{-\frac{1}{4}\frac{ba\varepsilon^{-1}-\omega_0^2}{1+a^2\varepsilon^{-1}}};$$

it is shown in Fig.3b.

The variation of the basic frequency allows to vary a resolution factor. For small ω_0 the individual spikes may be easily resolved. With increasing basic frequency, the number of oscillations of the wavelet within a typical window also increases. However, while the resolution of harmonics becomes better their the spatial localization worsen. We illustrate this for a set of periodic signals composed of Dirac δ–functions:

$$f(f) = \sum_n \delta(t - t_n), \quad w(a,b) = \sum_n \frac{C}{a\sqrt{2\pi}} e^{-i\omega_0 \frac{b-t_n}{a}} e^{-\frac{1}{2}\left(\frac{b-t_n}{a}\right)^2},$$

see Fig.4.

Note that we may clearly detect period of the sequence by the line only in Fig.4a, where $\omega_0 = \pi$. At $\omega_0 = 1.5\pi$ (Fig.4b) one can reveal the second line corresponding to the second harmonics in the Fourier–transform of the signal. This phenomenon is demonstrated in Fig.4c, where also we may see the appearance of high harmonics.

We wish to stress that the standard way of CWT calculation (used, for example, in [11]) includes FFT as an intermediate step. However, in spite of the algorithm simplicity and its hight effectiveness, the FFT suffers from certain disadvantages: the initial sample must have 2^N equidistant nodes, and the obtained data have the corresponding frequency distribution. Violation of this condition leads to a sufficient complications of calculation and/or to the

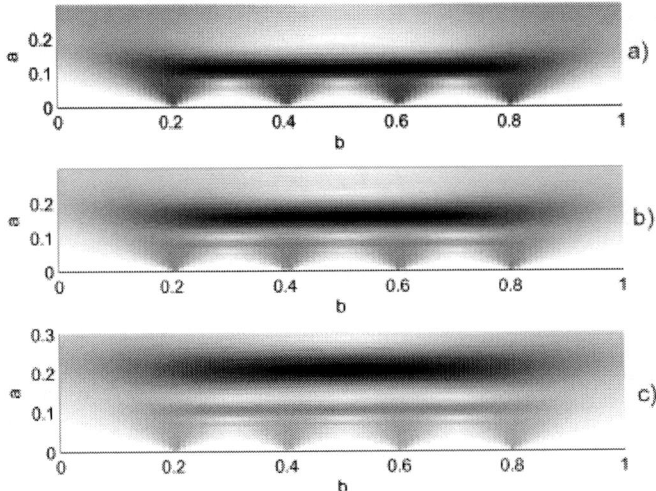

Figure 4. Transform of the sequence of impulses in the form of the Dirac δ–functions.

loss of accuracy. Here we attack the problem in a quite different way: We use the fact that the integral (5) has a form of the solution of a partial differential equation (PDE), where the wavelet ψ plays a role of the Green function. Here the variables a and b correspond to the "time" and "space" variable of this PDE. Let us consider this statement in more details.

It is known that the wavelet–image (5) with the Morlet wavelet (see [23]) is a solution of the following PDE:

$$\left(a\frac{\partial^2}{\partial b^2} - \frac{\partial}{\partial a} - i\omega_0\frac{\partial}{\partial b}\right) w(a, b) = 0. \tag{8}$$

In [23] this differential equation has been used to reveal local properties of a priori known wavelet–image. As a next step we recast the continuous transform (5) with the kernel (7) as:

$$w(a, b) = \int_{-\infty}^{+\infty} f(t)\frac{e^{-\frac{1}{2}\left(\frac{t-b}{a}-i\omega_0\right)^2}}{\sqrt{2\pi a^2}}dt. \tag{9}$$

This expression corresponds to the normalization (6) with $C = \exp(-\omega_0^2/2)$.

As is known, the integral (9) does not depends on the imaginary subtrahend in the exponent. Also, the transformation kernel turns to the Dirac delta-function in the limit $a \to 0$. Hence, $w(0, a) = f(t)$ is an initial value for the PDE (8). Since $\psi(a, b)$ is a solution of (8) with the initial value $f(t) = \delta(t)$ it affirms that wavelet in (9) is indeed a Green function for Eq. (8).

In practical calculations it is convenient to write the wavelet-image as a sum of real and the imaginary parts

$$w(a, b) = u(a, b) + iv(a, b).$$

With these notations Eqs.(8) reads,

$$\frac{\partial u}{\partial a} = a\frac{\partial^2 u}{\partial b^2} + \omega_0 \frac{\partial v}{\partial b}, \tag{10}$$

$$\frac{\partial v}{\partial a} = a\frac{\partial^2 v}{\partial b^2} - \omega_0 \frac{\partial u}{\partial b}. \tag{11}$$

with the initial conditions

$$u(0,b) = Re(f(b)),$$
$$v(0,b) = Im(f(b)).$$

Correspondingly, the modulus of the transform is

$$|w(a,b)| = \sqrt{u^2(a,b) + v^2(a,b)}.$$

From the applied point of view the proposed approach has the following advantages.

- There exist the stable finite–difference algorithms for the numerical solution of partial differential equations of the diffusion type (see, e.g. [24, 25]. The solver of the parabolic PDE realized in MATLAB (used in our examples) is based on this algorithms. In every point (a, b) for any a the value of the wavelet–transform $w(a, b)$ is determined by the solution of the equations (10). The wavelet–image obeys this system. Any given in advance the scale small step may be easily realized by a finite–difference scheme.

  Such an approach has advantages over the standard one. The matter is that, the method using FFT (see [3] allows to find the required transform only for the discrete set of scales. In the other points the value of the wavelet–transform can be calculated by the interpolation methods (say, parabolic). This means that it is not satisfied to the exact transform expression. This is especially evident for the cases of small scales. The minimal period resolution which is permitted by the sample of N points is N^{-1}. At the same time, the value of the transform at $a = 0$ is equal to zero in all points of the interval using the quadratic norm. This norm is required for the FFT. Thus, values in the small scale areas via the interpolation do not reveal the local properties of the signal. In addition, the passage from the obtained result to the amplitude norm is reduced to the division by $\sqrt{a}$ and obviously, it does not decrease the errors of the method. This is due to the fact that such an operation is ill–posed, i.e. it uses the division of two numbers close to zero. On the contrary, in the proposed approach expression $f(b) = w(0, b)$ is an initial value for the PDE. The solution in the small scale area possesses the maximal possible accuracy. Besides, the given method does not require equispaced sampling of the initial values. For more details and examples see [26].

- The finite–difference algorithms do not have difficulties with the sample periodization (the Gibbs phenomenon, ringing, and aliasing). In the case of inequality of the function at the beginning and at the end points of the interval, one should consider the function with the finite discontinuities in the countable set of points. For such functions the pointwise convergence of the Fourier series (the Gibbs phenomenon) does not hold true (see [27]). As a result, in the vicinity of the boundary points the artificial oscillations (as addition maxima in the area of small scales) appear. This is known as the ringing phenomenon. Secondly, in the cone of influence of the boundary points the value of errors is a quite large. This is similar to the cones in the points of delta-functions (see Fig. 4). Depending on the value of discontinuity their magnitude may be sufficiently more than the general level of the modulus of the wavelet–transform of the signal. In this case valuable information will be hidden.

At the signal periodization there is also the folding of high frequency components over a low frequency interval which is known as aliasing ([3]). At the inverse Fourier transform this leads to an additional low-frequency filtering of the wavelet image. At the same time, the algorithm based on the PDE solution by finite–difference schemes works with local values of the function (in the knots of the sample). For example, the standard finite–difference approximation of the second derivative requires the knowledge of the function only in the three points. Thus, the global characteristics of the sample (its length and the length of support of Fourier–image) does not influence on the local properties (except for large scales compared with the half length of the sample).

Besides, owing to the reduction of the problem to the PDE system, periodic boundary conditions are not unique. Moreover, as it was mentioned, these conditions are undesirable. In our approach the Dirichlet or Neuman boundary conditions are more adequate. Conditions at every of two ends of the interval may be chosen independently and adapted to the properties of the signal in the vicinity of this points.

3.2. Example: Detection of Typical Structures in Saturn Rings

Let us apply the described algorithm to the processing images of the Saturn rings obtained by the spacecraft "Cassini". We chosen images from the NASA/JPL/Space Science Institute collection [29]. These images does not allow a full quantitative analysis applicable for density estimation, etc since the brightness is calibrated only the public release purposes, but the determination of the wave/wake's properties are correct due to the fact that characteristic sizes of the ring areas may be quantified. Because this review is dedicated to the demonstration of principal advantages of an application of the wavelet transform to the ring's pictures processing, that is enough for our goal.

We have cut out a quite narrow stripe in the radial (across the rings) direction from every image such that we can neglect a certain curvature of the rings. We used the following images of A ring: PIA06099 (1022×20 pixels, Fig.5a), PIA06094 (891×23 pixels,Fig.6a), PIA06095 (902×23 pixels and Fig.7a); B ring: PIA06543 (1024×17 pixels, Fig.8) and C ring: PIA06537 (1024×15 pixels, Fig.9). For more deep visualization all images were stretched in the lateral direction.

Due to the fact that the signal is a real, we used initial conditions $u(0, b) = f(b)$ and $v(0, b) = 0$, where the function $f(b)$ can be obtained by the sample average (see Fig.5b – 9b). Because the signal has a bounded width the Cauchy problem for the equations (10)–(11) is to be replaced by the boundary problem. Accordingly, we have chosen the boundary conditions of the first kind: the value of the transform's real part is equal to the signal values, the transform's imaginary part is zero. We used the base frequency $\omega_0 = \pi$ since it

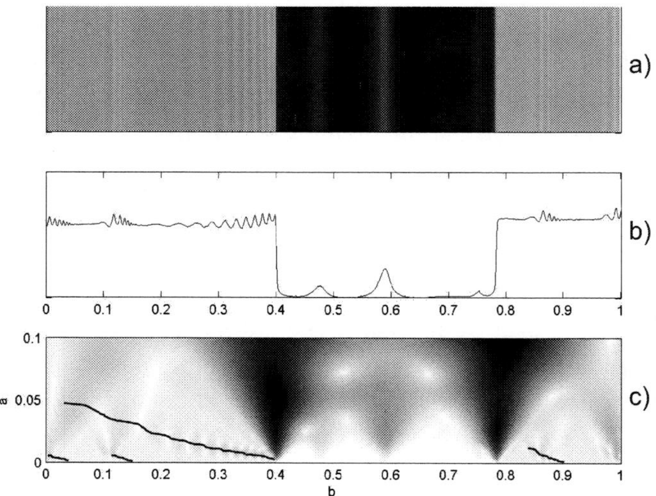

Figure 5. The Encke division. The left side of this picture is almost coincided with the position of the resonance 11:10 with Pandora. The next wave–like structure is generated by the resonance 15:14 with Prometheus. The first wave train following after this division is generated by the resonance 12:11 with Pandora.

allows to select clearly lines of the instant period, as it is shown above. In the diagrams of the module of wavelet–images the maxima lines are marked by contrast black lines.

An interesting fact found in the neighborhood in the Encke division, is the co-existence of resonance waves of certain periods. One can note that a large-scale evolution of a spiral wave excited by Pan from the edge division, which provides the growth of the value of its instant period, admits a continuous transformation to a maxima line corresponding to large-scale bursts. The typical size of such scales is about the length of the train of resonance waves generated by the resonances 11:10 with Pandora and 15:14 with Prometheus. Here there is a clear crossing of the maxima lines of various resonances, which does not initiate their interaction. We found also the effect of the crossing of the maxima line of large- and small-scale (forming by Pan) resonance wave structures in the outside of the Encke division (see Fig.6c).

The other type of the inhomogeneity which can be detected by the proposed wavelet–analysis is the waves with relatively stable period. Therewith, for A ring the presence of small-scale periodicity in interresonance areas is typical. For B and C rings we have long-periodic waves against a background of which the resonance peaks take place.

To carry out a detailed analysis of the A ring small-scale in the interresonance areas, consider the density waves formed by the resonances with Janus, Pandora and Prometheus. A typical staircase-like shape of their instant spatial period is shown in Fig.7c. obtained at the basis frequency π. To get more frequency resolution let us increase the basis frequency up to the value of $\omega_0 = 1.5$ (Fig.7d) and $\omega_0 = 2\pi$ (Fig. 7e). We apply the following criterium: we consider only the areas in the plane (b, a) where the condition

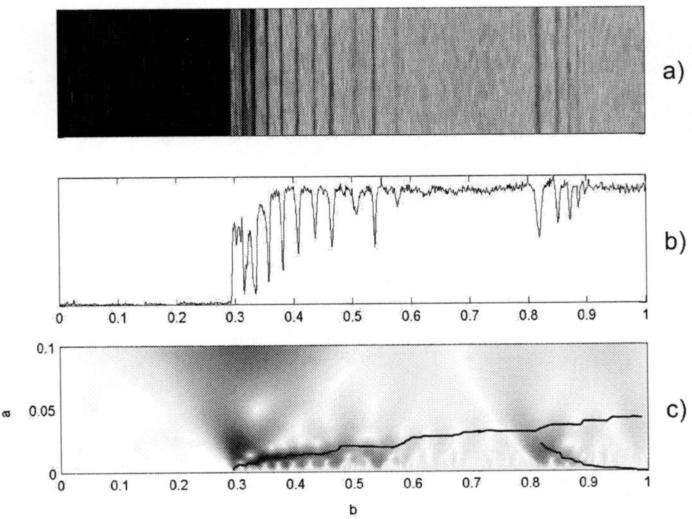

Figure 6. The edge of the Encke division far from Saturn.

$\exp\left(-(b-b_0)^2/2a^2\right) \geq 10^{-5}$ is satisfied, with $b_0 = 0$ or $b_0 = 1$.

To increase the sensitivity to the small amplitude module we use various gray tints for larger values. This method leads to the tailing of resonance lines but allows to separate (by the brightness lines) almost stable periodic signal in the interval $[0.35, 0.45]$ which couples the two first resonance wave trains. Its period is 4.5 ± 1 km. In the interval $[0.52, 0.67]$, between the second and the third resonances, a shortwave signal is also detected. However it has unstable spatial period which is changed in the limits $4.5 - 6.6 \pm 1$ km. Therewith, the first value coincides with the wave period between the resonances of Janus (4:3) and Pandora (5:6).

For B ring, in Fig.8c one can see that in the region with dimensionless coordinates form 0.2 to 0.65 there exists a smooth variation of period with 380 to 200 km, and the maxima line has a staircase-like shape with three plateaus. Also, one can see another part of the maxima line corresponding to period 115 km. With this background and with such resolution we may find only an inclined line of resonance in a neighborhood of the point with the coordinate 0.4. Small-scale structures with the ordered periodicity are practically lacking.

One can detect a large variety of structures in a part of C ring (see Fig.9). Here, besides a background density wave with period 460 km we may clearly recognize two classical resonance curves. They become apparent already in Fig.9c, being more clearly visible in Fig.9d, where the maxima lines are emphasized by contrast black and white curves. It should be noted that a flat grey line, which is below this white line, does not manifest an additional phenomena. It is just the consequence of a large basis frequency as is shown at the analysis of non-sinusoidal waves in the first part.

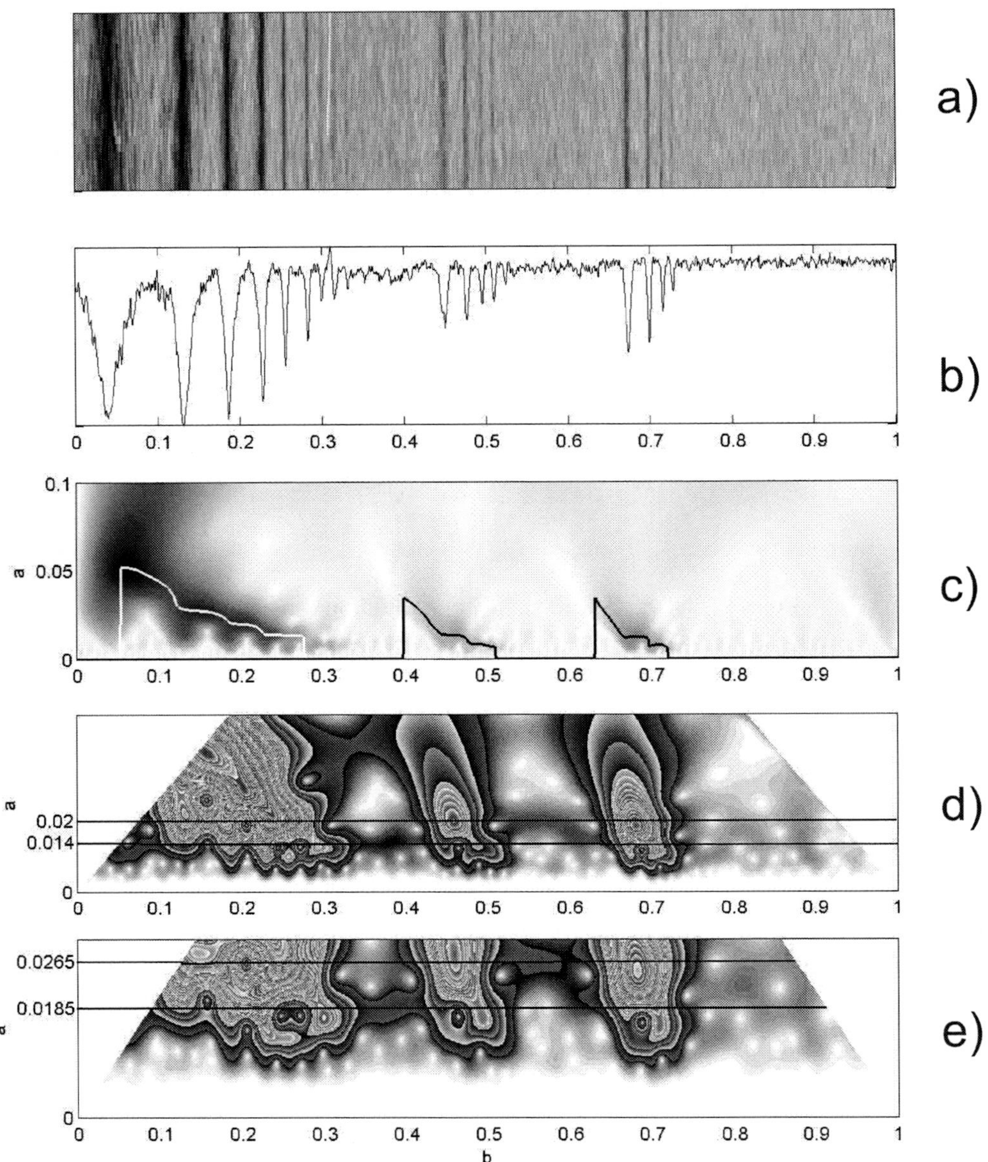

Figure 7. Outside of A ring containing the resonance 4:3 with Janus, 6:5 with Pandora and 7:6 with Prometheus.

4. Conclusion and Perspectives

Thus, we have shown in out study, that the continuous wavelet transform is an effective tool for the analysis of spatial radial structures of Saturn rings. It makes possible to resolve individual ringlats as well to observe the instant period evolution at various scales and detect

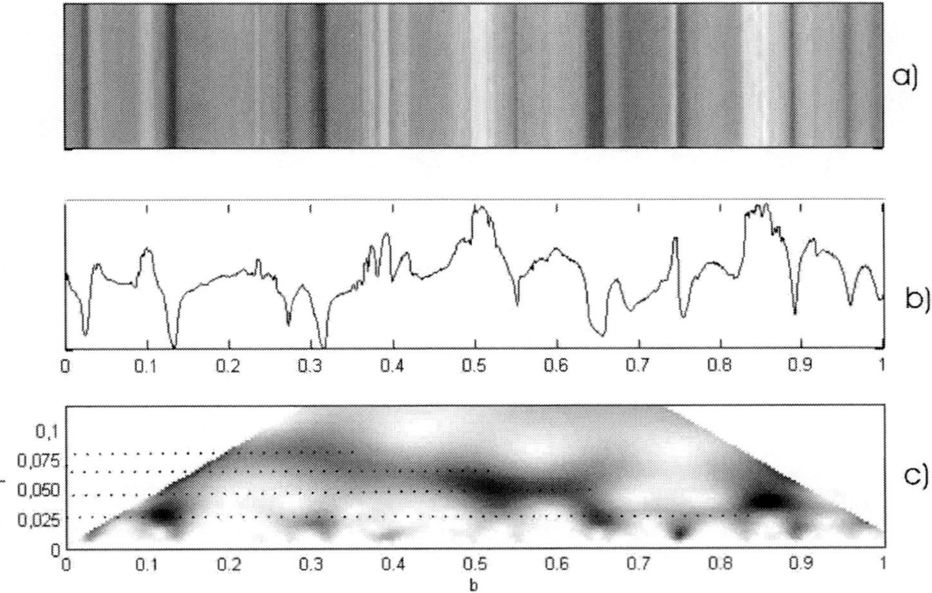

Figure 8. The central part of B ring.

the periodic and aperiodic bursts of different nature.

In the present study we develop a new approach for the evaluation of the complex continuous transform with the Morlet wavelet and show that it may be successively applied for the analysis of the images of Saturn's rings structures recently obtained during the "Cassini" mission. It allows to investigate in detail the resonance zones and reveal the coexistence of waves with stable periods and the wave trains with a variable instant period. It has a number of merits over the standard ways based on the Fast Fourier Transform (FFT) as an intermediate step. These abilities of the method steams essentially from the proposed algorithm, which allows to derive a continuous small steps for the scale variable. It is the direct consequence of the application of the algorithm of solving partial differential equations which replaces the integral transform.

Besides such a representation has the natural connection with the method of tracing of wavelet maxima and ridges offered in the paper [28]. Additionally, the method does not suffer, as the FFT from the artificial effects connected with the periodization of the sample.

We hope that its application will makes possible to find parameters which characterize the mass distribution as well as the other physical properties of the rings more preciously as it has been made before. The detailed analysis of the wave processes in Saturn rings should include a question concerning the interaction of long-wave parts of perturbations with small-scale wave trains generated by the resonance effect of satellites. Also, this analysis should take into account mechanisms of the formation of almost monochromatic small-scale waves in regions between the of resonance zones.

From the modulus diagram, we reveal that almost monochromatic spatial density oscil-

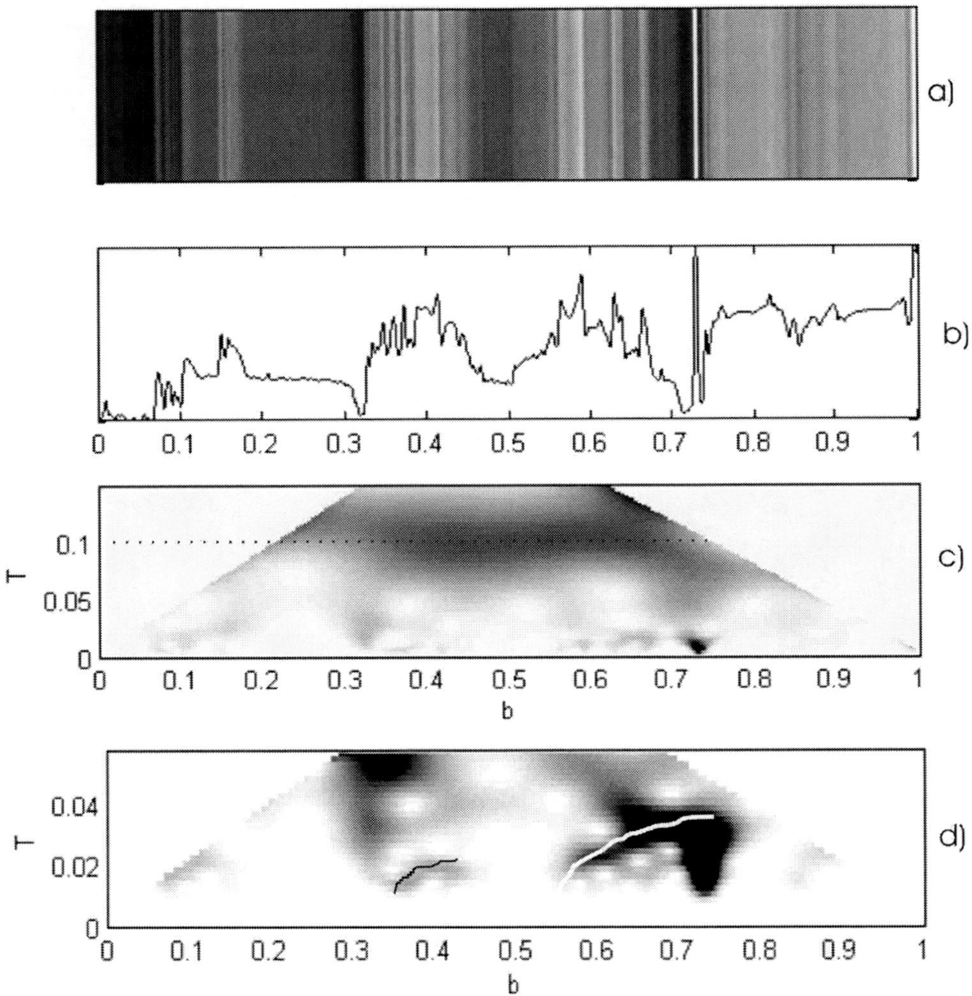

Figure 9. A part of C ring. The distance between the center of this image and Saturn is about 75000 km.

lations take place in such regions. They may be detected by the maxima lines which are drawn parallel to the coordinate axis. This line of maxima virtually connects the other lines of maxima corresponding to two resonance staircases (i.e. points that correspond to the minimal scale). We may propose a mechanism of such type of behavior: when the frequency of the resonance spatial swings of the matter is comparable with the typical frequency of viscously-unstable standing waves, this interaction induces visco-elastic oscillations.

The opportunity of the inverse transformation for the exact form of the Morlet basis makes it possible to cut from the radial signal and analyze the small-scale wave features, e.g. the resonance trails. By substituting the wavelet–image values chosen along the maxima line (and its small neighborhood) into the expression of an inverse wavelet–transform, we may obtain a set of purely resonance oscillations. Then, subtracted them from a total signal, we get a function diagram which allows us to analyze a small-scale structure. Localization of regions of the stable periodicity and detection of their period in a quasihydrodynamic model provides us a data about the density and composition of the ring matter of the required typical viscosity caused the structure with the given wavelength. It should be noted that the complex wavelet–transform is much better one for such an analysis than the other window transformation. This is due to the fact that, owing to the self-similarity property of the analyzing wavelet, in any frequency range one and the same typical oscillation numbers is packed on the window width. Moreover, elimination by the described method of a high-frequency component (this means near zero scales) allows to denoise a signal. Finally, aperiodic spots in the wavelet–image allows to detect structures in the form of narrow dense rings (i.e. ringlets).

References

[1] Esposito, L.W., 2002. *Rep Prog Phys*, 2002, 65, 1741–1783.

[2] Porco C.C. et al. *Science*, 307, 1226-1236.

[3] Mallat, S. A Wavelet Tour of Signal Processing; Academic Press, 1999.

[4] Poularikas, A.D. The Transforms and Applications Handbook; CRC Press, 2000.

[5] Michtchenko, T.A.; Nesvorný D. *Astron Astrophys*, 1996, 313, 674–678.

[6] Vela-Arevalo, L.V. Time-Frequency Analysis Based on Wavelets for Hamiltonian Systems. PhD thesis. Caltech, 2002.

[7] Bendjoya, Ph.; Petit, J.-M.; Spahn, F. *Icarus*, 1993, 105, 385-399.

[8] Postnikov, E.B.; Loskutov, A. http://www.arxiv.org/abs/astro-ph/0502375.

[9] Postnikov, E.B.; Loskutov, A. *J Exp Theor Phys*, 2005, 101, 646-652.

[10] Postnikov, E.B.; Loskutov, A. *J Exp Theor Phys*, 2007, 104, 417-422.

[11] Tiscareno, M.S.; Burns J.A., Nicholson P.D. et al. *Icarus*, 2007, 189, 14-34.

[12] http://pds-rings.seti.org/

[13] Shu, F.H., Cuzzi, J.N., Lissauer, J.J. *Icarus*, 1983, 53, 185-206.

[14] Shu, F.H.; Yuan, C.; Lissauer, J.J. *Astroph J*, 1985, 291, 356-376.

[15] Shu, F.H., Yuan, C., Lissauer, J.J., *Astroph J*, 299, 542-573.

[16] Spilker, L.J.; Pilorz. S; Lane, A.L et al. *Icarus*, 2004, 171, 373-390.

[17] Thiessenhusen, K.-U., Esposito, L.W., Kurths, J., Spahn, F. *Icarus*, 1995, 113, 206-212.

[18] Schmit, U.; Tscharnuter, W.M., *Icarus*, 2001, 138, 173–187.

[19] Griv, E.; Gedalin, M. *Planet Space Science*, 2003, 51, 899-927.

[20] Horn, L.J.; Cuzzi J.N. *Icarus*, 1996, 119, 285-310.

[21] http://pds.jpl.nasa.gov/

[22] Mallat, S.; Zhong, S. *IEEE T Pattern Anal*, 1992, 14, 710-732.

[23] Haase, M. In: Paradigms of Complexity. M.M.Novak, M.M., Ed.: World Scientific, 2000, pp. 287-288.

[24] Skeel, R.D.; Berzins, M. 1990. *SIAM J Sci Stat Comput*, 1990, 11, 1-32.

[25] Kiseliov, R.V.; Postnikov E.B. In: Proceedings of the XIX Session of the Russian Acoustical Society. Acoustical Measurements and Standartization. Nizhny Novgorod, 2007, pp. 233-235.

[26] Postnikov, E.B. *Comp. Math. and Math. Phys.*, 2006, 46, 73-78.

[27] Jerri, A. The Gibbs Phenomenon in Fourier Analysis, Splines and Wavelet Approximations. Kluwer Academic Publishers, 1998.

[28] Haase, M.; Widjajakusuma, J.; Bader, R. In: Emergent Nature. Novak, M.M., Ed.: World Scientic, 365-374.

[29] http://ciclops.org

In: Space Exploration Research
Editors: J.H. Denis and P.D. Aldridge
ISBN: 978-1-60692-264-4

Chapter 14

SPACE SCIENCE APPLICATIONS OF AEROGELS

Steven M. Jones*
Jet Propulsion Laboratory, California Institute of Technology, CA, USA

ABSTRACT

The unique physical properties of aerogels have provided enabling technologies to a variety of both flight and proposed space science missions. The extremely low values of the density and corresponding high values of the porosity of aerogels make them suitable for stopping high velocity particles, as highly efficient thermal barriers and as a porous medium for the containment of cryogenic fluids. The use of silica aerogel as a hypervelocity particle capture and return medium for the Stardust Mission, which launched in 1999 and returned to earth in 2006, has drawn the attention of the scientific community, as well as the public, to these fascinating materials. Aerogels are currently being used as the thermal insulation material in the 2003 Mars Exploration Rovers and will be used on the 2009 Mars Science Laboratory rover, as well. The SCIM (Sample Collection for the Investigation of Mars) and the STEP (Satellite Test of the Equivalence Principle) Missions are proposed scientific missions, in which the use of aerogel is critical to their overall design and success. Composite materials comprised of silica aerogel and oxide powders are currently under development for use in a new generation of thermoelectric devices that are planned for use in future mission designs. Work is ongoing in the development and production of non-silicate and composite aerogels to extend the range of useful physical properties, and thus, the applications of aerogels in future space science missions.

1. AEROGELS

Aerogels were first produced in the 1930's by Stephen Kistler at the College of the Pacific in Stockton, California. Kistler wanted to study the networks of gels, however to do so, he needed to find a way to dry the gels without collapsing the porous networks. He did this by raising the pressure of the gel and then heating it, so that the liquid within the gel

* Steven.M.Jones@jpl.nasa.gov

network was in the supercritical state [1]. The first aerogels he produced were silica aerogels, however, he went on the produce magnesia, chromia, tin oxide, thoria, alumina, and organic aerogels [2]. Later, Kistler went to work for the Monsanto Corporation, where he conducted research into the development and production of aerogels for new applications. Powdered silica aerogel was used as a thixotropic, a thickening agent, in toothpaste and cosmetics for many years. However, the use of silica aerogel died off with the development of fumed silica, which is significantly cheaper to produce.

In the 1980's, research into the production and physical properties of aerogels was revived and aerogel laboratories began to appear around the world. Major U.S. national laboratories, such as the Lawrence Berkeley and the Lawrence Livermore Laboratories in California, have been producing and studying aerogels for several decades. Various companies have been started to commercialize aerogels, although they have tended to fail after several years. More recently, Aspen Aerogels and Cabot Chemicals have initiated the production of aerogel-based thermal insulation materials, which are gradually establishing themselves in the thermal insulation market. Despite the fact that aerogels exhibit many useful physical properties, they are somewhat difficult to produce and to package for their eventual applications. Research is ongoing to produce low cost, durable aerogels that are more competitive with existing materials. Nevertheless, aerogels have proven to be extremely useful materials in many space science missions and hold the promise of enabling many more missions.

Aerogels are composed of micro-porous networks made up of interlocking nano-scale filaments. This results in a highly porous, open cell structure that imparts on aerogel a variety of unique physical properties. Despite the fact that typical aerogels are up to 99.9% by volume open pore space, they are still solid materials. Among the more potentially useful properties are the extreme low density, high porosity, low thermal conductivity, optical transparency, low refractive index and low dielectric constant. Aerogels are typically produced by the hydrolysis and condensation of either a metal salt or a metal alkoxide. This results in a continuous solid network that is filled with solvent. To remove the solvent without collapsing the fragile gel network, the gel is pressurized above the vapor pressure of the solvent, heated to a temperature above the supercritical temperature of the solvent, and then depressurized to withdraw the supercritical solvent from the interstitial space of the network. This process is known as high temperature supercritical solvent extraction. Aerogels can also be produced by exchanging the initial liquid solvent with supercritical carbon dioxide and then removing the carbon dioxide. There are many good reviews about the production and properties of aerogels [3 - 6].

The proposed applications of aerogel are many and varied, yet most remain unrealized. Commercial applications such as for thermal window insulation, acoustic insulation, optical coatings, capacitor electrodes, low dielectric constant layers in integrated circuits, piezoelectric transducers and catalytic supports have all been proposed, although little in the way of actual use has resulted [1, 7 - 14]. However, monolithic silica aerogel has been used extensively in high energy physics in Cerenkov radiation detectors [15 - 19].

Starting in the 1990's, NASA employed aerogel in three of its exploration missions, these being the Mars Pathfinder, the Stardust Missions and the 2003 Mars Exploration Rovers. These missions utilized the unique thermal and mechanical properties of aerogel and have sparked continued interest in the further use of aerogel for space science applications.

2. Hypervelocity Particle Capture

2.1. Initial on Orbit Studies

After conducting many ground-based experiments of the feasibility of using aerogels as hypervelocity particle capture media, aerogels were first flown on the Space Shuttle (STS) in 1992 to test their performance in space [20]. Aerogel panels were placed on the lids of Get-Away-Special (GAS) Canisters in the payload bay of the STS-42 [21]. Aerogel proved to be sufficiently durable to survive the launch and reentry environments and captured three particles during this initial flight. Since that initial flight, aerogels have been flown on the lids of many other GAS Canisters.

In the mid-1990's, aerogels were also flown on Spacehab 2 and the Mir Space Station. The Orbital Debris Collection Experiment consisted of 0.63 square meters of single density, 20 milligram per cubic centimeter silica aerogel mounted on Mir and stayed on orbit for eighteen months. After being returned to the Johnson Space Center for analysis, it was observed that a wide range of impacts had occurred during its time in space. Both man-made, e.g., paint flakes, solder, and naturally occurring, e.g., cosmic dust, samples were collected and analyzed [22].

2.2. The Stardust Mission

The Stardust Mission was designed around the fact that low density aerogel is an excellent hypervelocity particle capture medium [20]. The mission plan was to transport a grid of aerogel cells into space, rendezvous with a comet, capture material from the comet in the aerogel and return the collected samples to the earth [23]. The primary science requirement of the mission was that one thousand cometary particles, fifteen microns or larger in diameter, be captured and returned to earth.

The timeline for the mission was as follows:

Launch - February 1999
Interstellar Collection A - October 1999 through March 2000
Interstellar Collection B – May 2002 through October 2002
Comet Wild 2 Encounter - January 2004
Return to earth – January 2006

During the comet encounter the aerogel grid was deployed such that particles from the coma of the comet impacted with the aerogel and embedded themselves in the porous network of the aerogel (figure 1). Having captured the cometary particles, the aerogel collection grid was retracted into the Sample Return Capsule (SRC) and, after cruising through space for two years, returned to earth. Upon the spacecraft's return to earth, it released the SRC, which parachuted down to the Utah Testing and Training Range.

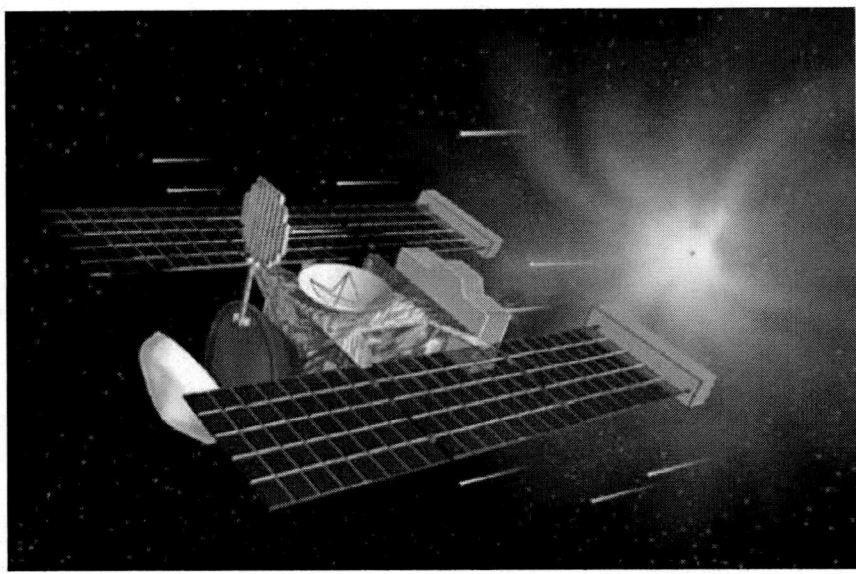

Figure 1. Artist's image of the Stardust spacecraft with the aerogel collector deployed for the cometary encounter. Image provided by NASA/JPL/Caltech.

Two grids of aerogel were assembled for the mission: one for the collection of the cometary particles and one for the collection of interstellar particles. The two grids were mounted back-to-back so that one could collect particles while the other was facing away from the particle stream. Each grid was composed of one hundred and thirty cells of gradient density, silica aerogel (figure 2). The cells for comet particle capture were approximately 4 by 2 by 3 centimeters, while those for the interstellar particles are 4 by 2 by 1 centimeters.

Figure 2. The cometary and interstellar particle collection grids each contained one hundred and thirty cells of gradient density silica aerogel.

Aerogel is an excellent hypervelocity particle capture medium due to the fact that it is a highly porous material whose microstructure is made up of nano-scale filaments (figure 3). Since the particles impacting the aerogel are microns in size, while the filaments of the aerogel are nanometers in size, the filaments yield to the force exerted by the particles. As the individual filaments are broken and destroyed, the kinetic energy of the particle is gradually transferred to the aerogel as it is converted to thermal and mechanical energy. Thus, the particle is slowed and finally stopped, largely intact (figure 4). When small dense particles are captured in low density aerogel they produce a characteristic long, gradually tapering penetration track. These are commonly known as a “carrot” track or stylus. Impact features known as pits that are relatively shallow and wide have also been observed in many aerogels returned from low earth orbit capture experiments, including the Stardust aerogels. The additional observation of blunt-nosed, cylindrical impact features suggests that there is an entire spectrum of the morphologies of such features [24]. Carrot tracks typically have a terminal particle at the end, whereas pits do not.

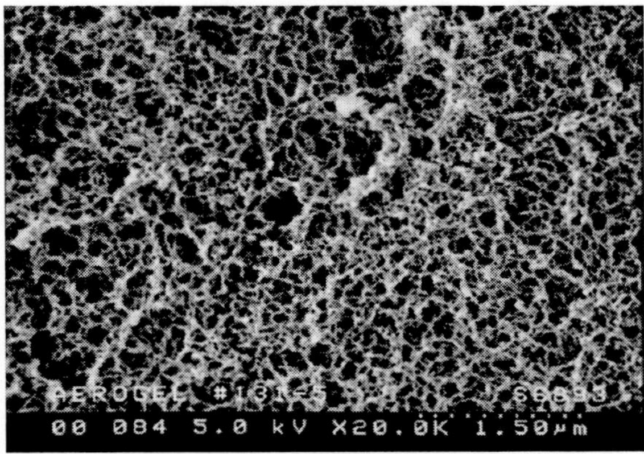

Figure 3. Scanning electron micrograph of filaments composing a silica aerogel network.

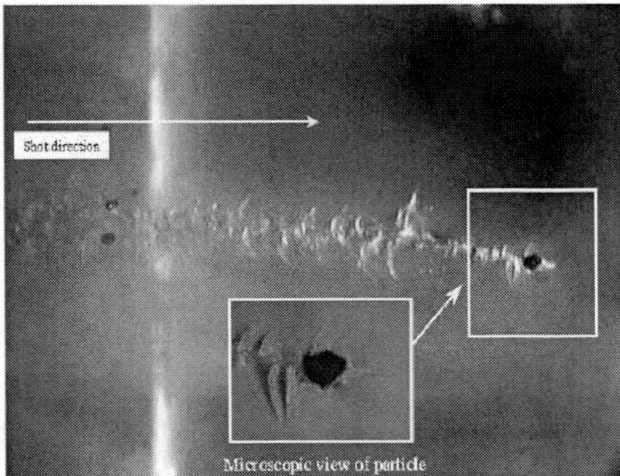

Figure 4. Terminal particle at the end of a hypervelocity impact penetration track.

Prior to the Stardust aerogel development work, hypervelocity particle capture experiments had been conducted with single density materials. Gradient density silica aerogel was developed and produced for the Stardust particle capture grids, since it was considered to have superior capture properties. When high velocity particles begin to penetrate the aerogel, they first encounter the low density material that begins the deceleration process. As the particle moves further into the aerogel, slowing as it progresses, it gradually encounters higher density material, which increases the resistance on the particle. The cometary aerogel has a gradient profile that changes from approximately10 milligrams per cubic centimeter at the impact surface to 50 milligrams per cubic centimeter in the lower half of the cell.

To produce the aerogel, a sol (a suspension of sub-micron sized particles in a liquid) was made from tetraethyl ortho silicate [25]. The sol was then mixed with acetonitrile, water and ammonium hydroxide to form the gel precursor. The density gradient was achieved by a method that gradually diluted the gel precursor and then pumped the mixture to the molds in which the aerogels were cast [26]. A mixture that would result in a high density aerogel was pumped initially to the base of the molds from a stock high density precursor. A precursor mixture that would result in a low density aerogel was then gradually pumped into the stock high density precursor, where they were stirred together. This resulted in a gradual dilution of the high density precursor. As the diluted mixture was pumped to the mold, the concentration of the hydrolyzed and partially condensed silica decreased and thus the final density of the aerogel across the profile decreased. Once the pumping was complete, the precursor was allowed to gel and it was then dried in a semi-automated supercritical solvent extraction system.

By varying the rate at which the precursors were combined and the rate at which the mixture was pumped to the mold, a variety of different gradients was produced. The density of the final aerogel was determined by using the Gladstone method, where the density is determined by measuring the refractive index of the aerogel. A laser is passed through the corner of a piece of aerogel, which acts as a prism, thus deflecting the laser beam. Since the density of the material and the refractive index are correlated, the density was then calculated from the value of the deflection.

While this method was developed for producing gradient density aerogel, it should be noted that the method could be used to produce materials with gradient thermal, optical, acoustic and dielectric properties, since these properties are density dependent. It can also be used to produce gradient oxide or dopant compositions. Novel materials such as these could also someday be used in future space exploration missions.

The requirements of the aerogel by the Stardust Mission were that it be capable of efficiently capturing hypervelocity particles, that it be the proper shape and size, that it be free of significant flaws, e.g., cracks, inclusions, that it be of reasonably high purity, and that it be capable of surviving launch and landing environments. Efficient hypervelocity particle capture was verified primarily by testing with light gas gun hypervelocity impact testing. Many tests were conducted in which small particles (10 to 125 microns in diameter) were fired into aerogels and then imaged and extracted. Suitable quartz molds and processing methods were developed to result in aerogel cells that were of the correct shape and size, were free of significant flaws and were of reasonable purity. The shape and size of the aerogel was crucial since the aerogel is held in the flight grid purely by compression. The aerogel cells were made to be slightly larger than the spaces machined out of the flight tray, and thus were compressed slightly as they were inserted into the tray. The launch and landing survivability of the aerogel was verified by conducting random shake and landing pyro shock tests, as well as by the previous launches and returns on the SST of experimental capture grids. During the launch, the aerogel also had to survive the transition from atmospheric pressure to the vacuum of space. Since aerogel is an open cell network, interstitial gases are able to flow through the network. Testing was conducted to demonstrate that the depressurization experienced during launch would not damage or disrupt the aerogel in any way.

The Stardust spacecraft was successfully launched in February of 1999 and encountered the comet Wild 2 in January of 2004. The encounter with Wild 2 went according to the mission design and in 2006 the samples were parachuted to earth successfully. The samples returned by Stardust are the first solid samples from a specified extraterrestrial body, other than the Moon. The analyses done thus far indicate that comets contain organic molecules not observed in other extraterrestrial materials, e.g., meteorites, and that the composition of comets are more complex than previously believed [27, 28]. Comparisons with existing data gathered from meteorites, micro-meteorites and interplanetary dust particles is ongoing and is anticipated to improve models of the origin of the solar system and of silicate materials in our solar system. The results drawn from the samples collected by Stardust have already indicated that there was significant radial mixing of the solar nebula [29]. This conclusion has been arrived at by the fact that high-temperature minerals have been identified in the samples. These minerals must have been formed relatively close to the sun and then been transported to the outer regions of the solar nebula and incorporated into comets.

The analysis of the aerogel grids that were deployed to collect interstellar dust during the cruise phase of the mission is also ongoing. Interstellar dust is a key component of the interstellar medium, since it is ubiquitous and interacts extensively with solar radiation. Identification and analyses of contemporary interstellar dust samples would provide valuable information regarding galactic chemical evolution and stellar nucleosynthesis. The search for captured interstellar grains is being done by first scanning the aerogel cells with an automated microscope, which will take digital images of minute portions of the cells. The images are being analyzed by thousands of volunteers over the internet and candidates of interstellar grains are currently being extracted for further study [30].

2.3. The Proposed SCIM Mission

The proposed SCIM (Sample Collection of the Investigation of Mars) Mission would fly a spacecraft through the upper Martian atmosphere, collect dust particles suspended in the atmosphere in aerogel and return the sample to earth (figure 5) [31, 32]. It would be the first sample return mission from another planet. The baseline proposed mission design is based largely on the technology developed and produced for the Stardust Mission [33]. SCIM would be a low cost, low risk mission for gathering and returning a sample from Mars to the Earth. Like the Stardust Mission, returning extraterrestrial samples to earth for study unleashes the entire spectrum of handling and analytical techniques available, rather than the extremely limited techniques that can be transported to another planet. One advantage of this scheme for sample collection over landing on the planetary surface is that since the dust in the Martian atmosphere represents a global sample of the surface of Mars, this mission will return a global sample rather than a sample primarily specific to one particular location. SCIM would also collect a sample of the Martian atmosphere, which would also be returned to earth for analysis.

The science goals of the mission are: 1. determine the extent of aqueous processing of Martian crustal materials, 2. quantitatively establish the chemical, isotopic and mineralogical composition of the Martian dust, 3. quantify the size and compositional distribution of dust in the Martian atmosphere, 4. provide ground truth for past and future remote sensing observations of Marian surface materials, 5. determine whether or not Martian SNC (shergottite, naklite and chassigny) meteorites originated on Mars, 6. find evidence relevant to Mars' original volatile chemical inventory and its subsequent evolution and 7. investigate current atmospheric chemical composition. Accomplishing these goals would make possible basic advances in our understanding of the geological and the climatic history of Mars and, ultimately, its habitability by humans.

The amount of dust suspended in the Martian atmosphere varies a great deal from season to season. The vertical and latitudinal distribution also varies greatly. Therefore, the SCIM Mission would be planned so that it would arrive during the period of greatest atmospheric dust content and at the best latitude and elevation for the dust collection. The presence of a significant dust storm on Mars can not be accurately predicted, however, a dust storm would only increase the atmospheric dust content and thus the amount of sample collected and returned.

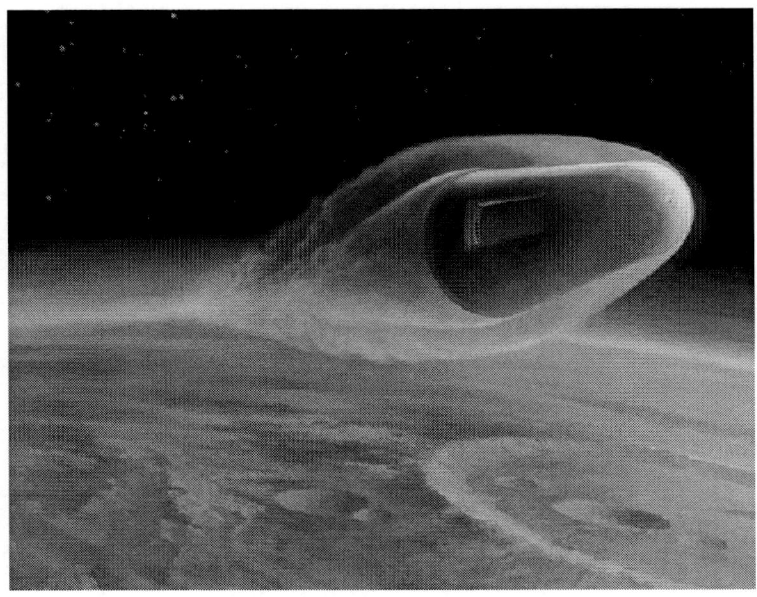

Figure 5. Artist's image of the SCIM spacecraft exiting the Martian atmosphere after the sample collection period.

Similar to the design of the Stardust Mission, SCIM would use aerogel to capture particles in aerogel and would return the aerogel to earth. Rather than a grid of aerogel cells, SCIM would use two rows of individual aerogel collection modules (figure 6). Each row of collector modules, located on opposing sides of the spacecraft, would be launched in place with ejectable covers protecting the aerogel. Prior to the aeropass through the Martian atmosphere, the covers would be ejected. After the aeropass is complete, the aerogel would be retracted into a sterilization oven, where they would be heated at 250°C for several hours. This would be done to satisfy Planetary Protection Protocols, ensuring that no live organisms that might possibly be present on the dust survive the return to earth. The aerogel cells would then be stowed in the Sample Return Capsule (SRC) for earth return. Based on the successful return of the Stardust samples, the SRC would be released as the spacecraft approached the earth and would parachute to the ground for retrieval. The aerogel collection modules would be taken to a specially designed curatorial facility, for preliminary analyses. The techniques developed for analysis of the Stardust samples and the lessons learned in employing those techniques would be drawn to bear on the returned Martian samples.

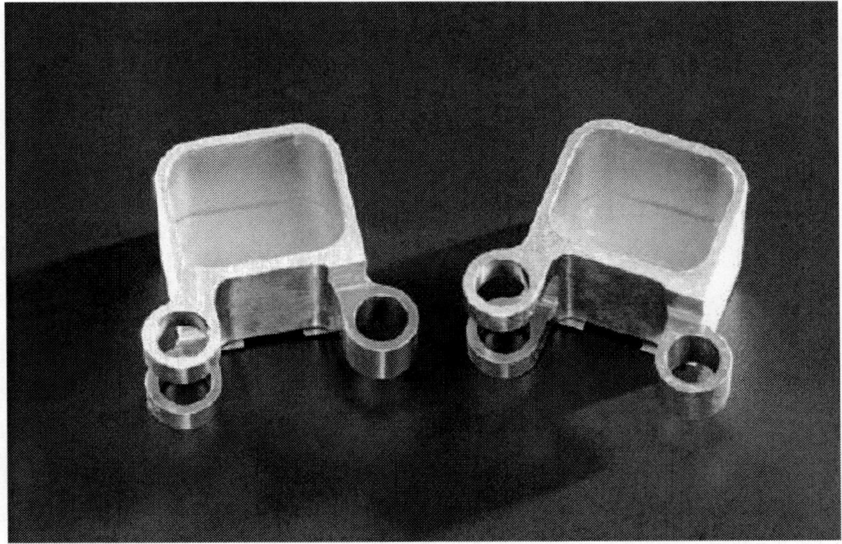

Figure 6. SCIM aerogel collection module with aerogel and 4 X 2 X 3 cm sample of silica aerogel.

However, since SCIM would be passing through an atmosphere at a velocity of several kilometers per second during the collection phase, the aerogel must survive significant heating during the collection period. Computer models have been developed to simulate the atmospheric gas flow field around spacecraft. The models were used to calculate the surface heating of the proposed aerogel collectors. Test assemblies were built of the aerogel collection modules and tested in the arc jet facility at the Ames Research Center Aerodynamic Heating Facility (AHF). The test assemblies were subjected to heating rates of between 1.5 W/cm^2 and 10 W/cm^2 for 120 seconds. Silica aerogel samples that were tested at heating rates of less than 6.5 W/cm^2 exhibited no significant surface melting. The computer models estimated that the aerogel collection modules would experience heating rates below 6.5 W/cm^2 during the Martian aeropass; therefore, it was determined that the aerogel would survive the rigorous conditions of the aeropass and perform as an efficient partice capture medium.

To ensure that the heating of the aerogel is less than 6.5 W/cm^2 the design of the spacecraft is such that the aerogel collection modules are housed in two bands on opposite sides of the aeroshell to prevent extensive heating of the aerogel cells (figure 7). Particles greater in diameter than 2 microns would travel in a straight path into the aerogel, while the atmosphere would flow past the sloped aeroshell and the aerogel. Extensive modeling of the dust content of the Martian atmosphere has indicated that even under poor collection conditions, SCIM would collect more than the targeted 1,000 particles.

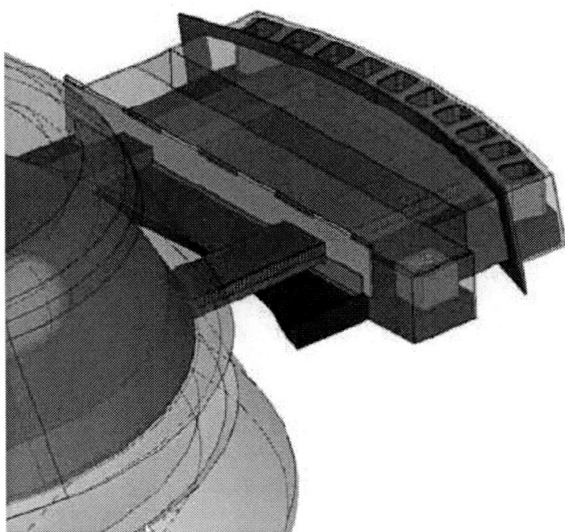

Figure 7. SCIM Dust Collection Experiment (DuCE) design. Aerogel collection modules seen as red squares at end of the retraction tray assembly. Image provided by NASA/Goddard Space Flight Center.

The SCIM Mission was proposed as a 2003 Mars Scout Mission, proceeded through a Phase A study period, but was not selected as the final flight project. Since sample return from Mars is considered a high priority goal for the Mars Exploration Program, a mission such as SCIM should draw further consideration. Now that Stardust has proven to be so highly successful, its technical successes can be applied to other sample capture and return missions. In fact, in the ten years since the delivery of the aerogel for the Stardust mission, the production and testing of non-silicate aerogel has advanced considerably. By using newer, more highly developed types of aerogels for the collectors in future missions, such as SCIM, the science return from these missions can be expanded.

2.4. Non-Silicate Aerogels

The science return from the Stardust Mission could have been enhanced if aerogels other than those made of silicon dioxide could have been used. This is due to the fact that since silicon was a major element in the capture medium and many of the Wild 2 particles fragmented and mixed on the submicron-scale with the aerogel, it was not possible to determine element, e.g., iron, nickel, to silicon ratios. Silicon is used as the standard for the presentation of elemental abundances in geochemistry and cosmochemistry since it is found throughout our solar system. Therefore, iron had to be used as the reference element for the Stardust geochemical analyses and comparison of Wild 2 analyses and existing reference data on other samples was further complicated. Using both silicate and non-silicate aerogels as the capture mediums would mitigate this problem. Unfortunately, at the time of the Stardust aerogel development and production, non-silicate aerogels had not been sufficiently developed to be used for this purpose.

Since the time of the development and production of the aerogel for the Stardust Mission, studies have been done to determine the suitability of non-silica aerogel as efficient hypervelocity particle capture medium [34]. Scanning electron microscope (SEM) images of the networks of various non-silicate aerogels indicate that they are similar in structure to silica aerogels (figure 8). However, the fact that the networks of the various aerogels are similar to that of silica aerogel does not verify that they are efficient hypervelocity capture media. This can only be proven by conducting hypervelocity impact tests.

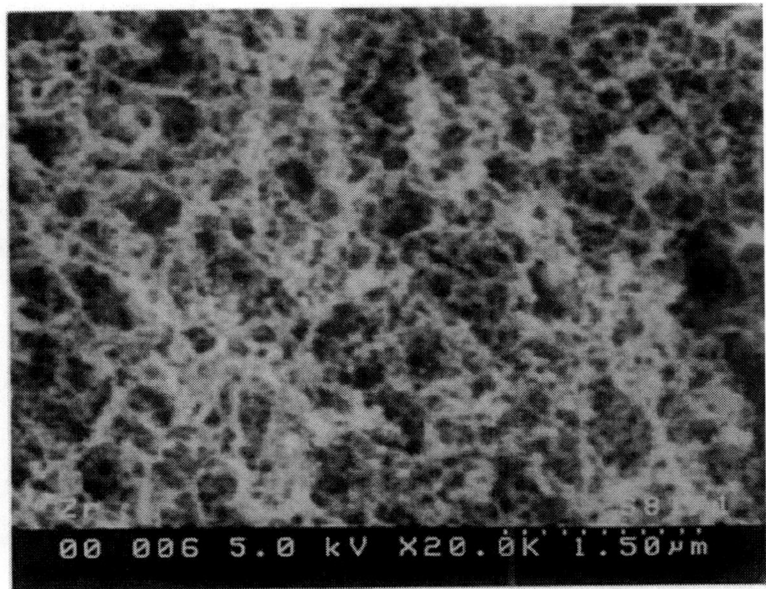

Figure 8. Scanning electron microscope image of zirconia aerogel. Note similarity to silica network shown in figure 3.

Initial hypervelocity impact tests on non-silicate aerogel were done using 20, 50 and 100 micron glass microspheres as the impactors. The microspheres were launched with a light gas gun and captured in different types of non-silicate aerogels. These tests indicated that resorcinol-formaldehyde (RF) (an organic polymeric aerogel), alumina and zirconia aerogels were the most promising candidate materials based on their structural robustness and their ability to efficiently capture particles intact.

Further testing was then conducted using meteoritic fragments as the projectiles. Meteoritic fragments were used because they are a reasonably good analog of the type of extraterrestrial materials that might be expected to be captured using non-silicate aerogels. To be able to compare the results of these tests to the analyses done on the returned Stardust comet samples, the same methods of in situ analysis and particle extraction were used. Synchrotron x-ray microprobe (SXRM) was used to locate and conduct in situ compositional analyses of the captured particles. The captured meteoritic particles were located by their high iron content and then chemical analyses were done of their compositions. The RF and alumina aerogels proved to be the materials best suited for this analytical technique, since the x-ray background scattering of these materials is low due to the fact that carbon and aluminum are low atomic number elements. While the smallest particles that could be analyzed in the zirconia aerogel were 22 microns in diameter, particles as small as 7 microns in diameter could be analyzed in the RF and alumina aerogels. These results are comparable to those obtained for silica aerogels.

The accepted technique for extracting particles captured in the Stardust aerogel is by micromachining the aerogel cells and removing the track and any terminal particles in a "keystone" [35]. A keystone was cut and removed from a sample of RF aerogel that had been impact tested (figure 9). The compositional texture of RF aerogel proved to be quite suitable for this technique. Having demonstrated that RF aerogel can capture hypervelocity particles efficiently, that the captured particles can be analyzed in situ with SXRM and that keystones can be cut and removed effectively, this material should be considered for future hypervelocity sample capture and return missions.

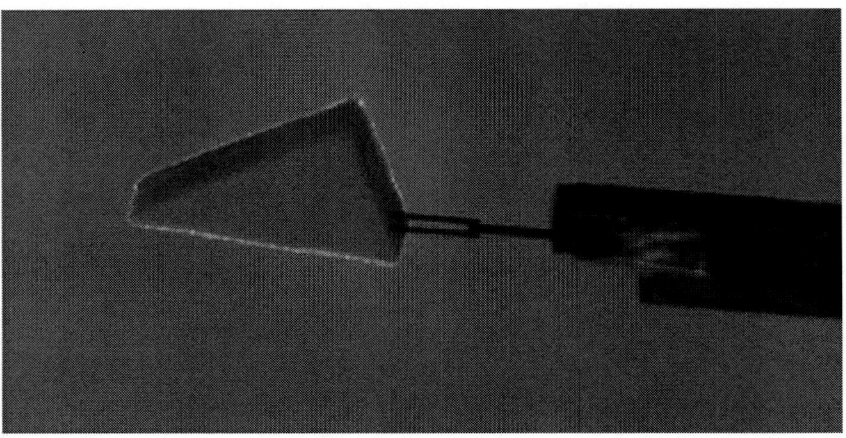

Figure 9. Keystone cut and removed from impact tested RF aerogel. Keystone is approximately 600 X 250 X 75 microns.

2.4. Calorimetric Aerogel

The capture and identification of interstellar grains is extremely important to current astronomical studies. However, due to the small size of these grains, typically less than 1 micron in diameter, identifying them once they have been captured in aerogel is extremely difficult. Recently, work has been done on the development of novel aerogel materials that could greatly facilitate the identification process. Silica and alumina/silica aerogels doped with transition metals, e.g., chromium, titanium, or lanthanide elements, e.g., terbium, gadolinium, have been produced and tested as capture media for interstellar grains [36, 37]. When a hypervelocity particle is captured in a porous material, e.g., low density aerogel, much of the kinetic energy is converted to heat. When this occurs in these new doped aerogels, the heat produced causes the aerogel immediately adjacent to the penetration track to undergo a permanent phase change. By correctly choosing the matrix and dopant materials, the resultant phase changed materials are highly fluorescent. Thus, the capture of a hypervelocity particle results in a distinct fluorescent signature that is easily identified with a low power fluorescent microscope (figure 10). Since the strength of the fluorescent signal is proportional to the kinetic energy of the particle the aerogel becomes a calorimeter. A variety of compositions of these types of aerogels have been produced and impact tested and proposals for their use in a flight detector are being developed.

New specialty aerogels such as these could greatly enhance the use of aerogel in astronomical research. For example, a grid of such aerogel could be flown in low earth orbit on the space station to capture extraterrestrial particles and orbital debris. Since interstellar grains have velocities much greater than anthropogenic orbital debris, the resultant fluorescent signatures they produce could be used to easily identify their impact locations. Since these materials are calorimetric, they can be used in a variety of future sample capture experiments to determine the velocities of different types of interplanetary and interstellar particles.

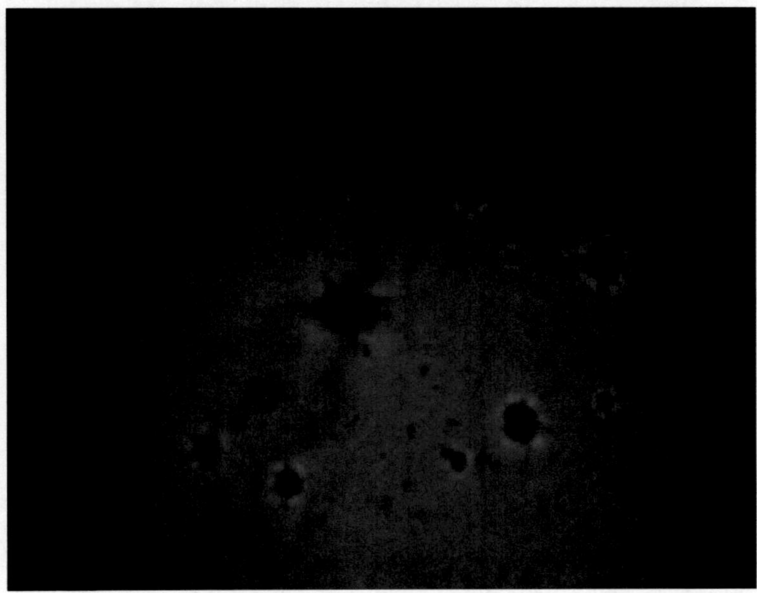

Figure 10. Fluorescent image of impacts of a hypervelocity particles into calorimetric aerogel. Particle velocity was 4.7 km/sec.

3. Thermal Insulation

3.1. 2003 Mars Exploration Rovers

The 2003 Mars Exploration Rovers, Spirit and Opportunity, were launched in May and June of 2003 (figure 11). They landed and deployed successfully on the surface of Mars in January of 2004. Their exploration missions began in February of 2004 and, as of February 2009, they are both still exploring the Martian landscape.

Figure 11. Artist image of a 2003 Mars Exploration Rover on the surface of Mars.

To meet the rigorous thermal requirements of the rovers, aerogel was used as the thermal insulation material in the Warm Electronics Boxes (WEB) of each [38]. The thermal insulation was used, in addition to the eight watts of heat from the Radioisotope Heater Units (RHU's) located near the battery, to help keep the rover electronics at a reatively steady temperature during the approximately 100°C variations in temperature during the changes from Martian day and night. Aerogel was used since it is low weight and yet is an extremely efficient thermal barrier.

Aerogel was first used as the thermal insulation material in the 1996 Mars Pathfinder rover, Sojourner. Since the aerogel used in the Sojourner rover was found to be quite successful, a similar aerogel was used in the 2003 rovers. However, to prevent irradiative thermal transport the aerogel was rendered opaque by adding 0.4% by weight graphite. Due to its high porosity and nano-scale silica network, silica aerogel exhibits extremely low conductive and convective thermal transport. However, since it is largely transparent to infrared radiation, it does not prevent irradiative thermal transport. By making the aerogel opaque, irradiative thermal transport is prevented by absorbing the thermal energy.

Since the exact shapes and sizes of the pieces of aerogel needed were not known when the aerogel production began, relatively large panels were produced and the final shapes and sizes were then cut and shaped from these panels (figure 12). A two-step process was used to produce the aerogel, in which a sol was first made by partially condensing tetraethyl ortho silicate and then forming a gel from the sol by adding acetonitrile, water, and ammonium hydroxide [25]. Since the graphite incorporated into the gel was found to significantly change the drying characteristics of the gels, changes in the processing of the sol were needed. It was found that to produce the panels without significant cracking and shrinking taking place during the supercritical solvent extraction process, the sol had to be refluxed for extended periods of time.

Figure 12. Silica aerogel opacified with graphite was cut and attached to the walls of the WEB prior to rover assembly completion.

The 2003 Mars Exploration Rovers were still traversing the surface of Mars, five years after landing there. In part, the longevity of the rovers is attributed to the successful performance of the aerogel. Since the aerogel has acted as such an efficient thermal insulator, providing power from the batteries to keep the WEB warm has been kept to a minimum.

3.2. Mars Science Laboratory

Silica aerogel, opacified with graphite, will also be used for thermal control on the Mars Science Laboratory (MSL). MSL is scheduled to launch from the Earth in 2011. However, the aerogel will not be used as the thermal insulation material in the WEB, as was done on the 2003 rovers. Since MSL is much larger than the 2003 rovers its thermal and power architecture is very different. The 2003 rovers used solar panels to generate electrical power for on-board instruments and to maintain thermal stability under cold conditions. MSL will generate electrical power from the heat produced by a radioisotope source. For MSL, the aerogel will be used as the thermal insulation material in the heat exchangers that extract heat from the Multi-Mission Radioisotope Thermoelectric Generator (MMRTG). Power generation is accomplished with thermoelectric materials that when placed in a thermal gradient produce electrical power. To maintain the thermal balance of the system waste heat must be rejected. This is done with heat exchangers, which absorb the waste heat on a hot side and reject the heat on a cool side. The rejected heat from the MMRTG will be used to keep MSL's instruments and electronics warm.

Aerogel based thermal insulation is being used to separate the hot and cold sides of the heat exchangers. Graphite doped silica aerogel is being used to fill the open spaces of a resin based honeycomb core. The honeycomb core provides structural support for the panels, while the aerogel acts as the thermal barrier between the hot and cold sides.

3.3. Radioisotope Thermoelectric Generators

The radioisotpoe thermoelectric generators (RTGs) that were launched in the two Voyager probes several decades ago are still functioning today. This demonstrates the reliability of these devices. RTG's are important to the continuing exploration of the outer solar system (beyond Mars) because for some space exploration missions there is not sufficient solar radiation for photovoltaic power generation to be used for energy conversion for long term missions. RTGs use a radioisotope for the source of their thermal energy and employ special materials to convert this thermal energy into electric energy [39]. Thermoelectric power conversion is based on the Seebeck effect, which describes the electromotive force that arises between certain metals, metal alloys or semiconductors when they experience a temperature gradient (figure 13). Since aerogels have such low thermal conductivity values, they are used to channel the thermal energy through the legs and to efficiently maintain the thermal gradient [40].

Using aerogel can simplify the manufacturing process of RTG modules since the aerogel can be formed in place (figure 14). Because the aerogel precursor is a liquid, it can fill the spaces between the legs and around the module. Once the gel has formed the entire assembly can be placed in an supercritical drying system and the gel can be dried under supercritical conditions. It is especially important that the aerogel maintain intimate contact with the legs so that there is no open path for the heat to follow around the legs. It has also been found that the aerogel acts as a sublimation barrier for certain elements within the metal alloys being used, e.g., skutterudites, thus making intimate contact even more important to the continuing working efficiency of the modules.

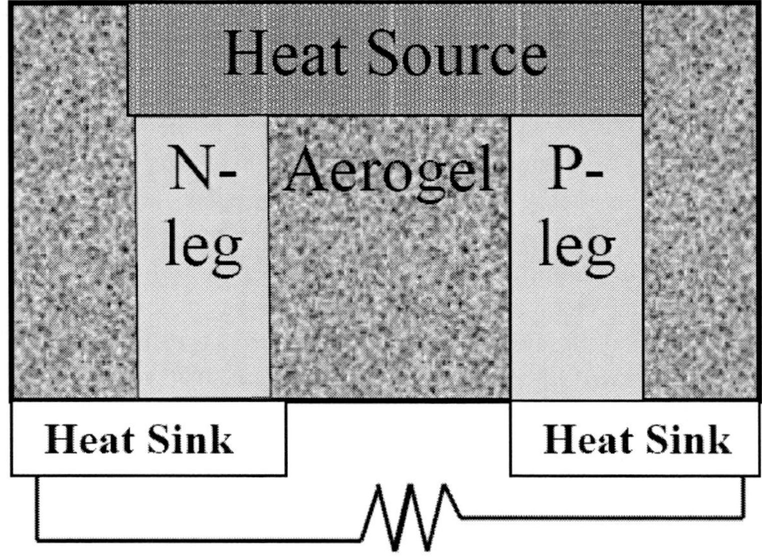

Figure 13. Schematic diagram of a simple radioisotope thermoelectric generator module using aerogel as the thermal insulation material.

Figure 14. Mock-up of a thermoelectric generator module with silica aerogel insulation cast in place.

Certain elements, such as antimony, tend to preferentially sublime out of the skutterudite materials at elevated temperatures. Tests have shown that aerogel significantly inhibits this sublimation process and thus maintains the integrity of the module. Thus far, aerogel has been used to demonstrate that the sublimation process can be decreased by several orders of magnitude.

The use of RTGs is considered to be of significant importance to the future exploration of our solar system. Future missions working where solar power is not practical, e.g., on the dark side of the moon, beyond the orbit of Mars, will require advanced power sources, such as RTGs. Aerogels might make it possible to produce highly efficient and more stable sources of power for these future missions.

3.4. Stirling Radioisotope Generators

Advanced aerogel composites have recently been developed and are being proposed for use in Stirling Radioisotope Generators (SRGs). Unlike RTGs, which generate electrical power from a temperature gradient by non-mechanical means, ASRGs generate electrical power from temperature gradients by the motion of pistons. Temperature gradients drive the motion of the pistons that in turn drive a generator, which produces electrical power. The sources of the thermal power are General Purpose Heat Sources (GPHS), which contain four iridium-clad plutonium 238 fuel pellets. SRGs produce roughly four times the power than RTGs from the same heat source and are therefore being considered for use in the designs of flagship space missions to the outer planets.

Composite aerogels will be used in thermally insulating the radioisotope heat source that will drive the stirling engine pistons [41]. By surrounding the heat source with a very low thermal conductivity material, the primary path for heat to escape the source will be through the stirling pistons. Aerogel composites have been found to have thermal conductivities the same as the best commercially available high temperature thermal insulation materials. However, they have lower bulk densities, which is extremely important when choosing materials for space missions where mass is of crucial importance.

Since the temperature of the heat source will be many hundreds of degrees Celsius, e.g., 700°C to 800°C, pure silica aerogel cannot be used as the thermal insulation material. At these temperatures silica aerogel shrinks and densifies significantly and is rendered a poor insulation material. To overcome this drawback of silica aerogel, a composite has been developed that combines silica aerogel, titania powder, silica powder and a very low density glass felt. The glass felt provides overall mechanical stability to the composite, while the titania and silica powders prevent irradiative thermal transport through the material. The powders also provide some mechanical stability. The aerogel composites are produced by combining a typical silica aerogel precursor, i.e., silica sol, water, catalyst and organic solvent, with a dispersion of titania and silica powders in the same organic solvent. This mixture is then poured into the appropriate molds over the glass fiber felt. The mixture forms a wet gel due to the hydrolysis and condensation of the silica particles in the sol and in so doing traps the oxide powder particles in the silica matrix formed. The wet gel is then dried by taking the solvent to the supercritical state and depressurizing the system at a temperature above the supercritical temperature. Since the initial mixture is a liquid the aerogel composites can be formed into any shape required by the overall system design. By producing many precisely shaped panels, an insulation package can be assembled to house a radioisotope heat source (figure 15).

Aerogel composites have the added working advantage that they shrink significantly at temperatures of over a thousand degrees Celsius. If a stirling piston should stop functioning, the heat from the GPHS will not be carried away from the source through the piston. Under this scenario, the temperature of the radioisotope source can reach well over one thousand degrees Celsius. At these temperatures the cladding on the radioisotope pellets becomes unstable and can disintegrate. To prevent the heat source from reaching these temperatures, the thermal insulation must fail and allow the excess heat to escape. By shrinking at these very elevated temperatures aerogel composites do exactly that. Tests conducted that increased the temperature of the thermal insulation assembly to 1300°C have demonstrated that the insulation panels shrink at these temperatures and allow heat to escape, thus reducing the temperature to acceptable levels. Due to the anisotropic nature in which the glass fiber felt is incorporated in the composites, the shrinkages are anisotropic. The felt is cut into panels and placed in the molds in layers. When the samples are heated to 1300°C for 30 minutes and then 1200°C for two hours, the shrinkage has been observed to be roughly 40% in the directions parallel to the glass fiber felt and roughly 50% in the directions perpendicular to the plane of the fiber felt.

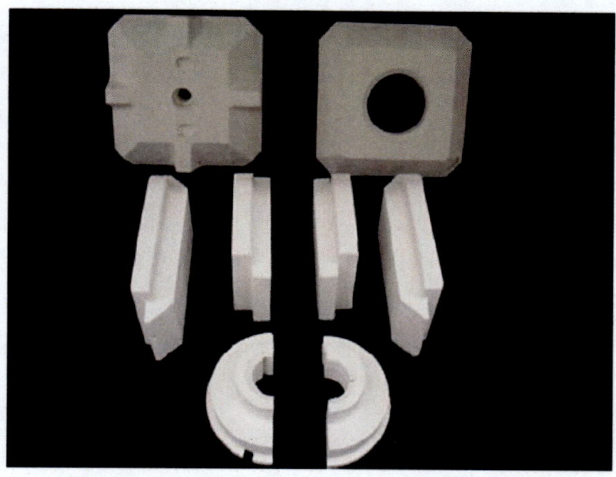

Figure 15. Aerogel composite panels for thermal insulation of a Stirling Radioisotope Generator heat source.

4. Cryogenic Fluid Containment

The STEP (Satellite Test of the Equivalence Principle) Mission proposes to place a spacecraft in earth orbit that is equipped to test the ratio of inertial and gravitational mass [42, 43]. Any difference found in the ratio of inertial and gravitational mass would violate Einstein's Theory of General Relativity. By measuring the constant acceleration of the masses while on orbit, the equivalence of inertial and gravitation mass can be determined using differential accelerometers. Since the spacecraft is to be held on a constant orbit around the Earth and since these conditions provide for a thoroughly isolated environment, the fundamental constant of gravity, G, could also be determined with very high precision. To maintain the stability of the test system and the detectors, they must be cooled to a few degrees Kelvin. To maintain these extremely low temperatures, the test masses and detectors would be located in the core of a Dewar filled with liquid helium (figure 16).

Figure 16. The STEP spacecraft is primarily a large Dewar (banded grey structure) filled with liquid helium which surrounds the detectors. Image provided by Stanford University.

However, the liquid helium required to establish and maintain this cryogenic environment poses a new technical challenge in that any bulk flow of the liquid would introduce inertial changes to the spacecraft that could disturb experimental results. To prevent this from occurring, the liquid helium must be contained in a small-scale network, thus locally confining the liquid helium. The network must have an open cell configuration so that the helium is allowed to flow into and subsequently out of the network. Silica aerogel is composed of a microscopic porous network that would act to prevent any bulk movement of the liquid helium and yet still allows for the helium to fill the network initially and then gradually bleed out from it as the mission proceeded. The helium being bled from the Dewar would be directed through spacecraft thrusters and would be used to maintain the constant orbit required for mission success.

The addition of silica aerogel to the Dewar posed various potential risks. Ground testing was conducted to mitigate these risks. The primary concerns regarding the introduction of the aerogel to the system were that the relatively large pieces of aerogel would not fill with liquid helium or that the aerogel could survive launch vibration environment while filled with liquid helium. To test whether large samples of silica aerogel would fill with liquid helium, an annular test cylinder of aerogel that was 50 mm tall, 150mm outer diameter and 38mm inner bore diameter was placed in a test Dewar and cooled with liquid helium. Visual observations indicated that the aerogel filled with liquid helium and a thermal probe measured a temperature of 2.6^{o}K. The liquid helium was then removed from the aerogel. To test whether silica aerogel would survive a launch vibration environment, a similar piece of aerogel was manually inserted into an aluminum sleeve and attached to the bottom of an experimental shake Dewar. After cooling the Dewar and filling the aerogel with liquid helium the entire assembly was subjected to the launch vibration levels of a Delta II launch vehicle. The aerogel was then removed from the assembly and observed visually. No signs of damage or degradation were observed after the testing.

The basic shape of the aerogel required for the mission is an annular cylinder, since the cylindrical Dewar has a central column running down the middle of it, which contains the instruments. Thus, the initial pieces of aerogel were cast in the form of annular cylinders (figure 17). These pieces were used to test the ingress and egress of the liquid helium through the aerogel network [44]. However, since an annular cylinder of aerogel large enough to fill the Dewar could not be produced, it was necessary to produce smaller, geometrically shaped pieces that could be assembled into the structure needed. The individual blocks were rounded trapezoids of approximately 2.5 liters in volume (figure 18).

Six of these units would be assembled into an annular ring. These six piece rings would then be stacked one on top of another the form an annular cylinder. Secondary rings, with larger radii of curvature, would then surround the first sets of rings to form an assembly that completely filled the Dewar. The individual rings of aerogel blocks would be held together by encircling them with lightweight, flexible bands, slightly compressing the blocks together. This work demonstrated that, despite the fact that very large pieces of aerogel cannot be produced, smaller precisely cast pieces can be produced and assembled into much larger structures.

The STEP Mission was not selected for flight when the final NASA 2003 Small Experiments Mission selection was made. However, work on the technology of the mission continues and with the success of Gravity Probe B, which employs similar technology, it is hoped that STEP will someday be using aerogel to conduct fundamental physics experiments in space.

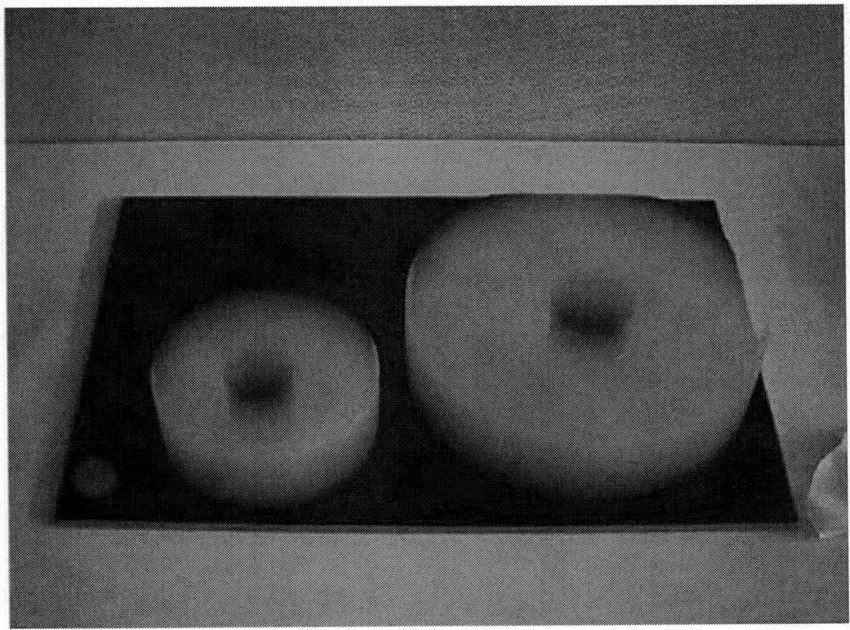

Figure 17. 3cm, 15cm and 25cm diameter annular discs of silica aerogel for testing.

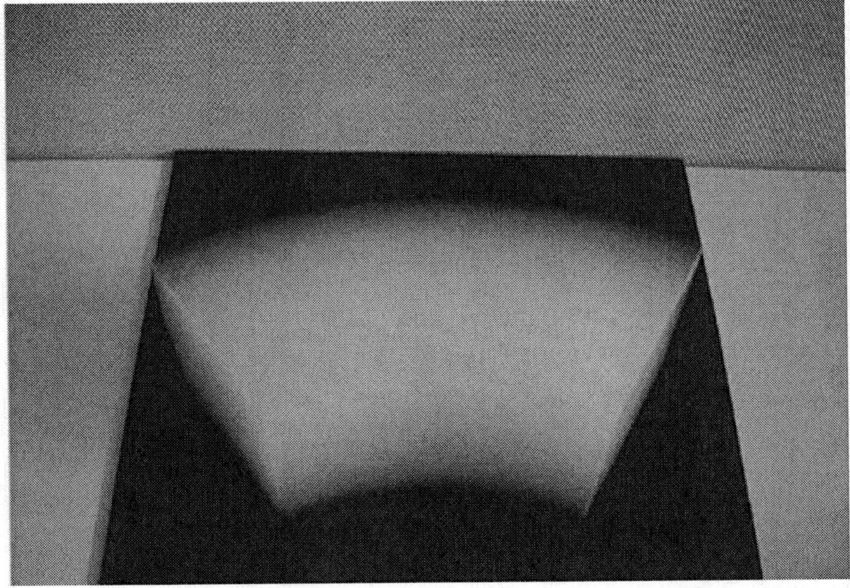

Figure 18. Rounded trapezoidal block of silica aerogel for assembly into larger ring.

5. CONCLUSION

The use of aerogel in space exploration began with its use in the Mars Pathfinder rover, Sojourner, in 1996. Since that time, several other missions have either used aerogel or have been planned around an aerogel based primary instrument. As the types of aerogel available expand beyond silica, e.g., carbon, non-silicate oxides and non-oxides, and the expertise in the production of these aerogels, as well as specially doped aerogels, become more extensive, the use of aerogel in space missions can expand. Many unique properties of aerogel have yet to be utilized in the quest to explore space. The fact that aerogels exhibit several unique physical properties concurrently has not even been explored yet, when considering the potential space science applications of aerogels. One commonly held idea, that the extreme fragility of low density aerogels prevents its use, can now be dispelled. If aerogel can stand up to the rigors of space flight testing and actual launches, then it can be deemed a viable material for almost any challenging environment.

ACKNOWLEDGEMENTS

The research described in this chapter was carried out at the Jet Propulsion Laboratory, California Institute of Technology, under a contract with the National Aeronautics and Space Administration.

REFERENCES

[1] S. S. Kistler, *J. Phys. Chem.,* 34 (1930) 52.
[2] S. S. Kistler, *US Patent.* 2,093,454.
[3] A. C. Pierre and G. M. Pajonk, *Chem. Rev.,* 102 (2002) 4243-4265.
[4] J. Livage and C. Sanchez, *J. Non-Cryst. Solids,* 145, 11 (1992).
[5] L. L. Hench and J. K. West, *Chem. Rev.*, 90(1), 33, 1990.
[6] C. J. Brinker and G. W. Scherer, *Sol-Gel Science: The Physics and Chemistry of Sol-Gel Processing.* (Academic Press, San Diego, 1990).
[7] Y. K. Akimov, *Instr. Exper. Techniques*. 46(3) (2003) 287-299.
[8] G. Herrmann, R. Iden,, M. Mielke, F. Teich, and B. Ziegler, *J. Non-Cryst. Solids,* 186, 380 (1995).
[9] L. W. Hrubesh, *J. Non-Cryst. Solids,* 225, 335 (1998).
[10] M. Schmidt and F. Schwertfeger, *J. Non-Cryst. Solids,* 225, 364 (1998).
[11] D. R. Ulrich, *J. Non-Cryst. Solids,* 121, 465 (1990).
[12] J. Fricke and T. Tillotson, *Thin Solid Films,* 297, 212 (1997).
[13] J. D. MacKenzie, *J. Non-Cryst. Solids,* 100, 162 (1988).
[14] A. C. Pierre and G. M. Pajonk, *Chem. Rev.,* 102, 4243, (2002).
[15] I. Adachi, T. Sumiyoshi, K. Hayashi, N. Iida, R. Enomoto, K. Tsukada, R. Suda, S. Matsumoto, K. Natori, M. Yokoyama, and H. Yokogawa, *Nucl. Instr. Meth. Phys. Res. A,* 355, 390 (1995).

[16] D. Asner, F. Butler, J. Dominick, V. Fadeyev, G. Masek, B. Nemati, P. Skubic and R. Strynowski, *Nucl. Instr. Meth. Phys. Res. A,* 374, 286 (1996).

[17] T. Sumiyoshi, I. Adachi, R. Enomotoi, T. Iijima, R. Suda, M. Yokoyama and H. Yokogawa, *J. Non-Cryst. Solids,* 225, 369, (1998).

[18] M. Ishino, J. Chiba, H. En'yo, H. Funahashi, A. Ichikawa, M. Ieiri, H. Kanda, A. Masaike, S. Mihara, T. Miyashita, T. Murakami, A. Nakamura, M. Naruki, R. Muto, K. Ozawa, H. D. Sato, M.Sekimoto, T. Tabaru, K. H. Tanaka, Y.Yoshimura, S. Yokkaichi, M. Yokoyama and H. Yokgawa, *Nucl. Instr. Meth. Phys. Res. A,* 457, 581 (2001).

[19] R. DeLeo, L. Lagamba, V. Manzari, E. Nappi, T. Scognetti, M. Alemi, H. Becker, R. Forty, I. Adachi, R. Suda, T. Sumiyoshi, A. Leone, R. Perrino, C. Matteuzzi, J. Seguinot, T. Ypsilantis, E. Cisbani, S. Frullani, F. Garibaldi, M. Iodice and G. M. Uriuoli, *Nucl. Instr. Meth. Phys Res. A,* 401, 187 (1997).

[20] P. Tsou, *J. Non-Cryst. Solids.* 186 (1995) 415-427.

[21] P. Tsou, D. E. Brownlee, S. A. Sandford, F. Horz and M. E. Zolensky, *J. Geophys. Res.* 108(E10) (2003) SRD3,1-21.

[22] F. Horz, M. E. Zolensky, R. P. Bernhard, T. H. See and J. L. Warren, *Icarus*, 147(2) (2000) 559-579.

[23] D. B. Brownlee, P. Tsou, K. L. Atkins, C-w., Yen, J. M. Vellinga, S. Price and B. C. Clark, *Acta Astronautic,* 39 (1- 4), 51 (1996).

[24] F. Horz, et al., *Science,* 314, 1716 (2006).

[25] T. M. Tillotson and L. W. Hrubesh, *J. Non-Cryst. Solids,* 145, 44 (1992).

[26] S. M. Jones, *J. Sol-Gel Sci. Technol.* 44 (2007) 255-258.

[27] S. A. Sandford, et al., *Science,* 314, 1720 (2006).

[28] M. E. Zolensky, et al., *Science,* 314, 1735 (2006).

[29] D. Brownlee, et al., *Science.* 314 (2007) 1711-1716.

[30] A. J. Westphal, A. L. Butterworth, C. J. Snead, N. Craig, D. Anderson, S. M. Jones, D. E. Brownlee, R. Farnsworth and M. E. Zolensky, *Lunar Planetary Sci. Con. XXXVI, Lunar Planetary Institue, Houston, TX.* (2005) Abstract 1908.

[31] L. A. Leshin, A. Yen, J. Bomba, B. Clarke, C. Epp, L. Fourney, T. Gamber, C. Graves, J. Hupp, S. Jones, A. J.G. Jurewicz, K. Oakman, J. Rea, M. Richardson, K. Romeo, T. Sharp, B. Sutter, M. Thiemens, J. Thornton, D. Vicker, W. Willcockson asnd M. Zolensky, *Sample Collection for Investigation of Mars (SCIM): An Early Mars Sample return Mission Through the Mars Scout Program,*Lunar Planetary Sci. Conf. XXXIII, Lunar Planetary Institute, Houston, TX. (2002) Abstract

[32] L. A. Leshin, B. C. Clark, L. Forney, S. M. Jones, A. J. G. Jurewicz, R. Greeley, H. Y. McSween, M. Richardson, T. Sharp, M. Thiemens, M. Wadhwa, R. C. Wiens, A. Yen, and M. Zolensky, *Scientific Benefit of a Mars Dust Sample Capture and Earth Return With SCIM* (Abstract), Lunar Planetary Sci. Conf.,XXXIV, Lunar Planetary Institute, Houston, TX, 2003.

[33] A. J. G. Jurewicz, L. Forney, J. Bomba, D. Vicker, S. Jones, A. Yen, B. Clark, T. Gamber, L. A. Leshin, M. Richardson, T. Sharpe, M. Thiemens, J. M. Thornton and M. Zolensky, *Investigating the Use of Aerogel Collectors for the SCIM Martian Dust Sample Return* (Abstract), Lunar Planetary Sci. Conf. XXXIII, Lunar Planetary Institute, Houston, TX., 2002.

[34] S.M. Jones, (in preparation).

[35] Westphal, A. J., et al., (2004) *Meteor. Planet. Sci.* 39(8) 1375-1386.

[36] G. Dominguez, A. Westphal, M. Phillips, and S. Jones, *Astrophysical Journal.* 592 (2003) 631.

[37] G. Dominquez, A. J. Westphal, S. M. Jones, and M. L. F. Phillips, *J. Non-Crystalline Solids.* 350 (2004) 383.

[38] K. S. Novak, c. J. Phillips, G. C. Burir, E. T. Sunada and M. T. Pauken, *Technology Applications International Forum.* 2003, Feb. 2 – 6, 2003.

[39] R. G. Lange and E. F. Mastal, in A Critical Review of Space Nuclear Power and Propulsion 1984-1993, Ed. M. S. El-Genk (American Institute of Physics Press, New York, 1994), p. 1.

[40] J.-A. Pak, J. Sakamoto and S. Jones, *NASA Tech. Briefs,* 30 (9), 26, 2006.

[41] J.-A. Paik, S. M. Jones and J. Sakamoto, Composite Aerogels for High Temperature Thermal Insulation (In preparation).

[42] P. Worden, R. Torii, J. C. Mester and C. W. F. Everitt, *Adv. Space Res.* 25(6) (2000) 1205 – 1208.

[43] J. Mester, R. Torii, P. Worden, N. Lockerbie, S. Vitale and C. W. F. Everitt, *Class. Quantum Grav.* 18 (2001) 2475 – 2486.

[44] S. Wang and R. Torii, S. Vitale, *Classical Quantum Gravity,* 18, 2551 (2001).

In: Space Exploration Research
Editors: J.H. Denis and P.D. Aldridge

ISBN: 978-1-60692-264-4

Chapter 15

Origin and Early Evolution of the Atmospheres and Oceans on the Terrestrial Planets

Lin-gun Liu
Department of Earth Sciences; The University of Hong Kong
Hong Kong, China

Abstract

The atmosphere of the Earth is very different from those of its two neighbors Venus and Mars, and furthermore, the oceans exist only on the Earth. We have attempted to explain these differences by undertaking a study of the origin and early evolution of the atmospheres and oceans of these planets during and right after accretion. It is conceivable that both dehydration and decarbonation of the primordial planetesimals and the surface of the growing planets had to take place when infalling materials impacted the growing planets that exceeded a certain mass. In addition, during accretion, the composition of the growing planets would become practically frozen (except for hydrogen escape) and the growing planets were covered by a "magma ocean" after the planets grew to certain masses. The proto-atmospheres, that formed after the composition became frozen, would envelope the growing planets and consist primarily of CO_2 (This CO_2 proto-atmosphere is preserved on both Venus and Mars today). At the same time, most of the H_2O would remain in the magma ocean due to the high solubility of H_2O in silicate melts at high pressures. The giant impact that formed the Moon is the most viable mechanism that drove out nearly all the H_2O from the partial melting zone of the Earth. This H_2O was then incorporated in Earth's CO_2 proto-atmosphere to form a supercritical H_2O-CO_2 mixture. When the Earth's surface further cooled down below 450~300 °C, the dense supercritical H_2O-CO_2 mixture would have precipitated to form the indigenous ocean. The "hot soda water" might have reacted quickly with feldspars to produce carbonate and clay minerals that precipitated on the bottom of the ocean, and effectively removed all CO_2 from the Earth's proto-atmosphere. The mantle of Venus is not yet completely solidified. Some parts of the original "magma ocean" may still be entrapped inside Venus, forming the so-called partial melting zone. Thus, unlike Earth and Mars, only a small amount of H_2O from the magma ocean was added to the Cytherean proto-atmosphere after accretion, and this H_2O has been lost due to hydrogen escape and the hot surface temperature. Mars should have been completely solidified during a certain stage of its life. Thus, H_2O should have been incorporated in the Martian atmosphere after complete solidification. In view of its small mass, there should not be a large amount of H_2O in the Martian atmosphere, and H_2O and CO_2 should act rather independently during cooling. The H_2O might have condensed to form an ocean when the Martian surface cooled to below 100 °C, but most of the H_2O from either the ocean or atmosphere should have been lost to outer-space throughout Martian life due to its small mass. The mass of the growing planets beyond which the loss of CO_2 to outer-space became negligible during accretion is calculated to be 6.4 x 10^{26} g. The CO_2 contents in the proto-atmosphere of the Earth are calculated to be 5.9 x 10^{23} g, which is equivalent to a partial pressure of 114 bar on the surface of the early Earth. The total CO_2 contents are estimated to be 5.1 x 10^{23} g for Venus, 6.3 x 10^{23} g for Earth, and 3.6 x 10^{22} g for Mars.

INTRODUCTION

Whilst Venus, Earth and Mars are the three consecutive neighboring terrestrial planets in the Solar system, the atmospheric composition of the Earth is very different from those of Venus and Mars which are made up of more than 95% CO_2. Even though the Earth is about 22% more massive than Venus, the atmospheric pressure on Venus is more than 90 times that on the Earth. Furthermore, the oceans are only present on Earth and Venus is known to be deficient in H_2O by a factor of 10^4 to 10^5, relative to Earth (e.g., Yung and DeMore, 1999). Therefore, all models or hypotheses attempting to deal with the origin and evolution of the atmospheres and oceans of the terrestrial planets must account for these facts. Here we present a model which assumes that all terrestrial planets, except Mercury, were accreted from primordial planetesimals having similar chemical compositions (i.e., containing both hydrous and carbonate minerals) and experienced similar physical processes of accretion.

H_2O and CO_2 should be the most abundant and important volatile species in/on the terrestrial planets Venus, Earth and Mars. For as long as the primordial planetesimals that accreted these terrestrial planets contained both hydrous and carbonate minerals, many aspects of the scenario that deals with the amounts and evolution of H_2O and CO_2 to be elucidated here should have taken place during and right after the completion of accretion of these planets. Certain aspects of the role of H_2O during Earth's accretion have been addressed by e.g., Holland (1984), Fukai and Suzuki (1986), and Liu (1987; 1988). Much less attention has been paid to the role of CO_2 during accretion, however. Some accounts of the role of CO_2, together with H_2O, during accretion and in the early life of the Earth were addressed recently by Liu (2004).

It is conceivable that both dehydration and decarbonation reactions had to take place during accretion, if primordial planetesimals containing hydrous and carbonate minerals impacted the planets that grew to a certain mass according to the results of shock-wave experimental studies of these minerals. The possibility that the surface of a growing planet may have been covered entirely by a "magma ocean" has been suggested by Hofmeister (1983), Abe and Matsui (1985) among others. The possible effects of such a magma ocean on the evolution of both H_2O and CO_2 during accretion have also been incorporated in the scenario. It is also quite feasible that an appreciable loss of gases to outer-space occurred but ceased (except hydrogen) when the growing planets reached over a certain mass during accretion.

DEHYDRATION AND DECARBONATION DURING ACCRETION

Lange and Ahrens (1984) and their earlier works have carried out shock-wave experimental studies of serpentine and other hydrous minerals. It has been found that dehydration reactions in hydrous minerals start generally at around 200 kbar, and complete dehydration occurs around 600 kbar. The results from similar studies on calcite ($CaCO_3$) and other carbonate rocks indicate that decarbonation reactions begin at about 100 kbar and are complete near 700 kbar (Boslough et al., 1982; Kotra et al., 1983; and Lange and Ahrens, 1986). Thus, it can be concluded that all H_2O and CO_2 contained in hydrates and carbonates had to be buried inside Venus, Earth and Mars (and probably all planets as well for as long as there were infalling planetesimals containing hydrates and carbonates) during the early stage of accretion before the impact pressure reached over 100~200 kbar. Even after dehydration and decarbonation started, there should still be some amount of both H_2O and CO_2 buried inside Venus, Earth and Mars up until the impact pressure exceeded 600~700 kbar.

Lange and Ahrens (1986) considered that the primordial materials may have a bulk composition as proposed by Ringwood (1979), which contained about 3 wt% H_2O. Fukai and Suzuki (1986) also proposed that the infalling planetesimals contained ~2 wt% H_2O. Liu (1987; 1988), on the other hand, argued that the H_2O contents in the primordial materials proposed by these authors are probably one order of magnitude too high. Assuming that the H_2O content in the infalling materials is 0.33 wt%, Liu (1988) calculated that the total H_2O buried inside Venus, Earth and Mars should be ~1.4×10^{24} g which is equivalent to the mass of today's oceans estimated by Holland (1984).

The CO_2 content in the primordial materials has not been properly appraised. Nevertheless, like H_2O, some appreciable amount of CO_2 must have also been buried inside Venus, Earth and Mars before the impact pressure reached over 700 kbar, though the exact amount cannot be estimated at the moment (see later discussion).

While some amount of both H_2O and CO_2 were buried inside these planets during the early stage of accretion, some amount of both H_2O and CO_2 due to dehydration and decarbonation should have escaped from the gravitational fields of the growing planets and should have been lost to outer-space when the impact pressures exceeded 100~200 kbar during accretion. According to Donahue (1986), however, loss of gas (except hydrogen) should occur so slowly from growing planets greater than about 10^{26} g, that the composition becomes practically frozen except for continuous growth.

This model becomes even more intricate if the presence of a "magma ocean" on the surface of the growing planets is also taken into consideration. Matsui and Abe (1986) suggested that the surface was melted once the growing Earth exceeded ~40% (a radius of ~2550 km) of its final radius. Once the "magma ocean" was formed, accretion then became the impact of solid planetesimals to the liquid "magma ocean". Therefore, the solid-solid impact models of both dehydration and decarbonation described earlier must be modified. This is not only because the results of a solid-liquid impact differ drastically from those of a solid-solid impact, but also because silicate melts are capable of dissolving an appreciable amount of H_2O at high pressures. High-pressure experimental studies indicate that at least 6 wt% H_2O can be dissolved in silicate melts at 3 kbar and the solubility increases with increasing pressure for all silicate melts known (see Liu, 1987). However, no appreciable amounts of CO_2 are known in silicate melts at high pressures (Solubility of CO_2 in some silicate melts is known to be at least two orders of magnitude less than that of H_2O (Y. Liu, personal communication, 2008).). So, the behaviors of dehydration and decarbonation are very different once the surface of the growing planets was covered by "magma ocean". When an impactor hit the magma ocean on a growing planet, instead of producing large quantities of impact-induced dusts and releasing volatiles to the proto-atmosphere as would be expected in a solid-solid impact, an impactor would penetrate into the magma ocean to greater depths. Nearly all the H_2O released during impacting and penetration would be dissolved in the magma ocean and most of the CO_2 released would escape from the magma ocean. In other words, even if the impact pressure exceeded 200 kbar, escape of H_2O from the growing planets and/or loss of H_2O to outer-space would not happen once the surface was covered by magma ocean. Escape and loss of CO_2 to outer-space, on the other hand, would take place once the impact pressure exceeded 100 kbar regardless of the formation of a magma ocean on the surface during accretion, but the released CO_2 from decarbonation due to impact had to be retained and formed the proto-atmosphere when the planets grew to over $\sim 10^{26}$ g.

The conclusion that nearly all the H_2O in the infalling planetesimals would be preserved in the silicate melts when they impacted on the magma ocean was also reached by Holland (1984) in his calculation of the solubility of various gases in the molten Earth. Fukai and Suzuki (1986) also concluded that 'nearly 100% of the accreted water was incorporated in the interior of the Earth, leaving only a very small proportion in the form of atmosphere (p. 9225)'. If these conclusions envisaged for the Earth are true, the same should also be applicable to the accretion of Venus and Mars in the present model which assumes that all the terrestrial planets accreted through similar physical processes and from materials having similar composition.

Thermal Models of the Planets

In order to gain some insights into the behavior of both H_2O and CO_2 in the terrestrial planets during and immediately after accretion, it is essential to know the temperature profile in the interiors of these planets (see figure 1). Although the thermal models are applicable to today's planetary interiors, they can shed light on the thermal properties of these planets right after accretion. Unfortunately, the temperature profile, even for Earth, is one of the most poorly constrained internal physical properties known. Liu (1987) adopted the thermal model of the Earth proposed by Hasebe et al. (1970) which has gained some support from Delany and Helgeson (1978) based on their detailed review. The characteristic feature of the model is that temperatures are relatively low compared with most others (see later discussion) and are even below the water-saturated solidus curve of peridotite determined by Kushiro et al. (1968).

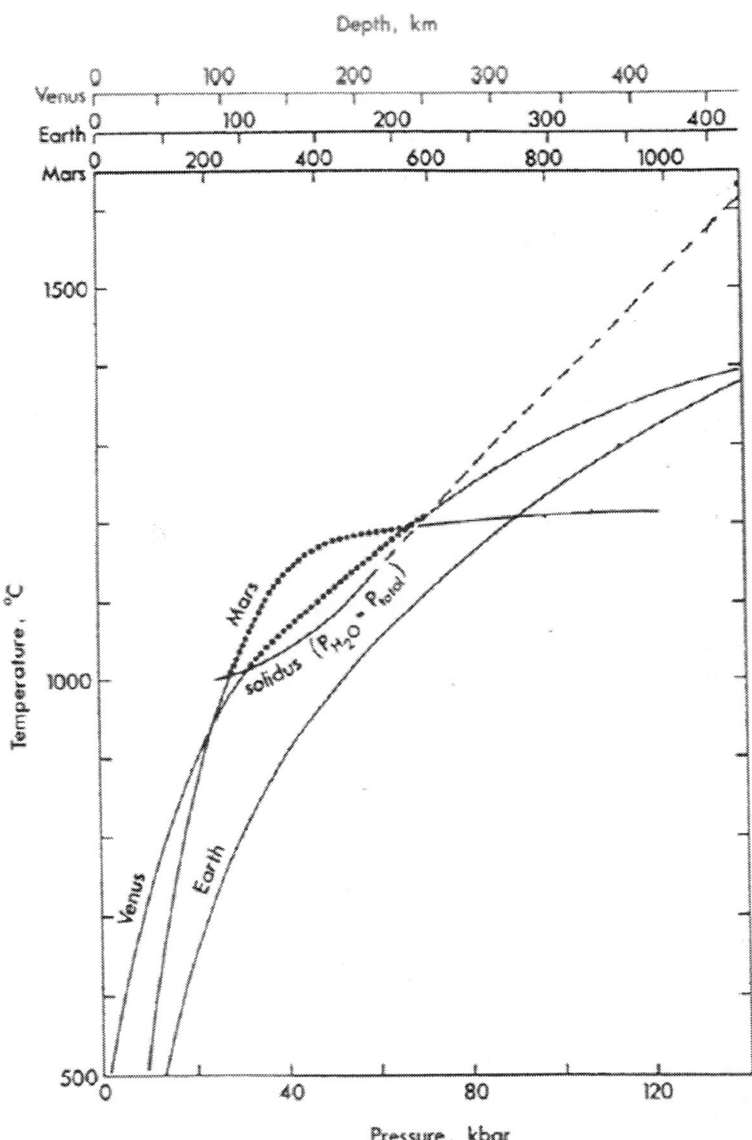

Figure 1. The thermal models for Venus and Mars respectively proposed by Fricker and Reynolds (1968) and Toksoz et al. (1978) are corrected against the Earth model of Hasebe et al. (1970). A water-saturated peridotite solidus curve is also shown for comparison.

There are several thermal models for the interiors of Venus (e.g., MacDonald, 1963; Fricker and Reynolds, 1968; Toksoz et al., 1978) and Mars (e.g., MacDonald, 1963; Toksoz et al., 1978) that are available in the literature. The pros and cons of these models have been discussed in detail by Liu (1988). Fricker and Reynolds (1968) have proposed thermal models for both Earth and Venus. For self-consistency and for a justified comparison, Liu (1988) assumed that the Earth model proposed by Fricker and Reynolds (1968) is identical to that of Hasebe et al. (1970), and then adjusted the internal temperatures of Venus proposed by Fricker and Reynolds (1968) accordingly. The thermal model of Venus proposed by Toksoz et al. (1978) is nearly the same as that proposed by Fricker and Reynolds (1968). So, one can also adjust the thermal model of Mars proposed by Toksoz et al. (1978) accordingly for a self-consistent comparison. It is interesting, after such a comparison, to find that the internal temperatures of Venus are greater than the water-saturated peridotite solidus curve at depths between 120 and 250 km (figure 1). To great surprise, even the internal temperatures of Mars are also greater than the solidus curve at depths between 230 and 570 km. If these thermal models are acceptable, there probably exists a partial melting zone below about 120 km in Venus and below about 230 km in Mars, but such a zone does not exist inside the Earth. The seismologically observed low-velocity zone in the upper mantle of the Earth has been interpreted to be due to the presence of a free water zone by Liu (1987).

H_2O AND CO_2 IN/ON THE TERRESTRIAL PLANETS

1. Venus

In considering the observed surface temperatures on Venus (~460 °C), it is quite conceivable that the internal mantle temperatures of Venus are most likely higher than those of the Earth, exceeding the peridotite solidus curve as shown in figure 1. If a large scale partial melting zone still exists inside Venus, it may suggest that solidification of Venus after accretion has not gone to completion. Thus, the proto-atmosphere of Venus during and immediately after accretion should be composed primarily of CO_2 that was released due to decarbonation after Venus grew to over ~10^{26} g. Some amount of H_2O should also later be added to Cytherean proto-atmosphere once solidification of a magma ocean had started. The amount of H_2O released due to solidification from the top 120 km of the Cytherean mantle throughout the life of Venus is probably rather small, because the large scale internal silicate melt would act like a sponge that absorbs all free H_2O. Thus, most of the Cytherean H_2O is still entrapped inside the partial melting zone between 120 and 250 km, and forms high-pressure hydrous silicates such as phases A, B, C and D below 250 km.

The small amount of H_2O that made its way to the Cytherean atmosphere throughout the life of Venus has probably been lost due to hydrogen escape. This leaves no appreciable amount of H_2O in the atmosphere of Venus today. Venus is known to be deficient in H_2O, relative to Earth, by a factor of 10^4 to 10^5. Although this deficiency may be accounted for by a speculative theory of hydrodynamic escape, Yung and DeMore (1999) favored a theory of planetary evolution, which suggested that most of the water in the magma ocean is still trapped in the thick partial melting zone inside the present Venus as envisaged by Liu (1988), and only CO_2 and a small amount of H_2O were outgassed in the Cytherean atmosphere.

According to the present scenario, all the CO_2 released during accretion when Venus grew to over ~10^{26} g should be retained in its proto-atmosphere and all that CO_2 should also be preserved in today's Cytherean atmosphere because there are no apparent mechanisms that have removed CO_2 from the Cytherean atmosphere. Consequently, there should not be much appreciable amount, or not at all, of both hydrous and carbonate minerals on the surface of Venus. Some amount of CO_2 should still be buried inside Venus as diamond and/or carbonates (most likely to be magnesite in the deep interior). Some CO_2 had to escape and was lost to outer-space when the impact pressure exceeded 100 kbar and before Venus grew to over ~10^{26} g during accretion.

2. Earth

Like Venus, the Earth's proto-atmosphere during and after accretion should also be composed primarily of CO_2. After accretion was completed, a small amount of H_2O was added to Earth's CO_2 proto-atmosphere when the Earth started to solidify. It was not until the late stage of solidification, however, that nearly all H_2O in the top several hundred kilometers of the Earth's mantle was trapped in the partial melting zone as in today's Venus. Then, suddenly, all the trapped H_2O was released to the proto-atmosphere when the Earth was completely solidified. The complete solidification of Earth should have occurred at rather an early stage after accretion because the geochemical evidence revealed from 4.4 Gyr old detrital zircons suggests the early growth of the oceans (Wilde et al., 2001 and Mojzsis et al., 2001). On the other hand, if indeed Venus has not been completely solidified today, complete solidification of the Earth at such an early stage of evolution is probably not a very viable and convincing hypothesis.

Because the formation of the Moon, as envisaged in the giant impact hypothesis (e.g., Benz et al., 1986; 1987), also occurred at a very early stage of the Earth's evolution, the fate of the Earth might be unique among the terrestrial planets. Thus, it is most likely that the Earth, like all other terrestrial planets, was not completely solidified before its capture of the Moon. It is the same giant impact process which formed the Moon that drove out nearly all the H_2O originally dissolved in the partial melting zone in the Earth's interior. The released H_2O was then incorporated in the Earth's CO_2 proto-atmosphere after the giant impact.

Liu (2004) has modeled the proto-atmosphere of the Earth at this stage and has determined that it comprised 560 bar of H_2O (twice the amount in the present oceans) and 100 bar of CO_2 (or some 5.2 x 10^{23} g). The latter figure appears to be reasonable in terms of the amount of CO_2 stored in near surface carbonates and all known organics (~3 x 10^{23} g). Today's Cytherean atmosphere is composed of 96.5% CO_2 that is equivalent to 4.7 x 10^{23} g of CO_2. The fact that the amount of CO_2 stored in the near surface carbonates of the Earth is about 2/3 that of the Cytherean atmosphere is not merely a coincidence. It hints that the amount of CO_2 in the Earth's proto-atmosphere would be at least that in the Cytherean atmosphere, and would most likely be greater (see later detailed estimation) because the Earth is more massive than Venus. The hot ~100 bar CO_2 proto-atmosphere of the Earth was also adopted by Zahnle et al. (2007) in their modeling of the Earth's early evolution after the giant impact.

Because of the relatively high pressure and temperature imposed on the Earth's surface by its proto-atmosphere, the H_2O released from the partial melting zone, after the giant impact, reacted with the existing CO_2 to form a supercritical H_2O-CO_2 mixture in the Earth's proto-atmosphere. When the Earth cooled down further and the surface temperature reached about 300~450 °C, the heavy supercritical H_2O-CO_2 mixture in the proto-atmosphere started to precipitate on the surface to form the indigenous ocean (figure 2). Thus, the indigenous ocean on the Earth was hot (300~450 °C) and was composed of a heavy supercritical H_2O-CO_2 mixture, or the "hot soda water". The hot supercritical H_2O-CO_2 mixture would then react rather quickly with the most abundant mineral, feldspar, to form carbonate and clay minerals at the bottom of the ocean, effectively removing all the CO_2 from the proto-atmosphere (Liu, 2004). Thus, the ocean had to be formed after the giant impact process that formed the Moon, and the atmosphere of the early Earth after the formation of the ocean should be composed primarily of H_2O. The removal of the CO_2 proto-atmosphere from the

Earth appears to require the presence of a large quantity of H_2O which would have formed a supercritical H_2O-CO_2 mixture at high pressure and high temperature (Liu, 2004). The complete removal of CO_2 from the Earth's early atmosphere might have helped the Earth to cool down much faster than the neighboring Venus even if their relative distances from the Sun are taken into consideration. Consequently, the internal temperatures of the Earth are much lower than those of Venus as shown in figure 1.

Like Venus, the remaining CO_2 retained inside the Earth should form diamond and magnesite. Because there is no large scale partial melting zone inside the Earth, the H_2O retained inside the Earth should have formed various high-pressure hydrous silicates as in Venus.

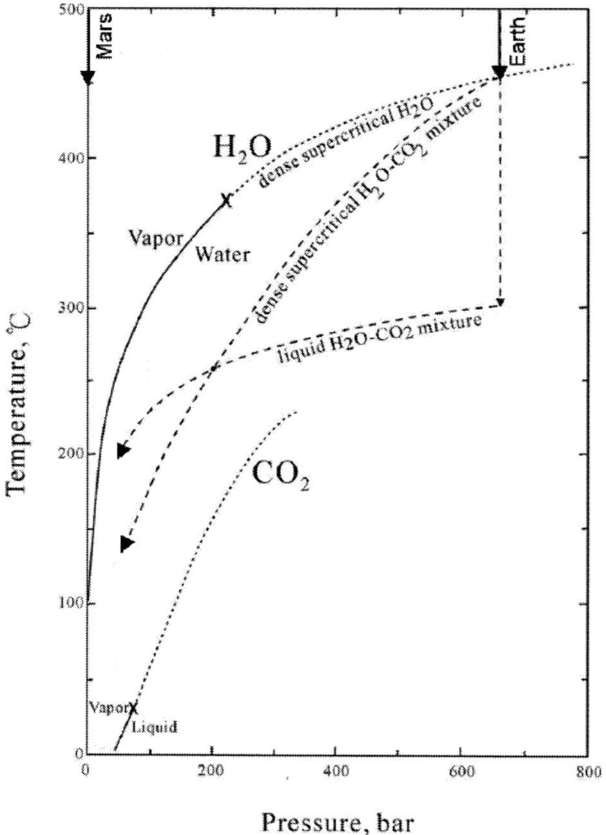

Figure 2. The vaporization temperature for both H_2O and CO_2 as a function of pressure (the solid curves). X marks the critical conditions of these materials. The short-dashed lines indicate the pressure-temperature conditions (beyond the critical points) at which the 'dense' supercritical H_2O and CO_2 exist. The long-dashed lines indicate the possible routes that dense supercritical H_2O-CO_2 mixture and/or liquid H_2O-CO_2 mixture may precipitate on the Earth's surface from its proto-atmosphere after accretion and the H_2O was driven out by the same giant impact process that formed the Moon. Mars total atmospheric pressure of both H_2O and CO_2 right after accretion and solidification is too small to shown in the present scale (indicated by an arrow along the vertical axis). At such a low pressure, H_2O and CO_2 would act rather independently except possibly for forming some HCO_3^-. Thus, H_2O in the Martian proto-atmosphere would firstly condense directly to form an ocean when Mars surface cooled down below 100 °C, leaving behind CO_2 in the proto-atmosphere.

3. Mars

According to the present model, like both Venus and Earth, the Martian proto-atmosphere should also be composed primarily of CO_2 right after accretion. From the size of Mars, it is indicative that the surface of the growing Mars should have also been covered by a magma ocean in which all the H_2O was dissolved during accretion. Based on the thermal models of these planets shown in figure 1, the roles of H_2O and CO_2 during and after accretion of Mars should be very similar to those of Venus, namely that nearly all the H_2O of Mars is still entrapped inside the partial melting zone 230 km below the surface and the Martian atmosphere is composed primarily of CO_2. While the latter appears to be true as observed today, the high mantle temperatures of Mars shown in figure 1 are very doubtful.

Mars is rather small as compared with Venus and Earth (0.13 and 0.11 respectively in mass ratio, see table 1) and is situated at the far side away from the Sun among all the terrestrial planets. If the Earth has already reached the stage of complete solidification as shown in figure 1, it would be reasonable to assume that Mars has also reached complete solidification at a certain stage during its evolution. After complete solidification, H_2O was suddenly released from the partial melting zone and added to the Martian proto-atmosphere if the surface temperature was higher than ~100 °C. Unlike the Earth, however, no supercritical H_2O-CO_2 mixture was formed in Martian proto-atmosphere because the amount of both H_2O and CO_2, and thus the atmospheric pressure, were rather small. Today's Martian atmosphere contains slightly more than 3.72×10^{19} g of CO_2 that is equivalent to 10 mbar atmospheric pressure on the Martian surface. So, in proportion to that on the Earth, there should not be more than a fraction of 1 bar (50~60 mbar) of H_2O in the Martian proto-atmosphere. According to figure 2, H_2O and CO_2 should act like two rather independent volatile species in the Martian proto-atmosphere, except possibly for forming some HCO_3^-. The small amount of H_2O in the Martian proto-atmosphere should have condensed firstly to form a Martian ocean when its surface temperature cooled down below 100 °C and should then have been totally lost to outer-space via hydrogen escape throughout the life of Mars, leaving CO_2 as the main volatile in today's Martian atmosphere (~95%). On the other hand, the released H_2O might have directly formed the Martian ocean without being incorporated in the proto-atmosphere if the surface temperature had already been cooled down below ~100 °C after complete solidification. Whatever the cases, the Martian ocean should be much less old than the Earth's ocean because it had to be derived after complete solidification of the Martian mantle.

Table 1. The mass of different regions and CO_2 quantity for Venus, Earth and Mars right after accretion calculated in this study

Planets	M (total mass)	M1 + M2	M3	CO2 (in air)	CO2 (total)
Venus	4.869×10^{27}	6.415×10^{26}	4.23×10^{27}	4.69×10^{23}	5.1×10^{23}
Earth	5.976×10^{27}	6.415×10^{26}	5.33×10^{27}	5.9×10^{23}	6.3×10^{23}
Mars	6.419×10^{26}	6.415×10^{26}	3.35×10^{23}	3.72×10^{19}	3.6×10^{22}

All units are in g. For M_1, M_2 and M_3, see Fig. 3 and the text. The values of CO_2 (in air) for Venus and Mars, or respectively CO_2 (V) and CO_2 (M), were assumed to be the same as in today's atmospheres, and CO_2 (E) should be the value in the proto-atmosphere of the Earth.

If the Martian ocean has ever been sustained for an appreciable long time during its life, a small amount of the atmospheric CO_2 might have dissolved in the ancient ocean and then reacted with feldspars to form carbonate and clay minerals on the Martian surface. The amount of both hydrous and carbonate minerals (the latter in particular) on Mars should be much less than those on the Earth, however. Like both Venus and Earth, some amount of both H_2O and CO_2 should still remain in the interior of Mars. The retained H_2O should have formed various hydrous silicates such as serpentine and amphibole, and the retained CO_2 should form various carbonates and possibly diamond because the Martian internal pressures are not so high.

ESTIMATE OF THE CO_2 INVENTORY

Following the above scenario, it can be concluded that today's atmospheres on both Venus and Mars should have retained all of the CO_2, in addition to other minor volatiles such as nitrogen, as in their proto-atmospheres, assuming there were few major interactions between the atmosphere and the solid planets throughout the history of Venus and Mars. This is particularly true for the former because there probably never exists an ocean on Venus.

It has been mentioned earlier that the dehydration and decarbonation started when the impact pressures exceeded 100~200 kbar during accretion, and that the impactor and the surface of the growing planets started to melt when they grew over to a radius of ~2550 km. In fact, the exact impacting pressures, the radius and/or the mass of the growing planets estimated by various investigators are not important here, but the concepts postulated are adopted in the present estimate. It is first assumed that the CO_2 of infalling planetesimals is all in the form of carbonate (some may exist as graphite), and that complete decarbonation

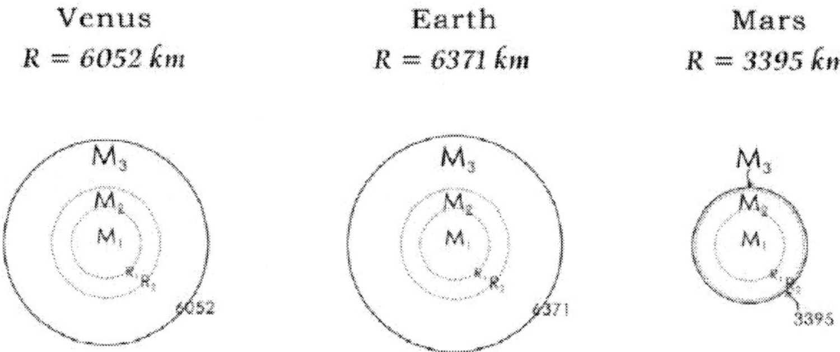

Figure 3. Imaginary internal structure of the terrestrial planets during and right after accretion. R_1 is the radius beyond which complete decarbonation of the infalling planetesimals and the surface of the growing planets occurred, and M_1 is the mass contained within R_1. R_2 is the radius beyond which the composition of the growing planets (except hydrogen) is frozen, and M_2 is the mass retained between R_1 and R_2. CO_2 in the proto-atmospheres was derived from mass M_3 and is also preserved in today's atmospheres of Venus and Mars. R_1 (V) = R_1 (E) = R_1 (M) and R_2 (V) = R_2 (E) = R_2 (M) before M_3 started to grow. Due to a self-contraction, R_1 (M) > R_1 (V) > R_1 (E) and R_2 (M) > R_2 (V) > R_2 (E) after accretion completed. V, E and M inside the parentheses stands for Venus, Earth and Mars, respectively.

due to impacting processes occurred both in infalling planetesimals and at the surface of the growing planets when Venus, Earth and Mars grew to a mass of M_1 (see figure 3). Complete loss of CO_2 into outer-space from these planets occurred when their radius grew from R_1 to R_2 (incipient CO_2 loss should occur even below R_1), and no loss of CO_2 should have occurred when these planets grew beyond R_2. In figure 3, the mass between R_1 and R_2 is marked as M_2, and the gravitational field of (M_1 + M_2) is large enough that loss of CO_2 to outer-space should not occur from the growing planets. Consequently, although both decarbonation and escape of CO_2 still took place when the planets grew beyond R_2, the escaping CO_2 would remain in the impact-induced proto-atmosphere during and after accretion. Notice that the formation of a magma ocean during accretion and the later solidification of magma ocean after accretion had no direct effect on the role of CO_2 in the proto-atmosphere.

The amount of CO_2 in the present atmospheres of Venus and Mars should have been derived from M_3 in figure 3 (the mass that grew beyond R_2). In other words, the mass of CO_2 in the present atmospheres of Venus and Mars divided by their respective masses of M_3 should be equal, i.e.

$$CO_2\,(V)/M_3(V) = CO_2\,(M)/M_3(M) \tag{1}$$

where V and M inside the parentheses represent Venus and Mars, respectively. The other boundary conditions are obvious as follows:

$$M_1(V) + M_2(V) + M_3(V) = M(V) \tag{2}$$

and

$$M_1(M) + M_2(M) + M_3(M) = M(M) \quad , \tag{3}$$

where M(V) and M(M) are respectively the total mass of Venus and Mars.

According to the present model, M_1 and M_2 should all be the same for Venus, Earth and Mars (see figure 3). Thus, Eqs. 2 and 3 reduce to

$$M_3(V) - M_3(M) = M(V) - M(M) \quad . \tag{4}$$

In principle, values of M(V), M(M), CO_2(V) and CO_2(M) should all be available in the literature (e.g., Yung and DeMore, 1999). The value of CO_2(M), however, is complicated by the fact that the atmospheric pressure of Mars fluctuates annually (over a Martian year) by more than 30%. On the basis of the data from Viking Landers 1 and 2, Tillman (1988) showed that the annual atmospheric pressure on the surface of Mars varies from 6.8 mbar to 10.0 mbar. The annual high atmospheric pressure was interpreted to be due to the sublimation of the solid CO_2 polar caps on Mars. It is reasonable to assume that not all of the CO_2 polar caps can be added to the Martian air during the seasons of high atmospheric pressure and that there is some solid CO_2 remaining in the polar caps at any given time on the present Mars. A 10.0-mbar atmosphere on Mars should yield $CO_2(M)$ = 3.72×10^{19} g, and the total CO_2

contents on the surface of Mars should be slightly greater than that. In addition, there is also a possibility that a small amount of atmospheric CO_2 may have been stored as carbonates on the Martian surface via its ancient ocean as elucidated earlier. However, because the value of CO_2(M) is many orders of magnitude smaller than that of M(V), M(M) and CO_2(V) (see table 1), the resolved values of M_3(V) and M_3(M) are not sensitive at all to the exact value of CO_2(M) used in the calculation. On the basis of the data listed in table 1 and Eqs. 1 and 4, one is able to obtain $M_3(V) = 4.23 \times 10^{27}$ g and $M_3(M) = 3.35 \times 10^{23}$ g. This gives

$$M_1 + M_2 = 6.415 \times 10^{26} \text{ g}.$$

It has been mentioned earlier on that Donahue (1986) has proposed a theoretical mass ($> \sim 10^{26}$ g) of a growing planet beyond which loss of gas (except hydrogen) to outer-space should occur so slowly that the composition becomes practically frozen except for the continuous growth. The mass of ($M_1 + M_2$) calculated for CO_2 in the present study matches with Donahue's prediction exactly. In this sense, it is lucky that the mass of Mars is just slightly greater than 6.415×10^{26} g. Otherwise, it is most likely that there would not exist an atmosphere and the possible ancient ocean on Mars, like the Moon (7.35×10^{25} g) and Mercury (3.30×10^{26} g).

Once the value of ($M_1 + M_2$) becomes available, one is able to calculate M_3 and CO_2 in the proto-atmosphere of the Earth, or M_3(E) and CO_2(E) in table 1. Thus, the partial pressure of CO_2 in the proto-atmosphere of the Earth is calculated to be 114 bar, which is close to the value (100 bar) adopted by Liu (2004) and Zahnle et al. (2007). This amount of CO_2 in the Earth's proto-atmosphere was removed to form carbonate and clay minerals via the formation of the oceans in the early history of the Earth (Liu, 2004).

It is obvious that one is still not able to calculate the total inventory of CO_2 for Venus, Earth and Mars in the present model. This is because the value of either M_1 or M_2 is not available in the literature and cannot be directly resolved in the present approach. In the present model, there must be some appreciable amount of CO_2 retained inside M_1 during and/or right after accretion (of course, the position of M_1 was most likely occupied by metallic cores in the very early life of Venus and Earth). During accretion, these deeply buried CO_2-containing materials (or carbonates) could be the sources of carbon that produced magnesite and/or the lower mantle diamonds inside the Earth (e.g., Liu, 1999). However, the approximate amount of deeply buried CO_2 in the present model can be estimated. First, one is able to calculate the CO_2 content in the infalling planetesimals using $CO_2(V)/M_3(V)$ or $CO_2(E)/M_3(E) = 1.1 \times 10^{-4}$. If M_1 approximates M_2, then the deeply buried CO_2 is calculated as 3.6×10^{22} g. This is most likely to be the maximum value, because M_1 does not contain the whole amount of the CO_2 in the infalling planetesimals. This value is more than one order of magnitude less than that in the present Cytherean atmosphere and in the Earth's proto-atmosphere. Thus, this amount of CO_2 should not be too significant to the nature and evolution of both Venus and Earth. However, it should be rather significant to Mars. Thus, the total inventory of CO_2 is estimated to be 5.1×10^{23} g for Venus, 6.3×10^{23} g for Earth, and 3.6×10^{22} g for Mars. The total inventory of CO_2 estimated for the Earth is about twice that ($\sim 3 \times 10^{23}$ g) stored in the near surface carbonates estimated by Ronov and Yaroshevsky

(1976) and Holland (1984). So, the remaining half of the inventory of CO_2 should be stored in the Earth's deep interior as diamond and/or magnesite.

The present model for the Earth should be complicated by the giant impact theory of the formation of the Moon after accretion. It bears at least two consequences that are directly relevant to the present model. First, after the giant impact, the total mass and/or composition of the Earth may not be the same as that right after accretion. Therefore, the boundary conditions such as Eqs. 2 and 3 may not be applicable to the Earth. Second, the amount of both H_2O and CO_2 in the proto-atmosphere and/or the indigenous ocean of the Earth might have been slightly modified after the giant impact. On the other hand, the giant impact hypothesis seems to affect the results relevant to the Earth only, because the CO_2 content in the infalling planetesimals and the mass of ($M_1 + M_2$) were not derived from any of the Earth data in the present calculation.

CONCLUSION

The present model was derived from the following assumptions:

1. Venus, Earth and Mars were formed from similar infalling materials via a similar accretion process, and the composition of primordial materials containing both hydrates and carbonates was assumed to be homogeneous throughout accretion.
2. Complete dehydration and decarbonation of the infalling planetesimals and the surface of the growing planets took place when terrestrial planets grew to a certain mass (e.g. M_1 for carbonates as shown in figure 3).
3. The composition of the growing planets became practically frozen (except for the escape of hydrogen) after the planets grew to a certain mass ($M_1 + M_2$ as shown in figure 3). According to a theoretical calculation by Donahue (1986), this mass should be greater than ~10^{26} g, which was also independently derived in the present study for CO_2 ($M_1 + M_2 = 6.4 \times 10^{26}$ g). During accretion, also, the growing planets were covered by a "magma ocean" after the planets grew to a certain mass. Matsui and Abe (1986) suggested that the surface was melted when the Earth grew over to ~40% (a radius of ~2550 km) of its final radius.
4. There were no major interactions between the CO_2 atmospheres and the solid planets throughout the life of Venus and Mars after accretion. Consequently, the present atmospheres of Venus and Mars have to retain nearly all their CO_2 from their proto-atmospheres.

Both assumptions 1 and 4 are intuitive, but reasonable. Assumptions 2 and 3 are substantiated by shock-wave experiments and theoretical calculations, respectively.

Immediately after the accretion, the proto-atmospheres of Venus, Earth and Mars were all composed primarily of CO_2. Nearly all the CO_2 have been preserved in today's atmospheres of Venus and Mars (the Martian poles always retain a small amount of solid CO_2 and the Martian surface may also possibly exist a small amount of carbonates that were derived from Martian atmosphere via an ancient ocean). The partial pressure of CO_2 in the proto-atmosphere of the Earth is calculated to be 114 bar and all that was removed by the formation of an ocean which was hot (300~450 °C) and composed of supercritical H_2O-CO_2 mixture. Most of the Cytherean H_2O is still entrapped inside the partial melting zone at depths between 120 and 250 km. The H_2O dissolved in the partial melting zone of the Earth was driven out by the same giant impact process that formed the Moon at the very early stage of the Earth's evolution. The amount of the H_2O released from Mars due to its solidification should not be large and should have condensed to form an ocean when the Martian surface cooled down below 100 °C. Most of the H_2O from either the ocean or atmosphere should have been lost to outer-space throughout Martian life due to its small mass. The maximum CO_2 contents are estimated to be 5.1×10^{23} g for Venus, 6.3×10^{23} g for Earth, and 3.6×10^{22} g for Mars.

Acknowledgments

I am indebted to J. C. Aitchison, M. H. Lee and T. P. Mernagh for critical comments on the manuscript.

References

Abe, Y. and Matsui, T. (1985) The formation of an impact-generated H_2O atmosphere and its implication for the early thermal history of the Earth, *J. Geophys. Res.* 90, C545-C559.

Benz, W., Slattery, W. L. and Cameron, A. G. W. (1986) The origin of the Moon and the single-impact hypothesis I, *Icarus.* 66, 515-535.

Benz, W., Slattery, W. L. and Cameron, A. G. W. (1987) The origin of the Moon and the single-impact hypothesis II, *Icarus.* 71, 30-45.

Boslough, M. B., Ahrens, T. J., Vizgirda, J., Becker, R. H. and Epstein, S. (1982) Shock-induced devolatilization of calcite, *Earth Planet. Sci. Lett. 61*, 166-170.

Delany, J. M. and Helgeson, H. C. (1978) Calculation of the thermodynamic consequences of dehydration in subducting oceanic crust to 100 kb and >800 °C, *Amer. J. Sci.* 278, 638-686.

Donahue, T. M. (1986) Fractionation of noble gases by thermal escape from accreting planetesimals, *Icarus.* 66, 195-210.

Fricker, P. E. and Reynolds, R. T. (1968) Development of the atmosphere of Venus, *Icarus.* 9, 221-230.

Fukai, Y. and Suzuki, T. (1986) Iron-water reaction under high pressure and its implication in the evolution of the Earth, *J. Geophys. Res.* 91, 9222-9230.

Hasebe, K., Fujii, N. and Uyeda, S. (1970) Thermal processes under island arcs, *Tectonophysics.* 10, 335-355.

Hofmeister, A. M. (1983) Effect of a Hadean terrestrial magma ocean on crust and mantle evolution, *J. Geophys. Res.* 88, 4963-4983.

Holland, H. D. (1984) *The Chemical Evolution of the Atmosphere and Oceans*, Princeton Univ. Press, Princeton, N. J.

Kotra, R. K., See, T. H., Gibson, E. K., Horz, F., Cintala, M. J. and Schmidt, R. S. (1983) Carbon dioxide loss in experimentally shocked calcite and limestone (abstract), *Lunar Planet. Sci.* 14, 401-402.

Kushiro, I., Syono, Y. and Akimoto, S. (1968) Melting of a peridotite nodule at high pressures and high water pressures, *J. Geophys. Res.* 73, 6023-6029.

Lange, M. A. and Ahrens, T. J. (1984) FeO and H_2O and the homogenous accretion of the Earth, *Earth Planet. Sci. Lett.* 71, 111-119.

Lange, M. A. and Ahrens, T. J. (1986) Shock-induced CO_2 loss from $CaCO_3$; implications for early planetary atmosphere, *Earth Planet. Sci. Lett.* 77, 409-418.

Liu, L. (1987) Effects of H_2O on the phase behaviour of the forsterite-enstatite system at high pressures and temperatures and implications for the Earth, *Phys. Earth Planet. Inter.* 49, 142-167.

Liu, L. (1988) Water in terrestrial planets and the Moon, *Icarus* 74, 98-107.

Liu, L. (1999) Genesis of diamonds in the lower mantle, *Contr. Mineral. Petrol.* 134, 170-173.

Liu, L. (2004) The inception of the oceans and CO_2 atmosphere in the early history of the Earth, *Earth Planet. Sci. Lett.* 227, 179-184.

MacDonald, G. J. F. (1963) The internal constitutions of the inner planets and the Moon, *Space. Sci. Rev.* 2, 473-557.

Matsui, T. and Abe, Y. (1986) Evolution of an impact-induced atmosphere and magma ocean on the accreting Earth, *Nature.* 319, 303-305.

Mojzsis, S. J., Harrison, T. M. and Pidgeon, R. T. (2001) Oxygen-isotope evidence from ancient zircons for liquid water at the Earth's surface 4,300 Myr ago, *Nature.* 409, 178-181.

Ronov A. B. and Yaroshevsky, A. A. (1976) A new model for the chemical structure of the Earth's crust, *Geochem. Int.* 13, 89-121.

Ringwood, A. E. (1979) *Origin of the Earth and Moon*, Springer Pub., N. Y.

Tillman, J. E. (1988) Mars global atmospheric oscillations--annually synchronized, transient normal-mode oscillations and the triggering of global dust storms, *J. Geophys. Res.* 93, 9433-9451.

Toksoz, M. A., Hsui, A. T. and Johnson, D. H. (1978) Thermal evolution of the terrestrial planets, *Moon Planets.* 18, 281-320.

Wilde, S. A., Valley, J. W., Peck, W. H. and Graham, C. M. (2001) Evidence from detrital zircons for the existence of continental crust and oceans on the Earth 4.4 Gyr ago, *Nature* 409, 175-178.

Yung Y. L. and DeMore, W. B. (1999) *Photochemistry of Planetary Atmospheres*, Oxford Univ. Press, N. Y.

Zahnle, K. , Arndt, N., Cockell, C., Halliday, A., Nisbet, E., Selsis, F. and Sleep, N. H. (2007) Emergence of a Habitable Planet, *Space Sci. Rev.* 129, 35-78.

Reviewed by

Dr. J. C. Aitchison and Dr. M. H. Lee, Department of Earth Sciences, The University of Hong Kong, Hong Kong, China

Dr. T. P. Mernagh, Minerals Division, Geosciences Australia, Canberra, A.C.T., 2601, Australia.

Reviewed by

Dr. J. C. Aitchison and Dr. M. H. Lee, Department of Earth Sciences, The University of Hong Kong, Hong Kong, China

Dr. T. P. Mernagh, Minerals Division, Geosciences Australia, Canberra, A.C.T., 2601, Australia.

In: Space Exploration Research
Editors: J.H. Denis and P.D. Aldridge

ISBN: 978-1-60692-264-4

Chapter 16

DYNAMICS AND DISRUPTION MECHANISMS OF ASTEROIDS: ORIGIN OF NEAR-EARTH OBJECTS AND CONSEQUENCES ON THEIR PHYSICAL PROPERTIES

Patrick Michel
Laboratoire Cassiopée, Observatoire de la Côte d'Azur, CNRS,
Université de Nice Sophia-Antipolis, France

Abstract

During their evolutions, the small bodies of our Solar System are affected by several mechanisms which can modify their properties. While dynamical mechanisms are at the origin of the orbital evolutions of asteroids, there are other mechanisms which can change their shape, spin, and their size or result in their disruption. These dynamical and disruption mechanisms have been identified and studied, both by analytical and numerical tools. It was thus found that a large fraction of the population of Near-Earth Objects (NEOs) come from several zones of the asteroid main belt where efficient dynamical mechanisms are present. Actually, most NEOs are now believed to be fragments of large parent bodies that are collisionally disrupted in the main belt. These fragments are more or less directly injected into zones where these dynamical mechanisms occur. Such mechanisms, called resonances, can increase the eccentricity of an asteroid such that it eventually crosses the Earth's orbit. The main disruption mechanisms of small bodies are collisional events, such as the ones at the origin of asteroid families and NEOs, tidal perturbations, and spin-ups. In the particular case of collisional disruption, it has been found that most large fragments generated by the disruption of at least kilometer-size bodies correspond to gravitational aggregates formed by smaller fragments which reaccumulate during their ejection from the parent body. Thus, since NEOs are likely to have undergone this process, a large fraction of them should be gravitational aggregates or rubble-piles. However, the efficiency of disruption mechanisms depends on the strength of the material constituing the small body. As there are several evidence that suggest that most asteroids larger than a few hundred meters in radius are rubble piles, i.e. cohesionless bodies, a fluid model is often used to represent them and to estimate parameters such as their tidal disruption (Roche) distance to a planet. However, even cohesionless, solid bodies do not behave like fluids. In particular, they are subjected to different failure criteria depending on the supposed strength model. This paper reviews the main studies that led us to understand the origin of NEOs and to estimate their orbital and size distribution. Then, several important aspects of material strengths that are believed to be adapted to Solar System small bodies are exposed and the most recent studies of the different disruption mechanisms of asteroids are reviewed. These studies rely on our poor understanding of the complex process of rock failure. While our knowledge of these mechanisms has improved, there is still a large debate on the appropriate strength models for Solar System small bodies and on their most likely physical properties. Current and future space missions to some of these bodies devoted to precise in-situ analysis and sample return will allow us to determine whether those models are appropriate or need to be revised.

1. Introduction

In our Solar System, there are several populations of small bodies, which differ both by their locations and by their physical properties. While most asteroids evolve in the main belt, a region located between the orbits of Mars and Jupiter, some of them cross the orbits of the terrestrial planets (the so-called Near-Earth Objects or NEOs), while another population evolves on the same orbit as Jupiter on the L4 and L5 lagrangian points (the so-called Trojan asteroids). At larger distances from the Sun, a population of small bodies called Kuiper Belt Objects (or KBOs) evolves beyond the orbit of Neptune and is at the origin of the Jupiter-Family Comets (JFCs). Finally the Long Period Comets (LPCs) come temporarily in the Solar System from an external location at a distance of about 50 000 Astronomical Units (AU), called the Oort Cloud.

Small bodies of our Solar System are all affected by planetary gravitational perturbations. Thus, their orbits are more or less stable, depending on their locations. It has thus been found that some populations of small bodies are at the origin of other populations. In particular, as we will explain in Sec. 2, in the inner Solar Sytem, the main fraction of the population of NEOs comes from the asteroid main belt (MB, hereafter). Asteroids evolving in the MB undergo collisions which can lead to a catastrophic disruption. During such an event, fragments are generated, and some of them can be more or less directly injected into an unstable zone. Such zones are numerous in the MB and correspond to mean motion or secular resonances with planets. When a small body is injected in a resonance, its orbital eccentricity increases such that its perihelion distance becomes eventually shorter than 1 AU. Depending on the strength of the resonance, the timescale of transport to Earth-crossing orbit is as short as a few Myr, e.g. [20], [10].

In addition to these changes in their trajectories, small bodies of our Solar System can also undergo dramatic changes in their physical properties due to different mechanisms. Lightcurves obtained by ground-based observations, and images obtained from space missions, all show that these bodies can have very irregular shapes and heavily cratered surfaces, indicating a quite intense collisional activity. Moreover, spin rates give important clues about the composition and strength of these bodies. Then, the presence of binary objects, which represent about 15% of the main belt and NEO populations, indicates that some processes are efficient to form such systems.

So, what are the mechanisms that can modify the physical properties of a small body? We know at least three mechanisms which can be effective enough to change the shape or disrupt a small body, depending on its strength. The first most intuitive one is the collisional process. It is well known that populations of small bodies evolve collisionally. Witnesses of these collisional events are for instance the asteroid families in the main belt. About 20 asteroid families have been identified, and each corresponds to a group of small bodies who share the same orbital and spectral properties. From these characteristics, reproduced recently by numerical simulations (see, e.g. [48] and references therein), it is now established that an asteroid family is the outcome of the disruption of a large asteroid due to an impact with another small asteroid. As a consequence, a large asteroid is transformed into a group of smaller bodies, and the shapes, sizes, spins and orbits of these objects depend on several parameters of the collision, one of them being the strength of the parent body. Our understanding of the origin of NEOs suggests that most of them originate from such a large event before being injected in a resonance that leads them to the near-Earth space, and such a birth event has great implications on their physical properties, as will be shown in this chapter. The second mechanism which can lead to a change of the physical properties of an object is the increase of its spin due to a thermal effect called the YORP effect ([75]). When a rotating body has an irregular shape, it can reemit the light received from the Sun in a different direction than the one from which it received it, and such difference in direction can lead to a change of its spin rate. Although acting on long timescales and only for small enough bodies, such effect has recently been observed ([36]). When an acceleration occurs, depending on the strength and internal properties of the object, the spin can reach the threshold above which the shape of the body is not in equilibrium anymore. Then, either the shape readjusts into another equilibrium or the body breaks up. The third mechanism which can produce similar effects is due to tidal encounters with a massive object (a planet). It is well known that below a certain distance limit, tidal forces can cause the deformation or the disruption of an object. This distance is known as the Roche limit for fluid bodies ([66]), but as we will see, it can take different values and the bodies can take a wide range of shapes at this distance when solid materials (with and without cohesion) are considered.

In the following, we first expose our current understanding of the origin of the population of NEOs (Sec. 2.). As already said, their origin has great implications on their physical properties. Since the efficiency of the different mechanisms described above which can affect the physical properties of small bodies relies at least partially on their assumed material strength, we expose important concepts of material strength (Sec. 3.). Then, we summarizes the most recent study on the spin limits of small bodies and what the observed spins tell us about the strength and internal structure of these objects (Sec. 4.). The latest results on the limit distances of small bodies to a planet as a function of their strength are then presented (Sec. 5.). Several reviews have already been devoted to our current understanding of the collisional disruption of small bodies based on numerical simulations (see., e.g., [47], [48]), therefore this problem is briefly discussed (Sec. 6.), concentrating only on some important issues and open areas. A conclusion is finally exposed (Sec. 7).

2. Origin and Characterization of the Near-Earth Object Population

Strong biases exist against the discovery of objects on some types of orbits, due to the limited portion of observable space from the ground. In particular, most observations are made toward the opposition and close to the ecliptic. Consequently, the observed orbital distribution of Near Earth Objects, which constitutes the main population of impactors in the inner Solar System, is not representative of the real distribution. However, using our knowledge on the origin of NEOs, it is possible to obtain an estimate of their complete orbital and size distributions.

Actually, two methods have been developed to obtain an estimate of the real NEO population from the observed one. The first method relies entirely on the data from observational surveys and tries to apply a correction for observational biases. This approach has been used on the largest detection sample size obtained by the LINEAR project ([72]). However, this direct de-biasing method requires using 1-D projections of absolute magnitude, semi-major axis a, eccentricity e and inclination i in order to beat down the small number statistics problem. As a consequence, it cannot capture any potential dependency of the distribution of an orbital element on another. For instance, if a difference exists between the inclination distribution at low semi-major axes and that at large semi-major axes, this method cannot capture it, which is a severe limitation. The other method uses theoretical orbital dynamical constraints in combination with the detections from observational programs with known biases. This method and its results are summarized in the following paragraphs.

From the results of numerical integrations, it is actually possible to estimate the steady state orbital distribution of the NEOs coming from each of the main source regions of these bodies. These regions of origin have actually been clearly identified in the last decade (see [52] for a complete review on this topic). In this approach, the key assumption is that the NEO population is currently in steady state. Such assumption is supported by the lunar and terrestrial crater records, which suggest that the impact flux has been relatively constant (within a factor of 2) during the last 3.6 Gyr ([71]). To compute the steady state orbital distribution of the NEOs coming from a given source, the following method is employed: first, the dynamical evolutions of a statistically significant number of particles, initially placed in the considered NEO source region(s), are numerically integrated. The particles that enter the NEO region are followed through a network of cells in the (a, e, i)-space during their entire dynamical lifetime. The mean time spent in each cell (called residence time hereafter) is computed. The resultant residence time distribution shows where the bodies from the source statistically spend their time in the NEO region. As it is well known in statistical mechanics, in a steady state scenario, the residence time distribution is equivalent to the relative orbital distribution of the NEOs that originated from the source. In other words, we can expect a larger number of NEOs in regions where the residence time of particles is greater.

This dynamical approach has been used with modern numerical integrations ([8], [10]). It allowed the computation of the steady-state orbital distributions of the NEOs coming from the three main sources: the ν_6 secular resonance, the $3:1$ mean motion resonance with Jupiter, and the Mars Crosser population.

The $3:1$ mean-motion resonance with Jupiter occurs at approximately 2.5 AU from the

Sun (see Fig. 1), and its effect is to increase the orbital eccentricity of main belt asteroids injected into it. When the eccentricity increases, the perihelion distance decreases and eventually, it becomes smaller than the orbital radius of the Earth, so that the originally main belt particle becomes an Earth-crossing one. For a population initially uniformly distributed inside the resonance, the median time required to cross the orbit of the Earth is about 1 Myr, the median lifetime is about 2 Myr, and the typical end-states are a collision with the Sun (70%) and an ejection on hyperbolic orbit (28%) ([20]). The mean time spent in the NEO region is 2.2 Myr ([10]), and the mean collision probability with the Earth, integrated over the lifetime in the Earth-crossing region, is 0.002 ([50]).

The ν_6 secular resonance occurs when the precession frequency of the asteroid's longitude of perihelion is equal to the sixth secular frequency of the planetary system. The latter can be identified with the mean precession frequency of Saturn's longitude of perihelion, but it is also relevant in the secular oscillation of the eccentricity of Jupiter. This secular resonance essentially marks the inner edge of the main belt and its effect depends on the location of the small body in this zone. In the powerful region the resonance causes a regular but large increase of the eccentricity of the asteroids. Earth- (or Venus-) crossing orbits can thus be reached, and in several cases the small bodies collide with the Sun, their perihelion distance becoming smaller than the solar radius. The median time required to become Earth-crosser, starting from a quasi-circular orbit, is about 0.5 Myr. Accounting also for the subsequent evolution in the NEO region, the median lifetime of bodies initially placed in the ν_6 resonance is about 2 Myr, the typical end-states being a collision with the Sun (80% of the cases) and an ejection on hyperbolic orbit (12%) ([20]). The mean time spent in the NEO region is 6.5 Myr ([10]), and the mean collision probability with the Earth, integrated over the lifetime in the Earth-crossing region, is about 0.01 ([50]).

In addition to these powerful resonances, the main belt is densely crossed by hundreds of thin resonances: high order mean motion resonances with Jupiter (where the orbital frequencies are in a ratio of large integer numbers), three-body resonances with Jupiter and Saturn (where an integer combination of the orbital frequencies of the asteroid, Jupiter and Saturn is equal to zero ([54], [57], [58]), and mean motion resonances with Mars ([51]). Many –if not most– main-belt asteroids have a chaotic evolution due to this dense presence of resonances. The magnitude of this chaotic effect remains very weak so that the time required to reach a planet-crossing orbit (Mars-crossing in the inner belt, Jupiter-crossing in the outer belt) ranges from several 10 Myr to some Gyr, depending on the resonance and on the starting eccentricity ([53]). The high rate of diffusion of asteroids from the inner belt can explain the existence of the population of numerous Mars-crossers. The population of Mars-crossers extends up to semi-major axes about 2.8 AU, suggesting that the phenomenon of chaotic diffusion from the main belt extends at least up to this threshold. To reach Earth-crossing orbits, the Mars-crossers random walk in semi-major axis under the effect of Martian encounters until they enter a resonance that is strong enough to further decrease their perihelion distance below 1.3 AU. For the main group of Mars-Crossers (see [42] for the complete characteristics of the Mars-Crosser population), which is called Intermediate Mars-Crossers (IMC) as they are just an intermediate population between the main belt and NEO ones, the median time required to become Earth-crosser is about 60 Myr; about 2 bodies larger than 5 km become NEOs every million year, consistent with the estimated supply rate from the main belt ([51]). The mean time spent in the NEO region

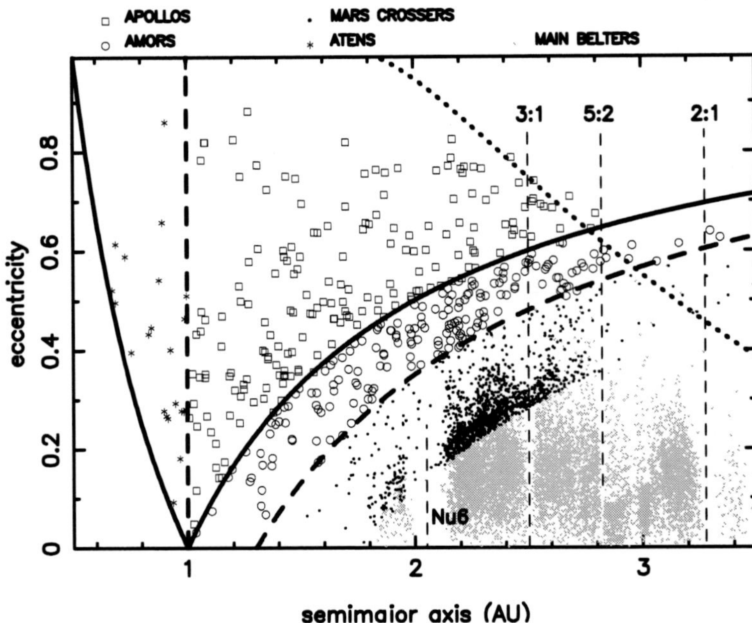

Figure 1. Source regions of the NEO population represented in the (semi-major axis a, eccentricity e) plane. The three NEO groups, namely Atens, Apollos and Amors, and the Mars-Crossers (see [42]) are also indicated, as well as the location of main mean motion resonances with Jupiter (indicated as $3:1$, $5:2$, $2:1$) and the secular resonance ν_6 (see text for details). The full lines represent the Earth-crossing lines (perihelion and aphelion equal to 1 AU), the curved dash line represents the Mars-crossing line at perihelion and the dotted line at the top right of the plot is the Jupiter-crossing line at the aphelion.

is 3.75 Myr ([10]). Fig. 1 shows the positions of the different sources in the (semi-major axis, eccentricity) plane.

The overall NEO orbital distribution has then been constructed as a linear combination of the distributions from these three different sources. The NEO magnitude distribution, assumed to be source-independent, was constructed so its shape could be manipulated using an additional parameter. The resulting NEO orbital-magnitude distribution was then "virtually" observed by applying on it the observational biases associated with the Spacewatch survey ([33]). This allowed us to determine a good combination of the three distributions, which resulted in a model distribution fitting appropriately the orbits and magnitudes of the NEOs discovered or accidentally re-discovered by Spacewatch. To have a better match with the observed population at large semi-major axes, the model has been extended by considering also the steady-state orbital distributions of the NEOs coming from the outer asteroid belt ($a > 2.8$ AU) and from the Jupiter Family Comets ([10]). The resulting best-fit model nicely matches the distribution of the NEOs observed by Spacewatch, without restriction on the semi-major axis (see Fig. 10 in [10]).

An important aspect of this model is that once the values of the parameters of the model are determined by best-fitting observations of a defined survey, the steady-state orbital-magnitude distribution of the entire NEO population is determined. This distribution is thus valid also in those regions of the orbital space not sampled by any survey because of extreme observational biases. This underlines the power of the dynamical approach for debiasing the NEO population.

From this model, the total NEO population is estimated to contain about 1 200 objects with absolute magnitude $H < 18$ and semi-major axis $a < 7.4$ AU. In January 2007, approximately 75% of these objects with $H < 18$ have been observed, as indicated by the number of discoveries provided by the system Horizons of the Jet Propulsion Laboratory. The NEO absolute magnitude distribution is of the type $N(< H) = C \times 100.35 \pm 0.02H$ in the range $13 < H < 22$, implying 29 400 $\pm$ 3 600 NEOs with $H < 22$. Assuming that the albedo distribution is not dependent on H, this magnitude distribution implies a power law cumulative size distribution with exponent -1.75 ± 0.1. This distribution is in perfect agreement with that obtained by directly debiasing the magnitude distribution observed by the NEAT survey ([61]). It is also consistent with the crater size distribution on the Moon (-2 exponent) when scaling laws are applied to derive the corresponding projectiles' size distribution.

The comparison between the debiased orbital-magnitude distribution of the NEOs with $H < 18$ and the observed distributions of discovered objects suggests that most of the undiscovered NEOs have H larger than 16, and semi-major axis in the range 1.5-2.5 AU. With this orbital distribution, and assuming random values for the argument of perihelion and the longitude of node, about 21% of the NEOs turn out to have a Minimal Orbital Intersection Distance (MOID) with the Earth smaller than 0.05 AU. The MOID is defined as the minimal distance between the osculating orbits of two objects. By definition, NEOs with MOID < 0.05 AU are classified as Potentially Hazardous Objects (PHOs), and the accurate orbital determination of these bodies is considered a top priority.

Thus, most of the NEOs come from the main asteroid belt where they are injected into efficient transport mechanism to the near-Earth space. A first mechanism needs to take place to produce this injection. Two such mechanisms, which are not necessarily exclusive, can be at the origin of their injection in a resonance: one is a slow diffusion in semi-major axis due to the Yarkovsky thermal effect, which eventually leads an object originally far from a resonance into it; the second is the catastrophic disruption of a large body in the proximity of a resonance, which generates a great number of fragments, some of them being eventually injected into the resonance as a direct result of their ejection from the parent body or later due to their slow diffusion under the Yarkovsky effect. This second scenario is the most likely one for NEOs. Indeed, the largest NEO is of the order of ten kilometer in size, which is much smaller than the sizes of largest components of the main belt population, which range from 100 to 1000 km for the largest one called (1) Ceres. Then, the shapes of NEOs are generally very irregular and/or ellongated. These characteristics strongly suggest that most NEOs, if not all, are fragments of a larger body that has been disrupted in the main belt, and these fragments have then been injected into a resonance that transported them in the near-Earth space. As we will see (Sec. 6.), a large majority of fragments resulting from such a collisional event consist of gravitational aggregates produced by the gravitational reaccumulation of smaller ones during their ejection from the parent body. Hence, the implication of the origin of NEOs is that most of them are likely to be gravitational aggregates or rubble piles, which is consistent with some observational data, such as the low bulk density ($1.9\ \mathrm{g/cm^3}$) of the 500 m-size NEO (25143) Itokawa vis-

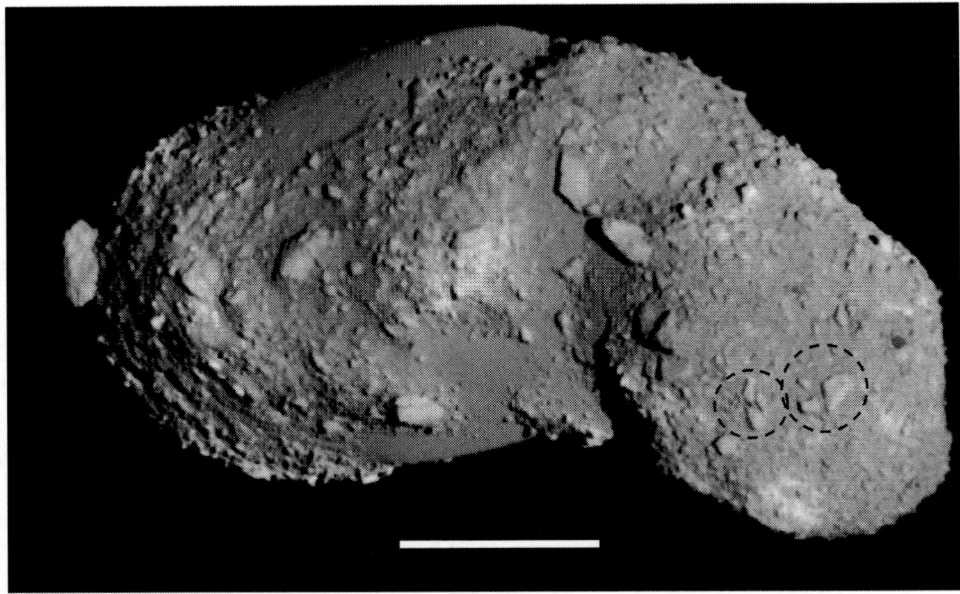

Figure 2. An overview of the asteroid (25143) Itokawa (ISAS/JAXA). The horizontal scale bar represents 100 meters. The boulder at the left end is the largest boulder called Yoshinodai. Boulders in circles look like if they were originated from a single larger boulder which underwent a cratering impact on its surface. (c) ISAS/JAXA

ited by the japanese JAXA/Hayabusa spacescraft in 2005 ([19]). A simple look to images of this asteroid, such as the one shown in Fig. 2 already suggests that this body is highly inhomogeneous with many debris reaccumulated on its surface.

Thus, NEOs and asteroids in general, are subjected to a great numbers of mechanisms during their evolutions. Some of them affect their dynamical properties but other ones can modify their physical ones. These last mechanisms highly depend on the original physical properties of these objects, one of them being their material strength. Because material strength can take different definitions, before describing the mechanisms that can affect the physical properties of these small bodies, we first expose some important concept that can help better understand the definition of strength of a material.

3. Material Strength of Asteroids: Basic Concepts

3.1. What Do We Mean by Strength?

The behavior of a small solid body subjected to different forces is a wide area of research, and the results depend at least partially on the definition used for the strength of the material. In this section, we expose some important concepts which can help better understand the meaning of strength of a small body.

There is no doubt that the term "strength" is often used in imprecise ways. Given the implications of this concept in different areas of study, we believe that it is important to present it in different places to ensure that a same language is spoken among researchers dealing with it. The description presented here is largely inspired from different works by Holsapple, and Holsapple and Michel ([26], [27] and [49]). Materials such as rocks, soils and ices, which are the main constituants of small bodies of our Solar System, are complex and characterized by several kinds of strength.

Generally, the concept of *strength* is a measure of an ability to withstand stress. But stress, as a tensor, can take on many different forms. One of the simplest is a uniaxial tension, for which one principal stress is positive and the two others are zero. The tensile strength, i.e. the value of this stress at which the specimen breaks, is often (mis)used to characterize material strength as a whole. Thus, while it is common to equate "zero tensile strength" to a fluid body, that is not correct. In fact, a body can both be solid and have no tensile strength. For instance, dry sand has no tensile strength. However, contrary to a fluid, dry sand and granular materials in general can withstand considerable shear stress when they are under pressure: that is why we can walk on dry sand but not on water. Here a second kind of strenght comes into play: the shear strength which measures the ability to withstand pure shear. The shear strength in a granular material under confining pressure comes from the fact that the interlocking particles must move apart to slide over one another, and that is resisted by the confining pressure. A third kind of strength, the compressive strength, governs the ability to withstand compressive uniaxial stress. Thus, in general, a material has tensile strength, shear strength at zero pressure (technically the "cohesion") and compressive strength. In geological materials such as soils and rocks, the failure stresses depend strongly on the confining pressure; as a result, these three strength values can be markedly different. Then, contrary to a common assumption, a cohesionless body is simply a solid body whose cohesion (shear strength at zero pressure) is null, but that does not mean it does not have any shear strength under confining pressure. For instance, there is strong evidence that probably most asteroids greater than a few kilometers in diameter are rubble piles or gravitational aggregates (see [64] for a definition of those terms). For such bodies, cohesion can be ignored but they should not be represented by a fluid. In their case, the confining pressure at the origin of the shear strength is played by their self-gravity. Hence, a body can be cohesionless but nevertheless solid.

3.2. Failure Criteria of Solid Bodies

Once the strength of a material has been defined consistently, a failure law is required to determine imminent failure states of stress. Failure criteria for geological materials parallel the yield criteria for metals. Recall that the maximum stress at which a load can be applied without causing any permanent deformation defines the *elastic limit*. It is also called the *yield point*, for it marks the initiation of plastic or irreversible deformation.

There are two common yield criteria for metals: the Tresca criterion and the von Mises criterion. The Tresca criterion states that yield occurs when the maximum shear stress on any plane reaches a critical value. The von Mises criterion replaces the shear stress with the square root of the second invariant J_2 of the deviatoric stress tensor (non-diagonal components of the stress tensor), which depends on all shear stresses. A common assumption of these criteria is that the average stress (pressure), given in terms of principal stresses σ_i as $P = (\sigma_1+\sigma_2+\sigma_3)/3$, has no effect. Then, in a plane in principal stress space perpendicular to the pure pressure axis, the von Mises criterion is a circle, while the Tresca criterion is a hexagon (see, for example, [15] for a good discussion of various yield and failure criteria).

For geological materials, failure can also be described by two such criteria, but with an important addition: because the allowable shear depends on the confining pressure, the size of either of the circle or hexagon depends on the pressure or normal stress. The Mohr-Coulomb criterion (MC) assumes that the maximum shear stress on any plane (τ_{max}) depends linearly on the normal stress (σ_n) on the plane:

$$\tau_{max} = Y - \sigma_n \tan(\phi) \quad (1)$$

where the constant of proportionality is the tangent of the *angle of friction* ϕ, and the constant Y is called the *cohesion* (shear strength at zero pressure); both are material constants determined by experiments. This defines an envelope (limit curve) of maximum shear stress. Thus, compressive stress (negative) increases the allowable shear. In three-dimensional principal stress space, this criterion defines a hexagonal cylinder that increases linearly in size for increasing pressure ([15]). The MC criterion can be considered a Tresca criterion generalized to account for the normal stress effect.

Another criterion called Drucker-Prager (DP) is also common model for geological materials. The DP criterion can be considered a modification of the von Mises criterion, which now assumes that the allowable shear stress depends linearly on the confining pressure. The shear stress magnitude is measured by the square root of the second invariant J_2 of the deviator stress (see Eq. 3). Thus, the DP criterion is similar to models for linear friction and is defined by two constants: one characterizes the "cohesion" (shear strength at zero pressure), and the second characterizes the dependence on the confining pressure and is related to the angle of friction. Those two constants determine the tensile and compressive strengths. When the cohesion is zero, so is the tensile strength, but not the compressive strength. Physically, the pressure dependence is, as already explained, the consequence of the interlocking of the granular particles and not the friction of the surfaces of the particles. In fact, a closely packed mass of uniform rigid *frictionless* spherical particles has an angle of friction about 23°. So the term angle of friction is somewhat a misnomer and angle of *interlocking* would be more correct. However, we will keep using the usual name angle of friction. Figure 3 gives a representation of the DP model. Using the three principal stresses σ_1, σ_2, σ_3 (positive in tension) of a general three-dimensional stress state, the pressure (positive in tension) is given as:

$$P = \frac{1}{3}(\sigma_1 + \sigma_2 + \sigma_3) \quad (2)$$

and the square root of the second invariant of the deviator stress is:

$$\sqrt{J_2} = \frac{1}{\sqrt{6}}\sqrt{[(\sigma_1 - \sigma_2)^2 + (\sigma_2 - \sigma_3)^2 + (\sigma_3 - \sigma_1)^2]} \quad (3)$$

Then, the DP failure criterion is generally given as:

$$\sqrt{J_2} \leq k - 3sP \quad (4)$$

which is illustrated as a straight line with slope $3s$ and intercept k on Fig. 1. Clearly negative pressure (compression) increases the allowable $\sqrt{J_2}$ when s is positive.

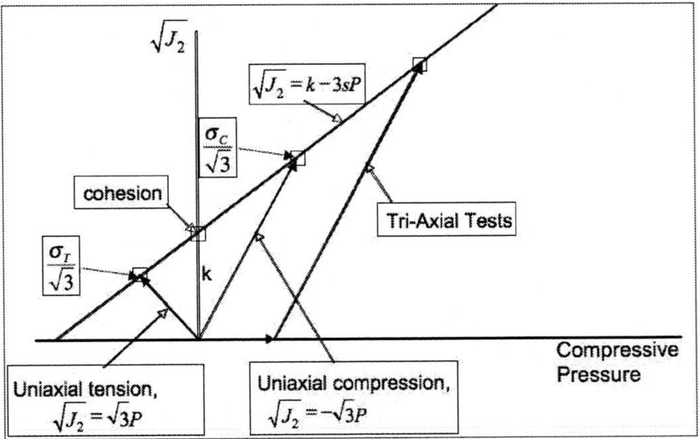

Figure 3. The Drucker-Prager failure criterion. The abscissa is positive in compression. The four small squares indicate the failure condition in, respectively from the left: tension, shear, compression and a confined compression or tri-axial test. The intercept at zero pressure at the value k is called the cohesion and the slope of the line passing through k is $3s$. From [27] and [49].

For the special case of a pure shear stress only, $\sqrt{J_2}$ is just that shear stress and the pressure P is zero. On Fig. 3, the uniaxial tension strength σ_T has $\sqrt{J_2} = 3^{-1/2}\sigma_T$ and $P = \sigma_T/3$. The uniaxial compression strength σ_c has $\sqrt{J_2} = -3^{-1/2}\sigma_C$ and $P = \sigma_C/3$.

The DP criterion can be made to match the MC one in all combinations of pressure plus uniaxial compression if the parameters s and k are related to the cohesion and the angle of friction ϕ used in the MC model. In particular, the slope s is related to the angle of friction ϕ of the MC model by:

$$s = \frac{2\sin\phi}{\sqrt{3}(3 - \sin\phi)}. \tag{5}$$

The intercept k of the DP model is also the shear stress τ for failure in pure shear. Technically the term "cohesion" means this intercept value of shear stress at zero pressure. When the cohesion is zero, so is the tensile strength, and vice versa; both cases would have the envelope starting at the origin on Fig. 3. For instance, an appropriate failure criterion for rubble piles is a criterion for which those two measures are zero, while the plot shows the more general case where they are non-zero.

The tensile stress σ_T for failure is located at the intersection of the tensile line shown on Fig. 3 sloping to the left with the straight line representing the criterion. Its value is given by:

$$\sigma_T = \frac{\sqrt{3}}{\sqrt{3}s + 1}k. \tag{6}$$

Similarly, the compressive failure occurs when the compression line intercepts the failure line, and the resulting compressive stress is given by:

$$\sigma_P = \frac{\sqrt{3}}{\sqrt{3}s + 1}k. \tag{7}$$

Hence, the ratio of compressive strength to tensile strength is given by:

$$\frac{\sigma_C}{\sigma_T} = \frac{\sqrt{3}s + 1}{\sqrt{3}s - 1} \tag{8}$$

which defines the slope of the DP criterion, i.e. the friction coefficient s.

To give an order of idea of the difference between those strengths, a common friction angle for rocks is 45°, so that $s = 0.356$, and the ratio of compressive strength to tensile strength is $-4.22 : 1$. In this case, from Eq. (6), the shear stress for failure k would be 0.93 of σ_T, i.e. the shear and tensile strengths are roughly equal.

This envelope is usually determined experimentally by a test known as a "confined compression" or a "tri-axial" test. In such a test, a uniform confining pressure is applied to a specimen in all directions, and the axial stress is increased in compression until failure occurs when the envelope is reached. This path is illustrated on Fig. 3.

Finally, from a practical point of view, the two criteria (MC and DP) can offer different advantages. The MC model defines a maximum shear stress directly, which is determined by the difference of the maximum and minimum principal stresses. As a consequence, to use this criterion in algebraic manipulations involving general stress states, one must first determine the principal stresses, then which is the largest and which is the smallest. The result is a difference in the algebra of the results in six different regimes, where the three principal stress components take on different orderings. An example of the six possible cases of the ordering of the stress magnitudes is given in Fig. 4 of [23]. Moreover, there are "corners" in the curves shown where the ordering of the principal stresses changes. In contrast, the DP criterion has a single algebraic relation for all stress states. Thus, although the algebraic form of that relation is more complicated than the MC criterion, there is no need to consider the six different possibilities of the ordering of the stress magnitudes. The algebraic complexity of the DP model is of little consequence when an algebraic manipulation program such as " ø¨ a ø¨ ø is used. For instance, Holsapple and Michel used the DP failure criterion to characterize the tidal disruption limit distance of a cohesionless ellipsoid to a planet, noting that the differences between the two models are small (see [26] and Sect. 5.).

3.3. Strength Dependence on Object's Size and Loading Rate

It is generally believed that the effective static cohesive and tensile strengths decrease with increasing body size. The origin of this assumption comes from indications that a distribution of incipient flaws is present within the volume of a solid body. Because larger bodies are more likely to contain larger natural flaws than smaller bodies, the strength is expected to decrease with the body's size. Thus, the use of a strength measure that decreases with size is now a common feature of the studies of disruption of small bodies by impacts (see, e.g. [24], [48]) and has even been demonstrated experimentally ([30]).

A common model for a distribution of incipient flaws in a solid body is a power-law Weibull distribution ([78]). Such a distribution is used in numerical simulations of catastrophic disruption of solid bodies (see Sect. 6.) to generate the initial flaws in the bodies involved. A two-parameter Weibull distribution is usually assumed to describe the network of incipient flaws in any material, expressed as:

$$N(\varepsilon) = k\varepsilon^m \tag{9}$$

where ε is the strain and N is the number density of flaws that activate (i.e. start their propagation) at or below this value of strain. The Weibull parameters m and k are material constants which have been measured for a number of geological and industrial materials, although data are quite scarce for some important rocks (see [39]). In particular the parameter k varies widely between various rock types and the exponent m ranges typically between 6 and 12, but can have a wider range of values. Recently it was measured for the first time for the same basalt material as the one used in some impact experiments, and its value was found to be around 17 in static loadings ([56]).

From the Weibull distribution, it is easily shown that the most probable static strength S of a specimen of volume V (diameter D) decreases with increasing size as:

$$S \propto V^{-1/m} \propto D^{-3/m}. \tag{10}$$

As explained above, such a decrease in strength is simply because larger specimens are more likely to have larger cracks.

The values of the Weibull parameters represent important material properties. Large values of m describe homogeneous rocks with uniform fracture threshold, while small values apply to rocks with widely varying flaw activation thresholds. The existence of incipient flaws within any rock is understood to originate from its cooling history and from crystal lattice imperfections. Due to the initial presence of these flaws, when a finite strain rate $\dot{\varepsilon}$ is applied, a stress increase occurs in time, which is compensated by the propagation of active flaws causing a stress release. Thus a competition takes place between the stress increase due to loading and the stress release due to flaw activation and propagation, until a temporary equilibrium is reached at the time of peak stress. Then, the stress decreases to zero as active flaws propagate rapidly through the rock.

From these explanations, it is obvious that the crack growth velocity c_g is an important parameter since it governs the stress release due to an active flaw. Experiments indicate that it relates to the speed c_l of longitudinal waves in a rock by $c_g \approx 0.4c_l$, and this is usually the value used in numerical simulations of fragmentation. Since cracks propagate at this fixed velocity, under moderate conditions, the weakest flaws (those which activate at lower values of ε) suffice to accomodate the growing stresses. Therefore, the peak stress at failure is low and fragments are relatively large (see Fig. 4). Conversely, more resistant flaws have time to activate at high strain rates. In this case, the peak failure stress is high and fragments are small. This process depends strongly on the assumed value of the crack growth velocity c_g.

In particular, fragment sizes scale with c_g. For instance, a higher velocity would enhance the efficiency at which a crack relieves stress, since stress release is proportional to crack length cubed. As a consequence, fewer flaws would be required to relieve a given increase in stress.

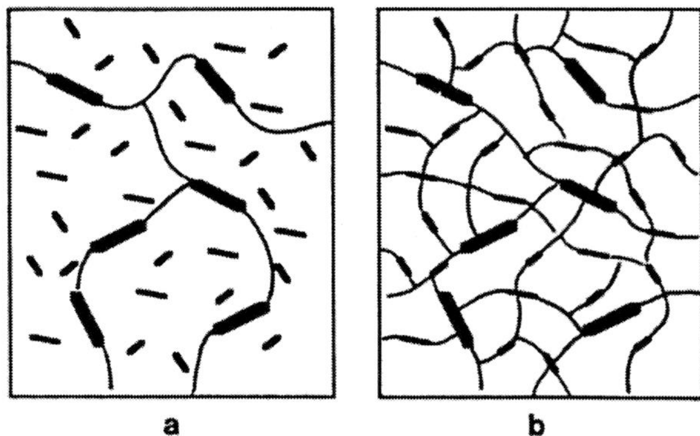

Figure 4. Dynamic fragmentation illustrated using low and high strain rates. Starting with a Weibull distribution of flaws (shown as lines, with the weakest being the longest), cracks propagate by the activation of these flaws according to their threshold strains. Stresses are then relieved and cracks coalesce to form fragments. (a) At low strain rates, the strongest and by far most numerous flaws are never activated because the weak flaws have enough time to relieve the stresses. The resulting fragments are thus large and the peak stress of failure is low; (b) at high strain rate, since the rate of crack growth (c_g) is finite, the weak flaws are not numerous enough to relieve the growing stresses and strong flaws can then be activated, leading to smaller fragment sizes and a higher failure stress. A relationship between dynamic tensile stress and strain rate has been derived from these properties, $\sigma \propto \dot{\varepsilon}^{\gamma}$ ($\gamma \approx 1/4$), and fragment size L and strain rate s are also linked by the relation $L \propto 1/s$. Kindly provided by E. Asphaug ([1])

Thus, the concept of material strength reaches another level of complexity as it can also depend on the dynamical context. From the explanations above, one may conclude that defining the material strength as *the stress at which sudden failure occurs* is not rigorously adequate. Material strength could rather be defined as the stress at which the first flaw begins to fail, thereby initiating an inelastic behavior characterized by irreversible deformation. But in practice, the adopted definition is the peak stress which the rock undergoes prior to failure. It is then not a material constant, since as explained above, the peak stress is a function of the loading history of the rock. This is the reason why a distinction is made between static strength and dynamic strength on the basis of the loading rate. For extremely small loading rates, elastic stresses increase in equilibrium until the onset of catastrophic failure. This occurs at loading rates that are typically smaller than $\approx 10^{-6}$ strains per second. Static tensile strength decreases with increasing size of the rock due to the greater probability of finding a weaker (larger) flaw. At high enough loading rates, stresses can continue to build while catastrophic rupture has begun. In this case, it is more appropriate to speak of dynamic failure. For most rocks, dynamic strain rates are of the order of 1 s^{-1} and decrease with increasing rock size. Therefore, the peak stress that the rock suffers prior to failure is rigorously called the material's dynamic strength at that strain rate. This dynamic strength increases with strain rate and is always greater than the static strength.

All hypervelocity impacts into small targets are in the dynamic regime, but some impacts on large bodies can still be in a regime close to the static one. In this case, part of the event, close to the impact point, can be dynamic, but some important aspects can also be understood in terms of quasi-static failure. Therefore, in all studies that are described in the following sections, apart from the problem of catastrophic disruption, the tensile strength will generally be the static one.

4. Rotation Rates and Implications on the Strength of Small Bodies

The spin rates of small bodies of the Solar System give an important clue about the composition and strength of those bodies. Indeed, the greatest spin that a body can take without being deformed or disrupted depends directly on those properties. For instance, a simple analysis based on the property of zero tensile stresses at the body's poles ([22]) led to the conclusion that an object whose assumed typical mass density is 2.5 g/cm^3 has a period limit of 2.1 h. This value is smaller than the measured rotation period of all large asteroids. Thus, it was suggested that most asteroids must be gravitational aggregates or rubble piles with no tensile strength. This value was later revised ([23], [25]) by a complete stress analysis of spinning, self-graviting, ellipsoidal bodies using the MC failure criterion for cohesionless solid bodies (see Sect. 3.2.). From this analysis, it was concluded that the spin limits are not determined by tensile failure, but by shear failure. Consequently, it was found that the spin limits depend on the angle of friction (see Sect. 3.2.) of the material of the body. A typical minimum period was found to be about 2.6 h, which is higher than the previous estimate ([22]), but still smaller than the rotation period of large asteroids. Numerical experiments of spinning rubble piles (modeled as hard spheres maintained together by gravity) found that such rubble-piles behave in a manner consistent with those last theoretical expectations ([65]).

It was thus tempting to conclude on this basis that most asteroids are rubble piles, because none of them was found to rotate faster than the limit above which a rubble pile would break, in principle. However, recent data for small asteroids indicated that some of them rotate at a rate which is much greater than those previous limits, which suggests that they have some cohesive and tensile strength. This raised the question whether the spin limits observed for large asteroids really rule out that their material is strengthless.

These questions have been addressed in a recent study ([27]), and this section summarizes its principle and main results. It is an extension of previous studies ([23], [25]) and considers spinning bodies with cohesive (and therefore tensile) strength. The fundamental approach consisted of calculating the internal stress state in an ellipsoidal spinning body as a function of size, shape and spin of the body, and comparing the stress state with the limit failure state (provided by the DP model; see Sect. 3.2.) to determine the spin limits at which the failure occurs. It must be noted that rather than solving for the stress state by assuming linear elasticity from some actual prior history of the material, the limit states are solved in the spirit of limit analyses of plasticity theories. Those limit states correspond

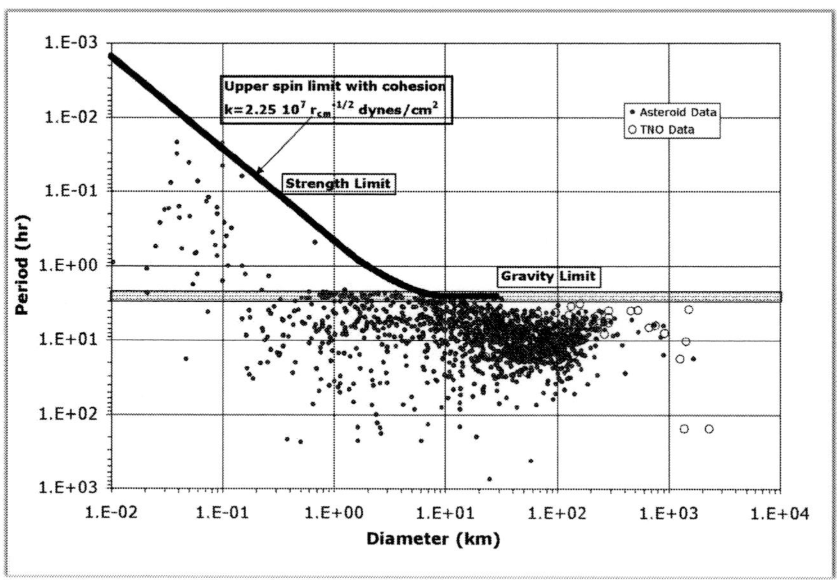

Figure 5. Spin limits and data for small Solar System bodies. The dark sloped line assumes a size dependent strength; it becomes asymptotic to the horizontal red band for materials without cohesion. On the left, the spin limit for cohesive bodies is determined by the cohesive/tensile strength and defines a strength regime. The horizontal asymptote on the right characterizes a gravity regime where tensile/cohesive strength is of no consequence. Gravity regime values do depend on shape and angle of friction, so average values have been assumed to represent them on the plot. The data in the upper left triangular region are the fast spinning near-Earth asteroids. The empty circles for the large diameter bodies on the right are trans-Neptunian objects (from [27] and [29]).

to the situation when the body has reached a final state at which collapse is imminent, and give the final and greatest loadings for failure. The great advantage of this approach is that the prior history of the material is not a factor, and the analyses give a limiting loading state with a greater spin than any found from first failure using a linear elastic approach. This is important because small bodies of our Solar System are formed by complex processes and undergo disruptions, reaccumulations, heating, cratering events, tidal forces and so on. Those processes inevitabily introduce residual stresses that cannot be known, so the assumption of linear elastic behavior from a virgin stress-free state is not reasonable given the history of these bodies. There exist several explicit examples (see [26]) that illustrate the important differences between stress analyses using elasticity theory, and those using the limit state approach. This limit state approach will also be used to determine the tidal disruption limit distances of solid bodies ([26], [29], [49]; see Sect. 5.).

A significant complexity is added once the cohesive terms are included in the strength measure. Indeed, for a cohesionless failure criterion, it was found ([25]) that the limit states had simultaneous failure at all points in the body and the algebra to find that state was rather simple. When the material has cohesion, failure is rather attained only at certain points and planes within the body, and the algebra that determines the stress state with certain failure locations seems insurmontable (even for a symbolic program such as ” ø¨ a ø¨ ø).

Therefore, the approach chosen by Holsapple in his study was to construct the volume-averaged stresses and to use those stresses to compare to the failure criterion ([27]). In other words, spin states are determined which, on average, cause stresses to equal the failure threshold (see Sec. 3.2.). It is clear that for a body having a certain degree of cohesion, more or less critical stress states than the average may exist at some locations within the body, so in reality failure may occur at a lower spin than that found with this approach. But in the particular case of a cohesionless material, failure occurs at all locations in the body at the spin limit. Therefore in this case, both the average and the exact methods give identical results. In other words, in the gravity regime (in which self-gravity dominates over strength, so that cohesion can be ignored) the results are exact, while in the strength regime (in which self-gravity is negligible) they are only approximate.

Once the average stresses have been expressed ([27]), they can be inserted into the DP criterion (Eq. (4)) to find the combinations of spin and shape that satisfy this criterion. Then for a given shape, represented by the aspect ratios of an ellipsoid, the spin at a given limit state can be found. Defining a scaled spin as:

$$\Omega^2 = \frac{\omega^2}{\pi \rho G}, \tag{11}$$

where ω is the actual spin, ρ is the bulk density and G is the gravitational constant, the results can be put into the form of relations for the scaled spin at failure states as functions of the cohesion, the angle of friction ϕ (see Sect. 3.2.), the average body radius $\bar{r} = (abc)^{1/3}$ (where a, b and c are the semi-axes of the ellipsoid representing the object with $a > b > c$) and the two aspect ratios $\alpha = c/a$ and $\beta = b/a$:

$$\Omega = F(k, \phi, \bar{r}, \alpha, \beta). \tag{12}$$

The spin limits are indicated on Fig. 5. Both constant strength and decreasing strength with body size have been considered. For the size-dependent strength (see Sect. 3.3.), the value assumed for the Weibull parameter m (Eq. (9)) is $m = 6$, which seems to fit many data for crack distributions in samples from micron to kilometer sizes ([30]). From Eq. (10), for $m = 6$, the strength of a body decreases with the body's diameter to the power of $-1/2$. Therefore, the strength (cohesion) expresses as:

$$k = \kappa \bar{r}^{-1/2} \tag{13}$$

where the strength coefficient κ is the strength of an object of 1 cm in radius (here k is used for the cohesion and should not be confused with the Weibull parameter having the same name). The spin limits represented on Fig. 5 have been estimated assuming $\kappa = 2.25 \times 10^7$ dynes/cm$^{3/2}$, which is one order of magnitude below the measurements of the tensile strength of Georgia Keystone granite specimens ([30]). The sloped line corresponding to this size-variable strength (Fig. 5) gives an extremely good upper envelope for the current data over the entire range of small body sizes. Measured strengths are still scarce, so it is not necessarily surprizing that the best fit is provided by the line corresponding to a strength value smaller than the measured one.

Note that these estimates of spin limits by Holsapple ([27]) assume that the bodies have ideal ellipsoidal shapes (with aspect ratios of 0.7 and prolate for the representation on the figure) and a fixed friction angle. The reality is obviously more complex, which may explain that the observed spin limits are smaller than the ones assuming the measured tensile strength. In particular, small bodies do not have ideal shapes, their actual friction angle is not known, they may have weaker materials and other non-ideal properties. Moreover, determinations of the observed body's sizes and shapes contain their own error bars which may shift the points on Figure 5 toward higher or lower values. Note finally that the magnitude of the dependence of the estimated spin limit on shape and friction angle is well within a factor two, except in extreme cases. Also, the average method used to make those estimates gives an upper bound to the spin limit in the strength regime.

Thus, a detailed investigation of the spin limits as a function of the mentioned parameters, including strength, has been performed recently by Holsapple ([27]) and the conclusion of this investigation is rich in implications. In particular, it is found that the presence of tensile and cohesive strength for a large body ($>$ 10 km in diameter) makes no difference in its spin limit. Therefore, the observed spin limit (also called the spin barrier) for large bodies cannot be interpreted as evidence of a zero-strength (cohesive /tensile) rubble pile structure. It is the gravity that limits the spin in those cases, even if they have some cohesion. So, large asteroids may be rubble piles, but not on the pure basis of the so-called spin barrier (however, other evidence may point to a rubble pile structure). On the other hand, the strength that allows the higher spins of the smaller and fast spinning km-sized bodies is only of the order of 10-100 kPa, which is very small compared to the strength of small terrestrial rocks. So, these small asteroids do not have to be very strong to be able to rotate so fast. They could be some kind of rubble piles that have accumulated slight bondings between constituents.

As a conclusion, the spin data of small bodies can give us some indications on the internal structure of these bodies. However, based on the current observed spins and our current understanding on the strength of large bodies, they are not sufficient to indicate whether these bodies are rubble-piles or monolithics. If some small asteroids were found to spin above the theoretical limits provided by the described approach ([27]), then this would be a first indication of the potential existence of strong (monolithic) rocky bodies in the Solar System.

5. Tidal Disruption of Small Bodies

When a small body has a close encounter with a planet, depending on the approach distance, it can be subjected to tidal forces that may change its shape or even disrupt it. Such a mechanism was at the origin of the observed disruption of Comet Shoemaker-Levy 9, which was fragmented into 21 pieces during a first passage close to Jupiter, and which collided with the giant planet during its next passage in 1994. Tidal disruption has often been proposed as a formation mechanism for binary asteroids, which represent 15% of the Near-Earth Object population, and as an explanation of crater chains and doublet craters on planetary surfaces. The strength of a small body is an important parameter in the determination of its limit distance to a planet for tidal disruption (or shape readjustment).

The investigation of the limit distance for tidal disruption started in the 19th century, using a fluid to represent the small body. This led to the concept of the Roche limit ([66]), which is still often used nowadays. A great number of studies followed until now, which accounted for important parameters in different manners from one study to the other. The last theoretical studies on this problem provided a continuum theory which allows the determination of this limit distance for cohesionless bodies, and a lower bound of this distance for small bodies with cohesion ([26], [49], [29]). We summarize here the theory and results provided by these studies, and the reader interested in other previous studies on these problems can refer to these last publications in which a history of previous works and their differences are well exposed.

Although the Roche limit ([66]) for tidal disruption of orbiting satellites assumes a fluid body, a length to diameter ratio of exactly $2.07 : 1$, and a particular orientation of the body, it is often used in studies of Solar System satellites and small asteroids or comets encountering a planet. Clearly, these bodies are neither fluid, nor generally are that elongated, so more appropriate theories are needed and have been developed since this first work. Recently, exact analytical results for the distortion and disruption limits of solid spinning ellipsoid bodies subjected to tidal forces, using the DP model with zero cohesion (see Sect. 3.2.) have been presented ([26]). The study used the same approach as the one exposed in Sect. 4. to study the spin limits for solid ellipsoidal bodies. It was followed by a study along the same lines, in which the cohesion of the small bodies was now considered, which, due to the added complexity, could not provide exact but only approximate results ([29]), for the same reasons as the ones already exposed in the case of spin limits (see Sect. 4.). Thus, a static theory was developed that predicts conditions for breakup and the nature of the deformation at the limit state, but it does not track the dynamics of the body as it comes apart. At the end of the section, we will briefly expose results from dynamical investigations.

In the case of cohesionless bodies, as already indicated in previous sections, the strength is essentially characterized by a single parameter associated with an angle of friction ranging from 0° to 90°. The case with a null angle of friction has no shear strength whatsoever, so it corresponds to the case of a fluid or a gas (and the limit distance corresponds to the Roche limit). The case of 90° represents a material that cannot fail in shear, but still has zero tensile strength. Typical dry soils have angles of friction of 30°- 40°. As most satellites are spin-locked with the planet around which they evolve, both the spin-locked case and the zero spin case, a possible case for a passing stray body, have been considered to characterize the limit distance.

The equilibrium problem of an ellipsoid body has been described by Holsapple ([23]). Three stress equilibrium equations must be satisfied by the stresses σ_{ij} in any body in static equilbrium with body forces b_i, which are given as (using repeated index summation convention):

$$\frac{\partial}{\partial x_j}\sigma_{ij} + \rho b_i = 0 \tag{14}$$

where ρ is the bulk density of the body. An (x, y, z) coordinate system aligned with the ordered principal axes of the ellipsoid is used. In the problems here, the body forces arise from mutual gravitational forces, centrifugal forces, and/or tidal forces; they all have the simple linear forms $b_x = k_x x$, $b_y = k_y y$, $b_z = k_z z$. The full expressions of k_x, k_y and k_z are explicitly presented by Holsapple and Michel ([26]). Then, for the limit states sought, the stresses must satisfy the DP failure criterion (see Sect. 3.2.) at all points x, y and z. Also, the surface tensions are zero on the surface points of the ellipsoidal body surface defined by: $\frac{x}{a}^2 + \frac{y}{b}^2 + \frac{z}{c}^2 - 1 = 0$. This problem has been solved ([23]), showing that the distribution of stresses in that limit state just at uniform global failure has the simple form:

$$
\begin{aligned}
\sigma_x &= -\rho k_x a^2[1 - \left(\frac{x}{a}\right)^2 - \left(\frac{y}{b}\right)^2 - \left(\frac{z}{c}\right)^2], \\
\sigma_y &= -\rho k_y b^2[1 - \left(\frac{x}{a}\right)^2 - \left(\frac{y}{b}\right)^2 - \left(\frac{z}{c}\right)^2], \\
\sigma_z &= -\rho k_z c^2[1 - \left(\frac{x}{a}\right)^2 - \left(\frac{y}{b}\right)^2 - \left(\frac{z}{c}\right)^2], \qquad (15)
\end{aligned}
$$

and the shear stresses in this coordinate system are all zero. The body force constants k_x, k_y, k_z depend on the body forces, so those forces must be such that the DP failure criterion is not violated. That condition determines the limit states. Putting the expressions of these components into the DP criterion (Eq. (4)), one can see that the common functional dependence $1 - (\frac{x}{a})^2 - (\frac{y}{b})^2 - (\frac{z}{c})^2$ will cancel out of the Eq. (15). That is because the limit stress state has simultaneous failure at all points. Thus, we can omit that functional dependence and focus on finding the combinations of the leading multipliers of the three terms of Eq. (15) that satisfy the failure criterion. We define the dimensionless distance by:

$$\delta = \left(\frac{\rho}{\rho_p}\right)^{1/3} \frac{d}{R} \qquad (16)$$

where ρ_p is the bulk density of the primary (the planet). Then, as detailed by Holsapple and Michel ([26]), failure will occur when:

$$\frac{1}{6}[(c_x - c_y)^2 + (c_y - c_z)^2 + (c_z - c_x)^2] = s^2[c_x + c_y + c_z]^2 \qquad (17)$$

where, for arbitrary spin and when the long axis points towards the Earth:

$$
\begin{aligned}
c_x &= \left(-A_x + \frac{1}{2}\Omega^2 + \frac{4}{3}\delta^{-3}\right), \\
c_y &= \beta^2\left(-A_y + \frac{1}{2}\Omega^2 - \frac{2}{3}\delta^{-3}\right), \\
c_z &= \alpha^2\left(-A_z - \frac{2}{3}\delta^{-3}\right) \qquad (18)
\end{aligned}
$$

and A_x, A_y and A_z are the components of the self-gravitational potential of a homogeneous ellipsoidal body of uniform mass density ρ in the body coordinate system expressed as: $U = \pi\rho G(A_0 + A_x x^2 + A_y y^2 + A_z z^2)$ (e.g. [14]). Ω is the scaled spin already expressed in Sect. 4.. A similar form when the long axis points along the trajectory at its closest approach can be obtained ([26]). The criterion expressed in these forms can then be used to solve for the dimensionless distances δ at the failure condition as a function of the aspect ratios α and β (which determine the A_x, A_y and A_z), the mass ratio p of the secondary to the primary, and for any value of the constant s related to the angle of friction. The solution always has the dimensionless form:

$$\delta = \frac{d}{R}\left(\frac{\rho}{\rho_p}\right)^{1/3} = F[\alpha, \beta, p, \phi, \Omega] \qquad (19)$$

so that the bulk density ratio only occurs with this cube root. Note that in the spin-locked case, the spin is given by:

$$\omega = \frac{G(M+m)}{d^3} \tag{20}$$

where M and m are the masses of the primary and secondary, respectively. The number of independent variables is then reduced by one when the scaled spin is zero or the spin-locked value, and by another one when $p = 0$, i.e. when the mass of the secondary is negligible compared to that of the primary, which is the case for an asteroid flying by a planet or a small satellite of a giant planet.

Note that this distance limit to the primary corresponds to the distance below which a secondary cannot exist with its assumed shape, because the failure criterion would be violated. However, it does not mean that below this distance, the secondary would disrupt. A flow rule is required to indicate the nature of any readjustment (or disruption). Then, if those changes lead to a new configuration that is within failure at the given distance, a shape change is indicated. Otherwise, if the new shape still violates the failure criterion, a global disruption is indicated. Such analysis has been done by Holsapple and Michel ([26] but goes beyond the scope of this review.

The results provided by this static theory show that a spin-locked spherical body can approach a planet as close as $d/R = 1.23168(\rho_p/\rho)^{1/3}$ if its angle of friction is 90°, and the orbit distance decreases smoothly as the angle of friction increases. For a generic rock value, say $\phi = 30°$, the closest orbit for a spherical satellite is about $d/R = 1.5(\rho_p/\rho)^{1/3}$. The fluid case with zero angle of friction has a distance of infinity, as there is no solution for a spherical body in this case. Other general ellipsoid shapes have then been fully investigated, and it was found that for each combination of aspect ratios α and β, there is a range of permissible orbital distances for any angle of friction $\phi > 0°$. For instance, a prolate body of negligible mass with aspect ratios of 0.8 and $\phi = 20°$ can orbit as close as $d/R = 1.78261(\rho_p/\rho)^{1/3}$, center to center, and if $\phi = 40°$, it can orbit as close as $d/R = 1.15141(\rho_p/\rho)^{1/3}$. Then, an elongated prolate body with $\alpha = 0.4$ and $\phi = 40°$ can orbit as close as $d/R = 1.92929(\rho_p/\rho)^{1/3}$. Figure 6 illustrates the same application to a stray body with zero spin, $p = 0$ and $\alpha = 0.8$.

Thus, all of these are noticeably closer than the fluid Roche limit of $d/R = 2.455(\rho_p/\rho)^{1/3}$, and for a solid, even cohesionless, the shapes are not limited to fluid shapes. This is why it is important to make no confusion between a fluid and a cohesionless bodies. This is particularly important in the context of the study of satellites of giant planets. In fact, many planetary satellites are inside their Roche limit, and do not have the aspect ratios required for a fluid at this limit (see e.g. Table 1 in [26]). The same holds true for stray (zero-spin) cohesionless bodies (the case of a fluid body in this configuration is called the Jean's problem, [32]).

The difference with a fluid body is even more striking when cohesion is added to the strength of a small body. This has been investigated recently by Holsapple and Michel ([29]) who extended their previous analysis of limit distances of cohesionless ellipsoids to the limit distances of ellipsoids with cohesion. Recall that when the cohesion term k is zero in the DP criterion (Eq. (4)), the general form of the stresses satisfying the equilibrium condition, the boundary values and the failure criterion at all points in the body can be found exactly in closed form with a quadratic dependence on the coordinates. Conversely,

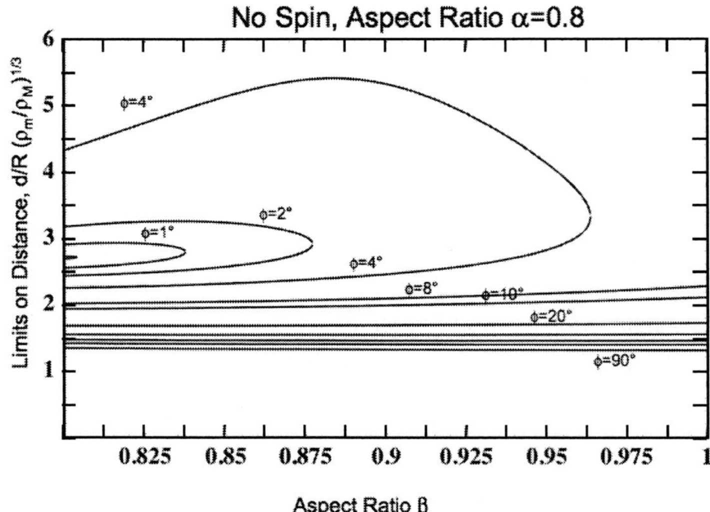

Figure 6. Limits on the possible distances for a stray body with no spin and for various aspect ratios α and β, and different angles of friction Φ. Note that for small angles of friction, the region of permissible distances is a closed curve. This means that the body cannot approach too close, because tidal forces would be too strong, nor range too far as some tidal forces are still necessary for its equilibrium. The limit case of such behavior is the Roche fluid case, for which there is a single point at the center of the closed curves, near the prolate bodies on the left. For typical angles of friction ($20°$ to $40°$), there is no large variation with the aspect ratios. As in the spin-locked situation, the miminum distances for cohesionless stray bodies are all well below the Roche limit, and a much wider range of aspect ratios can exist.

when the cohesion is not zero, there is no solution to this general equilibrium problem with simultaneous failure at all locations. Therefore, an exact answer cannot be determined, but an approximate solution can be found by averaging the stresses across the body, as explained in Sect. 4. for the spin limits. The difference here with the spin limit study is that the tidal terms are added. It has been proven that the loads for which the average stresses are at failure are equal to or greater than the actual limit loads ([28]). Thus, using average stresses to look for failure gives an upper bound on actual limit loads. In other words, the analysis provides a lower bound to the closest approach for collapse. But recall again that in the special cases with zero cohesion the results are exact. In fact, a study along the same lines devoted to tidal encounters of granular bodies has been done by Sharma et al. using the volume average approach and, although the calculations of distance limits were not investigated with the same level of details, it was consistent with the results above using the closed form for the stresses ([68]).

The distance limit for a small cohesive body can be expressed in terms of 7 non-dimensional parameters (one dependent, six independent variables), leading for each orientation of the body to the reduced form:

$$\delta = \frac{d}{R}\left(\frac{\rho}{\rho_p}\right)^{1/3} = G[\alpha, \beta, p, \Omega, \phi, \frac{k}{\rho^2 G r^2}] \quad (21)$$

using the previously defined aspect ratios α, β, mass ratio p, and a scaled cohesion term $k^* = k/(\rho^2 G r^2)$, where r is the body's average radius. This scaled term corresponds to the ratio of the cohesive strength to a gravity pressure. For any given value of these six parameters, numerical results can then easily be obtained, using appropriate programs (see [29] for significant examples and illustrations of important dependencies). In this analysis, the cohesion is also assumed to decrease with the body's size, and the same size dependency as in Sect. 4. is used.

For bodies larger than a few km in diameter, the distance limit is the one provided by the theory for cohesionless bodies, as the cohesion is so small that gravity is dominant (due to the decrease of cohesion with size). Thus, the new approach in which cohesion is included is only relevant for those bodies whose size is below a few kilometers. For them, the distance limit becomes much closer to the primary, and depending on the values of the six parameters (or on the density ratio) the distance can even be smaller than the primary's radius, which means that these small bodies cannot be disrupted by tidal forces. An interesting application relates to planetary satellites, and a preliminary analysis showed that some of them must have non zero friction angles or cohesion to evolve at the observed distance from their primary (see [29]). Thus, the approach developed to determine distance limits can provide some indirect indications on the possible internal structure of real objects, which is a very interesting aspect. Also, it can tell us whether some shape readjustment or disruption will occur during the close approach with the Earth of a real asteroid. For instance, if we assume that the asteroid Apophis (2004 MN4) is a cohesionless body, then the static theory developed by Holsapple and Michel indicates that its close approach with the Earth at 5.6 Earth's radii in 2029 will correspond to its tidal distance limit only if its bulk density is smaller than $0.25\ \text{g/cm}^3$ or unless it is highly elongated, or its angle of friction is less than about $5°$ ([26], [49]).

Thus, according to these recent studies, both cohesionless bodies and cohesive ones with expected properties of geological solids can exist in arbitrary ellipsoidal shapes and much closer to a primary than a fluid body. The main limitation of these studies is that they are not based on a dynamic approach but a static one, which is used to determine the onset of disruption or shape readjustment. Actually, in case of readjustment, the angular moment of inertia would change, so the spin would change. Then the new state would not be in the spin-locked or zero spin configuration as supposed here. The effects of tidal torques on the asteroid's spin have been investigated in details ([69], [70]), and a complete analysis should incorporate these results, thereby accounting for the change in the distance limit as the small body's spin rate changes during a fly-by. Such analysis will have to be performed in the future.

Also, when a small body goes through the distance limit, the resulting motions are affected by how the body changes its shape or breaks up, and by the resulting dynamics. That is a much more complicated problem, which is left for future studies. Some semi-analytical studies have been recently done to address this problem ([68]), but numerical simulations are probably the best tools. In a pioneering numerical study ([62]), the break-up of bodies encountering a planet was considered. Small bodies were modeled as granular aggregates comprised of 247 smooth spheres that interacted with each other only through inelastic collisions and were held together by gravity. Numerical simulations were then used to determine the motion of individual spheres. Various parameters, such as the initial angular velocity vector and encounter variables, were changed to explore the consequences of different close approach configurations. But this study was limited by the low resolution (number of smooth spheres) constrained by the computer performances and numerical codes at that time. More recently, numerical simulations of tidal disruption of rubble piles at higher resolutions using a sophisticated N-body code named *pkdgrav* have been performed ([76]) in order to determine whether tidal disruptions can explain the presence of 15% of binaries in the NEO population. The results show that tidal encounters with a planet (Earth) can form binaries. However, when a small body experiences a tidal encounter, it is likely to experience another such encounter in its close future, so that the binary which is first formed is often eventually disrupted by the next encounter. Thus, although tidal encounters are efficient to form temporary binaries, they cannot be at the origin of the high fraction of binaries observed in the NEO population. Another mechanism has been found recently which explains both the formation and observed properties of small binaries (see [77] for details). Also, these simulations only addressed the problem of tidal approaches of cohesionless bodies, and they will be extended to the case of cohesive bodies in the close future. In principle, the limit distances should be similar to the ones provided by the static approach (for some values of the angle of friction and cohesion), but a dynamical approach will also allow the determination of the behavior of the small body and its seperate pieces once it breaks up. The gravitational evolution of fragments from a disrupted body has already been studied but only in the specific case of the collisional disruption of a small body. This is the last desmechanism of disruption which is briefly summarized in the following section.

6. Collisional Disruption of Small Bodies

In this section, we just summarize the most important concepts and issues concerning the catastrophic disruption of an asteroid due to a collision. The reader interested in more details can refer to a few articles which expose the main recent results concerning this process (see, e.g., [47], [48]).

Collisional processes occur frequently between the small bodies of our Solar System. The best witnesses of those events are *asteroid families* in the main belt. Each family originates from the break-up of a large body, which is now represented by a group of asteroids sharing the same spectral and orbital properties (see, e.g., [31], [81]). Fig. 7 shows the distribution of main belt asteroids in a semimajor axis versus sine of inclination diagram where groups corresponding to asteroid families can be easily identified, and three major ones are indicated.

As in the case of spin limits (Sect. 4.), two regimes of collisional disruption have been defined: the *strength* regime, in which the fragmentation of the body is the only process determining the outcome (this is the case of impact experiments in laboratory on cm-size targets), and the *gravity* regime, in which not only the fragmentation but also the gravitational interactions of fragments have an influence on their final size and velocity distribu-

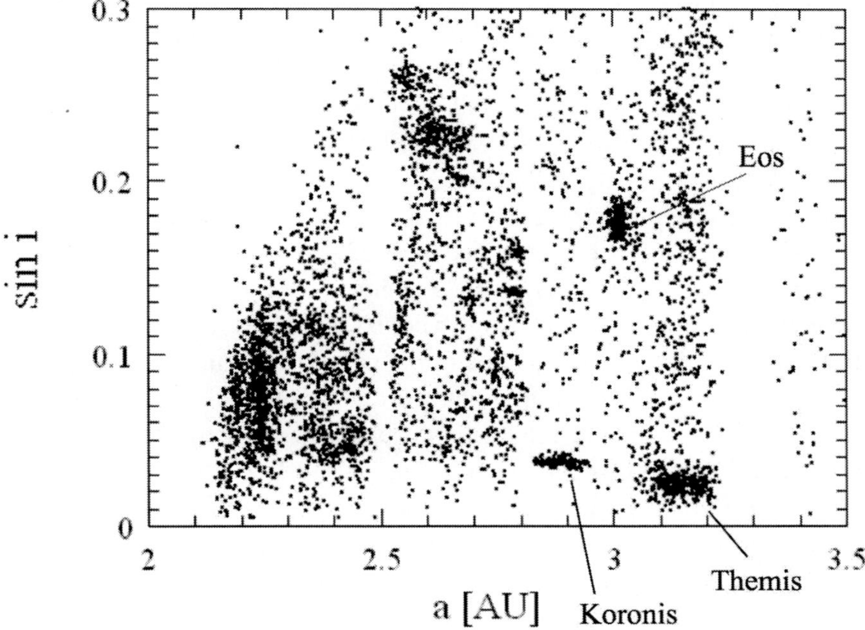

Figure 7. The distribution of 5335 numbered asteroids with respect to proper semimajor axis a and inclination i (see ([41]); data from PDS Small Body Node). The inner edge is determined by the ν_6 secular resonance. The gap at 2.5 AU corresponds to the location of the 3 : 1 mean motion resonance with Jupiter. Three known asteroid families, Eos, Koronis, and Themis, are also indicated.

tions. The transition between the two regimes has been found to occur for body sizes in the hundred meter range by numerical simulations ([6]), while the transition between the two regimes derived from the spin limits occurs at higher diameters (kilometer range; see Sect. 4. and [27]). Therefore, the size of the body at which the transition occurs is still a subject of debates and ranges between a few 100 m and several km.

The first numerical simulations which succesfully reproduced large scale events represented by asteroid families ([43], [44], [45], [46]), showed that when a large parent body (several tens of kilometers in diameter) is disrupted by a collision with a projectile, the generated fragments interact gravitationally during their ejections, and some of them reaccumulate to form aggregates. The final outcome of such a disruption is thus a distribution of fragments, most of the large ones being aggregates formed by gravitational reaccumulations of smaller ones. The implication of these results is that most large family members should be rubble piles and not monolithic bodies. Moreover, it was found that collisional disruptions form naturally binary systems and satellites ([43], [18]), although the timescale of their stability was not determined. Sec. 6.1. reviews briefly some of these results.

The physics of the gravitational phase during which generated fragments evolve under their mutual attractions relies on the fundamental laws of classical mechanics, which are well understood. However, the development of numerical simulations which account for all the processes that may occur during this phase is a difficult task. When a large body is fragmented, the number of generated sizeable fragments can be as large as a few millions. Therefore, the numerical difficulties come from the fact that the forces must be computed for a large number of particles, up to millions, and this requires the use of efficient numerical methods to reduce the CPU time to reasonable values. Moreover, during their evolutions, these fragments do not only evolve under distant interactions, they can also undergo physical collisions between them, which must also be handled properly. A numerical N-body code called *pkdgrav* has been developed (see [63]) to compute the evolutions of large numbers of particles. It is a parallel tree-code which is able to compute the gravitational evolutions of millions of particles and handles collisions between them. During the gravitational phase, such collisions are assumed to not cause fragmentation, but only mergers or bounces. This simplification is justified by the fact that the relative velocities between the ejected fragments are small enough that collisions between them are quite smooth. So, when collisions occur during this phase, depending on some velocity and spin criteria, the particles either merge into a single one whose mass is the sum of the particle masses and whose position is at the center of mass of the particles, or bounce with some coefficient of restitution to account for dissipation (see [44] for details). In this approximation, while the aggregates that are formed have "correct" masses, their shapes are all spherical because of the merging procedure. Of course, the final shape and spin distributions are also an important outcome of a disruption and a study is currently under way to improve the simulations: instead of merging into a single spherical particle, colliding particles will be able to stick together using rigid body approximations. Such improvement will allow the determination of the shapes and spins of aggregates formed during a catastrophic disruption.

The most poorly understood part of the collisional process is the fragmentation phase immediately following the impact of the projectile. It usually lasts twice the time for the shock wave to propagate through the whole target (a few seconds for a kilometer-size body). The process of rock fragmentation is still a widely open area of research, relying on a large number of assumptions based on a limited number of data (such as material strength; see Sec. 3.). Moreover, not only is the physics badly understood, the numerical techniques used to perform the computation are also confronted with some difficulties. Indeed, the fragmentation process in a rock involves two kinds of approaches, which are generally incompatible. A high-velocity impact on a rock generates a shock wave, followed by a rarefaction wave which will activate the crack propagation. Thus, the rock can be seen as a continuum for the shock treatment. On the other hand, a rock contains some discrete elements (the initial cracks). This mixture of continuum and discrete features makes the development of a numerical scheme difficult. A numerical code used to compute the fragmentation phase is generally called a *hydrocode*, which emphasizes the fact that this process involves the physics of hydrodynamics, although it occurs in a solid body. Indeed, the difference between a fluid and a solid is that the deviatoric (non-diagonal) part of the stress stensor is not null in the case of a solid, while in a fluid only the spherical (diagonal) part of the stress tensor representing the pressure plays a role (see Sect. 3. for a detailed explanation of the difference between a fluid and a solid). Thus, three kinds of waves (elastic, plastic and shock) propagate through a rock during an impact. Elastic waves are well known and determined by linear realtionships between the stress and strain tensors. Plastic waves begin to develop when the material strength changes with the wave amplitude. Then, at wave amplitudes that are high enough and associated to shock waves, the body is treated as a fluid. Being non-linear, the transitory behaviors between these kinds of waves are difficult to determine analytically from constitutive models, and this probably motivated the development of numerical algorithms. The process has thus been studied by implementing the bulk properties of a given rock in a numerical model of continuous medium (a *hydrocode*), including a yielding criterion and an equation of state for the appropriate material. The main power of this method is that no assumption on the form of the stress wave that drives the fragmentation is required since the initial conditions evolve numerically based on a rheological model and a failure criterion. The appropriate regime (elastic, plastic or shock) is determined by the computation.

The 3D Lagrangian hydrocode developed by Benz and Asphaug ([5]) represents the state-of-the-art in numerical computations of dynamical fracture of brittle solids. It uses the method called Smooth Particle Hydrodynamics (SPH) (see [4] for a review of this method). Basically, the values of the different hydrodynamics quantities are known at finite numbers of points which move with the flow. Starting from a spatial distribution of these points called particles, the SPH technique allows the computation of the spatial derivatives without the necessity of an underlying grid. The 3D SPH hydrocode is thus able to simulate consistently from statistical and hydrodynamical points of view the fragments that are smaller or larger than the chosen resolution (number of SPH particles). The resulting system has proven to predict successfully the sizes, positions and velocities of fragments measured in laboratory experiments, without requiring the adjustment of too many free parameters ([5]); moreover, associated with the N-body code *pkdgrav*, it has succesfully reproduced the main properties of asteroid families ([43], [44], [45], [46]).

For the sake of completeness, we recall here the basic equations that must be solved to compute the fragmentation process. Other important concepts, such as the equations of state, the model of brittle failure used to propagate damage and the method to distribute appropriately incipient flaws in the modeled rock with a Weibull distribution are not reproduced here, as they have been described several times (see [5], [48]).

The basic equations that must be solved to compute the process are the well-known conservation equations of hydrodynamics that can be found in standard textbooks. The first equation represents the mass conservation. Its expression is:

$$\frac{d\rho}{dt} + \rho \frac{\partial}{\partial x_\alpha} v_\alpha = 0 \tag{22}$$

where d/dt is the lagrangian time derivative. Other variables have their usual meaning (i.e. ρ is the bulk density, v is the velocity and x the position) and the usual summation rule over repeated indices is used. The second equation describes the momentum conservation (in absence of gravity):

$$\frac{dv_\alpha}{dt} = \frac{1}{\rho} \frac{\partial}{\partial x_\beta} \sigma_{\alpha\beta} \tag{23}$$

where $\sigma_{\alpha\beta}$ is the stress tensor given by:

$$\sigma_{\alpha\beta} = -P\delta_{\alpha\beta} + S_{\alpha\beta} \tag{24}$$

where P is the isotropic pressure and $S_{\alpha\beta}$ is the traceless deviatoric stress tensor.The energy conservation is then expressed by the equation:

$$\frac{du}{dt} = -\frac{P}{\rho}\frac{\partial}{\partial x_\alpha}v_\alpha + \frac{1}{\rho}S_{\alpha\beta}\dot{\varepsilon}_{\alpha\beta} \tag{25}$$

where $\dot{\varepsilon}_{\alpha\beta}$ is the strain rate tensor given by:

$$\dot{\varepsilon}_{\alpha\beta} = \frac{1}{2}\left(\frac{\partial}{\partial x_\beta}v_\alpha + \frac{\partial}{\partial x_\alpha}v_\beta\right). \tag{26}$$

This set of equations is still unsufficient in the case of a solid since the evolution in time of $S_{\alpha\beta}$ must be specified. The basic Hooke's law model is assumed in which the stress deviator rate is proportional to the strain rate:

$$\frac{dS_{\alpha\beta}}{dt} = 2\mu\left(\dot{\varepsilon}_{\alpha\beta} - \frac{1}{3}\delta_{\alpha\beta}\right) + S_{\alpha\gamma}R_{\beta\gamma} + S_{\beta\gamma}R_{\alpha\gamma} \tag{27}$$

where μ is the shear modulus and $R_{\alpha\beta}$ is the rotation rate tensor given by:

$$R_{\alpha\beta} = \frac{1}{2}\left(\frac{\partial}{\partial x_\beta}v_\alpha - \frac{\partial}{\partial x_\alpha}v_\beta\right). \tag{28}$$

This term allows the transformation of the stresses from the reference frame associated with the material to the laboratory reference frame in which the other equations are specified.

This set of equations can now be solved, provided an equation of state is specified, $P = P(\rho, u)$, linking the pressure P to the density ρ and internal energy u. The Tillotson equation of state for solid material ([74]) is generally used. Its expression and method of computation, as well as parameters for a wide variety of rocks are described in [38] (Appendix II). Other equations of states have been developed and all have different pros and cons and remain necessarily limited to materials studied in laboratory. This is one of the limits of any collisional model that necessarily relies on the behavior of known materials that do not necessarily represent the materials constituing an asteroid.

Perfectly elastic materials are well described by these equations. Plastic behavior beyond the Hugoniot elastic limit is introduced in these relations by using the von Mises yielding criterion (see Sect. 3.). This criterion limits the deviatoric stress tensor to:

$$S_{\alpha\beta} = fS_{\alpha\beta} \tag{29}$$

where f is computed from:

$$f = \min[\frac{Y_0^2}{3J_2}, 1], \tag{30}$$

where J_2 is the second invariant of the deviatoric stress tensor (see Sect. 3.) and Y_0 is a material dependent yielding stress which generally depends on temperature, density, etc. in such a way that it decreases with increasing temperature until it vanishes beyond the melting point.

The von Mises criterion is adapted to describe the failure of ductile media such as metals. Brittle materials like rocks do not undergo a plastic failure but rather "break" if the applied stresses exceed a given threshold. Conversely, the yielding beyond the Hugoniot elastic limit does not prescribe any permanent change in the constitution of the material, since once stresses are reduced the original material remains behind, possibly heated by the motion against the remaining stress, but otherwise not weakened. Therefore, it is not adapted to impacts into rocks, as any yielding beyond the elastic limit invariably involves irreversible damage, and one needs to know how the rock is permanently altered by the event. A realistic fracture model is then clearly required to study the disruption of a solid body. The Grady-Kipp model of brittle failure is generally the model implemented in numerical codes aimed at simulating fragmentation processes in solid bodies ([21]).

Despite the recent successes of impact simulations to reproduce some experiments and asteroid family properties, there are still many issues and uncertainties in the treatment of the fragmentation phase. In particular, one of the main limitations of all researches devoted to the fragmentation process comes from the uncertainties on the material properties of the objects involved in the event. For instance, ten material parameters describe the usually adopted Tillotson equation of state (see e.g. [38], Appendix II). Other sensitive material-dependent parameters are, for instance, the shear and bulk modulus, but the most problematic parameters are probably the two Weibull parameters m and k used to characterize the distribution of initial cracks in the target. In fact, as already mentioned (Sect. 3.3.), data are still scarce about these parameters, due to the experimental difficulty to determine their values. This is a crucial problem because so far, the validation of numerical simulations by confrontation to experiments has been done by choosing freely those missing values so as to match the experiments ([5]). This is not a totally satisfactory approach for an *øæ* ¨± method such as the one provided by SPH simulations. Unfortunately, this is often the only alternative which one has. A database including both the material parameters of targets and outcomes of impact experiments using these targets is thus required to perform a full validation of numerical codes. Such a project has started using the experimental expertise of japanese researchers from Kobe University, and the numerical expertise of french and swiss researchers from Côte d'Azur Observatory and Berne University. For instance, measurements of Weibull parameters of a Yakuno basalt used in impact experiments were made for this purpose ([56]). Another problem concerns the model of fragmentation that is assumed to be appropriate for small Solar System bodies. Up to now, all published simulations of impact disruption have been done using in general the Grady-Kipp model of brittle failure ([21]). In this model, damage increases as a result of crack activation, and microporosity effects (pore crushing, compaction) are not treated. However, several materials contain porosity at micro-scale (e.g. pumice, gypsum), and asteroids belonging to dark taxonomic types (e.g. C, D) are believed to contain a high level of microporosity. The behavior of a porous material subjected to an impact is likely to be different than the behavior of a non-porous one, as already indicated by some experiments (e.g. [30]). Therefore, a model for porous materials is required, in order to be able to address the problem of dark-type asteroid family formations, and to characterize the impact response of porous bodies in general (including porous planetesimals during the phase of planetary growth). Such models have been developed recently and inserted in numerical codes ([79], [7], [34]). The model by Juzi et al. ([34]) has then been recently validated by comparison to impact experiments on pumice targets ([35]).

Last but not least, so far all simulations of catastrophic disruption have been performed starting with a non-rotating target. However, in the real world, small bodies are spinning (see Sect. 4.), and the effect of the rotation on the fragmentation is totally unknown. Some preliminary experiments have been performed suggesting that, everything else being equal, a rotating target is easier to disrupt than a non-rotating one (K. Housen, private communication). If this is confirmed, this will be an important result as all models of collisional evolutions of small body populations use prescriptions that are provided by numerical simulations on non-rotating targets. In particular, the lifetime of small bodies may be shorter than expected if their rotation has an effect on their ability to survive collisions. It will thus be important to characterize the impact response of rotating bodies, both experimentally and numerically, although on both sides, starting with a rotating body poses several difficulties.

6.1. Formation of Asteroid Families

In the following, the most recent results obtained by numerically simulating the catastrophic disruptions of large asteroids are presented. The simulations rely on the theory of fragmentation of solid bodies and have been performed using the 3D SPH hydrocode and the gravitational N-body code mentioned in previous sections.

Recall that asteroid families represent a unique laboratory that Nature offers us to study large-scale collisional events. Indeed, observed asteroid families in the main asteroid belt are each composed of bodies which originally resulted from the break-up of a large parent body (e.g. [37]).

Interestingly, until recently, the theory of the collisional origin of asteroid families rested entirely on these similarities in dynamical and spectral properties and not on the detailed understanding of the collisional physics itself. Indeed, laboratory experiments on centimeter-scale targets, analytical scaling rules, or even complete numerical simulations of asteroid collisions were not able to reproduce the physical and dynamical properties of asteroid families (e.g. [67]). The extrapolation of laboratory experiments to asteroidal scales yields bodies much too weak to account for both the mass distribution and the dynamical properties of family members. More precisely, in a collision resulting in a mass distribution of fragments resembling a real family, which can contain many big members, the ejection velocities of individual fragments are much too small for them to overcome their own gravitational attraction. The parent body is merely shattered but not dispersed and therefore no family is created. Conversely, matching individual ejection velocities and deriving the necessary fragment distribution results in a mass distribution in which no big fragment is present, contrary to what is indicated by most real families (e.g. [16], [13]).

The collisional origin of asteroid families thus implies that not only has the parent body (up to several hundred kilometers in size) been shattered by the propagation of cracks but also that the fragments generated this way typically escape from the parent and reaccumulate elsewhere in groups in order to build up the most massive family members. Such a process had already been suggested (e.g. [12]), and the possibility that at least the largest fragment from a collision consists of a rubble pile had also been indicated later by means of numerical simulations ([6]). The effect of gravitational reaccumulation was then estimated by a procedure which consists of searching for the largest group of gravitationally bound debris immediately following the collision, and not in computing explicitly the gravitational interactions between the fragments. It is only very recently that the formation of many large family members by reaccumulation of smaller fragments has been demonstrated explicitly by Michel and his colleagues ([43], [44], [45], [46], [47]).

Such aggregates of gravitationally bound fragments, as the ones formed in the last simulations, are usually defined as *rubble-piles* in the asteroidal community, which means that they are *loose* aggregates of fragments held together by gravity. A detailed definition and a review of this topic are presented in [64]. Roughly, such bodies have little to no tensile strength, i.e. they can be torn apart easily by planetary tides. Only indirect evidence for such structures exist. Indeed, the structural properties of asteroids are difficult to establish since directly measurable quantities do not distinguish between solid bodies and rubble piles. Rubble piles have been invoked to explain, for instance, the low density of some observed bodies like the 50 km-size main belt asteroid (253) Mathilde whose measured density by the *NEAR* probe is $1.35\ \mathrm{g/cm}^3$ ([80]) and the 500 m-size NEO (25143) Itokawa whose bulk density is only $1.9\ \mathrm{g/cm}^3$ ([19]).

For the first time, Michel and his colleagues have simulated entirely and successfully the formation of asteroid families in two extreme regimes of impact energy leading to either a small or a large mass ratio of the largest remnant to the parent body M_{lr}/M_{pb} ([43]). Two well-identified families have been used for comparison with simulations: the Eunomia family, with a 284 km parent body and $M_{lr}/M_{pb} = 0.67$, has been used to represent the barely disruptive regime, whereas the Koronis family, with a 119 km parent body and $M_{lr}/M_{pb} = 0.04$, represented the highly catastrophic one. In these simulations, the collisional process was carried out to late times (typically several days), during which the gravitational interactions between the fragments could eventually lead to the formation of aggregates or rubble piles far from the largest remnant. It was first assumed somewhat unrealistically that particles colliding during the gravitational phase always stuck perfectly and merged regardless of relative velocity and mass. Michel and his colleagues then improved on this treatment by allowing for the dissipation of kinetic energy in such collisions and applying an energy based merging criterion ([44]). This improved treatment did not change the conclusion obtained with the more simplistic one, and the reaccumulation process remains at the origin of large family members.

6.1.1. Disruption of Monolithic and Pre-shattered Asteroids

The two studies mentioned above used as a starting condition a monolithic parent body represented by a basalt sphere. The projectile's parameters (diameter, velocity and impact angle) were defined such that the expected value of M_{lr}/M_{pb} was successfully obtained by the simulation (see [43], [44] for details). A first unexpected result was found in all the explored impact energy regimes: the fragmentation phase always leads to the complete pulverization of the parent body down to a fragment size corresponding to the resolution limit (of the order of 1 km). Then, the gravitational phase during which these fragments interact leads to many reaccumulations (see Fig. 8). Eventually, the fragment size distribution is dominated by aggregates formed by reaccumulation of smaller fragments and only the smallest size end of the distribution consists of individual (or *intact*) fragments. Moreover, for each family, the simulated distribution is qualitatively compatible with that of real

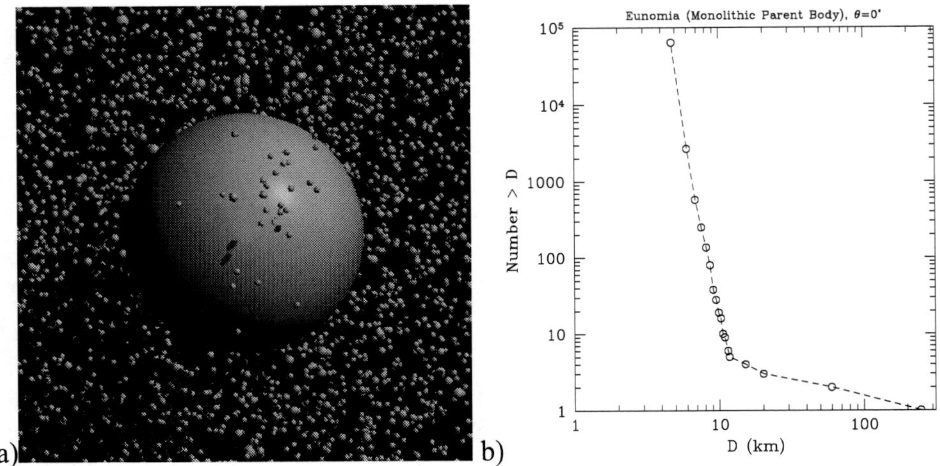

Figure 8. (a) Image taken from the simulation of the gravitational phase of the Eunomia family formation $\approx$ 84 minutes after fragmentation of the parent body, when the largest fragment of Eunomia is almost already formed; it is represented by the large sphere at the center of the image which is produced by the reaccumulations of smaller ones; (b) Cumulative diameter distribution (in km) of the fragments resulting from the simulation of the Eunomia family formation in a log–log plot; the simulation was performed using a target with diameter and bulk density corresponding to the parent body of the family. The number of particles used to define the target was set to 2×10^5. The impact angle was $\theta = 0°$ and the impact velocity of the projectile was 5 km s^{-1}. These results are described in the paper of Michel and his colleagues ([44])

family members, even though some large fragments always lack between the largest and smaller ones, compared to the observed distribution (this difference will disappear using pre-shattered parent bodies; see further). Another important result is that many binary systems are formed during the gravitational phase, so that the formation of asteroid satellites can be seen as a natural outcome of catastrophic collisions. However, the timescale of the simulations is too short and external perturbations (solar tides, etc...) should be included to determine their long-term stability and lifetime. Nevertheless, their occurence provides a possible explanation for the origin of some of the observed ones. The interested reader can find a detailed study of satellite formation during collisions in a recent paper ([18]).

As for the orbital dispersion of fragments obtained from these simulations, it is always smaller than the dispersion of real family members computed from their proper orbital elements ([81], [31]). However, the computation of the dispersion of simulated fragments is confronted with many unknowns such as the values of the true anomaly and perihelion argument of the parent body at the instant of impact, which are required by Gauss' formulae to convert ejection velocities into orbital elements (see e.g. [44]). Nevertheless, the compact orbital dispersions found in simulations are still consistent with the larger ones of real families. Indeed, it has been realized that the current orbital dispersion of family members may not represent the original one following the break-up of the parent body. The post-collisional evolution of family members is affected by dynamical mechanisms such as high order resonances and/or the thermal Yarkovsky effect which can both cause a diffusion of the orbital elements over long enough timescales (see e.g. [9], [11], [60]). When the estimated age of the considered family and the diffusion timescale of the orbital elements of family members due to these mechanisms are taken into account, the comparison between the compact dispersions obtained by the simulations and the larger one of real members leads to a good agreement as shown in Fig. 9.

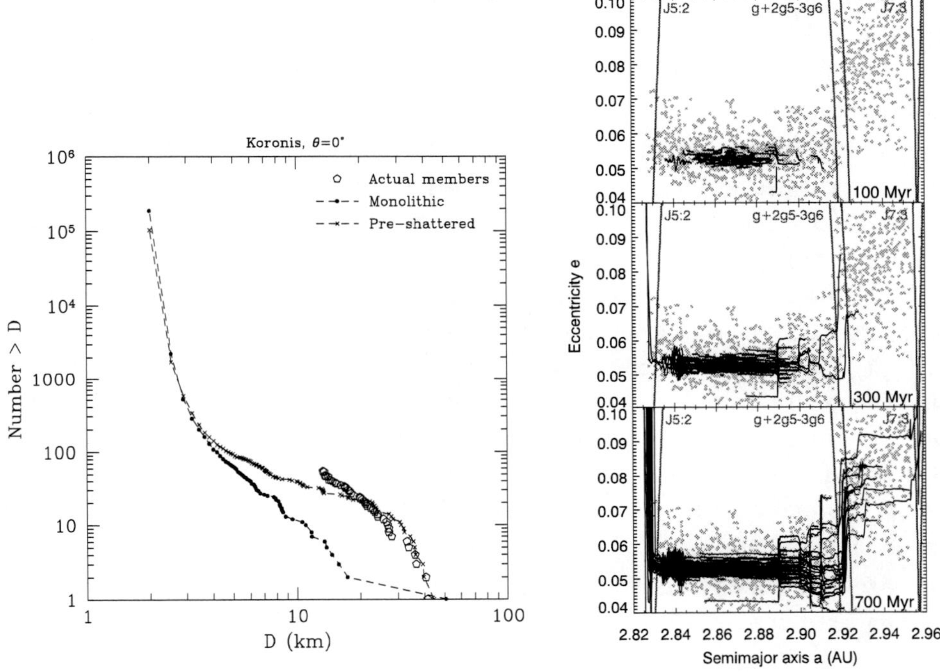

Figure 9. **Left:** Cumulative diameter distributions (in km) in a log-log plot for the fragments of the simulated Koronis family ([46]). An impact at 5 km s^{-1} with an angle of incidence of 0° gave rise to these distributions. Different symbols are used depending on the different parent body models used in the simulations as indicated on the plots. The estimated sizes of the actual members are also shown on the plot ([73]). Note that simulations of the disruption of a pre-shattered parent body are able to produce the four similar-sized largest members. **Right:** Evolution in the (semimajor axis a, eccentricity e) plane of 210 simulated Koronis family members under the influence of the Yarkovsky thermal effect and dynamical diffusion due to several resonances indicated on the plot. The test family members were started with a dispersion which is consistent with the ones obtained from our simulations of Koronis family formation. They were integrated over 700 Myr, which is still shorter than the estimated age of the family ($>$ 1 Gyr). However, this evolution shows that the current shape of the family cluster in the proper element space does not represent the original one from the collisional event but is well explained by its subsequent evolution (from [9]).

To start from monolithic family parent bodies already allowed to reproduce qualitatively the main properties of real family members. However, the cumulative size distribution of simulated fragments is quite systematically characterized by a lack of intermediate sized bodies and a very steep slope for the smaller ones (see plot b) of Fig. 8). Such characteristics are not observed in the size distribution of real family members, which generally looks rather continuous. It is then important to determine whether this systematic effect is a general failure of the simulations, or whether it depends on the initial parameters and assumptions used to start the simulations. In particular, instead of being monolithic, the parent bodies may be modelled with an internal structure composed of different zones of voids and fractures, as if they had first been pre-shattered during their collisional history before undergoing a major event leading to their disruption. If a better agreement could be obtained by playing with the original internal structure of the parent body, this would be an important result with many implications. For instance, simulations could then be used to constrain the internal structure of parent bodies at the origin of known asteroid families that allows us to produce the corresponding member properties.

The assumption that large parent bodies are pre-shattered before being disrupted is appropriate not just because it may potentially lead to a closer match with observed properties. The assumed pre-shattered state is thought to be a natural consequence of the collisional evolution of main belt asteroids. Indeed, several studies have indicated that for any asteroid, collisions at high impact energies leading to a disruption occur with a smaller frequency than collisions at lower impact energies leading to shattering effects only (see, e.g. [3], [17], [64]). Thus, in general, a typical asteroid gets battered over time until a major collision eventually disrupts it into smaller dispersed pieces ([40]). Consequently, since the formation of an asteroid family corresponds to the ultimate disruptive event of a large object, it is reasonable to think that the internal structure of this body has been modified from its primordial state by all the smaller collisional events that it has suffered over its lifetime in the belt.

The size and velocity distributions of family members are the major constraints that can be compared with results of simulations starting from different models of parent bodies. However, as already stated, these distributions can be modified over time by collisional erosion and dynamical diffusion, depending on the proximity of diffusion mechanisms to family members and on the family's age. Recently, a very young family, called Karin, has been identified thanks to the increasing database of asteroid proper elements. Its young age of 5 Myr has been estimated by integrating numerically the orbits of its members backward in time from their current state to the state where the heliocentric orbits of all cluster members and their orientations were nearly the same ([59]). The authors estimated that this convergence could not happen by chance. The Karin family is thus so young that it provides a unique opportunity to study a collisional outcome almost unaffected by erosion and dynamical diffusion.

Michel and his colleagues have recently reported on numerical simulations which showed how the collisional outcome depends on the internal structure of the parent body of a family ([45]). More precisely, two kinds of parent bodies, both spherical in shape, were considered: (1) purely monolithic, and (2) pre-fragmented (pre-shattered with damage zones but no internal voids or rubble pile with internal voids). The simulations then showed that the heterogeneities introduced in this way result in significant changes in the size spectrum of the final fragments left after the collision, even though in both cases the targets are completely shattered by the impact. In particular, whereas the fragment size distribution obtained from a monolithic parent body still lacks some fragments at intermediate sizes, disruptions involving a pre-fragmented (pre-shattered or rubble pile) parent body on the other hand result in a much more continuous cumulative size distribution of fragments.

Hence, the presence of intermediate sized bodies is most likely a direct consequence of the presence of large-scale fractures or big blocks within the parent body.

The authors then remarked that intermediate size fragments are actually present in most major asteroid families, implying that many parent bodies in the asteroid belt were probably pre-fragmented, and confirmed this interpretation by new simulations ([46]). More precisely, they remade their simulations of the Eunomia and Koronis family formations using pre-shattered parent bodies and compared the results with the ones obtained with a monotithic parent body. The best agreement was then found with pre-shattered parent bodies. In particular, in the case of the Koronis family, an interesting result has been obtained from these simulations, which may have important implications concerning the real family history. The size distribution obtained from the disruption of a pre-shattered parent body contains four largest fragments of approximately the same size (Fig. 9). This peculiar characteristic is shared by the real family, and has been a source of debate as it was assumed that a single collisional event cannot produce such a property (see [46] for a discussion). Moreover, the simulation using a monolithic parent body did not result in such a distribution. The authors then demonstrated numerically for the first time, by using a pre-shattered parent body, that these fragments can actually be produced by the original event, and therefore no subsequent mechanism needs to be invoked to form them, which would require a revision of the entire family history ([46]). According to those results showing that even old families may well have originated from pre-shattered parent bodies, the authors concluded that most large objects in the present-day asteroid belt may well be pre-shattered.

Another big interest of these simulations is that they can be used to determine the impact energy needed to produce a given degree of disruption as a function of the internal structure of the parent body. It is generally found that disrupting a pre-shattered target requires less energy per unit mass than disrupting a monolithic body with the same degree of disruption. This, at first glance a surprising result, is related to the fact that the fractures as modelled by the authors (no porosity and no material discontinuities except damage) do not affect shock waves but only tensile waves. Hence, fragments can be set in motion immediately upon being hit by the shock wave without having to wait for fracture to occur in a following tensile wave. Thus, transfer of momentum is more efficient and disruption facilitated. Note that the presence of large voids such as those in rubble piles affect the propagation of the shock wave, thus reversing this trend. A more detailed study of these properties, as well as the dependency of the impact strength as a function of macro- and micro-porosity, is in development, but it is already clear that the internal structure of an asteroid plays an important role in the determination of its response to impacts (see also [2]). This is not only relevant for estimating the collisional lifetime of a body in the asteroid belt, but also for developing strategies to deflect a potential Earth impactor.

7. Conclusion

More than two hundreds years have passed since the first asteroid (1) Ceres was discovered by G. Piazzi in Palermo in 1801, and our knowledge about asteroids, and small bodies of our Solar System in general, has changed quantitatively and continuously during the last decades. Our understanding of these solid bodies is very likely to be challenged again with future observational programs and space missions.

Most small bodies have been subjected to collisional processes, and this is particularly true for NEOs whose majority is believed to consist of fragments of large bodies which have been disrupted by a collision with another body in the main asteroid belt. This disruption mechanism is not the only one which can affect the physical properties of a small body and our understanding of the different disruption mechanisms has greatly improved in the last decades, thanks to the development of analytical theories and sophisticated simulations. However, there are still many uncertainties, and problems that need to be investigated.

Concerning the spin limits and tidal disruptions, the theories that have been developed so far are all static. Nevertheless, they allowed us to understand that the spin barrier observed for large asteroids does not imply necessarily that these bodies are pure rubble piles, in contrast with the usual interpretation. On the other hand, the small fast rotators do not need to have much cohesion to spin at such high rates (see Sect. 4.).

Then, these theories allowed us to revisit the concept of Roche limit for solid ellipsoidal bodies with and without cohesion. They showed that, contrary to a fluid, a solid body can come much closer to a planet or a primary and with a wide variety of shapes (see Sect. 5.). However, these theories rely on a model of material strength (Drucker-Prager or Mohr-Coulomb) which is not necessarily unique and they are limited to bodies whose shapes are idealized ellipsoids. Some other strength models or non-idealized shapes should certainly be considered. Then, a complete dynamical investigation of these problems is required to determine the outcome of rotational and tidal break-ups of small bodies.

Sophisticated numerical codes have been developed to study the process of impact disruption of a small body. The outcome of some impact experiments and the main properties of some asteroid families have been reproduced successfully with one of those codes, based on the Smooth Particle Hydrodynamics technique and the Grady-Kipp model of brittle failure or a model for porous materials. However, as discussed in Sect. 6., there are still many issues, and the road is still long before being able to characterize with high accuracy the impact response of a small body as a function of its material properties. This is a challenging topic which has many applications. Indeed, the collisional process plays a fundamental role in the different phases of the history of our Solar System, from the phase of planetary growth by collisional accretion to the current phase during which small bodies are catastrophically disrupted. Moreover, the determination of the impact response of a small body as a function of its physical properties is crucial in the definition of efficient mitigation strategies aimed at deflecting a potential threatening near-Earth asteroid whose trajectory leads it to the Earth.

Thus, researches devoted to these disruption mechanisms and to the concept of strength will certainly keep future generations of researchers busy, and they will also take advantage of future space missions devoted to in-situ investigations and sample returns from small bodies.

References

[1] Asphaug, E. (1993). Dynamic fragmentation in the solar system: applications of fracture mechanics and hydrodynamics to questions of planetary evolution. PhD Thesis, The University of Arizona.

[2] Asphaug, E., Ostro, S.J., Hudson, R.S., Scheeres, D.J. & Benz, W.(1998). Disruption of kilometer-size asteroids by energetic collisions. *Nature*, **393**, 437-440.

[3] Asphaug, E., Ryan, E.V. & Zuber, M.T. (2002). Asteroid Interiors. In W.F. Bottke, A. Cellino, P. Paolicchi, R.P. Binzel, & T. Gehrels (Eds.), *Asteroid III* (pp. 463-484). Tucson, USA: University of Arizona Press.

[4] Benz, W. (1990). Smooth Particle Hydrodynamics - A Review. In J.R. Buchler (Ed.), *Proceedings of the NATO Advanced Research Workshop on The Numerical Modelling of Nonlinear Stellar Pulsations Problems and Prospects* . Dordrecht, Germany: Kluwer Academic Publishers.

[5] Benz, W. & Asphaug, E. (1994). Impact simulations with fracture. I- Method and tests. *Icarus*, **107**, 98-116.

[6] Benz, W. & Asphaug, E. (1999). Catastrophic disruptions revisited. *Icarus*, **142**, 5-20.

[7] Benz, W. & Jutzi, M. (2007). Collision and impact simulations including porosity. In A. Milani, G.B. Valsecchi, & D. Vokrouhlicky (Eds.), *Near-Earth Objects, our celestial neighbors: Opportunity and Risks* , IAU Symposium 236 (pp. 223-231). Cambridge, UK: Cambridge University Press.

[8] Bottke, W.F., Jedicke, R., Morbidelli, A., Petit, J.M. & Gladman, B.J. (2000). Understanding the distribution of Near-Earth Asteroids. *Science*, **288**, 2190-2194.

[9] Bottke, W.F., Vokrouhlický, D., Borz, B., Nesvorný, D., & Morbidelli, A. (2001). Dynamical spreading of asteroid families via the Yarkovsky effect. *Science*, **294**, 1693-1696.

[10] Bottke, W.F., Morbidelli, A., Jedicke, R., Petit, J.M., Levison, H., Michel, P. & and Metcalfe, T.S. (2002). Debiased orbital and absolute magnitude distribution of the Near-Earth Object population. *Icarus*, **156**, 399-433.

[11] Bottke, W.F., Vokrouhlický, D., Rubincam, D.P. & Broz, M. (2002). The Effect of Yarkovsky Thermal Forces on the Dynamical Evolution of Asteroids and Meteoroids. In W.F. Bottke, A. Cellino, P. Paolicchi, R.P. Binzel, & T. Gehrels (Eds.), *Asteroid III* (pp. 501-515). Tucson, USA: University of Arizona Press.

[12] Chapman, C.R., Davis, D.R., & Greenberg R. (1982). Apollo asteroids: relationship to main belt asteroids and meteorites. *Meteoritics*, **17**, 193-194.

[13] Chapman, C.R., Paolicchi, P., Zappalà, V., Binzel, R.P. & Bell, J.F. (1989). Asteroid families: Physical properties and evolution. In R.P. Binzel, T. Gehrels, & M.S. Matthews (Eds.), *Asteroids II*, (pp. 386-415). Tucson, USA: University of Arizona Press.

[14] Chandrasekhar, S. (1969). *Ellipsoidal Figures of Equilibrium* . New-York: Dover.

[15] Chen, W.F., & Han, D.J. (1988). *Plasticity for Structural Engineers*. Berlin: Springer.

[16] Davis, D.R., Chapman, C.R., Weidenschilling, S.J., & Greenberg, R. (1985). Collisional history of asteroids: evidence from Vesta and Hirayama families. *Icarus*, **62**, 30-53.

[17] Davis, D.R., Durda, D.D., Marzari, F., Campo Bagatin, A. & Gil-Hutton, R. (2002). Collisional Evolutions of Small Body Populations. In W.F. Bottke, A. Cellino, P. Paolicchi, R.P. Binzel, & T. Gehrels (Eds.), *Asteroid III* (pp. 545-558). Tucson, USA: University of Arizona Press.

[18] Durda, D.D., Bottke, W.F., Enke, B.L., Merline, W.J., Asphaug, E., Richardson, D.C., & Leihnardt, Z.M. (2004). The formation of asteroid satellites in large impacts: results from numerical simulations. *Icarus*, **170**, 243-257.

[19] Fujiwara, A., et al. (2006). The rubble-pile asteroid Itokawa as observed by Hayabusa. *Science*, **312**, 1330-1334.

[20] Gladman, B.J., Migliorini, F., Morbidelli, A., Zappalà, V., Michel, P., Cellino, A., Froeschlé, Ch., Levison, H.F., Bailey, M., & Duncan, M. (1997). Dynamical lifetimes of objetcs injected into asteroid belt resonances. *Science*, **277**, 197-201.

[21] Grady, D.E. & Kipp, M.E. (1980). Continuum modeling of explosive fracture in oil shale. *Int. J. Rock Mech. Min. Sci. Geomech. Abstr.*, **17**, 147.

[22] Harris, A.W. (1996). The rotation rates of very small asteroids: Evidence for "rubble pile" structure. *Lunar Planet. Sci.*, **27**, 493-494.

[23] Holsapple, K.A. (2001). Equilibrium configurations of solid ellipsoidal cohesionless bodies. *Icarus*, **154**, 432-448.

[24] Holsapple, K.A., Giblin, I., Housen, K.R., Nakamura, A. & Ryan, E. (2002). Asteroid impacts: Laboratory experiments and scaling laws. In W.F. Bottke, A. Cellino, P. Paolicchi, R.P. Binzel, & T. Gehrels (Eds.), *Asteroid III* (pp. 443-462). Tucson, USA: University of Arizona Press.

[25] Holsapple, K.A. (2004). Equilibrium figures of spinning bodies with self-gravity. *Icarus*, **172**, 272-303.

[26] Holsapple, K.A. & Michel, P. (2006) Tidal disruptions: a continuum theory for solid bodies. *Icarus*, **183**, 331-348.

[27] Holsapple, K.A. (2007). Spin limits of Solar System bodies: From the small fast-rotators to 2003 EL61. *Icarus*, **187**, 500-509.

[28] Holsapple, K.A. (2007). Spinning rods, ellipticaldisks and solid ellipsoidal bodies: Elastic and plastic stresses and limit spins. submitted (2007)

[29] Holsapple, K.A., & Michel, P. (2007). Tidal disruption II: a continuum theory for solid bodies with strength, with applications to the satellites of our Solar System. *Icarus*, accepted.

[30] Housen, K.R. & Holsapple, K.A. (1999). Scale effects in strength-dominated collisions of rocky asteroids. *Icarus*, **142**, 21-33.

[31] Knezěvič, Z., Lemaître, A. & Milani, A. (2002). The determination of asteroid proper elements. In W.F. Bottke, A. Cellino, P. Paolicchi, R.P. Binzel, & T. Gehrels (Eds.), *Asteroid III* (pp. 603-612). Tucson, USA: University of Arizona Press.

[32] Jeans, J.H. (1917). The motion of tidally-distorded masses, with special reference to the theories of cosmogony. *Mem. Roy. Astron. Soc. London*, **62**, 1.

[33] Jedicke, R. (1996). Detection of Near-Earth Asteroids based upon their rates of motion. *Astron. J.*, **111**, 970-982.

[34] Jutzi, M., Benz, W., & Michel, P. (2008). Numerical simulations of impacts involving porous bodies I. Implementing sub-resolution porosity in a 3D SPH hydrocode. *Icarus*, **198**, 242-255.

[35] Jutzi, M., Michel, P., Hiraoka, K., Nakamura, A.M., & Benz, W. (2009). Numerical simulations of impacts involving porous bodies II. Confrontation with laboaratory experiments. *Icarus*, in press.

[36] Lowry, S.C., et al. (2007). Direct detection of the asteroidal YORP effect. *Science*, **316**, 272.

[37] Marzari, F., Davis, D., & Vanzani V. (1995). Collisional evolution of asteroid families. *Icarus*, **113**, 168-187. (1995)

[38] Melosh, H.J. (1989). *Impact cratering: a geologic process*. New-York: Oxford University Press.

[39] Melosh, H.J., Ryan, E.V., & Asphaug, E. (1992). Dynamic fragmentation in impacts - Hydrocode simulations of laboratory impacts. *J. Geophys. Res.*, **97**, 14735-14759.

[40] Melosh, H.J. & Ryan, E.V. (1997). NOTE: Asteroids: shattered but not dispersed. *Icarus*, **129**, 562-564.

[41] Milani, A., & Knezěvič, Z. (1994). Asteroid proper elements and the dynamical structure of the asteroid belt. *Icarus*, **107**, 219-254.

[42] Michel, P., Migliorini, F., Morbidelli, A. & Zappalà, V. (2000). The population of Mars-crossers: classification and dynamical evolution. *Icarus*, **145**, 332-347.

[43] Michel, P., Benz, W., Tanga, P. & Richardson, D.C. (2001). Collisions and gravitational reaccumulations: forming asteroid families and satellites. *Science*, **294**, 1696-1700.

[44] Michel, P., Benz, W., Tanga, P. & Richardson, D.C. (2002). Formation of asteroid families by catastrophic disruption: simulations with fragmentation and gravitational reaccumulation. *Icarus*, **160**, 10-23.

[45] Michel, P., Benz, W., & Richardson, D.C. (2003). Fragmented parent bodies as the origin of asteroid families. *Nature*, **421**, 608-611.

[46] Michel, P., Benz, W., & Richardson, D.C. (2004). Disruption of pre-shattered parent bodies. *Icarus*, **168**, 420-432.

[47] Michel, P., Benz, W., & Richardson, D.C. (2004). Catastrophic disruption and family formation: a review of numerical simulations including both fragmentation and gravitational reaccumulation. *Planet. Space Sci.*, **52**, 1109-1117.

[48] Michel, P. (2006). Modelling collisions between asteroids: from laboratory experiments to numerical simulations. In J. Souchay (Ed.), *Dynamics of Extended Celestial Bodies*, (Lecture Notes in Physics, pp. 117-143), Berlin, Germany: Springer.

[49] Michel, P. & Holsapple, K.A. (2007). Tidal disturbances of small cohesionless bodies: limit on planetary close approache distances. In A. Milani, G.B. Valsecchi, & D. Vokrouhlicky (Eds.), *Near-Earth Objects, our celestial neighbors: Opportunity and Risks*, IAU Symposium 236 (pp. 201-210). Cambridge, UK: Cambridge University Press.

[50] Morbidelli, A. & Gladman, B.J. (1998). Orbital and temporal distributions of meteorites originating in the asteroid belt. *Meteoritics and Planetary Science*, **33**, 999-1016.

[51] Morbidelli, A. & Nesvorný, D. (1999). Numerous weak resonances drive asteroids towards terrestrial planets orbits. *Icarus*, **139**, 295-308.

[52] Morbidelli, A., Bottke, W.F., Froeschlé, Ch. & Michel, P. (2002). Origin and evolution of Near-Earth Objects. In W.F. Bottke, A. Cellino, P. Paolicchi, R.P. Binzel, & T. Gehrels (Eds.), *Asteroid III* (pp. 409-422). Tucson, USA: University of Arizona Press.

[53] Murray, N., & Holman M. (1997). Diffusive chaos in the outer asteroid belt. *Astron. J.*, **114**, 1246-1259.

[54] Murray, N., Holman, M. & Potter, M. (1998). On the origin of chaos in the asteroid belt. *Astron. J.*, **116**, 2583-2589.

[55] Nakamura, A., & Fujiwara, A. (1991). Velocity distribution of fragments formed in a simulated collisional disruption. *Icarus*, **92**, 132-146.

[56] Nakamura, A.M., Michel, P., & Seto, M. (2007). Weibull parameters of Yakuno basalt targets used in documented high-velocity impact experiments. *J. Geophys. Res.*, **112**, E02001.

[57] Nesvorný, D., & Morbidelli, A. (1998). Three-body mean motion resonances and the chaotic structure of the asteroid belt. *Astron. J.*, **116**, 3029-3037.

[58] Nesvorný, D., & Morbidelli, A. (1999). An analytic model of the three-body mean motion resonances. *Celest. Mech. Dyn. Astron.*, **71**, 243-271.

[59] Nesvorný, D., Bottke, W.F., Dones, L., & Levison, H.F. (2002). The recent breakup of an asteroid in the mainbelt region. *Nature*, **417**, 720-722.

[60] Nesvorný, D., Ferraz-Mello, S., Holman, M. & Morbidelli, A. (2002). Regular and Chaotic Dynamics in the Mean-Motion Resonances: Implications for the Structure and Evolution of the Asteroid Belt. In W.F. Bottke, A. Cellino, P. Paolicchi, R.P. Binzel, & T. Gehrels (Eds.), *Asteroid III* (pp. 379-394). Tucson, USA: University of Arizona Press.

[61] Rabinowitz, D., Helin, E., Lawrence, K. & Pravdo, S. (2000). A reduced estimate of the number of kilometre size Near-Earth Asteroids. *Nature*, **403**, 165-166.

[62] Richardson, D.C., Bottke, W.F., & Love, S.G. (1998). Tidal distortion and disruption of Earth-crossing asteroids. *Icarus*, **134**, 47-76.

[63] Richardson, D.C., Quinn, T., Stadel, J. & Lake, G. (2000). Direct large-scale N-body simulations of planetesimal dynamics. *Icarus*, **143**, 45-59.

[64] Richardson, D.C., Leinhardt, Z.M., Bottke, W.F., Melosh, H.J. & Asphaug, E. (2002). Gravitational aggregates: Evidence and evolution. In W.F. Bottke, A. Cellino, P. Paolicchi, R.P. Binzel, & T. Gehrels (Eds.), *Asteroid III* (pp. 501-515). Tucson, USA: University of Arizona Press.

[65] Richardson, D.C., Elankumaran, P. & Sanderson, R. (2005) Numerical experiments with rubble piles: equilibrium shapes and spins. *Icarus*, **173**, 349-361.

[66] Roche, E.A. (1847). *Acad. Sci. Lett. Montpelier. Mem. Section Sci.*, **1**, 243.

[67] Ryan, E.V., & Melosh, H.J. (1998). Impact fragmentation: from the laboratory to asteroids. *Icarus*, **133**, 1-24.

[68] Sharma, I., Jenkins, J.T., & Burns, J.A. (2006). Tidal encounters of ellipsoidal granular asteroids with planets. *Icarus*, **183**, 312-330.

[69] Scheeres, D.J., Ostro, S.J., Werner, R.A., Asphaug, E. & Hudson, R.S. (2000). Effect of gravitational interactions on asteroid spin states. *Icarus*, **147**, 106-118.

[70] Scheeres, D.J. (2001). Changes in rotational angular momentum due to gravitational interactions between two finite bodies. *Celest. Mech. Dynam. Astron.*, **81**, 39-44.

[71] Shoemaker, E.M. (1998). Impact cratering through geological time. *Journal of the Royal Astronomical Society of Canada*, **92**, 297.

[72] Stuart, J.S. (2001). A Near-Earth Asteroid population estimate from the LINEAR survery. *Science*, **294**, 1691-1693.

[73] Tanga, P., Cellino, A., Michel, P., Zappalà, V., Paolicchi, P. & dell'Oro, A. (1999). On the size distribution of asteroid families: the role of geometry. *Icarus*, **141**, 65-78.

[74] Tillotson, J.H. (1962). Metallic equations of state for hypervelocity impact. *General Atomic Report*, **GA-3216**, July.

[75] Vokrouhlický, D. & Capek, D. (2002). YORP-induced long-term evolution of the spin state of small asteroids and meteoroids: Rubbincam's approximation. *Icarus*, **159**, 449-467.

[76] Walsh, K.J. & Richardson, D.C. (2006). Binary near-Earth asteroid formation: Rubble pile model of tidal disruptions. *Icarus*, **180**, 201-216.

[77] Walsh, K.J., Richardson, D.C., & Michel, P. (2008). Rotational break up as the origin of binary asteroids. *Nature*, **454**, 188-191.

[78] Weibull, W.A. (1939). *Ingvetensk. Akad. Handl.*, **151**, 5.

[79] Wuennemann, K., Collins, G.S., & Melosh, H.J. (2006). A strain-based porosity model for use in hydrocode simulations of impacts and implications for transient crater growth in porous targets. *Icarus*, **180**, 514-527.

[80] Yeomans, D.K., et al. (1997). Estimating the mass of asteroid 253 Mathild from tracking data during the NEAR flyby. *Science*, **278**, 2106.

[81] Zappalà, V., Cellino, A., Dell'Oro, A. & Paolicchi, P. (2002). Physical and dynamical properties of asteroid families. In W.F. Bottke, A. Cellino, P. Paolicchi, R.P. Binzel, & T. Gehrels (Eds.), *Asteroid III* (pp. 619-631). Tucson, USA: University of Arizona Press.

In: Space Exploration Research
Editors: J.H. Denis and P.D. Aldridge

ISBN: 978-1-60692-264-4

Chapter 17

HILDA ASTEROIDS IN THE JUPITER NEIGHBORHOOD

Romina P. Di Sisto and Adrián Brunini
Facultad de Ciencias Astronómicas y Geofísicas (UNLP)
and IALP - CCT La Plata - CONICET, Argentina

Abstract

Hilda asteroids are objects that orbit the Sun at a distance where the orbital period is exactly 2/3 of the orbital period of Jupiter. Previous analytical and numerical studies have focused their attention to the great dynamical stability within the orbital region where the Hilda asteroids are found. However the group is indeed dispersing. In this paper we are going to review the contribution of Hilda asteroids to the Jupiter neighborhood. Based on simulations of long term dynamical evolution of escaped Hilda asteroids, it is find that 8% of the particles leaving the resonance end up impacting Jupiter. Also they are the main source of small craters on the Jovian satellite system, overcoming the production rate by comets and Trojan asteroids. Almost all the escaped Hildas pass through the dynamical region occupied by JFCs, and the mean dynamical life time there is 1.4×10^6 years. $\sim 14\%$ of JFCs with $q > 2.5$ A.U. could be escaped Hilda asteroids. We analyzed also the possibility that the Shoemaker - Levy 9 was an Hilda asteroid.

1. Introduction

Asteroids are small solid bodies that orbit the Sun. In general they are rocky, some ones have a metal composition, and others are rich in water ice, other volatiles and organic materials. To this last group belong the Hilda asteroids. They are placed at $\sim$ 4 AU in the $3:2$ mean motion resonance with Jupiter, namely in the time that Jupiter completes two periods an Hilda asteroid completes three. It is a numerous group of asteroids that make evident the protecting mechanism present in this resonance. The angle $\varphi = 3\lambda_J - 2\lambda - w$, where λ_J and λ are the mean longitudes of Jupiter and the asteroid respectively and w is the longitude of perihelion of the asteroid, is called critical angle and librates around zero. Then, when the asteroid is in conjunction with Jupiter ($\lambda = \lambda_J$), it is near perihelion ($\lambda \sim w$) and far from the planet. The configuration of Hildas with respect to Jupiter, and the

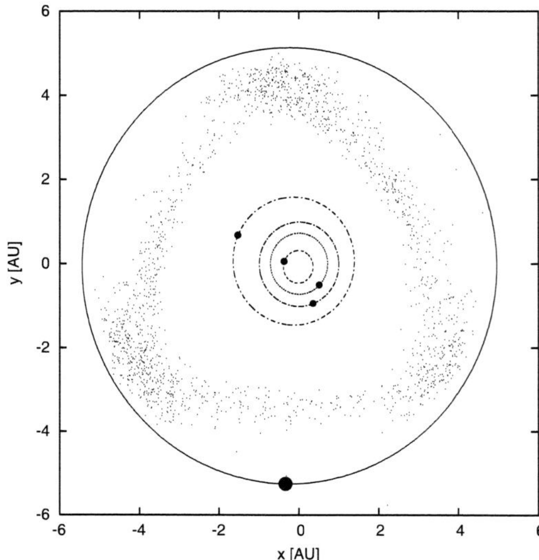

Figure 1. Instantaneous positions of Hilda asteroids (dots) in the ecliptic coordinate system, projected onto the ecliptic plane for the epoch UT 0 h 27 October 2007 . The orbits of Jupiter and the terrestrial planets are drawn for scale. The black points represent the location of the planets from Mercury to Jupiter, for the Hildas osculating epoch.

mentioned protecting mechanism can be seen in Fig. 1. There it is plotted the instantaneous positions of Hilda asteroids obtained from the ASTORB database of the osculating orbital elements for asteroids (maintained by E. G. Bowell of Lowell Observatory), projected onto the ecliptic plane. There it is also shown the location of the planets from Mercury to Jupiter for the Hildas osculating epoch, and their orbits.

The central zone of the 3:2 resonance has a great dynamical stability, where an asteroid can last for the age of the solar system (Nervorný et al. 1997, Ferraz Mello et al. 1998). However, the dynamical evolution of Hildas and mainly their collisional activity can remove them from this protected zone and they can quickly reach the unstable regions that surround the stable zone. Then they can reach a very chaotic boundary, where the characteristic permanence times are very short and they will be quickly ejected.

In this paper we are going to review the contribution of Hilda asteroids to the Jupiter neighborhood. We will see the dynamical evolution of Hilda asteroids that are recently removed from their stable orbits, their relation to Jupiter Family Comets (JFC) and the contribution to the cratering rate on the major satellites of Jupiter.

The general dynamics of Hildas recently removed from the resonance is addressed in section 3.

Comets, in particular Jupiter Family Comets (JFCs), are minor bodies whose dynamical evolution is mainly controlled by Jupiter and so they pass through the Jupiter's neighborhood. JFCs come from the Trans Neptunian region (Fernández, 1980), they have unstable orbits and a short lifetime. They are formed by a solid nucleus consisting of intimately mixed ices and dust. When they approach to the Sun, the ices sublimate forming the coma and the tail. Hilda asteroids are in the external zone of the asteroid belt and they could also contain volatile ices, like comets. In fact they have D and P taxonomic class that can be considered as spectral analogous of comets. However Hilda asteroids can't approach enough to the Sun to have the possibility of sublimate their volatiles. They could be dormant comets, this is a comet nucleus with no detectable activity, but if delivered to lower semi major axes, it could be reactivated, according to the definition by Hartmann et al. (1987). After escape from the resonance, Hildas follow dynamical routes that are controlled by Jupiter (Di Sisto et al. 2005), they can reach low perihelion distances and then they could have the possibility of sublimate their volatiles. JFCs are mainly found in the same zone of escaped Hildas (Di Sisto, et al. 2005). Also there are JFCs associated with the Hilda asteroid region (Toth, 2006) and with its boundaries (Kresák, 1979). This last group, called the quasi-Hilda group of comets, is very important because it could give the keys of the relation between asteroids and comets.

The relation between asteroids and comets is a vastly studied subject but not an already closed one. There are many questions that continue still open and in discussion. In particular the relation between comets, Hildas and quasi-Hildas could clarify the question. Toth (2006), makes an update of the inventory of this group. He found 11 new members of the quasi-Hilda cometary group of comets making a total of 15, and 23 outliers of the Hilda asteroid family in the Lagratian element space. He suspect that there could be some dormant or extinct comet nuclei among these asteroidal objects. If this were the case, the cometary nature of Hildas could be proved. We are going to attend the relation between Hildas and comets in section 4.

Evidence of impact cratering can be found in most of the solar system planets and satellites as well as on the population of the minor bodies. The collision among bodies was the fundamental mechanism during planetary accretion but also we have evidence of a present impact process at a slow rate. The collision of the Shoemaker-Levy 9 (SL9) comet with Jupiter in 1994 was a recent example of an impact between two celestial bodies at present. Craters are important markers in establishing sequences of geologic events though also, impact structure forms may have changed through geologic time. So the study of the distribution of impact craters and their surrounding terrain is very important in order to establish an age sequence. Then, the study of the origin of the craters of a celestial body provide important information of both, the source and the target. As we mentioned, Hilda asteroids that escaped from the resonance are under the control of Jupiter, so we could expect there would be collisions between Hildas and the satellites of Jupiter. In particular we can study the Hilda cratering contribution to the four Galilean satellites: Io, Europa, Ganymede and Callisto. We are going to analyze this topic in section 5.

2. Hilda Asteroid Group

As we mentioned, in the 3:2 mean motion resonance with Jupiter it is present an important population, the Hilda asteroid group. They librate in a thin fringe in semimajor axis of 0.1 AU wide centered at $\sim$ 3.97 AU. Up to the present there are 1744 Hildas cataloged in the ASTORB database.

They have a wide range of eccentricities but the great majority is clustered in the interval $0.1 < e < 0.3$, the inclinations extend up to $\sim 42°$. In Fig 2 we can see the distribution of semimajor axis and eccentricity of Hildas and their relative location to the Main Asteroid Belt.

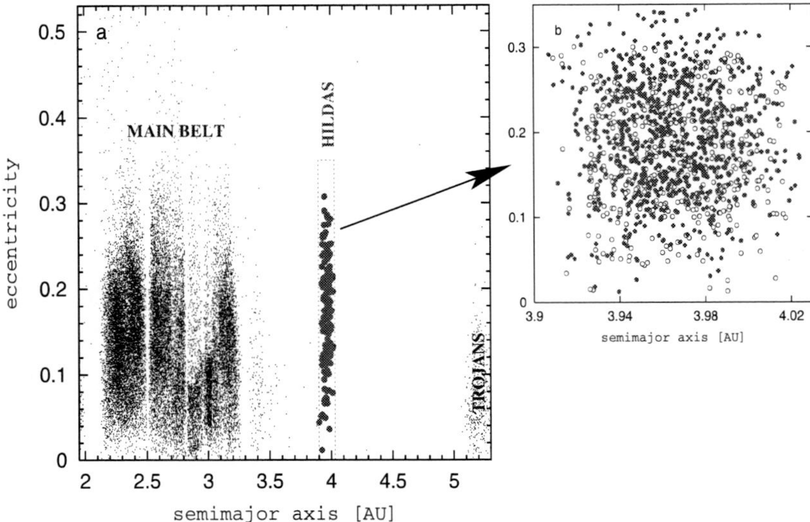

Figure 2. Asteroids in the a vs e plane and in detailed the Hilda zone where the red points are the real Hildas and the black empty circles are the fictitious Hildas generated in the simulation by Di Sisto et al. (2005).

This population is characterized by a central zone of a great stability (Nervorný et al. 1997, Ferraz Mello et al. 1998), but is surrounded by unstable regions. If an asteroid reaches these limit zone it can be quickly ejected from the resonance. Gil-Hutton and Brunini (2000) have shown that mutual collisions is the most efficient mechanism that, at present, is injecting Hilda asteroid fragments into the unstable regions beyond the boundaries of the resonance. The dynamical diffusion is very slow and then it should be taken into account the evaporation by mutual collisions to study the evolution of escaped Hildas (Brunini et al., 2003).

The biggest asteroid in the group is 153 Hilda with a diameter of 170.6 km. The whole population is characterized by a cumulative power low size distribution function in diameter (D), given by:

$$N(> D) \propto D^{2.11}, \tag{1}$$

The sample could be complete up to $D \sim 12$ km (Brunini et al. 2003). They are placed in the external zone of the asteroid belt, so they should have ices in their composition. Spectroscopic studies of Dahlgren and Lagerkvist (1995) and Dahlgren et al. (1997), reveal that Hilda asteroids are mainly of taxonomic classes D and P, which are associated to organic materials and ice enriched silicates. They also obtain a correlation between the taxonomic class and the size of Hildas. There are more D - type particles at smaller sizes.

However Hildas should have ices, they could be integrated or separated from the other material. If ices were integrated with the other material, sublimation could be more difficult and we could not detected it from Earth. But sublimation is also restricted by the formation of carbon mantles on the asteroid surface because their long exposure to the solar radiation in their stable resonance zone. So it is difficult from observations to really say the state of ices in Hilda asteroids. Only more dedicated observations could give light to this point.

3. Dynamical Evolution of Hilda Asteroids Removed from the Resonance

Di Sisto et al. (2005), studied the Hilda asteroid family as another probable source of JFCs. They performed a numerical integration of 500 fictitious Hilda asteroids under the gravitational influence of the Sun and the planets from Mercury to Neptune.

In Fig 2 we plot all asteroids in the semimajor axis vs eccentricity plane, and in detailed the Hilda zone, where the red points are the real Hildas and the black empty circles are our fictitious Hildas.

The sample was integrated for $1 \times 10^9 y$ which is an interval comparable to the age of the solar system. In this integration they followed the dynamical evolution of Hilda asteroids that escaped from the resonance. They obtained that 78.2% of them are ejected from the resonance and the rest, the 21.8% remains stable within it. In most of the cases, the asteroids left the resonance increasing first their eccentricity, and then their semimajor axis, in such a way that in the dynamical evolution they pass through the external solar system, having successive close encounters with the outer planets, mainly with Jupiter. The general dynamics of escaped Hildas is dominated by Jupiter's scattering. Therefore, at any time, most escaped Hildas may be found near Jupiter's orbit.

Being escaped Hildas under the gravitational control of Jupiter, one usual path of the dynamical evolution of them is the residence into a mean motion resonance with Jupiter (other than the 3:2) for some time. This behavior can be observed, as an example, in the dynamical evolutions of three particles of the simulation by Di Sisto et al. (2005) shown in Fig. 3. In particular in Fig 2 b and c the particles go through the 1:1 resonance with Jupiter. The first particle transits that state from 3.1305×10^8 years to 3.1317×10^8 years, this is an interval of ~ 120000 years. The second one transits the 1:1 resonance from 590000 years to 685000 years, this is an interval of ~ 95000 years. Other frequent state of escaped particles is the pass for a Kozai resonance. A Kozai resonance occurs when the argument of pericentre, w, librates about a constant value. For low inclinations it is possible for w to librate about $w = 0°$ and $w = 180°$, and for large inclinations about $w = 90°$ and $w = 270°$. The semimajor axis of the object remains constant but the eccentricity and the inclination of the orbit are coupled in such a way that e is a maximum when i is a minimum, and vice versa.

4. Relation with JFCs

Comets and asteroids represent different parts of the planetesimal spectrum formed and processed in the Primitive Solar Nebula. The distinction between asteroids and comets is based in their observational qualities and orbital characteristics, rather than in their different properties or chemical composition. Comets are characterized by their *coma* of sublimated gas and dust, that gives them their typical aspect in the sky, but asteroids appear like light points in the sky.

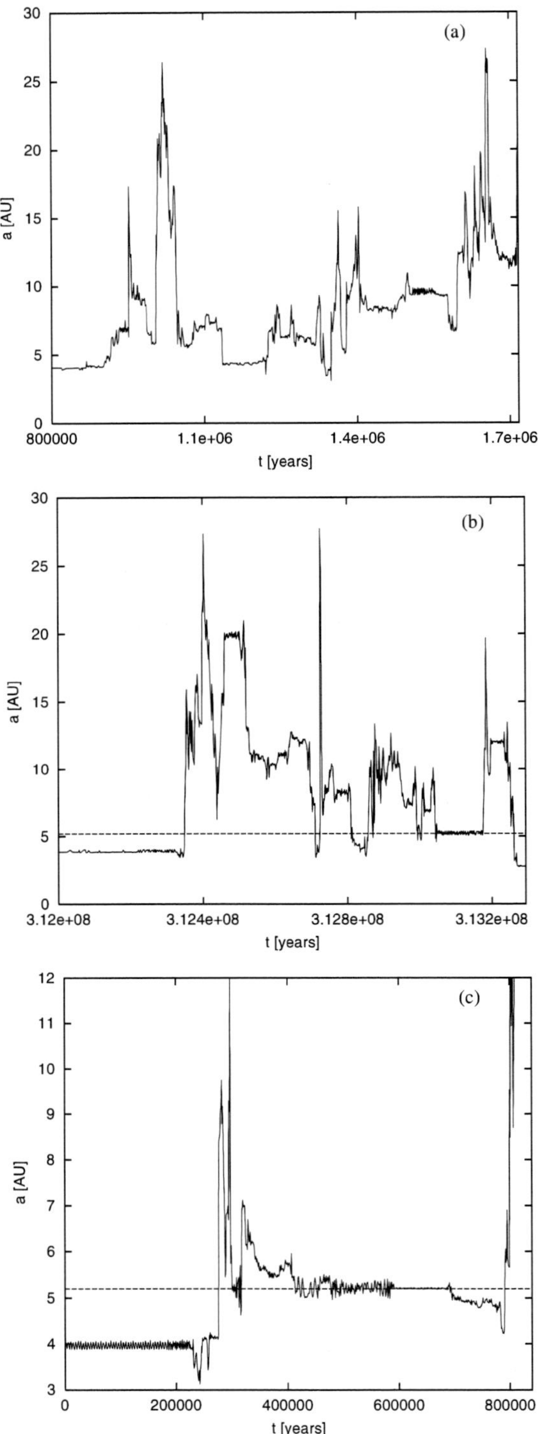

Figure 3. Dynamical evolution of three escaped Hildas. In fig 2 b and c the particles go through the 1:1 resonance with Jupiter at $a = 5.2$ AU for considerable intervals of time. This resonance is marked with the dashed line in both plots

Asteroids and JFC are indeed two dynamically different populations. Comets have highly eccentric orbits, being the great part of their orbits in the outer solar system, where their ices are preserved against depletion by solar heating. Asteroids have more circular orbits in proximity to the Sun and thus received much more solar heating over their lifetimes, making the survival of ices more difficult. The dynamical distinction is frequently defined by the Tisserand parameter with respect to Jupiter, T_j. The Tisserand parameter is an approximation of the Jacobi constant of motion in the circular restricted three body problem (CR3BP). In the problem with respect to Jupiter, the Tisserand parameter is given by

$$T_j = \frac{a_j}{a} + 2\sqrt{\frac{a}{a_j}\left(1 - e^2\right)}\ cos\, i, \tag{2}$$

where a, and e are the semimajor axis and eccentricity of the comet (or asteroid) and i is the inclination of the orbit with respect to the orbital plane of Jupiter.

The limit for the division between asteroids or comets is $T_j = 3$, since objects with $T_j < 3$ are dynamically coupled to Jupiter and then are considered cometary, while objects with $T_j > 3$ are decoupled to Jupiter and are considered asteroidal.

But although Hildas, in the external Main Belt, are in stable resonant motions and are protected from encountering Jupiter, once they escape from the resonance, they are dominated by the gravitation of Jupiter, and in this sense, they can resemble the dynamical state of a JFC.

Di Sisto et al. (2005) follow the evolution of escaped Hildas in order to determine if they pass some time like a JFC, assessing the population that, at present, contaminate the "genuine" population of JFC coming from the Trans Neptunian region. They define the JFC zone taking into account that JFCs pass through the Hill Sphere of Jupiter and have mostly semimajor axis between 2 and 9 AU and eccentricities less than 0.9. Besides, the comets are outside the 3:2 resonance. This criterion is almost equivalent to the range of Tisserand constant with respect to Jupiter $2 < T < 3$ proposed by Levison (1996) for the classification of JFCs. They obtained that from the bodies that left the resonance zone, 98.7% live at least 1000 years in the JFC region being their average residence time in this region of phase space 1.4×10^6 years. Also they calculated the number of escaped Hildas that are at present in the JFC region taking into account the collisional evaporation rate obtained in Brunini et al. (2003). Escaped Hildas mainly populate the JFC zone with $q > 2.5$ AU, and the contribution to the inner zone is negligible. It is expected that ~ 143 Hilda asteroids with diameter grater than 1 km are at present among the population of JFCs with $q > 2.5$ A.U., so they would be the $\sim 14\%$ of the JFC in that region. Although because of the physical life time of a comet is only 5 - 20 % of their dynamical lifetime, it could be that almost all the kilometer sized Hildas were dormant comets. They also estimate that the contribution of escaped Trojan asteroids to JFCs would be negligible.

Alvarez-Candal and Licandro (2006) studied the population of asteroids in cometary orbits (ACOs) in order to discriminate its cometary or asteroidal origin based on the comparison of cumulative size distributions of asteroids and comets. The ACOs are object with $T_j < 3$, so they would be cometary by the classification given above, but they do not present any signature of cometary activity. They divided the ACO population in two sub samples of ACOs in NEO (Near Earth Objects) orbits and non NEO orbits. They also analyzed the size distribution of Hildas and Trojans. They conclude that unstable Trojans and Hildas

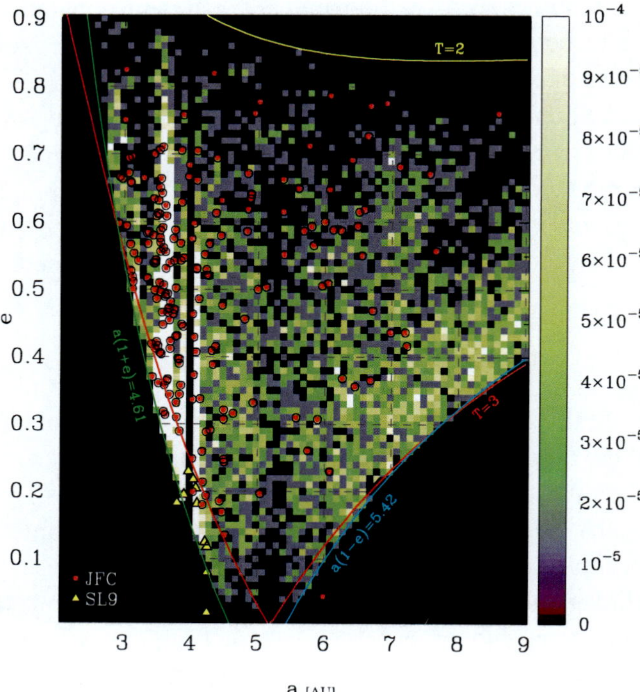

Figure 4. Time-weighted distribution of escaped Hildas in the a vs e space of the JFC zone. The red circles are the observed JFC. The yellow triangles are the pre-capture orbital elements of SL9 fragments (Chodas and Yeomans 1996)

have size distribution indices similar to that of the NEO sample of ACOs, suggesting that a fraction could be scattered objects from the outer main belt. However Di Sisto et al. (2005) found that the contribution of escaped Hildas to the NEO population is negligible. On the other hand Alvarez-Candal and Licandro (2006) analyzed the size distribution index of JFCs and found that it is similar to that of the non-NEO sample of ACOs, suggesting that JFCs could be an important source of the non-NEO population.

Di Sisto et al. (2005) also give a picture of how escaped Hilda asteroids populate the JFC zone and so the Jupiter's neighborhood. In its Fig. 4, that is reproduced here in Figs. 4 and 5, they plot the distribution of residence times or time-weighted distribution of escaped Hildas obtained in the simulation in the orbital element space. They are also plotted the observed JFCs. As we see most of the JFCs are in the denser regions of escaped Hildas. The zone near the boundaries of the resonance is where escaped Hildas spend more time, then they increase their eccentricities leaving the stable region of the resonance. The break of the protection mechanism (libration around $\varphi \sim 0$) makes it possible close encounters with Jupiter. This causes the increase of their semimajor axis, crossing Jupiter's orbit and leaving quickly the JFC region. It is also seen a depletion of Hildas in the neighborhood of Jupiter's orbit at semimajor axis of $a_J \sim 5.2$ AU. A particle that crosses the Jupiter' s orbit will have a close encounter with the planet that quickly removes it from the zone. This causes the depletion of asteroids with $a \sim a_J$

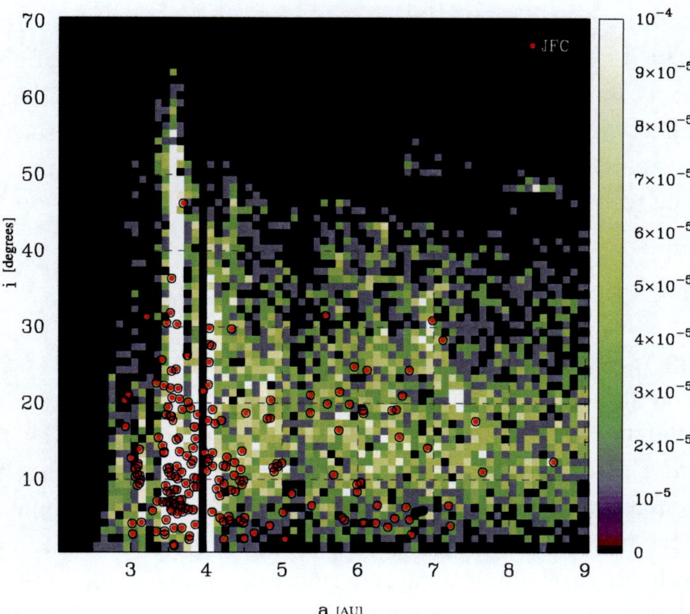

Figure 5. Time-weighted distribution of escaped Hildas in the a vs i space of the JFC zone. The red circles are the observed JFC.

4.1. The Relation with Quasi-Hildas and the SL9 Case

The study of the interrelation between the quasi Hilda cometary group of comets and Hilda asteroids could be broach from a dynamical point of view and also from a compositional point of view. The presence of comets inside or in the borders of the resonance is an evidence, both dynamical and physical, of the connection between them. As it is seen in Figs. 3 and 4 the probability of finding objects in the borders of the resonance is high and then is probable that comets observed in this zone be Hilda asteroids or fragments that have left the stable zone of the resonance. If they really were, the detection of activity would reveal the presence of ices in those asteroids and their comet nature. Although, the 3:2 resonance is a very stable zone, so Hildas have been there for a long time, and they could have slowly lost their surface ices by the continuous action of the Sun. Other phenomenon that is present is the formation of irradiation mantles on the asteroid surface. In any case the collisional activity produces fragments that could expose their trapped ices and then escape like potential comets. If we suppose that Hilda asteroids are really dormant comets, this population could be an important source of quasi-Hildas and also JFCs. They could be active or dormant, depending on their real composition and the time they have been exposed to the solar radiation.

A particular case is that of the Shoemaker-Levy 9 (SL9) comet that allow us to see live a collision of two celestial objects in the space. SL9 was tidally split into numerous fragments in 1992 after its perijove passage and then struck Jupiter in 1994. There was evidence of dust tails and out-gassing. Benner and Mc.Kinnon (1995), calculated the pre-capture orbital elements of SL9 and found two possible pre-capture orbits because of the chaoticity of the encounter. One of the possible pre-capture orbital element set overlaps the space of orbital elements of escaped Hildas and also the quasi-Hilda zone. This set of pre-capture orbital elements of SL9 is shown in Fig 4. So, at first, by virtue of this overlap we can say that an escaped Hilda could have been the parent of the fragments that struck Jupiter in 1994. Therefore, there is a chance that SL9 was an Hilda asteroid, and if it was, its pre-capture elements could be restricted.

5. Impact Cratering on the Galilean System

Impact cratering was a common process in the solar system as it is evident from the signatures left on the surfaces of most of its planets and satellites as well as on the population of the minor bodies. The collision among bodies was the fundamental mechanism during planetary accretion. Although the impact rate has been decreasing as the solar system was stabilizing, we have a considerable amount of evidence of an intense impact process in the past that continues at present at a much more slow rate. The determination of the present cratering rate on the surfaces of the solar system bodies furnishes one of the most fiducial ways of dating the geological time scale of activity.

The impact process of the inner solar system has been vastly studied, mainly, trough examinations of lunar cratering. To study the impact cratering history of bodies in the outer solar system the most natural and appropriate scenario are the four Galilean satellites: Io, Europa, Ganymede and Callisto. Io has no known impact craters. Europa, is an icy world crossed by a network of dark fractures. The existence of few impact craters suggest that geologic processes are active today, and that it has a young surface. The largest Galilean satellite, the icy moon Ganymede, has a dark, heavily cratered terrain with more recent brighter grooved terrain. It also shows evidence of geologic activity, but its surface is very old, dating from the Late Heavy Bombardment or earlier. Also Callisto is very old. Its surface is completely saturated by craters. One feature unique to Callisto is the remnant structures of numerous impacts.

At looking for populations capable of producing craters on the Galilean surfaces we should investigate those objects that may be found at the distance of Jupiter. Zahnle et al. (1998), studied a number of populations that produced the craters on the Galilean satellites. They discussed in detail the Ecliptic Comets, using the numerical study of Levison and Duncan (1997). Other sources, like Long Period Comets, Trojan asteroids and asteroids from the main belt were also considered. Brunini et al. (2003) have investigated the possible contribution of escapees from the 3:2 mean motion resonance with Jupiter to the present cratering rate on the Galilean satellites. They shown that 8% of the objects leaving the 3:2 resonance impact Jupiter. Therefore, it would be natural that a fraction of them could also impact some of the Galilean Satellites. The impact rate on the surfaces of the four Galilean satellites was computed by Brunini et al. (2003) by assuming that after escaping from the resonance by collisional evolution, Hilda fragments follow a dynamical evolution similar to the one obtained in their numerical simulation of escaped Hildas (Di Sisto et al. 2005). From the 8 % of the fragments that end up impacting Jupiter, a small fraction (from 10^{-4} to 10^{-5}) impacts a given Galilean satellite. They do so at a velocity ranging from ~ 15

Table 1. Mean interval between impacts on the Galilean satellites in years, capable to produce craters with $D > 10km$ km and $D > 4km$. For these calculations the radius of the minimum target considered to be member of the Hilda population was $0.4km$.

Cratering rate

Satellite	$\Delta T(D = 10km)$	$\Delta T(D = 4km)$
Io	$2.1 \times 10^7 y$	$2.0 \times 10^6 y$
Europe	$1.5 \times 10^7 y$	$1.8 \times 10^6 y$
Ganymede	$1.1 \times 10^7 y$	$1.3 \times 10^6 y$
Callisto	$2.6 \times 10^7 y$	$3.3 \times 10^6 y$

kms^{-1} for Callisto to $\sim$ 30 kms^{-1} for Io. These impacts produce craters on the Galilean satellite surfaces whose diameter depend on several factors (e.g. the impact velocity, the mass of the target and the projectile mass).

The mean interval between the production of craters of two different diameter D computed by Brunini et al. (2003) are shown in Table 1.

The other main source of craters are the Ecliptic comets. Zahnle et al. (1998) computed the mean intervals between craters produced by them. They are 2.6×10^7 years on Io, 1.5×10^7 years on Europe, 1.0×10^7 years on Ganymede and 2.6×10^7 years on Callisto. These timescales are comparable to the time scales for the production of craters of $D > 10km$ by Hilda asteroids, but are longer than the time scales for $D > 4km$. Therefore, Brunini et al. (2003) conclude that Hilda asteroids is the main source of small craters on the surfaces of the Galilean satellites.

Europa has only 27 known craters with $D > 4km$ (Moore et al. 2001). Although Bierhaus et al. 2001 states that the vast majority of small craters are secondaries, probably all craters larger than 2 km are primary craters. So, $D > 4km$ seems to be a safe election for primary craters.

One difference between the cratering rates from Hilda asteroids and Ecliptic comets is regarding the time-scales involved in the process. Brunini et al. (2003) shown that nearly 75% of the objects that impact Jupiter do so in less than 5×10^4 yr after escape from the resonance. It represents a very short time scale compared with the 72 My that typically spends a comet to travel from the Kuiper Belt to the Jupiter region (Di Sisto and Brunini 2007). We could speculate that after a catastrophic collision, many fragments escape from the Hilda region. A considerable number of them end-up colliding Jupiter ($\sim$ 8 %) during the relatively short time interval of some 10^4 yr. If some signature of this kind of process is found on the surfaces of the Gallilean satellites, it would furnish important clues regarding the past history of the Hildas.

6. Conclusion

In this paper we have review the contribution of Hilda asteroids to the Jupiter neighborhood. We have seen how escaped Hildas, recently removed from their stable orbits populate the zone of JFCs and its contribution to the cratering rate on the major satellites of Jupiter. We can point the following items:

- Hilda asteroids contribute with $\sim 14\%$ to the JFC population with $q > 2.5$ A.U.. Almost all the escaped Hildas pass through the dynamical region occupied by JFCs, and the mean dynamical life time there is 1.4×10^6 years. The more abundant taxonomic class in the Hilda group is D, with more D asteroids with less diameter. Then, as escaped Hildas are fragments resulting from a catastrophic collision, it is more likely that the escaped objects were D types. This is the more primitive class and the more similar to comets.

- Escaped Hildas spend more time in the zone near the boundaries of the resonance. If Hilda asteroids were really dormant comets, this population could be an important source for quasi-Hildas.

- One of the two sets of pre-capture orbital elements of SL9 obtain by Benner and Mc.Kinnon (1995) overlaps with the orbital elements of the Hildas in the JFC zone, supporting the idea that it would be an Hilda asteroid, and if it was, the pre-capture orbital elements would be defined.

- From numerical simulations of the dynamical evolution of fictitious Hilda asteroids, 8 % of the particles that leave the resonance hit Jupiter. Hilda asteroid population could be the main source of small impact craters on Jupiter satellites, overcoming the contribution of ecliptic comets calculated by Zahnle et al. (1998).

- There are some open issues in relation to the escaped Hildas that is convenient to point out. The existence of the quasi-Hilda comets in the borders of the 3:2 resonance could give the keys of the relation between asteroids and comets. It is needed more observations of Hilda asteroids directed to detect activity and study their surface compositions in order to understand the real nature of Hildas. On the other hand, there is a short time scale involved in the production of craters on the Galilean satellites from escaped Hildas. So this process must leave some signature, for example sets of small craters on the surfaces of the satellites produced by many escaped fragments of a catastrophic collision in the Hilda zone. These characteristics could be searched by dedicated spatial missions.

References

[1] Alvarez-Candal, A. and Licandro, J. The size distribution of asteroids in cometary orbits and related populations. *A&A* **458**, Issue 3, 1007-1011 (2006).

[2] Benner, L.A.M., Mc.Kinnon, W.B. On the orbital evolution and origin of Comet Shoemaker-Levy 9. *Icarus*. **118**, 155-168 (1995).

[3] Bierhaus, E. B., C. R. Chapman, W. J. Merline, S. M. Brooks and E. Asphaug. Pwyll Secondaries and Other Small Craters on Europa. *Icarus*. **153**, 264-276 (2001).

[4] Brunini, A., Di Sisto R.P., and Orellana R.B. Cratering rate on the jovian system: the contribution from Hilda asteroids. *Icarus*. **165**, 371-378 (2003).

[5] Dahlgren, M. and C.-I. Lagerkvist. A study of Hildas asteroids. I. CCD spectroscopy of Hilda asteroids. *Astronomy and Astrophysics*. **302**, 907-914 (1995).

[6] Dahlgren, M., C.-I. Lagerkvist, A. Fitzsimmons, I.P. Williams. and M. Gordon. A study of Hildas asteroids. II. Compositional implications from optical spectroscopy. *Astronomy and Astrophysics*. **323**, 606-619 (1997).

[7] Chodas, P., W. & Yeomans, D., K. *IAU Colloq. 156: The Collision of Comet Shoemaker-Levy 9 and Jupiter* (eds Noll, K. S., Feldman, P. D. & Weaver, H. A.), 1-30 (1996).

[8] Di Sisto R. P., A. Brunini, L. D. Dirani and R. B. Orellana. Hilda asteroids among Jupiter family comets. *Icarus* **174**, 81-89 (2005).

[9] Di Sisto R. P. and A. Brunini. The origin and distribution of the Centaur population *Icarus* **190**, Issue 1, 224-235 (2007).

[10] Fernández, J. A. On the existence of a comet belt beyond Neptune. *Mon. Not. R. Astron. Soc.* **192**,481-491 (1980).

[11] Ferraz-Mello, S., Michtchenko, T. A. Nesvorný, D., Roig, F. and Simula, A. The depletion of the Hecuba gap vs. the long-lasting Hilda group. *Planet. Space Sci.*. **46**, No 11/12 1425-1432 (1998).

[12] Gil-Hutton, R. & Brunini, A. Collisional Evolution of the Outer Asteroid Belt. *Icarus* **145**, 382-390 (2000).

[13] Hartmann, W.K., D.J. Tholen and D.P.Cruikshank. The Relationship of active Comets, "Extinct" Comets, and Dark Asteroids *Icarus* **69**, 33-50 (1987).

[14] Kresák, L. Dynamical interrelations among comets an asteroids. In Asteroaids (T. Gehrels, Ed.) pp. 289-309. Univ. of Arizona Press, Tucson (1979).

[15] Levison, H. Comet taxonomy. In *Completing the Inventory of the Solar System* (T.W. Retting and J. M. Hahn, Eds.), Astron. Soc. of Pacific Conference Proceedings, **107**, 233-244 (1996).

[16] Levison, H. and M. Duncan. From the Kuiper Belt to Jupiter-family comets: The spatial distribution of ecliptic comets. *Icarus* **127**, 13-32 (1997).

[17] Moore, J. M., and 25 collegues. Impact features on Europa: Results of the Galileo Europa Mission. *Icarus* **151**, 93-111 (2001).

[18] Nervorný, D. and Ferraz Mello, S. On the Asteroidal Population of the First-Order Jovian Resonances. *Icarus* **130**, 247-258. (1997).

[19] Toth, i. The quasi-Hilda subgroup of ecliptic comets - an update (Research Note) *A&A* **448**, 1191-1196 (2006).

[20] Zahnle, K., Dones, L. and H.F. Levison. Cratering Rates on the Galilean Satellites. *Icarus*. **136**, 202-222 (1998).

[11] Ferraz-Mello, S., Michtchenko, T. A. Nesvorný, D., Roig, F. and Simula, A. The depletion of the Hecuba gap vs. the long-lasting Hilda group. *Planet. Space Sci.*. **46**, No 11/12 1425-1432 (1998).

[12] Gil-Hutton, R. & Brunini, A. Collisional Evolution of the Outer Asteroid Belt. *Icarus* **145**, 382-390 (2000).

[13] Hartmann, W.K., D.J. Tholen and D.P.Cruikshank. The Relationship of active Comets, "Extinct" Comets, and Dark Asteroids *Icarus* **69**, 33-50 (1987).

[14] Kresák, L. Dynamical interrelations among comets an asteroids. In Asteroaids (T. Gehrels, Ed.) pp. 289-309. Univ. of Arizona Press, Tucson (1979).

[15] Levison, H. Comet taxonomy. In *Completing the Inventory of the Solar System* (T.W. Retting and J. M. Hahn, Eds.), Astron. Soc. of Pacific Conference Proceedings, **107**, 233-244 (1996).

[16] Levison, H. and M. Duncan. From the Kuiper Belt to Jupiter-family comets: The spatial distribution of ecliptic comets. *Icarus* **127**, 13-32 (1997).

[17] Moore, J. M., and 25 collegues. Impact features on Europa: Results of the Galileo Europa Mission. *Icarus* **151**, 93-111 (2001).

[18] Nervorný, D. and Ferraz Mello, S. On the Asteroidal Population of the First-Order Jovian Resonances. *Icarus* **130**, 247-258. (1997).

[19] Toth, i. The quasi-Hilda subgroup of ecliptic comets - an update (Research Note) *A&A* **448**, 1191-1196 (2006).

[20] Zahnle, K., Dones, L. and H.F. Levison. Cratering Rates on the Galilean Satellites. *Icarus*. **136**, 202-222 (1998).

In: Space Exploration Research
Editors: J.H. Denis and P.D. Aldridge

ISBN: 978-1-60692-264-4

Chapter 18

OBSERVATION OF CHAOTIC PROCESSES IN EARTH ROTATION

Valérie Frède *

ECCD-Octogone University of Toulouse, UFE Paris Observatory, France

ABSTRACT

Astronomy research is concerned with the precise description and understanding of Earth orientation parameters (polar motion and length of day), time series that are obtained from several terrestrial and space observational techniques. The Earth rotation time series has been extensively studied from a linear point of view, considering that the underlying dynamics were composed of well-defined oscillations above a level of noises. If this can apply to the main part of the signal, where periodic oscillations are observed, it becomes problematic when dealing with short-term fluctuations. Indeed, linear method would conclude that this part of the signal is only composed of noises. Nevertheless, as the main forcing processes of the short-term fluctuations of the Earth comes from atmospheric processes through angular momentum exchanges, and because it was demonstrated that their dynamics were mainly nonlinear, it appears necessary to investigate the research of nonlinear dynamical processes in Earth rotation data. This is a challenge that we have initiated and that needs to be integrated in future research.

We present in the chapter a review of several studies we have carried out on this topic and some perspectives of research. As an introductory part, we describe the main components of the Earth rotation signal and their astronomical or geophysical origin. In a second part, we explain the principles of nonlinear time series analysis and their theoretical links to the dynamical system theory. We demonstrate how these tools give the possibility to break with the traditional spectra analysis that fails to capture nonlinear processes. To illustrate those points, we present results from several studies (on the Earth's rotation itself and on the related atmospheric flows) where we have described chaotic components in a topological and a dynamical way. We explain why our results have important implications for future research in astronomy. Indeed, they show that deterministic processes can be found and have to be searched in the "noisy" part of the system. Additionally, the prediction of the Earth orientation parameters is still performed with linear methods. We encourage applying in the future nonlinear predicting tools for

* Dr. Valérie Frède, University of Toulouse, 56 avenue de l'URSS, 31078 Toulouse Cedex. France. valerie.frede@free.fr

the Earth's rotation. Moreover, the oceanic processes are also acting on the Earth and an analysis of oceanic time series can precisely measure how all of those dynamics interact together.

1. Introduction

The Earth is a complex body composed of parts whose chemical and dynamical properties are different and interact. Moreover, it is stressed to external gravitational constraints and possesses various oscillations. Each of the observed fluctuations of the Earth's rotation is of different origin and type. Their time duration goes from the Earth's age to several hours. Earth rotation was for a long time considered a stable process; for instance, length of day was a time reference until 1960. But thanks to the improvement in observation techniques, an important number of new fluctuations could be observed in the polar motion as well as in the length of day.

The study of the Earth orientation parameter variations (length of day and polar motion) concerns several research areas such as astronomy, geophysics, space research, and meteorology. The principal source of information regarding those parameters comes from observation that leads to time series. The study of the components of the time series gives information about the nature of the main oscillations. Those time series are either very simple (one dimensional series) or very complex (they are the results of several processes interacting). The difficulty lies in extracting the maximum of relevant information from them. Some parts of the fluctuations observed in the Earth orientation parameters remain unknown or partially explained. Those limits can come either from the noise level added to the data, the imprecision of some models, or from the employed methods in data analysis. Indeed, even if nonlinearities are dominant in nature, it is usual to apply linearization.

How do people analyse the time series in a linear way? They first of all evaluate the Fourier spectra and isolate the noise from the signal. The noise is characterised by continuous spectra while the signal is considered as a sum of independent oscillations. This approach excludes totally any nonlinear component even if it exists. If we consider that geophysical forces are mainly nonlinear, we can imagine the interest in using nonlinear methods to detect a nonlinear component in the Earth rotation time series.

Nonlinear time series analysis is thus different from Fourier transform. It gives the possibility to construct the asymptotic dynamic of a physical system whose information source is only one or several time series. It consists first in identifying the deterministic signal by separating in the continuous part of the spectra stochastic noises from deterministic processes. The analysis and description of the signal are not in the frequency domain anymore but rather in a multidimensional space called pseudo phase space. It is not the time series that is studied but a set of vectors whose components are the points of time series delayed in time. They form then a geometrical figure, called *attractor,* whose properties characterise the system's asymptotic behaviour. The invariant quantities are not the frequencies but the Lyapunov exponents (measure of the orbit's stability) and the dimensions of the attractor.

We will present in the next section a brief description of the Earth rotation time series components and the knowledge of their main forcing at the time we performed our nonlinear studies (Frede, 1999). Then, in section 3, we will highlight the limitation of the linear and usual methods. Section 4 is devoted to present the principles of nonlinear time series analysis.

In the last sections, the results of the application of those methods to Earth orientation parameters and linked phenomena such as atmospheric time series will be presented and discussed.

2. The Earth's Rotation: Synthetic Description

Various observational techniques give access to the Earth orientation parameters (EOP) under the form of time series. We will focus on the polar coordinates: PMX (polar motion in the x direction) and PMY (polar motion in the y direction) (Gross, 1992) and on the length of day (LOD).

The Polar Motion

The polar motion time series can be divided into several components: a tendency, a free oscillation (Chandler oscillation), seasonal terms (annual and semi-annual) and a short period term, principally considered as dominated by noise.

The long period variations concern the secular and decennial variations. The secular variation might come from the melting ice effect 10,000 years ago and from the slow inelastic answer of the Earth. It is called the "post-glacial rebound" (Gross, 1990; Ming and Danam, 1988). The decennial variations (Markowitz oscillation) have a period range of 20–30 years. The origin of this oscillation could be caused by terrestrial behaviours: transfer of mass due to the sea level, underground water and ices.

The free wobble of the Earth (whose period is about 433 days) presents unexplained amplitude and phase variations. Are they caused by stochastic fluctuations or are they systematic? The origin of the damping is unexplained (it can come from the inelasticity of the mantle and oceanic dissipation). Indeed, Chandler wobble is damped and an excitation source is necessary to maintain it. Various candidates have been proposed, but none of them nor their combined effect were enough to explain the oscillation: we can mention, for instance, the air mass fluctuations (atmosphere), the fluid mass movements (underground), the Earthquakes, and the electromagnetic or topographic torques between mantle and core due to the presence of a liquid core. Nevertheless, it appears that atmospheric pressure variations are the most important source of excitation of the Chandler wobble. Wahr (1983) showed that the atmosphere explained 25% of the total excitation. Thirty percent of the wobble is explained by the combined effect of atmospheric pressure and underground waters (Kuehne and Wilson, 1991). Earthquakes are too small to have a significant effect on the Chandler wobble (Gross, 1986). The study of the mantle-core torques showed that the pressure variations at the border could be of enough magnitude to excite the wobble (Runcorn et al., 1990). Observational proofs are still needed.

The seasonal terms (annual and semiannual oscillation of the polar motion) are very stable. The causes of annual and semiannual variations are mainly attributed to the atmospheric, oceanic and underground water variations.

In contrast, the short-term fluctuation part of the signal is difficult to predict because it is most of time assimilated to noise, more precisely geophysical noise or red noise. There are no apparent periods in the signal, and the observed oscillations are rapid. Nevertheless, the main excitation source of the rapid fluctuations is coming from the atmosphere (Eubanks et al., 1988). Indeed, a strong correlation between the short-term fluctuations and the atmospheric times series is observed. For periods from two weeks to several months the correlation is about 50–60%. The missing part is thought to come mainly from the ocean (Ponte, 1990).

The Length of Day

We saw that the observed variations in the polar motion are mainly due to mass redistributions in the Earth and its fluid parts. Concerning length of day, variations are mainly caused by wind friction and oceanic currents at the Earth surface. As for the polar motion, we can divide the signal into several time range components. Concerning the long period variations, the speed of rotation has a secular decrease caused by dissipation during the tides. The length of day is thus increased of about 1.4 ms per century. There are also decennial variations coming from electromagnetic and topographic torques at the core mantle boundary.

Gravitational torques create terrestrial and oceanic tides on Earth. We observe them under the form of periodic variations of length of day that we usually don't consider in the geophysical studies of length of day. The inter-annual variations (1–10 years) observed in the length of day are mainly correlated to the Southern Oscillation (SO) and to the Quasi Biennal Oscillation (QBO). Those two oscillations are quasi periodic oscillations appearing in atmospheric and oceanic parameters. SO manifests under the form of 2–6 years' variations in the pressure and sea lever measures. In particular the El Nino event is a hot water movement in the Pacific during several months. El Nino is known as the oceanic part of the SO oscillation and we refer to these phenomena under the name of ENSO (El Nino Southern Oscillation). ENSO is explained by the results of nonlinear interaction between air and water (Barnett et al., 1991). ENSO episodes are not periodic; a stochastic approach was initially considered and explained the ENSO variations as an external stochastic forcing on a linear model of Gaussian correlated noise. More recently ENSO was considered through a nonlinear approach and could be explained by the result of a deterministic chaos of low dimension caused by nonlinear interactions in the oceanic and atmospheric system in the Pacific. Stone et al. (1998) proposed a coupled model by examining the effect of a stochastic forcing on a nonlinear deterministic model of ENSO. To summarise, in this period range, atmospheric processes explain about 50% of the variance of the inter-annual variations of the length of day. The missing excitation sources are to be searched in the ocean mainly.

The seasonal variations are composed of annual variations and semi annual variations. They are produced mainly by the atmosphere (zonal and stratospheric winds). The oceanic contribution could be associated to the semi annual variations (Hopfner, 1998). We call "short period variations", the variations of the length of day that are under 100 days. They are used to be considered as stochastic noise without a real geophysical cause. It is the same thing for the short-term periods of the wind term of atmospheric and oceanic time series. The short-term variations of length of day are nevertheless mainly attributed to angular momentum exchanges between solid Earth and atmosphere (Rosen and Salstein, 1983; Eubanks et al., 1985).

3. LIMITATIONS OF THE METHOD OF ANALYSIS USED

In the studies based on Earth rotation time series and their forcing, a stochastic and linear approach was adopted, whereas the main processes in the geophysical fluids are nonlinear. For the main part of the signal it is not a problem (periodic oscillations) but for the short-term fluctuations, it is not convincing. Indeed, those dynamics are mainly caused by nonlinear processes (atmospheric and oceanic forcing). Thus, if nonlinear responses are present, the adopted methods can not observe and characterise them. This has a consequence on Earth parameters prediction, for instance. Indeed capturing the exact dynamics of the Earth will help improving the models of prediction of the EOP (Earth orientation parameters).

In Earth orientation parameters studies, it is usual to split the signal into several independent periodical components and the short-term fluctuations are described as coloured noises thanks to Fourier analysis. But working only in the frequency domain can lead to misinterpretation of the EOP dynamics. Indeed, it was shown that a Fourier transform can not distinguish between a coloured noise and a low dimensional deterministic chaotic process. From linear methods we can only detect linear signals. We will show how the use of nonlinear methods can lead to detect all the components of the signal (either linear or nonlinear) in a complementary way to usual methods. Figures 1a, 1b and 1c present the Fourier spectra of the short term (1–100 days) EOP parameters (respectively, PMX, PMY and LOD). We observe that the spectra are continuous; this could lead to the conclusion that EOP short- term fluctuations are stochastic processes, similar to coloured noises.

To illustrate the limitation of the Fourier analysis, we present also some examples of Fourier spectra of a chaotic series (obtained from Lorenz attractor, (Lorenz, 1963)), (figure 2a), a white noise (figure 2b), a coloured noise (figure 2c) and a biperiodic series (figure 2d). Except for the periodic series, where the Fourier method is very wall adapted, all the spectra are continuous and it is difficult to observe from the Fourier results that the first series (Lorenz attractor, figure 2a) is a deterministic and predictable one and not a stochastic noise (even if its spectra is continuous).

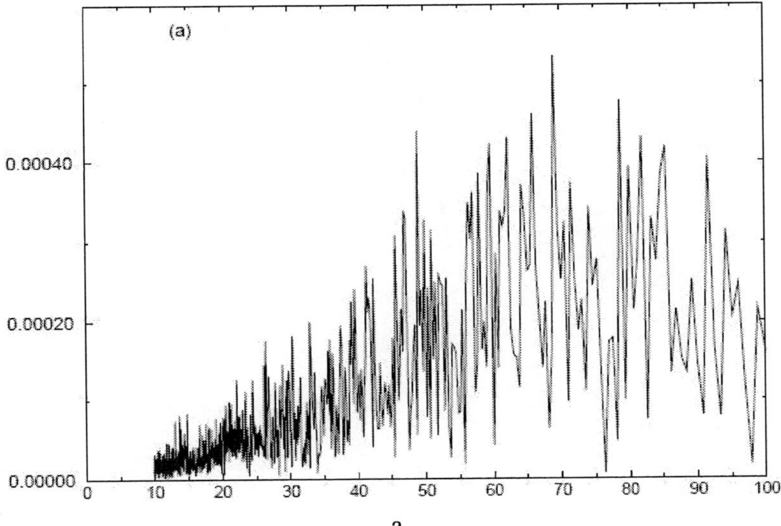

a

Figure 1. (Continued).

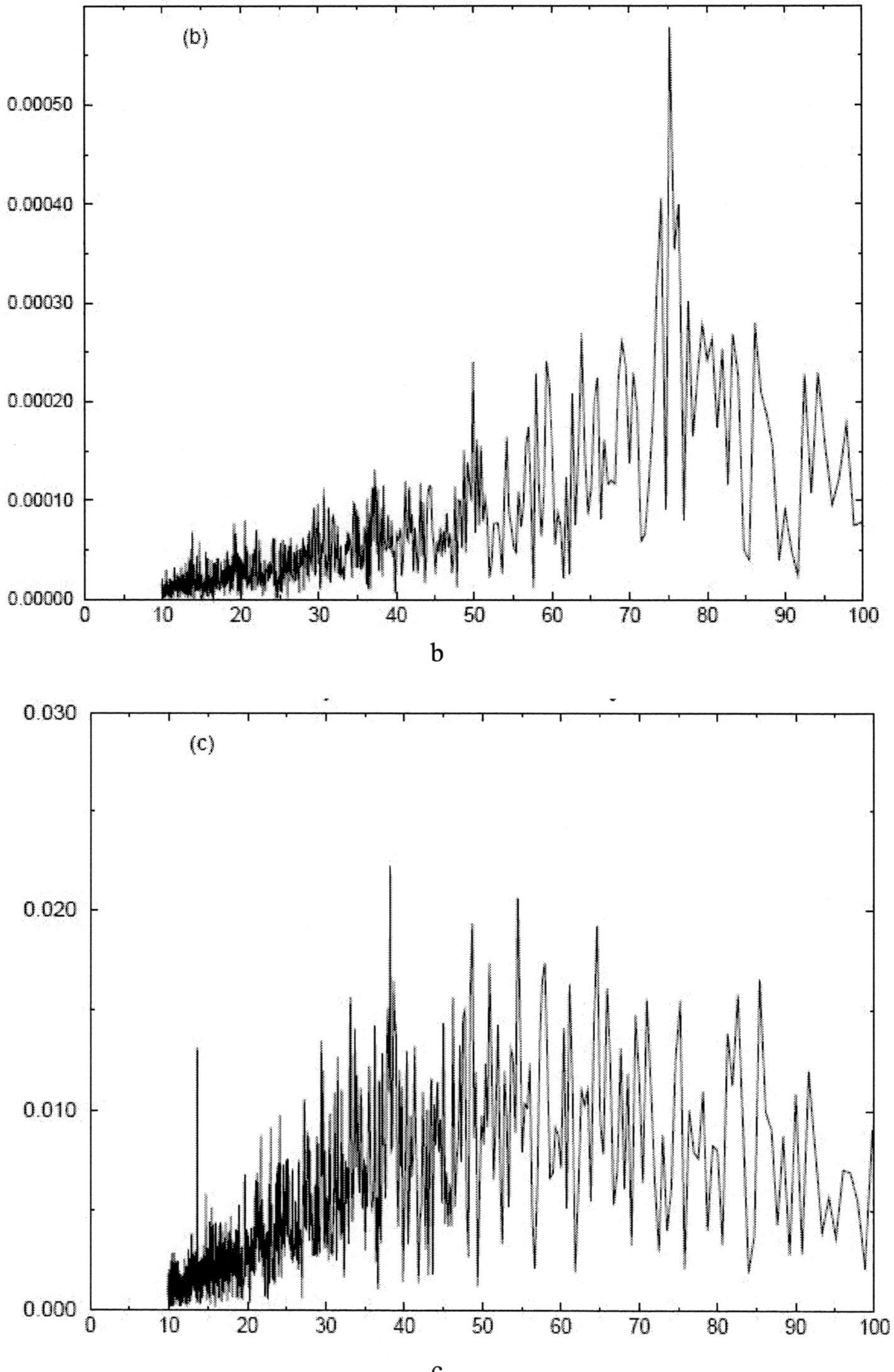

Figure 1 a,b,c. EOP spectra obtained from Fourier transform from the PMX (a, in mas), PMY (b, in mas) and LOD (c, in ms) short-term fluctuations [1–100 days] time series.

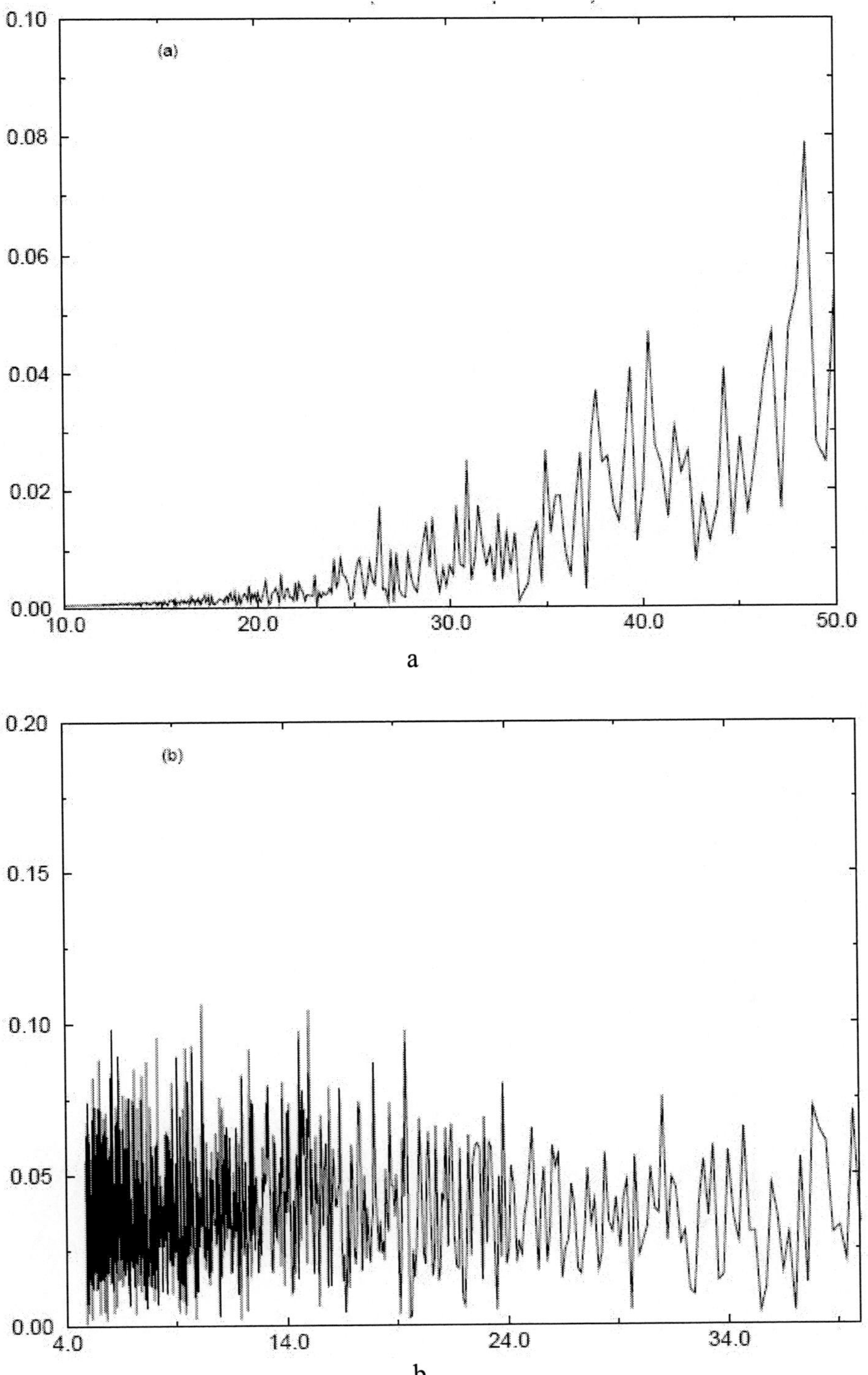

Figure 2. (Continued).

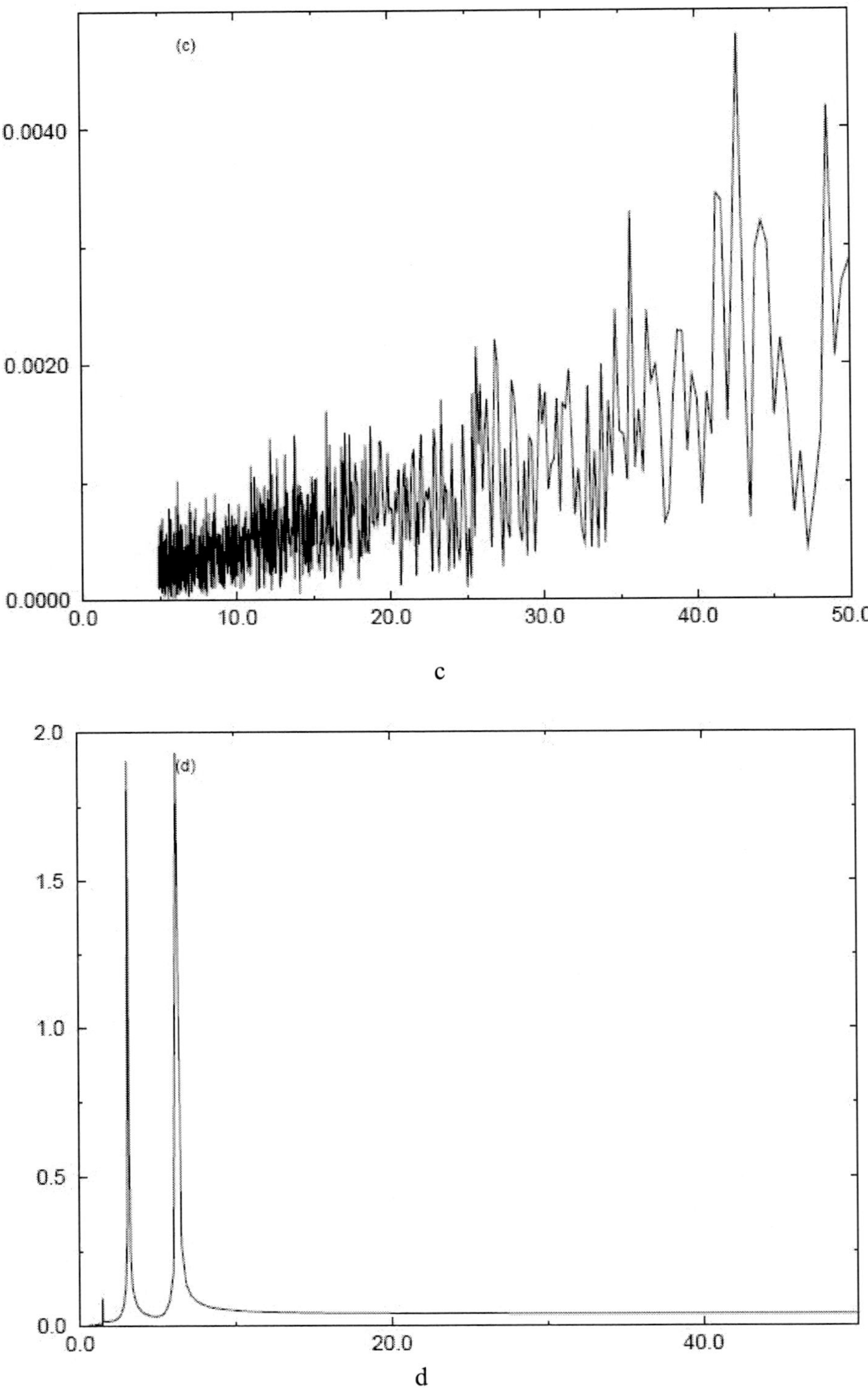

Figure 2a,b,c,d. Various Fourier spectra obtained from various time series: for a chaotic system (a), a stochastic white noise (b), a stochastic coloured noise (c) and a biperiodic series (d). The first three spectra are continuous.

4. Nonlinear Time Series Analysis: Overview, Synthetic Description

For most of the observational geophysical systems, no equations are available. The knowledge and characterisation of the processes depend only on data analysis from time series analysis. As, we said earlier, usual methods consist in Fourier spectra analysis, autocorrelation evaluation and modelling residuals by autoregressive or stochastic models. Nevertheless, some complex systems can have an irregular behaviour without being stochastic. Non linear time series analysis gives the possibility to identify those systems from stochastic ones. The methods are mainly geometrical and come from the dynamical system theory. They have been used in several domains (biology, medicine, physics, economy, music,…). The nonlinear time series analysis gives the possibility to analyse in a dynamical way, signals whose original equations are not available. The vocabulary, we use, relies on dynamical system theory and we will define briefly the main principles before describing the methods.

4.1. General Dynamical Systems Definitions

Dynamical system theory gives the possibility to study temporal evolution of physical systems described by ordinary differential equations through a geometrical and topological characterisation of the orbits drawn by the variables in the solution space, called the phase space. Those orbits can be very different, the solution of differential equations can be either fixed point, periodic or quasi periodic cycles, or turbulent (non periodic). Study of dynamical systems gave a strong change in the turbulence study. Indeed, until 1971, it was believed that to obtain a non periodic behaviour, a great number of frequencies were needed. In contrast, Ruelle and Takens (1971) proposed a "chaotic" definition of the turbulence introducing the concept of strange attractor. Only a limited number of bifurcations were necessary to obtain turbulence. Lorenz in 1963 had already shown that from only three nonlinear coupled equations, an aperiodic movement could be observed.

A dynamical system is determinist if it can be described by a finite system of autonomous ordinary differential equations of order n. The temporal evolution of a deterministic system is determined in a unique way from the specification of its initial conditions. As said before, the space of the solution of the equations is called the phase space. The dimension of the phase space is the number of state variables of the system. We consider dissipative systems and thus are interested in the asymptotic permanent behaviour of the system, which means the description of its limit sets.

The stability of the system is studied by measuring the effect of a small perturbation applied on the orbit. In a general manner, from a linearization of the flow around the trajectory, we determine the eigenvalues of the linearised matrix. It can be obtained from the Lyapunov exponents that are directly linked to the eigenvalues and that apply to all kind of dynamics. A negative Lyapunov exponent is associated to a contracting direction (stable) of the flow, a null exponent is associated to a neutral direction of the flow and a positive exponent is associated to an unstable direction of the flow. The orbits of a chaotic system have a double property: a small perturbation will have an important effect in time (strong divergence of orbits linked to the existence of a positive Lyapunov exponent) but the system being deterministic, the attraction still remains (contracting direction associated to the negative exponents).

An attractor is defined as a stable limit set, all the trajectories will converge to it. Different kind of attractor can exist : a fixed point, (the dimension of such an attractor is zero) a periodic system (dimension of the attractor is 1), quasi periodic system (r independent frequencies), dimension is r (for instance biperiodic, the 2-tore is the attractor of dimension 2). For a chaotic system the dimension of the attractor can be non integer (we call it "strange attractor"). It is because a chaotic attractor possesses at least one unstable element. As the attractor is stable and dissipative by definition, this results in a sort of layer folding of the orbits to link contraction and dilatation of orbits. Thus a chaotic attractor needs at least 3 dimensions and three exponents (one positive, one negative and one null along the trajectory). If we consider the Lorenz attractor, the topological dimension is 2 (continuous part of the attractor) while the attractor itself has a 2.06 dimension (fractal dimension due to the layer folding of the orbits in the third direction) (figure 3).

We can also define the kind of attractor of a system from the sign of its Lyapunov exponents. A fixed point is stable if all its Lyapunov exponents are negative. A periodic solution has a limit cycle stable if one exponent is null and the others negative. A positive exponent reveals a divergence of the trajectories and the existence of an unstable manifold. It is characteristic of chaotic attractors.

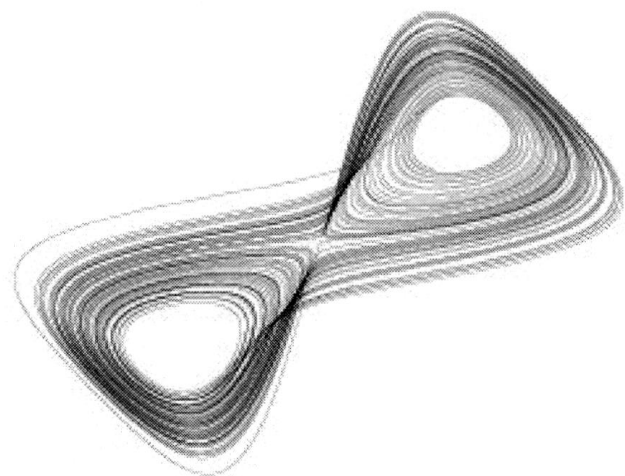

Figure 3. Lorenz attractor. Chaotic attractor of dimension 2.06.

We will see now how we can retrieve all this not from differential equations but from observations only when equations are not available. It is called nonlinear time series analysis.

4.2. Application: Nonlinear Time Series Analysis

Embedding, Phase Space Reconstruction

In order to analyse a time series in a nonlinear way, we must embed the data in a multidimensional phase space, usually called pseudo phase space. Those data will construct a geometrical figure in this space (different from the time space) from which a number of characteristics belonging to the underlying system can be highlighted. The observable we have is thus a time series x(t). This time series represents the observation part of a more complex system involving several unknown and may be coupled variables. We usually do not know the differential equations underlying the system and we cannot observe those initial variables nor find the phase space dimension. At best x(t) is a nonlinear combination of several "real" variables.

The embedding method gives the possibility to construct a pseudo phase space directly from the observable data x(t) which provide information on the unknown dynamical system. It consists in the transformation of the observed x(t) into pseudo orbits y(t) embedded in the pseudo phase space. The behaviour and the underlying dynamical properties of the y(t) is similar to the one of the initial variables.

Takens (1981) gives the way to construct the pseudo phase space easily. It is based on the evaluation of a time delay (T). The vectors, y(t), can be constructed as:

$$y(t)=[x(t), x(t+T),\ldots,x(t+De^*-1)T)]$$

where De^* is the dimension of the pseudo phase space.

Even if, for a dynamical system, there is a unique phase space dimension, different time series measurements x(t) and x'(t) both arising from the same source may lead to different embedding dimensions. De*, the dimension of the pseudo phase space is thus the lower limit of the real phase space dimension. With this dimension we can embed the whole dynamics of the system in an optimal way.

The determination of the time delay T is important because if it is taken too small then two measures x(t) and x(t+T) will not be independent enough and we will get no more information from x(t+T) than from x(t). It will result in a reconstruction of y(t) along a diagonal line. If the time delay is chosen too high, we haven't any link between the measurements anymore and we can not reconstruct a coherent figure in the pseudo phase space.

There exists two ways to determine the time delay, from the autocorrelation function of the data and from the average mutual information function. From the first method, the time delay is read as the first zero of the correlation function. But this method was discussed as it could give false results when applied to nonlinear processes (Fraser and Swinney, 1986). Indeed, nonlinear processes can have spatial dependence but any temporal links. The other method which is much relevant when dealing with possible nonlinear processes is based on the probability distribution of events in the data series. It gives the possibility to detect similar sequences in phase space that are not necessarily successive sequences in the time domain. It gives a more global measure of the dependence. Given an observation x(i), we look for how many additional information is given from observation x(j). The time delay is then the first minimum of the average mutual information function (Fraser and Swinney, 1986, Abarbanel, 1996).

Figure 4 illustrates the principle of the embedding: a dynamical system possesses a dynamic described by z(n) in a Rk dimensional space (phase space). Its attractor is called A. We can not have a direct access to this complete dynamics but only to a time series x(n), that contains a coupling of the z(n). From this x(n) we can construct another attractor B in a Rl dimensional space (pseudo phase space). The orbits y(n) that compose the attractor B have similar properties than the original system.

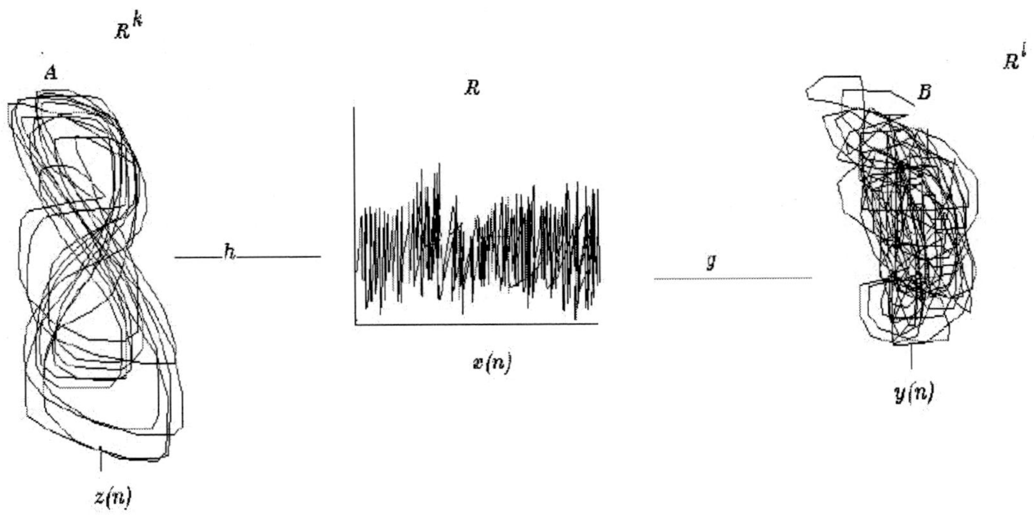

Figure 4. Embedding principles.

Determination of the Embedding Dimension

The Method of Global False Neighbours

The time series available is one dimensional. It results from the projection of the whole dynamics on one axis. After the determination of multidimensional vectors from the time delay, we need to increase the dimension of the embedded space in order to "open" the projected structure. The relevant embedding dimension will be the one when the orbits are totally "open". The method is based on the false neighbours (Kennel et al., 1992) and consists in measuring the distances between points. If neighbours remain neighbours when increasing the dimension, they are "true neighbours", otherwise, they are just close because of the projection. The dimension is reached when there are not any false neighbours anymore and thus when increasing the dimension has no effect on the distances between neighbours. Figure 5 shows how the optimal dimension is reached for the Lorenz system. Dimension $De^{*}=1$ is too small all points are projected on an axis. Dimension $De^{*}=2$ is too small, there are still false neighbours (distances are changing when increasing the dimension), while dimension $De^{*}=3$ is the smallest optimal dimension. All the dynamics is open, there are no false neighbours anymore.

For the complete description of the equations and criteria, see Abarbanel, 1996 or Frede and Mazzega, 1999a.

This method is relevant for deterministic systems. Indeed, if a noise is tested, the embedding dimension can't be found and is infinite. A white noise spread in all the dimensions and the number of false neighbours never reaches zero. For a red noise, results are less convincing even if the process is not deterministic and of infinite dimension. Indeed the percentage of false neighbours becomes very low. This false result comes from the fractal nature of the curve and not from the dynamics (Frede and Mazzega, 1999a).

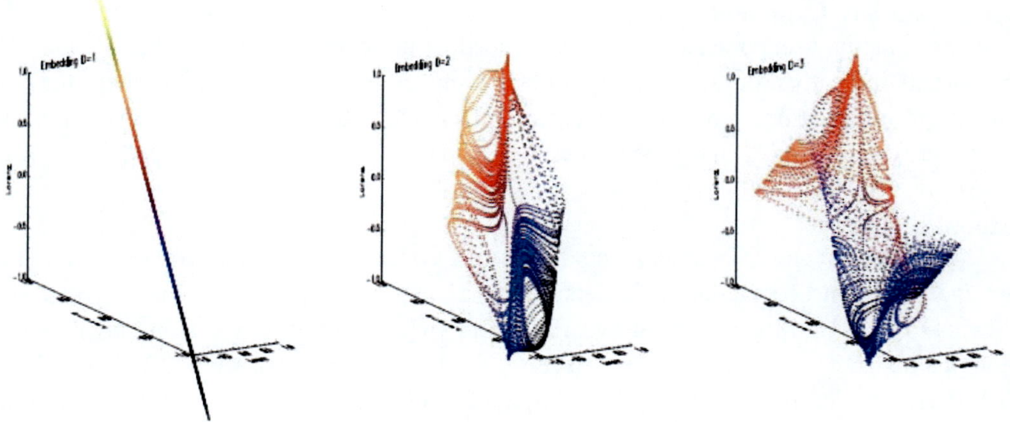

Figure 5. Embedding of the Lorenz attractor, in 1, 2 and 3 dimensions for the pseudo phase space.

The Method of the Correlation Dimension

There exists another way to determine the embedding dimension. This method is called the correlation dimension or the Grassberger and Procaccia algorithm (1983). It was initially developed to determine the dimension of the attractor of a system. It consists in defining the Euclidian distance between every pairs of vectors in the phase space and then to calculate the probability density of points on the attractor in hyper spheres or r radius. Indeed, when orbits draw an attractor of DA dimension, the correlation dimension C(r) behaves in r^{DA} when r tends to 0. It depends also on the embedding dimension and DA is read as the slope value on the curves logC(r)/log(r) in $D=De^*$. De^* is reached when the slopes remain constant; more precisely, when the density of points on the attractor does not depend anymore on the dimension of the space it is embedded. This method is very performing for long data set and data with very little noise. Nevertheless, it is more relevant, with noisy data and quite short time series, to use the geometrical method of false neighbours.

The Dynamics Characteristics

The Dynamical Dimension

The embedding dimension is a global system dimension. It gives the possibility to embed the attractor but the dimension needed to describe the system can be lower. For instance a 2-torus (biperiodic system) needs an embedding space of 3 to be recovered but locally only 2 dimensions are active (it can be understood as the dimension of the hyper surface tangent to the attractor). The local dimension gives also the number of differential equations needed to describe the system, it remains invariant whatever the data time series it is found from. It is always lower or equal to the embedding dimension. Moreover, it is directly linked to the Lyapunov exponents. Indeed, the local dimension gives the number of Lyapunov exponents of the system. To evaluate the local dimension, we look at the number of principal directions on the attractor along which the points are preferably distributed. From the determination of the local covariate matrix, we evaluate the eigenvalues. The local dimension corresponds to the number of significant eigenvalues. Those eigenvalues indicate the main directions in phase space.

The Lyapunov Exponents

The Lyapunov exponents are directly linked to the eigenvalues described earlier. They are strong indicators of the stability of a dynamical system. As said earlier, they give important information about the diverging nature of the orbits and thus are useful to characterise and identify chaotic systems.

Global Exponents

To define the global Lyapunov exponents of a system from a time series, we need to find the Jacobian matrix from measurements. Several methods have been proposed and tested (Eckmann et al., 1986; Bryant et al., 1990; Brown et al., 1991). We won't explicit those methods here, the interested reader can refer, for instance, to Abarbanel, 1996. The idea of their determination is to define the neighbours of a point y(i) in the phase space and to follow the temporal evolution of the neighbours. The important thing to understand is that a null exponent means that there is a flow direction, a negative exponent means there is a contracting direction, there is attraction, and the existence of at least one positive exponent means that the system has an unstable direction, and is chaotic. If the sum of the Lyapunov exponents is negative, the system is dissipative. Figure 6 shows the global Lyapunov exponents for the Lorenz time series (figure 6a) and for a biperiodic time series (figure 6b). For the Lorenz system, there are three exponents: one null, one negative and one positive. The observed dissipative deterministic system is thus chaotic. For the biperiodic system, there are two exponents: one null and one negative. The observed dissipative deterministic system has a 2-torus attractor.

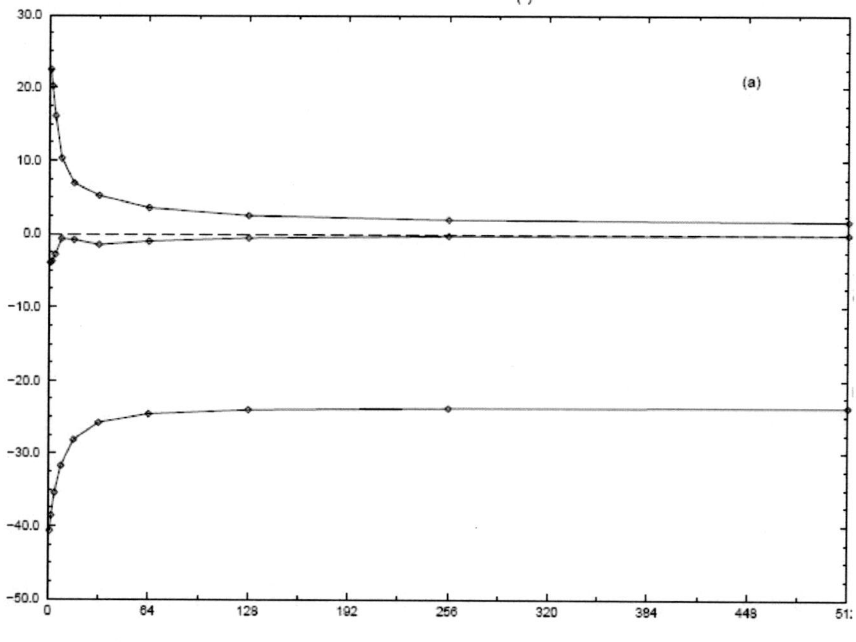

(a)

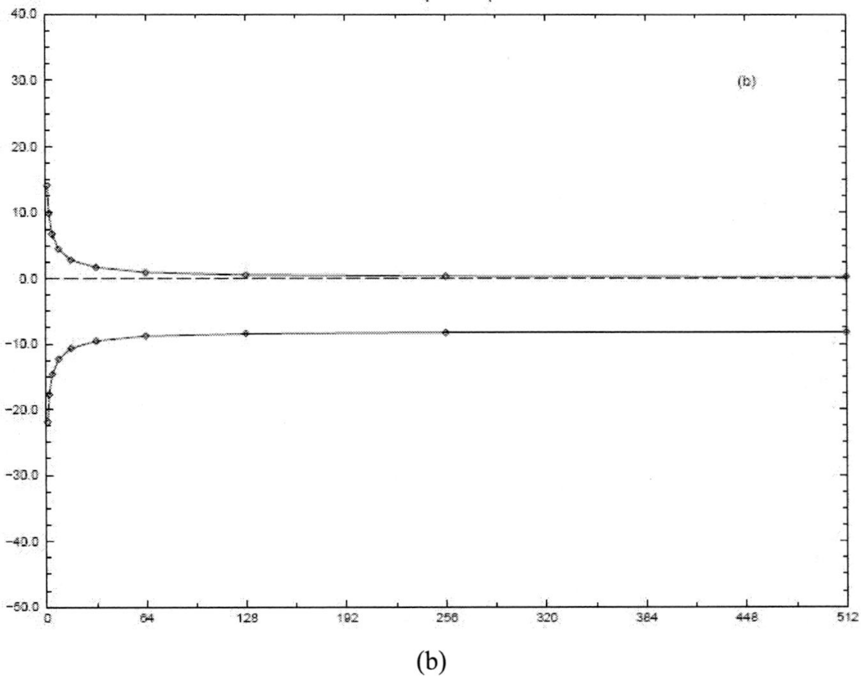

(b)

Figure 6. Global Lyapunov exponents for the Lorenz time series (a) and for a biperiodic time series (b). The exponents are evaluated on a window of length L=500 (L (given in abscissa) Jacobian matrices are computed on all the time series).

Local Lyapunov Exponents

The global exponents describe the global stability of the system. But there can exist local stability variations depending on the localisation on the attractor (Abarbanel and Kennel, 1993). Those variations are important for the system dynamics understanding, as they are associated to the horizon of prediction of the system. The principal Lyapunov exponent (the highest positive eigenvalue) gives a theoretical limit to the horizon of prediction of the system. Its variation can be observed thanks to the local analysis and gives indication of stability loss during the system evolution. The method is similar to the one for the global exponents but the difference is that the local exponents are evaluated on a finite and short orbit length. Practically we are interested in the variations of the principal local exponent λ_1.

The horizon of prediction H_p is given by the formulae:

$$H_p = 3.9/\lambda_1$$

where λ_1 is the principal Lyapunov exponent (the highest positive exponent).

Dimension of the Attractor

The dimension DA of the attractor can be found from the correlation integral as described earlier. This method has been largely applied to various systems (Nicolis and Nicolis, 1984; Essex et al., 1987; Tsonis and Elsner, 1988…). But it was shown that this method was not very performing when time series are quite short and noisy and can give spurious results. Another dimension which is more robust is the Lyapunov dimension, directly linked to the global Lyapunov exponents. It was initially defined by Kaplan and Yorke, 1979 as:

$$D_{Lyap}=K+\frac{\sum_{a=1}^{K}\lambda_a}{\left|\lambda_{K+1}\right|} \quad \text{with} \quad \sum_{a=1}^{K}\lambda_a>0 \text{ and } \sum_{a=1}^{K+1}\lambda_a<0$$

where K is an integer lower than DL and $\lambda_{k+1}<0$. D_{Lyap} is a superior limit to DA.

In Figure 7, we have reported some examples of attractors projected in 2D and when it was possible, their dimensions. We observe that depending on the nature of the underlying systems (stochastic, chaotic, periodic), the embedded attractors have well defined characteristics in phase space.

Figure 7a shows a 2-torus attractor associated to a biperiodic time series. The dimension of such an attractor is 2, the orbits are continuous. Figure 7b shows the attractor associated to a white noise process. No embedding dimension can be found, it tends to infinity and the points are distributed in all directions. There exists no attractor for this system, which is not determinist but stochastic. Figure 7c shows the reconstruction of a red noise process. Even if an embedding dimension can be found because of limitation of the algorithm, the curve is not continuous at all but follows a fractal drawing. Figure 7d shows again the Lorenz attractor of dimension 2.06, the orbits are continuous. Figure 7e is the attractor of a periodic time series (dimension 1, limit cycle).

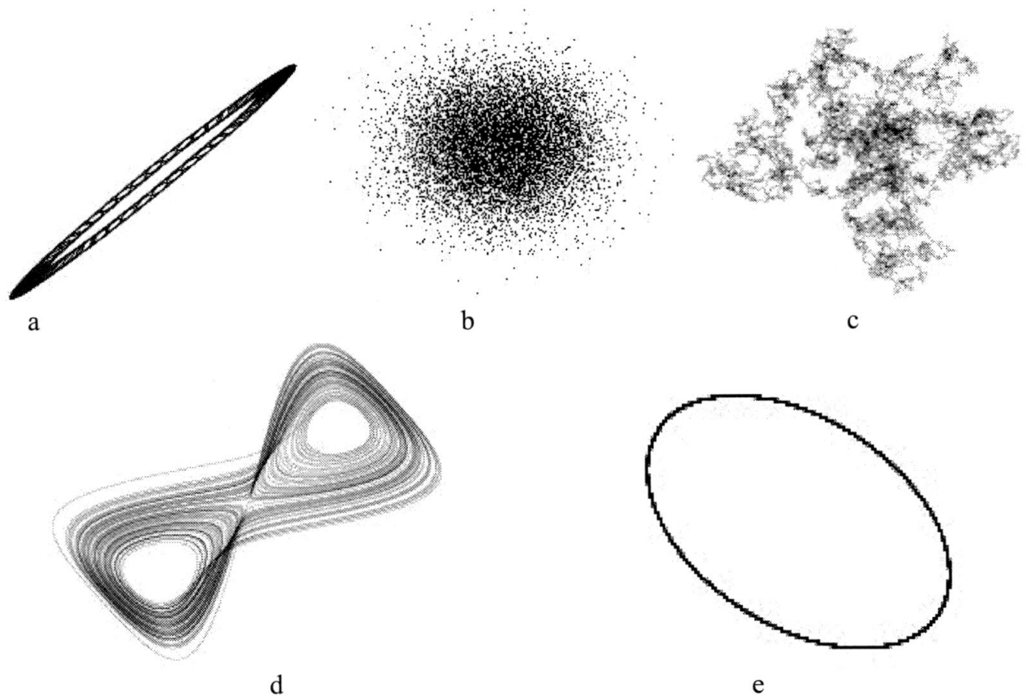

Figure 7. Examples of attractors projected in 2D : torus (a), white noise (b), red noise (c), Lorenz (d), periodic system (e).

5. APPLICATION TO EARTH ROTATION STUDIES

We will describe the studies we performed on several Earth variations and atmospheric associated fluctuations and show that nonlinear methods were relevant to that part of the signal (Frede, 1999).

5.1. EOP Short-Term Fluctuations: Polar Motion and Length of Day

Before performing our study, we had an initial question: Is it possible to detect a nonlinear deterministic process of low dimension in EOP time series short-term fluctuation?

Our motivations were that for periods under 100 days, the origins of the fluctuations observed in the Earth's rotation are mainly coming from atmospheric and oceanic forcing (Eubanks et al., 1988; Rosen and Salstein, 1983) and that the Fourier spectra of the EOP at that time scale are continuous. Moreover, the dynamics underlying those forcing (ocean and atmosphere) are complex and show several behaviour from a fixed point to a turbulent movement. Most of the circulation models in geophysics fluids possess a nonlinear coupling of their variables and have either stable or unstable solutions. The application of the nonlinear time series analysis to the short-term fluctuations (1–100 days) of the EOP time series gives the following results (various tests of significance have been performed before choosing those values see Frede and Mazzega, 1999a and 1999b for details). The time delay obtained from the average mutual information function is of 15 days for PMX (polar motion in x), 18 days for PMY (polar motion in y) and 10 days for LOD (length of day). The embedding dimension was taken as 5 or 6 for the three time series.

The computation of the local dimension gave the value of 5, we will thus compute 5 global Lyapunov exponents and the underlying system could be modelled by 5 ordinary autonomous differential equations. The sum of the 5 Lyapunov exponents is negative, the system is thus dissipative. Nevertheless there exist 2 positive exponents, a null exponent (obtained with 5% of precision), and two negative exponents. The principal Lyapunov exponent for each series gives the horizon of prediction which is around 10 days (Frede and Mazzega, 1999a).

The local temporal behaviour of the principal exponent for the three time series gives interesting results. The most important one, concerns the length of day process. It is shown that there exists a stability loss around 1983 where the horizon of prediction is reduced by a factor 2 or 3. It is very interesting to remark that at that date a major El Nino event took place (Frede and Mazzega, 1999b).

How can we conclude about the EOP short-term fluctuations time series? The Fourier analysis gave us continuous spectra that could be interpreted as red noises processes. The nonlinear time series analysis puts into light several arguments in favour of the existence of a low dimensional deterministic process underlying the EOP time series. In order to identify the real dynamics and to check if the EOP time series are the sum of a low dimensional deterministic process and stochastic noise or are only stochastic processes similar to red noises, we have additionally represented the orbits of the PMX, PMY and LOD embedded in a 6 dimensional pseudo phase space, in a 3D projections (Figure 8). We obtain continuous trajectories and not fractal curves. Thus the red noise process can be discarded. There exists a low dimensional deterministic process in the EOP short-term time series. The existence of 2 positive exponents is another strong argument to the fact that the system is chaotic and that it could not be predicted over than about 10 days.

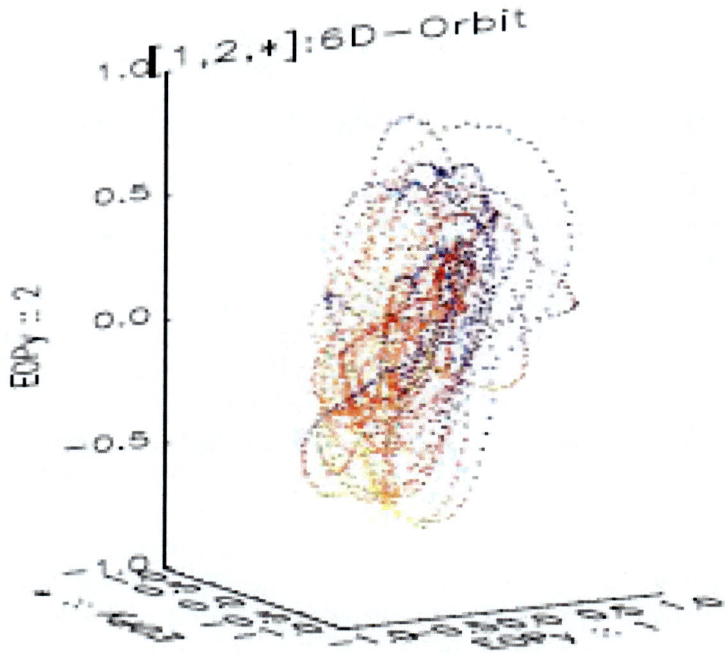

Figure 8. 3D projection of the PMY time series embedded in a 6-dimensional pseudo phase space.

5.2. Angular Momentum Short-Term Fluctuations Time Series: Comparison with the Earth

As the main forcing of the Earth is atmospheric, we were interested in applying the same methodology to EAAM (Effective Atmospheric Angular Momentum) short-term time series (Kalnay et al. 1996). The description of the time series and the topological and dynamical properties are described in Frede, 2000. As we have detected and described a low dimensional chaotic process in the short-term fluctuations EOP, we were interested in the nature of the dynamics of the main forcing of the EOP fluctuations. After isolating the short-term fluctuations of the EAAM time series, we applied classical Fourier analysis that revealed a continuous spectra for the three times series (we worked with the x1p, and x2p (pressure terms) and x3w (wind term) time series. The embedding dimension was 5 and two positive exponents were retrieved. The representation of the embedded attractor in 2D has been reported in Figure 9 for the wind term of the EAAM time series. The orbits are continuous and thus do not behave as a red noise. The associated horizon of prediction was found around 6 days which is in coherence with the values given by the atmospheric models (Bell et al. 1991). The differences observed between the chaotic dynamics of the Earth and the chaotic dynamics of the atmosphere show that another process is acting and could be directly linked to the oceanic turbulent dynamics.

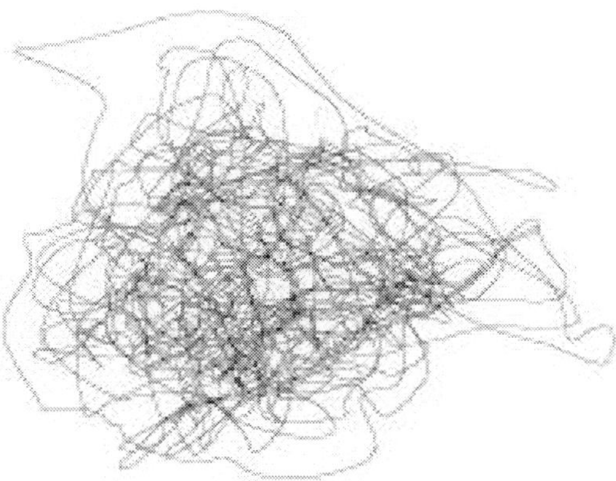

Figure 9. Attractor associated to x3w time series (2D projection).

We have thus identified ant characterised a new nonlinear deterministic atmospheric signal. Chaos has been observed in various atmospheric time series at various time scales (Essex et al., 1987; Tsonis and Elsner, 1988). With our study, we complete this knowledge by determining chaos in the short term fluctuations (1–100 days) of the angular momentum (EAAM) time series. As said before and as expected, the characteristic of this chaos are quite similar to the one observed in the EOP time series for the same time scales. We would also need 5 ordinary differential equations to model the system with two unstable directions on the attractor.

5.3. The Chandler Wobble

The Chandler wobble is a resonant mode of the rotation axis of the Earth around its figure axis, whose mechanism of excitation remains unknown. The free wobble was initially evaluated by Euler in 1765, for a rigid Earth, at 306 days. In 1891, Chandler has observed that this free wobble had a period of about 14 months (430 days). The fact that this period was higher than the computed one was explained by the elasticity of the Earth. Without any source of excitation to maintain this oscillation, the Chandler wobble, will disappear by dissipation. A forcing source is needed. As seen in section 2, several candidates have been proposed but the exact and complete nature of the forcing remains unknown. As all studies have considered the Chandler wobble analyse in a linear way and none of them has managed to describe completely the signal and its oscillations, we were interested in applying nonlinear methods to see if a nonlinear deterministic signal could exist and help explanation.

We have considered the Chandler wobble time series in X and Y (IERS, 1997). The complete description of our work is described in Frede and Mazzega (2000). The time delay needed to embed the time series is about 105 days for the Y component and 115 days for the Y component. The time series could be embedded in a pseudo phase space of dimension 3 or 4 (the presence of noise has the effect to increase the embedding dimension). From the representation of the orbit in the pseudo phase space, we observe a perturbed 1-Tore (Figure 10). The modulations around the 1-Tore can reach about 100 mas. Three Lyapunov exponents are found. One positive, one null and one negative. The existence of a positive exponent shows that instabilities can develop around the 1-Tore (manifold of 1 dimension).

Figure 10. 2D- representation of the attractor from the X component of the Chandler wobble.

The horizon of prediction is around 368 days for X and 276 days for Y. The Lyapunov dimension gives a dimension of 2.5 for the attractor. The two previous results : sensitivity of initial conditions (existence of an unstable manifold on the attractor) and fractal dimension (strange attractor) give clues for the characterisation of a low dimensional nonlinear deterministic process whose dynamic is chaotic. Additionally, the study of the local variations of the principal Lyapunov exponent shows important fluctuations around 1860 and 1940 for the X-component. The possible candidates to these stability losses have still to be searched.

In 2001, Maneville made a reanalysis of the EOP time series and Chandler wobble time series by varying the time scales. He chose to develop the analysis for the interannual time scales. He completed our studies by focusing on quantitative descriptions and confirmed a strong correlation between length of day and El Nino event. He also discussed the differences between the dynamical system approach and conventional signal analysis.

Conclusion

The contribution of nonlinear time series analysis to geophysical study of the Earth's rotation is important. It appeared relevant to work in other domains than the frequency domain. Indeed, we have shown that some terrestrial components could be modelled by a low dimensional deterministic nonlinear process in a multidimensional phase space. We could identify dynamical invariants and theoretical horizons of prediction. It becomes now possible to compare dynamical properties of various processes acting on the Earth. We should not limit our comparisons to time series correlations or frequencies, in order to identify the complete forcing acting on the Earth.

We have also observed that a good way to distinguish deterministic process from stochastic process, is to represent the embedded attractor, in a multidimensional phase space. The continuous properties of the orbits are fundamentally different. For instance, creating a synthetic stochastic signal with similar statistic properties (spectra) than the studied and observed signal, and then comparing their attractors, is a method that appears very efficient. From this technique, we saw that the asymptotic behaviour of the orbits for EOP and EAAM short-term fluctuations was fundamentally different from the one of a random walk (even if they have similar Fourier spectra).

We have found a nonlinear deterministic process in the EOP short-term fluctuations where previous studies concluded with "red noise". This process can be described in a five-dimensional space, with fractal attractor with two unstable directions. Moreover, the deterministic and chaotic nature of the process indicates a limit of prediction of about 10 days. We have also put into light a new atmospheric attractor from EAAM time series, whose fractal dimension is between 4 and 5. The associated theoretical horizon of prediction is about five days. This result is in concordance with the horizons of prediction given by the atmospheric models (Bell et al. 1991). Moreover, the similar values found in the EOP and EAAM dynamics show that the main factor accountable for the Earth's rotation variations is the atmosphere. It is now necessary to study the oceanic part of this excitation by applying in the same way a nonlinear analysis of the oceanic time series (Frede and Bizouard, 2003).

How must we interpret the chaotic attractor found in the short-term fluctuations of the Earth's rotation? We believe that the Earth system is not "chaotic". This chaotic component does not belong to the Earth's own dynamics but rather is induced by atmospheric forcing, which is mainly nonlinear. This was confirmed by the EAAM short-term time series analysis. In other words, the nonlinear study of the time series is a sort of inverse problem to learn about the forcing properties.

Regarding the Chandler term, our study (Frede and Mazzega, 2000) and the study of Manneville (2001) were the first to consider it through a nonlinear point of view. We have shown that dynamically the Chandler wobble could be described as a dissipative oscillator with an unstable manifold induced by forcing and excitations. This unstable manifold implies a theoretical horizon of prediction of about one year. The excitation and the damping of the Chandler wobble are not well understood. Thus, from the nonlinear analysis, the dissipative rate (whose inferior limit value is given by the sum of the Lyapunov exponents) should be compared to the quality factor Q of the Chandler term. The quality factor value is directly linked to the damping of the oscillation. A comparative study could help in determining more precisely this term. Moreover, we have shown that the forcing of the Chandler wobble is not stochastic in nature. A more complete characterisation of the dynamics would help identifying them, maybe from a complete analysis of the local variation of the principal Lyapunov exponent.

A strong perspective to these studies would be to develop a nonlinear prediction method for the Earth rotation time series. Up to now, one uses linear method to predict irregular and nonlinear parts of the signal of the Earth's rotation. Indeed, the EOP short-term fluctuations (1–100 days) and the Chandler wobble are mainly predicted by autoregressive methods. Classical linear methods work sequentially in time, while nonlinear methods can exploit spatial correlations and are based on the neighbours' notion. They appear then more relevant for those nonlinear parts of the signal. Moreover, we know that, in prediction, the error's behaviour depends on the nature of the process. For a noncorrelated Gaussian noise (white noise), the error remains stable in time. For a red noise it decreases rapidly, as for a chaotic system. Additionally, by studying the correlation coefficient between the predicted values and the real values, it is possible to distinguish a red noise from a chaotic system. This method was used by Tsonis and Elsner (1992) to confirm from a Southern Oscillation Index (SOI) time series the chaoticity of ENSO event.

To conclude, our results show that the study of the rotation of the Earth can't be limited to linear description only. If we want to overcome the descriptive step from nonlinear time series analysis and try to model quantitatively the underlying dynamics, we have to develop a method for recovering differential equations from observation and from time series. Indeed, as explained in this paper, the nonlinear time series analyses give the possibility to describe the topological and geometrical properties of the system dynamics (phase space, attractor, local dimension...) but don't allow a mathematical formalism. The next step would be to recover the differential equations that can describe exactly the observed system. It would consist in the construction of a system of autonomous differential equations by finding the general laws of the orbit evolution on the attractor. The difficulty is to work globally on the attractor; local analysis is easier to perform. For instance, for the EOP short-term fluctuations, we would need exactly five differential equations to describe the system and the attractor. Several methods have been proposed to construct such system of equations (i.e., Crutchfield and Mcnamara, 1987; Gouestbet, 1991) but none of them was used and appeared relevant with observational time series of the Earth orientation parameters.

REFERENCES

Abarbanel, H.D.I. (1996). Analysis of Observed chaotic data. Springler verlag, Berlin.

Abarbanel, H.D.I. and Kennel, M.B. (1993). Local false nearest neighbors and dynamical dimension from observed chaotic data. *Phys. Rev. E., 47*, 3057-3068.

Barnett, T.P., Latif M., Kirk E. and Roeckner E. (1991). On ENSO. *Physics J. Climate, 4,* 487-513.

Bell, R.S., Hide, R. and Sakellarides, G. (1991). Atmospheric angular momentum forecasts as novel test of global numerical weather prediction models. *Phil. Trans. R. Soc. Lond. A 334*, 55-92.

Brown, R., Bryant, P. and Abarbanel, H.D.I. (1991). Computing the Lyapunov spectrum of a dynamical system from an observed time series. *Phys. Rev. A., 43(6),* 2787-2806.

Bryant, P., Brown, R. and Abarbanel, H.D.I. (1990). Lyapunov exponents from observed time series. *Phys. Rev. Lett., 65(13)*, 1523-1526.

Crutchfield, J. and Mc Namara, B. (1987). Equations of Motion from a data series. *Complex systems, 1,* 417.

Eckman, J.P., Kamphorst, S.O., Ruelle, D. and Ciliberto S. (1986). Liapunov exponents from time series. *Phys. Rev. A, 34(6),* 4971-4979.

Eubanks, T.M., Steppe, J.A., Dickey, J.O. and Callahan P.S. (1985). A spectral analysis of the Earth's angular momentum Budget. *Journal Geophys. Res., 90, B7,* 5385-5404.

Eubanks, T.M., Steppe, J.A. and Dickey, J.O. (1988). The atmospheric excitation of rapid polar motions. In : Bancock and Wilkins (Eds), *The Earth's Rotation and reference Frames for Geodesy and Geodynamics,* 365-371, Kluwer, Dordecht.

Essex, C., Lookman, T. and Nerenberg, M.A.H. (1987). The climate attractor over short timescales. *Nature, 326*, 64-66.

Fraser, A.M. and Swinney H.L. (1986). Independent coordinates for strange attractors from mutual information. *Phys. Rev. A, 33(2),* 1134-1140.

Frède, V. (1999). Apport de l'analyse non linéaire à l'étude géophysique de la rotation de la terre. PhD manuscript (in French). Paris Observatory.

Frède, V. (2000). Dynamical analysis of the atmospheric angular momentum short term fluctuations. Comparison with the Earth. *Journal of Geodesy 73:* 660-670.

Frède, V. and Mazzega, P. (1999a). Detectability of Deterministic Nonlinear Processes in Earth Rotation Time Series : I-Embedding. *Geophysical Journal International 137, 2* :551-564.

Frède, V. and Mazzega, P. (1999b). Detectability of Deterministic Nonlinear Processes in Earth Rotation Time Series : II-Dynamics. *Geophysical Journal International 137,2* :565-580.

Frède, V. and Mazzega, P. (2000). A preliminary Nonlinear Analysis of the Earth's Chandler Wobble. *Discrete Chaotic Dynamics in Nature and Society Journal vol 4/1*. 39-53.

Frède, V. and Bizouard, C. (2003). Effect of the nonlinear behaviour of atmospheric forcing on the Earth rotation. *Poster presented in the European geophysical Society General Assembly*, Nice (France).

Gouesbet, G. (1991). Reconstruction of the vector fields of continuous dynamical systems from numerical scalar time series. *Phys. rev. A., 43,* 5321.

Grassberger, P. and Procaccia, I. (1983). Characterization of strange attractors. *Phys. Rev. Lett., 50,* 346-349.

Gross, R.S. (1986). The influence of Earthquakes on the Chandler wobble during 1977-1983. *G.J.R. ast. Soc, 85*, 161-177.

Gross, R.S. (1990). The secular Drift of the rotation Pole. In: Boucher and Wilkins (Eds), *Earth rotation and coordinates reference frames, IAG Symposium 105,* 146-153.

Gross, R.S. (1992). Correspondance between theory and observations of polar motion. *Geophysical Journal International. 109,* 162-170.

Hopfner, J. (1998). Seasonal variations in length of days and atmosphereic angular momentum. *Geophys. J. Int., 135*, 407-437.

International Earth Rotation Service, 1997, IERS 1996 Annual report. Central Bureau of IERS, Observatoire de Paris, Paris.

Kalnay, E., Kanamistu, M., Kistler, R., Collins, W., Deaven, D., Gandin, L., Iredell, M., Saha, S., White, G., Woollen, J., Zhu, Y., Chelliah, M., Ebisuzaki, M., Higgins, W., Janowiak, J., Mo, K.C., Ropelwski, C., Wang, J., Leetmaa, A., Reynolds, R., Roy, J. Dennis, J. (1996). The NCEP/NCAR 40 years reanalysis Project. *Bulletin of the American Meteorological Society, 77, 3*, 437-471.

Kaplan, J.L. and Yorke, J.A. (1979). Chaotic behaviour in multidimensional difference equations. In : Peitgen. H.O. and H.O. Walthers. (Eds), *Functional Differential Equations and Approximations of Fixed Points*. Springler. Berlin pp 204-227.

Kennel, M.B., Brown R. and Abarbanel H.D.I. (1992). Determining embedding dimension for phase space reconstruction using a geometrical construction. *Phys. Rev. A, 45, 6,* 3403-3411.

Kuehne, J.K. and Wilson, C.R. (1991). Terrestrial Water Storage and Polar Motion. *Journal Geophys. res., vol 96, n° B3,* 4337-4345.

Lorenz, E.N. (1963). Deterministic Nonperiodic Flow. *Journal of the Atmospheric Sciences, 20, 2,* 130-141.

Manneville, P., (2001). Time-Delay Study of Earth Orientation Parameters at Inter-Annual Time Scales. *Nonlinear Phenomena in Complex system, 4:1,* 94-105.

Ming, Z. and Danan, D. (1988). A new Research for the secular Polar Motion in this Century. In : Bancock and Wilkins (Eds), *The Earth's rotation and reference Frames for Geodesy and Geodynamics,* 385-392, Kluwer, Dordecht.

Nicolis, C. and Nicolis, G. (1984). Is there a climatic attractor? *Nature, 311,* 529-532.

Ponte, R.M. (1990). Barotropic Motions and the Exchange of Angular Momentum Between the Oceans and Solid Earth. *Journal Geophys. Res., 95, C7,* 11 369-11 374.

Rosen, R.D. and Salstein D.A. (1983). Variations in Atmospheric Angular momentum on Global and regional scales and the length of day. *Journal Geophys. Res, 88, C9*, 5451-5470.

Ruelle, D. and Takens, F. (1971). On the nature of turbulence. *Communications in Mathematical Physics, 20,* 167.

Runcorn, S.K., Wilkins, G.A., Groten, E., Lenhardt, H., Campbell, J., Hide, R., Chao, B.F., Souriau, A., Hinderer, J., Legros, H., Le Mouel, J.L. and Feissel, M. (1990). The excitation of the Chandler Wobble. *Surv. Geophys., 9*, 419-449.

Stone, L., Saparin, P.I., Huppert, A. and Price, C. (1998). El Nino Chaos : the role of noise and stochastic resonance on the ENSO cycle. *Geophys. Res. Lett., 25, 2,* 175-178.

Takens, F. (1981). Detecting strange attractors in turbulence. *Lecture Notes in Mathematics. n° 898,* Springler-Verlag.

Tsonis, A.A. and Elsner, J.B. (1988). The weather attractor over very short timescales. *Nature, 333,* 545-547.

Tsonis, A.A. and Elsner, J.B. (1992). Non linear prediction as a way of distinguishing chaos from random fractal sequences. *Nature, 358*, 217-220.

Wahr, J.M. (1983). The effects of the atmosphere and oceans on the Earth's wobble and on the seasonal variations in the length of day. II-Results. *Geophysical J.R. Astr. Soc., 74,* 451-487.

In: Space Exploration Research
Editors: J.H. Denis and P.D. Aldridge
ISBN: 978-1-60692-264-4

Chapter 19

AGRICULTURE ON EARTH AND ON MARS

Hidenori Wada, Masamichi Yamashita, Naomi Katayama, Jun Mitsuhashi, Hiroshi Takeda and Hirofumi Hashimoto
Pharmacy and Life Science, Hachigi, Tokyo, University of Tokyo, Japan

ABSTRACT

Since the end of the final glaciation, agriculture on Earth has been developed in several areas by modifying various ecosystems. We have replaced natural producers with crop plants and made humans the final consumer in each ecosystem. Though terrestrial agriculture has been established after a process of trial and error, agriculture on Mars should be prepared rapidly and firmly without any failure, even under severe natural conditions completely different from those on Earth. In addition, the agriculture on Mars should supply clean water and air, as well as foods, fibers and timber necessary for sustaining human life. This engineering target can be achieved by a strategy consisting of two stages. During the first stage, air, water and regolith are brought into a pressurized dome and are regulated somewhat similar to those on Earth. In the second stage, sustainable agriculture will be managed on Mars, mainly by utilizing on-site resources and actions taken by members of the ecosystem inside the dome.

INTRODUCTION

Agriculture on Mars (Martian agriculture) is indispensable for some dozens of emigrants on long-duration missions to that remote planet, where at least 2.5 years are necessary for making the return trip. Prior to discussion about Martian agriculture, we will reconsider the significance of Earth's agriculture (terrestrial agriculture) and ecosystem. Then, we will carefully examine differences between terrestrial agriculture and Martian agriculture in terms of the ecosystem. Based on these considerations, we will explain a strategy to establish agriculture on Mars.

Agriculture can be broadly defined as a human activity that produces and supplies organisms and their products to us for sustaining our life and for assisting with the development of human society. The agricultural organisms are various, including crops, livestock, and pets. The products of agriculture are not limited to edible biomass for human foods, fibers and timber but also include revitalized air and water.

An ecosystem is composed of living organisms and their environment, consisting of both biotic and abiotic members, such as air, water, and soil. Bio-elements recycle in the ecosystem through a concerted process of all its member components, and are driven by solar radiation. Photosynthetic plants (Producers) produce organic substances containing bio-elements. Animals (Consumers) consume a part of the organic substances of the producer, either directly (Herbivores) or indirectly (Carnivores). Herbivores are usually prey to some carnivores, which are often prey to other carnivores. Thus plants, herbivores and carnivores form a food web in the terrestrial ecosystem. Microorganisms (Decomposers) decompose carcasses and excrements of animals and the remaining organic substances of plants to liberate bio-elements from these organic compounds. The liberated bio-elements are taken up by Producers, leading to completion of the recycling of the bio-elements in the ecosystem.

CHARACTERISTICS OF AGRICULTURE ON EARTH AND ON MARS

Terrestrial Agriculture

Terrestrial agriculture is a revolutionary new technology in human history, introduced at several places in different ecosystems after the end of the last glaciation. Agriculture has secured the supply of foods to humans, and helped humans to take off from other living creatures. It enabled humans to increase in population and provided us spare time for deep thinking, which created advanced culture, religion, civilization, and other new technologies in various fields. In this sense, the invention of terrestrial agriculture is called the first revolution (agrarian revolution) in the history of humans.

The supply of foods necessary for developing human society is considered the main role of terrestrial agriculture. Actually, agriculture also provides fiber, timber and even pet animals, ornamental flowers and plants. In addition, agriculture forms an artificial ecosystem, which is a stable, friendly environment for humans: people feel this artificial ecosystem is an indispensable and heart-warming part of their hometown.

The traditional strategy of terrestrial agriculture is to modify a natural ecosystem by replacing natural wild producers with crop plants, and making human the sole final consumer. Furthermore, revitalization of air and water has been trusted mainly to the intact natural ecosystems. Terrestrial agriculture has been developed through trial and error through the long history of humans. Nevertheless, several fundamental agricultural technologies have been invented since its early stage, such as domestication of wild animals, breeding of livestock and crops, reclamation of grassland and forest, irrigation of dry land, drainage of wetland, application of organic waste or compost, and weeding. After the industrial revolution, industry became dominant in developed countries, and terrestrial agriculture was greatly advanced by adopting scientific knowledge and new technologies: agricultural work has become large in scale and easier in operation. In recent years, modern agriculture, thus industrialized, has taken full advantage of new scientific technologies, including biotechnologies, and has been rather freely utilizing chemical fertilizers, synthetic agricultural chemicals, heavy agricultural machines and gene-engineered organisms. This has resulted in rapid and extensive destruction of precious natural ecosystems, which are characterized by their high degree of biodiversity.

Accordingly, the strategy of terrestrial agriculture, together with those of industry and urbanization, causes severe environmental problems, such as pollution of water and air, a decrease in biodiversity, desertification and global warming.

Martian Agriculture

Martian agriculture will need to be initiated within a short period of time and operated confidently without any catastrophic failure. It is required to be in operation under conditions completely different from those on Earth. As well, Martian agriculture must revitalize air and water, in addition to supplying ordinary agricultural products like foods, fibers and timber. Recycled use of materials is a major function of agriculture on Mars. Martian agriculture is also characterized by limited species in the ecosystem, contained inside a small dome constructed on either the surface or subsurface of Mars. An outline of Martian agriculture is shown in figure 1.

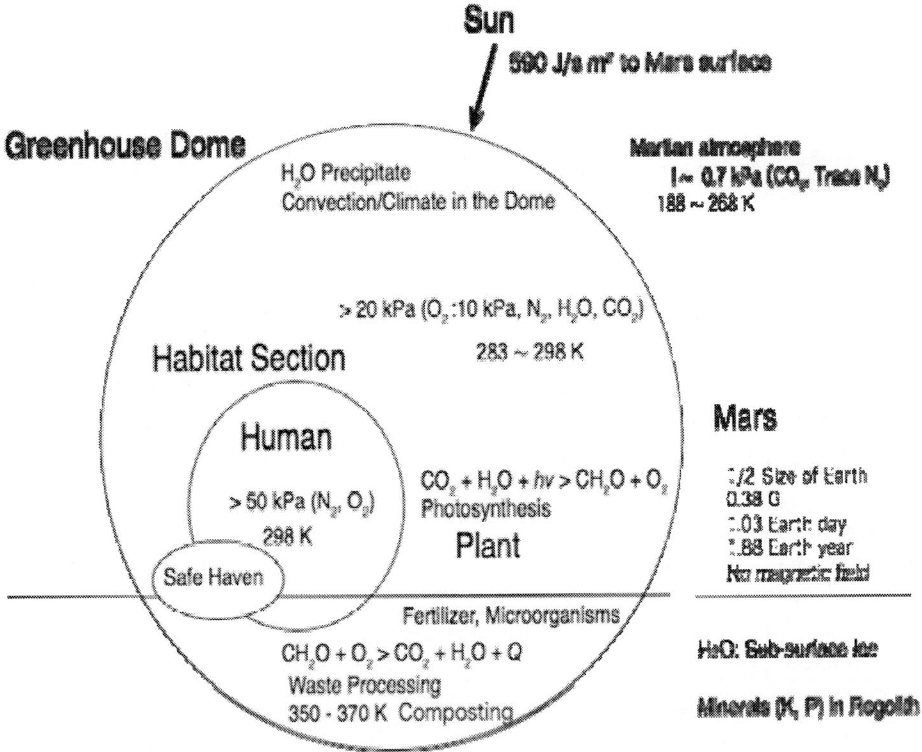

Figure 1. Outline of Martian agriculture.

Creation of Agriculture on Mars

Present Conditions of Mars for Agriculture

Mars orbits the Sun with a mean radius of 228 x 10^6 km. This distance of Mars from the Sun is about 1.5 times longer than that of Earth. Accordingly, solar constant for Mars is about 2.25 times lower than that for Earth. However, solar radiation energy deposited at the ground surface of Mars is about 2/3 that of Earth due to the lower albedo of Mars than of Earth (Salisbury et al. 2002). In spite of this, the surface temperature of Mars (avg. –55°C) is much lower than that of Earth (avg. +15°C). This is because the balance between deposited solar energy and radiative heat outgoing from the surface determines the surface temperature; the loss of heat is much larger from Mars than from Earth due to a much weaker greenhouse effect on Mars than on Earth.

Composition of air is remarkably different between Mars and Earth, as shown in table 1. Gravity on Mars is 0.38 of that on Earth. This low gravity is responsible for the low atmospheric pressure, through loss of gas molecules with low escape velocity.

Table 1. Composition of planetary atmosphere

	Mars	Earth
CO_2	95.30%	0.03%
N_2	2.70%	79%
O_2	0.13%	21%
CO	0.07%	0.12 ppm
H_2O	0.03%	2%
Ar	1.60%	0.90%
CH_4	< 10 ppb	1.7 ppm

(Kieffer et al. 1992, Formisano et al. 2004).

Information about Martian rocks and regolith, obtained by searches of Mars and analysis of Martian meteorites, is insufficient and somewhat contradictory, in terms of resource evaluation for agriculture. Some information pertinent to agriculture is:

1. Regolith and rock of varying sizes are seen on the surface of Mars in the images sent from spacecrafts. However, texture (particle size distribution) of regolith has not been directly measured.
2. Large-scale dust storms of fine regolith sometimes occur.
3. All the regolith is red in color due to a covering of iron oxides (e.g. hematite).
4. Most rocks are judged to be basaltic, because all the meteorites from Mars are basalt, and olivine ($(Fe,Mg)_2SiO_4$) is found everywhere in the sediment of Martian regolith.
5. Although the presence of clay has been reported, the mineralogical species of clay minerals (phyllosilicates) have not yet been determined. However, some scientists still insist that the main phyllosilicates are likely smectite.
6. Jarosite ($KFe_3(OH)_6(SO_4)_2$ or $NaFe_3(OH)_6(SO_4)_2$), sulfates and halides are found.

7. A thick layer of limestone has not been found as yet, despite much searching. Instead, a small amount (~2 to 5 weight %) of carbonates, especially $MgCO_3$, has recently been found almost ubiquitously in all the sediments of fine regolith (Bandfield et al. 2003).
8. Frozen water exists as icy lakes or frozen layers of regolith with ice inside their pores.
9. Surface layers of regolith are completely free from organic matter.
10. Signs of any living organisms have not been confirmed on the surface layers of regolith.

According to a recent hypothesis (Halevy et al. 2007), in the early era, Mars retained liquid water by being wrapped with SO_2-enriched air, which has a strong greenhouse effect. Thus, sulfate was formed, but thick layers of carbonate could not be formed under such acidic conditions. Another hypothesis is that the ubiquitous existence of carbonates, especially $MgCO_3$, in the present sediment of regolith, which may be deposited from circulated water, is formed by a reaction between CO_2 in the air and basic minerals of regolith. This hypothesis appears conformable to the ubiquitous existence of olivine in the present sediment.

It is probable that SO_2 in the air was oxidized and dissolved in the liquid water to form an acid sea or acid lake. The acid sea or lake was neutralized by reacting with minerals in the bed rocks. This resulted in the release of cations (Na^+, K^+, Ca^{2+}, Mg^{2+}, Fe^{2+}, Al^{3+}, etc.) from the minerals. The disappearance of SO_2 from the air caused a decrease in Martian surface temperature and the newly formed regolith was left more or less alkaline. This theory is reminiscent of a hypothesis proposed over 50 years ago (Rubey 1951), which was later doubted when the plate tectonics theory became popular. Rubey hypothesized that volatile substances (e.g. H_2O, CO_2, SO_2, HCl) were released directly or indirectly from magma, when the Earth was still very hot, and the ocean became acidic by dissolving these acid substances. After the Earth cooled, the acidic ocean was neutralized with bedrocks, leaving the present-day salty ocean. Ca^{2+} released to the salty ocean by the decomposition of plagioclase with the acid is deposited as carbonate in large amounts. This hypothesis of Rubey's may be still tenable for Mars, where the plate tectonics are less active and the cooling down and the loss of gaseous components is much quicker and more remarkable than for Earth.

On Mars, after neutralization of strong acids such as H_2SO_4 and HCl, the alkalinity of the newly formed minerals by volcanic activity is partly neutralized by reacting with CO_2 in the air, as mentioned above. The remaining alkalinity of the partially neutralized minerals is further neutralized when released Fe^{2+} is oxidized to iron oxides. This may be the reason that the pH of the regolith, measured by the Viking labeled release experiment, is estimated to be about 7.2+/-0.1 (Plumb et al. 1993).

The geological history of Mars is considered the cause of a big spatial variation in pH of the present Martian regolith. This is because a quantitative balance between acid substances and alkaline substances has changed through Martian history and by local geological events such as volcanic activity.

Presently on Mars, the dominance of CO_2 in the thin atmosphere may be caused by both the absence of liquid water and photosynthetic organisms, both of which are responsible for composition of terrestrial atmosphere.

The presence of dust storms on Mars suggests that 1) fine regolith widely covers the ground surface of Mars, 2) the thickness of the cover varies according to topography and air flow conditions, and 3) the particle size of wind-blown dust on Mars is much finer than that on Earth due to lower bearing capacity of the wind on Mars than on Earth.

Strategy for Creation of Agriculture on Mars

The selection of a dwelling site is an important early step in the initiation of human habitation on Mars. Criteria used in the selection of habitation sites on Earth could be applied to the Martian case, with certain modifications. Humans usually dwell in sunny places where both a watering spot and a shelter are within easy distance. This means that detailed topographic maps of Mars are very helpful for selecting the dwelling site. If the regolith at the selected site is sandy loam or loamy sand in texture and its sedimentation thickness is about 1 meter, the emigrants are fortuitous, because other textures of regolith or thinner sediment of regolith are considered unsuitable for agriculture, as discussed below.

Because the main target of Martian exploration is astrobiology research to search for extraterrestrial life or biotic substances, physical isolation barriers or a minimum distance should be secured between the site of scientific exploration and the dwelling site. Furthermore, careful unmanned exploration should be conducted prior to a manned mission and agriculture on Mars (Hashimoto et al. 2006).

To initiate appropriate agriculture at the selected dwelling site under the harsh Martian conditions described earlier, we must adopt a strategy different from that used to establish terrestrial agriculture. We propose a strategy with two stages. During the first stage, air and water on Mars will be brought into a pressurized dome, rather than making their recycling loops closed. Regolith, air and water inside the dome will be regulated to approximate their levels on Earth, principally by artificial means, without reliance on the autonomous control in an ecosystem. During the second stage, Martian agriculture will support human life by sufficiently supplying the necessities of subsistence mainly through recycling bio-elements inside the dome. In addition, Martian agriculture will gradually expand in scale by taking on-site resources available on Mars into the recycling loop of materials.

The First Stage in the Initiation of Martian Agriculture

Soon after landing on Mars, the emigrants should set up domes and deploy numerous solar panels on the ground surface. Extensive areas are available for placing a huge number of solar panels that can generate the necessary electrical energy for the emigrants during the whole period of their stay. The solar panels can work for many years, as evidenced by the Mars rovers. However, generation of energy by the solar panels is often disrupted due to the interception of solar radiation at night, and even during daytime by dust storms. Consequently, some facilities, such as batteries, are required to store the generated electric energy. For the safety of the dwelling site on Mars, it might be equipped with a nuclear power unit in order to make the power system redundant.

The structure of the domes and the materials that sheath them must be tough enough to endure the inner human-friendly pressure of the domes surrounded by the outer ambient Martian atmosphere, which is as low as 1/100 of terrestrial atmosphere. A multi-layered or -celled structure for the domes is desirable to reduce the mechanical load applied to the domes through dividing the pressure. This type of structure is also effective for insulating heat. The filmy sheathing materials themselves should be optically transparent to admit solar radiation necessary for human life and photosynthesis of plants, but opaque to the harmful ultraviolet part of the solar radiation, and should effectively shield cosmic rays.

The configuration of the domes consists of two units roughly divided from each other. One is a living quarter for the emigrants and is placed near a shelter. The other unit is for agriculture and should be placed near a watering spot.

A tubular or spherical shape should be chosen for the pressurized domes of the agricultural unit, rather than the ordinary shape of a terrestrial green house, which has only an upper structure placed on the ground. That is, in order to make the domes air tight, a tubular or spherical form might be preferable when considering the reduced magnitude of gravity on Mars (Nakamura 2005).

Another option is to place the dome sub-surface, where direct exposure to ultraviolet solar radiation is eliminated and cosmic rays can be shielded. Instead of direct introduction of solar radiation, an array of solar light collectors can be set up on the ground surface to guide light into the subsurface dome, after eliminating short wave-length radiation by the color aberration action of an optical lens (Tanatsugu et al. 1985). The use of electric lamps powered by the solar panel is another way of providing light to both the agriculture and living quarters.

Modification of Air and Water

The Martian air outside the domes is pumped into the dome by compressors driven by the generated electricity. Appropriate total pressure and partial pressure of O_2 and other gas species in the agriculture quarter is determined by considering the growth and yield of cultivated plants. A trace amount of N_2 in the outside air is selectively collected and supplemented to the air in the domes. In the living quarter, the environment is maintained for keeping the emigrants healthy and pleasant. At reduced pressure under reduced gravity on Mars, heat and mass transport phenomena in the domes differ from those on Earth (Yamashita et al. 2006). This fact should be taken into account when ducts for transport of gas, steam and water are installed in the domes. Another concerns are plant physiology dominated by transport phenomena, such as transpiration of leaf and thermal environment for plants. Effects of reduced pressure and gravity on plant biomass production should be carefully examined on ground and possibly verified in lunar base.

Water can be obtained by melting frozen lake water or frozen regolith mined from subsurface layers. The water is electrolyzed to generate O_2 and H_2. The H_2 is stored in a hydrogen adsorbing material until needed for electric power generation by fuel cells. Alkaline water, a byproduct of the electrolysis, is used to absorb CO_2.

The water obtained from frozen regolith may be contaminated with NaCl, Na_2SO_4, $NaHCO_3$, Na_2CO_3, $CaCl_2$, $CaSO_4$, or $Ca(HCO_3)_2$, because the regolith probably contains these salts. Such a saline solution is unsuitable for both drinking and irrigation and can be purified by distillation if necessary.

Some organisms can help with the modification of Martian air and water. For instance, some calcareous algae grow in saline water, and deposit $CaCO_3$ to its body. Spirulina, photosynthetic bacteria and tilapia may flourish in the resulting saline alkaline water, as evidenced at alkaline and/or salty lakes in Africa. Spirulina converts CO_2 to O_2 and biomass by photosynthesis. Calcareous algae do more than photosynthesis, and cover their algal body with $CaCO_3$. This inorganic deposit can be used by the humans as a material for construction, or for neutralization of acid substances. Spirulina, tilapia and photosynthetic bacteria are also nutritious foods for humans during the first stage.

Hyper-thermophilic aerobic bacteria can also help modify Martian air and water. If the organic waste and the waste water are kept in the spacecraft in order to prevent contamination of exploration targets on Mars, and if O_2 can be generated in excess of life support demand in the spacecraft, the organic waste can be quickly composted at temperatures as high as 80-100°C with hyper-thermophilic aerobic bacteria (Kanazawa et al. 2008) at a special corner inside the domes. During this composting, any harmful organisms in the organic waste are killed by natural autoclaving. Heat generated during the composting can be used for heating the air inside the domes, melting ice at the watering spot and obtaining distilled water after the cooled steam is deodorized. If the organic waste and waste water include human excreta rich in N, the hot steam would be contaminated with NH_3. This NH_3 can be collected by peat moss and used as an N-fertilizer. NO_3 made from NH_3 by nitrification can be used as another N-fertilizer and may turn to N_2 after denitrification.

Modification of Regolith (Formation of Martian Soil)

Regolith is modified in the agricultural quarter to make it habitable for plant roots and soil microorganisms. This effort can be called artificial soil formation on Mars. Unfortunately, information about the regolith is still fragmentary, and somewhat contradictory in certain parts. We shall explore Mars to map and collect comprehensive data on the regolith, which is required to design space agriculture and select the habitation site.

The regolith on Mars is widely covered by hematite. This indicates that ferruginous minerals such as olivine, pyroxene and hornblende have been weathered to release ferrous iron, which is oxidized by mechanisms not yet fully understood. In addition, the presence of $CaSO_4$ and jarosite indicates that even Ca-, Na-, and K-rich minerals such as feldspars have been weathered. However, the common presence of olivine indicates that the degree of the weathering is generally low. Furthermore, the presence of jarosite would mean the regolith is acidic, because jarosite is formed and kept intact under acidic conditions. While on the contrary, the presence of $MgCO_3$ or $CaCO_3$ would mean the regolith is somewhat alkaline.

In any case, the following stepwise efforts are necessary in order to form Martian soil:

Desalinization: Martian regolith usually contains water-soluble salts. If their concentration is so high as to harm most organisms except for halo-tolerant or halophilous organisms, desalinization (removal of the water-soluble salts) is imperative. Desalinization can be achieved by leaching the salt-affected regolith with non-saline water. The leaching water enriched with the salts (e.g. NaCl, $NaSO_4$, $CaCl_2$, $CaSO_4$) can be used to strengthen regolith bricks for use in the construction of space habitation.

Neutralization: The pH of the regolith may vary from place to place. pH should be very low in places dominated by jarosite and should be high in places dominated by $NaHCO_3$ or Na_2CO_3. However, in most places, pH may be somewhat high where $CaCO_3$ or $MgCO_3$ is found. Common plants are damaged by soil with too low or too high pH. Accordingly, we must prepare methods of neutralization for both too acidic and too alkaline regolith. The application of calcareous algae rich in $CaCO_3$, or fine regolith rich in olivine, is effective for increasing the pH of acid regolith. For alkaline regolith, the application of peat moss with low pH may solve the problem.

Improvement of physical properties: From a soil chemist's standpoint, the sediment of fine textured regolith is considered to be favorable for agriculture, because its cation exchange capability (CEC), mineral nutrient content, and ability to retain organic matter are expected to be high. However, from a soil physicist's standpoint, fine textured sediment is problematic because of poor aeration and poor drainage, which result in O_2-deficiency for plant roots and microorganisms living in the sediment, and the difficulty in desalinization by leaching, respectively. When water is added to the sediment of fine textured regolith, a large portion of the pores in the sediment are liable to be completely filled with water by capillary force (capillary water) working against gravity. Accordingly, loamy textured soils have been considered to be desirable for common upland crops in terrestrial agriculture. This suggests that a sandy loam or loamy sand texture of Martian regolith is desirable for Martian agriculture under low gravity. However, the aeration and drainage of soil is controlled not only by texture but also by a mode of aggregation of soil particles. For instance, drainage of terrestrial soils with sandy loam or loamy sand texture is often poor. This problem with the physical properties of terrestrial soil is usually overcome by the presence of aggregates of suitable sizes (e.g. a few mm), which are naturally formed by a concerted action of biotic and abiotic processes and can be artificially formed by applying synthetic polymers or some materials with coarse pores. Pores among the aggregates of suitable sizes are coarse enough for desirable aeration and drainage. Accordingly, the artificial aggregation of Martian regolith should improve its physical properties. A promising candidate synthetic polymer is polyvinyl alcohol (PVA) (Dejbhimon and Wada 2005). Another feasible method to improve the physical properties of Martian regolith is to apply peat moss or an equivalent processed plant material produced in agriculture on Mars. Both are rich in coarse pores for aeration and drainage.

Another possible problem with the physical properties of the Martian regolith is the presence of hard impervious layers cemented with $CaCO_3$ or $CaSO_4$. Such layers are recognized in the arid regions of Earth and are called petrocalcic horizon and petrogypsic horizon (Soil Survey Staff 2006a), respectively. These layers are formed by thick precipitation of $CaCO_3$ or $CaSO_4$. If these layers are near the surface at some places in the selected dwelling site, these places should not be used for agriculture, because these layers are obstacles for tillage and percolation of water. However, such places may be suitable for the living quarters of the emigrants. In other words, these hard layers, if present, should be taken into account when selecting a dwelling site and dividing the dome into two areas.

The leachate from the desalinized fields should be carefully managed, because the supply of water enriched with these salts can lead to new formation of these impermeable layers at the place of desalinization. This may be troublesome if it occurs at a shallow depth in the agricultural area. However, it may be utilized to increase the hardness of dwelling basements.

Nutrients: Most of the plant nutrients, except N and P, are expected to be available on Mars, because they may take forms accessible to plant roots after they are released from minerals by weathering. N is not a component of any minerals and accordingly is utterly absent in the regolith. Application of some N fertilizer (NH_3, NO_3^-, urea, excreta of animals) or cultivation of some green manure crop plants capable of N-fixation is inevitable. However, the N-fixation is not active if the amount of available P is far short of the demand of the green manure crop plants.

P is actually contained in the minerals of the Martian regolith, and may be released by weathering in a similar way as other nutrients. However, the released P is considered to be unavailable for plant roots, due to its strong adsorption (P-fixation) on $CaSO_4$, $CaCO_3$ and iron oxides. Thus, P fertilizers should be applied by three appropriate measures to avoid this

P-fixation. The first is to apply P fertilizers mixed with peat moss (Kawakami et al. 2007). The second is the removal of $CaSO_4$ by leaching if the main P-adsorbent is this substance. The third is the adjustment of pH at about neutral where P-adsorption with both $CaCO_3$ and iron oxides is most suppressed if these two substances are dominant P-adsorbents.

Excess amounts of several elements are harmful to common terrestrial crop plants, and this problem may occur on Mars. For instance, if the regolith is enriched with olivine, excess Mg, Ni and Cr may damage common crop plants. The application of $CaCO_3$ may cure excess Mg, while the application of peat moss, which possesses a strong capability to adsorb Ni and Cr, may effectively mitigate the harm caused by excess Ni and Cr.

Organic matter: Organic matter contributes to many important functions of the soil, two of which may be essential in forming desirable Martian soil. One function is to form and maintain aggregates of suitable sizes in the fine regolith. The other is to store nutrients (bio-elements), especially N. The nutrients stored in organic matter are released when the organic matter is microbiologically decomposed. Therefore, accumulation of organic matter containing N in the regolith is imperative to establish space agriculture. This can be achieved by applying compost and excreta of animals.

Microorganisms: Microorganisms inhabiting the soil are responsible for the most of the functions of the soil, such as decomposition of various organic waste, release of nutrients stored in the organic matter, nitrification, denitrification and N-fixation. To endow the regolith with these functions, the Martian regolith should be inoculated with desired microorganisms. For this purpose, it is promising to use peat moss, which contains the desired microorganisms.

The processes going on during the first stage of agriculture on mars are summarized in figure 2.

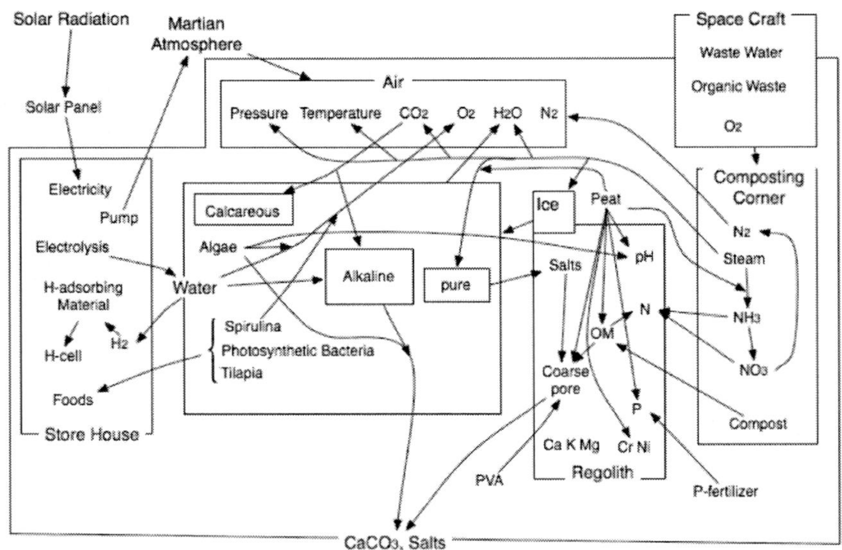

Peat: Peet moss containing desired microorganisma, OM: Organic Matter + microorganisms, PVA: Polyvinyl alcohol.

Figure 2. Proceses in the First Stage.

The Second Stage in the Initiation of Martian Agriculture

When the goal of the first stage is almost attained, the second stage of space agriculture should be started. The main objectives in the second stage are to establish a sustainable agriculture that can continuously supply to the emigrants clean air and water, as well as foods, fiber and timber, mainly by recycling substances as shown in figure 3. Reservoirs of the recycling flows of substances are not drawn in figure 3, but they are necessary for buffering fluctuation of the flows.

During the second stage, most of the tasks in the first stage should continue to work to keep the conditions inside the domes favorable for recycling substances in the second stage.

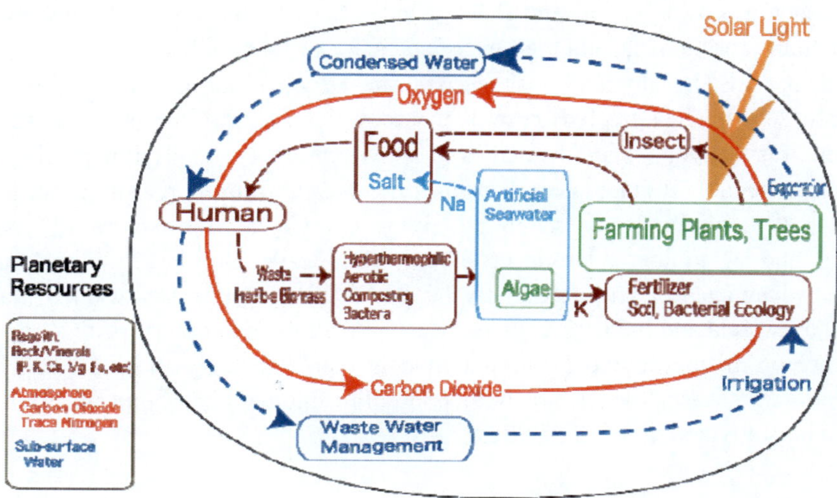

Figure 3. Materials recycle in Martian agriculture.

Supply of Foods, Fiber and Timber

Foods should be not only sufficient in quantity and high in quality, but also tasty enough for the emigrants to enjoy. In addition, the farming of high yielding plants, which can supply storable foods, is desirable. This is especially true for staple foods. For meeting these requirements, the selection of planting organisms is an important issue.

Carbohydrate (Staple Food)

Cereals and potato are promising candidates for the carbohydrate source in space agriculture. Among various cereals, rice (*Oryza sativa*) is most favorable, because the yield of rice is higher than that of other cereals. Paddy fields for cultivating rice can be used to co-culture *Azolla*, an aquatic fern with symbiotic blue green algae capable of N-fixation (Katayama et al. 2007), and rear loach (*Misgumus anguillicaudatus*) in the flood water. The paddy field on Mars will be similar to a "vinyl paddy field," in which a polyvinyl chloride sheet liner is filled with a mass of soil. This technology was introduced about 50 years ago in Japan.

Potato (*Solanum tuberosum*) is expected to be cultivable on Martian soil, because the cultivation of potato originated in Inca, on cold infertile soils in the high plateaus of the

Andean Mountains. The natives in the region freeze and dry the harvested potato to make it storable for a long period. This method can be easily applied on Mars.

Sweet potato (*Ipomoea batatas*) is another promising high-yielding crop plant, which can grow in infertile soil. Unfortunately, sweet potato has low cold resistance and storability. However, its cold resistance could be increased by breeding or genetic engineering, and it can be changed into storable foods by several simple processing methods.

Animal Food (Amino Acids)

Food of animal origin is important for the emigrants to eat, as it supplies certain amino acids, lipids, and vitamins, which are hard to get from plant food. Insects are recommended to be reared on Mars (Katayama et al. 2008) as a source of animal food, instead of common livestock, such as cows, horses, and sheep. This is because insects can be fed on substances that are inedible for humans, and their excrements are easily handled.

Promising insect candidates are silkworm (*Bobyx mori*), termites and fly larva (Mitsuhashi 2007). They are fed on mulberry leaf, plant debris rich in cellulose and lignin, and excreta, respectively. The advantages of silkworm are 1) its physiology, life history and genetics have been well elucidated, 2) it does not escape from its rearing room to become a nuisance to the emigrants staying in the living quarter, 3) it produces silk cocoon as a byproduct, and 4) mulberry produces edible fruit, leaves for healthful tea and wooden materials. Advantages of termite and fly larva are to convert waste wooden materials and carcasses, or excreta and industrial organic waste, to edible biomass, respectively.

Another promising source of animal food is fish, especially those feeding on organic debris or organic waste, which are both non-edible biomass for humans. Some fish (e.g. loach) can be cultivated in the paddy field.

Lipid

Appropriate combinations of the above-mentioned foods can fulfill all the essential lipids required for maintaining human health.

Minerals and Vitamins

Nutritional analysis of the appropriate combinations of the above-mentioned foods revealed that they contain almost all minerals and vitamins necessary for human health. However, some menus consisting of these foods may be deficient in some minerals and vitamins, especially vitamin D. In such cases, the deficiency can be overcome by taking other foods, such as supplements, mushrooms cultivated on waste timber, or fish (e.g. loach), rich in vitamin D. In addition, leaf vegetables soybean, sesame and radish are desirable for preparing tasty and healthy meals.

Na-K Enigma

Na and K are somewhat different from other minerals and nutrients. Both Na and K are water-soluble and have no organic forms. In addition, humans need a much higher Na/K ratio than common crop plants, though both humans and the common crops need Na and K in large amounts. Furthermore, humans daily excrete urine containing Na. This implies that the water discharged from the living quarter into the quarter for agriculture contains Na that will damage those crops sensitive to salt (Yamashita et al. 1985). One conceivable method to solve this problem is to lead composted urine through a channel to an agricultural section for halophytes, which can grow in the salt-affected soil and accumulate Na in the edible parts of the plant. Candidate halophytes are the ice plant (*Mesembryanthemum crystallinum*) (Adams et al. 1998) and saltwort (*Salicornia herbacea* L.) (Shimizu 2007).

Menu

Several menus using a combination of the above-mentioned foods are proposed. All of them are tasty and sufficiently nutritious for the emigrants. As an example, figure 4 shows a menu composition using silkworm pupa as an animal food, and figure 5 illustrates an entire day's menu.

Rice 300g, Soybean 100g, Sweet Potato 200g, Green Vegetable 300g, Silkworm Pupa 50g, Loach fish 120g, Salt 3g per person per day.

Figure 4. Food composition for Martian menu.

Figure 5. (Continued).

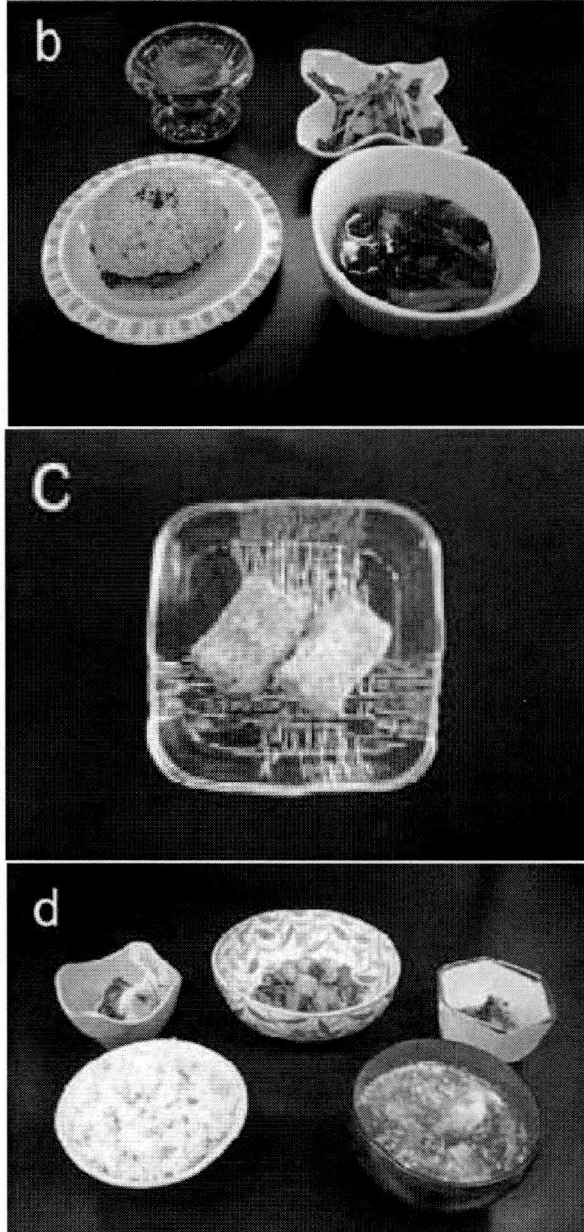

Figure 5. Menu for one day on Mars. Ingredients are listed clockwise from bottom left. a) Breakfast: Boiled unpolished rice, baked thin rice cake with powdered silkworm pupa, green vegetable and loach wrapped with deep-fried thinly sliced soy-bean curd, miso (fermented soy bean paste) soup containing Komatsuna (green-yellow vegetable) and diced tofu (soy bean curd). b) Lunch: Rice burger with a filling of loach and silkworm pupa, soft rice cake with soy milk and silkworm pupa powder, fresh radish sprout salad, loach soup with green vegetable and diced tofu. c) Snack: Soft rice paste cake coated with roasted soy bean powder and sprinkled with silkworm pupa. d) Supper: Boiled unpolished rice, steamed diced sweet potato topped with roasted sesame cream flavored with silkworm pupa powder, boiled diced sweet potato with green yellow vegetable seasoned by silkworm pupa powder, boiled green vegetable with vinegared miso dressing, and miso soup with loach and green vegetable.

Composting

Organic waste emerges from both the living quarter and the agriculture quarter. Its small part can be directly applied to the soil. However, its large part should be composted before application to the farming soil. The organic waste is principally composted before application to the farming soil. Some manageable organic waste like insect excreta can be directly applied to the soil. This is a main process of recycling of materials in Martian agriculture, because this is the transfer from consumer to decomposer in the Martian ecosystem. One problem with the composting is consumption of the precious O_2 in the air. This problem may be solved at least partly using *Geobacter*, which can oxidize organic substances with the iron oxides commonly available on Mars and generate electricity (Lovely 2006).

The hyper-thermophilic aerobic composting discussed above can also be used in the second stage when quick treatment of a large amount of organic waste is urgently needed.

Maintenance of Soil Conditions

The soil formed from the regolith in the first stage should be maintained to be productive and manageable. In other words, the application of a suitable amount of compost and/or organic waste is not only necessary for material recycling, but also necessary for maintaining soil productivity. Furthermore, accumulation of organic matter in the soil helps to ensure high O_2 content in the air and acts as a reservoir for bio-elements.

In addition, careful irrigation is necessary for the following three reasons. The first is to save water. The second is to avoid accumulation of salts in the soil by keeping the evapotranspiration rate lower than the irrigation rate. The third is to avoid the damage of dikes and other constructions by turbation and the formation of deep cracks in the soil. Such a problem is known in clayey terrestrial soils rich in smectite, which are classified as Vertisols (Soil Survey Staff 2006b). Vertisols swell when moist, and shrink when dry, leading to cracks and turbation of the clayey soils. If such problems are anticipated in Martian soil, the soil should not be exposed to distinct alternation of wetting and drying by improper irrigation.

Supply of Clean Water and Air

Water is principally recycled through irrigation, evapotranspiration and condensation. This cycling cannot work smoothly without reservoirs, because the evapotranspiration from the arable places changes with both the growth stage of plants and the daily and seasonal changes in soil temperature.

Concentrations of O_2, CO_2, N_2+Ar and H_2O in the air are partly controlled by photosynthesis of plants, respiration of organisms, microbiological decomposition of organic debris, N-fixation, nitrification and denitrification. However, this control is insufficient to keep the concentrations of each gaseous component at their desired levels. Among these components of the air, a decrease in O_2 concentration is most critical for life support of the emigrants. Accordingly, O_2 concentration should be kept sufficiently high by electrolysis of water, when photosynthetic activity of plants is insufficient.

Ark for Initiating Martian Agriculture

We must bring organisms and materials from Earth to Mars to initiate agriculture on Mars. We must minimize both the variety and the quantity because of the limited capability of inter-planet transportation. The above-referred organisms and materials directly relevant to the Martian agriculture are listed below:

For the First Stage

Organisms* (calcareous algae, spirulina, tilapia, green manure crops)

Useful microorganisms* (photosynthetic bacteria, rhizobia, nitrifying bacteria, denitrifying bacteria, heterotrophic bacteria, hyper-thermophilic aerobic bacteria)

Fertilizers (P-fertilizers, urea)

Soil conditioners (PVA, peat moss)

Hydrogen absorbing materials

For the Second Stage

Cereals *(paddy rice)

Corm *(potato, sweet potato)

Vegetables *(*Azolla*, mushroom, ice plant, salt wort, mulberry, leaf vegetables, soy bean, sesame, radish)

Insects *(silkworm, termite, fly)

Fish* (loach)

Microorganisms* (heterotrophic microorganisms, hyper-thermophilic aerobic bacteria, *Geobacter*)

Plastic sheet for “vinyl paddy field”

Note: Organisms with * are genetically engineered to be adaptable to Martian agriculture. Microorganisms liable to escape from the dwelling site to the site for scientific exploration should be genetically marked for easy identification of their Earth origin.

Conclusion

A conceivable plan is proposed to realize Martian agriculture on the lifeless barren regolith under hostile conditions, based on the knowledge and experiences of terrestrial ecosystems and agriculture. In this plan, the selection of both the dwelling site on Mars and the organisms for Martian agriculture is the essential issue. To test the feasibility of the plan, many terrestrial experiments should be carried out to examine each part of the plan under simulated Martian conditions, starting with the first step. Then, a computer simulation should be conducted using the data obtained through these terrestrial experiments in order to confirm feasibility and validate the stability of the system’s operation as a whole.

Currently, many terrestrial ecosystems are being quickly destroyed, leading to a decrease in biodiversity. At the same time, terrestrial agriculture in many places has been losing its productivity, principally due to soil damage. This situation with terrestrial agriculture somewhat resembles the conditions for Martian agriculture. This suggests that the success of

Martian agriculture may be helpful for reviving destructed terrestrial ecosystems and renewing damaged terrestrial agriculture. This may be the next revolution of our human history.

REFERENCES

Adams, P., Nelson, D.E., Yamada, S., Chmara, W., and Jensen, R.G. (1998). Tansley Review No. 97 Growth and Development of *Mesembryanthemum crystallinum* (Aizoaceae). *New Phytologist*, 138, 171-190.

Bandfield, J.L., Glotch, T.D. and Christensen, P.R. (2003). Spectroscopic identification of carbonate minerals in the Martian dust. *Science,* 301, no. 5636, 1084-1087.

Dejbhimon, K., Wada, H. (2005) Improvement of the saline soil in Northeast Thailand using polyvinyl alcohol (PVA). In: *Proceedings of "Management of Tropical Sandy Soils for Sustainable Agriculture.* A holistic approach for sustainable development of problem soils in the tropics" 27th November-2nd December 2005, Khon Kaen, Thailand. 123-128.

Formisano, V., Atreya, S., Encrenaz, T., Ignatiev, N., and Giuranna, M. (2004) Detection of Methane in the Atmosphere of Mars. *Science*, 306,1758 – 1761.

Hashimoto, H., Koike, J., Yamashita, M., Oshima, T. (2006). Space Agriculture Saloon; Proposal for Extension of Planetary Protection Policy to Avoid Biological Contamination of Mars with Manned Exploration Supported by Space Agriculture, *Viva Origino*, 34, 86-89.

Helevy, I., Zuber M.T. and Schrag. D.P. (2007). A sulfur dioxide climate feedback on early Mars. *Science*, 318. 1903-1907.

Kanazawa, S., Ishikawa, Y., Timita-Yokotani, K., Hahimoto, H., Kitaya, Y., Yamashita, M., Nagatomo, M., Oshima, T., Wada H. and Space Agriculture Task Force. (2008). Space agriculture for habitation on Mars with hyperthermophilic aerobic composting bacteria. *Adv. Space Res.* 41(5), 696-700.

Katayama, N., Ishikawa, Y., Takaoki, M., Yamashita, M., Nakayama, S., Kiguchi, K., Kok, R., Wada, H., Mitsuhashi J. and Space Agriculture Task Force. (2008). Entomophagy: a key to space agriculture. *Adv. Space Res*, 41(5), 701-705.

Katayama, N., Yamashita, M., Kishida, K., Liu, C.C., Watanabe, I., Wada H. and Space Agriculture Task Force. (2008). *Azolla* as a dish of space diet for habitation on Mars. *Acta Astronautica*, accepted.

Kawakami, H., Niijima, Y., Ota, Wada, H., Lan, Z., Wang, Z., Li, S. (2007). Increase in growth and yield of peanut through improvement of the saline alkali soil by applying a domestic peat. *J. Japan Peat Society*, 4(1), 22-32.

Kieffer, H.H., Jakosky, B.M., Snyder, C.W., and Matthews, M.S. (1992). *Mars*, Univ Arizona Press.

Loveley, D.R. (2006). Bug juice: harvesting electricity with microorganisms. *Nature Reviews, Microbiology*, 4, 497-508.

Mitsuhashi, J. (2007). Use of Insects in Closed Space Environment. (in Japanese) *Biol. Sci. Space*, 21, in print .

Nakamura, Y. (2005). Engineering of pressurized structure and fire safety for space habitation on Mars - Needs of fundamental fire researches for space enclosure system- . *J. Space Technol. Sci.*, 21(2), 39-48.

Plumb, R.C., Bishop, J.L. and Edwards, J.O. (1993). The pH of Mars. In: Lunar and Planetary Inst. "Mars: Past, Present and Future. Results from the MSATT Program, part 1" 40-41.

Rubey, W.W. (1951). Geological history of seawater. *Geological Society of American Bulletin*, 62,1111-1147.

Salibury, F.B., Dempster, W.F., Allen, J.P., Alling, A., Bubenheim, D., Nelson, M. and Silverstone, S. (2002). Light, plants, and power for life support on Mars. Light Support and Biosphere Science, Vol.8, *Cognizant. Comm. Corp.*161-177.

Shimizu, K. (2000). Effect of salt treatments on the production and chemical composition of salt wort (*Salicornia herbacea* L), rodesgrass and alfalfa. *Jpn. J. Trop. Agr.* 44 (1):61-67.

Soil Survey Staff. (2006a). Keys to Soil Taxonomy, 10th ed. USDA-Natural Resources Conservation Service, Washington, DC. USA, 13

Soil Survey Staff. (2006b). Keys to Soil Taxonomy, 10th ed. USDA-Natural Resources Conservation Service, Washington, DC. USA, 283-294.

Tanatsugu, N., Yamashita, M., Mori, K. (1985). A Conceptual Design of a Solar-Ray Supply System in the Space Station, *Space Solar Power Review*, 5, 221-230.

Yamashita, M., Ohya, H., Nitta K. and Yatazawa. M. (1985). Sodium and potassium recycle in closed ecosystem-space agriculture (in Japanese). *J. Jpn. Soc. Aeron. Space Sci.*, 376, 288-296.

Yamashita, M., Ishikawa, Y., Kitaya, Y., Goto, E., Arai, M., Hashimoto, H., Tomita-Yokotani, K., Hirafuji, M., Omori, K., Shiraishi, A., Tani, A., Toki, K., Yokota, H., Fujita, O. (2006). An Overview of Challenges in Modeling Heat and Mass Transfer for Living on Mars, *Ann. N.Y. Acad. Science*, 1077, 232-243.

In: Space Exploration Research
Editors: J.H. Denis and P.D. Aldridge

ISBN: 978-1-60692-264-4

Chapter 20

THE GEOLOGY OF THE VENUSIAN PLAINS

Alvaro Ignacio Martín-Herrero*[1]*, Beatriz Gómez-Izquierdo*[1,2]*,
Guillermo Caravantes*[1,2]*, Esther Velasco*[3]*, Javier Ruiz*[3]
***and Francisco Anguita*[4]**

[1] Seminario de Ciencias Planetarias, Facultad de Ciencias Geológicas,
Universidad Complutense de Madrid, 28040 Madrid, Spain
[2] Volcano Dynamics Group, The Open University,
MK7 6AA, United Kingdom
[3] Centro de Biología Molecular,
CSIC-Universidad Autónoma de Madrid,
28049 Cantoblanco, Madrid, Spain
[4] Departamento de Petrología y Geoquímica,
Universidad Complutense de Madrid,
28040 Madrid, Spain

ABSTRACT

A large fraction of the venusian surface is formed by rolling plains, whose altitude is typically 1 to 3 kilometers under the planet mean radius. The plains' most interesting features are volcanic structures, winkle ridges and tectonic belts. The volcanic activity has generated huge lava fields that have inundated almost all the planet's surface concealing any previous features that possessed. This activity has resulted in the formation of the plains, volcanic buildings whose morphologies are very similar to terrestrial analogs and that usually appear in groups of several individuals, although sometimes they are isolated; may well find a shield volcanoes, slag and tephra cones and domes among others. Wrinkle ridges are interpreted as resulting from failure of the crust under compression. Tectonic belts are linear zones of concentrated deformation hundreds of kilometers long, tens of kilometers wide, which include thrusted anticlines, grabens, tight folds, and strike-slip fractures. The combination of these structures makes up belts with different kinematics: extensional belts mainly characterized by grabens, thrust and fold contractional belts, and strike-slip shear belts usually defined by the en echelon disposition of folds (transpression) or grabens (transtension). Another important characteristic of the ridged plains, indicative of a crust excess density, is that the majority of them exhibits a geoid negative anomaly and are far from isostatic equilibrium.

Whereas tectonic belts have received an ample attention, there is not consensus on their origin has not been reached. Several hypotheses have been proposed, for example excess density (deduced from geoid anomaly) causing compression through the drag of vertical traction and inducing plains downwelling, push of the geoid highs over low-standing terrains, or crustal delamination. Thus, venusian plains are useful keys for the understanding of the evolution of Venus.

1. INTRODUCTION

Impact crater counts allow us to estimate the approximate age of the surface of a solid planetary body, the greater the number of craters, the older the surface will be. When we observe Venus, we will quickly realize about the almost non existence of impact craters on its surface. This fact indicates that its surface has suffered rejuvenation processes in a not too distant past or even at present. In the Earth there are currently two types of crust rejuvenation processes. The first and most appreciable by human eye is the erosion, due mainly to atmospheric processes. The second, less evident to us but not for that less important are the tectonic processes originated by the movement of tectonic plates. Therefore, one should expect that on Venus, a planet with an atmosphere 90 times denser than ours, the erosion processes will be of equal or greater magnitude than on Earth, but this is not the case.

In 1970, the Russian Lander Venera 7, landed on the surface of our neighbor. The acquired data during the short period of time in which remained operational were amazing. Planet surface temperature was approximately 450 degrees centigrade. Wind speeds close to 3 km per hour were far from the 500 km per hour reached in the upper zones of the atmosphere. There is also a total lack of liquid water, although there are signs of its presence millions of years ago. Taking these data into account, the rejuvenation of the surface of the planet would be controlled by similar tectonic processes that those occurring on Earth. However, Venus lacks tectonic plates, whose movement drifting on the venusian mantle would produce earthquakes or generate ridges as they collide or suffer friction against each other. This is due to the relative calm in which the interior of the planet is, because while the Earth rotates on itself once every 24 hours, Venus instead does it once every 243 terrestrial days.

Thus, if the rejuvenation of the planet's surface is not due to erosion or tectonic processes similar to those of the Earth, how can this process happen? Today we know that the internal heat of a rocky planet like ours is produced by two main sources. The first one is the residual heat that occurred during the formation of the planet, due to the collisions between the planetesimals that constituted the protoplanet, as well as subsequent additions in the form of meteorites. The second one is the disintegration of radioactive elements (such as thorium, uranium or potassium) in the interior of the planet. On Earth, we know that this heat is the engine that puts into operation the convective flows in the mantle which trigger movement of the lithospheric plates, break them, make them collide and generate open ways for the internal heat to escape through faults and volcanism. In Venus, as has been said, there are no plate tectonics as on Earth. It is believed that the lithosphere of Venus, much thicker than that of the Earth, isolates the heat inside the planet, warming the mantle to temperatures as high as in the core. This heat accumulation would lead to a sudden release of energy in certain moments when a critical point is reached; provoking huge collapses of the inner crust and favoring a planetary scale volcanism with enormous amounts of melts flooding the surface and erasing any possible previous structure. It is has been estimated that several hundreds of million years ago Venus suffered one of these critical stages, and the materials expelled flooded practically the entire surface of the planet, especially the so-called "tectonic plains", from which a high proportion of the lower areas of the planet is made up.

According to what has been said, Venus would present cycles of absolute calm alternating with tectonic and volcanic catastrophic activity periods, which remodel the entire surface of the planet every several hundred million years. However, in 1992, the Magellan probe reached our neighbor, starting a mission that would last little more than two years until its loss in the venusian atmosphere. During these 2 years, Magellan conducted, among other projects, a high resolution mapping of the planet's surface through radar images that covered approximately the 98% of it. This cartography showed us an active planet, with high elevations, enormous deformation belts and wide plains (figure 1).

Figure 1. This image is a composite Global View Centered at 180 Degrees East Longitude obtained by the Magellan mission. It shows craters on the surface of about 300 to 500 million years old. (Image font: www.jpl.nasa.gov).

An important percentage (65-70%) of the surface of Venus is covered by the so-called Venusian plains. With an altitude that goes from 1 to 3 km below the mean planetary radius, plains consist of extensive sheets of flood lavas hundreds of kilometers across with a remarkably flat topography. The origin of these volcanic plains is thought to be related to a period of extensive volcanic activity between 300 and 750 million years ago (Basilevsky and Head 1995; Head et al. 1996; Basilevsky et al. 1997; McKinnon et al., 1997), although other authors have proposed a more gradual origin for the formation of the observed geology of these areas (Guest and Stofan, 1999). The appearance in radar images of these regions is mostly dark, indicating a smooth surface, but they can be also bright when corresponding to higher levels of deformation. The main features found on the surface of the plains are lava flows, small volcanoes, wrinkle-ridges and tectonic belts.

In the present chapter we will try to make a revision of the current knowledge about the geology of the venusian plains, its volcanic and tectonic features, and the relation between them. Different models have been proposed for the formation of the ridged plains: for example, compression through the drag of vertical traction and inducing plains downwelling (e.g., Bindschadler et al., 1992; Magee and Head, 1995), or push of the geoid highs over low-standing terrains (Sandwell et al., 1996). The depiction of the sequence of volcanic and tectonic events can be performed by detailed cartographic works, and could eventually lead to the development of a model for the formation of the current morphology of the plains.

2. Volcanism

The data obtained by the Magellan's SAR (Synthetic Aperture Radar) showed that the reflective properties of the materials, which constituted the volcanic plains, to the radar pulses varied significantly from regions with a very low signal return (black) to areas with a high rate of pulses return (white). The amount of the pulses that bounced back to the source depends on a diverse number of details such as the incidence angle of the pulses as well as the roughness and electrical properties of the surface where bounced. For the radar data from the Magellan, the local variations of brightness were controlled by changes in the roughness of materials, always within the scale of the wavelength of the SAR pulses (12.6 cm).

Thus, the dark areas would indicate materials with soft and smooth surfaces, whereas materials with white color could be interpreted as rough surfaces within a decimetric to metric scale (figure 2). To demonstrate the validity of these appreciations, these data were compared to those obtained trough similar systems on Earth that could be checked in situ (figure 3). After this check, data showed that the presented materials possessed similar roughness to those of the aa type lavas here on Earth; seeing that the young materials can be both, bright or dark, depending on their surface textures, while the oldest materials tend to be extremely dark, maybe due to slow metamorphic processes under the high pressures and temperatures of the planet or by smoothing of their surfaces by the deposit of debris by wind action.

The origin of the lava flows on the Venusian plains may not be always identified. When it is possible to recognize their origin, they come from volcanic edifices, surface fractures, etc. They can have very different horizontal extensions, from a few to hundreds of kilometers, and present on their surface lava channels, levees, pressure ridges and flow margins; features which provide information about lava flow direction and formation style. Small volcanoes on the surface of Venus are considered to have typically less than 20 kilometers in diameter (Slyuta and Kreslavsky, 1990; Head et al., 1992). They are the predominant volcanic construction type on the plains, and can be subdivided into several categories: cones, which are features with a central pit, steep slopes and a circular shape and usually occur in clusters; shields, with shallow slopes, sometimes diffuse outer limits and a central pit (groups of small shields are common and often associated with fracture belts); and domes.

Previous works review the volcanism of the venusian plains (e.g., Guest and Bulmer, 1992; Roberts and Guest, 1992; Saunders, 1999). Below, we will describe the main volcanic features on the plains.

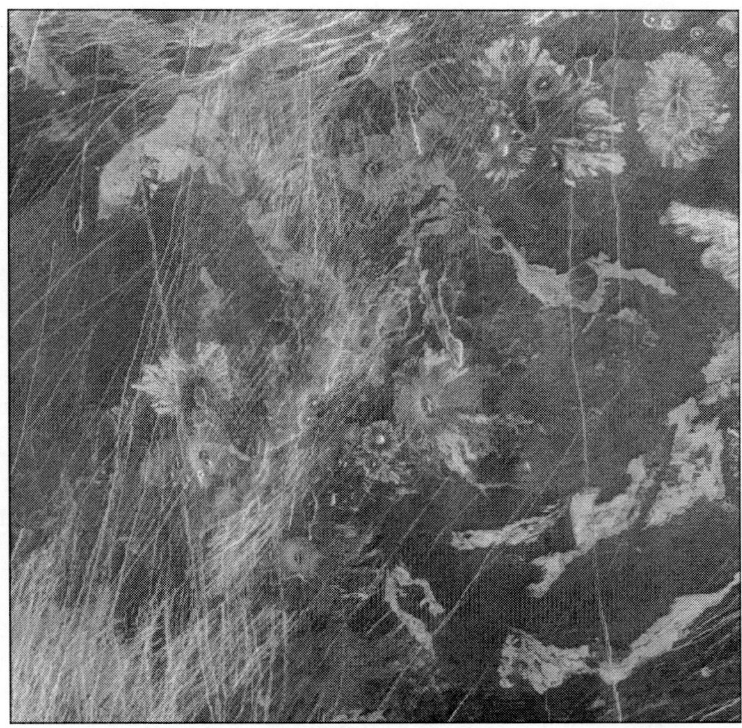

Figure 2. Volcanic and tectonic features in Atla Regio (Image font: www.jpl.nasa.gov).

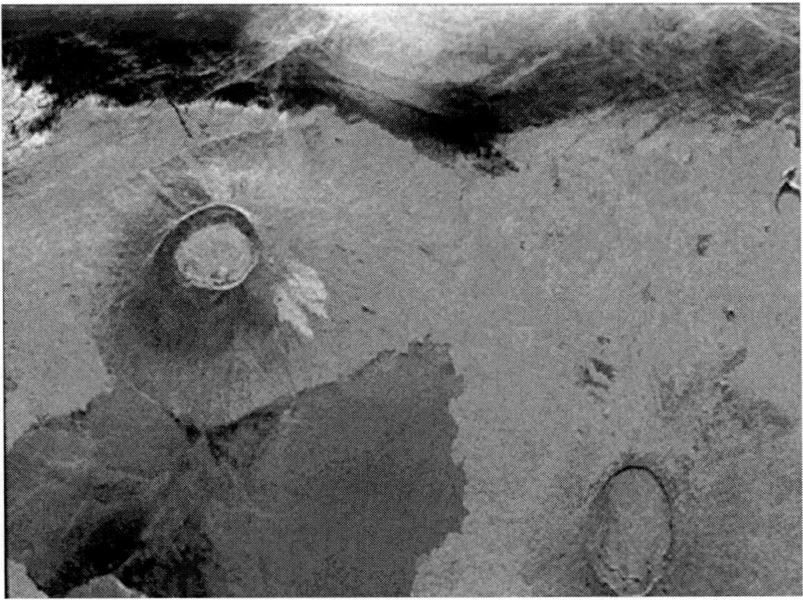

Figure 3. Image showing part of Isla Isabela in the western Galapagos Islands. The western Galapagos Islands have six active volcanoes similar to the volcanoes found in Hawaii. Alcedo and Sierra Negra volcanoes shows the rougher lava flows as bright features, while ash deposits and smooth pahoehoe lava flows appear dark. (Image font: www.jpl.nasa.gov).

Flood Lavas

Before the arrival of the Magellan space probe, the Russian's Venera probes compiled data about the composition of Venus surface, mainly chemicals, some even obtained images (figure 5). The data indicates basaltic composition rocks and the images showed flattened slabby rocks, separated by minor amounts of soil; reminding the basaltic flows present in our planet. Some extensive flows present uniform radar characteristics, while others are mottled at different scales; some show a gradual change from the proximal regions to the distal or from the center to the margins. Many of the plains appear to be formed by extensive layers of basalt flooding, though some of the largest as Mylitta Fluctus, consist of a varied number of long and relatively narrow units of flow, shaping a complex field of lava flow. For most of these flows small evidences of channeling exist, although others, such as those of Alta Regio, clearly show channels and flow characteristics. In these cases, the crust of the bright materials seems to have been broken just before it is transformed into broad slabs of tens of kilometers. Among those slabs there is dark material that represents younger lavas that did flow between them.

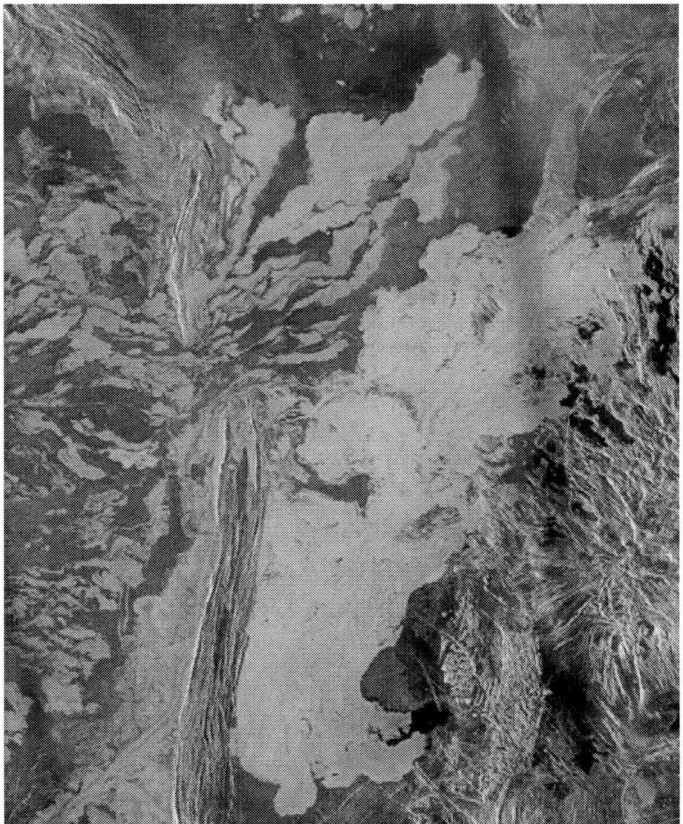

Figure 4. Radar image that shows a ridge belt, direction N-S and a lava field. More or less centred in the image, it is shown that some of lava flows cover part of the ridge belt. The meaning of this is that lava deposits is younger than the belt. Colour changes of the lava flows due to different rough textures. (Image font: www.jpl.nasa.gov).

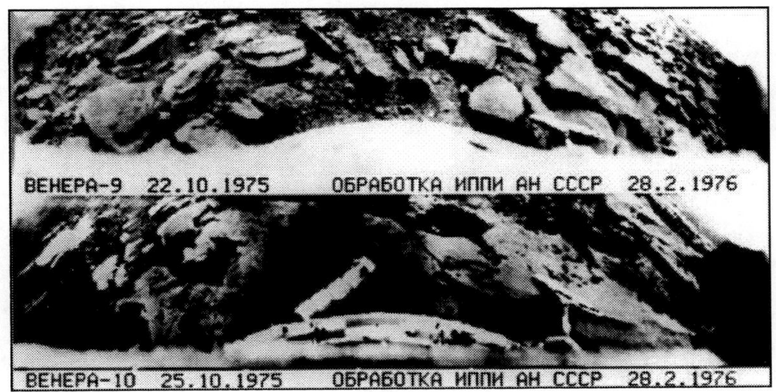

Figure 5. Images taken by the Russian spacecraft, Venera 9 (top) and 10 (bottom). The lander portions of Venera 9 and 10 parachuted to the surface of Venus and took the first black and white photos of the surface of Venus. (Image font: www.jpl.nasa.gov).

The volumes of individual lava fields are variable, but some have been estimated in 8000 km^3, very large compared with terrestrial standards. The lavas appear to be highly fluid, with very high emission rates, similar to those estimated in the Columbia River basin. Flood basalts with high volumes, as much as 3000 km^3, are estimated to have been emitted during a series of eruptions within a lapse that could last several days or weeks. Normally, source areas on Venus are not always identifiable, but in many cases there are large faults as is the case of similar lavas here on Earth. Indeed, in the Venusians plains there is evidence of the presence of subsurface dikes hundreds of kilometers long. The morphology of the lavas is consistent with the morphology of the basaltic flows as we know them, and in the same way, a basaltic composition is also consistent with the analyses by the Venera and Vega probes, although there is the possibility for them to have had komatiitic or even carbonate-rich lava composition (carbonatites). Many of the candidates for this type of extreme compositions are those flows associated with long channels that have maintained a constant thickness for thousands of kilometers, formed by lava erosion.

Pyroclastic Deposits

The recognition of pyroclastic deposits through remote sensing techniques in planetary bodies is difficult. In Mars for example, the identification of this type of deposits depends largely on the ease to be eroded presented by some units under the fluvial action. In addition, pyroclastic fall deposits are difficult to distinguish from lava flows in many cases, because both types of deposits are often formed from dense non Newtonian fluids.

The Magellan probe was capable of mapping some of these deposits. They were often presented as a zone of dark material with the shape of a thin mantle, and a pattern of reticular fractures that cut the underlying lavas (clearly visible through the deposit); not being its distribution affected by the topography in their margins. Some of these deposits, like the one found in Guinevere Planitia, are connected with a group of about 20 small craters, each of them surrounded by a circular area of rough material with a radius of approximately 10 km. Apparently, these deposits thin away from a particular crater but form a thick embankment around it. All these features suggest air pyroclastic fall for its origin. The material which constituted them seems very fragmented, with grain sizes easily transportable by wind. This interpretation would imply an eruption capable of generating a large column to disperse away the materials up to 10 km from the source; similar to the terrestrial Plinian-type eruptions. Estrombolian-type eruptions, which are more likely to occur given venusian conditions, would produce deposits of limited sizes compared to the terrestrial ones. Thus, the size of the eruption needed to produce this type of deposits is intriguing because the content of gases in the magmas is very similar to the terrestrial ones, but the highly explosive activity would be inhibited under the venusian atmospheric pressure, particularly in the present low levels of the plains. Thus, it is believed that the magmas generated under the Venusians plains present a gas concentration far higher than that observed on Earth, but also depending of course on the type of gases present in magmas.

One alternative explanation for this type of deposits is pyroclastic flow as a source, a more likely process under Venus present conditions. But this explanation does not seem right for two reasons: the margins of these deposits would be controlled by topography; and pyroclastic flows tend to increase its thickness away from the source, while the deposits discussed tend to do the opposite, as they lose thickness at increasing distance from the source.

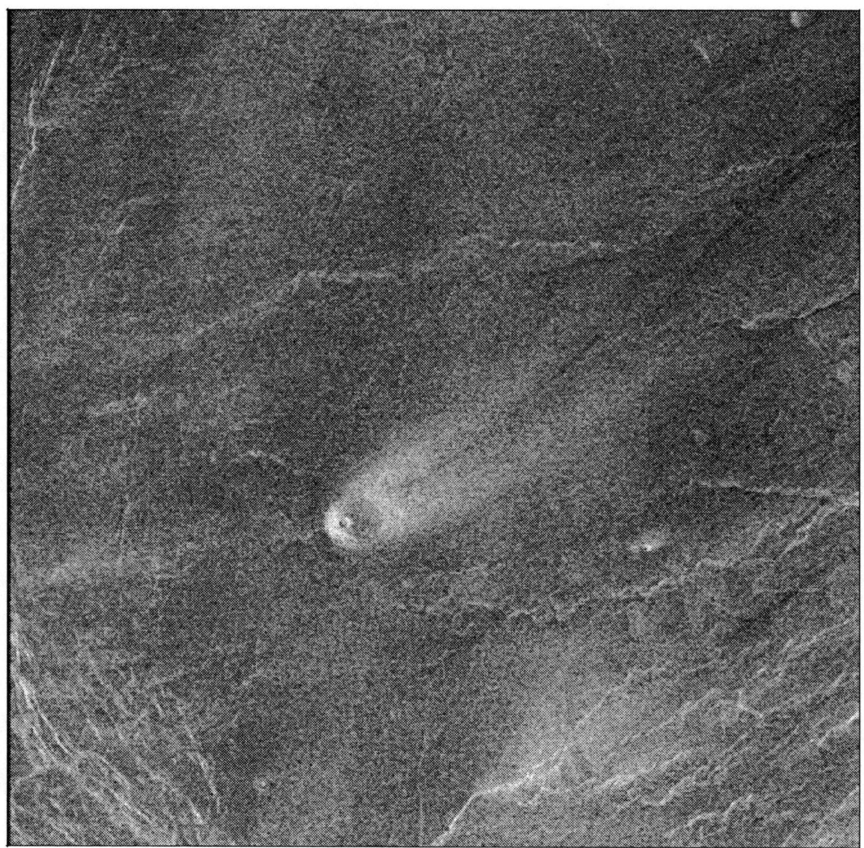

Figure 6. Radar image of isolated volcanic structure located a the western end of Parga Chasma at 9.4°S – 245°E. It is shown a comet like tail due to wind action that swepts the volcanic materials as dust and pyroclastic ashes. (Image font: www.jpl.nasa.gov).

Small Edifices

As we know, volcanism on Venus has been able to generate enormous buildings like coronae, always related to the great deformations belts, but within the plains, buildings with more modest dimensions are predominant; having been identified through quantitative measures of height, slope and diameter, three types of small buildings: shields, cones and domes. As a rule, these buildings do not exceed 20 km in diameter, although some can exceed them. It should be noted that although they are characteristic of the plains, can be found associated with big constructions as well.

Shields

This kind of buildings has been defined through surveys of reflectivity to the radar waves, but it always remains quite difficult to study due to its low slopes. Many present circular and planimetric forms, with smooth slopes a few degrees of inclination, being comparable with the terrestrial shield volcanoes. Most show convex slopes, ending in a small crater approximately 1 km in diameter. The smaller buildings do not show it, probably due to the resolution of the methods employed. Less frequently, some craters over 1 km in diameter exist. The circular form may be due to eruptions occurred at surface level.

Other buildings present a different morphology. The exterior slopes form a broad ramp around the building, with little inclination. The center of the building is a kind of flattened platform, surrounded by steep edges and generally with a central crater. They have been interpreted as buildings with a caldera collapse, surrounded by a spatter embankment and padded to the edge by a lake of lava. For this type of buildings, the surface of the lava lake is dark radar, which involves a relatively flat surface.

Those buildings that are totally dark radar seem not to possess bright areas at all and their margins are very sensitive to small variations in topography. The way in which the lava flows are channeled into the margins through relatively narrow fractures, suggests fluid lava types in the moment of emplacement. The low response to the radar of these surfaces means that they are much smoother than most of the other types of shield buildings, maybe similar to the pahoehoe lavas with vitreous surface. This kind of building possesses a central sunk and occasionally processes of caldera collapse with diameters between 3-4 km have been identified.

One type of distinctive building is the specifically related to linear fractures hundreds of km long, interpreted as surface manifestation of dykes in depth. They are buildings low in altitude but some tens of kilometers long. Although some are mainly circular in shape, others present a long axis parallel to the direction of the subjacent dyke with a complex caldera, elongated in the same direction. In contrast to other shield buildings, this type of building possesses well defined flows, dark near the crater, and becoming brighter in distal areas, suggesting slabby or blocky edges. This kind of buildings is comparable to the typical shield volcanoes of Iceland both in size and form, as well as to the so-called dorsal building type in the Canary Islands (Ancochea et al., 1990) (figure 7). Lava flows for this type of building can be as long as for the other types. The difference in response to the radar between proximal and distal zones of the same flows is certainly due to differences in the surface developed as the lava flows cool down away from the source. Normally, several buildings of the same type appear in a region, showing a variety of sizes and with separations about 200 km.

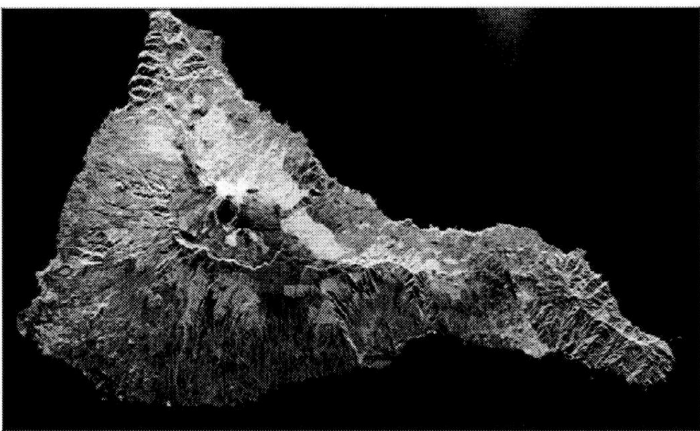

Figure 7. Satellite image of Tenerife island, Spain. The origin of this volcanic isle was three volcanic vents, located at the three corners, that subsequently they were joined with the "Teide- Pico Viejo" volcano (center-east of the image) by the so-called dorsal building type. (Image font: www.jpl.nasa.gov).

Cones

This kind of buildings possesses baseline diameters close to 15 km and tends to be bright radar. As in shield volcanoes, in this type of buildings it is difficult to discern individual lava flows and similarly, tend to be grouped in clusters. In some cases, the flanks are furrowed, presumably due to slope fail with a consequent collapse. It is assumed that this type of buildings are formed by accumulation of thin lava flows from a central opening; or that they are similar to the terrestrial cinder cones, where cinder or tefra layers alternates with lava flows.

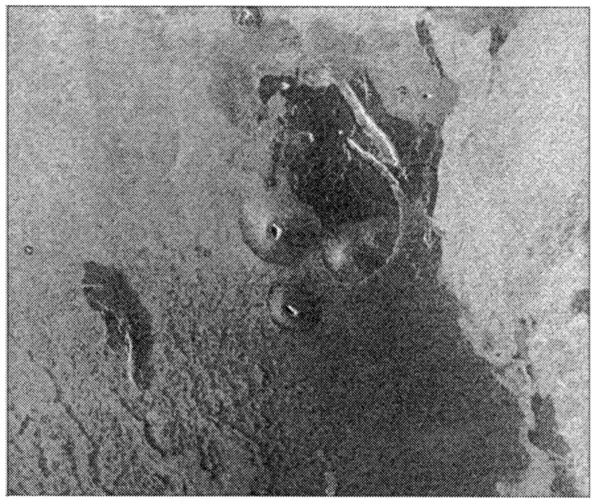

Figure 8. Volcanic Cones and lava flows in East Ovda Regio. (Image font: www.jpl.nasa.gov).

Domes

Studies have shown that in the venusian plains we can find similar domical structures analogous to the terrestrial domes produced by emission of lavas with effective viscosities similar to those of riolites. In this way we can differentiate three different types of morphologies for these domes. All of them present a relatively steep perimeter. They often present a flattened or slightly convex top, although others present concave tops, presumably owing to the retraction suffered at the end of the eruption. The margins often present a high radar pulse return, showing furrowing possibly related with mass movements of collapse breccia.

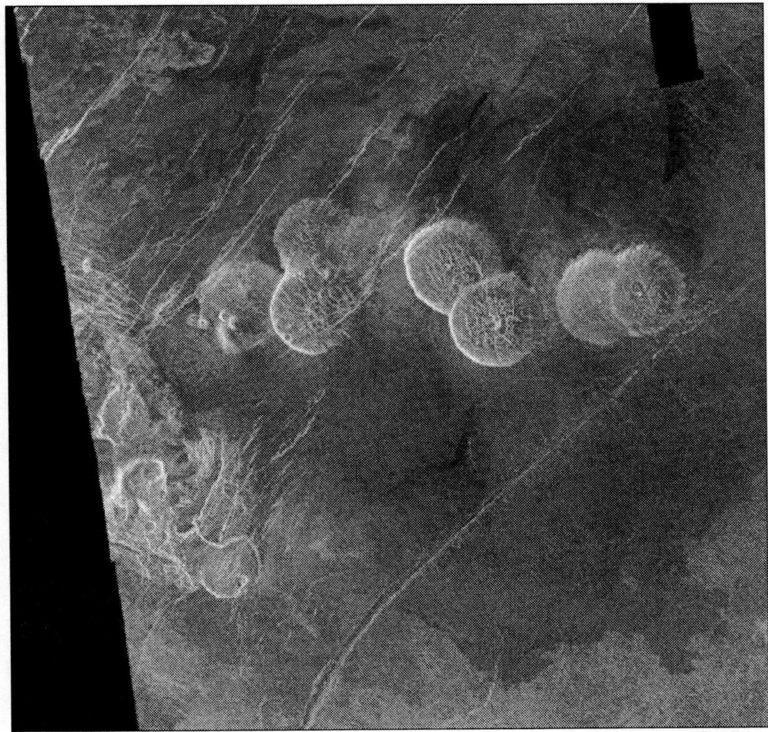

Figure 9. Radar image of Alpha Regio showing a cluster of domes with its top heavily fractured by both radial fractures as circulars. In more detail can be seen with their edges furrowed and even one of them seems to have suffered a collapse in its northern slope. (Image font: www.jpl.nasa.gov).

Compared with the terrestrial, venusian domes are generally large. They have often between 20 and 30 km in diameter and thicknesses close to 1 kilometer, with a usually flattened or slightly convex top, and a steep outer perimeter (24° approximately). They generally present circular forms, which implies that the ascent conduit can be found in its central zone and the surface in which they are generated is flattened and open, allowing the expelled materials to equally extend in all directions. Many present concentric radial fractures, indicative of inflation of the dome that produces the fracture of the exterior layers already cooled. The exterior flanks are cut equally by radial fractures that generate a kind of groove in the margins, possibly due to the development of collapse breccia.

This kind of buildings is often related to linear fracture belts, and where a group of them is relatively nearby, the oldest tend to be cut by parallel fractures. Some can be formed in isolation or as loose groups and even occasionally overlap occurs between them. When this happens, a curious feature appears, being the form of the overriding dome not substantially modified by the relief of the underlying dome. This suggests that both domes began to form at the same time, joining by their edges. Subsequently one stopped its growth while the other continued, producing a little overriding and giving the impression of being younger.

The presence of small craters is fairly common in these buildings, and can be found anywhere on their surface. It is believed that these craters are produced as a result of small explosions in the final stages of the eruption. These breakages can occur in the flanks, due to fails of the outer layer, allowing lava flows to emerge or potential pyroclastic flows to extend over the adjacent plain. Domes with a central crater surrounded by steep edges, suggest a minor explosive activity at the end of this eruptive process, producing an embankment around the conduit.

The volumes of magma involved in the creation of one of these buildings have been estimated at hundreds of cubic kilometers. Eruptions of this magnitude in Earth are due to siliciclastic magmas, generating extensive and thick ignimbrite layers. As this type of eruptions is believed not to be possible on Venus due to the high atmospheric pressure, this type of lavas is emitted in a way that generates domes.

There is a type of buildings with similar characteristics to the domes, including circular forms and a relatively flatted top, but without traces of a crater. They have small thickness values compared to their diameters. Their margins are abrupt and have often between 2 and 5 km in diameter. It is not common to find them forming groups, but several of them can be found in isolation in the same area. All of them show similar characteristics in terms of return of radar signals, which is very high, making easier to distinguish them due to its brightness. This kind of buildings is probably the product of eruptions from a central vent of lavas with a moderate effective viscosity, over flat surfaces. Another possibility is its origin as large lakes of lava as those observed in Hawaii. The outer abrupt perimeter would be a kind of retaining wall built in the edge of a growing lava lake.

Scalloped Margin Domes

While most buildings of venusian volcanic plains have a terrestrial counterpart, at least in form if not in scale, there is a kind of morphology also positive that seems not to have a direct terrestrial equivalent. This morphology presents a concave or convex top flat; and abrupt margins with furrowed or scalloped aspect, what is considered as an evidence of failures in the slopes of the exterior flanks of some domes. Some present central craters and even calderas. Most of these buildings are located in the plains, usually associated geographically with domes which have not suffered landslides. Some were found at the top of large shield buildings as Sapas Mons.

In some cases, a mantle of debris with a different texture compared to the margins of the dome can be distinguished. There are several examples in which part of the dome seems to have collapsed allowing lava flows to emerge and sometimes transport parts of the original crust. The fact that these mantles of debris cannot always be visible is mainly due to the similar response to the radar pulses that their surfaces present compared to that of the surrounding materials. The debris may also form an unconsolidated layer that can be crossed easily by radar pulses, down into the lower units, thus resulting invisible to radar. In addition, these buildings often present concentric fractures generated along with the collapse of the dome that may mask these mantles of debris. These fractures, which seem to be steeply inclined towards the dome, may be associated with normal faults.

The collapse of the edges of domes on Earth can be caused by their steepening and the subsequent explosions, which generate avalanches of blocks and ash. However, breakdowns in Venus are of a scale several orders of magnitude higher than on Earth, having been estimated slides of up to 200 km^3. One possible explanation is that due to high temperatures, the crust of solid rock that can be formed in these huge lava flows is very thin and weak. The steepening of the front of these flows can produce failures as well as seismic activity, perhaps related to the development of systems of fractures surrounding this type of domes.

Figure 10. Radar image of scalloped margin domes with a top flat morphology and abrupt margins with furrowed or scalloped aspect. It is shown a relationship with the fracture belts in Guinevere Planitia. (Image font: www.jpl.nasa.gov).

Small Edifice Clusters

Most of the small size volcanic buildings on the venusian plains are grouped. It has been estimated that there are nearly 600 groups of shield type buildings on Venus. Typically found in areas of average elevation within the plains, they are much less frequent in high and low altitudes; and not very common in areas of strong deformation (belts). However, they are often associated with the linear fracture belts that cut the plains.

The comparative analysis made between different groups (Guest et al., 1992) shows that they are formed by a predominant type of building, shields or cones. This can be easily appreciated at a global scale, although within each group, the morphology varies. Their relationship with the fracture belts is significant because some buildings are cut by fractures, thus more ancient, while others overlap with them. It is believed that these groups are associated with a single magma chamber, whose size would be given by the extension of the group of buildings, each building representing an eruption.

A clearly similar clustering can be found in the Snake River Plain volcanic region in Idaho. This volcanic region covers an arc of approximately 550 km long and 100 km wide, occupying a structural depression partially limited by faults, and with an extension close to 49,200 km^2 (comparable with the extension of many of the groups of buildings in Venus). Another terrestrial analogue can be found in the bottom of our oceans, where groups of volcanic buildings arise near the mid-ocean ridges, in many cases with elongated forms and structures dominated by fissures; and where bigger buildings have also been identified, with sources of emission in its central zone, being interpreted as produced by volcanic eruptions from a single magma chamber (figure 11).

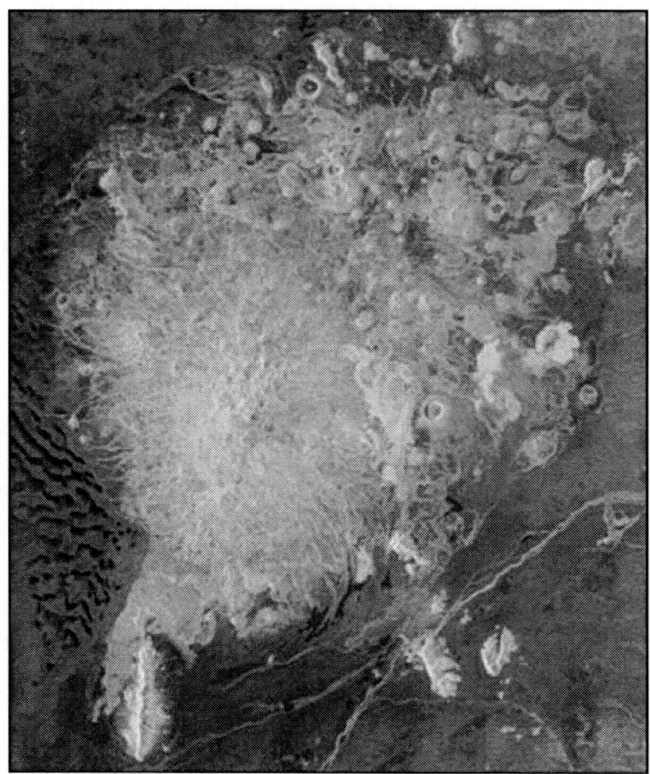

Figure 11. This spaceborne radar image shows the Pinacate Volcanic Field in the state of Sonora, Mexico. More than 300 volcanic vents occur in the Pinacate field, including cinder cones, the larger circular craters seen in the image are a type of volcano known as a "maar". The highest elevations in the volcanic field, about 1200 meters, occur in the "shield volcano" structure shown in bright white, occupying most of the left half of the image. (Image font: www.jpl.nasa.gov).

3. Tectonism

As for the rest of the Venus surface, neither SAR images nor topography measurements have been able to reveal similar features to the terrestrial expressions of plate tectonics in the venusian plains. The images from the Magellan were unable to show evidences such as similar features to high scale subduction trenches (only in chasmata, whose origin is still discussed, Sandwell and Schubert 1992a, Solomon et al 1992, Hansen and Phillips 1993, Brown and Grimm 1997), or spreading ridges (McKenzie et al 1992a, Solomon et al 1992). Although there are similar rift valleys to those on Earth, the rifting appears to be related to convective upwelling rather than plate motions (McGill et al 1981, Foster and Nimmo 1996). The same is true of plains deformation, which can be coherent over thousands of square kilometres (Solomon et al 1992).

The ridged plains, which are usual in the venusian lowlands, make more than 60% of the surface of Venus, are formed by volcanic material deformed by wrinkle ridges after their emplacement (e.g., Basilevsky and Head, 1996, 1998), and should have covered a larger area when were formed, because younger plains have partially overprinted them subsequently. Wrinkle ridges are long, narrow, sinuous and almost parallel between themselves, with an average spacing of few tens of kilometers and a nearly constant orientation along thousands of kilometers, and usually considered as the surface expression of low-amplitude compressive anticlines, which are associated to blind thrust faults (e.g., Bilotti and Suppe, 1998; Schultz, 2000b). They tend to be bright on Magellan SAR images compared to the surrounding terrains, not being this brightness contrast related to the orientation of the ridges with respect to radar look direction (McGill 1993), but to wavelength-scale surface roughness (Banerdt et al., 1996). They usually appear forming sets (e.g. Plescia ond Golombek, 1986), two or three of which can be found sometimes overlapped, what gives us the opportunity to study cross-cutting relationships.

Because winkle ridges are so widespread and abundant on Venus they provide us with global clues on the orientation of shallow crustal stresses (McGill, 1993), being one of the few effective methods to study Venus crust, given the lack of detailed seismic and gravity data during many years after the launch of the Magellan probe. Most models explaining the formation of the wrinkle ridges imply compression normal to the longest ridge axis. From crater counting, geologic relations, and climatic considerations, it can be deduced that a high proportion of the wrinkle ridges formed during a relatively short span of time (millions of years) after the formation of the plains (McGill, 2004). However, some of them are thought to have formed during longer time lapses, coevally to the formation of other stratigraphic units. In any case, due to the younger ages of wrinkle ridges compared to those of the plains, we cannot employ them to constrain and define different stratigraphic units on the plains (McGill, 2004).

Tectonic belts are linear zones of concentrated deformation hundreds of kilometers long, tens of kilometers wide, usually elevated above the surrounding plains, which include thrusted anticlines, grabens, tight folds, and strike-slip fractures (e.g., Young and Hansen, 2004). In them, long-wavelength and short-wavelength deformation are spatially independent, with the first one temporally predating regional deformation (e.g. wrinkle ridges). Short-wavelength structures in the structural core of the belt could possibly represent a local tectonic regime as the long-wavelength warp could be formed in response to later regional processes (Young and Hansen, 2004). These belts are a common feature of Venus tectonics, presenting complex geometries as dividing and coalescing branches. They also record complex temporal evolution, and are an evidence of crustal shortening and mobility of the uppermost lithosphere as well. Many hypotheses have been formulated to explain their origin, which still remains controversial, implying compressional, extensional and even strike-slip forces; but in any case a valid hypothesis must be able to explain the presence of two potential conflicting processes as volcanism and contractional movements. The relation between belts and wrinkle ridges can be of major importance to understand the dynamics and tectonic evolution of Venusian plains. At least three different types of major belts are formed as a result of the combination of these features: extensional belts, thrust and fold contractional belts, and strike-slip shear belts.

Several models try to explain the tectonics of venusian plains. Early models supposed that the lower crust could be weakened by the Venus' high surface temperature, which predicted simultaneous development of two length scales of deformation (Zuber, 1987). The apparent periodic occurrence of deformation, the prominence of warp-like morphologies at various scales, and the great length of some arrays of deformation belts, could have formed as harmonic buckling instabilities driven by regional compresion (Zuber, 1987). This model has unresolved problems as the evidence for localized volcanic activity within deformation belts (Young and Hansen, 2004). Moreover, the apparently basaltic nature of surface rocks and an extremely dry atmosphere should imply that Venus' crust is very strong (e.g., Brown and Grimm, 1997; Phillips et al., 1997; Mackwell et al., 1998).

Nowadays, there are two main models to explain the tectonics of the venusian plains, both based on vertical tectonic forces. The downwelling model (e.g., Bindschadler et al., 1992; Magee and Head, 1995) attributes volcanism and rifts areas on Venus to deep-seated mantle plume activity linked to upwelling and decompression of mantle material. It implies a volcanism prior to the tectonic activity that produced background plains which were deformed by subsidence of the basin. In this model the deformation of the venusian plains is due to compression and crust thickening, originating wrinkle ridges and thrust faults, above mantle downwelling. This process would cause compressional hoop and peripheral extensional radial strains, formed by buckling of strong upper crust in response to subsidence. Finally, subsequent volcanism would have produced extensive volcanic plains between and among the deformation belts apparently from sources within the basin.

The swell-push model is based in two principal assumptions. First: the topography on Venus is locally compensated somewhere within the lithosphere or supported dynamically by poloidal flow in a nearly uniform viscosity mantle (Sandwell et al., 1996), where the swell-push body force acts on the lithosphere in a proportional way of the gradient of the geoid height. Second: the lithosphere of Venus is well approximated by an elastic shell of uniform thickness, which means to assume that the magnitude swell-push force is proportional to geoid height, causing extensional stress on geoid highs and compressional stress on geoid lows. The model predicts mainly that the direction of the principal stresses would follow the curvature of the geoid for each point. The strong point of this theory would be the agreement between the strain predicted from geoid height and the strain inferred from rift zones and wrinkle ridges. The regions where the prediction does not match the evidence in terms of wrinkle ridges spatial disposition, as well as the zones with two or more sets of non-parallel wrinkle ridges would be explained as changes in the driving forces over geologic time (Sandwell et al., 1996).

All these models explain venusian belts as compressional tectonic areas. But ongoing research of Lavinia Planitia and Velamo Planitia is finding evidences that these belts show not only thrusted anticlines and tight folds, but also grabens and strike-slip fractures (F. Anguita, personal communication). Therefore, these models are not able to totally explain the complex tectonics of Venusian plains.

4. Conclusion

The currently accepted vision for Venus geology does not include the existence of plate tectonics, at least as we understand it on the Earth. Indeed, the models proposed to describe the tectonic of Venus explain the observed structures from a verticalist point of view. Nevertheless, compressive belts show not only compressive structures, but also extensional and transcurrent structures, which can not been easily explained with these verticalist models.

It is thought that venusian volcanism could be quite similar to the Hawaiian type of terrestrial volcanism, as well as to the fissural volcanism or basaltic flooding on Earth. Notwithstanding, in some few cases, it could be very close to the Strombolian type of terrestrial volcanism. But the characteristics of these kind of volcanism on Venus are different due to environmental conditions (high atmospheric pressures and surface temperatures) which cause low degasification and slow cooling of magmas, inhibiting large explosions in different eruptions and allowing that the lava flows reach large extension both wide and length.

The main remaining question is the relationships existing between tectonic and volcanism on Venus. Several times extensional features are observed join to lava flows or even volcanic buildings. Sometimes the volcanic buildings cover the fractures and other times the fractures cut the buildings. So, the question is if the tectonic causes the volcanism, the volcanism causes the tectonic or both are simultaneous. Another important question is if the volcanism is a continuous process or it happens in a cyclical periods where most of the crust are renewed when a crytical point is reached that leads to renewal of the entire planet surface, erasing all previous features.

References

Banerdt, W. B., McGill, G. E., and Zuber, M. T., 1997. Plains tectonics on Venus. In Venus II: Geology, Geophysics, Atmosphere, and Solar Wind Environment, Bougher, S.W., Hunten, D.M., Phillips, R.J. (Eds.), Univ. of Ariz. Press, Tucson, pp. 901–930.

Basilevsky, A.T., Head J.W., 1995. Global stratigraphic survey of Venus: Analysis of a random sample of thirty-six test areas. *Earth Moon Planets*. 66, 285–336.

Basilevsky A.T., Head J.W., 1996. Evidence for rapid and widespread emplacement of volcanic plains on Venus: stratigraphic studies in the Baltis Vallis region. *Geophys. Res. Lett.* 23, 497–500.

Basilevsky, A.T., Head, J.W., Schaber, G.G., Strom, R.G., 1997. The resurfacing history of Venus. In: In Venus II: Geology, Geophysics, Atmosphere, and Solar Wind Environment, Bougher, S.W., Hunten, D.M., Phillips, R.J. (Eds.), Univ. of Arizona Press, Tucson, pp. 1047–1084.

Basilevsky, A.T., Head III, J.W., 1998. The geologic history of Venus: a stratigraphic view. *J. Geophys. Res.* 103, 8531–8544.

Bilotti, F., Suppe, J., 1999, The Global Distribution of Wrinkle Ridges on Venus. *Icarus.* 139, 137-157.

Bindschadler, D.L., Schubert, G., Kaula, W.M., 1992. Coldspot and hotspots: global tectonics and mantle dynamic of Venus. *J. Geophys. Res.* 97, 13,495-13,532.

Brown, C.D., Grimm, R.E., 1997. Tessera deformation and the contemporaneous thermal state of the plateau highlands, *Venus. Earth Planet. Sci. Lett.* 147, 1-10.

Foster, A., Nimmo, F., 1996. Comparisons between the rift valleys of East Africa, Earth and Beta Regio, *Venus. Earth Planet. Sci. Lett.* 143:183–96.

Guest, J.E., Bulmer, M.H., 1992. Small Volcanic Edifices and Volcanism in the Plains of Venus. *J. Geophys. Res.* 97, 15,949-15,966.

Guest, J.E., Stofan, E. R., 1999. A new view of the stratigraphic history of Venus. *Icarus.* 139, 55–66.

Hansen, V.L., Phillips, R.J., 1993. Tectonics and volcanism of Eastern Aphrodite Terra, Venus: no subduction, no spreading. *Science.* 260:526–30

Head, J.W., Crumpler, L.S., Aubele, Jayne C., Guest, J.E., Saunders, R.S., 1992. Venus volcanism: Classification of volcanic features and structures, associations, and global distribution from Magellan data. *J. Geophys. Res.* 97, 13,153-13,197.

Head, J.W., Basilevsky, A.T., Wilson, L., Hess, P.C., 1996. Evolution of volcanic styles on Venus: Change but not Noachian? *Lunar Planet Sci.* 27, 525–526.

Mackwell, S.J., Zimmerman, M.E., Kohlstedt, D.L., 1998. High-temperature deformation of dry diabase with application to tectonics on Venus. *J. Geophys. Res.* 103, 975-984.

Magee, K.P., Head, J.W., 1995. The role of rifting in the generation of melt: Implications for the origin and evolution of the Lada Terra-Lavinia Planitia region of Venus. *J. Geophys. Res.* 100, 1527-1552.

McGill, G.E., Warner, J.L., Steenstrup, S.J., Barton, C., Ford, P.G., 1981. Continental rifting and the origin of Beta Regio, Venus. *Geophys. Res. Lett.* 8:737–40.

McGill, G.E., 1993. Wrinkle ridges, stress domains, and kinematics of venusian plains. *Geophys. Res. Lett.* 20, 2407–2410.

McGill, G.E., 2004. Geologic map of the Bereghinya Planitia quadrangle (V-8), Venus. U.S. *Geol. Surv., Geol. Inv. Series,* Map I-2794.

McKenzie, D., Ford, P.G., Johnson, C., Parsons, B., Sandwell, D., et al., 1992a. Features on Venus generated by plate boundary processes. *J. Geophys. Res.* 97:13533–44.

McKinnon, W.B., Zhanle, K.J., Ivanov, B.A., Melosh, J.H., 1997. Cratering on Venus: models and observations. In: Bougher, S.W., Hunten, D.M., Phillips, R.J. (Eds.), Venus II. Univ. of Arizona Press, Tucson. pp. 969-1014.

Plescia, J. B., Golombek, M. P., 1986. Origin of planetary wrinkle ridges based on the study of terrestrial analogs. *Geological Society of America Bulletin,* v. 97, p. 1289-1299.

Phillips, R.J., Johnson, C.L., Mackwell, S.J., Morgan, P., Sandwell, D.T., Zuber, M.T., 1997. Lithospheric mechanics and dynamics of Venus. In: Bougher, S.W., Hunten, D.M., Phillips, R.J. (Eds.), Venus II. Univ. of Arizona Press, Tucson. pp. 1163-1204.

Roberts, K.M., Guest, J.E., 1992. Mylitta Fluctus, Venus: Rift-Related, Centralized Volcanism and the Emplacement of Large-Volume Flow Units. *J. Geophys. Res.* 97, 15,991-16,015.

Sandwell, D.T., Johnson, C.L., Bilotti, F., Suppe, J., 1996. Driving Forces for Limited Tectonics on Venus. *Icarus.* 129, 232-244.

Sandwell, DT, Schubert, G., 1992. Flexural Ridges, Trenches, and Outer Rises Around Coronae on Venus, *J. Geophys. Res.,* 97(E10), 16,069–16,083.

Saunders, R.S., 1999. Venus. In: The New Solar System, Fourth Edition, Beatty, J.K., Petersen, C.C., Chaikins, A. (Eds.), Cambridge University press & sky publishing Corp., pp. 97-110.

Schultz, R.A., 2000b. Localization of bedding-plane slip and backthrust faults above blind thrust faults: keys to wrinkle ridge structure. *J. Geophys. Res.* 105, 12035–12052.

Slyuta, E.N., Kreslavsky, M.A., 1990. *Intermediate (20-100 km) sized volcanic edifices on Venus Lunar Planet Sci.* 21, abstract.

Solomon, S.C., Smrekar, S.E., Bindschadler, D.L., Grimm, R.E., Kaula, W.M., et al., 1992. Venus

tectonics: an overview of Magellan observations. *J. Geophys. Res.* 97:13199–13256.

Squyres, S.W., Jankowski, D.G., Simons, M., Solomon, S.C., Hager, B.H., McGill, G.E., 1992. Plains Tectonism on Venus: The Deformation Belts of Lavinia Planitia. *J. Geophys. Res.* 97, 13,579- 13,599.

Young, D.A., Hansen, V.L., 2005. Poludnista Dorsa, Venus: History and context of a deformation belt. *J. Geophys. Res.* 110, E030D1, doi: 10.1029/2004J001180.

Zuber, M.T., 1987. Constraints on the lithospheric structure of Venus from mechanical models and tectonic surface features. *J.Geophysic Research.* 92, 341-551.

In: Space Exploration Research
Editors: J.H. Denis and P.D. Aldridge

ISBN: 978-1-60692-264-4

Chapter 21

EUROPA: NEW HORIZONS FOR ASTROBIOLOGY

Esther Velasco[1], Guillermo Caravantes[2,3], Beatriz Gómez-Izquierdo[2,3], Alberto Rodriguez[3] and Javier Ruiz[1]

[1] Centro de Biología Molecular,
CSIC-Universidad Autónoma de Madrid,
28049 Cantoblanco, Madrid, Spain

[2] Volcano Dynamics Group, The Open University,
MK7 6AA, United Kingdom

[3] Seminario de Ciencias Planetarias, Facultad de Ciencias Geológicas,
Universidad Complutense de Madrid, 28040 Madrid, Spain

ABSTRACT

Europa is the smallest of the Galilean satellites of Jupiter, although it is however a remarkable body and the primary focus of astrobiological interest in the Jovian system. Geological and magnetic evidences, as well as theoretical considerations, strongly suggest that Europa has a thin ice shell, maybe a few kilometers or a few tens of kilometers thick, which floats on an internal ocean of liquid water. Some anomalies have been observed in the huge magnetic field of Jupiter whenever Europa goes through it, which seems to indicate that this ocean is made of conductive water, due to a probable high content of salts. Moreover,the low density of impact craters suggest an surface age of about 60 Myr in average, which is very young in geological terms, indicating that internal activity, driven by tidal heating, is resurfacing the moon at the present time. So, Europa shows an overprint of crisscrossing ridges and linear markings, some extending for thousand of kilometers, which are locally disrupted for chaotic terrains, building a colorful and exotic textured surface. All of these features make of Europa a theoretically good niche for life. Biological organisms might exist in the salty internal ocean, in the icy shell (active or latent) at places such as fractures, veins or liquid reservoirs, on the internal sea floor in hydrothermal vents or submarine volcanoes, or even within the rocky interior.

INTRODUCTION

Europa, the fourth largest Jovian satellite, included in the four Galilean satellites, has about the same size (and slightly lower density) as the Earth's Moon, but its surface is between the brightest in the Solar System as consequence of sunlight reflecting from an icy shell. Talking about geological structures, the surface of Europa shows a wide variety of geologic terrains and features (Lucchita and Soderblom, 1982; Malin and Pieri, 1986; Greeley et al., 1998, 2000), with local prominent relieves of up to 1 km (Schenk, 2002). From the scarcity of impact craters, an age of ~10^7 years has been suggested for the surface (Zahnle et al., 2003). However, after the formation of these craters, almost no tectonics has deformed them, what may mean a decrease in tectonic activity in a moment in a recent past.

Moreover, there are solid evidences for the existence of an internal ocean of liquid water below the icy shell of this satellite of Jupiter. The more strong evidence comes from magnetic data acquired by the Galileo spacecraft, which recorded magnetic anomalies in the huge magnetic field of Jupiter whenever Europa goes through it (Kivelson et al., 2000). These observations indicates that there is a global conductive electric layer locate ~20 km below the surface, and it could be explained by an internal ocean made of salty water (Kivelson et al., 2000; Schilling et al., 2004) maybe compositionally similar to that of the Earth's oceans. This internal ocean could be habitable (e.g., Reynolds et al. 1983; Chyba 2000; Chyba and Phillips 2001, 2002; Schulze-Makuch and Irwin 2002; Marion et al. 2003). Similarly, biological organisms may exist, active or latent, inside the icy shell above this internal ocean, at places such as fractures, veins and liquid reservoirs (e.g., Greenberg et al., 2000; Marion et al., 2003; Lipps and Rieboldt, 2005), and inside and around warm ice diapirs (Price, 2003; Ruiz et al., 2007).

In order to understand the biologic potential of Europa and facilitate the search for possible ecological niches, we must review the present knowledge of the ice shell, and its structure and chemical-physical conditions, due to its key role as a direct or indirect communication between ocean and surface, and as a potential marker of the ocean's composition.

EUROPA'S SURFACE: A COMPLEX GEOLOGIC RECORD

Europa seems to be a geologically young world, with a mean age about 60 Myr (Zahnle et al., 2003), and may therefore continue to be active today. This conclusion is based on the relative low density of impact craters observed on its surface. The surface shows, mainly, three kinds of terrains: bright plains, mottled terrains and disrupted regions (For a review of the geology of Europa see Greeley, 2004). They are (more or less) mixed on the Europa surface, which exhibits an overprint of crisscrossing ridges and linear markings, some extending for thousand of kilometers. These near linear features are locally disrupted for chaotic terrains, building a colorful and exotic textured surface.

The bright plains seem to be the basic unit of Europa's surface, and many of the rest of units can be derived, mainly, from that one. These plains show plateaus up to 10 km wide and tens of meters high with, in some places, multiple sets of ridges and grooves with a length of a few tens meters wide. This structure could be formed by repeatedly tectonic processes, which deform a relative low-density ice of the icy shell. The mottled terrains show an irregular patchwork of darker zones, of a range in size from ~50 km to ~500 km across or more, with irregular margins grading into the bright plains. This terrain could be formed in response to endogenic forces, which have broken bright plains into plates with intervening areas of chaotic terrain. This low-lying chaotic terrain is darker than disrupted plates and probably contains a greater proportion of non-ice material. So, this terrain is one of the most important places to looking for biosignatures. Other important places on the surface are the so-dubbed "lenticulae", which are round, elliptic or irregular shaped features, including dark spots, pits, domes, microchaos areas and are considered an evidence of the younger geological (endogenic) activity recorded on Europa (Pappalardo et al., 1998; Prockter et al., 1999; Figueredo and Greeley, 2000); their size goes from less than ten to several tens of kilometers, and in some areas are highly concentrated (Greeley et al., 1998; Pappalardo et al., 1998).

The differences of color and spectral properties between terrains cannot be explained in a simple way. The shell composition may include silicate minerals, magnesium and sodium sulfate hydrates, sulfuric acid, and so on (McCord et al. 1998; Carlson et al. 1999, Kargel et al. 2000; Spencer et al. 2006), and these most likely are of internal origin. The abundance of biologic elements on Europa is still unknown.

Thickness of the Icy Shell

One of the main topics regarding the existence of life is the ice shell thickness, because it determines the model and dynamics of the ice shell and therefore constraint the physical conditions in the inner ocean. Two models have been proposed to explain the observed features and physical parameters, and constitute the famous controversy thin shell vs. thick shell.

For the thin shell model (Greenberg et al., 1999, 2000; Hoppa et al., 1999) some features, such as the so-called cycloids, would seem to suggest the presence of an ocean a few kilometers below the surface, which would be locally exposed in the surface due to the melt-through of a thermally conductive ice shell. It would be warmed from below, with a surface temperature of about 100 K and a maximum thickness of a few kilometers. It would be required significant tidal heating in the rocky interior to maintain such thin shell, maybe including a molten silicate layer. On the other hand, the thick shell model proposes a shell a few tens of kilometers thick. After this model, the icy shell can be subdivided into a upper, cold and inmovil, stagnant lid, and a thick layer of ductile, maybe convective, ice (Pappalardo et al., 1999a,b; Pappalardo and Head, 2001).

The detailed analysis of size and depth of some of the major impact structures provides a thickness of ~19-25 km for the ice shell (Schenk, 2002), and the important amplitude of surface topography also supports the thick shell model (Prockter and Schenk, 2005). Moreover, the most accepted hypothesis to explain the origin of the lenticulae proposes that these features reflect the effects on the near surface crust of convective upwelling plumes ("thermal diapirs"); in this case, the lenticulae would be showing us the recent activity of the convective system, and the spacing between them would indicate that the thickness of the convective layer is ~10-20 km thick (Pappalardo et al., 1998; Spaun et al., 2004).

So, the evidences favoring a, at least relatively, thin shell, imply that the communication between the surface and the internal ocean would be greatly restricted to thermal diapirism (linked to the convection), and maybe cryovolcanism and meteoritic impacts.

Thermal State of the Icy Shell

There is scarce information about the heat flow of Europa, although some estimation have been proposed based on relation between geological features and the mechanical and thermal properties of the lithosphere. For Europa, it is usual to put the brittle-ductile transition 2 km deep at most (e.g., Pappalardo et al., 1999b). For example, the undulations found in Astypalaea Linea, interpreted as folds (Prockter and Pappalardo, 2000), have a wavelength of ~25 km, which would imply a brittle-ductile transition ~2 km deep (Dombard and McKinnon, 2006); undulations with a similar spacing have also been observed in the leading hemisphere of Europa (Figueredo and Greeley, 2000). On the other hand, troughs, possibly grabens, in the Callanish and Tyre multiring impact structures are up to ~2 km wide: if these troughs are interpreted as grabens then its width implies a depth of faulting of ~2 km, which could correspond to the brittle-ductile transition (Ruiz, 2005; Lichtenberg et al., 2006). For this depth the vertical heat flows must be close to 100 mWm^{-2} (Ruiz and Tejero, 2000; Ruiz, 2005; Dombard and McKinnon, 2006; Lichtenberg et al., 2006).

On the other hand, the effective elastic thickness of the lithosphere varies depending on the region. Available estimations range from ~0.4 below ridges and domes to ~2.9 km below a plateau in the crater Cilix region. The lower values correspond to geological setting where tidal heating can be differentially enhanced, and here not representative for average Europa. For the Cilix region the estimated heat flow would be ~25-35mWm^{-2} (Nimmo et al., 2003; Ruiz, 2005), value lower than those deduced from the depth brittle-ductile transition, which suggest the existence of local variations in the thermal state of the ice shell, maybe related to variations in the vigor of the convection (Ruiz et al., 2005).

These indications obtain heat flow high values when compared to the component produced in the rocky fraction by radiogenic heat at the present, which is ~6-8 mWm^{-2} (Cassen et al., 1982; Spohn and Schubert, 2002). This implies an important role for tidal heating, what should happen mainly in the warm inner part of a convective layer (McKinnon, 1999; Ruiz and Tejero, 2003). Indeed, some theoretical studies of tidally heated convection in the ice shell of Europa obtain heat flows of ~100 mW mWm^{-2} (Nimmo and Manga, 2002; Ruiz and Tejero, 2003; Moore, 2006). Because convection could start in the shell for heat flows under 10-45 mW m^{-2} (the exact value depends on the particular model; e.g., Ruiz and Tejero, 2003), most of the surface heat flow would be the result of tidal heating in the convective layer. On the other hand, a conductive ice shell at least ~19-25 km thick (Schenk 2002) implies that the surface heat flow should be at most ~20-30 mW m^{-2} (Ruiz, et al., 2005). This implies a heat flow of at most ~30-45 mW m^{-2} reaching the shell base (or equivalently, the sea floor), which limit the possibility of hydrothermal circulation in the rocky portion of Europa.

LIMITING FACTORS FOR LIFE IN EUROPA

We have not found any definitive probe of life outside the Earth. However, on the Earth, life can resist conditions that we thought life would not be able to resist. Also, the conditions that we think they are optimal for life (293 K, pH = 7, and so on) are an environment where some kinds of life are not able to live. In fact, Europa has some places where life (as we know life), would have able to take place and develop in a different way with respect to Earth. So, a review the limiting factors for Earth's life is necessary for an assessment of the possibility of life on Europa.

Temperature

On Earth, temperature tolerate for life usually going to -20° to 121° C (Marion et al., 2003). However, recently has been discussed evidence for microbial metabolic activity in Antarctic glacial ice at -40°C (Price and Sowers, 2004). It is important to know why the temperature is a limiting for life. 100° C is the temperature of the normal boiling point of water, but in hydrothermal vents and the subterranean deep biosphere, the high pressures impede water to boil, because the water molecules are kept sufficiently together for the chemical bonds are joining enough to keep the liquid state. So, in these cases, life can resist up to 110-121°C.

There are several problems for life under the freezing point. Firstly micro-crystals of ice are formed inside the cells, which broke cellular structures, not only for the effect of the micro-crystals themselves, but because water in solid state has more volume than water in liquid state. Secondly, cells need water movement through its membrane to equilibrate the ion's internal concentration; so, life cannot use icy water and have dehydration problems even in presence of abundant water in form of ice. To avoid this problem, Earth's life could have adopted strategies to reduce the water freezing point, both in the extra- and intra-cellular environments, by using cryoprotectant substances, for example through the accumulation of compatible solutes with the intracellular metabolism, as trehalose, glycerol or glycine betaine, substances that are efficient cryoprotectants (e.g., Welsh, 2000; Yancey, 2005). Moreover, terrestrial life at low temperature is adapted to low water activity in the same way as it is adapted to high osmolarity or desiccation conditions (e.g., Potts, 1994; Thomas and Dieckmann, 2002), implying that a decrease in temperature would be equivalent to an increase in osmolarity or desiccation. For this reason, it has been suggested (Ruiz et al., 2007) that the synthesis of compatible solutes can be universal metabolic strategy for hypersaline, desiccation or low temperature conditions, and for any organism living within the ice shell of Europa.

Moreover, life could be possible as long as liquid water is available. So, the term of "eutectophiles" has been used (Deming, 2002) for these hypothetical organisms living at liquid veins or reservoirs close to the eutectic point of water containing impurities.

This increases the potential habitats for life inside the icy shell of Europa.

Salinity

On Earth, we can find a lot of kind of hypersaline habitats (Madigan, 1997; Marion, 2003). Microorganisms adapted to high salinity are called halophiles, or even extreme halophiles organisms, when, not only can resist high salinity, but also, they need a high salinity to be alive. These extreme halophiles, mainly inside the group of unicellular microorganisms called archaeas, can have metabolic activity in salinity close to saturation (Madigan, 1997). So, the briny ocean of Europa could be a habitable one.

Acidity

On Earth, exist examples of extreme pH tolerances (e.g., Bachofen, 1986; Zhilina and Zarwarzin, 1994; Johnson, 1998; Scheleper et al., 1995; Marion 2003), from < 1 to > 11. For this reason, the pH value seems not a great limiting factor for Earth's life, and all the proposed compositions for the Europa's ocean (Kargel et al., 2000; Marion, 2001, 2002) could, in theory, support life.

Radiation

Although there are organism on Earth that could survive under ultraviolet and ionizing radiation (e.g., Bacholen 1986; Dose et al 2001; Rothschild, 2001), which could be the great pitfall for the possibility of Europan life near surface. The resistance to this kind of radiations lies in the capacity of the organism to repair DNA quickly and not to an immune skill of the leaving beings.

Pressure

Some authors think that pressure, or the relation between temperature and pressure, is a limiting factor for organisms on Earth. It could be true in giant gas planet, where the high temperature exceed amply the known range of temperature for life, or underground, some kilometers of dip inside the crust. Nevertheless, in aquatic environment, pressure are not a problem for life in general (some organism could be very sensitive to changes of pressure, others not, but it is not a problem for life), because the water has the next property, the heaviest or the most dense water has a temperature of ~4° C. Warmer water floats to upper parts in the water column, and colder water and ice floats on the surface. So, 4° C is a temperature inside the range of life temperature, and there are lots of organisms which can live in that temperature. And only thinking in pressure effects, it could be a problem for life if there were high variations of pressure in aerial, or partially aerial, ecosystems. However, in aquatic systems, it would not have to be a problem, because water has a constant volume with pressure variations, and the pressure is the same outside and inside the cell or organism. Then, all the ocean's floor could be a possibly habitat for life.

WHERE LOOK FOR?

Europa is a world with a special fascination and interest that may help us to understand very aspect hat we don't know yet, for example: what are the limits inside which Life, such as we know and understand it, can exist? Is it possible that Life is a more frequent process than we thought? Earth contains the unique biosphere in the Solar system?

There are different ecosystems considered to be potentially viable for support life on Europa. Chyba and Hand (2001) and Lipps and Rieboldt (2005), have correctly argued that a photosynthetically-driven ecosystem is not impossible, but doubtful. Photosynthesis could be possible close to the surface, for organisms living at liquid veins or reservoirs inside the icy shell. But sunlight is not as accessible as on the Earth, and for this reason it is more probable the existence of ecosystems based in other modes of energy supply. Zolotov and Shock (2004), have proposed several possible biogeochemistry models for Europa, mainly based on sulfur, carbon and iron reduced compounds.

On Earth, the best studied non-photosynthetic ecosystems are marine hydrothermal vents, where chemoautotrophic organisms use mainly the metabolism of H_2S and FeS to obtain nourishment through the oxidation of these inorganic chemical compounds. So, the internal sea floor, on the possible hydrothermal vents or submarine volcanoes, could be a good candidate for an Europan niche, although it is not clear if hydrothermal activity exists. Other possibilities for habitats are the Europa's rocky interior (similarly to Earth underground ecosystems), the water column and the water-icy shell interface. The icy shell can harbor life, inside of liquid veins or reservoirs, or close to thermal diapirs (Ruiz et al., 2007). Even possible habitats on the Europa's surface have been described (Lipps and Reiboldt, 2005), but the superficial conditions are really extreme in that habitats, at least for life as we know.

CONCLUSION

Although several possible niches which could potentially maintain Europan life, this life would be subject to extreme conditions, at least on the basis of what we considerate extreme conditions on our planet. But, life can exist on Earth under most of these conditions.

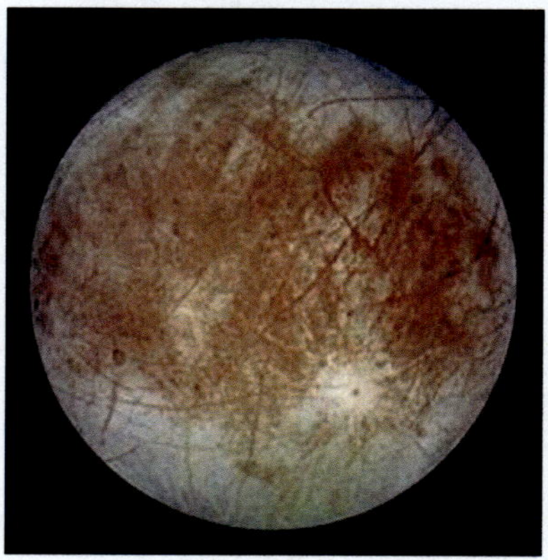

Figure 1. Global view of Europa. This image reveal the three basic elements of the geology of Europa: bright plains, mottled terrain, and linear features.

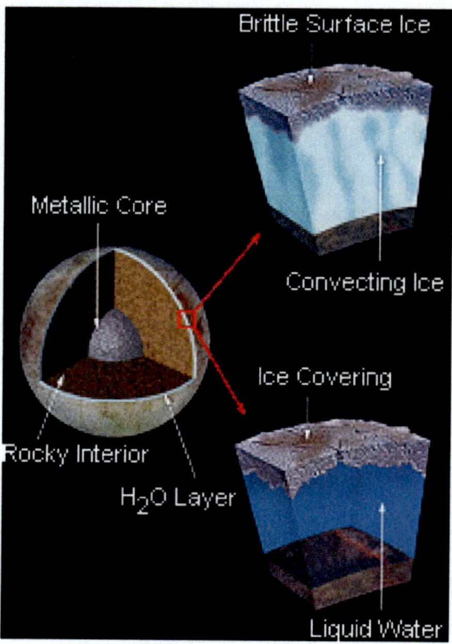

Figure 2. Europa's Interior Models: the upper one considers that there is a convecting ice layer under a brittle ice cover, and the lower one considers that there is a liquid ocean under the ice covering. The second one is supported by magnetic anomalies data (source: www.jpl.nasa.gov).

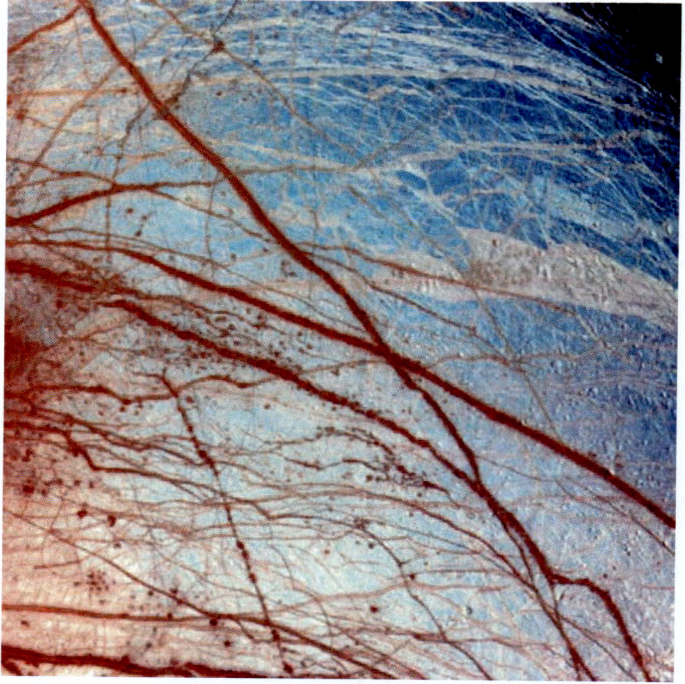

Figure 3. False-color composite of visible and infrared images from the Galileo spacecraft. The image is centered at 45° N- 221° W and shown an area 1,260 km across.

References

Bachofen, R. 1986. Microorganisms in extreme environments: introduction. *Experientia.* 42, 1179–1182.

Carlson, R. W., Johnson, R. E. and Anderson, M. S.: 1999, Sulfuric acid on Europa and the radiolytic sulfur cycle, *Science.* 286, 97-99.

Cassen, P.M., Peale, S.J., Reynolds, R.T., 1982. Structure and thermal evolution of the Galilean satellites. In: Morrison, D. (Ed.), Satellites of Jupiter. Univ. of Arizona Press, Tucson, pp. 93–128.

Chyba, C.F., 2000. Energy for microbial life on Europa. *Nature.* 403, 381-382.

Chyba, C.F., Phillips, C.B., 2001. Possible ecosystems and the search for life on Europa. *Proc. Nat. Acad. Sci.* 98, 801-804.

Chyba, C.F., Phillips, C.B., 2002. Europa as an above of life. Origin Life Evol. *Biosphere.* 32, 47-68.

Deming, J.W., 2002. Psychrophiles and Polar regions. *Curr. Opin. Microbiol.* 5, 301-309.

Dombard, A.J., McKinnon, W.B., 2006. Folding of Europa's icy lithosphere: An analysis of viscous-plastic buckling and subsequent topographic relaxation. *J. Struct. Geol.* 28, 2259–2269.

Dose, K., Bieger-Dose, A., Ernst, B., Feister, U., Gomez-Silva, B., Klein, A., Risi, S., Stridde, C.,2001. Survival of microorganisms under the extreme conditions of the Atacama Desert. *Orig. Life Evol. Biosph.* 31, 287–303.

Figueredo, P.H., Greeley, R., 2000. Geologic mapping of the northern hemisphere of Europa from Galileo solid-state imaging data. *J. Geophys. Res.* 105, 22629-22646.

Greeley, R., et al., 1998. Europa: initial Galileo geological observations. Icarus 135, 4–24.

Greeley, R., Chyba, C.F., Head, J.W., McCord, T.B., McKinnon, W.B., Pappalardo, R.T., Figueredo, P.H., 2004. Geology of Europa, in Bagenal, F., et al. (Eds.), *Jupiter: The Planet, Satellites and Magnetosphere,* Cambridge Univ. Press, Cambridge, pp. 329-362.

Greenberg, R., Hoppa, G.V., Tufts, B.R., Geissler, P., Riley, J., Kadel, S., 1999. Chaos on Europa. *Icarus.* 141, 263-286.

Greenberg, R., Geissler, P., Tufts, B.R., Hoppa, G.V., 2000. Habitability of Europa's crust: The role of tidal-tectonic processes. *J. Geophys. Res.* 105, 17551-17562.

Hoppa, G.V., Tufts, B.R., Greenberg, R., Geissler, P.E., 1999. Formation of cycloidal features on Europa. *Science.* 285, 1899-1902.

Johnson, D.B., 1998. Biodiversity and ecology of acidophilic microorganisms. *FEMS Microbiol. Ecol.* 27, 307–317.

Kargel, J.S., Kaye, J.Z., Head, J.W., Marion, G.M., Sassen, R., Crowley, J.K., Ballesteros, O.P., Grant, S.A., Hogenboom, D.L., 2000. Europa's crust and ocean: Origin, composition, and the prospects for life. *Icarus.* 148, 226-265.

Kivelson, M.G., Khurana, K.K., Russell, C.T., Volwerk, M., Walker, R.J., Zimmer, C., 2000. Galileo magnetometer measurements: a stronger case for a subsurface ocean at Europa. *Science.* 289, 1340-1343.

Lichtenberg, K.A., McKinnon, W.B., Barr, A.C., 2006. Heat flux from impact ring graben on Europa. *Lunar Planet. Sci.* 37, abstract 2399.

Lipps, J.H., Rieboldt, S., 2005. Habitats and and taphonomy of Europa. *Icarus.* 177, 515-527.

Lucchita, B.K., Soderblom, L.A., 1982. The geology of Europa, in: Morrison, D. (Ed.), Satellites of Jupiter, Univ.of Arizona Press, *Tucson,* pp. 521-555.

Malin, M.C., Pieri, D.C., 1986. Europa, in: Burns, J.A., Matthews, M.S. (Eds.), Satellites, Univ. Arizona Press, T*ucson,* pp. 689-717.

Madigan, M.T, Martinko, J.M, Parker, J., 1997. Brock Biology of Microorganisms. Prentice Hall.

Marion, G.M., 2001. Carbonate mineral solubility at low temperatures in the Na-K-Mg-Ca-H-Cl-SO4-OH-HCO3-CO3-CO2-H2O system. *Geochim. Cosmochim. Acta.* 65, 1883-1896.

Marion, G.M., 2002. A molal-based model for strong acid chemistry at low temperatures (< 200 to 298 K). *Geochim. Cosmochim. Acta.* 66, 2499-2516.

Marion, G.M., Fritsen, C.H., Eicken, H., Payne, M.C., 2003. The Search for Life on Europa: Limiting Environmental Factors, Potential Habitats, and Earth Analogues. *Astrobiology.* 3, 785-811.

McCord, T.B., Hansen, G.B., Fanale, F.P., Carlson, R.W., Matson, D.L., Johnson, T.V., Smythe, W.D., Crowley, J.K., Martin, P.D., Ocampo, A.R., Hibbitts, C.A., Granahan, J.C., 1998. Salts on Europa's surface detected by Galileo's Near-Infrared Mapping Spectrometer. *Science.* 280, 1242-1245.

McKinnon, W.B., 1999. Convective instability in Europa's floating ice shell. *Geophys. Res. Lett.* 26, 951-954.

Moore, W.B., 2006. Thermal equilibrium in europa's shell. *Icarus.* 180, 141-146.

Nimmo, F., Manga, N., 2002. Causes, characteristics and consequences of convective diapirism on Europa. *Geophys. Res. Lett.* 29, 2109, 10.1029/2002GL015754.

Nimmo, F., Giese, B., Pappalardo, R.T., 2003. Estimates of Europa's ice shell thickness from elastically-supported topography. *Geophys. Res. Lett.* 30, 1233, 10.1029/2002GL016660.

Pappalardo, R.T., et al., 1998. Geological evidence for solid-state convection in Europa's ice shell. *Nature.* 391, 365-368.

Pappalardo, R.T., Head, J.W., the Galileo Imaging Team, 1999a. Europa: role of the ductile layer. *Lunar Planet. Sci.* XXX, 1967 abstract, [CD-ROM].

Pappalardo, R.T., et al., 1999b. Does Europa have a subsurface ocean? Evaluation of the geological evidence. *J. Geophys. Res.* 104, 24015-24055.

Pappalardo, R.T., Head, J.W., 2001. The thick-shell model of Europa's geology: Implications for crustal processes. *Proc. Lunar Sci. Conf.* 32. Abstract 1866.

Potts, M., 1994. Desiccation Tolerance of Prokaryotes. *Microbiol. Rev.* **58,** 755-805.

Price, P.B., 2003. Life in solid ice on Earth and other planetary bodies. In: Norris, R. and Stootman, F. (Eds.), *Bioastronomy.* 2002: Life among the stars, IAU Symposium Series No. 213, pp. 363-366.

Price, P.B., Sowers, T., 2004. Temperature dependence of metabolic rates for microbial growth, maintenance, and survival. *Proc. Natl. Acad. Sci. USA.* 101, 4631-4636.

Prockter, L.M., Schenk, P., 2005. Origin and evolution of Castalia Macula, an anomalous young depression on Europa. *Icarus.* 177, 305-326.

Prockter, L.M., Antman, A.M., Pappalardo, R.T., Head, J.W., Collins, G.C., 1999. Europa: Stratigraphy and geological history of the anti-jovian region from Galileo E14 solid-state imaging data. *J. Geophys. Res.* 104, 16531-16540.

Reynolds, R.T., Squyres, S.W., Colburn, D.S. McKay, C.P., 1983. On the habitability of Europa. *Icarus.* 56, 246–254.

Rothschild, L.J., Mancinelli, R.L., 2001. Life in extreme environments. *Nature.* 409, 1092-1101.

Ruiz, J., 2005. The heat flow of Europa. Icarus 177, 438–446.

Ruiz, J., Tejero, R., 2000. Heat flows through the ice lithosphere of Europa. *J. Geophys. Res.* 105, 23283–23289.

Ruiz, J., Tejero, R., 2003. Heat flow, lenticulae spacing, and possibility of convection in the ice shell of Europa. *Icarus.* 162, 362–373.

Ruiz, J., Montoya, L., López, V., Amils, R., 2007. Thermal diapirism and the habitability of the icy shell of Europa. *Orig. Life Evol. Biosph.* 37, 287-295.

Scheleper, C., Pühler, G., Kühlmorgen, B., Zilling, W., 1995. Life at extremely low pH. *Nature.* 375, 741-742.

Schenk, P.M., 2002. Thickness constraints on the icy shells of Galilean satellites from a comparison of crater shapes. N*ature.* 417, 419-421.

Schilling, N., Khurana, K.K., Kivelson, M.G., 2004. Limits on an intrinsic dipole moment in Europa. *J. Geophys. Res.* 109, E05006, 10.1029/2003JE002166.

Schulze-Makuch, D., Irwin, L.N., 2002. Energy cycling and hypothetical organisms in Europa's ocean. *Astrobiology.* 2, 105-121.

Spaun, N.A., Head, J.W., Pappalardo, R.T., 2004. Europan chaos and lenticulae: a synthesis os size, spacing, and areal density analyses. *Lunar Planet. Sci. Conf.* 35, abstract 1409.

Spencer, J.R., Grundy, W.M., Dumas, C., Carlson, R.W., McCord, T.B., Hansen, G.B., Terrile, R.J., 2006. The nature of Europa's dark non-ice surface material: Spatially-resolved high spectral resolution spectroscopy from the Keck telescope. *Icarus.* 182, 202-210.

Spohn, T., Schubert, G., 2003. Oceans in the icy Galilean satellites? *Icarus.* 161, 456–467.

Thomas, D.N., Dieckmann, G.S., 2002. Antarctic Sea Ice-A Habitat for Extremophiles. *Science.* 295, 641-644.

Welsh, D.T., 2000. Ecological significance of compatible solute accumulation by micro-organisms: from single cells to global climate. *FEMS Microbiol. Rev.* 24, 263-290.

Yancey, P.H., 2005. Organic osmolytes as compatible, metabolic and counteracting cytoprotectants in high osmolarity and other stresses. *J. Exp. Biol.* 208, 2819-2830.

Zahnle, K., Schenk, P., Levison, H., Dones, L., 2003. Cratering rates in the outer Solar System. *Icarus.* 163, 263-289.

Zhilina, T.N., Zavarzin, G.A., 1994. Alkaliphilic anaerobic community at pH 10. *Curr. Microbiol.* 29, 109–112.

Zolotov, M.Y., Shock, E.L., 2004. A model for low-temperature biogeochemistry of sulfur, carbon, and iron on Europa. *J. Geophys. Res.* 109, E06003, 10.1029/2003JE002194.

In: Space Exploration Research
Editors: J.H. Denis and P.D. Aldridge
ISBN: 978-1-60692-264-4

Short Communication

NEUTRON DOSIMETRY INSIDE THE INTERNATIONAL SPACE STATION

Hideki Koshiishi, Haruhisa Matsumoto, Kazuhiro Terasawa, Kiyokazu Koga and Tateo Goka
Japan Aerospace Exploration Agency, Japan

ABSTRACT

This article presents descriptions of instruments and measurement results of neutron dosimetry for astronaut safety inside the International Space Station (ISS). Most neutrons inside the ISS are the secondary particles, and, especially for E > 100 keV, are one of the major contributors to the radiation dose received by astronauts because of their high radiation quality factor. The Bonner Ball Neutron Detector (BBND) experiment was conducted onboard the US Laboratory Module of the ISS as part of the Human Research Facility project of NASA over an eight-month period in 2001, corresponding to the maximum period of solar-activity variation, in order to evaluate the neutron radiation environment in the energy range from thermal up to 15 MeV inside the ISS. The BBND experiment is the first active measurement of neutrons inside the ISS, providing new knowledge about astronaut safety concerning the radiation environment as well as important inputs for further radiation environment model calculations. However, the BBND experiment can be considered as only half of the necessary evaluation of the neutron radiation environment inside the ISS since neutrons of 10 MeV < E < several hundreds MeV are considered to make a comparable contribution to the radiation dose to that in the BBND measurement energy range. A scintillation fiber neutron detector with a sensitive energy range higher than 10 MeV has been developed, which is complementary to the BBND instrument. Wide-energy range measurements of neutrons by these detectors shall permit evaluation of total contribution from neutrons to the radiation dose received by astronauts.

INTRODUCTION

The neutron radiation environment inside the International Space Station (ISS) is of two types. One is secondary neutrons; albedo neutrons that are produced by the interaction of primary particles (mostly galactic cosmic rays (GCRs) and trapped protons) with the atmosphere, and local neutrons that are produced by the interaction of primary particles with the mass of the ISS. The other is primary solar neutrons that are considered to be accelerated in the solar atmosphere, and can pass through the ISS walls. Nearly all neutrons inside the ISS are secondary particles, which, especially for E > 100 keV, are one of the major contributors to the radiation dose received by astronauts because of their high radiation quality factor.

Therefore, measurements of neutrons inside the ISS in the middle- to the high-energy range over extended time periods are important from the perspective of astronaut safety. Passive detectors have been used many times to evaluate the neutron radiation environment [e.g. Benton et al., 2001; Reitz, 2001]. However, few investigations have been conducted using active detectors over extended time periods because of various technical difficulties [e.g. Armstrong and Colborn, 2001; Badhwar et al., 2001; Lyagushin et al., 2001]. NASA has been developing the Human Research Facility project to evaluate the effects of long-duration space flight on astronaut health. The Bonner Ball Neutron Detector (BBND) experiment was conducted as part of this project by the Japan Aerospace Exploration Agency (JAXA) in order to evaluate the neutron radiation environment inside the ISS. The BBND experiment was carried out over an eight-month period in 2001, corresponding to the maximum period of solar-activity variation.

EXPERIMENTS

Bonner Ball Neutron Detector

The BBND instrument has six sensors. Figure 1 illustrates the cross-sectional view of each sensor. A 2-inch-diameter spherical ^{3}He proportional counter that is sensitive to thermal neutrons is located at the center of each sensor, and covered with two types of neutron moderators [Bramblett et al., 1960]. One moderator is comprised of polyethylene spheres with thicknesses of 1.5, 3, 5, and 9 cm for four of the sensors to provide each sensor with a different energy-response function for incident neutrons. The other moderator consists of a 1 mm-thick gadolinium cover over two of the sensors to eliminate thermal neutrons. The sensors are situated in the BBND instrument as illustrated in figure 2.

The energy-response functions for each sensor were calibrated in individual irradiation experiments using several monoenergetic neutron fields. Figure 3 depicts the energy-response functions for each sensor in the assembled configuration, which were obtained by the Monte Carlo calculations for 0.025 eV < E < 100 MeV using interaction cross-section libraries based on the calibration results of each sensor. The effective measurement energy range is from thermal up to 15 MeV.

The gain of each sensor was adjusted to produce a peak signal from thermal neutrons in a specific output channel in order to discriminate charged particles such as protons. The pulse-height distribution of the output signal obtained by each sensor in the BBND experiment presented a sharp peak signal from thermal neutrons, which appeared separately from the signal caused by background protons. Thus, the BBND instrument can readily distinguish neutrons in mixed charged-particle/neutron fields [Matsumoto et al., 2001].

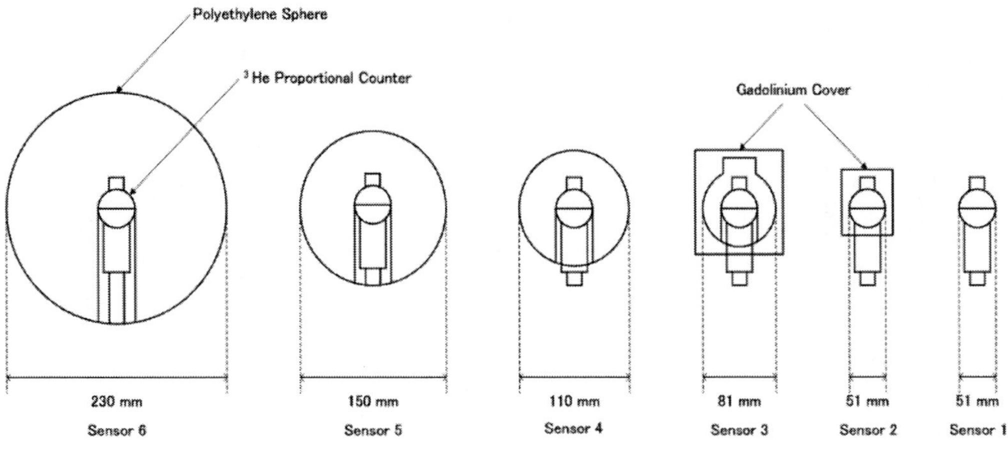

Figure 1. Cross-sectional view of the BBND sensors.

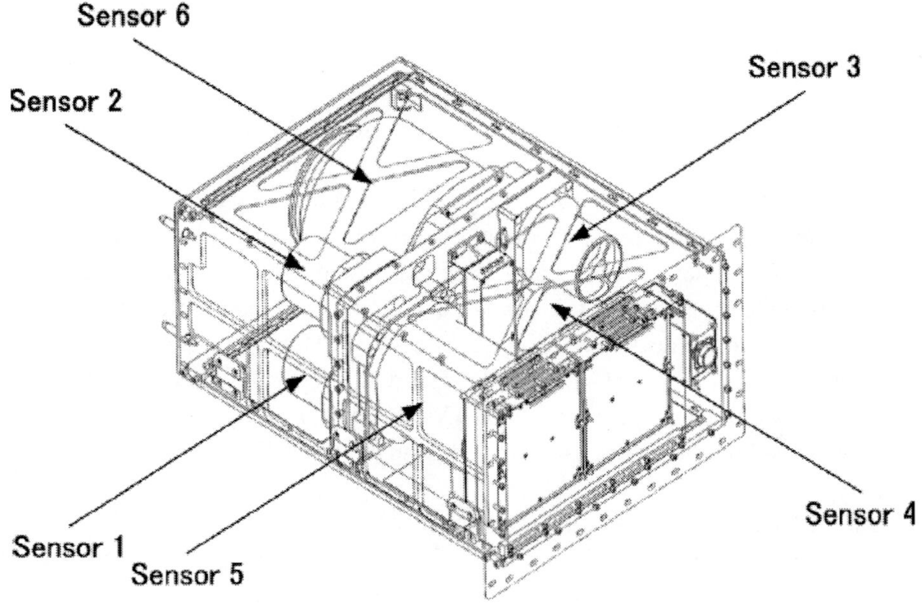

Figure 2. Assembled configuration of the BBND sensors.

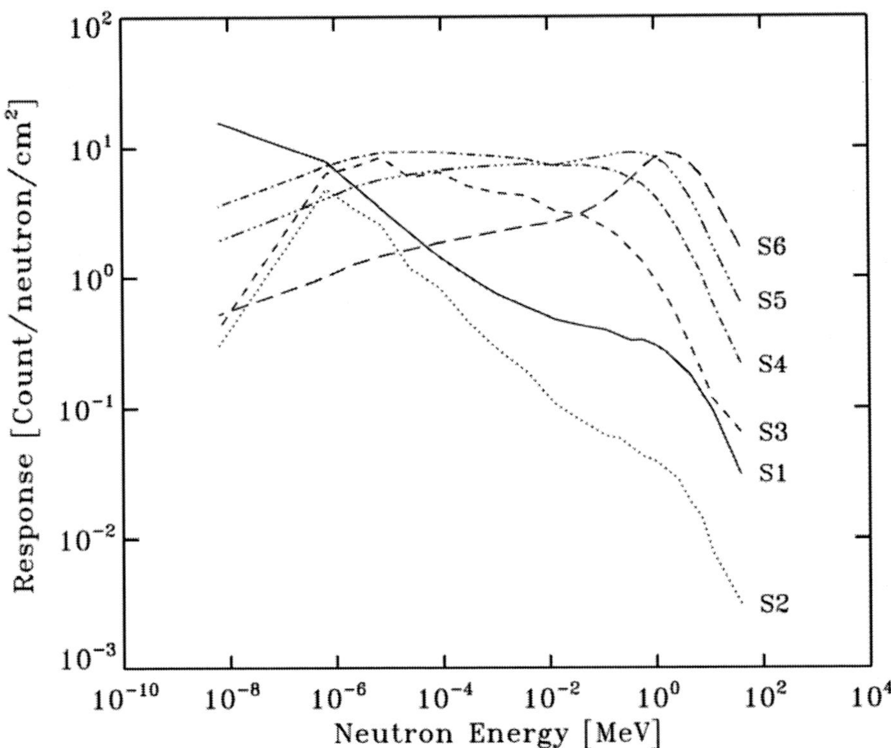

Figure 3. The energy-response functions for the BBND sensors in the assembled configuration.

Measurements

The BBND instrument was launched onboard the STS-102 Space Shuttle flight on 8 March 2001, and installed inside the US Laboratory Module of the ISS on 14 March. The BBND experiment was carried out from 23 March through 14 November 2001, corresponding to the maximum period of solar-activity variation. Useful data were collected for a total period of 219 days, 4 hours, and 36 minutes with a temporal resolution of 1 minute. The BBND instrument was relocated from a corner on the port side of the module to the center of the module at the deck inside the US Laboratory Module on 9 August as indicated in figure 4. Data were retrieved by the STS-104 and the STS-108 Space Shuttle flight, and transmitted to the Tsukuba Space Center of JAXA. During the BBND experiment, the ISS altitude varied between 368 km and 415 km, with an average altitude of 394 km [Koshiishi et al.,2007].

The same BBND sensors were also previously mounted inside the STS-89 Space Shuttle flight with 51.6 degree orbital inclination and 387 km orbital altitude as a pilot experiment from 24 through 27 January 1998, corresponding to the rising phase of solar-activity variation [Matsumoto et al., 2001].

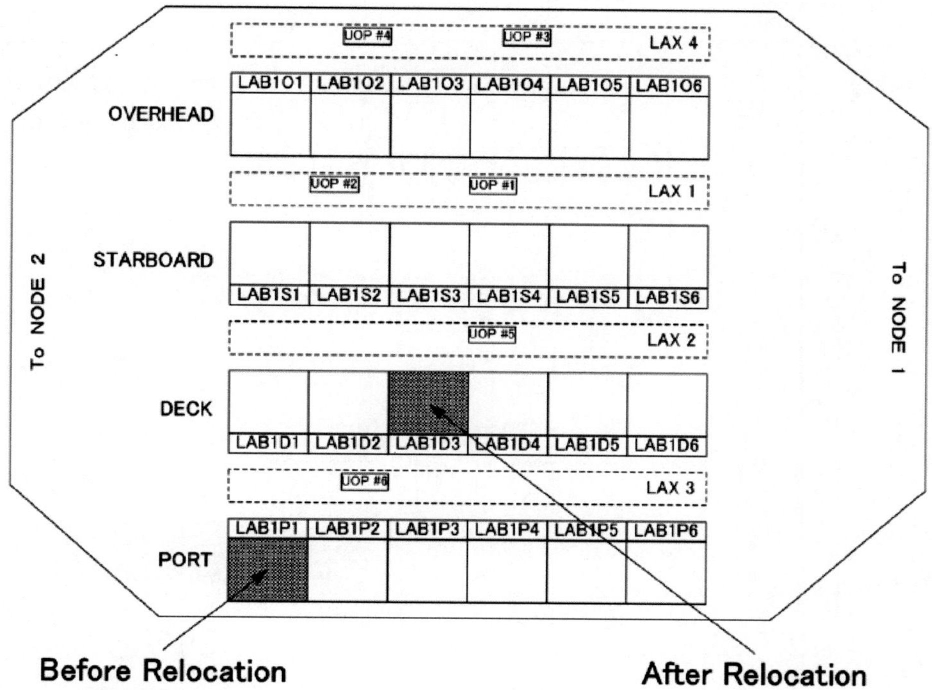

Figure 4. Locations of the BBND instrument inside the US Laboratory Module of the ISS.

RESULTS

Differential-Energy Spectrum

The neutron differential-energy spectrum is evaluated by applying the energy-response functions in figure 3 to the data obtained in the BBND experiment using the NEUPAC-83 unfolding code [Taniguchi et al., 1983]. The model neutron spectrum predicted for the inside of the ISS for the maximum period of solar-activity variation reported by Armstrong and Colborn [1998; 2001] is used as an initial guess in the unfolding procedure, together with the assumption that the effective shielding thickness inside the ISS is 20 g/cm^2 of aluminum [Armstrong and Colborn, 1998; 2001].

Figure 5 and table 1 show the orbit-averaged neutron spectra obtained inside the ISS before and after the relocation of the BBND instrument, and the orbit-averaged neutron spectrum obtained inside the STS-89. The neutron spectra have a power-law shape of neutron energy except for 10 keV $< E <$ 1 MeV. In this energy range, the neutron spectra show two local maxima. The first maximum at $E \sim$ 100 keV can be explained by the contribution from albedo neutrons as suggested by Lyagushin et al. (2001). The second maximum at $E \sim$ 1 MeV is supported by the intranuclear cascade-evaporation model (e.g. Armstrong, 1980).

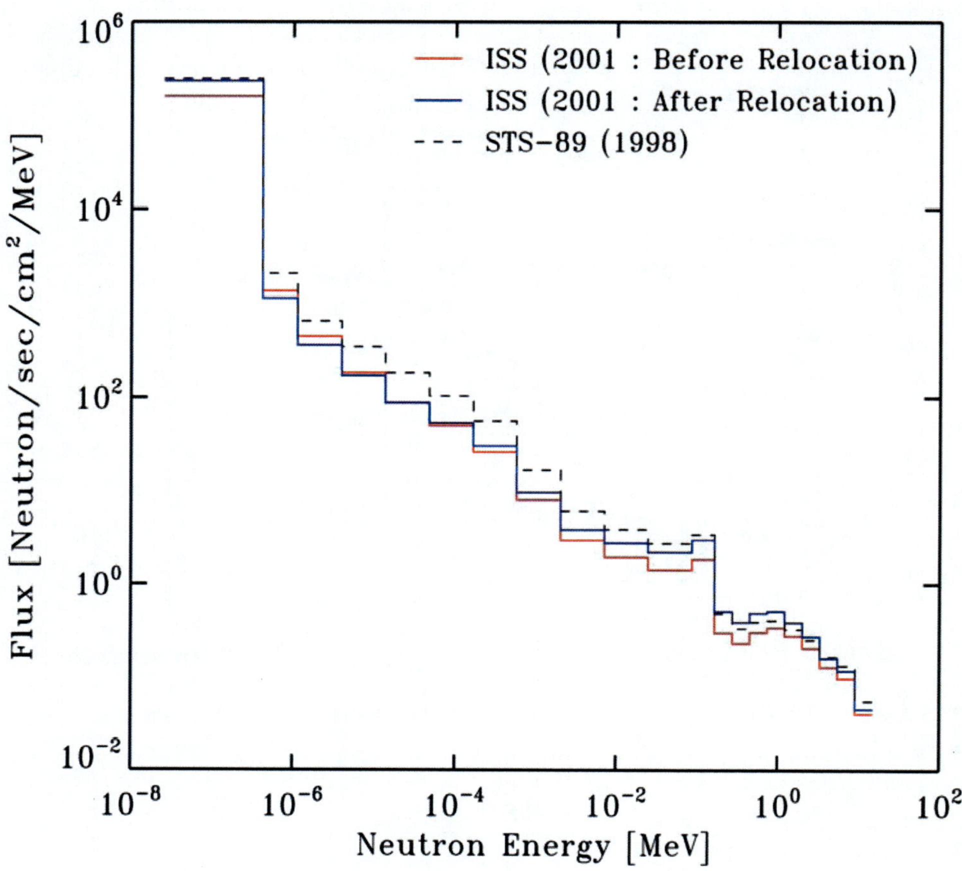

Figure 5. The orbit-averaged neutron spectra obtained inside the ISS from 23 March through 9 August 2001 (before the relocation of the BBND instrument), from 9 August through 14 November 2001 (after the relocation of the BBND instrument), and the orbit-averaged neutron spectrum obtained inside the STS-89 from 24 through 27 January 1998.

The neutron flux inside the ISS after the relocation shows a harder spectrum than that before the relocation. The location of the BBND instrument after the relocation in figure 4 was surrounded by a greater amount of mass than the original location. Thicker shield leads to a reduction of low-energy neutrons, while producing more high-energy neutrons, resulting in a harder spectrum. The neutron flux inside the ISS also shows a harder spectrum than that inside the STS-89, suggesting that the shielding thickness surrounding the BBND instrument inside the ISS was greater than that inside the STS-89. On the other hand, the neutron spectrum inside the STS-89 has higher flux than those inside the ISS, which is mainly due to solar-activity variation, namely, the anti-correlation of GCR flux and trapped-proton flux with solar-activity [e.g. Dachev et al., 1999].

Table 1. The orbit-averaged neutron spectra obtained inside the ISS from 23 March through 9 August 2001 (before the relocation of the BBND instrument), from 9 August through 14 November 2001 (after the relocation of the BBND instrument), and the orbit-averaged neutron spectrum obtained inside the STS-89 from 24 through 27 January 1998.

Neutron Energy [MeV]	Flux [Neutron/sec/cm^2/MeV]		
	ISS Before Relocation	ISS After Relocation	STS-89
$2.5 \times 10^{-8} - 4.1 \times 10^{-7}$	1.6×10^{5}	2.4×10^{5}	2.5×10^{5}
$4.1 \times 10^{-7} - 1.1 \times 10^{-6}$	1.4×10^{3}	1.1×10^{3}	2.1×10^{3}
$1.1 \times 10^{-6} - 3.9 \times 10^{-6}$	4.4×10^{2}	3.6×10^{2}	6.5×10^{2}
$3.9 \times 10^{-6} - 1.4 \times 10^{-5}$	1.8×10^{2}	1.7×10^{2}	3.5×10^{2}
$1.4 \times 10^{-5} - 4.8 \times 10^{-5}$	8.8×10^{1}	8.8×10^{1}	1.8×10^{2}
$4.8 \times 10^{-5} - 1.7 \times 10^{-4}$	5.0×10^{1}	5.3×10^{1}	1.0×10^{2}
$1.7 \times 10^{-4} - 5.8 \times 10^{-4}$	2.6×10^{1}	3.0×10^{1}	5.6×10^{1}
$5.8 \times 10^{-4} - 2.0 \times 10^{-3}$	8.0×10^{0}	9.6×10^{0}	1.7×10^{1}
$2.0 \times 10^{-3} - 7.1 \times 10^{-3}$	3.0×10^{0}	3.8×10^{0}	6.0×10^{0}
$7.1 \times 10^{-3} - 2.5 \times 10^{-2}$	1.9×10^{0}	2.7×10^{0}	3.9×10^{0}
$2.5 \times 10^{-2} - 8.7 \times 10^{-2}$	1.4×10^{0}	2.2×10^{0}	2.7×10^{0}
$8.7 \times 10^{-2} - 1.6 \times 10^{-1}$	1.8×10^{0}	3.0×10^{0}	3.4×10^{0}
$1.6 \times 10^{-1} - 2.7 \times 10^{-1}$	3.0×10^{-1}	5.1×10^{-1}	4.9×10^{-1}
$2.7 \times 10^{-1} - 4.5 \times 10^{-1}$	2.3×10^{-1}	3.9×10^{-1}	3.3×10^{-1}
$4.5 \times 10^{-1} - 7.4 \times 10^{-1}$	3.0×10^{-1}	4.9×10^{-1}	3.9×10^{-1}
$7.4 \times 10^{-1} - 1.2 \times 10^{0}$	3.4×10^{-1}	5.1×10^{-1}	4.1×10^{-1}
$1.2 \times 10^{0} - 2.0 \times 10^{0}$	2.8×10^{-1}	3.9×10^{-1}	3.3×10^{-1}
$2.0 \times 10^{0} - 3.3 \times 10^{0}$	2.1×10^{-1}	2.7×10^{-1}	2.5×10^{-1}
$3.3 \times 10^{0} - 5.5 \times 10^{0}$	1.3×10^{-1}	1.6×10^{-1}	1.7×10^{-1}
$5.5 \times 10^{0} - 9.0 \times 10^{0}$	9.8×10^{-2}	1.2×10^{-1}	1.3×10^{-1}
$9.0 \times 10^{0} - 1.5 \times 10^{1}$	4.1×10^{-2}	4.6×10^{-2}	5.7×10^{-2}

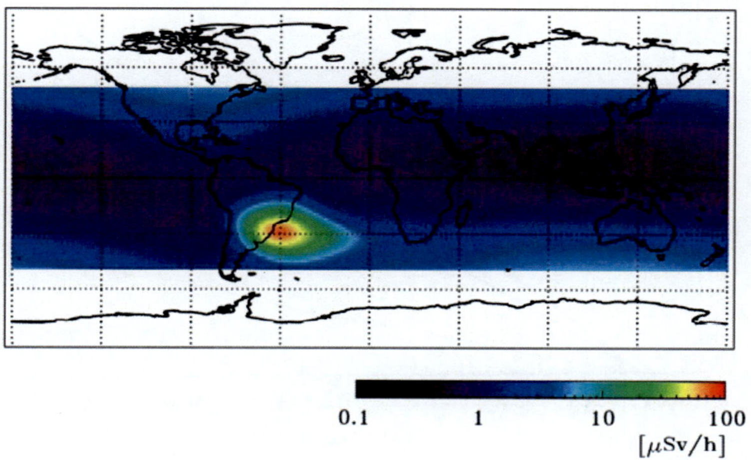

Figure 6. The smoothed geographic neutron dose equivalent rate distribution obtained inside the ISS.

DOSE EQUIVALENT

The neutron dose equivalent is evaluated by applying the ICRP 74 conversion coefficients [ICRP 74, 1996] to the neutron differential-energy spectrum obtained in the BBND experiment. Figure 6 illustrates the smoothed geographic neutron dose equivalent rate distribution obtained inside the ISS. The neutron dose equivalent rate distribution is determined by the geomagnetic cut-off rigidity distribution except in the SAA region where trapped protons dominate. The peak neutron dose equivalent rates before and after the relocation are 62 microSv/h and 72 microSv/h at 46W 29S using a geographic Gaussian-fit distribution applied to the central SAA region.

Figure 7 depicts the contribution from neutrons produced by GCRs, the contribution from neutrons produced by trapped protons, and the total cumulative neutron dose equivalent during the BBND experiment. The neutron dose equivalent that occurs in the SAA region is defined as the contribution from trapped-proton-induced neutrons. The remainder of the neutron dose equivalent is considered to be from GCR-induced neutrons. The ratio of the neutron dose equivalent contributed from GCR-induced neutrons to that from trapped-proton-induced neutrons is 3.0 before and 3.3 after the relocation. For the minimum period of solar-activity variation, the ratio is expected to be smaller, 2.0 for the STS-89 pilot experiment, because trapped-proton-induced neutrons are influenced by solar-activity variation greater than GCR-induced neutrons.

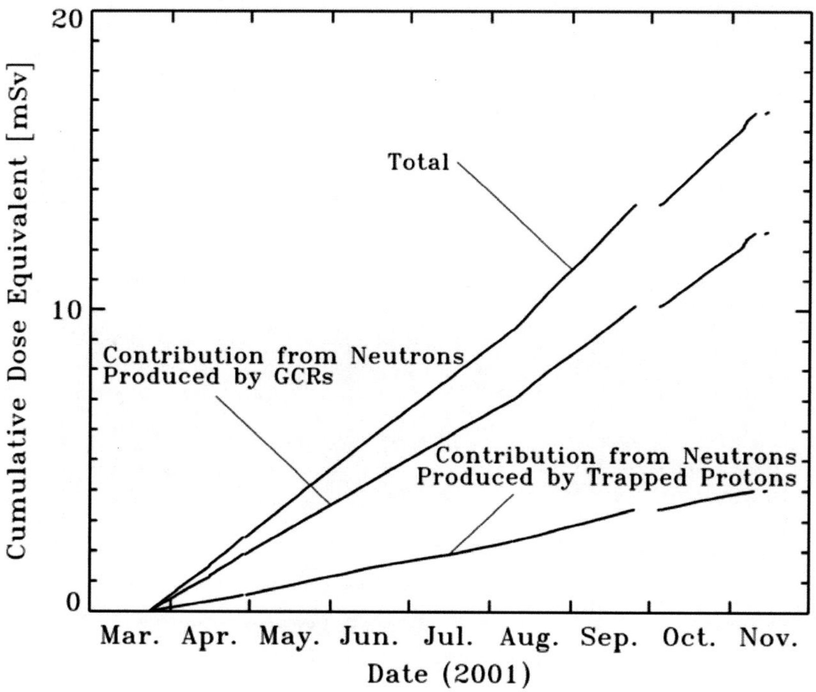

Figure 7. The contribution from neutrons produced by GCRs, the contribution from neutrons produced by trapped protons, and the total cumulative neutron dose equivalent obtained inside the ISS during the BBND experiment.

Figure 8 depicts the variation in the 1-hour averaged neutron dose equivalent rate during the BBND experiment, with average rates of 69 microSv/day before and 88 microSv/day after the relocation. Even though the relocation was carried out inside the same pressurized module, there is a 30 % increase in the average neutron dose equivalent rate due to the difference in the localized shielding environment.

During the BBND experiment, the radiation environment in the ISS orbit was sometimes strongly disturbed by solar and geomagnetic events, which are seen as high neutron dose equivalent rate spikes, especially in middle April and early November 2001. The most significant event was a X1.0/3B solar flare that occurred on 4 November 2001, and was followed by a coronal mass ejection (e.g. Nitta et al., 2003). The disturbance of the radiation environment due to the solar event appeared in the high-latitude region, while the radiation environment in the low-latitude and the SAA regions remained relatively unchanged. The increase in the total cumulative neutron dose equivalent due to the solar event is 0.15 mSv, which is seen in figure 7 as a rapid increase in the contribution from GCR-induced neutrons. Given that the annual neutron dose equivalent estimated from the BBND experiment is 28 mSv, the solar event contributes less than 1 % to the annual neutron dose equivalent.

The average neutron dose equivalent rate during the STS-89 pilot experiment is 90 microSv/day, which is higher than those during the BBND experiment. In comparison with the results obtained in the experiments previously conducted in 1990's onboard the MIR and the Space Shuttle flights at the same orbital inclination [Dudkin et al., 1996; Badhwar et al., 1997; Ing, 2001; Lyagushin et al., 2001], the highest neutron dose equivalent rate was measured in 1995, corresponding to the minimum period of solar-activity variation. There is, at most, a factor of 2 variation among these experiments.

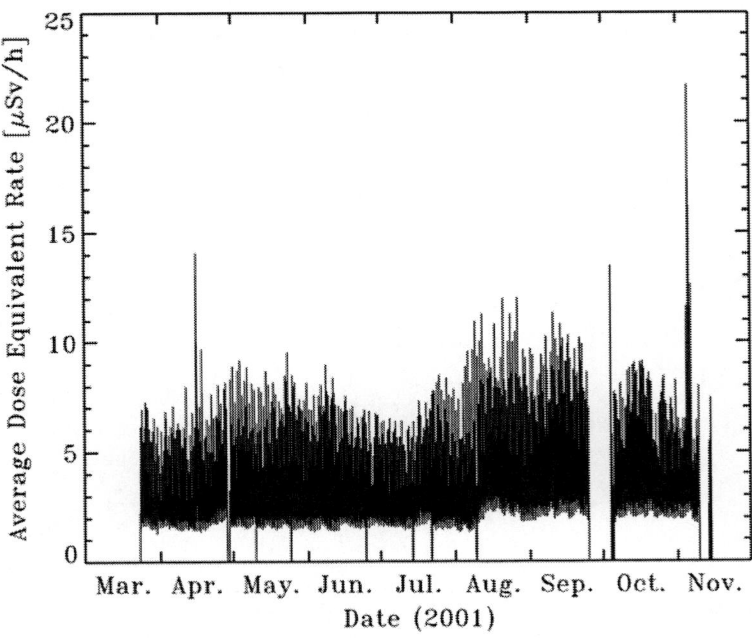

Figure 8. The variation in the 1-hour averaged neutron dose equivalent rate obtained inside the ISS during the BBND experiment.

Influence of Altitude Variation

In the estimation of the influence of the ISS altitude variation on the neutron dose equivalent rate, the contribution from GCR-induced neutrons and the contribution from trapped-proton-induced neutrons are independently related to the ISS altitude variation. The influence of the ISS altitude variation on the contribution from GCR-induced neutrons can be determined by the geomagnetic cut-off rigidity distribution calculated using the CREME86 code [Adams, 1986]. The influence of the ISS altitude variation on the contribution from trapped-proton-induced neutrons can be dominated by the proton integrated flux distribution in the energy range of 10 MeV - 400 MeV calculated using the AP-8 MAX code [Sawyer and Vette, 1976] incorporating the secular drift of the SAA region.

By using the combination of these relationships, together with the geomagnetic cut-off rigidity and the proton integrated flux distributions at each altitude, the influence of the ISS altitude variation on the neutron dose equivalent rate over the ISS altitude variation of 300 km - 500 km is estimated as shown in figure 9. The contribution from GCR-induced neutrons remains almost constant over this altitude range. In contrast, the contribution from trapped-proton-induced neutrons increases rapidly with increase in the altitude. Thus, the variation in the neutron dose equivalent rate with the altitude is mainly due to the contribution from trapped-proton-induced neutrons. This result appears in figure 8 as the variation in the 1-hour averaged neutron dose equivalent rate due to the ISS altitude variation in the SAA region. The neutron dose equivalent rate increases by a factor of 2 over the ISS altitude variation of 300 km - 500 km. For the minimum period of solar-activity variation, the factor is expected to be greater because the contribution from trapped-proton-induced neutrons to the neutron dose equivalent becomes larger.

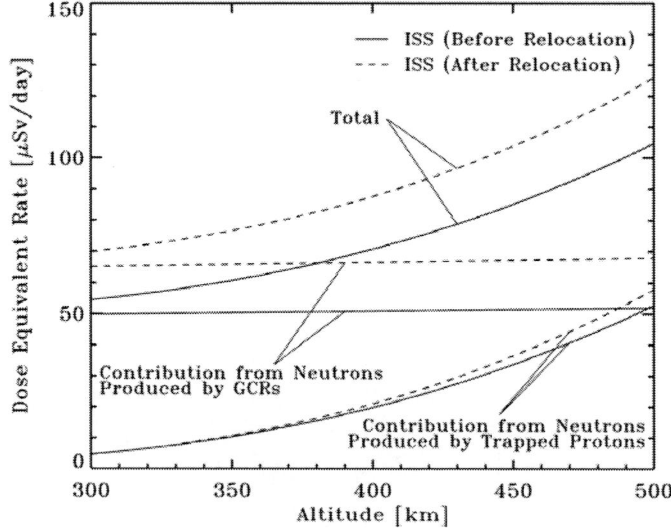

Figure 9. Estimation of the influence of the ISS altitude variation on the contribution from neutrons produced by GCRs, the contribution from neutrons produced by trapped protons, and the total neutron dose equivalent rate obtained inside the ISS before and after the relocation of the BBND instrument over the ISS altitude variation of 300 km - 500 km.

FUTURE PLAN FOR NEUTRON DOSIMETRY IN SPACE

The BBND experiment can be considered as only half of the necessary evaluation of the neutron radiation environment inside the ISS since neutrons of 10 MeV < E < several hundreds MeV are considered to make a comparable contribution to the radiation dose to that in the BBND measurement energy range [Armstrong and Colborn, 1998; Badhwar et al., 2001; Lyagushin et al., 2001]. A scintillation fiber neutron detector with a sensitive energy range higher than 10 MeV has been developed, which is complementary to the BBND instrument [Terasawa et al., 2001a]. Wide-energy range measurements of neutrons by these detectors shall permit evaluation of total contribution from neutrons to the radiation dose received by astronauts.

Scintillation Fiber Neutron Detector

The scintillation fiber neutron detector is a type of 3-dimensional track detector that is comprised of a scintillation fiber stack and two image intensifiers. Figure 10 illustrates a schematic image of the scintillation fiber stack. The scintillation fiber stack consists of layers of scintillation fiber sheets that are alternatively stacked to be perpendicular to each other. The scintillation fibers are connected to the image intensifiers to view two 2-dimensional projections of a 3-dimensional track, which are used to construct 3-dimensional trajectories of incident particles. Figure 11 shows a schematic image of the principle to identify charged particles such as protons, gamma rays, and neutrons by the scintillation fiber neutron detector. High-energy neutrons collide elastically with hydrogens in the scintillation fiber, and produce recoil protons that cause scintillation emissions, which appear differently from those caused by protons and gamma rays. Thus, the scintillation fiber neutron detector can readily distinguish linear energy transfer of neutrons, which is essential to evaluate neutron dose equivalent, by analyzing individual 3-dimensional particle trajectory.

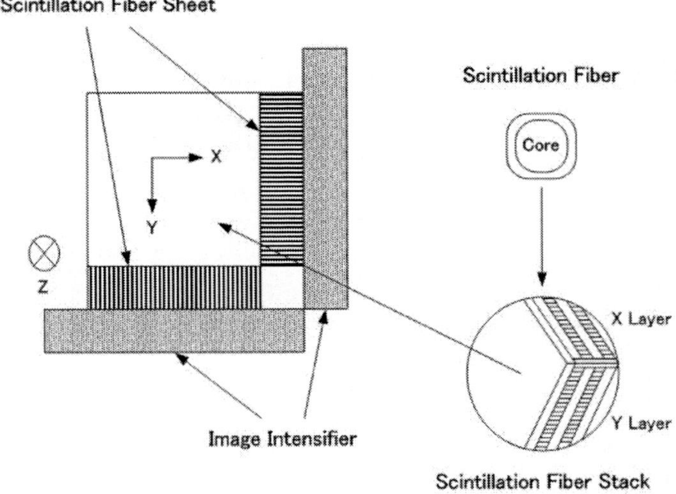

Figure 10. Schematic image of the scintillation fiber stack.

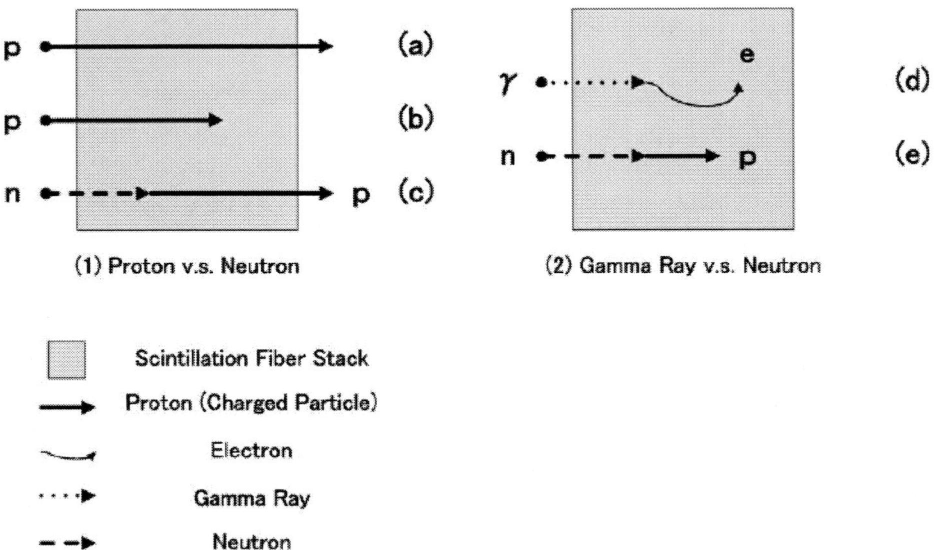

Figure 11. Schematic image of the principle to identify protons (charged particles), gamma rays, and neutrons by the scintillation fiber neutron detector. (a) Track of an incident proton passing through the stack. (b) Track of an incident proton stopping inside the stack. (c) Track of an incident neutron and a recoil proton moving outside the stack. (d) Track of an incident gamma ray and a produced electron. (e) Track of an incident neutron and a recoil proton stopping inside the stack. Solid lines and curves (protons and electrons) show particle trajectories with scintillation emissions that can be detected by the scintillation fiber neutron detector, while dashed lines (gamma rays and neutrons) show particle trajectories without scintillation emissions because of no interactions with the scintillation fibers. (1) If the beginning and/or the end of track are on the edge of the stack, the incident particle can be a proton or a neutron. The proton in case (a) can be discriminated because scintillation emissions yield on both edges of the stack. The proton in case (b) can also be discriminated because the Bragg peak of scintillation emissions yields at the end of track inside the stack. Otherwise, the incident particle is identified to be a neutron in case (c). (2) If the beginning and the end of track are inside the stack, the incident particle can be a gamma ray or a neutron. Scintillation emissions per unit length in case (e) yield brighter than those in case (d) so that the incident particle can be identified to be a neutron in case (e).

The scintillation fiber stack for the Neutron Monitor (NEM) experiment described below is 96 x 96 x 96 mm^3, which includes 512 fibers (16 fibers in a sheet x 16 layers of sheets for one direction x 2 perpendicular directions) having a 3 x 6 mm^2 cross section for each fiber. The scintillation fiber stack has been improved, and is 50 x 50 x 50 mm^3, which includes 40000 fibers (200 fibers in a sheet x 100 layers of sheets for one direction x 2 perpendicular directions) having a 0.25 x 0.25 mm^2 cross section for each fiber in the latest version [Terasawa et al., 2001b; Matsumoto et al., 2006].

Neutron Monitor Onboard JEM

The NEM instrument that has both the Bonner Ball and the scintillation fiber neutron detectors is planed to be onboard the Exposed Facility of the Japanese Experiment Module (JEM : KIBO) of the ISS in 2009, as one of the instruments of the Space Environment Data Acquisition Equipment (JEM SEDA) experiment [Imaida et al., 1999; Koga et al., 2001]. The

NEM experiment has two major objectives; evaluation of the neutron radiation environment outside the ISS for astronaut safety during extravehicular activity, and direct observation of solar neutrons that are considered to be accelerated in the solar atmosphere [e.g. Watanabe et al., 2003]. The JEM SEDA will be operated in space for three years.

CONCLUSION

The BBND experiment was conducted in order to evaluate the neutron radiation environment in the energy range from thermal up to 15 MeV inside the ISS for the maximum period of solar-activity variation. The BBND experiment is the first active measurement of neutrons inside the ISS, providing new knowledge about astronaut safety concerning the radiation environment as well as important inputs for further radiation environment model calculations. However, the BBND experiment can be considered as only half of the necessary evaluation of the neutron radiation environment inside the ISS since neutrons of 10 MeV < E < several hundreds MeV are considered to make a comparable contribution to the radiation dose to that in the BBND measurement energy range. A scintillation fiber neutron detector with a sensitive energy range higher than 10 MeV has been developed, which is complementary to the BBND instrument. Wide-energy range measurements of neutrons by these detectors shall permit evaluation of total contribution from neutrons to the radiation dose received by astronauts.

REFERENCES

Adams Jr., J.H. *NRL Memorandum Report.* 1986, 5901.

Armstrong, T.W. In *Computer Techniques in Radiation Transport and Dosimetry*; Nelson, W.R.; Jenkins, T.M.; 1980, 311-322; Plenum Press, New York.

Armstrong, T.W.; Colborn, B.L. *SAIC Report.* 1998, 98042R.

Armstrong, T.W.; Colborn, B.L. *Radiation Measurements.* 2001, 33, 229-234.

Badhwar, G.D.; Atwell, W.; Cash, B.; Weyland, M.; Petrov, V.M.; Tchernykh, I.V.; Akatov, Yu.A.; Shurshakov, V.A.; Arkhangelsky, V.V.; Kushin, V.V.; Klyachin, N.A.; Benton, E.V.; Frank, A.L.; Benton, E.R.; Frigo, L.A.; Dudkin, V.E.; Potapov, Yu.V.; Vana, N.; Schoner, W.; Fugger, M. *Radiation Measurements.* 1997, 26, 901-916.

Badhwar, G.D.; Keith, J.E.; Cleghorn, T.F. *Radiation Measurements.* 2001, 33, 235-241.

Benton, E.R.; Benton, E.V.; Frank, A.L. *Radiation Measurements.* 2001, 33, 255-263.

Bramblett, R.L.; Ewing, R.I.; Bonner, T.W. *Nuclear Instruments and Methods.* 1960, 9, 1-12.

Dachev, Ts.P.; Tomov, B.T.; Matviichuk, Yu.N.; Koleva, R.T.; Semkova, J.V.; Petrov, V.M.; Benghin, V.V.; Ivanov, Yu.V.; Shurshakov, V.A.; Lemaire, J.F. *Radiation Measurements.* 1999, 30, 269-274.

Dudkin, V.E.; Potapov, Yu.V.; Akopova, A.B.; Melkumyan, L.V.; Bogdanov, V.G.; Zacharov, V.I.; Plyuschev, V.A.; Lobakov, A.P.; Lyagyshin, V.I. *Radiation Measurements.* 1996, 26, 535-539.

ICRP 74; *ICRP Publication.* 1996, 74; Pergamon Press, Oxford.

Imaida, I.; Muraki, Y.; Matsubara, Y.; Masuda, K.; Tsuchiya, H.; Hoshida, T.; Sako, T.; Koi, T.; Ramanamurthy, P.V.; Goka, T.; Matsumoto, H.; Omoto, T.; Takase, A.; Taguchi, K.; Tanaka, I.; Nakazawa, M.; Fujii, M.; Kohno, T.; Ikeda, H. *Nuclear Instruments and Methods*. 1999, A421, 99-112.

Ing, H. *Radiation Measurements*. 2001, 33, 275-286.

Koga, K.; Goka, T.; Matsumoto, H.; Muraki, Y.; Masuda, K.; Matsubara, Y. *Radiation Measurements*. 2001, 33, 287-291.

Koshiishi, H.; Matsumoto, H.; Chishiki, A.; Goka, T.; Omodaka, T. *Radiation Measurements*. 2007, 42, 1510-1520.

Lyagushin, V.I.; Dudkin, V.E.; Potapov, Yu.V.; Sevastianov, V.D. *Radiation Measurements*. 2001, 33, 313-319.

Matsumoto, H.; Goka, T.; Koga, K.; Iwai, S.; Uehara, T.; Sato, O.; Takagi, S. *Radiation Measurements*. 2001, 33, 321-333.

Matsumoto, H.; Kawakami, H.; Terasawa, K.; Doke, T.; Goka, T.; Zhao, Z.; Song, Q.; Ye, B.; Ni, H.; Ferguson, I.; Howorth, J.; Nozaki, T.; Uchihori, Y. In *Infrared and Photoelectronic Imagers and Detector Devices 2*; Longshore, R.E.; Sood, A.; 2006, 6294; The Society of Photo-Optical Instrumentation Engineers, Washington.

Nitta, N.V.; Cliver, E.W.; Tylka, A.J. *Astrophysical Journal*. 2003, 586, L103-L106.

Reitz, G. *Radiation Measurements*. 2001, 33, 341-346.

Sawyer, D.M.; Vette, J.I. *National Space Science Data Center Report*. 1976, 76-06.

Taniguchi, T.; Ueda, N.; Nakazawa, M.; Sekiguchi, A. *NEUT Research Report*. 1983, 83-10.

Terasawa, K.; Doke, T.; Hasebe, N.; Kikuchi, J.; Kudo, K.; Murakami, T.; Takeda, N.; Tamura, T.; Torii, S.; Yamashita, M.; Yoshihira, E. *Nuclear Instruments and Methods*. 2001a, A457, 499-508.

Terasawa, K.; Doke, T.; Hara, K.; Hasebe, N.; Kikuchi, J.; Kudo, K.; Takeda, N.; Yoshihira, E. *IEEE Transactions on Nuclear Science*. 2001b, 48, 1118-1121.

Watanabe, K.; Muraki, Y.; Matsubara, Y.; Murakami, K.; Sako, T.; Tsuchiya, H.; Masuda, S.; Yoshimori, M.; Ohmori, N.; Miranda, P.; Martinic, N.; Ticona, R.; Velarde, A.; Kakimoto, F.; Ogio, S.; Tsunesada, Y.; Tokuno, H.; Shirasaki, Y. *Astrophysical Journal*. 2003, 592, 590-596.

INDEX

A

D

E

F

G

H

I

J

K

L

M

N

Q

R

S

T

U

V

W

X

Y

Z